Bioinformatics in Aquaculture

Bioinformatics in Aquaculture

Principles and Methods

Edited by Zhanjiang (John) Liu

Registered Offices
John Wiley & Sons Ltd, The Atrium, Southern Gate, Chichester, West Sussex, PO19 8SQ, UK

Editorial Office
111 River Street, Hoboken, NJ 07030, USA
9600 Garsington Road, Oxford, OX4 2DQ, UK
The Atrium, Southern Gate, Chichester, West Sussex, PO19 8SQ, UK
Boschstr. 12, 69469 Weinheim, Germany

For details of our global editorial offices, customer services, and more information about Wiley products, visit us at www.wiley.com.

Wiley also publishes its books in a variety of electronic formats and by print-on-demand. Some content that appears in standard print versions of this book may not be available in other formats.

Library of Congress Cataloging-in-Publication Data

Names: Liu, Zhanjiang, editor.
Title: Bioinformatics in aquaculture : principles and methods / edited by Zhanjiang (John) Liu.
Description: Hoboken, NJ : John Wiley & Sons, 2017. | Includes bibliographical references and index.
Identifiers: LCCN 2016045878 (print) | LCCN 2016057071 (ebook) | ISBN 9781118782354 (cloth : alk. paper) | ISBN 9781118782385 (Adobe PDF) | ISBN 9781118782378 (ePub)
Subjects: LCSH: Bioinformatics. | Aquaculture.
Classification: LCC QH324.2 B5488 2017 (print) | LCC QH324.2 (ebook) | DDC 572/.330285–dc23
LC record available at https://lccn.loc.gov/2016045878

Cover image: Jack fish © wildestanimal/Getty Images, Inc.;
Digital DNA strands © deliormanli/iStockphoto;
DNA illustration © enot-poloskun/iStockphoto
Cover design: Wiley

Set in 10/12pt WarnockPro by SPi Global, Chennai, India
Printed in Singapore by C.O.S. Printers Pte Ltd

10 9 8 7 6 5 4 3 2 1

Contents

About the Editor

Zhanjiang (John) Liu is currently the associate provost and associate vice president for research at Auburn University, and a professor in the School of Fisheries, Aquaculture and Aquatic Sciences. He received his BS in 1981 from the Northwest Agricultural University (Yangling, China), and both his MS in 1985 and PhD in 1989 from the University of Minnesota (Minnesota, United States). Liu is a fellow of the American Association for the Advancement of Science (AAAS). He is presently serving as the aquaculture coordinator for the USDA National Animal Genome Project; the editor for *Marine Biotechnology*; associate editor for *BMC Genomics*; and associate editor for *BMC Genetics*. He has also served on the editorial board for a number of journals, including *Aquaculture, Animal Biotechnology, Reviews in Aquaculture*, and *Frontiers of Agricultural Science and Engineering*. Liu has also served in over 100 graduate committees, including as a major professor for over 50 PhD students. He has trained over 50 postdoctoral fellows and visiting scholars from all over the world. Liu has published over 300 peer-reviewed journal articles and book chapters, and this book is his fourth after *Aquaculture Genome Technologies* (2007), *Next Generation Sequencing and Whole Genome Selection in Aquaculture* (2011), and *Functional Genomics in Aquaculture* (2012), all published by Wiley and Blackwell.

List of Contributors

Asher Baltzell
Arizona Biological and Biomedical Sciences
University of Arizona
Tucson, Arizona
United States

Lisui Bao
The Fish Molecular Genetics and Biotechnology Laboratory
School of Fisheries, Aquaculture and Aquatic Sciences and Program of Cell and Molecular Biosciences
Auburn University
Alabama
United States

Zhenmin Bao
Key Lab of Marine Genetics and Breeding
College of Marine Life Science
Ocean University of China
Qingdao
China

Matt Bomhoff
The School of Plant Sciences
iPlant Collaborative
University of Arizona
Tucson, Arizona
United States

Ailu Chen
The Fish Molecular Genetics and Biotechnology Laboratory
School of Fisheries, Aquaculture and Aquatic Sciences and Program of Cell and Molecular Biosciences
Auburn University
Alabama
United States

Jinzhuang Dou
Key Lab of Marine Genetics and Breeding
College of Marine Life Science
Ocean University of China
Qingdao
China

Qiang Fu
The Fish Molecular Genetics and Biotechnology Laboratory, School of Fisheries, Aquaculture and Aquatic Sciences and Program of Cell and Molecular Biosciences
Auburn University
Alabama
United States

Sen Gao
The Fish Molecular Genetics and Biotechnology Laboratory, School of Fisheries, Aquaculture and Aquatic Sciences and Program of Cell and Molecular Biosciences
Auburn University
Alabama
United States

Xin Geng
The Fish Molecular Genetics and Biotechnology Laboratory, School of Fisheries, Aquaculture and Aquatic Sciences and Program of Cell and Molecular Biosciences
Auburn University
Alabama
United States

Alejandro P. Gutierrez
The Roslin Institute, and the Royal (Dick) School of Veterinary Studies
University of Edinburgh
Edinburgh
United Kingdom

Yanghua He
Department of Animal & Avian Sciences
University of Maryland
College Park, Maryland
United States

Ross D. Houston
The Roslin Institute, and the Royal (Dick) School of Veterinary Studies
University of Edinburgh
Edinburgh
United Kingdom

Chen Jiang
The Fish Molecular Genetics and Biotechnology Laboratory, School of Fisheries, Aquaculture and Aquatic Sciences and Program of Cell and Molecular Biosciences
Auburn University
Alabama
United States

Yanliang Jiang
CAFS Key Laboratory of Aquatic Genomics and Beijing Key Laboratory of Fishery Biotechnology, Centre for Applied Aquatic Genomics
Chinese Academy of Fishery Sciences
Beijing
China

Yulin Jin
The Fish Molecular Genetics and Biotechnology Laboratory, School of Fisheries, Aquaculture and Aquatic Sciences and Program of Cell and Molecular Biosciences
Auburn University
Alabama
United States

Blake Joyce
The School of Plant Sciences, iPlant Collaborative
University of Arizona
Tucson, Arizona
United States

Mehar S. Khatkar
Faculty of Veterinary Science
University of Sydney
New South Wales
Australia

Chao Li
College of Marine Sciences and Technology
Qingdao Agricultural University
Qingdao
China

Jiongtang Li
CAFS Key Laboratory of Aquatic Genomics and Beijing Key Laboratory of Fishery Biotechnology, Centre for Applied Aquatic Genomics
Chinese Academy of Fishery Sciences
Beijing
China

Ning Li
The Fish Molecular Genetics and Biotechnology Laboratory, School of Fisheries, Aquaculture and Aquatic Sciences and Program of Cell and Molecular Biosciences
Auburn University
Alabama
United States

Yun Li
The Fish Molecular Genetics and Biotechnology Laboratory, School of Fisheries, Aquaculture and Aquatic Sciences and Program of Cell and Molecular Biosciences
Auburn University
Alabama
United States

Shikai Liu
The Fish Molecular Genetics and Biotechnology Laboratory, School of Fisheries, Aquaculture and Aquatic Sciences and Program of Cell and Molecular Biosciences
Auburn University
Alabama
United States

Zhanjiang Liu
The Fish Molecular Genetics and Biotechnology Laboratory, School of Fisheries, Aquaculture and Aquatic Sciences and Program of Cell and Molecular Biosciences
Auburn University
Alabama
United States

Qianyun Lu
Key Lab of Marine Genetics and Breeding, College of Marine Life Science
Ocean University of China
Qingdao
China

Jia Lv
Key Lab of Marine Genetics and Breeding, College of Marine Life Science
Ocean University of China
Qingdao
China

Eric Lyons
The School of Plant Sciences, iPlant Collaborative
University of Arizona
Tucson, Arizona
United States

Fiona McCarthy
Department of Veterinary Science and Microbiology
University of Arizona
Tucson, Arizona
United States

Zhenkui Qin
The Fish Molecular Genetics and Biotechnology Laboratory, School of Fisheries, Aquaculture and Aquatic Sciences and Program of Cell and Molecular Biosciences
Auburn University
Alabama
United States

Jiuzhou Song
Department of Animal & Avian Sciences
University of Maryland
College Park, Maryland
United States

Luyang Sun
The Fish Molecular Genetics and Biotechnology Laboratory, School of Fisheries, Aquaculture and Aquatic Sciences and Program of Cell and Molecular Biosciences
Auburn University
Alabama
United States

Xiaowen Sun
CAFS Key Laboratory of Aquatic Genomics and Beijing Key Laboratory of Fishery Biotechnology, Centre for Applied Aquatic Genomics
Chinese Academy of Fishery Sciences
Beijing
China

Suxu Tan
The Fish Molecular Genetics and Biotechnology Laboratory, School of Fisheries, Aquaculture and Aquatic Sciences and Program of Cell and Molecular Biosciences
Auburn University
Alabama
United States

Ruijia Wang
Ministry of Education Key Laboratory of Marine Genetics and Breeding, College of Marine Life Sciences
Ocean University of China
Qingdao
China

Shaolin Wang
Beijing Advanced Innovation Center for Food Nutrition and Human Health, College of Veterinary Medicine
China Agricultural University
Beijing
China

Shi Wang
Key Lab of Marine Genetics and Breeding, College of Marine Life Science
Ocean University of China
Qingdao
China

Xiaozhu Wang
The Fish Molecular Genetics and Biotechnology Laboratory, School of Fisheries, Aquaculture and Aquatic Sciences and Program of Cell and Molecular Biosciences
Auburn University
Alabama
United States

Peng Xu
CAFS Key Laboratory of Aquatic Genomics and Beijing Key Laboratory of Fishery Biotechnology, Centre for Applied Aquatic Genomics
Chinese Academy of Fishery Sciences
Beijing
China

Yujia Yang
The Fish Molecular Genetics and Biotechnology Laboratory, School of Fisheries, Aquaculture and Aquatic Sciences and Program of Cell and Molecular Biosciences
Auburn University
Alabama
United States

Jun Yao
The Fish Molecular Genetics and Biotechnology Laboratory, School of Fisheries, Aquaculture and Aquatic Sciences and Program of Cell and Molecular Biosciences
Auburn University
Alabama
United States

Zihao Yuan
The Fish Molecular Genetics and Biotechnology Laboratory, School of Fisheries, Aquaculture and Aquatic Sciences and Program of Cell and Molecular Biosciences
Auburn University
Alabama
United States

Peng Zeng
Department of Mathematics and Statistics Auburn University
Alabama
United States

Qifan Zeng
The Fish Molecular Genetics and Biotechnology Laboratory, School of Fisheries, Aquaculture and Aquatic Sciences and Program of Cell and Molecular Biosciences
Auburn University
Alabama
United States

Jiaren Zhang
The Fish Molecular Genetics and Biotechnology Laboratory, School of Fisheries, Aquaculture and Aquatic Sciences and Program of Cell and Molecular Biosciences
Auburn University
Alabama
United States

Lingling Zhang
Key Lab of Marine Genetics and Breeding, College of Marine Life Science
Ocean University of China
Qingdao
China

Degui Zhi
School of Biomedical Informatics and School of Public Health the University of Texas Health Science Center at Houston
Texas
United States

Tao Zhou
The Fish Molecular Genetics and Biotechnology Laboratory, School of Fisheries, Aquaculture and Aquatic Sciences and Program of Cell and Molecular Biosciences
Auburn University
Alabama
United States

Preface

Genomic sciences have made drastic advances in the last 10 years, largely because of the application of next-generation sequencing technologies. It is not just the high throughput that has revolutionized the way science is conducted; the rapidly reducing cost of sequencing has made these technologies applicable to all aspects of molecular biological research, as well as to all organisms, including aquaculture and fisheries species. About 20 years ago, Francis S. Collins, currently the director of the National Institutes of Health, had a vision of achieving the sequencing of one genome for US$1000, and we are almost there now. From the billion-dollar human genome project, to those genome projects of livestock with a budget of about US$1 million (down from US$10 million just a few years ago), to the current cost level of just tens of thousands of dollars for a *de novo* sequencing project, the potential for research using genomic approaches has become unlimited. Today, commercial services are available worldwide for projects, whether they are new sequencing projects for a species, or re-sequencing projects for many individuals. The key issue is to achieve a balanced of quality and quantity with minimal costs.

The rapid technological advances provide huge opportunities to apply modern genomics to enhance aquaculture production and performance traits. However, we are facing a number of new challenges, especially in the area of bioinformatics. This challenge may be paramount for aquaculture researchers and educators. Aquaculture students may be well acquainted with aquaculture, but may have no background in computer science or be sophisticated enough for bioinformatics analysis of the large datasets. The large datasets (in tera-scales) themselves pose great computational challenges. Therefore, new ways of thinking in terms of the education and training of the next generation of scientists is required. For instance, a few laboratories may be sufficient for the worldwide production of data, but several orders of magnitude more numbers of laboratories may be required for the data analysis or bioinformatics data mining required to link the data with biology. In the last several years, we have provided training with special problem-solving approaches on various bioinformatics topics. However, I find that the training of graduate students by special topics is no longer efficient enough. All graduate students in the life sciences need some levels of bioinformatics training. This book is an expansion of those training materials, and has been designed to provide the basic principles as well as hands-on experience of bioinformatics analysis. While the book is titled *Bioinformatics in Aquaculture*, it is not the intention of the editor or the book chapter contributors to provide bioinformatics guidance on topics such as programming. Rather, the focus is on providing a basic framework about the need for informatics analysis, and then to provide guidance on the practical applications of existing bioinformatics tools for aquaculture problems.

This book has 28 chapters, arranged in five parts. Part 1 focuses on issues of dealing with DNA sequences: basic command lines (Chapter 1); how to determine sequence identities (Chapter 2); how to assemble short read sequences into contigs and scaffolds (Chapter 3); how to annotate genome sequences (Chapter 4); how to analyze repetitive sequences (Chapter 5); how to analyze duplicated genes (Chapter 6); and how to deal with complex genomes such as tetraploid fish genomes (Chapter 7). Part 2 focuses on the issues involved in dealing with RNA sequences: how to assemble short reads of RNA-Seq into transcriptome sequences (Chapter 8); how to identify differentially expressed genes and co-regulated genes (Chapter 9); how to characterize results from RNA-Seq analysis using gene ontology, enrichment analysis, and gene pathways (Chapter 10); how to use RNA-Seq for genetic analysis (Chapter 11); analysis of long non-coding RNAs (Chapter 12); analysis of microRNAs and their target genes (Chapter 13); determination of allele-specific gene expression (Chapter 14); and epigenetic analysis (Chapter 15). Part 3 focuses on the issues involved in the discovery and application of molecular markers: microsatellites (Chapter 16); single-nucleotide polymorphisms (SNPs) (Chapter 17); SNP arrays (Chapter 18); genotyping by sequencing (Chapter 19); genetic linkage analysis (Chapter 20); genome selection (Chapter 21); QTL mapping (Chapter 22); GWAS (Chapter 23); and gene pathway analysis in GWAS (Chapter 24). Part 4 focuses on the issues involved in comparative genome analysis: comparative genomics using CoGe (Chapter 25). The last part, Part 5, introduces bioinformatics resources, databases, and genome browsers useful for aquaculture, such as NCBI resources and tools (Chapter 26); Ensembl resources and tools (Chapter 27); and the iAnimal bioinformatics infrastructures (Chapter 28).

This book was written to illustrate both principles and detailed methods. It should be useful to academic professionals, research scientists, graduate students and college students in agriculture, as well as students of aquaculture and fisheries. In particular, this book should be a good textbook for graduate training classes. I am grateful to all the contributors for their inputs; it is their great experience and efforts that made this book possible. In addition, I am grateful to the postdoctoral fellows and graduate students in my laboratory at Auburn University for recognizing the need for and inspiring the production of such a "manual-like" book, but with sufficient background for beginner-level graduate students. Also, I have had a pleasant experience interacting with Kevin Metthews (senior project editor) and Ramya Raghavan (project editor) of Wiley-Blackwell Publishing.

During the course of writing and editing this book, I have worked extremely hard to fulfill my responsibilities as the associate provost and associate vice president for research, while performing my duty and passion as a professor and graduate advisor. As a consequence, I have fallen short of fulfilling my responsibility as a father to my three lovely daughters—Elise, Lisa, and Lena Liu—and even more so to my granddaughter Evelyn Wong. I wish to express my appreciation for their independence and great progress.

Finally, this book is a product of the encouragement I received from my lovely wife, Dongya Gao. Her constant inspiration to rise above mediocrity has been a driving force for me to pile additional duties on my already very full plate. This book, therefore, is dedicated to my extremely supportive wife.

Zhanjiang (John) Liu

Part I

Bioinformatics Analysis of Genomic Sequences

1

Introduction to Linux and Command Line Tools for Bioinformatics

Shikai Liu and Zhanjiang Liu

Introduction

Dealing with huge omics datasets in the genomics era, bioinformatics is essential for the transformation of raw sequence data into meaningful biological information for all branches of life sciences, including aquaculture. Most tasks of bioinformatics are processed using the Linux operating system (OS). Linux is a stable, multi-user, and multi-tasking system for servers, desktops, and laptops. It is particularly suited to working with large text files. Many of the Linux commands can be combined in various ways to amplify the power of command lines. Moreover, Linux provides the greatest level of flexibility for development of bioinformatics applications. The majority of bioinformatics programs and packages are developed on the Linux OS. Although most programs can be compiled to run on Microsoft Windows systems, it is generally more convenient to install and use the programs on Linux systems. Therefore, familiarity with and understanding of basic Linux command lines is essential for bioinformatic analysis. In this chapter, we provide an introduction to the Linux OS and its basic command line tools.

An operating system (OS) is basically a suite of programs that make the computer work. It manages computer hardware and software resources and provides common services for computer programs. Examples of popular modern OSs include Microsoft Windows, Linux, macOS, iOS, BSD, Android, BlackBerry OS, and Chrome OS. All these examples share the root of a UNIX base, except for Microsoft Windows.

The UNIX OS was developed in the late 1960s and first released in 1971 by AT&T Bell Labs. It has been under continuous development ever since. UNIX is proprietary, however, which hindered its wide academic use. Researchers at University of California-Berkeley developed an alternative to AT&T Bell Labs' UNIX OS, called the Berkeley Software Distribution (BSD. BSD is an influential operation system, from which several notable OSs such as Sun's SunOS and Apple Inc's macOS system are derived. In the 1990s, Linus Torvalds developed a non-commercial replacement for UNIX, which eventually became the Linux OS. Linux was released as free open source software, with its underlying source code publicly available, freely distributed, and freely modified. Linux is now used in numerous areas, from embedded systems to supercomputers. It is the most common OS powering web servers around the world. Many Linux distributions have been developed, such as Red Hat, Fedora, Debian, SUSE, and Ubuntu. Each distribution has the Linux kernel at its core, but builds on top

Bioinformatics in Aquaculture: Principles and Methods, First Edition. Edited by Zhanjiang (John) Liu.

of that with its own selection of other components, depending on the target users of the distribution. From the perspective of end users, there is no big difference between Linux and UNIX. Both use the same shell (e.g., bash, ksh, csh) and other development tools such as Perl, PHP, Python, and GNU C/C++ compilers. However, because of the freeware nature of the Linux OS, it has the most active support community.

Linux is well known for its command line interface (CLI), while it also has a graphical user interface (GUI). Similar to Microsoft Windows, the GUI provides the user an easy-to-use environment. Currently, the most common way to interact with a Linux OS is via a GUI. In general, the GUI is powered by a derivative of the X11 Window System, commonly referred to as "X11." A desktop manager runs in the X11 Window System and supplies the menus, icons, and windows to interact with the system. The KDE (the default desktop for openSUSE) and GNOME (the default desktop for Ubuntu) are two of the most popular desktop environments. On the modern Linux OS, although the GUI provides the graphical "user-friendliness," the "unhandy" text-based CLI is where the true power resides. In the field of bioinformatics, almost all applications are executed with CLI.

Linux is a stable, multi-user, and multi-tasking system for servers, desktops, and laptops. It is particularly suited to working with large text files because it has a large number of powerful commands that specialize in processing text files. Most of these commands can be further combined in various ways to amplify the power of command lines. In the genomics era, with sequencing data being explosively accumulated, bioinformatics has become a scientific discipline of its own. Bioinformatics relies heavily on the Linux OS because it mostly works with text files containing nucleotide and amino acid sequences. Moreover, Linux provides the greatest level of flexibility for the development of bioinformatics applications. The majority of bioinformatics programs and packages are developed on Linux-based systems. Although most bioinformatics programs can be compiled to run on Microsoft Windows systems, it is more convenient to install and use the program on Linux-based systems.

In this chapter, we introduce the Linux OS and its basic command lines. All commands introduced in Linux are valid for UNIX or any UNIX-like OSs. This chapter functions as a boot camp of Linux command lines to assist bioinformatics beginners in going through with the commands and packages discussed in the remaining chapters of this book. Readers who are already familiar with Linux and its command lines can skip this chapter.

Overview of Linux

The Linux OS is made up of three parts: the kernel, the shell, and the program (Figure 1.1). The kernel is the hub of the OS, which allocates time and memory to programs, and handles the file system and communications in response to system calls. The shell and the kernel work together. As an illustration, let us suppose a user types in a command line `ls myDirectory`. The `ls` command is used to list the contents of a directory. In this process, the shell will search the file system for the file containing the program `ls`, and then request the kernel, through system calls, to execute the program (`ls`) to list the contents of the directory (`myDirectory`).

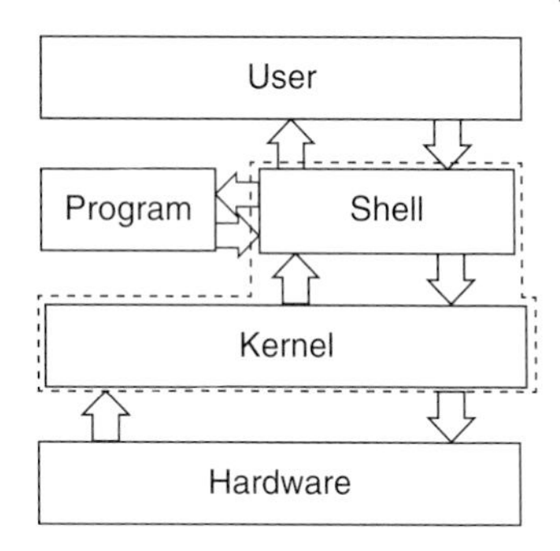

Figure 1.1 An illustration of the Linux operation system.

The shell acts as an interface between the user and the kernel. When a user logs in, the *login* program checks the username and password, and then starts another program called *shell*. The *shell* is a command line interpreter, which interprets the commands that the user types in and passes them to the OS to perform. The shell can be customized by users, and different shells can be used on the same machine. The most influential shells include the Bourne shell (sh) and the C shell (csh). The Bourne shell was written by Stephen Bourne at AT&T as the original UNIX command line interpreter, which introduced the basic features common to all UNIX shells. Every UNIX-like system has at least one shell compatible with the Bourne shell. The C shell was developed by Bill Joy for Berkeley Software Distribution, which was originally derived from the UNIX shell with its syntax modeled after the C programming language. The C shell is primarily for interactive terminal use, and less frequently for scripting and OS control. Bourne-Again shell (bash) is a free software replacement for the Bourne shell, which is written as a part of the GNU Project. Bourne-Again shell is distributed widely as the shell for GNU OSs and as a default interactive shell for users on most GNU/Linux and macOS systems.

The users interact with the shell through terminals—that is, programs called terminal emulators. A bunch of different terminal emulators are available. Most Linux distributions supply several, such as gnome-terminal, konsole, xterm, rxvt, kvt, nxterm, and eterm. Although many different terminal emulators exist, they all do the same thing: open a window and give users access to a shell session. After opening a terminal, the shell will give a prompt (e.g., $) to request commands from the user. When the current command terminates, the shell gives another prompt.

A computer program is a list of instructions passed to a computer to perform a specific task or a series of tasks. Linux commands are themselves programs. A command can take options, which change the behavior of the command. Manual pages are available for each command, to provide detailed information on which options it can take, and how each option modifies the behavior of the command.

Directories, Files, and Processes

Everything in Linux is either a file/directory or a process. A *process* is an executing program identified by a unique process identifier. A *file* is a collection of data such as a document (e.g., report and essay), a text of a program written in some high-level programming language (e.g., a shell script), a collection of binary digits (e.g., a binary executable file), or a directory. All the files are grouped together in the directory structure.

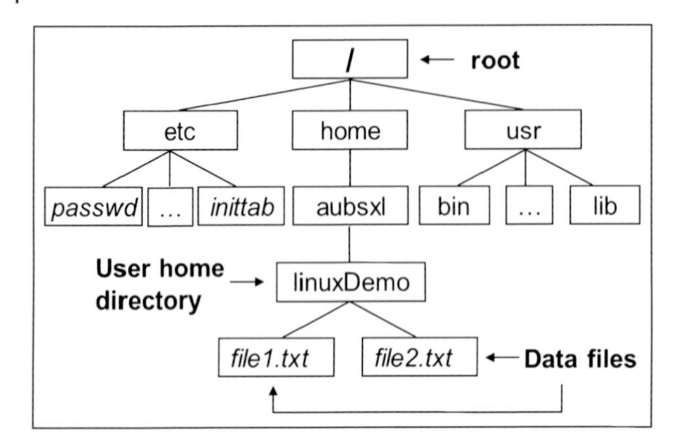

Figure 1.2 An illustration of the Linux directory structure.

Directory Structure

Linux files are arranged in a single-rooted, hierarchical structure, like an inverted tree (Figure 1.2). The top of the hierarchy is traditionally called the *root* (written as a slash—"/"). As shown in Figure 1.2, the home directory ("home") contains a user home directory ("aubsxl"). The user home directory contains a subdirectory ("linuxDemo") that has two files ("file1.txt" and "file2.txt"). The full path of the "file1.txt" is "/home/aubsxl/linuxDemo/file1.txt".

Filename Conventions

In Linux, files are named conventionally, starting with a lower-case letter and ending with a dot, followed by a group of letters indicating the contents of the file. For instance, a file consisting of C code is named with the ending ".c", such as "prog1.c". A good way to name a file is to use only alphanumeric characters (i.e., letters and numbers) together with underscores (_) and dots (.). Characters with special meanings—such as /, *, &, %, and spaces—should be avoided. A *directory* is merely a special type of file (like "a container for files"); therefore, the rules and conventions for naming files apply to directories as well.

Wildcards

Wildcards are commonly used in Linux shell commands, and also in regular expressions and programming languages. Wildcards are characters that are used to substitute for other characters, increasing the flexibility and efficiency of running commands. Three types of wildcards are widely used: `*`, `?`, and `[]`. The star (`*`) is the most frequently used wildcard. It matches against one or more character(s) in the name of a file (or directory). For instance, in the "linuxDemo" directory, type

```
$ ls file*
```

This will list all files that have names starting with the string "file" in the current directory. Similarly, type

```
$ ls *.txt
```

This will list all files that have names ending with ".txt" in the current directory.

The question mark (?) is another wildcard, which matches exactly one character. For instance,

```
$ ls file?.txt
```

This will list both "file1.txt" and "file2.txt", but will not list the file if it is named "file_1.txt".

The third type of wildcard is a pair of square brackets ([]), which represents a range of characters (or numbers) enclosed in the brackets. For instance, the following command line will list files with names starting with any letter from a to z:

```
$ ls [a-z]*.txt
```

File Permission

Each file (and directory) has associated access rights, which can be shown by typing "ls -l" in the terminal (Figure 1.3). Also, "ls -lg" gives additional information as to which group owns the file (e.g., "file1.txt" is owned by the group named "aubfish" in the figure).

The left-hand column in Figure 1.3 is a 10-symbol string that consists of symbols, including d, l, r, w, x, and -. If d is present, it will be at the left-hand end of the string, and will indicate a directory; otherwise - will be the starting symbol of the string indicating a file. The symbol of l is used to indicate the links of a file or directory.

The nine remaining symbols indicate the permissions, or access rights, and are taken as three groups of three (Figure 1.3).

1) The left group of three gives the file permissions for the user that owns the file (or directory) (i.e., "aubsxl" in the figure).
2) The middle group of three gives the permissions for the group of people who own the file (or directory) (i.e., "aubfish" in the figure).
3) The rightmost group of three gives the permissions for all other users.

The symbols have slightly different meanings, depending on whether they refer to a file or to a directory. For a file, the *r* (or -) indicates the presence or absence of permission to read and copy the file; *w* (or -) indicates the permission (or otherwise) to write (change) a file; and *x* (or -) indicates the permission (or otherwise) to execute a file. For a directory, the *r* allows users to list files in the directory; *w* allows users to delete files from the directory or move files into it; and *x* allows users to access files in the directory.

Change File Permission

The owner of a file can change the file permissions using the chmod command. The options of chmod are listed in Table 1.1. For instance, to remove read, write, and execute permissions on the file "file1.txt" for the group and others, type

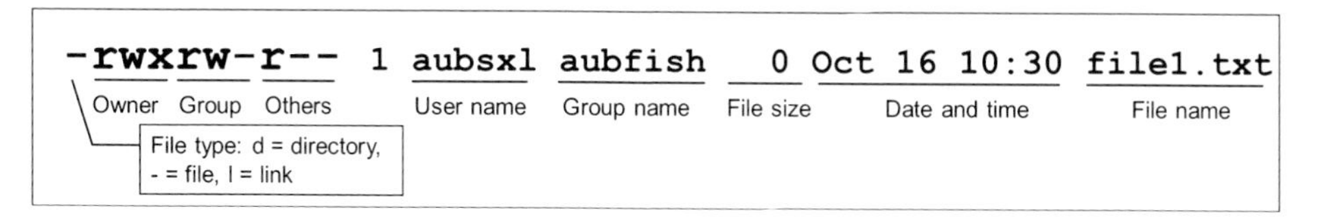

Figure 1.3 An illustration of file permission.

Table 1.1 The options of chmod command.

Option	Function
u	user
g	group
o	other
a	all
r	read
w	write
x	execute
+	add permission
-	take away permission

Table 1.2 List of octal numbers for file permissions.

Number	Permission
0	No permission
1	Execute only
2	Write only
3	Write and execute (1 + 2)
4	Read only
5	Read and execute (4 + 1)
6	Read and write (4 + 2)
7	Read, write, and execute (4 + 2 + 1)

```
$ chmod go-rwx file1.txt
```

To give read and write permissions on the file "file1.txt" to all, type

```
$ chmod a+rw file1.txt
```

The file permissions can also be encoded as octal numbers (Table 1.2), which can be used in the `chmod` command. For instance, to give all permissions on the file "file1.txt" to the owner, read and execute permission to the group, and no permission to others, type

```
$ chmod 750 file1.txt
```

Environment Variables

Each Linux process runs in a specific environment. An environment consists of a table of environment variables, each with an assigned value. When the user logs in, certain login files are executed, which initializes the table holding the environment variables for the process. The table becomes accessible to the shell once the login files pass the

Table 1.3 A list of examples of environment variables.

Environment variables	Value
USER	The login name
HOME	The path name of the home directory
HOST	The name of the computer
ARCH	The architecture of the computer's processor
DISPLAY	The name of the computer screen to display X11 windows
PRINTER	The default printer to send print jobs to
PATH	The directories the shell should search to find a command
PWD	The current working directory
BASH_VERSION	The version of bash being executed
BASHOPTS	The list of options that were used when bash was executed
HISTSIZE	Number of lines of command history allowed in memory
HISTFILESIZE	Number of lines of command history stored to a file
HOSTNAME	The hostname of the computer at this time
PS1	The primary command prompt definition
SHELLOPTS	Shell options that can be set with the set option
UID	The UID of the current user

process to the shell. When a parent process starts up a child process, it will give a copy of the parent's table to the child process.

Environment variables are used to pass information from the shell to programs that are being executed. Programs look "in the environment" for particular variables, and if they find the variables, they will use the stored values. Some frequently used environment variables are listed in Table 1.3. Standard Linux OS has two categories of environment variables: global environment variables and local environment variables.

Global Environment Variable

Global environment variables are visible from the shell session and from any subshells. An example of an environment variable is the HOME variable. The value of this variable is the path name of the home directory. To view global environment variables, the `env` or `printenv` command can be used. For instance, type

```
$ printenv
```

This command will display all the environment variables in the system. To display the value of an individual environment variable, only the `printenv` command can be used:

```
$ printenv HOME
```

This command line will display the path name of the home directory.

The `echo` command can also be used to display the value of a variable. However, when the environment variables are referred in this way, a dollar sign (`$`) needs to be placed before the variable name.

```
$ echo $HOME
```

Local Environment Variable

The shell also maintains a set of internal variables known as *local environment variables* that define the shell to work in a particular way. Local environment variables are available only in the shell where they are defined, and are not available to the parent or child shell. Even though they are local, they are as important as global environment variables. Linux systems define standard local environment variables by default. Users can also define their own local variables. There is no specific command to only display the local variables. To view local variables, the *set* command can be used, which displays all variables defined for a specific process, including local and global environment variables and user-defined local variables.

```
$ set
```

The output of the `set` command includes all global environment variables as displayed using the `env` or `printenv` command. The remaining variables are the local environment and user-defined variables.

Setting Environment Variables

A local variable can be set by assigning either a numeric or a string value to the variable using the equal sign.

```
$ myVariable=Hello
```

To view the new variable,

```
$ echo $myVariable
```

If the variable value contains spaces, a single or double quotation mark should be used to delineate the beginning and end of the string.

```
$ myVariable="Hello World"
```

The local variables set in the preceding example are available only for use with the current shell process, and are not available in any other child shell. To create a global environment variable that is visible from any child shell processes created by the parent shell process, a local variable needs to be created and then exported to the global environment. This can be done using the `export` command:

```
$ myVariable="Hello World"
$ export myVariable
```

After defining and exporting the local variable "myVariable", the child shell is able to properly display the variable's value.

When defining variables, spaces should be avoided among the variable name, the equal sign, and the assigned value. Moreover, in the standard bash shell, all environment variable names use uppercase letters by convention. It is advisable to use lowercase letters for the names of user-defined local variables to avoid the risk of redefining a system environment variable.

To remove an existing environment variable, the *unset* command can be used.

```
$ unset myVariable
```

Setting the PATH Environment Variable

When an external command is entered in the shell CLI, the shell will first search the system to locate the program. The PATH environment variable defines the directories in which the shell will look to find the command that the user entered. If the system returns a message saying "command: Command not found", this indicates that either the command does not exist on the system or it is simply not in your path. To run a program, the user either needs to directly specify the absolute path of the program, or has to have the directory containing the program in the path.

The PATH environment variables can be displayed by typing:

```
$ echo $PATH
```

The individual directories listed in the PATH are separated by colons. The program path (e.g., "/home/aubsxl/linuxDemo") can be added to the end of the existing path (the `$PATH` represents this) by issuing the command:

```
$ PATH=$PATH:/home/aubsxl/linuxDemo
```

To add this path permanently, add the preceding line to the .bashrc file after the list of other commands.

Basic Linux Commands

A typical Linux command line consists of a command name, followed by options and arguments. For instance,

```
$ wc -i FILE
```

The "`$`" is the prompt from the shell, requesting for the user's command; "`wc`" is the name of a command that the shell will locate and execute; "`-i`" is one of the options that modify the behavior of the command; and "`FILE`" is an argument specifying the data file that the command `wc` should read and process. Manual pages can be accessed by using the `man` command to provide information on the options that a particular command can take, and how each option modifies the behavior of the command. To look up the manual page of the `wc` command, type

```
$ man wc
```

In Linux shell, the [Tab] key is a useful shortcut to complete the names of commands and files. By typing part of the name of a command, filename, or directory, and pressing the [Tab] key, the shell can automatically complete the rest of the name. If more than one command name begins with those typed letters, the shell will beep and prompt the user to type a few more letters before pressing the [Tab] key again.

Here, we introduce a set of the most frequently used Linux commands. For documentation on the full usage of these commands, the readers are referred to the manual pages of each command.

List Directory and File

The `ls` command is used to list the contents of a directory. By default, `ls` only lists files whose names do not begin with a dot (.). Files beginning with a dot (.) are known as *hidden files*, and they usually contain important program configuration information. To list all files including hidden files, the -a option can be used.

```
$ ls -a
```

This command line will list all contents including hidden files in the current working directory.

```
$ ls -l
```

With the use of the -l option, this command line will list contents in the "long" format, providing additional information on the files.

```
$ ls -t
```

This command will show the files sorted based on the modification time.

Create Directory and File

The mkdir command is used to create new directories. For instance, to create a directory called "linuxDemo" in the current working directory, type

```
$ mkdir linuxDemo
```

A file can be created using the touch command. To create a text file named "linuxDemo.txt" in the current working directory, type

```
$ touch linuxDemo.txt
```

Files can also be created and modified using text file editors such as nano, vi, and vim. To create a file in nano, a simple text editor, type

```
$ nano filename.txt
```

In nano, text can be entered or edited. To write the file out, press the keys [Ctrl] and [O]. To exit the application, press the keys [Ctrl] and [X].

vi and vim are advanced text editors. To create a file using vim, type

```
$ vim linuxDemo.txt
```

vim has two different editing modes: *insert* mode and *command* mode. Insert mode can be initiated by pressing the key [I] to insert text. To return to command mode, press [ESC]. In command mode, press [Shift] and [:] to enter the command. To exit and write out the file, press [Shift] and [:], then type in wq and press [Enter] to save. To quit without saving changes, type in: q! and press [Enter].

Change to a Directory

The cd command is used to change from the current working directory to other directories. For instance, to change to the "linuxDemo" directory, type

```
$ cd linuxDemo
```

To find the absolute pathname of current working directory, the pwd command can be used, type

```
$ pwd
```

This will print out the absolute pathname of the working directory, for example, "/home/aubsxl/linuxDemo"

In Linux, there are several shortcuts for working with directories. For instance, the dot (.) represents the current directory, and the double-dot (..) represents the parent of the current directory. Home directory can be represented by the tilde character (~), which is often used to specify paths starting at the home directory. For instance, the path "/home/aubsxl/linuxDemo" is equivalent to "~/linuxDemo".

```
$ cd .
```

This will stay in the current directory.

```
$ cd ..
```

This will change to one directory level above the current directory.

```
$ cd ~
```

This will go to the home directory. Moreover, typing `cd` with no argument will also lead to the home directory.

```
$ cd
```

Manipulate Directory and File

The `cp` command is used to copy a file/directory.

```
$ cp file1 file2
```

This command will make a copy of "file1" in the current working directory and call it "file2".

```
$ cp file1 file2 myDirectory
```

This command line will copy "file1" and "file2" to the directory called "myDirectory".

The `mv` command can be used to move a file from one place to another. For instance,

```
$ mv file1 file2 myDirectory
```

This command line will move, rather than copy (no longer existing in the original directory), "file1" and "file2" to the directory called "myDirectory".

The `mv` command can also be used to rename a file when used without indications of a directory.

```
$ mv file1 file2
```

This command line will rename "file1" as "file2".

The `rm` command can be used to delete (remove) a file.

```
$ rm file1
```

This command will remove the file named "file1".

To delete (remove) a directory, the `rmdir` command should be used.

```
$ rmdir old.dir
```

Only an empty directory can be removed or deleted by the `rmdir` command. If a directory is not empty, the files within the directory should first be removed.

The `ln` command is used to create links between files.

```
$ ln file1 linkName
```

This command line will create a link to "file1" with the name "linkName". If "linkName" is not provided, a link to "file1" is created in the current directory using the name of "file1" as the "linkName". The `ln` command creates hard links by default, and creates symbolic links if the `-s` option is specified.

Access File Content

The command `cat` is used to concatenate the files. It can also be used to display the contents of a file on screen. If the file is longer than the size of the window, it will scroll past, making it unreadable. To display long files, the `less` command can be used. The `less` command writes the contents of a file onto the screen, one page at a time. Press the [Space bar] to see the next page, and type [Q] to quit reading. Using `less`, one can search through a text file for a keyword (pattern), by typing forward slash (/) followed by the keyword. For instance, to search through "linuxDemo.txt" for the word "linux", type

```
$ less linuxDemo.txt
```

Then, still in `less`, type a forward slash (/) followed by the word to be searched: "/linux". The `less` command will find and highlight the keyword. Type [N] to search for the next occurrence of the word.

The `head` command is used to display the first *N* lines of the file. By default, it writes the first 10 lines of a file to the screen. With more than one file, it displays contents of each file and precedes each output with a header giving the file name. When using the `-n` option, it prints the first *N* lines instead of the first 10. With the leading `-`, it prints all but the last *N* lines of each file. For instance,

```
$ head file1
```

This will print the first 10 lines of "file1".

```
$ head -n 50 file1
```

This will print the first 50 lines of "file1".

```
$ head -n -50 file1
```

This will print all but the last 50 lines of "file1".

Similarly, the `tail` command is used to write the last *N* lines of a file. Similar options can be used as those in head command.

Query File Content

The `sort` command is used to sort the contents of a text file line by line. By default, lines starting with a number will appear before lines starting with a letter; and lines starting with a lowercase letter will appear before lines starting with the same letter in uppercase. The sorting rules can be changed by providing the `-r` option. For instance,

```
$ sort months.txt
```

This will sort the file "months.txt" by default sorting rules, based on the first column.

```
$ sort -r months.txt
```

This will sort the file in the reverse order, based on the first column.

```
$ sort -k 2 months.txt
```

This will sort the file "months.txt" based on the second column.

```
$ sort -k 2n months.txt
```

This will sort the file based on the second column by numerical value. By default, the file will be sorted in ascending order; to sort in reverse order, use the `-r` option:

```
$ sort -k 2nr months.txt
```

The sort can be performed based on multiple lines. To sort the file first based on the third column, and then sort based on the second column in numerical value, type

```
$ sort -k 3 -k 2n months.txt
```

The `cut` command is used to select sections of text from each line of files. It can be used to select fields or columns from a line by specifying a delimiter. This command looks for the "tab" delimiter by default; otherwise, the `-d` option should be used to define the delimiter. For instance,

```
$ cut -f1 months.txt
```

This will cut the first column of the file.

```
$ cut -f1,2 months.txt
```

This will cut the first and second columns.

```
$ cut -f1-3 months.txt
```

This will cut the first to the third columns.

```
$ cut -d ' ' -f3 months.txt > seasons
```

This will cut the third column based on "spaces" as delimiters.

The `uniq` command is used to report and filter out repeated lines in a file. It only detects adjacent repeated lines, and therefore the file usually needs to be sorted before using `uniq`.

```
$ uniq months.txt
```

This will print lines with duplicated lines merged to the first occurrence.

```
$ uniq -c months.txt
```

This will print out lines prefixed with a number representing how many times they occur, with duplicated lines merged to the first occurrence.

```
$ uniq -d months.txt
```

This will only print duplicated lines.

```
$ uniq -u months.txt
```

This will only print unique lines.

The `split` command is used to split a file into several. It outputs fixed-sized pieces of input files to files named "PREFIXaa", "PREFIXab", etc.

```
$ split myfile.txt
```

This will, by default, split "myfile.txt" into several files, each containing 1000 lines, and prefixed with "x".

```
$ split -l 2000 myfile.txt myfile
```

This will split "myfile.txt" into several files, each containing 2000 lines, and prefixed with "myfile".

```
$ split -b 100 myfile.txt new
```

This will split the file "myfile.txt" into separate files called "newaa", "newab", "newac", etc., with each file containing 100 bytes of data.

The `grep` command is one of many standard UNIX utilities that can be used to search files for specified words or patterns. To print out each line containing the word "linux", type

```
$ grep linux linuxDemo.txt
```

The grep command is case sensitive, meaning that it distinguishes between "Linux" and "linux". To ignore upper/lower case distinctions, use the `-i` option.

```
$ grep -i linux linuxDemo.txt
```

To search for a phrase or pattern, the phrase or pattern should be enclosed in a pair of single quotes. For instance, to search for "Linux system", type

```
$ grep -i 'Linux system' linuxDemo.txt
```

Some of the other frequently used options of grep are:

`-v` to display those lines that do NOT match
`-n` to precede each matching line with the line number
`-c` to print only the total count of matched lines

More than one option can be used at a time. To print out the number of lines without the words "linux" and "Linux", type

```
$ grep -ivc linux linuxDemo.txt
```

The `wc` command can be used to query the file content for word count. To do a word count on "linuxDemo.txt", type

```
$ wc -w linuxDemo.txt
```

To find out how many lines the file has, type

```
$ wc -l linuxDemo.txt
```

Edit File Content

Files can be manually edited using text editors such as nano, vi, and vim. To automatically edit files, `sed`, a stream editor, can be used. `sed` is mostly used to replace text, but can also be used for many other things. Here, a few examples are provided to illustrate the use of sed:

1) *Common usage:* To replace or substitute a string in a file, type

```
$ sed 's/unix/linux/' linuxDemo.txt
```

This command will replace the word "unix" with "linux" in the file. Here, the "s" specifies the substitution operation, and "/" is a delimiter. The word "unix" is the searching pattern, and the word "linux" is the replacement string. By default, sed command only replaces the first occurrence of the pattern in each line.
To replace the *n*th occurrence of a pattern in a line, the /1, /2, ... , /*n* flags can be used. For instance, the following command replaces the second occurrence of the word "unix" with "linux" in a line.

```
$ sed 's/unix/linux/2' linuxDemo.txt
```

To replace all the occurrence of the pattern in a line, the substitute flag /g (global replacement) can be used. For instance,

```
$ sed 's/unix/linux/g' linuxDemo.txt
```

To replace the text from the *n*th occurrence to all the occurrences in a line, the combination of /1, /2, etc., and /g can be used. For instance,

```
$ sed 's/unix/linux/3g' linuxDemo.txt
```

This sed command will replace the word "unix" with "linux" starting from the third occurrence to all the occurrences.

2) *Replacing on specific lines:* The sed command can be restricted to replace the string on a specific line number. An example is

```
$ sed '3 s/unix/linux/' linuxDemo.txt
```

This sed command replaces the string only on the third line. To replace the string on several lines, a range of line numbers can be specified. For instance,

```
$ sed '1,3 s/unix/linux/' linuxDemo.txt
```

This sed command replaces the lines in the range of 1–3. Another example is

```
$ sed '2,$ s/unix/linux/' linuxDemo.txt
```

This sed command replaces the text from the second line to the last line in the file. The "$" indicates the last line in the file.
To replace only on lines that match a pattern, the pattern can be specified to the sed command. If a pattern match occurs, the sed command looks for the string to be replaced, and then replaces the string.

```
$ sed '/linux/ s/unix/centos/' linuxDemo.txt
```

This sed command will first look for the lines that have the word "linux", and then replace the word "unix" with "centos" on those lines.

3) *Delete, add, and change lines:* The sed command can be used to delete the lines in a file by specifying the line number, or a range of line numbers. For instance,

```
$ sed '2 d' linuxDemo.txt
```

This command will delete the second line.

```
$ sed '5,$ d' linuxDemo.txt
```

This command will delete lines starting from the fifth line to the end of the file.

To add a line after line(s) in which a pattern match is found, the "a" command can be used. For instance,

```
$ sed '/unix/ a "Add a new line"' linuxDemo.txt
```

This command will add the string "Add a new line" after each line containing the word "unix".

Similarly, using the "i" command, the sed command can add a new line before a pattern match is found.

```
$ sed '/unix/ i "Add a new line"' linuxDemo.txt
```

This command will add the string "Add a new line" before each line containing the word "unix".

The sed command can be used to replace an entire line with a new line using the "c" command.

```
$ sed '/unix/ c "Change line"' linuxDemo.txt
```

This sed command will replace each line containing the word "unix" with the string "Change line".

4) *Run multiple* sed *commands:* To run multiple sed commands, the output of one sed command can be piped as input to another sed command.

```
$ sed 's/unix/linux/' linuxDemo.txt | sed 's/os/system/'
```

This command line will first replace the word "unix" with "linux", and then replace the word "os" with "system". Alternatively, sed provides the -e option to run multiple sed commands. The preceding output can be achieved in a single sed command, as shown in the following:

```
$ sed -e 's/unix/linux/' -e 's/os/system/' linuxDemo.txt
```

Redirect Content

Most processes initiated by Linux commands take their input from the standard input (the keyboard) and write to the standard output (the terminal screen). By default, the processes write their error messages to the terminal screen. In Linux, both the input and output of commands can be redirected, using > to redirect the standard output into a file, and using < to redirect the input file. For instance, to create a file named "fish.names" that contains a list of fish names, type

```
$ cat > fish.names
```

Then type in the names of some fish. Press [Enter] after each one.

catfish
zebrafish

carp
stickleback
tetraodon
fugu
medaka
^D (press [Ctrl] and [D] to stop)

In this process, the `cat` command reads the standard input (the keyboard) and redirects (>) the output into a file called "fish.names". To read the contents of the file, type

```
$ cat fish.names
```

The form ≫ appends standard output to a file. To add more items to the file "fish.names", type

```
$ cat >> fish.names
```

Then type in the names of more fish

seabass
croaker
^D ([Ctrl] and [D] to stop)

The redirect > is often used with the cat command to join (concatenate) files. For instance, to join "file1" and "file2" into a new file called "file3", type

```
$ cat list1 list2 > file3
```

This command line will read the contents of "file1" and "file2" sequentially, and then output the text to the file "file3".

Similarly, the redirects apply to other commands. For instance,

```
$ sed -e 's/unix/linux/' -e 's/os/system/' linuxDemo.txt >
   linuxDemo_edit.txt
```

This command line will perform substitutions, and output to the new file "linux Demo_edit.txt" instead of the terminal screen.

The *pipe* (|) is used to redirect the output of one command as the input of another command. For instance, to find out how many users are logged on, type

```
$ who | wc -l
```

The output of the `who` command is redirected as the input of the `wc` command. Similarly, to find out how many files are present in the directory, type

```
$ ls | wc -l
```

The output of the `ls` command is redirected as the input of the `wc` command.

Compare File Content

The `diff` command compares the contents of two files and displays the differences. Suppose we have a file called "file1", and its updated version named "file2". To find the differences between the two files, type

```
$ diff file1 file2
```

In the output, the lines beginning with < denotes "file1", while lines beginning with > denotes "file2".

The comm command is used to compare two sorted files line-by-line. To compare sorted files "file1" and "file2", type

```
$ comm file1 file2
```

With no options, comm produces a three-column output. The first column contains lines unique to "file1", the second column contains lines unique to "file2", and the third column contains lines common to both files. Each of these columns can be suppressed individually with options.

```
$ comm -3 file1 file2
```

This command line will show the lines in both files.

```
$ comm -1 file1 file2
```

This command line will show the lines only in "file1".

```
$ comm -2 file1 file2
```

This command line will show the lines only in "file2".

Compress and Archive Files and Directories

1) zip is a compression tool that is available on most OSs such as Linux/UNIX, macOS, and Microsoft Windows. To zip individual files (e.g., "file1" and "file2") into a zip archive, type

   ```
   $ zip abc.zip file1 file2
   ```

   To extract files from a zip folder, use unzip

   ```
   $ unzip abc.zip
   ```

   To extract to a specific directory, use the -d option.

   ```
   $ unzip abc.zip -d /tmp
   ```

2) The gzip command can be used to archive and compress files. For example, to compress "linuxDemo.txt", type

   ```
   $ gzip linuxDemo.txt
   ```

   This will compress the file and place it in a file called "linuxDemo.txt.gz".
   To decompress files created by gzip, use the gunzip command.

   ```
   $ gunzip linuxDemo.txt.gz
   ```

3) bzip2 compresses and decompresses files with a high rate of compression together with reasonably fast speed. Most files can be compressed to a smaller file size with bzip2 than with the more traditional gzip and zip programs. bzip2 can be used without any options. Any number of files can be compressed simultaneously by merely listing their names as arguments. For instance, to compress the three files named "file1", "file2", and "file3", type

Table 1.4 A list of frequently used `tar` options.

Options	Description
-A	Append tar files to an archive
-c	Create a new archive
-d	Find differences between archive and file system
-r	Append files to the end of an archive
-t	List the contents of an archive
-u	Only append files that are newer than those existing in archive
-x	Extract files from an archive
-f	Use archive file or device F (default "-", meaning stdin/stdout)
-j	Use to decompress .bz2 files
-v	Verbosely list files processed
-z	Use to decompress .gz files

```
$ bzip2 file1 file2 file3
```

`bunzip2` (or `bzip2 -d`) decompresses all specified files. Files that are not created by `bzip2` will be detected and ignored, and a warning will be issued.

```
$ bunzip2 abc.tar.bz2
```

4) `tar` is an archiving program designed to store and extract files from an archive file known as a *tarfile*. The first argument to `tar` must be one of the options A, c, d, r, t, u, x (Table 1.4), followed by any optional functions. The final arguments to tar are the names of the files or directories that should be archived.

To create a tar archive named "abc.tar" by compressing three files, type

```
$ tar -cvf abc.tar file1 file2 file3
```

To create a gzipped tar archive named "abc.tar.gz" by compressing three files, type

```
$ tar -czvf abc.tar.gz file1 file2 file3
```

To extract files from the tar archive "abc.tar", type

```
$ tar -xvf abc.tar
```

To extract files from the tar archive "abc.tar.gz", type

```
$ tar -xvzf abc.tar.gz
```

Access Remote Files

Two programs (`wget` and `curl`) are widely used to retrieve files from websites via the command-line interface. For instance, to download the BLAST program "ncbi-blast-2.2.31 + -x64- linux.tar.gz" from NCBI ftp site using `curl`, type the following:

```
$ curl ftp://ftp.ncbi.nlm.nih.gov/blast/executables/blast+/
  LATEST/ncbi-blast-2.2.31+-x64-linux.tar.gz > ncbi-blast-
  2.2.31+-x64-linux.tar.gz
```

Alternatively, this can be done using `wget` as following:

```
$ wget ftp://ftp.ncbi.nlm.nih.gov/blast/executables/blast+/
  LATEST/ncbi-blast-2.2.31+-x64-linux.tar.gz
```

In addition, the program `scp` (e.g., secure copy) can be used to copy files in a secure fashion between UNIX/Linux computers, as following:

To send a file to a remote computer,

```
$ scp file1 aubsxl@dmc.asc.edu:/home/aubsxl/linuxDemo
```

To retrieve a file from a remote computer,

```
$ scp aubsxl@dmc.asc.edu:/home/aubsxl/linuxDemo/file1
   LocalFile
```

Check Process and Job

A process is an executing program identified by a unique PID (process identifier). The `ps` command provides a report of the current processes. To see information about the processes with their associated PIDs and status, type

```
$ ps
```

The `top` command provides an ongoing look at processor activity in real time. It displays a list of the most CPU-intensive processes on the system, and can provide an interactive interface for manipulating processes. It can sort the tasks by CPU usage, memory usage, and runtime. To display top CPU processes, type

```
$ top
```

A process may be in the foreground, in the background, or suspended. In general, the shell does not return the Linux prompt until the current process has finished executing. Some processes take a long time to run and hold up the terminal. Backgrounding a long process allows for the immediate return of the Linux prompt, enabling other tasks to be carried out while the original process continues executing. To background a process, type an & at the end of the command line. The & runs the job in the background and returns the prompt straight away, allowing the user to run other programs while waiting for that process to finish. Backgrounding is useful for jobs that will take a long time to complete.

When a process is running, backgrounded, or suspended, it will be entered into a list along with a job number. To examine this list, type

```
$ jobs
```

To restart (foreground) a suspended processes, type

```
$ fg jobnumber
```

For instance, to restart the first job, type

```
$ fg 1
```

Typing `fg` with no job number will foreground the last suspended process.

To kill a job running in the foreground, type ^C ([Ctrl] and [C]). To kill a suspended or background process, type

```
$ kill jobnumber
```

Other Useful Command Lines

quota

The quota command is used to check current quota and how much of it has been used.

```
$ quota -v
```

df

The df command reports on the space left on the file system. To find out how much space is left on the current file system, type

```
$ df .
```

du

The du command outputs the number of kilobytes used by each subdirectory. It is useful to find out which directory takes up the most space. In the directory, type

```
$ du -s *
```

The -s flag will display only a summary (total size), and the * indicates all files and directories.

free

The free command displays information on the available random-access memory (RAM) in a Linux machine. To display the RAM details, type

```
$ free
```

zcat

The zcat command can read gzipped files without decompression. For instance, to read the gzipped file "abc.txt.gz", type

```
$ zcat abc.txt.gz
```

For text with large size, the zcat output can be piped through the less command.

```
$ zcat abc.txt.gz | less
```

file

The file command classifies the named files according to the type of data, such as text, pictures, and compressed data. To report on all files in the home directory, type

```
$ file *
```

find

The find command searches through the directories for files and directories with a given name, date, size, or any other specified attribute. This is different from grep, which finds contents within files. To use find to search for all files with the extension of ".txt", starting at the current directory (.) and working through all sub-directories, and then to print the name of the file to the screen, type

```
$ find . -name "*.txt" -print
```

To find files over 1 MB in size, and to display the result as a long listing, type

```
$ find . -size +1M -ls
```

history

The `history` command can display a list of commands the user has typed in. Each command is given a number according to the order it is entered. To repeat a command, the user can either use the cursor keys to scroll up and down the list or type `history` for a list of previous commands.

```
$ history
```

Getting Help

There are manuals within the Linux system that give detailed information about most commands. The manual pages tell which options a particular command can take, and how each option modifies the behavior of the command. The `man` command can be used to read the manual page for a particular command. For instance, to find more information about the `wc` (word count) command, type

```
$ man wc
```

Alternatively, the `whatis` command can be used; type

```
$ whatis wc
```

This command line will give a one-line description of the command, but omit any information about options, etc.

When the user is not sure of the exact name of a command, the `apropos` command can be used.

```
$ apropos copy
```

This command line will list all the commands with the keyword "copy" in their manual page headers.

Installing Software Packages

Although each OS is preinstalled with a large number of commonly used commands, specific programs often need to be installed to perform specific tasks. On Microsoft Windows systems, every program has a simple "Setup.exe" or "program.zip" file. The installation can be done by simply clicking the "Setup.exe" file, which is followed by on-screen instructions. In the world of bioinformatics, software packages are commonly distributed in the form of either precompiled executables or source codes. To install precompiled executables, they can be simply put to the environment path, while software most often needs to be installed from the source code in which the compiling process is required.

Installing Packages from a Configured Repository

The standard Linux package format is RPM. The RPM packaging system was originally developed by Red Hat and is widely used in the Linux community. Distributions using it include Fedora, Mandriva, Red Hat (naturally), and SUSE. An RPM package file is normally named in the pattern "program-version-other.rpm". Another popular package format is DEB, the Debian software package. Debian packages and the Advanced Packaging Tool (APT) provide several advanced features that are now commonly used, such as automatic dependency resolution and signed packages. Debian packages are used by Debian GNU/Linux and distributions based on it, including Ubuntu, Knoppix, and Mepis. A Debian package file is normally named in the pattern "program-version-other.deb".

To install software from the configured repository, the user must become a SuperUser. A broad array of tools is available to work with DEB packages, and `apt-get` is commonly used. The use of `apt-get` is straightforward because it not only keeps track of what packages are installed, but also what other packages are available. To install packages using `apt-get`, type

```
$ sudo apt-get install packagename
```

To remove the software, type

```
$ sudo apt-get remove packagename
```

`yum` does for RPM packages roughly what `apt-get` does for Debian packages. As with `apt-get`, yum can download and install packages from a configured repository.

```
$ sudo yum install packagename
```

To remove the software, type

```
$ sudo yum remove packagename
```

Installing Software from Source Code

Software packages delivered in tarballs are mostly in source code, which must be compiled before installation. A number of steps are required to install the software package from source code: (1) locate and download the source code, which is usually compressed; (2) unpack the source code; (3) compile the code; (4) install the resulting executable; and (5) set paths to the installation directory. Of these steps, the most difficult is the compilation part.

Compiling source code is the process that converts high-level human-readable language code into a form that the computer can understand. For instance, C language source code is converted into a lower-level language called *assembly language*. The assembly language code is then further converted into object code, which the computer can directly understand. The final stage in compiling a program involves linking the object code to code libraries that contain certain built-in functions. The final stage produces an executable program.

As the number of UNIX-based OSs increases, it becomes difficult to write programs that could run on all derivative systems. The characteristics of some systems change from version to version, and developers frequently do not have access to every system.

A number of utilities and tools have been developed for programmers and end users to conduct compiling steps. The GNU *configure and build* system simplifies the building of programs distributed as source code. All programs are built using a simple, standardized, two-step process. The program builder does not need to install any special tools in order to build the program. Therefore, building a program is normally as simple as running `configure` followed by make.

The `configure` command is a shell script that is used to check the details of the system in which a software is going to be installed. This script checks for dependences required by the particular software to work properly in the system. If any of the major requirements is missing, the `configure` script exits, and installation is not able to proceed until those required dependences are installed. The `configure` script supports a wide variety of options. The `--help` option can be used to get a list of options for a particular configure script. Two frequently used generic options are the `--prefix` and `--exec-prefix`, which are used to specify the installation directories. The directory defined by the `--prefix` option holds machine-independent files such as documentation, data, and configuration files. The directory defined by the `--exec-prefix` option is normally a subdirectory of the `--prefix` directory, which holds machine-dependent files such as executables. The main job of the configure script is to create a "Makefile", which contains various steps that need to be taken when compiling the software, depending on the results of checking performed by the configure script.

The `make` program is a utility available on almost all UNIX systems. It depends on the Makefile, which instructs on how to compile the software and where to install the finished compiled binaries (executables), manual pages, data files, dependent library files, configuration files, etc.

Compiling a Package

To install a package, one needs to carefully read the README and INSTALL text files, which contain important information on how to compile and run the software. Generally, the shell commands `./configure`; `make`; and `make install` should configure, build, and install the package, respectively.

The simplest way to compile a package is:

1) `cd` to the directory containing the package's source code. Type `./configure` to configure the package for values of system-dependent variables. If `configure` has run correctly, it will create a Makefile with all the necessary options.
2) Type `make` to compile the package. After this, the executables will be created. To check if everything is compiled successfully, type `make check` to run any self-tests that come with the package, generally using the just-built uninstalled binaries.
3) Type `make install` to install the programs and any data files and documentation. By default, `make install` installs the package's commands under "/usr/local/bin", and includes files under "/usr/local/include". The installation prefix can be specified other than "/usr/local" by giving `configure` the option `--prefix = PREFIX`, where `PREFIX` must be an absolute file name. Separate installation prefixes can be specified for architecture-specific files and architecture-independent files. If the option `--exec-prefix = PREFIX` is passed to `configure`, the package

uses `PREFIX` as the prefix for installing programs and libraries. Documentation and other data files still use the regular prefix.

4) Optionally, to remove the program binaries and object files from the source code directory, the `make clean` command can be executed.

Accessing a Remote Linux Supercomputer System

Many ways are available for users to access a Linux system, such as installing Linux on a personal computer, running a Linux virtual machine, and using a live CD to run a Linux system. However, to run most bioinformatics applications, users need to access remote Linux machines, such as supercomputer clusters that provide much larger computing resources.

To gain access to a remote Linux-based system, a user name, password, and hostname (or IP address) are required. Once the account information is available, remote access can be done from Linux, macOS, and Microsoft Windows systems.

Access Remote Linux from Local Linux System

To connect to a remote UNIX/Linux computer securely, the program `ssh` (e.g., secure shell) can be used as below:

```
$ ssh user@hostname
```

The "user name" and "host name" need to be replaced with your user name and machine name before running this command. After running this command, a prompt will show up to request for the password of the account that is being connected to. After typing in the password and pressing [Enter], the remote computer is accessed via the CLI. Once done, all operations can be performed as if you were sitting in front of the supercomputer. The disconnection can be done by typing "exit" and pressing [Enter]. It should be noted that `ssh` will only work if "Remote Login and File Sharing" is enabled on the computer that is being connected to.

Access Remote Linux from macOS

macOS includes an application called "Terminal", which is located in the "Applications >> Utilities" folder. As in the Linux system, the remote Linux computer can be connected by using Terminal in macOS. To access a remote Linux system, launch Terminal and type

```
$ ssh user@hostname
```

Replacing the "user" and "hostname" with your user name and machine name, press [Enter], then type in the password to establish the connection.

Access Remote Linux from Microsoft Windows

On Microsoft Windows systems, a variety of third-party tools can be used to connect to a remote Linux system. One of the popularly used tool is PuTTY, which is a free program and can be installed by downloading the executable from the PuTTY website (http://www.chiark.greenend.org.uk/~sgtatham/putty/). Launching PuTTY will open a configuration window. Click "Session" in the left pane, and then enter user@hostname in the text box "Host Name (or IP address)", replacing the "user" and "hostname" with your user name and machine name. Click "Open" to establish a connection with the remote Linux system.

Demonstration of Command Lines

Here, we discuss a number of frequently used Linux commands to illustrate how to use Linux command lines and how to install Linux programs from scratch using source code. The introduction and usage of BLAST programs are not detailed here, and are provided in Chapter 2.

The step-by-step demonstration of command lines:

1) Login the remote Linux account using ssh

   ```
   $ ssh aubsxl@dmc.asc.edu
   ```

2) Create a directory named "linuxDemo"

   ```
   $ mkdir linuxDemo
   ```

3) Go to the directory "linuxDemo"

   ```
   $ cd linuxDemo
   ```

4) Create a subdirectory named "blast"

   ```
   $ mkdir blast
   ```

5) Download the BLAST source code from the NCBI FTP site

   ```
   $ wget ftp://ftp.ncbi.nlm.nih.gov/blast/executables/
     blast+/LATEST/ncbi-blast-2.2.31+-src.tar.gz
   ```

6) Decompress the tarball package

   ```
   $ tar -xvzf ncbi-blast-2.2.31+-src.tar.gz
   ```

7) Go to the new folder ("ncbi-blast-2.2.31+-src") just created after decompression

   ```
   $ cd ncbi-blast-2.2.31+-src
   ```

8) Go to subdirectory c++

   ```
   $ cd c++
   ```

9) Configure the software package

   ```
   $ ./configure --prefix=/home/aubsxl/linuxDemo/blast
   ```

10) Compile the software package

    ```
    $ make
    ```

11) Install the binary executables to the defined directory

    ```
    $ make install
    ```

12) Attach the path of compiled BLAST executables to the environment path

```
$ echo 'export PATH=$PATH:/home/aubsxl/linuxDemo/blast/
  bin' >> ~/.bashrc
```

Further Reading

A large number of useful books and web tutorials are available for beginners to learn UNIX/Linux. Some of the excellent ones are listed here:

1) *Linux in a Nutshell*: http://shop.oreilly.com/product/9780596154493.do. A classical Linux book that thoroughly covers programming tools, system and network administration tools, the shell, editors, and LILO and GRUB boot loaders.
2) *UNIX Power Tools*: http://shop.oreilly.com/product/9780596003302.do. One of the best-selling books that provides vital information on Linux, Darwin, and BSD, and offers coverage of bash, zsh, and other new shells, along with discussions about modern utilities and applications.
3) *Linux Command Line and Shell Scripting Bible*: http://www.wiley.com/WileyCDA/WileyTitle/productCd-111898384X.html. An essential Linux guide with detailed instructions and abundant examples.
4) The UNIX tutorial from the Tutorialspoint: http://www.tutorialspoint.com/unix/index.htm.
5) UNIX Tutorial for Beginners: http://www.ee.surrey.ac.uk/Teaching/Unix/.
6) UNIX & Perl Primer for Biologists by Keith Bradnam and Ian Korf: http://korflab.ucdavis.edu/unix_and_Perl/. An online course written in a fun and accessible style, providing step-by-step guidance to inspire and inform non-programmers about the essential aspects of UNIX and Perl.

2

Determining Sequence Identities: BLAST, Phylogenetic Analysis, and Syntenic Analyses

Sen Gao, Zihao Yuan, Ning Li, Jiaren Zhang and Zhanjiang Liu

Introduction

For molecular biologists, the first task is most often to determine the identities of a DNA sequence. This task is becoming increasingly easier as databases grow with known gene sequences. The easiest and most convenient way for the determination of sequence identities is to conduct a Basic Local Alignment Search Tool (BLAST) analysis. However, because many sequences share similar levels of identities, BLAST analysis alone may not be sufficient. Additional analysis may be required for determining sequence identities. Phylogenetic analysis is often very useful to determine the relationships of many similar sequences. Even with phylogenetic analysis, sometimes it can still be difficult to draw concrete conclusions as to what exactly the gene is. Additional analysis such as syntenic analysis using conserved syntenic blocks may be required for determining the sequence identities. In this chapter, we will sequentially introduce approaches for determining sequence identities with BLAST, phylogenetic analysis, and syntenic analysis.

Determining Sequence Identities through BLAST Searches

BLAST was first introduced in 1990 (Altschull *et al.*, 1990). It is the most popular program used for molecular biology. The success of BLAST lies in its ability to search for similarities of the input sequence(s) (the query) to existing sequences in the databases, as specified by the researcher. Using nucleotide or protein sequences as queries, scientists can search against public or customized nucleotide or protein databases using web-based BLAST or UNIX-based BLAST. Web-based BLAST provides a user-friendly interface. Only copying and pasting the query sequences to a dialog box of BLAST programs or directly uploading your sequences saved in local computers as files to the NCBI server are required. However, the web-based BLAST is not practical when dealing with large numbers of sequences. UNIX-based BLAST, in contrast, is less user-friendly, but is more flexible and more powerful as any sequence files can be used as subject database. Actually, sequence files used as subject database in one study can be used as query in another, and vice versa.

Also, an "expect value" (E-value) to estimate matches that occurred by chances was provided to the users for evaluating the pairwise sequence alignment, given the

Bioinformatics in Aquaculture: Principles and Methods, First Edition. Edited by Zhanjiang (John) Liu.

statistical significance. Other parameters including, but not limited to, IDs of query and subject sequences, percentage of identical bases, numbers of mismatches, numbers of open gaps, start and end positions on query and subject sequences, respectively, and bit scores are all provided in the outputs of BLAST searches. Optimal combinations of parameter settings can be determined when using UNIX-based BLAST. The results of BLAST using various parameters, along with other analysis, helps the users to determine if the alignment parameters are suitable for their project, or if biological relationships revealed in the analysis is reliable. For instance, after taking considerations of identical alignment as well as alignment gaps, bit score was interpreted in a way that the higher the bit score, the better the alignment. Tabular, XML, and ASN.1, along with other formats available for downstream analysis, are provided. The tabular outputs of UNIX-based BLAST, which is generated using the "`-outfmt 6`" option on the command line, is the first choice for general analysis of BLAST results, because it is easy for parsing and is compatible with Microsoft Excel, which provides more choices for handling analysis. The XML format, for most situations, is used as inputs for Gene Ontology and enrichment analysis (Conesa *et al.*, 2005, see Chapter 10). The XML format can be easily transformed to a tabular format. Conveniently, the ASN.1 format is highly flexible, as it can be transformed into various other formats for downstream analysis.

Finally, different BLAST programs can be used, based on the nature of the query and target sequences (Table 2.1). For example, when using contigs assembled from RNA-Seq reads query against NCBI non-redundant database, BLASTX should be used. In BLASTX, nucleotide sequences are first translated into amino acid sequences in all six possible reading frames, and then sequence comparison and alignments are made.

Web-based BLAST

As web-based BLAST (http://blast.ncbi.nlm.nih.gov/Blast.cgi) is self-explanatory, we will only make a few comments here. First, select proper BLAST based on your query sequences (Table 2.1). When running a BLAST search, select the desired database. One can exclude or limit your search to a specific organism. Second, for coding sequences, protein databases are more useful. However, BLASTN can be quite useful when analyzing for short conserved regulatory sequences such as promoters and/or enhancers, or a binding site for a specific transcriptional factor.

After finishing BLAST, alignment distribution of all BLAST hits on the query sequences will be shown up using different colors representing alignment scores, with red color indicating the best alignments. Determination of proper cutoff values is

Table 2.1 Basic types of BLAST.

Name	Query	Subject
blastn	Nucleotide	Nucleotide
blastp	Protein	Protein
blastx	Translated nucleotide	Protein
tblastn	Protein	Translated nucleotide
tblastx	Translated nucleotide	Translated nucleotide

important, but is dependent on the purposes. To be able to include distantly related proteins at the first stages of analysis, we tend to set the cutoff E values high, for example, $E = 10^{-5}$, such that no sequence is excluded before further analysis. However, to conclude that the gene is the equivalent in your species to a known gene, a low E value should be desirable, for example, $E \leq 10^{-30}$.

Generally, web-based BLAST provides more graphical results than UNIX-based BLAST, and is easier to understand for beginners. Unlike UNIX-based BLAST, almost every Internet browser based on different operation systems can be used for web-based BLAST. However, considering the running speed of web-based BLAST (about 3~5 minutes per sequence), it is not suitable for the analysis of large datasets (e.g., *de novo* assembly software Trinity usually generates hundreds of thousands of sequences). Furthermore, subject databases cannot be customized based on the researcher's needs. Thus, we highly recommend using web-based BLAST only for querying small datasets against public databases.

UNIX-based BLAST

For genomics projects, tens of thousands of sequences from newly studied species need to be annotated via BLAST searches through homology-based annotation. It is very difficult to execute these types of tasks using web-based BLAST. UNIX-based supercomputers are an ideal solution for this challenge. For instance, it takes several days for us to perform BLAST searches using ~20000 sequences against a ~47 GB NCBI non-redundant database using a high-speed supercomputer with 128 allocated CPUs to our project. However, users need to be familiar with the basic UNIX command lines (see Chapter 1) and the queue system used by the HPC management to schedule and execute the computational jobs submitted by researchers. Here, we present the step-by-step analysis processes, from downloading public databases to parsing the output results, for new users who wish to conduct BLAST searches using UNIX-based supercomputer clusters.

Download the Needed Databases

Among all the public databases, NCBI non-redundant database and Uni-prot TrEMBL database are the two most comprehensive ones for BLAST searches. They can be downloaded by using these command lines:

```
wget ftp://ftp.ncbi.nlm.nih.gov/blast/db/FASTA/nr.gz
wget ftp://ftp.uniprot.org/pub/databases/uniprot/current_
  release/knowledgebase/complete/uniprot_trembl.fasta.gz
```

Setup Databases

The downloaded databases are compressed. They need to be decompressed before conducting BLAST searches. After downloading, use `gunzip` to decompress:

```
gunzip nr.gz
gunzip uniprot_tremble.fasta.gz
```

Before running local BLAST, the databases need to be formatted using these commands:

```
/path/to/makeblastdb -in nr -dbtype prot -out nr
/path/to/makeblastdb -in uniprot_tremble.fasta -dbtype
   prot -out uniprot_tremble
```

Considering the sizes of these two databases, they are submitted to the queue system of HPC (e.g., our HPC aforementioned, use NCBI nr database as example):

```
bsub /path/to/makeblastdb -in nr -dbtype prot -out
```

Execute BLAST Searches

Here, we use `blastp` as an example:

```
~/path/to/blastp -db ~/path/to/nr -query /path/to/query.fa
   -out /path/to/results -evalue 0.00001 -outfmt 6
   -max_target_seqs 1
```

Option description: "`-db`" means database to BLAST; "`-query`" means input query file name; "`-out`" means output results file name; "`-evalue`" means expectation value thresholds, here $E = 10^{-5}$; "`-outfmt 6`" means tabular outputs format; and "`-max_target_seqs 1`" means maximum number of aligned sequences to keep (for this example, only top one sequence with hit was kept).

For users who want to use more CPUs of the HPC, the following options can be used (e.g., 128 CPUs are requested):

```
-num_threads 128
```

This will allow you to use 128 CPUs to execute the job if the management allocates the available CPUs to your project.

Parsing the BLAST Results

For tabular outputs, there will be 12 columns, each representing one parameter in the following order: query sequence id, subject sequence id, percentage of identities, length of the matched sequences, number of mismatches, number of gaps, starting position of query sequences, ending position of query sequences, starting position of subject sequences, ending position of subject sequences, expectation value, and bitscore. These columns can be removed or kept using this command:

```
cut -f1,2,11,12 results
```

The numbers in the command represent each corresponding column in the output file. This example command would delete all the other columns but keep columns 1, 2, 11, and 12. For additional information, readers are referred to NCBI handbooks and BLAST help: http://www.ncbi.nlm.nih.gov/books/NBK143764/ and http://www.ncbi.nlm.nih.gov/books/NBK52640/.

Determining Sequence Identities through Phylogenetic Analysis

BLAST searches provide a quick assessment of the sequence identities. For the most part, if the gene of interest is a single copy gene without being a member of gene families,

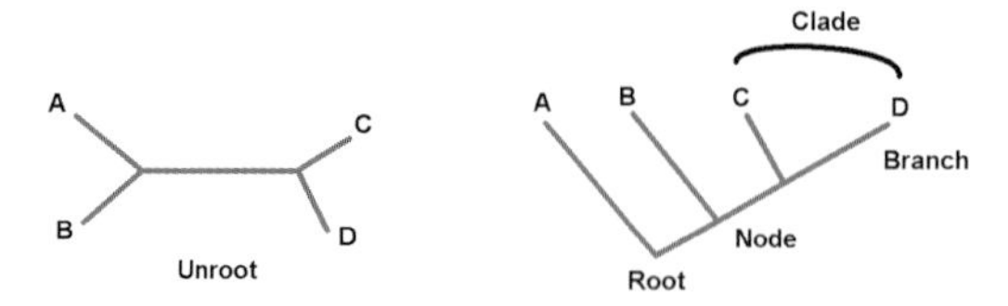

Figure 2.1 The unrooted tree (left) and rooted tree (right). The unrooted tree only shows the phylogenetic relationships, while the rooted tree shows the order formation of the species and common ancestors. The basic components of rooted phylogenetic trees include: branches (lines within the tree and equivalent to divergence); nodes (points where branches connect, or the tips of branches); roots (the basal node of the rooted tree); and clades (the group of the most recent common ancestor and all of its descendants).

or there is no duplication of the gene, BLAST searches are the most rapid approach for determination of the sequence identities. However, in most cases, especially with teleost fish, more than one homologous gene is present in the organism. As a result, BLAST is incapable of providing the identity to the sequence. Additional analysis is required. Phylogenetic analysis provides an alternative approach for a greater level of analysis of sequence identities, with their evolution relationships with all the known sequences existing in the databases.

Phylogenetic trees can be divided into unrooted trees and rooted trees, based on their topology (Figure 2.1). Unrooted trees illustrate only the relatedness of each species, while rooted trees reveal not only the relatedness of the species but also the most recent putative common ancestor; and the tree branch length represents the rough estimation of evolution time. For the root of the tree, it represents the ancestral lineage, and the tips of the tree represent the descendants of that ancestor. Moving from the root to the tips indicates moving forward in time. Splitting of a lineage signals a speciation event that separates the ancestral lineage to two or more daughter lineages. Horizontal lines represent evolutionary lineages changing over time; and the longer the branch, the larger is the degree of change. Nodes of each cluster can be separated into two types: external nodes, called "tips," and internal nodes. The external nodes usually display the gene or species names, whereas internal nodes represent putative ancestors for this cluster of descendants.

Procedures of Phylogenetic Analysis

Four steps are involved in a typical phylogenetic analysis (Figure 2.2). Notably, each of the steps is pivotal to the phylogeny analysis and should be given equal emphasis. Selection of different strategies at each step will lead to the generation of phylogenetic trees with distinct topologies.

Collecting Sequences

The first step of phylogenetic analysis is to collect homologous sequences to be used for the analysis. This is most often achieved through BLAST searches. The number of homologous sequences for the query can be very large after BLAST search. Therefore, representative sequences can be collected. For instance, several homologous sequences may exist for any given species, and all these sequences may be of interest. Homologous sequences may be present in a very large number of taxa that do not need to be all used

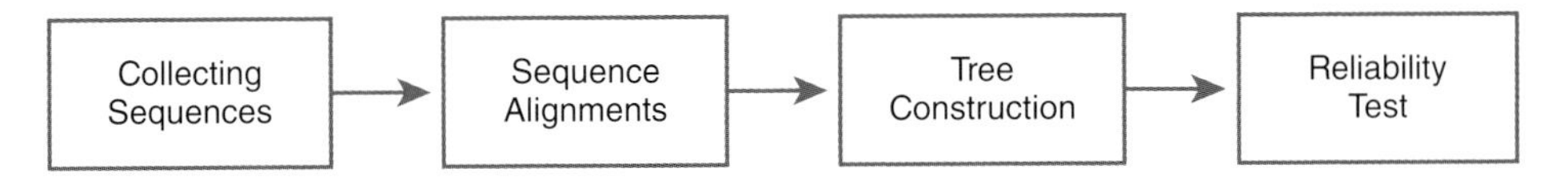

Figure 2.2 Four steps of phylogenetic tree construction.

for analysis. Rather, representative species from major taxonomic groups can first be chosen. For instance, from mammals, sequences from human, mouse, bovine, and swine species should be sufficient; from the bird group, sequences of chicken, turkey species may be sufficient; similarly, sequences from various other groups such as amphibians, reptiles, teleost fish, etc., can be selected. Because most aquaculture species are either fish (vertebrates) or shellfish (invertebrates), their relevant species for phylogenetic analysis depend on the species under study. Here, we will use fish as an example. Because the sequence of interest is from fish, known sequences from various fish species should also be included. For instance, homologous sequences from zebrafish, medaka, stickleback, catfish, rainbow trout, Atlantic salmon, etc., should be included for phylogenetic analysis. All the selected sequences can be retrieved from NCBI databases. The sequences can be either nucleotide or protein sequences. In most cases, protein sequences are used because they are more conserved.

Multiple Sequences Alignments

After the collecting of sequences used for phylogenetic analysis, the next step is to conduct multiple sequence analysis. This can be achieved by using any of the software currently available for multiple sequence alignments such as Clustal, Kalign2, MAFFT, MUSCLE, T-COFFEE, and Web-PRANK (Table 2.2). The most popular used software for multiple sequence alignments is probably Clustal. Clustal has been integrated into some Windows-based packages such as MEGA that is very user-friendly.

The format of input protein sequences is FASTA. Click on the "align" icon and select "edit/build alignment"; then select the "create a new alignment". After that, the software will ask for the input file format. After selecting the button of protein, an Alignment Explorer will show up; press the "open FASTA format" button to open protein FASTA file. After that, the software will show all the sequences colored by amino acids. To execute sequence alignment, users can directly click "align by Clustal" and use default parameters. To save the alignment results, users can click on "Data", then select "Export Alignment"; after that, three formats can be selected—"MEGA", "FASTA", and "PAUP"—before the construction of a phylogeny tree.

Tree Construction

The most common software for phylogenetic tree construction is MEGA (molecular evolutionary genetics analysis). It is a phylogenetic analysis software developed by the University of Pennsylvania. The current release version is Version 6 (Tamura *et al.*, 2013), and it is freely available at its website at: http://www.megasoftware.net. It carries out distance-matrix and character-based methods for nucleic acid sequences and protein sequences with Neighbor-Joining, Minimum Evolution, Maximum Likelihood method, UPGMA, and Maximum Parsimony method. In order to construct a phylogenetic tree, we can select "Phylogeny" and use the NJ tree as an example: for amino acid analysis, we

Table 2.2 Commonly used multiple sequence alignment software.

Name	Description	Official website and references
Clustal Omega	Suitable for medium-to-large alignments and divergent sequences. Makes use of multiple processors	http://www.clustal.org/omega/ (Sievers *et al.*, 2011)
Kalign2	Allows for external sequence annotation to be included. Fast, memory-efficient, and accurate for aligning large numbers of sequences	http://msa.sbc.su.se/ (Lassmann, Frings & Sonnhammer, 2009)
MAFFT	Implemented with two methods (FFT-NS-i, FFT-NS-2). Suitable for medium number of sequences	http://mafft.cbrc.jp/alignment/server/ (Katoh *et al.*, 2002)
MUSCLE	Algorithm includes fast distance estimation using k-mer counting, alignment using the log expectation score, refinement using tree-dependent restricted partitioning	http://www.drive5.com/muscle/ (Edgar, 2004)
T-Coffee	Combine results from several sources such as sequence alignment, structure alignment, threading, manual alignment, motifs, and specific constraints	http://www.tcoffee.org/ (Notredame, Higgins & Heringa, 2000)
WebPRANK	Easy-to-use interface. Visualization and supporting post-processing of the results. Makes use of evolution information to place insertions and deletions	http://www.ebi.ac.uk/goldman-srv/webprank/ (Löytynoja & Goldman, 2010)

usually select the Poisson model, and the number of bootstraps is usually set as 1000; then click "compute" to initiate the analysis. After the analysis is completed, the software will construct the phylogenetic tree. To export this tree, we can click on the "Image" button and choose to copy to the clipboard or save as a picture file. To emphasize the gene you are interested in, you can click on "Option" and select the labels you prefer. We can also select different tree branch styles by clicking on "Tree/Branch style". Several types of phylogeny trees can be viewed using MEGA, including rectangular, circular, and radiation. In fact, each cluster can be swapped round at internal node and topology of the tree. An example of a rectangular phylogenetic tree is shown in Figure 2.3.

Selection of a proper model for the construction of a phylogenetic tree depends on the type of data. The construction of the phylogenetic tree can be divided into two basic categories: distance-matrix methods and character-based methods. Fixed pairwise distances are computed for derivation of trees, and distribution of actual data patterns are optimized for the construction of phylogenetic trees. The most applied distance-based methods include unweighted pair group method with arithmetic mean (UPGMA), neighbor joining (NJ), minimum evolution (ME) method, and the Fitch-Margoliash method, while maximum parsimony and maximum likelihood are most often used for character-based methods (for references and characteristics of each of these models, see Table 2.3).

For distance-matrix methods, at least one out-group sequence should be included in the analysis. The out-group can be seen as a type of control group. The out-group should

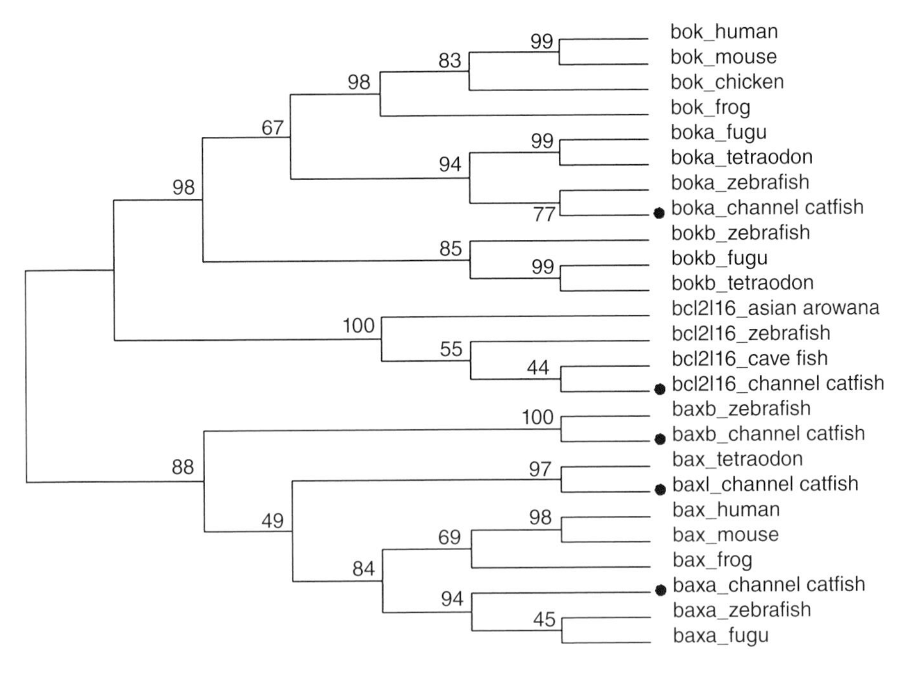

Figure 2.3 An example of a phylogenetic tree. Numbers above the nodes are bootstrapping values.

Table 2.3 The phylogenetic tree construction method.

	Name	Brief description and reference
Distance-matrix methods	Unweighted pair group method with arithmetic mean (UPGMA)	First and simplest phylogeny analysis algorithm (Sneath & Sokal, 1973)
	Neighbor joining (NJ)	Fast and accurate (Saitou & Nei, 1987)
Character-based methods	Maximum parsimony (MP)	Highly homologous sequences required (Farris, 2008; 1983)
	Maximum likelihood (ML)	Accurate, calculates the probability of each topology for fixed sequence order
	Bayesian methods	Accurate, calculates the probability of each OTU order for fixed topology

be the sequences that are distantly related to the query sequences. The out-group should have a longer branch length than other sequences, and should appear at the root of the rooted tree (Mount, 2001).

Reliability Test

To test the reliability of the topology of the phylogenetic tree, the bootstrapping method is a reasonable method and is commonly used. Since its introduction in 1979 (Efron, 1979), it has been widely applied to statistical analysis. The basic principle is to choose some of the sequence in the query and test if the same nodes can be recovered in iterated analysis. This is done through much iteration (e.g., 1000). Bootstrapping values are displayed on the nodes, and they are the percentages of the inferred trees that are the same as the trees built with all query sequences. For example, if the same nodes can be generated in 95 out of 100 iterations by resampling your tree, the node is well supported (the bootstrap value in that case would be 95).

Other Software Available for Phylogenetic Analysis

Many software packages are available for phylogenetic analysis. Many of these are listed on the web link http://evolution.genetics.washington.edu/phylip/software.html. Currently, 392 phylogeny packages and 54 free web servers are included in this web link, and they are classified by methods, supporting systems, the types of data they can process, etc.

Determining Sequence Identities through Synthetic Analysis

In most cases, sequence identities can be determined with BLAST analysis or through phylogenetic analysis. However, in some cases, even after phylogenetic analysis, the exact identity of a gene still cannot be determined, especially when dealing with genes with various types of duplications. Syntenic analysis is useful as one additional approach, especially for the determination of orthologies.

Synteny (Greek: Along with) is a term describing the phenomenon that different species possess common chromosomal sequences. The modern definition of synteny is: the physical co-localization relationship between two genetic loci represented on the same chromosomal pair that are being compared with each other, or (for haploid chromosomes) on the same chromosome within an individual or species. Synteny is important in comparative genomics for understanding genome evolution. It is widely used to identify closely related genes, especially for the analysis of orthologous and paralogous relationships. Although sometimes the terms *synteny* and *collinear* are mixed up, they have different meanings. Synteny describes a set of loci from different species located on the same chromosome, and the loci are not necessarily in the same order; collinear describes a set of loci from different species not only located on the same chromosome but also the loci are in the same order. In other words, the collinear is a special state of synteny where the loci are in the same order (Tang *et al.*, 2008). The differences are shown in Figure 2.4. In the literature, conserved syntenic blocks are widely used.

Based on the extent of conserved regions, synteny can be divided into *macrosynteny* and *microsynteny*. As reflected in the names, macrosynteny is the form of synteny

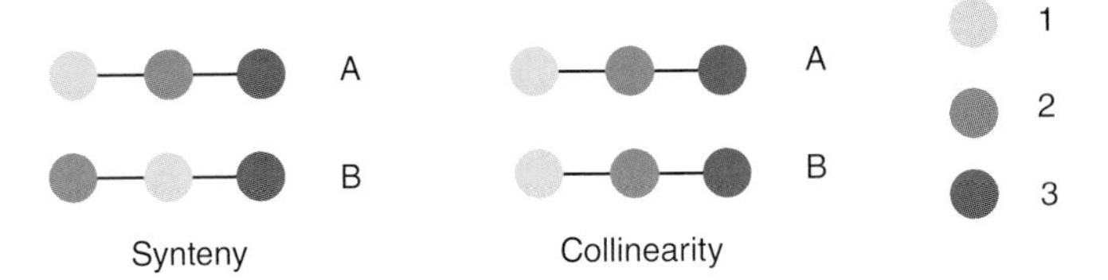

Figure 2.4 The comparison between synteny and collinearity. For species A and B, as long as gene 1, 2, 3 are on the same genomic neighborhood of a chromosome, they are called *synteny*. However, they have to be in the same order to be called *collinearity*.

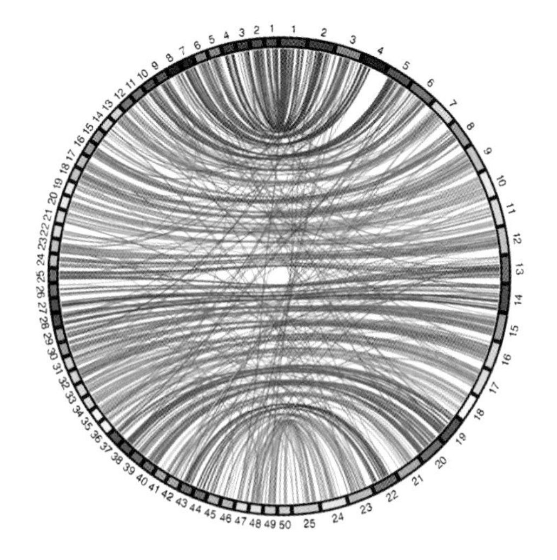

Figure 2.5 Macrosynteny blocks between common carp and zebrafish. The synteny analysis between 50 common carp chromosomes (left) 25 zebrafish chromosomes (right) demonstrates highly conserved syntenies between common carp and zebrafish, and the whole genome duplication of the common carp genome (Xu *et al.*, 2014).

involving a large number of genes in an extended chromosomal region. In contrast, microsynteny involves only a short segment of the chromosome region with limited number of conserved genes between/among species. Macrosynteny analysis is useful for an understanding of the entire genome organization. For example, the common carp (*Cyprinus carpio*) has 50 chromosomes, twice as many as those in zebrafish, making genome structure difficult to be analyzed (Postlethwait *et al.*, 2000; Xu *et al.*, 2014). Macrosynteny made it possible to study the common carp genome based on the known zebrafish genome. The macrosynteny analysis between carp and zebrafish chromosomes suggested an additional round of whole-genome duplication of the common carp genome (Xu *et al.*, 2014; Figure 2.5). In contrast, microsynteny analysis is widely used for the identification of genes, especially for the analysis of orthologies.

Procedures for Synteny Analysis

The purpose of syntenic analysis is to provide evidence for a gene that is orthologous to another gene. In this context, if the neighboring genes are all conserved between the gene under study and a known gene, that would provide very strong evidence that the gene under study is orthologous to the known gene. Here, we will illustrate the procedures of syntenic analysis using the catfish bcl2l12 (B-cell lymphoma 2-like 12) gene as an example.

First, the chromosome regions containing the gene under study need to be identified from your species and from the other species that will be compared. This can be

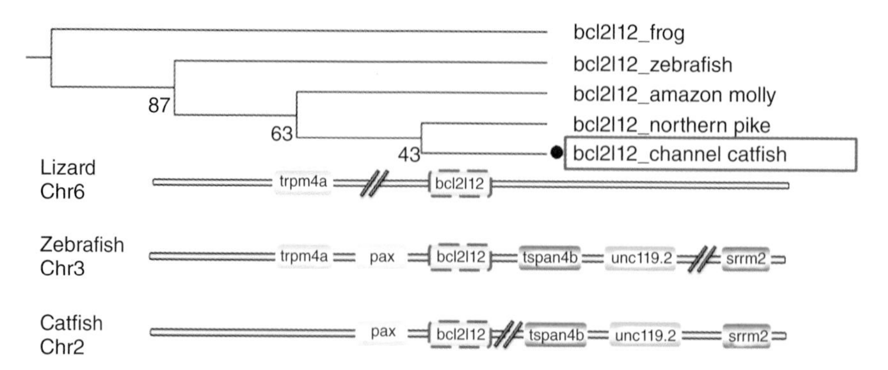

Figure 2.6 The phylogenetic and synteny analysis of bcl2l12 in channel catfish, zebrafish, and lizard. In order to identify a novel gene bcl2l12 in channel catfish (in red box), the phylogenetic tree of the bcl2l12 genes is made. The channel catfish has the closest relationship with zebrafish. However, in the phylogenetic tree, the channel catfish bcl2l12 and zebrafish bcl2l12 are not in the same cluster as expected. The synteny analysis shows that bcl2l12 in channel catfish and bcl2l12 in zebrafish share some same-neighbor genes both upstream and downstream of the gene, indicating the channel catfish bcl2l12 and the zebrafish bcl2l12 are orthologous.

accomplished by simple BLAST search against the reference genomic sequences. Second, the neighboring genes from these species need to be identified. This can be easily achieved by extending a distance, for example, 1 Mb, on both sides of the gene along the chromosome. Third, gene orders and orientations are determined. In the example here, the chromosome location of the catfish bcl2l12 gene can be determined by BLAST search using "bcl2l12 gene" as a query. It is located on chromosome 2. Its neighbor genes were identified to include trpm4a, pax, tspan4b, and unc119.2 (Figure 2.6). For zebrafish, the location of bcl2l12 and its neighbor genes can be directly visualized from ENSEMBL. More conveniently, through the web-based Genomicus (http://www.genomicus.biologie.ens.fr/genomicus-82.01/cgi-bin/search.pl), simply type in the gene name, and select the species in the windows of Genomicus, and the neighboring genes are exhibited in graphical views. Once the neighboring genes from the species of interest are all known, determination of the syntenic relationship can be made. Conservation of all or most of the genes neighboring the gene of interest suggests conserved syntenies or orthologies (Figure 2.6).

For macrosynteny analysis, MCScanX can be used. Once again, we will demonstrate this process using an example. For instance, putative homologous chromosomal regions can be identified between zebrafish and stickleback using this software. Before running MCScanX, BLAST all-versus-all should be conducted. This should be a process that includes searching all genes of zebrafish against all genes of stickleback and vice versa, searching zebrafish against itself, and searching stickleback against itself. Amino acid sequences are recommended, and all the aforementioned processes can be done using `blastp`. Other required input files are the gene coordinates files, which can be directly downloaded from the ENSEMBL database. Newly released profiles can be found at: ftp://ftp.ensembl.org/pub/release-83/gff3. Finally, BLAST result files and gene coordinates files should be put in the same directory, and the following command line should be used to run MCScanX:

```
/path/to/MCScanX /path/to/directory
```

Table 2.4 Commonly used synteny analysis software and their characteristics.

Software and references	Feature
SynMap (Lyons *et al.*, 2008; 2011)	• A tool from Coge (Comparative Genomics) • Generate syntenic dot plot of syntenic blocks between two organisms • Identify putative genes or homologous regions between two genomes • Calculate synonymous and non-synonymous site mutation data for syntenic protein-coding genes
Cinteny (Sinha & Meller, 2007)	• Identify syntenic regions across genomes • Assess extent of genome rearrangement • Measure evolutionary relationships of genomes • User can upload sequence or select genome
MCScanx (Multiple Collinearity Scan toolkit) (Wang *et al.*, 2012)	• Scan multiple genomes or subgenomes • The distance between genes are calculated in terms of gene differences in addition to base positions, or user can set distances between genes • Duplication depth at each locus and tandem genes are marked in the synteny result
Symap (Synteny Mapping and Analysis Program) (Soderlund, Bomhoff & Nelson, 2011; Soderlund *et al.*, 2006)	• Detect, display, and query syntenic relationships between sequenced genome sequences in a graphic user interface • Define the orientation of draft sequence by synteny • Multiple display modes (dot plot, circular, side-by-side, closeup, 3D) • Analyze cross-species gene families • Align fingerprinted contigs maps to sequenced genomes

This can be finished within minutes, and the format of the output file is html. There will be one file generated for each chromosome of species studied, and it can be viewed using Google Chrome. MCScanX is suitable for large gene families, for example, those with more than 50 members. Detailed instructions can be found at: http://chibba.pgml.uga.edu/mcscan2/.

In addition to MCScanX, several other software packages are available for the analysis of macrosyntenies. These include SynMap, Cinteny, and Symap. Their characteristics and references are included in Table 2.4.

References

Altschull, S.F., Gish, W., Miller, W. *et al.* (1990) Basic local alignment search tool. *Journal of Molecular Biology*, **215**, 403–410.

Conesa, A., Götz, S., García-Gómez, J.M. *et al.* (2005) Blast2GO: a universal tool for annotation, visualization and analysis in functional genomics research. *Bioinformatics*, **21**, 3674–3676.

Edgar, R.C. (2004) MUSCLE: multiple sequence alignment with high accuracy and high throughput. *Nucleic Acids Research*, **32**, 1792–1797.

Efron, B. (1979) Bootstrap methods: another look at the jackknife. *The Annals of Statistics*, **7**, 1–26.

Farris, J. (1983) The logical basis of phylogenetic analysis. *Advances in Cladistics*, **2**, 7–36.
Farris, J.S. (2008) Parsimony and explanatory power. *Cladistics*, **24**, 825–847.
Katoh, K., Misawa, K., Kuma, K.i. and Miyata, T. (2002) MAFFT: a novel method for rapid multiple sequence alignment based on fast Fourier transform. *Nucleic Acids Research*, **30**, 3059–3066.
Lassmann, T., Frings, O. and Sonnhammer, E.L.L. (2009) Kalign2: high-performance multiple alignment of protein and nucleotide sequences allowing external features. *Nucleic Acids Research*, **37**, 858–865.
Löytynoja, A. and Goldman, N. (2010) webPRANK: a phylogeny-aware multiple sequence aligner with interactive alignment browser. *BMC Bioinformatics*, **11**, 579.
Lyons, E., Freeling, M., Kustu, S. and Inwood, W. (2011) Using genomic sequencing for classical genetics in E. coli K12. *PLoS One*, **6**, e16717–e16717.
Lyons, E., Pedersen, B., Kane, J. and Freeling, M. (2008) The value of nonmodel genomes and an example using SynMap within CoGe to dissect the hexaploidy that predates the rosids. *Tropical Plant Biology*, **1**, 181–190.
Mount, D.W. (2001) *Bioinformatics: sequence and genome analysis*, vol. **2**, Cold spring harbor laboratory press, New York.
Notredame, C., Higgins, D.G. and Heringa, J. (2000) T-Coffee: A novel method for fast and accurate multiple sequence alignment. *Journal of Molecular Biology*, **302**, 205–217.
Postlethwait, J.H., Woods, I.G., Ngo-Hazelett, P. *et al.* (2000) Zebrafish comparative genomics and the origins of vertebrate chromosomes. *Genome Research*, **10**, 1890–1902.
Saitou, N. and Nei, M. (1987) The neighbor-joining method: a new method for reconstructing phylogenetic trees. *Molecular Biology and Evolution*, **4**, 406–425.
Sievers, F., Wilm, A., Dineen, D. *et al.* (2011) Fast, scalable generation of high-quality protein multiple sequence alignments using Clustal Omega. *Molecular Systems Biology*, **7**, 539.
Sinha, A.U. and Meller, J. (2007) Cinteny: flexible analysis and visualization of synteny and genome rearrangements in multiple organisms. *BMC Bioinformatics*, **8**, 82.
Sneath, P.H.A. and Sokal, R.R. (1973) Unweighted pair group method with arithmetic mean, in Numerical Taxonomy, W. H. Freeman, San Francisco, pp. 230–234.
Soderlund, C., Bomhoff, M. and Nelson, W.M. (2011) SyMAP v3. 4: a turnkey synteny system with application to plant genomes. *Nucleic Acids Research*, **39** (10), e68.
Soderlund, C., Nelson, W., Shoemaker, A. and Paterson, A. (2006) SyMAP: a system for discovering and viewing syntenic regions of FPC maps. *Genome Research*, **16**, 1159–1168.
Tamura, K., Stecher, G., Peterson, D. *et al.* (2013) MEGA6: molecular evolutionary genetics analysis version 6.0. *Molecular Biology and Evolution*, **30**, 2725–2729.
Tang, H., Bowers, J.E., Wang, X. *et al.* (2008) Synteny and collinearity in plant genomes. *Science*, **320**, 486–488.
Wang, Y., Tang, H., DeBarry, J.D. *et al.* (2012) MCScanX: a toolkit for detection and evolutionary analysis of gene synteny and collinearity. *Nucleic Acids Research*, **40**, e49.
Xu, P., Zhang, X., Wang, X. *et al.* (2014) Genome sequence and genetic diversity of the common carp. *Cyprinus carpio. Nature Genetics*, **46**, 1212–1219.

3

Next-Generation Sequencing Technologies and the Assembly of Short Reads into Reference Genome Sequences

Ning Li, Xiaozhu Wang and Zhanjiang Liu

Introduction

In this chapter, the various sequencing technologies are first introduced, and then the methods for the assembly of sequences generated using various sequencing platforms are presented.

Understanding of DNA Sequencing Technologies

The start of the Human Genome Project (HGP) 30 years ago marks the start of a new era in genomics. For the most part, the genomics revolution has been driven by advances in DNA sequencing technologies, including the development of automated DNA sequencers and powerful data-processing algorithms. Sanger sequencing (Sanger *et al.*, 1977), considered the “first-generation” DNA sequencing technology, dominated the field for almost 20 years till the mid-1990s. Since then, in the last 15–20 years, rapid progress was made in sequencing technologies. The new set of sequencing technologies since has been collectively known as the “next-generation” sequencing (NGS) technologies. Three platforms of NGS were initially available in the market: the Roche/454 FLX, the Illumina sequencers, and the Applied Biosystems SOLiD System. With years of competition in the sequencing market, the Illumina platform is now among the most popular platforms. In addition, several additional sequencing platforms were introduced to the market, such as Helicos Heliscope, Pacific Biosciences SMRT, Ion Torrent Personal Genome Machine (PGM), and Oxford Nanopore MinION sequencers.

Compared to Sanger sequencing, next-generation sequencing produces massive amounts of data rapidly and cheaply (Mardis, 2013). A typical cloning step in Sanger method is replaced in next-generation sequencing by adding universal adapters to the DNA fragments (Mardis, 2011). There is also no need to perform sequencing reactions in microtiter plate wells. Instead, the library fragments are amplified *in situ* on a solid surface, which is covalently derivatized with adapter sequences. Another difference is that next-generation sequencing conducts sequencing and detection simultaneously (Dolník, 1999). These steps allow hundreds of millions to billions of reaction loci to

Bioinformatics in Aquaculture: Principles and Methods, First Edition. Edited by Zhanjiang (John) Liu.

be sequenced per instrument run. Due to the signal-to-noise ratio, the read length of next-generation sequencing is generally short. However, as detailed in the following text, long reads can now be achieved with the third generation of sequencers such as PacBio. The shorter read length is a major shortcoming of next-generation sequencing for large and complex genome sequencing (Alkan, Sajjadian & Eichler, 2011). To overcome this disadvantage, various tactics such as paired-end and mate-pair sequencing can be applied, which help the assembly of short sequences into contigs and scaffolds, as detailed in the following text.

Based on the chemistry, next-generation sequencing technologies can be classified into five groups (Shendure *et al.*, 2004): pyrosequencing (454 Life Sciences), sequencing by synthesis (Illumina), sequencing by ligation (SOLiD), semiconductor sequencing (Ion Torrent), and single-molecule real-time sequencing (PacBio, Helicos and Oxford Nanopore). Here, we provide some introduction of these platforms because their characteristics are related to the design of a genome-sequencing project, and to the sequence assembly strategies.

454 Life Sciences

Roche 454 sequencing system was the first next-generation sequencing platform available in the market (Margulies *et al.*, 2005). It was founded by Jonathan Rothberg in 2005, and bought by Roche Diagnostics in 2007. In this platform, libraries are prepared either paired-end or mate-pair to obtain a mixture of short, adapter-flanked fragments. The major characteristic of the 454 system is emulsion PCR (Dressman *et al.*, 2003), which is used to generate clonal sequencing features with amplicons captured to the surface of 28 μm beads. After amplification, the emulsion shell is broken, and successfully amplified beads are enriched through binding with streptavidin magnetic beads. A sequencing primer is hybridized to the universal adapter at the appropriate position and orientation, that is, immediately adjacent to the start of unknown sequences. The Pico Titer Plate is loaded with one fragment carrying bead per well and smaller beads with the enzymes necessary for sequencing and for packing. Sequencing is performed by the pyrosequencing method (Ronaghi *et al.*, 1996). When ATP drives the luciferase-mediated conversion of luciferin to oxyluciferin, a burst of light is generated in amounts that are proportional to the amounts of ATP. Light is captured by CCD camera and recorded as a peak "proportional" to the number of nucleotides incorporated.

Relative to other next-generation platforms, the key advantage of the 454 platform is its long read length (Metzker, 2010). For example, the read length generated from a 454 instrument can reach up to 700 bp (base pairs). A major limitation of the 454 technology relates to homopolymer repeats (multiple bases of the same identity). Because there is no terminating moiety preventing multiple consecutive incorporations at a given cycle, the length of all homopolymer repeats must be inferred from the signal intensity. As a consequence, the dominant error type for the 454 system is insertion/deletion, rather than base substitution (Fox *et al.*, 2014). Other disadvantages include relatively low sequence coverage and relatively high price. In October 2013, 454 Life Sciences announced the decision to phase out their pyrosequencing-based instruments (the GS Junior and GS FLX+) in mid-2016 (http://www.biocompare.com/Editorial-Articles/155411-Next-Gen-Sequencing-2014-Update/).

Illumina Genome Analyzer

Previously known as "the Solexa", the Illumina platform was originally introduced by Turcatti and colleagues (Fedurco *et al.*, 2006; Turcatti *et al.*, 2008). This platform has been the most popular next-generation sequencing platform in the market since it took over the market from 454 Life Science in 2012. Illumina Genome Analyzer is a sequencing-by-synthesis-based platform (Mardis, 2013). Libraries can be constructed by any method as long as a mixture of adapter-flanked fragments up to 100–300 bp in length can be obtained. Sequences are amplified through cluster generation, which is performed on the Illumina Cluster Station (or Cbot) (Fedurco *et al.*, 2006). In brief, DNA fragments are individually attached to the flow cell by hybridizing to oligos on its surface complementary to the ligated adapters. DNA molecules are amplified by bridge PCR, and then reverse strands are cleaved and washed away in the end. After cluster generation, the amplicons are single-stranded (linearization), and a sequencing primer is hybridized to a universal sequence flanking the region of interest. Clusters are sequenced simultaneously, and each cycle of sequence interrogation consists of single-base extension with a modified DNA polymerase and a mixture of four nucleotides. After each synthesis step, the clusters are excited by a laser, which causes fluorescence of the last incorporated base fluorescence label. The fluorescence signal is captured by a built-in camera, producing images of the flow cell. Then the blocking groups are removed, allowing addition of the next base. This process continues for every base to be sequenced. The read length has increased from the original 25 bp (single-end reads) to the current 150 bp (paired-end reads) using the Illumina HiSeq 2500 instrument. The dominant error type is base substitution, rather than insertion/deletion as those in 454 Life Science (Fox *et al.*, 2014). The average raw error rate of most Illumina reads is approximately 0.5% at best, but higher accuracy bases with error rates of 0.1% or less can be identified through quality metrics associated with each base calling (Mardis, 2013). Low expense, high throughput, and high coverage have made Illumina secure roughly 60% of the next-generation sequencing market. However, its major disadvantage is the relatively short read length, making it less effective for complex *de novo* assembly projects (Metzker, 2010). HiSeq 3000/4000 instruments were introduced to the market in early 2015. The dual flow-cell HiSeq 4000 can generate up to 1.5 TB of 150 bp pair-end reads (750 GB for HiSeq 3000). The major difference between the HiSeq 4000/3000 and earlier models is their flow-cell design. Instead of using non-patterned flow cell in which sequencing clusters could form anywhere on the surface, the HiSeq 4000/3000 instruments use a patterned flow cell in which sequencing clusters are restricted to 400 nm wells spaced 700 nm apart (http://www.biocompare.com/Editorial-Articles/171872-Next-Gen-DNA-Sequencing-2015-Update/).

SOLiD

The SOLiD platform is based on sequencing by oligo ligation. It was invented by Shendure and colleagues at Church's lab and later acquired by Applied Biosystems in 2006 (now owned by Thermo Fisher Scientific). Same as the 454 platform, clonal sequencing features are generated by emulsion PCR, with amplicons captured to the surface of 1 μM paramagnetic beads (Shendure & Ji, 2008). After emulsion PCR, beads bearing templates are selectively recovered, and then immobilized to a solid surface to generate a dense, disordered array. Sequencing is driven by a DNA ligase: first, a universal primer is hybridized and ligated to the array of amplicon-bearing beads, and then a set of four fluorescently labeled di-base probes compete for ligation to the sequencing primer (Housby & Southern, 1998; Macevicz, 1998; McKernan *et al.*, 2012; Shendure *et al.*, 2005). Each cycle of sequencing involves the ligation of a degenerate population of fluorescently labeled octamers. After the first ligation cycle, the extended primer is denatured to reset the system with a primer complementary to the $n - 1$ position for a second round of ligation cycles. Multiple cycles of ligation, detection, and cleavage are performed, and the number of cycles determines the eventual read length.

The efficiency and accuracy of SOLiD is enhanced by the use of two-base encoding, which is an error-correction scheme in which two adjacent bases, rather than a single base, are associated with one color (McKernan *et al.*, 2012). Each base position is then queried twice, so that miscalls can be more readily identified. High accuracy up to 99.99% can be achieved through this step. Similar to Illumina, the dominant error type is base substitution (Fox *et al.*, 2014). With its high levels of accuracy, SOLiD is the best for genome resequencing or SNP capture. Short read length is the major disadvantage for this platform. An additional disadvantage, such as that with the Roche 454 system, is that emulsion PCR can be cumbersome and technically challenging.

Helicos

The Helicos sequencer (Harris *et al.*, 2008) is based on the work of Quake and colleagues (Braslavsky *et al.*, 2003). It is the first commercial machine to sequence a single DNA molecule rather than one copied many times. Libraries are prepared by random fragmentation and poly-A tailing, then captured by hybridization to surface-tethered poly-T oligomers to yield a disordered array of primed single-molecule sequencing templates. DNA polymerase and a single species of fluorescently labeled nucleotide are added in each cycle, resulting in template-dependent extension of the surface-immobilized primer–template duplexes. A highly sensitive fluorescence detection system is used to directly interrogate individual DNA molecules via sequencing by synthesis. After collection of images tiling the full array, chemical cleavage and release of the fluorescent label allow the subsequent cycle of extension and imaging (Shendure & Ji, 2008).

Homopolymer repeat runs are the major problem of Helicos, almost like the situation of the Roche 454 system. This is because, with Helicos, there is no terminating moiety present on the labeled nucleotides. However, the problem is less serious as compared to Roche 454 because individual molecules are being sequenced, and the problem can be mitigated by limiting the rate of incorporation events. The dominant error type is insertion/deletion due to the incorporation of contaminated, unlabeled, or non-emitting bases (Fox *et al.*, 2014).

Ion Torrent

Ion Torrent, founded by Rothberg and collogues, was commercialized in 2010 and later purchased by Life Technologies™ Corp, which in turn was acquired by Thermo Fisher Scientific in February 2014. Thus, Thermo Fisher Scientific now owns both SOLiD and Ion Torrent. The Ion Torrent platform is based on the detection of the release of hydrogen ions, a by-product of nucleotide incorporation, as quantitated by changes in pH through a novel coupled silicon detector (Rothberg *et al.*, 2011; Mardis, 2013). Genomic DNA is fragmented, ligated to adapters, and then clonally amplified onto beads using emulsion PCR. After emulsion breaking, template-bearing beads are enriched through a magnetic-bead-based process and primed for sequencing. This mixture is then deposited into the wells of an Ion Chip, a specialized silicon chip designed to detect pH changes within individual wells of the sequencer as the reaction progresses stepwise. The upper surface of the Ion Chip is designed as a microfluidic conduit to deliver the reactants needed for the sequencing reaction. The lower surface of the Ion Chip interfaces directly with a hydrogen ion detector to translate the released hydrogen ions from each well into a quantitative readout of nucleotide bases in each reaction step.

Same as 454, Helicos, and PacBio, the major error type of Ion Torrent sequencing is caused by insertion/deletion due to homopolymer repeats, although base substitution does occur, but at a low frequency (Fox *et al.*, 2014). Overall, the error rate of Ion Torrent on a per-read basis averages approximately 1% (Mardis, 2013). The average read length obtained by the Ion Torrent has increased from originally 50 bp to 400 bp, produced as single-end reads. Throughput has increased from 10 MB per run to 1 GB per run, on average. Reaction volume miniaturization and mass production of the Ion Chips make this platform relatively fast and inexpensive, and hence ideal for small-scale applications or small laboratories that do not require large datasets.

PacBio

PacBio was commercialized by Pacific Biosciences in 2010. It is based on single-molecule sequencing, and generally regarded as third-generation technology. It is a successful combination of nanotechnology with molecular biology and highly sensitive fluorescence detection to achieve single-molecule DNA sequencing (Eid *et al.*, 2009). On this platform, by applying a special nanotechnology called *zero-mode waveguide* (ZMW), tens to thousands of ZMWs can be fitted in a light-focusing structure as a regular array (Mardis, 2013). With fragmented genomic DNA, after polishing their ends, the hairpin adapters are ligated onto the ends of these fragments. Once ligated and denatured, these molecules can form DNA circles, and at the same time reaction by-products should be removed before binding library fragments with DNA polymerase molecules through a primer complementary to the adapter. Then, these mixed molecules are deposited onto the surface of the ZMW chip (SMRT Cell). The instrument excitation/detection optics can be trained on the bottom of each ZMW, where the polymerase attaches. Once fluorescent nucleotides are added to the chip surface, the instrument optics collect the sequencing data by actively monitoring each ZMW to record the incorporating fluorescent molecules dwelling in the active site as identified by their emission wavelength. Here, the highly sensitive fluorescence detection is aimed at single-molecule detection of fluorescently labeled nucleotide incorporation events in real time, as the DNA polymerase is copying the template.

Depending on the size of the library fragments and the time of data collection, this platform produces some of the longest average read lengths available in the market. The average read length is currently up to 10 KB, while some reads can reach 85 KB using a new sequencing chemistry called P6-C4. However, an overall high error rate is the major disadvantage of PacBio. On a per-read basis, the error rate is approximately 10–15% (Mardis, 2013). The dominant error type is insertion/deletion, although a small proportion of base substitution does occur (Fox *et al.*, 2014). This platform is mainly used in assembling bacterial genomes and gap filling for sequencing projects with large genomes.

Oxford Nanopore

Oxford Nanopore platforms (the GridION and MinION), first announced by a UK startup Oxford Nanopore Technologies in 2012 (Eisenstein, 2012), are label-free and light-free exonuclease sequencing platforms. Library preparation is similar to that for other next-generation sequencing applications, including DNA shearing, end repair, adapter ligation, and size selection. The library must be conditioned by addition of a motor protein, and then mixed with buffer and a "fuel mix" and loaded directly into the sequencer. As the sequencer runs, DNA is chewed up base by base by exonuclease tethered to a nanopore, and each base generates a different electric current pattern. Base calling takes place in real time using Oxford Nanopore's Metrichor cloud service. Data can be analyzed using either 1D or 2D workflows. The ingenious 2D workflow uses a hairpin adapter, which links the top and bottom strands of double-stranded DNA into one strand. The base caller recognizes the hairpin sequence and aligns both strands of the template molecule, with the goal of improving sequencing accuracy. The first device, MinION, a USB device, was available through the MinION Access Programme (MAP) (Loman & Watson, 2015). PromethION and GridION are still under development (http://www.biocompare.com/Editorial-Articles/171872-Next-Gen-DNA-Sequencing-2015-Update/). For MinION, with continuing updates of this instrument, especially the so-called "2D" reads, more and more positive publications have emerged to give credit to the accuracy and satisfying read length of this platform (Ammar *et al.*, 2015; Jain *et al.*, 2015). Its success in the future awaits validation of large sequencing projects.

Preprocessing of Sequences

During the library preparation and sequencing process, a variety of sequence artifacts, including adapter/primer contamination, base calling errors, and low-quality sequences, will negatively affect the quality of raw data for downstream analyses. Therefore, these quality issues necessitate better programs for quality control and preprocessing of all the raw data (Patel & Jain, 2012; Schmieder & Edwards, 2011; Trivedi *et al.*, 2014).

Data Types

Single-end data: Single-end sequencing only sequences DNA fragments from one direction, so all the fragments in the library are only sequenced one end, which means there is only one data output file from each sequencing template.

Paired end data: Paired-end sequencing sequences the same DNA fragment from both directions, which produces a pair of reads for each fragment. The genomic distance between these two reads was chosen during the size selection process and is used to constrain assembly solutions (Nagarajan & Pop, 2013). As a result, there are two output files for each of the sequencing templates in the form of paired-end data.

Mate pair: Mate pair sequencing requires long-insert-size paired-end libraries that are useful for a number of sequencing applications, such as *de novo* assembly and genome finishing. With large inserts, library preparation is more complicated than with short-insert libraries. Genomic DNA is fragmented to an approximate size range, usually 3–5 KB, 6–8 KB, or 7–10 KB. Fragments are then end-repaired with biotinylated nucleotides and go through size selection more specifically. After circularization, non-circularized DNA is removed by digestion, while those circularized fragments are then subjected to a second round of fragmentation, and the labeled fragments (corresponding to the ends of the original DNA ligated together) are purified by biotin–streptavidin cleanup. Purified fragments are end-repaired and ligated to sequencing adapters (http://www.illumina.com/technology/next-generation-sequencing/mate-pair-sequencing_assay.html).

The length of the initial size selection determines the mate pair gap size expected during the scaffolding of contiguous sequences (contigs). Combining mate pair data with data from short-insert-size paired-end reads provides a powerful tool for good assembly and high sequencing coverage across the genome. However, the ligation process may create chimeric molecules, which is a major source for artificial structural variations.

Quality Control

The quality control process of sequencing data typically involves assessing the quality metrics of the raw reads. One of the most popular programs for performing this is `FastQC` (http://www.bioinformatics.babraham.ac.uk/projects/fastqc/). It is a useful tool with a graphical user interface (GUI), and is widely used as an initial checkpoint to ensure the overall quality of the sequence datasets. Use:

```
fastqc PATH_TO_SEQUENCE_FILE
```

The evaluating result is named "FILENAME_fastqc.html", which can be opened in a browser. There are 12 tabs displaying 12 quality metrics, with the exclamatory mark representing "warning" and the cross representing "failure".

Some important quality metrics are: the *basic statistics*, which helps in the selection of assembly programs; the *per-base sequence quality*, which shows the overall quality distribution and average quality from the first base to the last base, helping to set the quality threshold for quality trimming; and the *per-base sequence content*, which may imply the existence of primers or adapters based on the non-random-sequence content distribution at the beginning. Furthermore, the last three tabs also function closely to display the primers or adapters.

Trimming

Based on the results from `FastQC`, adapter/primer trimming and quality trimming should be carried out before downstream analyses. Take mapping programs as an example: most popular mapping tools use global matching algorithms, trying to map the whole length of reads to reference genomes if the number of mismatches is below the user-defined threshold. If the adapter/primer sequences are not removed completely, the reads containing the adapter/primer sequences may not be mapped to reference genomes. Similarly, quality trimming is also important (Kong, 2011; Lindgreen, 2012). Besides these two trimming types, there are also ambiguity trimming (to remove stretches of Ns), length trimming (to remove reads shorter or longer than a specified length), and base trimming (to remove a specified number of bases at either 5' or 3' end of the reads). The choices of trimming types and trimming threshold depend on the needs of further analyses.

A number of trimming programs have been developed, such as Skewer (Jiang *et al.*, 2014), Trimmomatic (Bolger, Lohse & Usadel, 2014), Btrim (Kong, 2011), Adapter-Removal (Lindgreen, 2012), and cutadapt (Martin, 2011). Here, we will describe the detailed usage of cutadapt (https://pypi.python.org/pypi/cutadapt).

Base Trimming The parameter `-u` is used to remove a specified number of bases from the beginning or end of each read. If the value is positive, the bases are trimmed from the beginning; and if the value is negative, the bases are trimmed from the end. Base trimming is always performed before adapter trimming. For example, the following command removes five bases from the end of each read. The trimmed output is stored in the "output.fastq" file.

```
cutadapt -u -5 -o output.fastq input.fastq
```

Quality Trimming The parameter `-q` is used to trim low-quality ends from each read before adapter trimming. This step achieves the same goal with base trimming, so they usually do not appear in the same command.

```
cutadapt -q 10,15 -o output.fastq input.fastq
```

In this command, the value before the comma is the cutoff for the 5' end of each read, and the one after the comma is for the 3' end. If you only want to trim the 3' end, then use `-q` *15*. If you only want to trim the 5' end, then use 0 for the 3' end; for example, `-q` *10,0*.

Adapter Trimming Adapter trimming is complex due to the location of the adapters. The parameter `-a` is set for trimming the 3' adapter, while `-g` is used for trimming the 5' adapter. If it is possible for adapters to appear in both ends of the reads, then `-b` will fix this kind of problem. No matter what parameter is used for trimming, it is followed by the adapter sequences.

For the 3' adapter, `-a` will remove the whole or partial matches, and also any sequence that may follow. For example, the sequence of interest is `SEQUENCE` and the adapter is `ADAPTER`. The reads before trimming and the trimmed reads may look like this:

```
SEQUEN                          -       SEQUEN
SEQUENCEADAP                    -       SEQUENCE
SEQUENCEADAPTER                 -       SEQUENCE
SEQUENCEADAPTERSOMETHING        -       SEQUENCE
ADAPTERSOMETHING                -       (all trimmed)
```

For the 5' adapter, `-g` will remove the whole or partial matches, and also the sequences preceding the adapter. Again, assume the sequence of interest is `SEQUENCE` and the adapter is `ADAPTER`. The reads and the trimmed reads may look like this:

```
ADAPTERSEQUENCE                 -       SEQUENCE
PTERSEQUENCE                    -       SEQUENCE
SOMETHINGADAPTERSEQUENCE        -       SEQUENCE
SOMETHINGADAPTER                -       (all trimmed)
```

For the 5' or 3' adapter, `-b` has a rule for deciding which part of the read to remove: if there is at least one base before the adapter, then the adapter itself and everything that follows it are removed. Otherwise, only the adapter at the beginning is removed, and everything that follows it remains.

Paired-end Reads Trimming The parameter `-p` is used to trim paired-end reads separately at the same time, avoiding running `cutadapt` twice. But it functions properly only when the number of reads is the same in both files and the read names before the first space match. Moreover, both of them are processed in the meantime to make sure they are synchronized. For example, if one read is discarded from one file, it will also be removed from the other file. Assume paired-end files "read_1.fastq" and "read_2.fastq" have adapters *ADAPTER_1* and *ADAPTER_2* separately. The trimming command may look like this:

```
cutadapt -a ADAPTER_1 -A ADAPTER_2 -o out_1.fastq -p
   out_2.fastq read_1.fastq read_2.fastq
```

The parameter `-A` is used to specify the adapter that `cutadapt` should remove in the second file. There are also the parameters `-G` and `-B`. All of them only work in the paired-end reads trimming.

Length Trimming This process only functions after all the other types of trimming are completed. The trimmed reads that are shorter than a specified length are discarded using `-m`, while the ones that are longer than a specified length are discarded using `-M`.

Different kinds of parameters could be combined in a single command to perform a complete trimming process. Here is an example:

```
cutadapt -q 20,20 -a ADAPTER_1 -A ADAPTER_2 -g ADAPTER_3 -G
   ADAPTER_4 -m 30 -o out_1.fastq -p out_2.fastq
   read_1.fastq read_2.fastq
```

All the reads are first quality trimmed from both ends using 20 as the quality threshold, and then the 3' adapter *ADAPTER_1* and 5' adapter *ADAPTER_3* are trimmed from the first file "read_1.fastq". At the same time, the 3' adapter *ADAPTER_2* and 5' adapter *ADAPTER_4* are trimmed from the second file "read_2.fastq". After quality trimming and adapter trimming, the reads whose lengths are shorter than 30 are discarded in both "read_1.fastq" and "read_2.fastq". The final trimmed reads are kept in the "out_1.fastq" and "out_2.fastq" files.

A trimming report is printed after `cutadapt` has finished processing the reads. And then `FastQC` (described in the previous section) could be used to check the overall quality of the trimmed reads again to make sure all the issues have been addressed.

Error-Correction

After quality trimming, all the reads are above the quality threshold, but the good quality value does not mean the read is totally error free. That is the reason why we need an error-correction step prior to sequence assembly. Although many assembly programs have their error-correction modules in the packages, such as ALLPATHS-LG (Gnerre *et al.*, 2011), many stand-alone error correction programs have been developed as well, such as `Quake` (Kelley, Schatz & Salzberg, 2010), `BLESS` (Heo *et al.*, 2014), and `Lighter` (Song, Florea & Langmead, 2014). As errors are random and infrequent, the reads containing an error in a specific position can be corrected by the majority of reads containing the correct base (Yang, Chockalingam & Aluru, 2013).

Here, we will introduce one stand-alone error-correction tool, `Lighter` (https://github.com/mourisl/Lighter) (Song *et al.*, 2014). It is a fast and memory-efficient sequencing error-correction program without counting *k*-mers. Running this program is simple, without too many parameters. Assume we have a species with a genome size of 1 GB, and paired-end sequencing reads files "read_1.fastq" and "read_2.fastq" yielding a genome coverage of 70. Then the error-correcting command simply looks like:

```
lighter -r read_1.fastq -r read_2.fastq -k 17 1000000000
   0.1 -t 10 -od output
```

Here, `-r` is a required parameter to exhibit the input files. If the data set is a single-end read file, then only one `-r` is used; `-k` is followed by *k*-mer length, genome size, and alpha (alpha = 7/coverage, here 70); and the genome size does not need to be accurate. Instead, `-K` (capital K) can be used, followed by *k*-mer length and genome size. However, for `-K`, the genome size should be more accurate, because `Lighter` will go through the calculation to acquire the value of alpha. Considering running in parallel, `-t` could be used to set the number of threads. For each input file, `Lighter` will generate an error-corrected file in the directory "output", created by the parameter `-od`. In this case, the error-corrected files will be named "read_1.cor.fastq" and "read_2.cor.fastq". As for the determination of *k*-mer length, the performance of `Lighter` is not overly sensitive to it (Song *et al.*, 2014).

Sequence Assembly

The sequence assembly process is more like solving a jigsaw puzzle (Pop, 2009). Assembling small DNA fragments into whole genome sequences is similar to assembling small puzzle pieces into a whole picture. For jigsaw puzzles, complexity increases with the number of pieces, the size of the picture, and especially the ambiguous but similarly colored pieces. Similarly, the difficulty of the assembly process increases with the number of reads, the size of genomes, and the content of repetitive sequences.

Reference-Guided Assembly

After error-correction, the reads could be aligned to a known reference genome sequence or assembled *de novo*. Mapping reads to a known reference genome is also called *reference-guided assembly*. Two different aligners, BWA (Li & Durbin, 2009; 2010) and Stampy (Lunter & Goodson, 2011), are often utilized to do the mapping work. BWA consists of three algorithms: BWA-backtrack, BWA-SW, and BWA-MEM. The first one is designed for Illumina sequence reads of up to 100 bp, while the other two are for longer sequences ranging from 70 bp to 1 MB (http://bio-bwa.sourceforge.net/bwa.shtml). Stampy allows variation during mapping (particularly insertion/deletions), and thus can be used if you do not have a closely related reference genome sequence. In any case, reference genome sequences are required for mapping, and therefore it is not applied to numerous species assemblies without references. For most aquaculture species, reference genome sequences are not yet available, and, therefore, we will focus on *de novo* assembly.

De Novo Assembly

"*De novo*" literally means "from the beginning" in Latin. For genome projects, *de novo* assembly refers to assembly entirely based on reads generated from sequencing instruments. In other words, no prior assembled sequence data is utilized during assembly, such as pre-assembled sequences from genomes, transcripts, and proteins (Rodríguez-Ezpeleta, Hackenberg & Aransay, 2012). The software packages designed for sequence assembly are called *assemblers*. All assemblers rely on the simple assumption that highly similar DNA fragments originate from the same position within a genome (Nagarajan & Pop, 2013).

There are mainly three categories of assemblers: greedy graph-based assemblers, overlap-layout-consensus (OLC) assemblers, and de Bruijn graph (DBG) based assemblers.

Graph

A *graph* is the representation of a set of objects (also called *vertices* or *nodes*) where some pairs of objects are connected by links (also called *edges* or *arcs*). If the edges are only traversed in one direction between its connected nodes, they are called *directed edges*. Each directed edge connects a source node and a sink node. In a path where edges visit nodes in some order, the sink node of one edge is also the source node for any subsequent node. However, if the edges may be traversed in either direction, then they are called *undirected edges*. A graph can be conceptualized as a set of balls representing its nodes with connecting arrows representing its edges (Miller, Koren & Sutton, 2010).

When building graphs from real data, there are several complicated situations (Miller *et al.*, 2010). Spurs are short, dead-end subpaths diverging from the main path, which is shown in Figure 3.1A. They are mainly induced by sequencing error toward one end of a read. In a frayed-rope pattern shown in Figure 3.1B, paths converge and then diverge, which are induced by repeat sequences. Paths that diverge and then converge as shown in Figure 3.1C form bubbles. They are induced by sequencing error in the middle of a read and polymorphism in the genome. Cycles occur when paths converge on themselves, which are also induced by repeat elements in the genome.

Greedy Assemblers

Early assemblers were based on the greedy algorithm, such as phrap and TIGR Assembler. The term "greedy" refers to the fact that the decisions made by the algorithm optimize a local objective function (Pop, 2009), which means the assembler always joins the reads that overlap best and end until no more reads or contigs can be joined. The scoring function for "overlap best" commonly measures the number of matches in the overlap and level of identity (percentage of base pairs shared by the two reads) between the reads within the overlapping region (Pop, 2009). As an optimization, the greedy algorithm may utilize just one overlap for each read end and then may discard each overlap immediately after the highest-scoring reads were recruited to extend the contig. Consequently, it can get stuck at local maxima if the current contig is extended by reads that may have helped other contig extensions (Miller *et al.*, 2010).

Since overlaps induced by repetitive sequences may score higher than true overlaps, the greedy algorithm needs mechanisms to avoid the incorporation of false-positive

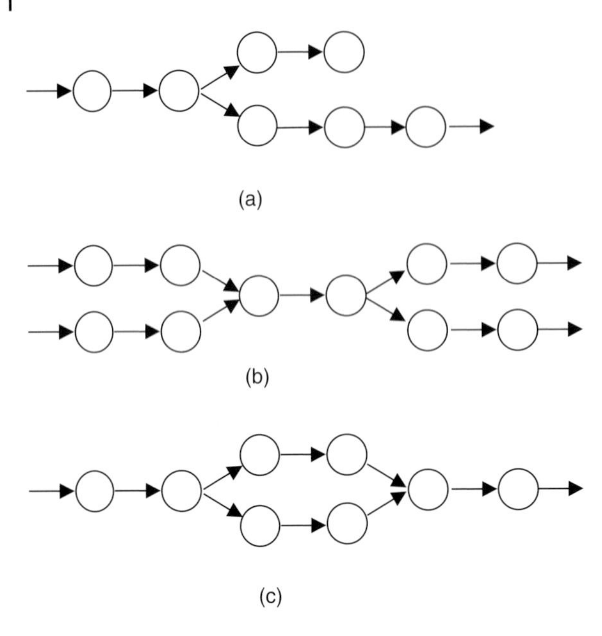

Figure 3.1 Several different types of graph complications are shown using balls representing nodes with connecting arrows representing edges: (1) spur, (2) frayed-rope, and (3) bubble (adopted from Rodríguez-Ezpeleta *et al.*, 2012).

overlaps into contigs (Rodríguez-Ezpeleta *et al.*, 2012). Otherwise, the assembler may join unrelated reads to either side of a repeat to create mis-assemblies. The greedy algorithm is not widely used any more (Nagarajan & Pop, 2013) due to the inherently local assembly process that does not use global information (such as mate-pair reads) to resolve repeats.

SSAKE SSAKE (Warren *et al.*, 2007) was the first short-read assembler, which was designed for unpaired short reads of uniform length (Miller *et al.*, 2010). It is written in PERL and runs on Linux. The algorithm assembles 25–300 bp reads from viral, bacterial, and fungal genomes (http://www.bcgsc.ca/platform/bioinfo/software/ssake). It is also the core of many other assemblers (such as VCAKE, QSRA, and SHARCGS).

SSAKE stands for short sequence assembly by progressive K-mer search and 3' read extension program, populating a hash table that is keyed by unique reads with values representing the number of occurrences in a single multi-fasta file (Warren *et al.*, 2007). The sequence reads are sorted by decreasing values to display coverage and identify reads that may contain errors. The first 11 of the 5' end bases are memorized in a prefix tree, and each unassembled read (r) is utilized to start an assembly. SSAKE uses the prefix tree to progressively perform perfect matches of each possible k-mer with another read (r'). Once perfect match is determined, r is extended by the 3' end bases of r', and r' is discarded from the hash table and the prefix tree. The output files are one log file with run information, one fasta file containing contig sequences, and one fasta file containing the unassembled reads.

One way to avoid mis-assemblies is to terminate the extension when a k-mer matches the 5' end bases of more than one read (`-s 1`), which could cause shorter contigs. The other way to control the stringency is to terminate the extension when a k-mer length is shorter than a user-set minimum word length (m).

VCAKE VCAKE (Jeck *et al.*, 2007) (verified consensus assembly by K-mer extension) is another iterative extension algorithm. Its assembly process is nearly the same as SSAKE, where sequence reads are searched from a hash table. However, compared to SSAKE, the main improvement of VCAKE is its ability to process imperfect matches during contig extension. Specifically, when extending contigs, each read offers a "vote" for the first overhanging base called by the contig (Jeck *et al.*, 2007). The votes are summed, and the base that exceeds a threshold ($-c$, which is the ratio of the most represented base required to extend assembly) is used to extend the contig. In this way, error-containing reads could be used for extension by adding one base at a time until no matches are found or the contig is stopped because of assumed repeat sequences.

QSRA QSRA (Bryant, Wong & Mockler, 2009) (quality-value-guided short read assembler) builds directly upon the VCAKE algorithm, where voting method is utilized to extend contigs one base at a time. To lengthen contigs past low coverage regions, one of the main improvements of QSRA is the usage of quality values in the contig extension. As long as the bases match the contig suffix or reach (or exceed) a minimum user-defined q-value ($-m$) at the overhanging region, the current assembly process continues. Another improvement of QSRA is the availability of suspect repeat sequences in a separate fasta file for repeat-related analysis.

Overlap-Layout-Consensus (OLC) Assemblers

Unlike the inherently local assembly in greedy algorithm, OLC assemblers use three stages to enable a global analysis of the relationships between the reads (Pop, 2009). The first two stages are to compute all-against-all comparison of reads and then construct an overlap graph. The overlap graph contains each read as a node, and an edge connects two reads that overlap each other. Overlaps are only taken into consideration between reads when they share a segment of sequences of length k, which is referred to as a k-mer. Therefore, the choice of k-mer minimum overlap length and percentage of identity required for an overlap are key factors that would affect discovery of overlaps and construction of graphs (Miller *et al.*, 2010). Larger parameter values increase accuracy of contigs but decrease the length. After these two steps, the optimized overlap graph produces an approximate read layout.

The last step is to determine the precise layout and then the consensus sequence by computing multiple sequence alignment (MSA) among the reads, which correspond to a single path traversing each node in the graph at most once. Full information of each read needs to be loaded into memory for base calls, but it can run in parallel, and be partitioned by contigs (Miller *et al.*, 2010).

Celera Assembler The Celera Assembler (Myers *et al.*, 2000) was originally a Sanger OLC assembler, and then was modified for combinations of Sanger and 454 data in a pipeline called CABOG (Celera Assembler with the Best Overlap Graph) (Miller *et al.*, 2008), which is robust to homopolymer run length uncertainty, heterogeneous read lengths, and high read coverage.

In order to exploit the increased expected read coverage, CABOG performs an overlap-based trimming step first. After that, it uses exact-match seeds (k-mer) to detect candidate overlapping reads quickly. To avoid highly repetitive k-mers, it precomputes a threshold M (k-mer with more than M occurrences accounts for at most 1% of all k-mer occurrences) for determining overlap seeds range. After this "anchors and overlaps" stage, directed overlaps with each overlap region spanning at least 40 bases at two-read ends and 94% or higher sequence identity are output for the next stage. Based on overlapping information, the best overlapping graph (BOG) is built by loading the best edge (i.e., best overlap alignment) connecting each node (read). Furthermore, any cycle is eliminated by random deletion of one edge. Then CABOG sorts the reads by their scores, which are counted by the number of other reads reachable from them in BOG paths. Starting from higher-scoring reads, unitigs (uniquely assemblable contigs) are constructed by following paths until reaching a path end or a read already being used for building some unitigs. On paths with a single intersection, CABOG always chooses the longer one first, while multiple intersections would lead to shorter unitigs. By incorporating paired-end constraints, contigs and scaffolds are created from unitigs, and, finally, CABOG derives consensus sequences through computing multiple sequence alignments from scaffold layouts and read sequences. One strength of Celera Assembler is its ability to handle various types of data, including those from sequencing by synthesis platforms (i.e., Illumina HiSeq) and single-module sequencing platforms (i.e., Pacific Biosciences) (Berlin *et al.*, 2015).

Edena Most OLC assemblers focus on assembly on Sanger and 454 data, while Edena (Hernandez *et al.*, 2008) (Exact DE Novo Assembler) applies the OLC approach to homogeneous-length short reads from the SOLiD and Illumina platforms. In order to improve assembly of such short reads, Edena employs two features: exact matching, and detection of spurious reads. First, it removes duplicate reads and finds all overlaps of a minimum length to build an overlap graph; and then the graph is cleaned by the removal of redundant overlaps and spurious edges and the resolution of ambiguity. Finally, all unambiguous paths in the graph are output as assembled contigs.

De Bruijn Graph (DBG) Approach

The DBG approach is also known as the k-mer graph approach or Eulerian path approach (Pevzner, Tang & Waterman, 2001). DBG-based assemblers are widely applied to short reads from the SOLiD and Illumina platforms. A whole set of programs have been developed using this approach, including Velvet (Zerbino & Birney, 2008), ABySS (Simpson *et al.*, 2009), SOAPdenovo (Li *et al.*, 2010; Luo *et al.*, 2012), and ALLPATHS (Butler *et al.*, 2008; Gnerre *et al.*, 2011; Maccallum *et al.*, 2009). Unlike the OLC approach, the DBG approach does not require all-against-all overlap search, and consequently, does not need to store reads or their overlaps, which makes it memory efficient for small genome assembly. Still, for large genomes, a massive amount of memory is still necessary for containing actual sequence and the graph.

During the construction of DBG, the reads are chopped up into a set of k-mers. Similar to the OLC approach, k-mers are represented as the nodes in the graph, and the edges connect the adjacent k-mers whose overlaps are exactly $k - 1$ letters (for instance, the 4-mers ATCG and TCGA share exactly three letters, "TCG"). And then, all the k-mers are stored in a hash table for graph construction. Hence, the assembly process is simply changed to finding a path that visits each node in the graph at most once.

In most cases, the DBG assemblers attempt to construct contigs originating from unambiguous and unbranching paths in the graph. Unfortunately, three factors in the real sequencing data complicate this process (Miller *et al.*, 2010). The first is that DNA is double-stranded. The forward sequence of a read may overlap the forward sequence or reverse complement sequence of another read. Currently, different implementations have been developed to handle this forward–reverse overlap. The second problem is sequencing error. However, many errors are easily recognized during graph construction. For example, errors in the middle of a read lead to bubbles in the path, and errors at the end of a read usually result in a single k-mer in the graph. Some assemblers preprocess the reads to remove errors by correcting erroneous bases or discard error-containing reads. Some discard those paths that are not supported by a high number of k-mers. Others convert paths to sequences for alignment algorithms and then collapse those nearly identical paths. The last factor is the repeats in the genome. Repeat regions always create branches in the paths, which means paths converge in the repeats and then diverge. Successful assembly usually collapses the repeats to avoid mis-assembly. On the other hand, mate-pair information is always used to resolve repeat-induced complications.

ABySS ABySS (Simpson *et al.*, 2009) (Assembly By Short Sequences) was originally developed for assembly of large data sets from individual human genome sequencing. In order to achieve that goal, ABySS applies a distributed representation of a de Bruijn graph, which allows parallel computation of the algorithm across a set of computers to utilize their combined large memory. It is implemented in C++, and the MPI (Message Passing Interface) protocol is employed for communication between nodes (an object on a network). The assembly process is mainly performed in two steps. First, paired-end information is applied to contigs extension until no unambiguous bases could be added or the path goes to a blunt end because of low coverage. Paired-end information is then applied to resolving ambiguities and merging contigs. Here, the usage of ABySS on Linux will be briefly discussed, including installation of the program, basic commands for assembly process, and simple analysis of assembly results.

The installation of ABySS requires several dependencies: Boost, Open MPI, and sparsehash. With all these libraries installed, the program can be downloaded (http://www.bcgsc.ca/platform/bioinfo/software/abyss). The next step is to compile this program from source (http://seqanswers.com/wiki/ABySS).

Compiling ABySS from Source

The following command line should compile and install ABySS:

```
./configure && make && make install
```

To install ABySS in a specified directory (i.e., ABySS), type:

```
./configure --prefix=PATH_TO_ABySS
```

As described earlier, ABySS uses the MPI protocol, so if you wish to build the parallel assembler with MPI support, type:

```
./configure --with-mpi=PATH_TO_openmpi
```

The Google sparsehash library should be used for internal hash table construction to reduce memory usage:

```
./configure --CPPFLAGS=-IPATH_TO_sparsehash/include
```

The default maximum *k*-mer size is 64 and may be decreased to reduce memory usage or increased during compiling. The value must be a multiple of 32 (i.e., 32, 64, 96, etc.):

```
./configure --enable-maxk=96
```

If so desired by the users, the preceding command lines can be combined into one command for compiling. To run ABySS, its executables should be found in the `PATH`:

```
PATH=$PATH:PATH_TO_ABySS/bin
```

Assembling Reads from Different Libraries

After compiling, ABySS can be utilized to assemble reads from different libraries. For example, we are going to assemble six data sets with three different fragment libraries, including two mate-pair reads, two paired-end reads, and two single-end reads.

Library pe180 has reads in two files, "pe180_1.fa" and "pe180_2.fa"
Library pe250 has reads in two files, "pe250_1.fa" and "pe250_2.fa"
Library mp3k has reads in two files, "mp3k_1.fa" and "mp3k_2.fa"
Library mp5k has reads in two files, "mp5k_1.fa" and "mp5k_2.fa"
Single-end reads are stored in two files, "se1.fa" and "se2.fa"

To assemble this dataset into contigs in a file named "human-contigs.fa" and scaffolds in a file named "human-scaffolds.fa", run the command:

```
abyss-pe -C PATH_TO_RESULTS k=64 name=human lib='pe180
   pe250' mp='mp3k mp5k' \
pe180='pe180_1.fa pe180_2.fa' pe250='pe250_1.fa
   pe250_2.fa' \
mp3k='mp3k_1.fa mp3k_2.fa' mp5k='mp5k_1.fa mp5k_2.fa'
   se='se1.fa se2.fa'
```

Where `-C` defines the specified directory for storing results; *k* is the *k*-mer length; *name* is the prefix for all the filenames in the results directory; *lib* is for fragment library; and *mp* is for mate-pair library. However, mate-pair libraries are used only for scaffolding (join disconnected contigs together), and they do not contribute to building contigs, *per se*.

Optimizing *k*-mer Length

In order to compute the best value for the parameter *k*, run multiple assemblies and then check the assembly statistics to determine the best *k*-mer length. The following command example would assemble every *k*-mer length from 40 to 70:

```
export k
for k in {40..70}; do
mkdir k$k
abyss-pe -C k$k name=human lib='pe180' pe180='pe180_1.fa
   pe180_2.fa'
done
abyss-fac k*/human-contigs.fa
```

`mkdir` is used to create new directory for storing results, corresponding to parameter `-C`. "abyss-fac" is used for analysis of assembly results, which can help determine the best *k*-mer value. The results of "abyss-fac" will be shown in the file named "human-stats.csv".

ALLPATHS ALLPATHS is a DBG assembler that can generate high-quality assemblies from short reads. It was originally published as an assembler based on simulated data (Butler *et al.*, 2008), and then revised for real data (Maccallum *et al.*, 2009).

First, ALLPATHS removes all read pairs containing 90% or more "A" bases, which are nearly always artifacts of the Illumina sequencing process. It then begins an error-correcting preprocessor by identifying putatively correct (or "trusted") *k*-mers in the reads based on high frequency and high quality (where each base must be confirmed by a minimum number of base calls with a quality score above a threshold). With lower-quality reads, they could be kept under two circumstances—(1) where up to two base substitutions to low-quality base calls make its *k*-mers trusted, or (2) if the read is essential for building a path. Another preprocessor step is called *unipath creation*. This step starts with the calculation of perfect read overlaps seeded by *k*-mers. An index is assigned to each *k*-mer, and a compact searchable database is built for constructing unipaths.

After the initial unipath creation, those with branching paths shorter than 20 *k*-mers, which are always caused by read errors, are removed to provide longer alternative branches. This step is also called "unipath graph shaving". Now, with the unipaths and read pairs in hand, those unipaths around which genome assembly will be constructed are picked as "seeds" (the perfect seeds being long and of low copy number). The genomic region containing the seed and extending 10 KB on both sides is called the "neighborhood" of the seed. The goal is to assemble the neighborhood with help of unipaths and paired-end reads.

In order to resolve repeat regions, ALLPATHS assembles the locally non-repetitive sequences first. This step runs in parallel. Once it ends, the gluing process is applied to iteratively joining together the output graphs that have long end-to-end overlaps. Next, a series of editing steps are carried out to remove spurs and ambiguous regions to yield the final assembly. Those unipaths that are not represented in the assembly are extended unambiguously where possible and then added to the final assembly results.

ALLPATHS-LG (Gnerre *et al.*, 2011) is an updated version of ALLPATHS, which is designed for large genomes from massively parallel sequence data. It requires construction of two different paired reads: a fragment library and a jumping library. The fragment library has a short insert size that is less than twice the read length, so that the two paired reads may overlap to produce a single longer "read" (i.e., read length is 100 bp and insert size is 180 bp, then these two reads have an overlap of 20 bp). The jumping library generates long-range mate-pair reads. This new version also makes extensive improvements in handling repetitive sequences and low coverage region, error-correction, and memory usage. Here, we will describe the detailed usage of the ALLPATHS-LG program.

Requirements for ALLPATHS-LG

ALLPATHS has a number of requirements, including:

1) A minimum of two paired-end libraries as described earlier: one fragment library whose insert size is less than twice the read length, and one jumping library.
2) Memory peak usage is around 1.7 bytes per read base. To compile and run ALLPATHS-LG on a Linux/UNIX system, a minimum of 32 GB memory for small genomes and 512 GB memory for large genomes are suggested.
3) The g++ compiler from GCC (https://gcc.gnu.org/).

4) The GMP library (https://gmplib.org/) compiled with C++ interface, which may already be included in GCC.
5) The Picard command-line utilities for SAM file manipulation (http://broadinstitute.github.io/picard/).

ALLPATHS Installation

The ALLPATHS-LG source code is available for download (http://www.broadinstitute.org/software/allpaths-lg/blog/?page_id=12). After the latest version of the program is downloaded, it can be decompressed using *tar*. Then, it can simply be compiled with *./configure* and *make*, and, at the same time, the *--prefix* parameter can be used to define the installation directory. All the source code is in the unpacked directory called "allpathslg- < revision>".

```
tar xzf allpathslg-<revision>.tar
cd allpathslg-<revision>
./configure --prefix=PATH_TO_INSTALLATION_DIRECTORY
make
make install
```

To run ALLPATHS-LG, its executables should be added to the *PATH*.

ALLPATHS-LG Pipeline Directory Structure

The assembly pipeline uses the following directory structure to store its inputs and outputs. If the directory does not exist, it can be automatically created by the pipeline. The names shown here are only for descriptions of the structure, and they are determined by the actual command line.

"PRE / REFERENCE / DATA / RUN / ASSEMBLIES / SUBDIR"

The "PRE" directory is the root for the assemblies, and there could be a number of "REFERENCE" directories in it. The "REFERENCE" directory is also called *organism directory*; in other words, it is used to separate assembly projects by organism. The "DATA" directory consists of the data whose formats are changed by the ALLPATHS-LG pipeline. The "RUN" directory contains intermediate files generated from the read data in preparation for the final assembly process and intermediate files used for evaluation. The "ASSEMBLIES" directory is the actual assembly directory, and its name is fixed. The "SUBDIR" directory (its name is "test", and it is also fixed) is where the localized assembly is generated.

Prepare the Data

Before assembly, original data (BAM, FASTQ, FASTA, or FASTB) needs to be prepared and imported into the ALLPATHS-LG pipeline. This step is completed by adding two comma-separated-values (csv) metadata files to describe the locations and library information.

"in_groups.csv"
"in_libs.csv"

Each line in "in_groups.csv" provides the following information:

group_name: a unique name for the data set
library_name: the library to which the data set belongs

`file_name`: the absolute path to the data set. Wildcards "*" and "?" are accepted for specifying multiple files (i.e., two paired files)

Here is one example for the "in_groups.csv" file:

```
group_name,        library_name,     file_name
1,                 paired-end,       /illumina/reads1*.fastq
2,                 mate-pair,        /illumina/reads2*.fastq
```

Each line in "in_libs.csv" provides the following information:

`library_name`: matches the same field in "in_groups.csv"
`project_name`: a string naming the project
`organism_name`: the name of the organism
`type`: fragment, jumping, etc.; field not necessary
`paired`: "0" for unpaired reads and "1" for paired reads
`frag_size`: estimated average fragment size
`frag_stddev`: estimated fragment size standard deviation
`insert_size`: estimated jumping average insert size
`insert_stddev`: estimated jumping insert size standard deviation
`read_orientation`: *inward* for fragment reads and fosmid jumping reads, *outward* for jumping reads
`genomic_start`: index of the first genomic base in the reads; if non-zero, trim all the bases before `genomic_start`
`genomic_end`: index of the last genomic base in the reads; if non-zero, trim all bases after `genomic_end`; if both `genomic_start` and `genomic_end` are blank or zero, keep all bases

Here is an example for the "in_libs.csv" file (Note: There should be three lines for this example, in which "..." means "to be continued".):

```
library_name,project_name,organism_name,type,paired,
  frag_size,...
paired-end,genome,human,fragment,1,180,...
mate-pair,genome,human,jumping,1, ,...
frag_stddev,insert_size,insert_stddev,read_orientation,...
20, , ,inward,...
  ,2000,200,outward,...
genomic_start,genomic_end
  ,
  ,
```

With these two description files, a Perl script is required to convert the original reads. "DATA_DIR" is the location of the DATA directory where the converted reads are stored. "PICARD_TOOLS_DIR" is the path to the Picard tools for reads conversion, if the data is in BAM format. "PLOIDY" indicates the genome, with "1" for haploid genomes and "2" for diploid genomes. Here is an example:

```
PrepareAllpathsInput.pl
   DATA_DIR=PATH_TO_DATA_DIRECTORY
   IN_GROUPS_CSV=PATH_TO_<in_groups.csv>
```

```
    IN_LIBS_CSV=PATH_TO_<in_libs.csv>
    PLOIDY=<1 or 2>
    PICARD_TOOLS_DIR=PATH_TO_<Picard_tools>
    HOSTS=<list of hosts to be used in parallel>
```

After a successful run of "PrepareAllpathsInput.pl", the necessary input files are ready for the assembly stage.

Assembling

There are a series of modules in ALLPATHS-LG. Each module performs one step of the assembly process. However, a single module called *RunAllpathsLG* controls the entire pipeline, assigning different modules to run efficiently at the proper time. This step specifically names the directories that were described in the directory structure part. Here is a basic example:

```
RunAllpathsLG
    PRE=allpaths
    REFERENCE_NAME=reference
    DATA_SUBDIR=data
    RUN=run
```

These command lines will create (if it does not exist) the following directory structure: "allpaths / reference / data / run / ASSEMBLIES / test".

Assembly Results

In the preceding example, the assembly results are in the "test" directory. Both files ("final.assembly.fasta" and "final.assembly.efasta") are scaffold results, where the *N*s represent the gaps between contigs. In the fasta file, an ambiguous base is represented by an *N* within each contig. For example, a C/G SNP is shown as "ATCNTGC". The efasta is "enhanced" fasta, which is generated and used by ALLPATHS-LG. In the efasta file, ambiguity is represented by listing all the possible bases within a pair of braces. For example, a C/G SNP is shown as "ATC{C,G}TGC". The options in the braces are ordered by decreasing likelihood. Hence, the fasta file can be easily generated from the efasta file by picking the first option for every ambiguity. Moreover, statistics for assembly results could be found in the "assembly.report" file.

PacBio Long Reads and Their Applications

Despite many advances in genome assembly programs, generation of complete and accurate assembly of genomes using only short reads data is still difficult, largely due to repetitive sequences (Alkan *et al.*, 2011; Kingsford, Schatz & Pop, 2010). Repetitive sequences can be properly assembled only when the read length is long enough to span the whole repetitive sequences or the read pair is uniquely anchored on both sides of the repeats. In 2011, Pacific Biosciences released their commercial "third-generation" sequencing instrument PacBio RS (Eid *et al.*, 2009). According to the information on the PacBio website (http://www.pacb.com/smrt-science/smrt-sequencing/read-lengths/), the average read length could reach 10 KB. Unfortunately, the long reads also come with relatively low nucleotide accuracy (~85%), compared with the 98% accuracy from short reads. However, short reads can be utilized to correct the long reads, allowing potential use of the long reads for the assembly.

Hybrid Assembly Short reads can be used in combination with long reads to improve the completeness and correctness of an assembly, making it a hybrid assembly strategy. Several programs have been developed to take advantage of both short and long reads. A PBcR (PacBio corrected Reads) algorithm (Koren *et al.*, 2012), implemented in the Celera Assembler, trims and corrects individual long reads by simultaneously mapping high-identity short reads to them and computing a highly accurate consensus sequence for each of them. This "hybrid" correction could achieve >99.9% base-call accuracy, and then the corrected reads may be *de novo* assembled or co-assembled with other data types using OLC assembly techniques. This method allows for the assembly of large genomes, but requires a large amount of running time for the correction process.

Cerulean (Deshpande *et al.*, 2013), unlike other hybrid assembly programs, does not use the short reads directly. Instead, it starts with a graph from ABySS (Simpson *et al.*, 2009) and extends contigs by resolving bubbles in the graph using PacBio long reads. The advantages of this method are that it requires minimal computational resources, and results in a highly accurate assembly. But the efficiency of cerulean when applied to large genomes remains to be verified.

ALLPATHS-LG (Gnerre *et al.*, 2011; Ribeiro *et al.*, 2012) is another program that can deal with PacBio long reads in combination with short reads from Illumina sequencing platforms. As described in the preceding section, it requires one fragment library and also a jumping library. These three different data types, whose accuracy, coverage, and range of repeat resolution are compensatory, thus enable satisfying assembly results. However, good performance requires much more efforts and money to construct the libraries and larger amounts of time to compute the data, which limits its application to large genomes. The assembly algorithm is implemented as parts of ALLPATHS-LG and is automatically invoked when PacBio data are provided as input to the algorithm.

Here is an example of descriptions for "in_groups.csv" and "in_libs.csv" when PacBio data is utilized for assembly:

For "in_groups.csv":

```
group_name, library_name, file_name
1,          paired-end,   /illumina/reads1*.fastq
2,          mate-pair,    /illumina/reads2*.fastq
3,          pacbio,       /pacbio/reads.fastq
```

For "in_libs.csv" (Note: There should be four lines for this example, in which "..." means "to be continued".):

```
library_name,project_name,organism_name,type,paired,
  frag_size,...
paired-end,genome,human,fragment,1,180,...
mate-pair,genome,human,jumping,1, ,...
pacbio,genome,human,long,0, ,...
frag_stddev,insert_size,insert_stddev,read_orientation,...
20, , ,inward,...
  ,2000,200,outward,...
  , , , ,...
genomic_start,genomic_end,
   ,
   ,
   ,
```

PacBio-only Assembly In order to avoid using different sequencing libraries and multiple sequencing platforms, a homogeneous workflow requiring only PacBio reads is favored to increase efficiency. The Hierarchical Genome Assembly Process (HGAP) (Chin *et al.*, 2013) was developed for high-quality *de novo* microbial genome assemblies using only PacBio long reads. This method uses the longest reads as seeds to recruit all the other reads to map to them, resulting in long and highly accurate preassembled reads. Next, quality trimming of preassembled reads is performed to improve the ability to detect sequence overlaps. OLC assemblers are then used for *de novo* assembly of multi-KB trimmed reads. To significantly reduce the remaining errors in the draft assembly, a quality-aware consensus algorithm implemented in HGAP is computed to derive a highly accurate consensus for the final assembly (99.999% accuracy). Undoubtedly, this method is successful with such high completeness and correctness. However, overlapping steps still consume large amounts of time, making it less computationally efficient for large genomes.

Scaffolding

The output of most assemblers is a large collection of fragmented contigs. However, downstream analyses such as analyzing gene synteny and carrying out comparative or functional genomics rely heavily on an assembly with good continuity (Hunt *et al.*, 2014). Hence, it is necessary to join disconnected contigs into continuous genome sequences in the correct orientation and order. This process is also called *scaffolding*, and it includes contig orientation, contig ordering, and contig distancing (Barton & Barton, 2012).

Contig Orientation Each strand of DNA could be sequenced to generate reads, and therefore the constructed contigs from assemblers could represent either strand. When scaffolding, contig orientation needs to be performed to keep contigs in the same direction, which requires reverse complementing sequences where necessary (Barton & Barton, 2012).

Contig Ordering Contigs are ordered within each scaffold, reflecting their spatial placement in the true genome sequences (Barton & Barton, 2012).

Contig Distancing Given the correct orientation and order, the distance is determined by putting certain numbers of "*N*s", which represent unknown nucleotides, between contigs. The size of each inter-contig gap corresponds to the number of *N*s between them. The total length of known and unknown regions is an estimate of the complete genome size (Barton & Barton, 2012).

Most commonly, the process of scaffolding is achieved from mate-pair information. Two contigs may be scaffolded together if one end of a mate-pair is assembled within the first contig, and the second contig contains the other end of the mate-pair. And the insert size of the mate-pair library provides an estimate for how far apart each read should appear in the final assembly (Paszkiewicz and Studholme, 2010). Furthermore, scaffolding information can also be obtained from whole-genome mapping data (Pop, 2009), which can be achieved from the program SOMA (Nagarajan, Read & Pop, 2008).

Several assemblers also contain a scaffolding module, such as Celera Assembler (Myers *et al.*, 2000), ABySS (Simpson *et al.*, 2009), SOAPdenovo (Li *et al.*, 2010; Luo *et al.*, 2012), and ALLPATHS-LG (Gnerre *et al.*, 2011). Stand-alone scaffolders are also available, such as GRASS (Gritsenko *et al.*, 2012), Scaffolder (Barton & Barton, 2012), SSPACE (Boetzer *et al.*, 2011), and SOPRA (Dayarian, Michael & Sengupta, 2010). Due to genomic complexity and missing data, scaffolding may finally produce a collection of scaffolds instead of a completed sequence, but a wise choice of scaffolder still can yield a decent scaffolding result. A comprehensive evaluation of scaffolding tools was performed to test their overall performance (Hunt *et al.*, 2014). Scaffolding tools vary in their ease of use, speed, and the number of correct and missed joins between contigs. The quality of the results depends highly on the choice of mapping tools and genome complexity. In general, SGA's scaffolding module (Simpson & Durbin, 2012), SOPRA (Dayarian *et al.*, 2010), and SSPACE (Boetzer *et al.*, 2011) perform better on the tested data sets (Hunt *et al.*, 2014).

Here, we introduce SSPACE and its detailed usage for scaffolding. SSPACE starts with the longest contig to build the first scaffold and continues to make joins with the majority of read pairs supporting it (Hunt *et al.*, 2014). It uses Bowtie (Langmead *et al.*, 2009) to map all the reads to the contigs, and, if desired, unmapped reads are used for contig extension. It requires limited computation resources and is able to handle large genomes in a reasonable amount of time (Boetzer *et al.*, 2011).

To apply for the basic version of SSPACE (free for academics), one needs to register on its website (http://www.baseclear.com/genomics/bioinformatics/basetools/SSPACE). Then, a library file containing information about each library needs to be established. All the columns are separated by spaces. Here is an example:

```
lib1 bowtie file1_1.fasta file1_2.fasta 200 0.25 FR
lib2 bowtie file2_1.fastq file2_2.fastq 2000 0.25 RF
unpaired bwa unpaired_reads.fasta
```

Column 1 Name of the library. Libraries of the same name are considered to possess the same insert size and deviation (columns 5 and 6). Moreover, they are used for the same scaffolding iteration. Libraries should be sorted by insert size (column 5). The first library is always scaffolded first, followed by the next libraries. For unpaired reads, the library name is "unpaired", as shown in the example.

Column 2 Name of the read aligner (bowtie, bwa, or bwasw).

Columns 3 and 4 Fasta or fastq files for datasets. For paired reads, the first read file is in column 3, and the second read file in column 4. They are always on the same line. For unpaired reads, only column 3 is required.

Columns 5 and 6 Column 5 represents the mean distance between paired reads, which is 200 for *lib1* in the example. Column 6 represents the minimum allowed deviation of the mean distance, which is 0.25 for *lib1*. This means the distance for paired reads in *lib1* is between 150 and 250.

Column 7 Column 7 indicates the orientation of paired reads. *F* means forward, and *R* means reverse. Usually for the case of mate-pair data, it is *RF*.

Before running SSPACE, its executables should be added to the *PATH*. There is one contig extension option (*-x*) whose values indicate whether to do extension using unmapped reads. Based on this option, different parameters are used in the commands.

Here is one example for scaffolding contigs without extending them:

```
perl SSPACE_Standard_v3.0.pl -l library_file -s contig_file
    -k 5 -a 0.7 -x 0 -b scaffolds_no_extension
```

-l is followed by the library file, and *-s* is followed by contig sequences file. To run SSPACE without extension, *-x* should be set to 0. *-k* and *-a* are parameters for controlling the scaffolding process. *-k* is the minimum number of read pairs (links) to compute scaffolds, and *-a* is the maximum link ratio between the two best contig pairs. Their default values are 5 and 0.7, which do not have to be listed if not reset. The final results are generated in the folder named "scaffolds_no_extension". The file "scaffolds_no_extension.summaryfile.txt" contains the statistics for the final scaffolds. The file "scaffolds_no_extension.final.scaffolds.fasta" contains the resulting scaffolds.

To run SSPACE with contig extension, *-x* should be set to 1. *-k* and *-a* use default values. Here, the minimal overlap of the reads with the contig during overhang consensus build-up is at least 32 bases (*-m*). The minimum number of reads required to extend a contig is 20 (*-o*). These two numbers are also default values for SSPACE version 3.0. The file "scaffolds_extension.summary.txt" in the folder "scaffolds_extension" contains the information on extended contigs. The sequences of the extended contigs are kept in the "intermediate_results" folder.

```
perl SSPACE_Standard_v3.0.pl -l library_file -s contig_file
    -x 1 -m 32 -o 20 -b scaffolds_extension
```

Considering that PacBio long reads are playing an increasingly important role in scaffolding genome assemblies in a cost-effective manner, SSPACE-LongRead (Boetzer and Pirovano, 2014) was developed to scaffold pre-assembled contigs using long reads as a backbone. This may enhance the quality of incomplete and inaccurate genomes constructed from next-generation sequencing data.

Gap Filling (Gap Closing)

In scaffolds construction, contigs with certain distance relationships are connected with *N*s. The majority of these gaps are composed of repetitive and low-coverage regions, which may contain essential sequences for the downstream analyses and need to be filled. The standard strategy to address this problem usually involves the design of specific primers to undergo target Sanger sequencing at contig ends (Tsai, Otto & Berriman, 2010). Without doubt, this method is expensive, labor intensive, and time consuming. At present, the gap-filling process can be made easier by using the original read-pair information (Ekblom and Wolf, 2014) with programs such as IMAGE (Tsai *et al.*, 2010), GapCloser (a module for SOAPdenovo2) (Luo *et al.*, 2012), FGAP (Piro *et al.*, 2014), GapFiller (Boetzer and Pirovano, 2012), and PBJelly (English *et al.*, 2012).

IMAGE (Iterative Mapping and Assembly for Gap Elimination) uses Illumina sequence data to improve draft genome assemblies by aligning paired-reads against contig ends and performing local assemblies to produce gap-spanning contigs (Tsai *et al.*, 2010). Similarly, this method also functions in GapCloser and GapFiller. But the improvement in GapFiller is that it takes into account the estimated gap size prior to closure and closes the gap only if the size of the sequence insertion corresponds closely to the estimated size. This will ensure that the local assembly results may reflect the true genomic situation. It also takes into account the contig edges, which is often a source of mis-assemblies, and trims them prior to gap filling (Boetzer and Pirovano, 2012). PBJelly was designed for genome finishing using long reads from the Pacific Biosciences RS platform (English *et al.*, 2012).

Here, we will introduce the detailed usage of GapFiller. It was developed by the same group as SSPACE. You need to register for downloading the program (free for academics) (http://www.baseclear.com/genomics/bioinformatics/basetools/gapfiller). The library information file from SSPACE is also applied to GapFiller. Before running, its executables should be added to the `PATH`. Here is an example of the gap-filling command:

```
perl GapFiller.pl -l library.txt -s scaffolds.fa -m 29 -o 2
   -r 0.7 -d 50 -n 10 -t 10 -T 1 -i 10 -b gapfiller
```

This command uses the paired reads in the "library.txt" to fill the scaffolds of "scaffolds.fa". The `-s` parameter should be followed by a fasta format file. `-m` is the minimum number of overlapping bases of the reads with the edge of the gap. It is suggested to take a value close to the read length. For example, it is suggested to use 30–35 as the `-m` value for a library with 36 bp reads. `-o` is the minimum number of reads required to extend the edge during gap filling. `-r` is the minimum base ratio needed to accept an overhang consensus base. Higher values of these three values lead to more accurate gap filling. `-d` is the maximum difference between the estimated gap size and the number of gap-closed sequences. Extension is stopped if the difference exceeds 50 bases for this case. `-n` is the minimum overlap for merging two sequences. `-t` is the number of nucleotides to be trimmed from the sequence edges of the gap. This parameter may discard low-quality and misassembled sequences to ensure correct overlap and extension. `-T` indicates the number of threads for reading in the files and mapping of the reads. `-i` is the number of iterations used to fill the gaps. GapFiller keeps using original reads and mapping them against the scaffold gaps until no more reads are used for gap filling or until the specified number of iterations is reached. `-b` gives the name for the output folder. In this case, all the gap-filling results are stored in the "gapfiller" folder, and the gapclosed sequences are in "gapfiller.gapfilled.final.fa", with the statistics information in "gapfiller.summaryfile.final.txt". The assembly process *in silico* is now considered to be finished.

Evaluation of Assembly Quality

Evaluation of assembly quality is of great importance. Considering the large number of operations and uncertain factors in the assembly processes, this evaluation can be complex and challenging. It is less difficult for the evaluation of reference assembly sequences. Since an available reference genome provides a comparable ground truth, we can assess the quality of a new assembly by simply comparing it to the reference genome. However, the lack of a reference genome makes the evaluation of *de novo* assembly a lot more complicated, and there are still no well-accepted measures to date. Before we get into the specific evaluation protocols, it should be very helpful to know the potential errors that may exist in the assembled sequences. Potential mis-assemblies (Phillippy, Schatz & Pop, 2008) may include: (1) repeat collapse regions where an assembler incorrectly joins distinct repeat copies into a single unit, resulting in increased coverage of the read density and sometimes generating "orphan" regions; (2) repeat expansion regions where the addition of copies occurs and the read density drops below normal coverage; and (3) rearrangement regions where the order of multiple repeat copies shuffles leading to the misplacement of the unique sequences, in a special case, inversions may occur when two repeat copies are oppositely oriented. Clearly, most of these mis-assemblies have always been caused by repeats, and the longer the reads, the fewer the repeats that cause errors in the assembly process.

The quality of assembled sequences is basically assessed through two indexes: size statistics and correctness scores. In the case of size statistics, commonly considered statistical values include maximum contig/scaffold length, minimum contig/scaffold length, mean contig/scaffold length, genomic coverage, and N50 size. In particular, the N50 size is the most widely used statistic, which is the minimum size of the contig/scaffold making the sum of the lengths equal to or greater than 50% of the entire genome size after sorting all contigs/scaffolds from longest to shortest. The N50 size can be used to compare different programs only if it is measured with the same combined length value.

The correctness of an assembly is very difficult to measure. Scientists tend to focus on contiguity and ignore the accuracy. To detect mis-assemblies, one measurement relies on comparing assemblies with other independently generated data such as finished BAC sequences (Istrail *et al.*, 2004), mapping data (Nagarajan *et al.*, 2008), transcriptome data (Zimin *et al.*, 2009), or the genomes of closely related organisms (Gnerre *et al.*, 2009). Another measurement uses intrinsic consistency to identify mis-assemblies. One case is the detection of regions with unusual depth of coverage (too high or too low); as we mentioned earlier in error types, repeat collapse or expansion may cause changes in the read density.

In addition, some computational technologies provide tools to evaluate assembly results, including: (1) AMOSvalidat (Phillippy *et al.*, 2008), a tool collected with several detecting mis-assembly mechanisms; (2) GAV (Choi *et al.*, 2008), an approach combined with multiple accuracy measures; (3) CGAL (Rahman & Pachter, 2013), a measurement considering both genome coverage and assembly accuracy with no need of reference sequences; (4) GAGE (Salzberg *et al.*, 2012), an evaluation of the very latest large-scale genome assembly algorithms; (5) REAPR (Hunt *et al.*, 2013), an evaluation tool providing a positional error call metric and identifying potential collapsed repeats; (6) SuRankCo (Kuhring *et al.*, 2015), a machine learning approach to predict quality scores and ranking for contigs; and (7) the validation scripts used in Assemblathon (Bradnam *et al.*, 2013; Earl *et al.*, 2011), which is a competition aiming to improve methods of genome assembly.

Even though there are so many methods out there evaluating assembly results, they only aim to display the issues existing in the sequences. Fixing them by sequencing technologies and varieties of programs is key to acquiring a complete and high-quality genome sequence as the ultimate resource for further genomic approaches. However, in our experience, validation using an independent method such as high-density linkage mapping is quite effective for the assessment of the sequence assembly. The order of the

SNP markers along the linkage map and along the whole-genome reference sequence should be the same, if the whole genome is assembled correctly.

References

Alkan, C., Sajjadian, S. and Eichler, E.E. (2011) Limitations of next-generation genome sequence assembly. *Nature Methods*, **8**, 61–65.

Ammar, R., Paton, T.A., Torti, D., Shlien, A. and Bader, G.D. (2015). Long read nanopore sequencing for detection of HLA and CYP2D6 variants and haplotypes. *F1000Research*, **4**, 17.

Barton, M.D. and Barton, H.A. (2012) Scaffolder – software for manual genome scaffolding. *Source Code for Biology and Medicine*, **7**, 4.

Berlin, K., Koren, S., Chin, C.-S. *et al.* (2015) Assembling large genomes with single-molecule sequencing and locality-sensitive hashing. *Nature Biotechnology*, **33**, 623–630.

Boetzer, M., Henkel, C.V., Jansen, H.J. *et al.* (2011) Scaffolding pre-assembled contigs using SSPACE. *Bioinformatics*, **27**, 578–579.

Boetzer, M. and Pirovano, W. (2012) Toward almost closed genomes with GapFiller. *Genome Biology*, **13**, R56.

Boetzer, M. and Pirovano, W. (2014) SSPACE-LongRead: scaffolding bacterial draft genomes using long read sequence information. *BMC Bioinformatics*, **15**, 211.

Bolger, A.M., Lohse, M. and Usadel, B. (2014) Trimmomatic: a flexible trimmer for Illumina sequence data. *Bioinformatics*, **30**, 2114–2120.

Bradnam, K.R., Fass, J.N., Alexandrov, A. *et al.* (2013) Assemblathon 2: evaluating *de novo* methods of genome assembly in three vertebrate species. *GigaScience*, **2**, 10.

Braslavsky, I., Hebert, B., Kartalov, E. and Quake, S.R. (2003) Sequence information can be obtained from single DNA molecules. *Proceedings of the National Academy of Sciences of the United States of America*, **100**, 3960–3964.

Bryant, D.W. Jr.,, Wong, W.K. and Mockler, T.C. (2009) QSRA: a quality-value guided *de novo* short read assembler. *BMC Bioinformatics*, **10**, 69.

Butler, J., MacCallum, I., Kleber, M. *et al.* (2008) ALLPATHS: *De novo* assembly of whole-genome shotgun microreads. *Genome Research*, **18**, 810–820.

Chin, C.-S., Alexander, D.H., Marks, P. *et al.* (2013) Nonhybrid, finished microbial genome assemblies from long-read SMRT sequencing data. *Nature Methods*, **10**, 563–569.

Choi, J.-H., Kim, S., Tang, H. *et al.* (2008) A machine-learning approach to combined evidence validation of genome assemblies. *Bioinformatics*, **24**, 744–750.

Dayarian, A., Michael, T. and Sengupta, A. (2010) SOPRA: scaffolding algorithm for paired reads via statistical optimization. *BMC Bioinformatics*, **11**, 345.

Deshpande, V., Fung, E.K., Pham, S. and Bafna, V. (2013) Cerulean: a hybrid assembly using high throughput short and long reads, in *Algorithms in Bioinformatics* (eds A. Darling and J. Stoye), Springer Berlin, Heidelberg, pp. 349–363.

Dolnık, V. (1999) DNA sequencing by capillary electrophoresis (review). *Journal of Biochemical and Biophysical Methods*, **41**, 103–119.

Dressman, D., Yan, H., Traverso, G. *et al.* (2003) Transforming single DNA molecules into fluorescent magnetic particles for detection and enumeration of genetic variations.

Proceedings of the National Academy of Sciences of the United States of America, **100**, 8817–8822.
Earl, D., Bradnam, K., St. John, J. *et al.* (2011) Assemblathon 1: a competitive assessment of *de novo* short read assembly methods. *Genome Research*, **21**, 2224–2241.
Eid, J., Fehr, A., Gray, J. *et al.* (2009) Real-time DNA sequencing from single polymerase molecules. *Science*, **323**, 133–138.
Eisenstein, M. (2012) Oxford nanopore announcement sets sequencing sector abuzz. *Nature Biotechnology*, **30**, 295–296.
Ekblom, R. and Wolf, J.B. (2014) A field guide to whole-genome sequencing, assembly and annotation. *Evolutionary Applications*, **7**, 1026–1042.
English, A.C., Richards, S., Han, Y., Wang, M., Vee, V., Qu, J. *et al.* (2012). Mind the gap: upgrading genomes with pacific biosciences RS long-read sequencing technology. *PLoS ONE*, **7**, e47768.
Fedurco, M., Romieu, A., Williams, S. *et al.* (2006) BTA, a novel reagent for DNA attachment on glass and efficient generation of solid-phase amplified DNA colonies. *Nucleic Acids Research*, **34**, e22.
Fox, E.J., Reid-Bayliss, K.S., Emond, M.J. and Loeb, L.A. (2014) Accuracy of next generation sequencing platforms. *Journal of Next Generation, Sequencing & Applications*, **1**, 1000106.
Gnerre, S., Lander, E.S., Lindblad-Toh, K. and Jaffe, D.B. (2009) Assisted assembly: how to improve a *de novo* genome assembly by using related species. *Genome Biology*, **10**, R88.
Gnerre, S., MacCallum, I., Przybylski, D. *et al.* (2011) High-quality draft assemblies of mammalian genomes from massively parallel sequence data. *Proceedings of the National Academy of Sciences*, **108**, 1513–1518.
Gritsenko, A.A., Nijkamp, J.F., Reinders, M.J.T. and de Ridder, D. (2012) GRASS: a generic algorithm for scaffolding next-generation sequencing assemblies. *Bioinformatics*, **28**, 1429–1437.
Harris, T.D., Buzby, P.R., Babcock, H. *et al.* (2008) Single-molecule DNA sequencing of a viral genome. *Science*, **320**, 106–109.
Heo, Y., Wu, X.-L., Chen, D. *et al.* (2014) BLESS: Bloom filter-based error correction solution for high-throughput sequencing reads. *Bioinformatics*, **30**, 1354–1362.
Hernandez, D., François, P., Farinelli, L. *et al.* (2008) *De novo* bacterial genome sequencing: Millions of very short reads assembled on a desktop computer. *Genome Research*, **18**, 802–809.
Housby, J.N. and Southern, E.M. (1998) Fidelity of DNA ligation: a novel experimental approach based on the polymerisation of libraries of oligonucleotides. *Nucleic Acids Research*, **26**, 4259–4266.
Hunt, M., Kikuchi, T., Sanders, M. *et al.* (2013) REAPR: a universal tool for genome assembly evaluation. *Genome Biology*, **14**, R47.
Hunt, M., Newbold, C., Berriman, M. and Otto, T. (2014) A comprehensive evaluation of assembly scaffolding tools. *Genome Biology*, **15**, R42.
Istrail, S., Sutton, G.G., Florea, L. *et al.* (2004) Whole-genome shotgun assembly and comparison of human genome assemblies. *Proceedings of the National Academy of Sciences of the United States of America*, **101**, 1916–1921.
Jain, M., Fiddes, I.T., Miga, K.H. *et al.* (2015) Improved data analysis for the MinION nanopore sequencer. *Nature Methods*, **12**, 351–356.

Jeck, W.R., Reinhardt, J.A., Baltrus, D.A. *et al.* (2007) Extending assembly of short DNA sequences to handle error. *Bioinformatics*, **23**, 2942–2944.

Jiang, H., Lei, R., Ding, S.-W. and Zhu, S. (2014) Skewer: a fast and accurate adapter trimmer for next-generation sequencing paired-end reads. *BMC Bioinformatics*, **15**, 182.

Kelley, D., Schatz, M. and Salzberg, S. (2010) Quake: quality-aware detection and correction of sequencing errors. *Genome Biology*, **11**, R116.

Kingsford, C., Schatz, M. and Pop, M. (2010) Assembly complexity of prokaryotic genomes using short reads. *BMC Bioinformatics*, **11**, 21.

Kong, Y. (2011) Btrim: a fast, lightweight adapter and quality trimming program for next-generation sequencing technologies. *Genomics*, **98**, 152–153.

Koren, S., Schatz, M.C., Walenz, B.P. *et al.* (2012) Hybrid error correction and *de novo* assembly of single-molecule sequencing reads. *Nature Biotechnology*, **30**, 693–700.

Kuhring, M., Dabrowski, P.W., Piro, V.C. *et al.* (2015) SuRankCo: supervised ranking of contigs in *de novo* assemblies. *BMC Bioinformatics*, **16**, 240.

Langmead, B., Trapnell, C., Pop, M. and Salzberg, S. (2009) Ultrafast and memory-efficient alignment of short DNA sequences to the human genome. *Genome Biology*, **10**, R25.

Li, H. and Durbin, R. (2009) Fast and accurate short read alignment with Burrows–Wheeler transform. *Bioinformatics*, **25**, 1754–1760.

Li, H. and Durbin, R. (2010) Fast and accurate long-read alignment with Burrows–Wheeler transform. *Bioinformatics*, **26**, 589–595.

Li, R., Zhu, H., Ruan, J. *et al.* (2010) *De novo* assembly of human genomes with massively parallel short read sequencing. *Genome Research*, **20**, 265–272.

Lindgreen, S. (2012) AdapterRemoval: easy cleaning of next-generation sequencing reads. *BMC Research Notes*, **5**, 337.

Loman, N.J. and Watson, M. (2015) Successful test launch for nanopore sequencing. *Nature Methods*, **12**, 303–304.

Lunter, G. and Goodson, M. (2011) Stampy: a statistical algorithm for sensitive and fast mapping of Illumina sequence reads. *Genome Research*, **21**, 936–939.

Luo, R., Liu, B., Xie, Y. *et al.* (2012) SOAPdenovo2: an empirically improved memory-efficient short-read *de novo* assembler. *GigaScience*, **1**, 18.

Maccallum, I., Przybylski, D., Gnerre, S. *et al.* (2009) ALLPATHS 2: small genomes assembled accurately and with high continuity from short paired reads. *Genome Biology*, **10**, R103.

Macevicz, S. (1998) DNA sequencing by parallel oligonucleotide extensions. U.S. patent No. 5,750,341. Washington, DC: U.S. Patent and Trademark Office.

Mardis, E.R. (2011) A decade's perspective on DNA sequencing technology. *Nature*, **470**, 198–203.

Mardis, E.R. (2013) Next-generation sequencing platforms. *Annual Review of Analytical Chemistry*, **6**, 287–303.

Margulies, M., Egholm, M., Altman, W.E. *et al.* (2005) Genome sequencing in microfabricated high-density picolitre reactors. *Nature*, **437**, 376–380.

Martin, M. (2011) Cutadapt removes adapter sequences from high-throughput sequencing reads. *EMBnet Journal*, **17**, 10–12.

McKernan, K., Blanchard, A., Kotler, L. and Costa, G. (2012) Reagents, methods, and libraries for bead-based sequencing. U.S. Patent No. 8,329,404. Washington, DC: U.S. Patent and Trademark Office.

Metzker, M.L. (2010) Sequencing technologies – the next generation. *Nature Reviews Genetics*, **11**, 31–46.
Miller, J.R., Delcher, A.L., Koren, S. *et al.* (2008) Aggressive assembly of pyrosequencing reads with mates. *Bioinformatics*, **24**, 2818–2824.
Miller, J.R., Koren, S. and Sutton, G. (2010) Assembly algorithms for next-generation sequencing data. *Genomics*, **95**, 315–327.
Myers, E.W., Sutton, G.G., Delcher, A.L. *et al.* (2000) A whole-genome assembly of Drosophila. *Science*, **287**, 2196–2204.
Nagarajan, N. and Pop, M. (2013) Sequence assembly demystified. *Nature Reviews Genetics*, **14**, 157–167.
Nagarajan, N., Read, T.D. and Pop, M. (2008) Scaffolding and validation of bacterial genome assemblies using optical restriction maps. *Bioinformatics*, **24**, 1229–1235.
Paszkiewicz, K. and Studholme, D.J. (2010) *De novo* assembly of short sequence reads. *Briefings in Bioinformatics*, **11**, 457–472.
Patel, R.K. and Jain, M. (2012). NGS QC Toolkit: a toolkit for quality control of next generation sequencing data. *PloS one*, **7**, e30619.
Pevzner, P.A., Tang, H. and Waterman, M.S. (2001) An Eulerian path approach to DNA fragment assembly. *Proceedings of the National Academy of Sciences*, **98**, 9748–9753.
Phillippy, A.M., Schatz, M.C. and Pop, M. (2008) Genome assembly forensics: finding the elusive mis-assembly. *Genome Biology*, **9**, R55.
Piro, V., Faoro, H., Weiss, V. *et al.* (2014) FGAP: an automated gap closing tool. *BMC Research Notes*, **7**, 371.
Pop, M. (2009) Genome assembly reborn: recent computational challenges. *Briefings in Bioinformatics*, **10**, 354–366.
Rahman, A. and Pachter, L. (2013) CGAL: computing genome assembly likelihoods. *Genome Biology*, **14**, R8.
Rodríguez-Ezpeleta, N., Hackenberg, M. and Aransay, A.M. (2012) *Bioinformatics for high throughput sequencing*, Springer, New York.
Ribeiro, F.J., Przybylski, D., Yin, S. *et al.* (2012) Finished bacterial genomes from shotgun sequence data. *Genome Research*, **22**, 2270–2277.
Ronaghi, M., Karamohamed, S., Pettersson, B. *et al.* (1996) Real-time DNA sequencing using detection of pyrophosphate release. *Analytical Biochemistry*, **242**, 84–89.
Rothberg, J.M., Hinz, W., Rearick, T.M. *et al.* (2011) An integrated semiconductor device enabling non-optical genome sequencing. *Nature*, **475**, 348–352.
Salzberg, S.L., Phillippy, A.M., Zimin, A. *et al.* (2012) GAGE: a critical evaluation of genome assemblies and assembly algorithms. *Genome Research*, **22**, 557–567.
Sanger, F., Air, G.M., Barrell, B.G. *et al.* (1977) Nucleotide sequence of bacteriophage [phi]X174 DNA. *Nature*, **265**, 687–695.
Schmieder, R. and Edwards, R. (2011) Quality control and preprocessing of metagenomic datasets. *Bioinformatics*, **27**, 863–864.
Shendure, J. and Ji, H. (2008) Next-generation DNA sequencing. *Nature Biotechnology*, **26**, 1135–1145.
Shendure, J., Mitra, R.D., Varma, C. and Church, G.M. (2004) Advanced sequencing technologies: methods and goals. *Nature Reviews Genetics*, **5**, 335–344.
Shendure, J., Porreca, G.J., Reppas, N.B. *et al.* (2005) Accurate multiplex polony sequencing of an evolved bacterial genome. *Science*, **309**, 1728–1732.

Simpson, J.T. and Durbin, R. (2012) Efficient *de novo* assembly of large genomes using compressed data structures. *Genome Research*, **22**, 549–556.

Simpson, J.T., Wong, K., Jackman, S.D. *et al.* (2009) ABySS: a parallel assembler for short read sequence data. *Genome Research*, **19**, 1117–1123.

Song, L., Florea, L. and Langmead, B. (2014) Lighter: fast and memory-efficient sequencing error correction without counting. *Genome Biology*, **15**, 509.

Sutton, G.G., White, O., Adams, M.D. *et al.* (1995) TIGR assembler: a new tool for assembling large shotgun sequencing projects. *Genome Science & Technology*, **1**, 9–19.

Trivedi, U.H., Cézard, T., Bridgett, S. *et al.* (2014) Quality control of next-generation sequencing data without a reference. *Frontiers in genetics*, **5**, 111.

Tsai, I.J., Otto, T.D. and Berriman, M. (2010) Improving draft assemblies by iterative mapping and assembly of short reads to eliminate gaps. *Genome Biology*, **11**, R41.

Turcatti, G., Romieu, A., Fedurco, M. and Tairi, A.P. (2008) A new class of cleavable fluorescent nucleotides: synthesis and optimization as reversible terminators for DNA sequencing by synthesis. *Nucleic Acids Research*, **36**, e25.

Warren, R.L., Sutton, G.G., Jones, S.J.M. and Holt, R.A. (2007) Assembling millions of short DNA sequences using SSAKE. *Bioinformatics*, **23**, 500–501.

Yang, X., Chockalingam, S.P. and Aluru, S. (2013) A survey of error-correction methods for next-generation sequencing. *Briefings in Bioinformatics*, **14**, 56–66.

Zerbino, D.R. and Birney, E. (2008) Velvet: algorithms for *de novo* short read assembly using de Bruijn graphs. *Genome Research*, **18**, 821–829.

Zimin, A.V., Delcher, A.L., Florea, L. *et al.* (2009) A whole-genome assembly of the domestic cow. *Bos taurus. Genome Biology*, **10**, R42.

4

Genome Annotation: Determination of the Coding Potential of the Genome

Ruijia Wang, Lisui Bao, Shikai Liu and Zhanjiang Liu

Introduction

With "raw" genomic sequences, the sequences themselves are meaningless, with just the four nucleotides, A, T, G, and C. Genome sequences become meaningful and practically much more useful when they are annotated, especially with protein-encoding genes. In this chapter, we will provide an overview for genome annotation, with some demonstration of the processes using examples. Although genome annotation pipelines differ in their details, they share a core set of features. Generally, existing bioinformatics software use two strategies: (1) evidence-based gene prediction using expressed sequence tags (ESTs), and/or proteins of target genes; and (2) *ab initio* gene prediction, which uses the genomic DNA sequence itself as the only input. We will review both of these methods, including their principles, algorithm, advantages, and disadvantages. The two most popular pipelines used for genome annotation, AUGUSTUS and FGENESH, will be discussed in detail. DNA sequences of zebrafish chromosome 1 and its corresponding cDNA sequences are used as inputs to demonstrate the processes.

The term "gene" is derived from the Greek words "genesis" or "genos", which represents birth or origin, respectively (Pearson, 2006). As the most important vocabulary in molecular biology, its meaning has been evolving. Biologically, it represents the basic unit of heredity in living organisms. It is widely used in molecular biology as a name given to stretches of DNA that encode for a polypeptide or for a functional RNA (Pennisi, 2007). While the concept of proteins or polypeptides is relatively stable, the concept of functional RNA has been evolving. At early phase of molecular biology, the functional RNAs were limited to the basic concept of messenger RNA (mRNA), the structural ribosomal RNA (rRNA), and the transfer RNA (tRNA). Therefore, DNA sequences coding for mRNA, rRNA, and tRNA were regarded as genes (Pennisi, 2007). However, in recent years, particularly after the discovery of small RNA and the rapidly evolving importance of non-coding RNAs, the concept of gene has become even more complex (Mattick & Makunin, 2006).

With the development of next-generation sequencing and genomics, the modern definition of a gene is "a locatable region of genomic sequence, corresponding to a unit of inheritance, which is associated with regulatory regions, transcribed regions, and/or other functional sequence regions" (Pearson, 2006). However, for the purpose of this chapter, which will focus on the informatic perspectives of the coding potential of a

Bioinformatics in Aquaculture: Principles and Methods, First Edition. Edited by Zhanjiang (John) Liu.

genome, we will use the narrow sense of the gene, that is, genomic sequences that encode for proteins. Readers interested in long non-coding RNAs or microRNAs are referred to Chapter 12 and Chapter 13.

The genome sizes of aquaculture species vary greatly, ranging from hundreds of millions of base pairs to several billions (Chen *et al.*, 2014; Xu *et al.*, 2014; Zhang *et al.*, 2012). The variation of the number of active protein-encoding genes, however, is relatively small. For instance, the invertebrate species can harbor 15000–20000 unique genes (Zhang *et al.*, 2012), while the diploid teleost fish generally harbor no more than 30000 unique protein-encoding genes (Chen *et al.*, 2014). However, with the teleost fish having gone through one more round of whole-genome duplication, and in some cases two more rounds, the total number of genes can be quite high in some tetraploid fish species such as common carp and Atlantic salmon (Chen *et al.*, 2014; Davidson *et al.*, 2010). For instance, the common carp has over 52610 genes. Even in this situation, the vast majority of genome sequences are not genes, but just intergenic sequences. Predicting the genes bioinformatically, therefore, is very important for the understanding and annotation of the genome.

At the beginning, gene discoveries relied on painstaking experimentation. With rapid advances in genome and transcriptome research, gene discoveries are now becoming largely a computational issue. The core component of genome annotation is the bioinformatic determination of the protein-coding genes. In addition, genome annotation certainly also includes prediction using bioinformatics for functional RNA-coding genes (other than mRNA), as well as prediction of other functional elements such as regulatory regions. Predicting the function of a gene and confirming the gene functions still call for in vivo experimentation (Sleator, 2010) through gene knockout and other assays, although bioinformatics research is making it increasingly possible to predict the function of a gene based on its sequences alone (Overbeek *et al.*, 2005). Many aspects of structural gene prediction are based on current understanding of underlying biochemical processes in the cell such as gene transcription, splicing, polyadenylation, translation, protein–protein interactions, and regulation processes, which are subjects of active research in the various Omics fields such as transcriptomics, proteomics, metabolomics, and, more generally, structural and functional genomics. Here in this chapter, we will cover the bioinformatics methods for the determination of the coding potential of a genome, that is, bioinformatics predictions of genes based on existing knowledge, information, and technologies.

Methods Used in Gene Prediction

Genome annotation is a key step following sequence assembly. To achieve the goals of genome annotation, genome sequences need to be repeatedly masked before annotation (Korf, 2004). Gene prediction is realized through two major methods, homolog-based methods and *ab initio* methods. Practically, the combined approaches integrating both empirical and *ab initio* methods are usually the optimal choice to obtain a decent annotation for genomes (Do & Choi, 2006).

Homolog-based Methods

Empirical method is also known as homolog- or evidence-based gene prediction method. The core principle of the empirical method of gene prediction is to search against the target sequences that are similar to extrinsic evidence in the form of known mRNA, ESTs, RNA-Seq reads, and protein products. For a single mRNA sequence, homolog-based methods are used to retrieve the unique genomic DNA sequence from which it is transcribed, whereas, for a protein sequence, a cluster of potential coding DNA sequences can be predicted through reverse translation of the genetic codons. After the determination of candidate DNA sequences, it is relatively straightforward to efficiently align the target genome sequences. Given a sequence, local alignment algorithms, such as BLAST, look for regions of similarity between the target sequences and candidate matches, which can be complete or partial matches, and perfect or imperfect matches. However, this approach relies heavily on the sequence alignments, and therefore it is limited by the contents and accuracy of the sequence databases.

With the homolog-based methods, the signal of a genome sequence with protein-coding potential is the high similarity to a known mRNA or protein product. However, the feasibility of this approach systemically depends on the extensive sequencing of mRNA and protein from the target species itself or its closed species. With multi-cellular eukaryotic organisms, only a subset of all genes of the genome are expressed at a given time, in a specific tissue, or under a specific environment, leading to the lack of extrinsic evidence (Saha *et al.*, 2002). Therefore, to collect extrinsic features for the genes as complete as possible in the whole genome of a complex organism, various RNA-Seq are required, which itself presents both expenditure and bioinformatics difficulties (Saha *et al.*, 2002).

In spite of these difficulties, more and more transcript and protein sequence databases are generated for many species. The most complete transcriptome resources are available for human as well as a number of model species in biology, such as mice and zebrafish. However, as a result of the application of next-generation RNA-Seq, transcriptome resources are now widely available for many species, including aquaculture species.

Ab initio methods

In most aquaculture species, the abundance of introns and long intergenic regions makes it difficult to collect extrinsic evidence for the homolog-based methods described earlier. As a result, gene prediction for the genome sequences of such species usually starts with *ab initio* methods, in which the conserved landmark features of protein-coding genes are searched in the genomic DNA sequence alone (Zhu, Lomsadze & Borodovsky, 2010). The *ab initio* approach is a computational gene prediction method based on the detection of certain features in the DNA (Do & Choi, 2006). These features can be either "signals, specific sequences that indicate the presence of a gene nearby, or statistical properties of protein-coding sequence itself" (Yandell & Ence, 2012). Because *ab initio* gene prediction relies only on the statistical parameters in the DNA sequence for gene identification, its result is perhaps more preliminary than that of the empirical gene prediction method (Zhu *et al.* 2010).

Ab initio gene prediction in aquaculture species, especially the tetraploid fish species (Xu *et al.*, 2014) or the less characterized species such as oysters (Zhang *et al.*, 2012), is considerably more challenging for several reasons. For instance, the promoter and other regulatory features in the genomes of eukaryotes are more complex than in prokaryotes, increasing the difficulty for accurate gene prediction. In addition, the alternative splicing mechanisms lead to various exons (Modrek & Lee, 2002). Since the redundant splice sites are themselves quite complex features, the eukaryotic gene finders need to be specifically set up to identify signals (Koonin & Galperin, 2003). A typical protein-coding gene in zebrafish might be split into a dozen exons, each less than 200 base pairs in length, and some as short as 20–30 base pairs (Howe *et al.*, 2014). It is therefore much more difficult to detect periodicities and other known content properties of protein-coding DNA in eukaryotes. In the cases of tetraploid organisms such as common carp, the confusion caused by paralogues itself is a great challenge (Xu *et al.*, 2014).

The complex probabilistic models combining information from a variety of different signal and content measurements, such as hidden Markov models (HMMs), are widely used in eukaryotic genome annotations with the advance of gene finders (Meyer & Durbin, 2002). A number of gene predictors are now publicly available with similar performance on gene prediction from a single sequence or short genomic region (Yandell & Ence, 2012)(Table 4.1). However, for gene prediction at the whole-genome level, only a few pipelines are provided. Several *ab initio* gene finders are useful for eukaryotic species. The notable examples are GENSCAN, FGENESH, and AUGUSTUS, which are frequently used with animal genomes (Burge, 1997; Foissac *et al.*, 2008; Salamov and Solovyev, 2000; Stanke *et al.*, 2004).

Table 4.1 A list of gene prediction software packages, modified from http://en.wikipedia.org/wiki/List_of_gene_prediction_software.

Name	Links	References
AUGUSTUS	http://augustus.gobics.de/	Stanke *et al.*, 2004
BGF	http://bgf.genomics.org.cn/	Zhao *et al.*, 2004
ChemGenome	http://bgf.genomics.org.cn/	Khandelwal & Bhyravabhotla, 2010
CRITICA	http://www.ttaxus.com/software.html	Henderson, Salzberg & Fasman, 1997
DNA SUBWAY	http://dnasubway.iplantcollaborative.org/	Micklos, Lauter & Nisselle, 2011
ExoniPhy	https://cgwb.nci.nih.gov/cgi-bin/hgTrackUi?hgsid=94955&c=chr6&g=exoniphy	Gross & Brent, 2005
EUGENE	http://eugene.toulouse.inra.fr/	Schiex, Moisan & Rouzé, 2001
FGENESH	http://linux1.softberry.com/berry.phtml?topic=fgenesh&group=programs&subgroup=gfind	Salamov & Solovyev, 2000
FRAMED	http://tata.toulouse.inra.fr/apps/FrameD/FD	Schiex *et al.*, 2003
GRAIL	https://www.broadinstitute.org/mpg/grail/	Burge & Karlin, 1997
GeneID	http://genome.crg.es/software/geneid/geneid.html	Blanco, Parra & Guigó, 2007
GeneMark	http://topaz.gatech.edu/GeneMark	Lukashin & Borodovsky, 1998
GeneTack	http://topaz.gatech.edu/GeneTack	Albert *et al.*, 2008

Table 4.1 (Continued)

Name	Links	References
GeneWise	http://www.ebi.ac.uk/Tools/psa/genewise/	Birney & Durbin, 2000
GeneZilla(TIGRscan)	http://www.genezilla.org/	Allen *et al.*, 2006
GENOMESCAN	http://genes.mit.edu/genomescan.html	Yeh, Lim & Burge, 2001
GenomeThreader	http://www.genomethreader.org/	Gremme *et al.*, 2005
GENSCAN	http://genes.mit.edu/GENSCAN.html	Burge & Karlin, 1997
GLIMMER	http://genes.mit.edu/GENSCAN.html	Delcher *et al.*, 1999
GLIMMERHMM	http://cbcb.umd.edu/software/glimmerhmm/	Allen *et al.*, 2006
GrailEXP	http://compbio.ornl.gov/grailexp/	Uberbacher, Hyatt & Shah, 2004
HMMgene	http://www.cbs.dtu.dk/services/HMMgene/hmmgene1_1.php	Krogh, 2000
JIGSAW(Combiner)	http://www.cbcb.umd.edu/software/jigsaw/	Allen *et al.*, 2006
mGene	http://bioweb.me/mgene/	Schweikert *et al.*, 2009
MORGAN	http://www.cbcb.umd.edu/~salzberg/morgan.html	Salzberg *et al.*, 1998
ORF FINDER	http://www.ncbi.nlm.nih.gov/gorf/gorf.html	Rombel *et al.*, 2002
PROCRUSTES	http://www-hto.usc.edu/software/procrustes/	Gelfand, Mironov & Pevzner, 1996
SGP2	http://genome.crg.es/software/sgp2/index.html	Parra *et al.*, 2003
SLAM	http://baboon.math.berkeley.edu/~syntenic/slam.html	Alexandersson *et al.*, 2003
TWINSCAN	http://mblab.wustl.edu/software.html	Wu *et al.*, 2004
VEIL	http://www.cs.jhu.edu/~genomics/Veil/veil.html	Henderson *et al.*, 1997

A number of factors need to be considered for the selection of an optimal whole-genome gene predictor for the target species. The most important factor is accuracy, which can be measured as sensitivity (Sn) and specificity (Sp). For the prediction of coding bases, exons, transcripts, and genes, *sensitivity* is defined as the number of correctly predicted features divided by the number of annotated features; and *specificity* is the number of correctly predicted features divided by the number of predicted features (http://augustus.gobics.de/accuracy). An exonic prediction is considered accurate when both splice sites at the annotated position of an exon are correct. Similar to exonic prediction, a predicted transcript is considered correct only when all exons are successfully predicted but without extra annotated exons. A predicted gene is considered correct when any of its transcripts are correct, that is, at least one isoform of the gene is exactly as annotated in the reference annotation (http://augustus.gobics.de/accuracy).

Among the available whole-genome gene predictors or annotators, the two most popular pipelines are FGENESH and AUGUSTUS. FGENESH is a commercially available,

Table 4.2 Accuracy of annotation of *C. elegans* using various gene-prediction programs. Accuracy values are from Coghlan *et al.* (2008), including sensitivity (Sn) and specificity (Sp) for the prediction of base, exon, transcript, and gene.

Programs	base		exon		transcript		gene	
	Sn	Sp	Sn	Sp	Sn	Sp	Sn	Sp
AUGUSTUS	97	89	86.1	72.6	50.1	28.7	61.1	38.4
FGENESH	98.2	87.1	86.4	73.6	47.1	34.6	57.8	35.4
GeneMark.hmm	98.3	83.1	83.2	65.6	37.7	24	46.3	24.5
GeneID	93.9	88.2	77	68.6	36.2	22.8	44.4	25.1
Agene	93.8	83.4	68.9	61.1	9.8	13.1	12	14.1
EUGENE	94	89.5	80.3	73	49.1	28.8	60.2	30.2
ExonHunter	95.4	86	72.6	62.5	15.5	18.6	19.1	19.2
GlimmerHMM	97.6	87.6	84.4	71.4	47.3	29.3	58	30.6
SNAP	94	84.5	74.6	61.3	32.6	18.6	40	19.1

user-friendly, interface-based, super-fast (50–100 times faster than other classic gene predictor, e.g., GENESCAN), and accurate gene finder that has been integrated into the comprehensive genomic analysis package MolQuest (http://molquest.com/). Its good performance was demonstrated by the rice genome sequencing projects, in which it was cited as "the most successful (gene prediction) program" (Yu *et al.*, 2002), and was used to produce 87% of all high-evidence predicted genes (Goff *et al.*, 2002).

AUGUSTUS is the most widely used command-line-based genome-level gene finder that has been successfully used in various species, including several genome projects of aquaculture fish species. The performance of AUGUSTUS was reported in the *C. elegans* genome annotation project (Coghlan *et al.*, 2008)(Table 4.2).

The advantage of FGENESH and AUGUSTUS genomic gene finders is not only their high prediction accuracy, but also the capability of their extra modules to integrate homology-based methods. For instance, in the homology-based module of FGENESH, FGENESH+, the accuracy of gene prediction can be up to 100% in mice. The detailed usage of FGENESH and AUGUSTUS will be described in the next section using the sequence of chromosome 1 of zebrafish as an example.

Case Study: Genome Annotation Examples: Gene Annotation of Chromosome 1 of Zebrafish using FGENESH and AUGUSTUS

Sample data access:

1) zebrafish_chr1.fa: chromosome 1 of zebrafish genome V9 (ftp://ftp.ensembl.org/pub/release-78/fasta/danio_rerio/dna/Danio_rerio.Zv9.dna.chromosome.1.fa.gz)
2) zebrafish.cdna.all.fa: a set of 48435 ESTs from zebrafish transcriptome (ftp://ftp.ensembl.org/pub/release-78/fasta/danio_rerio/cdna/Danio_rerio.Zv9.cdna.all.fa.gz)

Pipeline Installations

FGENESH

Getting FGENESH (Microsoft Windows based): download from http://molquest.com/cgi-bin/download.pl?file=molquest-win-2.4.5-trial

Installation: double-click "MolQuest-2.4.5-trial-1135-5.exe" and follow the interface installation instructions.

AUGUSTUS

Pre-requisite:

a) Creating gene structures: CEGMA
b) ESTs alignment: BLAT/Bowtie/Tophat

Getting AUGUSTUS: type `wget` http://bioinf.uni-greifswald.de/augustus/binaries/augustus-3.2.tar.gz

Installation:

1) Unpack

```
>tar -xzf augustus-3.2.tar.gz
```

2) Compile (if not already compiled)
   Install dependencies: to turn on the optional support of gzip-compressed input files
   1) Edit "common.mk" and uncomment the line "ZIPINPUT = true"
   2) Install these dependencies:
      1. Boost C++ libraries: libboost-iostreams-dev (on Ubuntu: `sudo apt-get install libboost-iostreams-dev`)
         zlib library for compression methods. Download from http://www.zlib.net/ or install via package manager.

      ```
      > make
      ```

      Use this command to install the pipeline.
      After compilation has finished, the command `bin/augustus` must be executable and should print a usage message.
   3) As a normal user, add the directory of the executables to the `PATH` environment variable. For example, issue

      ```
      PATH=$PATH:~/augustus/bin:~/augustus/scripts
      ```

      As an administrator, globally install AUGUSTUS by typing

      ```
      make install
      ```

Prepare the Input Files

Genome structure file (if EST or cDNA sequences are not available)

```
$ cegma -g zebrafish_chr1.fa -v -o zebrafish_chr1
```

`$ path-to-augustus/scripts/cegma2gff.pl` (transfer the result to a format that can be recognized by AUGUSTUS)

```
zebrafish_chr1.cegma.gff > augustus_chr1_training.gff
```

Generate Hint file from EST (can also be automatically done by autoAugTrain.pl)
Align the ESTs against chr1 using BLAT.

```
$ blat -minIdentity=92 -minIdentity=92 zebrafish_chr1.fa
   zebrafish.cdna.all.fa test_chr1.est.psl
```

Further filter the alignments and keep the best alignment for each query with at least 80% of the query length

```
$ cat test_chr1.est.psl | filterPSL.pl --best --minCover=80
   > cat test_chr1.est.f.psl
$ cat test_chr1.est.f.psl | sort -n -k 16,16 | sort -s -k
   14,14 > test_chr1.est.fs.psl
$ blat2hints.pl --nomult --in=test_chr1.est.fs.psl
   --out=hints.gff
```

Gene Prediction Using FGENESH

Parameter Setting in FGENESH

The analyst needs to set the path of input files; path and name of the output results; and the closest species model used for genome prediction. Some additional options, including print mRNA and exons, splicing prediction, and promotor prediction, can also be set in this step. After the parameters are set, the analyst can simply start the analysis by clicking the “run” button.

Gene Prediction Using AUGUSTUS

AUGUSTUS requires the genome sequences and at least one additional sequence/ structure file from either genome structure, cDNA, or protein level. The permissible combination of these files is listed as the following:

```
Genome file+cDNA file
```

In this case, the cDNA file is used to create training gene structures with PASA. After the parameters are trained using the so-created gene structure file, the cDNAs will additionally be used to create hints. Hints are extrinsic evidence for gene structures that are used during gene prediction with hints. And then, AUGUSTUS is used to predict genes in the genome file *ab initio* and with hints.

```
Genome file+protein file
```

In this case, the protein file is used to create a training gene set using Scipio. After parameter optimization, AUGUSTUS is used to predict genes in the genome sequence *ab initio*.

```
Genome file+gene structure file
```

In this case, the gene structure file is used as a training gene set. Gene structure files can be provided in two different formats: genbank format and gff format. Training gene sequences are extracted from the genome file prior to parameter optimization. Finally, AUGUSTUS is used for *ab initio* gene prediction.

```
Genome file+cDNA file+protein file
```

In this case, the protein file will be used to create a training gene set using Scipio.

`Genome file+cDNA file+gene structure file`

In this case, the gene structure file is used as a training gene set. cDNA sequences will be used as evidence for prediction in the form of hints that are generated using BLAT. Finally, AUGUSTUS is used to predict genes in the genome file *ab initio* and with hints.

Command-line (Genome file+cDNA file)
First step: initially training the genome of your species

```
$ path-to-augustus/scripts/autoAug.pl
   --species=zebrafish_chr1 --genome=zebrafish_chr1.fa
   -cdna=zebrafish.cdna.all.fa --pasa
$ cd autoAug/autoAugPred_abinitio/shells
$ ./shellForAug
```

Second step: continue to predict genome structure with AUGUSTUS without hints, no UTR

```
$ path-to-augustus/scripts/autoAug.pl
   --species=zebrafish_chr1 --genome=autoAug/seq/
   genome_clean.fa -cdna=zebrafish.cdna.all.fa
   --useexisting -v -v  --index=1
$ cd autoAug/autoAugPred_hints/shells
$ ./shellForAug
```

Note: if the analyst is utilizing RNA-seq sequences to generate the hint file other than ESTs, please manually edit “extrinsic.E.cfg” to “extrinsic.M.RM.E.W.cfg” in all the aug* scripts

Third step (optional): continue to predict genome structure with AUGUSTUS with hints and UTR (if possible)

```
$ path-to-augustus/scripts/autoAug.pl
   --species=zebrafish_chr1 --genome= autoAug/seq/
   genome_clean.fa -cdna=zebrafish.cdna.all.fa
   --useexisting -v -v -v --index=2
```

Output from AUGUSTUS
`augustus.aa`: the amino acid sequence of the genes predicted from the genome sequences

`augustus.codingseq`: the coding sequence of the genes predicted from the genome sequences

`augustus.gff`: the gene annotation file for the genome sequences in gff format

`augustus.gtf`: the gene annotation file for the genome sequences in gtf format

`augustus.gbrowse`: the gene annotation file used for gbrowse visualization

Discussion

For genome-level gene prediction, the first and most important part is to choose the suitable gene predictor for your species. Unfortunately, there are no shortcuts for this step except testing various gene predictors. Even if one is told that some software is better, it may not really suit the species under consideration. Some factors are helpful for predictor selection, including accuracy, running time, computational resource requirement, and also the updating speed for fixing emerging bugs. The analyst needs to evaluate the performance of different gene predictors according to these factors to choose the best one for the genome of the species under study. Another challenge is to make a choice between commercial and open-source gene predictors. The advantage of commercial gene predictors is obvious; they are usually faster, have fewer bugs, and are easier to use. As such, commercial packages do not require a high level of bioinformatics background for the analyst. However, these advantages are built on a high financial cost for the packages, usually ~US$5000 per year (e.g., FGENESH). On the other hand, open-source gene predictors are slower, and difficult to use and learn, but they have the potential capability to give you the best prediction by the combination of different level parameters and datasets without imposing any financial burdens. However, the time required for training analysts can be quite extensive, and, therefore, analysts need to compromise between the cost and their bioinformatics skills and make the decision.

References

Albert, I., Wachi, S., Jiang, C. and Pugh, B.F. (2008) GeneTrack – a genomic data processing and visualization framework. *Bioinformatics*, **24**, 1305–1306.

Alexandersson, M., Cawley, S. and Pachter, L. (2003) SLAM: cross-species gene finding and alignment with a generalized pair hidden Markov model. *Genome Research*, **13**, 496–502.

Allen, J.E., Majoros, W.H., Pertea, M. and Salzberg, S.L. (2006) JIGSAW, GeneZilla, and GlimmerHMM: puzzling out the features of human genes in the ENCODE regions. *Genome Biology*, **7**, S9.

Birney, E. and Durbin, R. (2000) Using GeneWise in the Drosophila annotation experiment. *Genome Research*, **10**, 547–548.

Blanco, E., Parra, G. and Guigó, R. (2007) Using GeneID to identify genes. *Current Protocols in Bioinformatics*, vol. 1, unit 4.3, pp. 1–28.

Burge, C. and Karlin, S. (1997) Prediction of complete gene structures in human genomic DNA. *Journal of Molecular Biology*, **268**, 78–94.

Burge, C.B. (1997). *Identification of Genes in Human Genomic DNA.* Department of Mathematics. Stanford, CA: Stanford University.

Chen, S., Zhang, G., Shao, C. *et al.* (2014) Whole-genome sequence of a flatfish provides insights into ZW sex chromosome evolution and adaptation to a benthic lifestyle. *Nature Genetics*, **46**, 253–260.

Coghlan, A., Fiedler, T.J., McKay, S.J. *et al.* (2008) nGASP—the nematode genome annotation assessment project. *BMC Bioinformatics*, **9**, 549.

Davidson, W.S., Koop, B.F., Jones, S.J. *et al.* (2010) Sequencing the genome of the Atlantic salmon (*Salmo salar*). *Genome Biology*, **11**, 403.

Delcher, A.L., Harmon, D., Kasif, S. *et al.* (1999) Improved microbial gene identification with GLIMMER. *Nucleic Acids Research*, **27**, 4636–4641.

Do, J.H. and Choi, D. (2006) Computational approaches to gene prediction. *Journal of Microbiology*, **44**, 137.

Foissac, S., Gouzy, J., Rombauts, S. *et al.* (2008) Genome annotation in plants and fungi: EuGene as a model platform. *Current Bioinformatics*, **3**, 87–97.

Gelfand, M.S., Mironov, A.A. and Pevzner, P.A. (1996) Gene recognition via spliced sequence alignment. *Proceedings of the National Academy of Sciences*, **93**, 9061–9066.

Goff, S.A., Ricke, D., Lan, T.-H. *et al.* (2002) A draft sequence of the rice genome (*Oryza sativa* L. ssp. japonica). *Science*, **296**, 92–100.

Gremme, G., Brendel, V., Sparks, M.E. and Kurtz, S. (2005) Engineering a software tool for gene structure prediction in higher organisms. *Information and Software Technology*, **47**, 965–978.

Gross, S.S. and Brent, M.R. (2005) Using multiple alignments to improve gene prediction. *Journal of Computational Biology*, **13**(2), 379–393.

Henderson, J., Salzberg, S. and Fasman, K.H. (1997) Finding genes in DNA with a hidden Markov model. *Journal of Computational Biology*, **4**, 127–141.

Howe, K., Clark, M.D., Torroja, C.F. *et al.* (2014) Corrigendum: the zebrafish reference genome sequence and its relationship to the human genome. *Nature*, **505**, 498–503.

Khandelwal, G. and Bhyravabhotla, J. (2010). A phenomenological model for predicting melting temperatures of DNA sequences. *PLoS one*, **5**(8), e12433.

Koonin, E.V. and Galperin, M.Y. (2003) Principles and methods of sequence analysis, in *Sequence—Evolution—Function*, US, Springer, pp. 111–192.

Korf, I. (2004) Gene finding in novel genomes. *BMC Bioinformatics*, **5**, 59.

Krogh, A. (2000) Using database matches with HMMGene for automated gene detection in Drosophila. *Genome Research*, **10**, 523–528.

Lukashin, A.V. and Borodovsky, M. (1998) GeneMark.hmm: new solutions for gene finding. *Nucleic Acids Research*, **26**, 1107–1115.

Mattick, J.S. and Makunin, I.V. (2006) Non-coding RNA. *Human Molecular Genetics*, **15**, R17–R29.

Meyer, I.M. and Durbin, R. (2002) Comparative *ab initio* prediction of gene structures using pair HMMs. *Bioinformatics*, **18**, 1309–1318.

Micklos, D., Lauter, S. and Nisselle, A. (2011) Lessons from a science education portal. *Science*, **334**, 1657–1658.

Modrek, B. and Lee, C. (2002) A genomic view of alternative splicing. *Nature Genetics*, **30**, 13–19.

Overbeek, R., Begley, T., Butler, R.M. *et al.* (2005) The subsystems approach to genome annotation and its use in the project to annotate 1000 genomes. *Nucleic Acids Research*, **33**, 5691–5702.

Parra, G., Agarwal, P., Abril, J.F. *et al.* (2003) Comparative gene prediction in human and mouse. *Genome Research*, **13**, 108–117.

Pearson, H. (2006) Genetics: hat is a gene? *Nature*, **441**, 398–401.

Pennisi, E. (2007) DNA study forces rethink of what it means to be a gene. *Science*, **316**, 1556–1557.

Rombel, I.T., Sykes, K.F., Rayner, S. and Johnston, S.A. (2002) ORF-FINDER: a vector for high-throughput gene identification. *Gene*, **282**, 33–41.

Saha, S., Sparks, A.B., Rago, C. *et al.* (2002) Using the transcriptome to annotate the genome. *Nature Biotechnology*, **20**, 508–512.

Salamov, A.A. and Solovyev, V.V. (2000) *Ab initio* gene finding in Drosophila genomic DNA. *Genome Research*, **10**, 516–522.

Salzberg, S., Delcher, A.L., Fasman, K.H. and Henderson, J. (1998) A decision tree system for finding genes in DNA. *Journal of Computational Biology*, **5**, 667–680.
Schiex, T., Gouzy, J., Moisan, A. and de Oliveira, Y. (2003) FrameD: a flexible program for quality check and gene prediction in prokaryotic genomes and noisy matured eukaryotic sequences. *Nucleic Acids Research*, **31**, 3738–3741.
Schiex, T., Moisan, A. and Rouzé, P. (2001) EuGene: an eukaryotic gene finder that combines several sources of evidence, in *Computational Biology*, Springer, Berlin, Heidelberg, pp. 111–125.
Schweikert, G., Zien, A., Zeller, G. *et al.* (2009) mGene: accurate SVM-based gene finding with an application to nematode genomes. *Genome Research*, **19**, 2133–2143.
Sleator, R.D. (2010) An overview of the current status of eukaryote gene prediction strategies. *Gene*, **461**, 1–4.
Stanke, M., Steinkamp, R., Waack, S. and Morgenstern, B. (2004) AUGUSTUS: a web server for gene finding in eukaryotes. *Nucleic Acids Research*, **32**, W309–W312.
Uberbacher, E.C., Hyatt, D. and Shah, M. (2004) GrailEXP and genome analysis pipeline for genome annotation. *Current Protocols in Human Genetics*, chapter 4, unit 4.9, pp. 1–15.
Wu, J.Q., Shteynberg, D., Arumugam, M. *et al.* (2004) Identification of rat genes by TWINSCAN gene prediction, RT–PCR, and direct sequencing. *Genome Research*, **14**, 665–671.
Xu, P., Zhang, X., Wang, X. *et al.* (2014) Genome sequence and genetic diversity of the common carp, *Cyprinus carpio*. *Nature Genetics*, **46**, 1212–1219.
Yandell, M. and Ence, D. (2012) A beginner's guide to eukaryotic genome annotation. *Nature Reviews Genetics*, **13**, 329–342.
Yeh, R.-F., Lim, L.P. and Burge, C.B. (2001) Computational inference of homologous gene structures in the human genome. *Genome Research*, **11**, 803–816.
Yu, J., Hu, S., Wang, J. *et al.* (2002) A draft sequence of the rice genome (*Oryza sativa* L. ssp. indica). *Science*, **296**, 79–92.
Zhang, G., Fang, X., Guo, X. *et al.* (2012) The oyster genome reveals stress adaptation and complexity of shell formation. *Nature*, **490**, 49–54.
Zhao, W., Wang, J., He, X. *et al.* (2004) BGI-RIS: an integrated information resource and comparative analysis workbench for rice genomics. *Nucleic Acids Research*, **32**, D377–D382.
Zhu, W., Lomsadze, A. and Borodovsky, M. (2010) *Ab initio* gene identification in metagenomic sequences. *Nucleic Acids Research*, **38**, e132–e132.

5

Analysis of Repetitive Elements in the Genome

Lisui Bao and Zhanjiang Liu

Introduction

Repetitive DNA makes up major portions of most eukaryotic genomes. Repetitive elements, especially transposable elements (TEs), have been a major force in genome evolution by introducing profound modifications to genome size, function, and structure. Prior to genome annotation in the whole-genome sequencing study, it is usually the first step to filter out repeats including TEs and/or low-complexity regions, because abundant repeats would complicate genome annotation and confuse the gene prediction programs. Meanwhile, the repetitive elements themselves remain the least-characterized genome component in most eukaryotic species studied so far. Therefore, repeats identification and characterization is a crucial step both for genome studies and for evolution studies. In this chapter, we will provide an overview of methods for the analysis of repetitive elements, a discussion of currently used software, and a practical demonstration of repeat analysis using selected software as examples.

Repetitive elements, also known as sequence repeats, are DNA sequences that occur in multiple copies under certain patterns throughout the genome (Benson, 1999). They are highly abundant in eukaryotic genomes, particularly in mammalian genomes. For instance, more than 50% of the human genome is made up of TE-derived sequences (Richard, Kerrest & Dujon, 2008). Based on their arrangements and distribution in the genome, repetitive elements can be categorized into two major groups: (1) tandem repeats, which are sequences that contain many copies of the same or similar short sequences, ranging from a single base to several mega bases, such as microsatellites, minisatellites, and satellites (Ahmed & Liang, 2012); and (2) interspersed repeats, which are the same or similar sequences that are dispersed in the genome, such as DNA transposons, long interspersed nuclear elements (LINEs), and short interspersed nuclear elements (SINEs) (Gorni *et al.*, 2005; Jjingo *et al.*, 2014).

Repetitive elements, once regarded as "junk" DNA sequences, are now recognized as the driving force of genome evolution (Kazazian, 2004). Examples of evolutionary events in which repetitive elements have been implicated include genome rearrangements (Kazazian, 2004), genome segmental duplications (Bailey, Liu & Eichler, 2003), and random drift to new biological functions (Brosius, 2003). Repeats often insert within other sequences such as into genes. When they are inserted in or adjacent to a functional gene, especially within the open reading frame (ORF), they can be

Bioinformatics in Aquaculture: Principles and Methods, First Edition. Edited by Zhanjiang (John) Liu.

identified as extra exons. In such cases, they could misguide the gene prediction program and hamper the genome annotation. Therefore, a good understanding of the repeat structures in a species can reduce much of the complexities involved in genome studies (Liu, 2007). However, in spite of the rapid progress accelerated by the recent advances in next-generation sequencing, bioinformatics tools used for the analysis of repeats are still not as optimized as those used for the analysis of protein-coding genes (Novak *et al.*, 2013; Yandell & Ence, 2012). Thus, analyses of repetitive elements and their roles in evolution are still key areas in genome biology research (Zhi *et al.*, 2006). In this chapter, we will provide an overview of the methods and software used for the analysis of repetitive elements, and then provide a step-by-step demonstration of the analysis of repetitive elements using a specific set of software packages.

Methods Used in Repeat Analysis

The task of computational identification of repeats in the genome of a given species can be divided into two steps: (1) construction of a representative repeat library of the species; and (2) categorization of the repeat sequences into different classes and super-families. The construction of the repeat library includes discovering all the repetitive sequences within the genome, removing redundant sequences, dividing related repeats into families, and rebuilding a consensus sequence for each family (Stojanovic, 2007). The classification of the repeat library involves inspecting the consensus sequences with a variety of criteria to allow the assignment of the family to a particular class and/or superfamily (Stojanovic, 2007). In this section, we will give an overview of the principles for repeat identification and the bioinformatics tools currently used for the identification and classification of repeats within whole-genome sequences.

Two distinct methods have been applied to the identification of repeats. The first is homology-based, which relies on detecting similarities in the genome sequences with known repetitive sequences. The second approach is the de novo identification of repeats with only the repetitive sequences themselves (Yandell & Ence, 2012).

Homology-Based Identification of Repeats

Homology-based identification of repeats is traditionally achieved using basic local alignment search tools (BLAST) (Johnson *et al.*, 2008). For instance, the widely used alignment program BLASTN is used to search the target sequences against the non-redundant database at GenBank. However, since this popular database contains most of the described repeats, it would be hard for the subsequent classification, and will bring redundancy into the analysis. Therefore, homology-based repeat identification can be made more sensitive by using more specific libraries of manually curated repeat sequences (Stojanovic, 2007). Such repeat databases are generally a collection of well-characterized repeats from years of TE research and literature. The most comprehensive reference libraries are the Repbase libraries (Jurka *et al.*, 2005). Currently, Repbase offers reference libraries for more than 800 eukaryotic species (Bao, Kojima & Kohany, 2015). As the number of curated repeat databases increases, this approach will become increasingly efficient for identifying the most typical repeats.

The major limitation of the homology-based approach is that only sequences with significant similarity to repeats previously described in the reference libraries will be identified (Stojanovic, 2007). Since repetitive elements usually diverge quickly, the utility of homology-based repeat identification often fails due to low similarity of the repeat sequences (Smith *et al.*, 2007).

De Novo Identification of Repeats

Two basic approaches have been employed for the de novo identification of repeats. The first approach starts with a self-comparison with a sequence similarity detection method to identify repeated sequences and then cluster similar sequences into families, based on a pre-defined threshold of similarity and alignment length (Bao & Eddy, 2002; Stojanovic, 2007). RepeatFinder (Volfovsky, Haas & Salzberg, 2001) and RECON (Bao & Eddy, 2002) are examples of programs using this approach. One disadvantage of this method is that all-vs-all alignments are computationally intensive, requiring large memory and CPU time. To improve this algorithm, an alternative and increasingly popular repeat identification method was introduced using word counting and seed extension (Singh *et al.*, 2010). This set of methods builds a set of repeat families starting with short sequence strings representing repeats in the genome. These strings are progressively extended into longer consensus sequences through multiple-iteration dynamic alignments within the query sequence. RepeatScout (Price, Jones & Pevzner, 2005) and ReAS (Li *et al.*, 2005) are two programs developed using this algorithm.

De novo repeat identification methods are powerful because they can discover repeats present within a given pool of sequences without a priori knowledge of any characterized repeats. One major drawback with de novo methods is the ambiguity in defining the biological boundaries of the repeats. In addition, de novo tools identify not only repetitive elements, but also those highly duplicated protein-coding genes such as histones and tubulins. Therefore, caution needs to be exercised to carefully remove protein-coding genes with the outputs from such programs (Yandell & Ence, 2012).

Software for Repeat Identification

Although software for repeats detection vary in repeat definition, search algorithm, and filtering method, they do share a core set of features. In general, a repeat finder program consists of three components: a detection unit, a filter unit and an output unit (Figure 5.1). The detection unit, harboring the search algorithm, is the core component of the program. Based on certain statistical selection criteria, it detects patterns of the input sequences under the customized parameters. Then, the identified candidate repeats will go through a filtering step to eliminate redundant sequences. Outputs and utilities can vary widely between programs, that is, including detailed information on the individual repeat, summary statistics, or even additional modules for subsequent analysis (Merkel & Gemmell, 2008).

Conventionally, the identification of highly tandem repeats is conducted by digesting genomic DNA with restriction enzymes. Tandem repeats tend to be similar in size per unit of the repeats, and, therefore, analysis of restriction-digested DNA by gel electrophoresis tends to generate a "bands on smear" type of image. Using this approach, a family of A/T-rich Xba elements was identified that were arranged in tandem, and

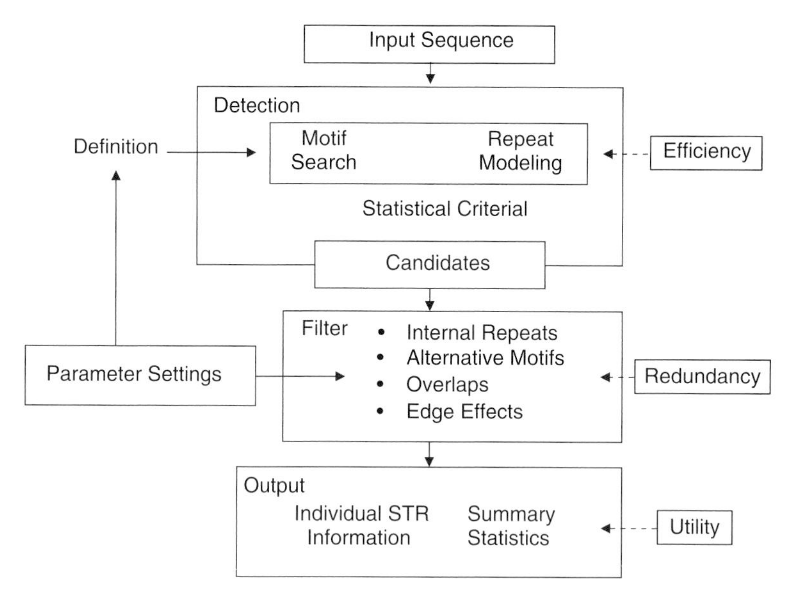

Figure 5.1 Workflow of a repeat finder program. Adapted from Merkel and Gemmell (2008).

they accounted for 5–6% of the catfish genome (Liu, Li & Dunham, 1998). However, it is labor-intensive to identify repeats through experimentation. Over the last 15 years, there have been numerous programs developed to identify tandem repeats by utilizing whole-genome shotgun (WGS) sequence data (Lim *et al.*, 2013). According to the survey conducted by Merkel and Gemmell (2008), more than 20 bioinformatics tools have been developed for detecting short tandem repeats (STR) (Table 5.1). Among these programs, Tandem Repeats Finder (TRF) is probably the most widely used software for finding tandem repeats, and has provided the basis for many other such tools to locate and display tandem repeats in DNA sequences (Benson, 1999; Boeva *et al.*, 2006). TRF requires no a priori knowledge of the pattern, pattern size, or number of copies. This program uses a probabilistic model of tandem repeats and a collection of statistical criteria to reduce the full-scale alignment matrix computations (Benson, 1999). Likewise, other repeat masking tools, such as RepeatMasker (Smit, Hubley & Green, 1996) or DUST (Morgulis *et al.*, 2006), are now standard components of sequence similarity search tools to efficiently reduce redundancy and speed up genome-wide pattern match searches. In addition, several repeat-specific databases have been established to serve as references for such diverging objectives as studying model organisms, for example, TandemRepeatDatabase (Gelfand, Rodriguez & Benson, 2007) and EuMicrosatdb (Aishwarya, Grover & Sharma, 2007). While the identification of tandem repeats is relatively straightforward, characterization of interspersed repeats is a great challenge (Liu, 2007), since they have more complex structures, including terminal repeats, polymerase (POL), envelope (ENV), and reverse transcriptase (RT) genes. Identification of repeat boundaries and classification of repeat families is more difficult (Pevzner, Tang & Tesler, 2004; Smith *et al.*, 2007). Fortunately, with the increasing number and diversity of sequences being generated in genome projects, there are many software packages that have been developed for the analysis of interspersed repeats (Table 5.2)

Table 5.1 Search tools used for STR detection. Modified from Merkel and Gemmell (2008).

Program	Operating System	Source	Reference
RepeatMasker	Linux	http://www.repeatmasker.org/	Smit, Hubley & Green, 1996
HighSSR	Linux	http://code.google.com/p/highssr/	Churbanov *et al.*, 2012
Misa	Linux	http://pgrc.ipk-gatersleben.de/misa/	Thiel *et al.*, 2003
SSRIT	Linux	http://www.gramene.org/db/searches/ssrtool	Temnykh *et al.*, 2001
Sputnik I	Windows, Linux	http://wheat.pw.usda.gov/ITMI/EST-SSR/LaRota/	Morgante, Hanafey & Powell, 2002
Sputnik II	Windows, Linux	http://cbi.labri.u-bordeaux.fr/outils/Pise/sputnik.html	La Rota *et al.*, 2005
Poly	Linux	http://bioinformatics.org/poly/	Bizzaro & Marx, 2003
TRF	Windows, Linux, Mac	http://tandem.bu.edu/trf/trf.html	Benson, 1999
ATRhunter	Windows, Linux	http://bioinfo.cs.technion.ac.il/atrhunter/ATRHunter.htm	Wexler *et al.*, 2005
TandemSWAN	Windows, Linux	http://strand.imb.ac.ru/swan/	Boeva *et al.*, 2006
Mreps	Windows, Linux, Mac	http://bioinfo.lifl.fr/mreps/	Kolpakov, Bana & Kucherov, 2003
STAR	Linux, Mac, Windows	http://atgc.lirmm.fr/star/	Delgrange & Rivals, 2004
STRING	Linux, Windows, Mac	http://www.caspur.it/~castri/STRING/index.htm	Parisi, De Fonzo & Aluffi-Pentini, 2003
TROLL	Linux	http://finder.sourceforge.net	Castelo, Martins & Gao, 2002
IMEx	Linux	http://203.197.254.154/IMEX/index.html	Mudunuri & Nagarajaram, 2007
SciRoko	Linux, Mac	http://www.kofler.or.at/bioinformatics/SciRoKo/index.html	Kofler, Schlotterer & Lelley, 2007
lobSTR	Linux, Mac	http://melissagymrek.com/lobstr-code/index.html	Gymrek *et al.*, 2012
Repbase	Eukaryotes	http://www.girinst.org/repbase/update/index.html	Jurka *et al.*, 2005
TRDB	Eukaryotes	http://tandem.bu.edu/cgi-bin/trdb/trdb.exe	Gelfand *et al.*, 2007
EuMicroSatdb	Eukaryotes	http://veenuash.info/web/index.htm	Aishwarya *et al.*, 2007
InSatdb	Insects	http://210.212.212.8/PHP/INSATDB/home.php	Archak *et al.*, 2007
Trbase	Human	http://trbase.ex.ac.uk/	Boby, Patch & Aves, 2005
VNTRfinder and PolyPredictR	Various	http://www.bioinformatics.rcsi.ie/vntrfinder/	O'Dushlaine & Shields, 2006

Table 5.2 Summary of programs for finding interspersed repeats. Modified from Saha *et al.* (2008b).

Program	Platform	Approach	Source	Reference
CENSOR	Linux/Mac, web	Library	http://www.girinst.org/censor/download.php	Jurka *et al.*, 1996
Dotter	Linux/Mac, Windows	*Ab initio*	ftp://ftp.cgb.ki.se/pub/esr/dotter/	Sonnhammer & Durbin, 1995
FINDMITE	Linux/Mac	Signature	jaketu.biochem.vt.edu/4download/MITE/	Tu, 2001
FORRepeats	Linux/Mac	*Ab initio*	al.jalix.org/FORRepeats/	Lefebvre *et al.*, 2003
HomologMiner	Linux/Mac	*Ab initio*	http://www.bx.psu.edu/miller_lab	Hou *et al.*, 2007
Inverted Repeats Finder	Linux/Mac, Windows	Signature	http://tandem.bu.edu	Warburton *et al.*, 2004
LTR_STRUC	Windows	Signature	http://www.genetics.uga.edu/retrolab/data/LTR_Struc.html	McCarthy & McDonald, 2003
MAK	Linux/Mac, Windows, web	Library/signature	http://wesslercluster.plantbio.uga.edu/mak06.html	Yang & Hall, 2003
MUMmer	Linux/Mac	*Ab initio*	http://mummer.sourceforge.net/	Delcher *et al.*, 2002
OMWSA	Linux/Mac, Windows	*Ab initio*	http://www.hy8.com/~tec/sw0Linux/Mac/omwsa0Linux/	Du, Zhou and Yan, 2007
PatternHunter I and II	Linux/Mac	*Ab initio*	http://www.bioinformaticssolutions.com	Ma, Tromp & Li, 2002
PILER	Linux/Mac	*Ab initio*	http://www.drive5.com/piler	Edgar & Myers, 2005
PILER-CR	Linux/Mac	*Ab initio*	http://www.drive5.com/pilercr	Edgar, 2007
ReAS	Linux/Mac	*Ab initio*	ftp://ftp.genomics.org.cn/pub/ReAS/software	Li *et al.*, 2005
Recon	Linux/Mac	*Ab initio*	http://selab.janelia.org/recon.html	Levitsky, 2004

(Saha *et al.*, 2008a). Based upon the core algorithms behind each software, these tools can be further subdivided into library-based techniques, signature-based techniques, and *ab initio* techniques (Saha *et al.*, 2008a). Among all of these programs, RepeatMasker is the predominant library-based tool in the repeat identification world (Tempel, 2012). It was designed to identify and annotate repetitive elements by searching the input sequence against a set of reference repeat sequences and masking them for further analysis. The repetitive elements, including tandem repeats and interspersed repeats, are annotated and replaced by Ns, Xs, or lowercase letters in the corresponding positions of the DNA sequence. The tool includes a set of statistically optimal scoring matrices calculated for a range of background GC levels permitting estimation of the divergence of query sequences compared to a curated repeat library (Jurka *et al.*, 2005). RepeatMasker can fully utilize multi-processor systems, which along with the simple database search approach makes it one of the most effective repeat finders available. Sequence search engines such as Cross_Match, RMBlast, or WUBlast/ABBlast are supported in the comparison process. To construct one's own customized repeat library, RepeatModeler (Smit, Hubley & Green, 2008) is an integrated de novo repeat identification and modeling package that collects results from three other programs—RECON (Bao & Eddy, 2002), RepeatScout (Price *et al.*, 2005), and TRF—to build a repeat library for RepeatMasker. RepeatScout is good for finding highly conserved repetitive elements, while RECON is more suited to find less highly conserved elements (Saha *et al.*, 2008b). The detailed usage of RepeatModeler will be demonstrated in the following section.

Using the Command-line Version of RepeatModeler to Identify Repetitive Elements in Genomic Sequences

Prerequisites

1) *Perl*: Available at http://www.perl.org/get.html. Developed and tested with version 5.8.8.
2) *RepeatMasker and libraries*: Developed and tested with open-4.0.5. This version must be used with open-4.0.5 or greater of RepeatMasker. The program is available at http://www.repeatmasker.org/RMDownload.html, and the libraries are at http://www.girinst.org.
3) *RECON—De Novo Repeat Finder*: The program is available at http://www.repeatmasker.org/RECON-1.08.tar.gz.
4) *RepeatScout—De Novo Repeat Finder*: This program is now available at http://repeatscout.bioprojects.org.
5) *TRF—Tandem Repeat Finder*: The program is available at http://tandem.bu.edu/trf/trf.html. RepeatModeler was developed using 4.0.4.
6) *RMBlast*: A modified version of NCBI Blast for use with RepeatMasker and RepeatModeler. Precompiled binaries and source can be found at http://www.repeatmasker.org/RMBlast.html
7) *FASTA file of the sequences interested*: Name this file as "mysequence.fa" in this tutorial.

RepeatModeler Installation

1) *Download RepeatModeler*: type `wget http://www.repeatmasker.org/RepeatModeler-open-1-0-8.tar.gz`
2) *Unpack distribution*: Unpack the distribution in your home directory or in a location where it may be shared with other users of your system (i.e., /usr/local/).

```
cp RepeatModeler-open-1-0-8.tar.gz /usr/local
cd /usr/local
gunzip RepeatModeler-open-1-0-8.tar.gz
tar xvf RepeatModeler-open-1-0-8.tar.gz
```

3) *Compile*: Run the "configure" script contained in the RepeatModeler distribution as: `perl ./configure`

Example Run

1) *Create a Database for RepeatModeler*:

```
[RepeatModelerPath]/BuildDatabase -name mysequence
   -engine ncbi mysequence.fa
```

Running `BuildDatabase` without any options can provide the full documentation on this utility. *Run RepeatModeler*:

```
[RepeatModelerPath]/RepeatModeler -engine ncbi -pa 3
   -database mysequence
```

The `-pa` parameter specifies how many searches should be run in parallel.
2) *Run RepeatMasker*: In the working directory, the final product, a file named "consensi.fa.classified", can be found. This is formatted as a RepeatMasker library and can be used directly with RepeatMasker as:

```
[RepeatMaskerPath]/RepeatMasker -lib consensi.fa.
  classified mysequence.fa
```

3) *Recover from a failed run*: If for some reason RepeatModeler fails in a run, you may restart this analysis from the last round it was working on. The `-recoverDir [ResultDir]` option allows you to specify a directory where a previous run of RepeatModeler was working, and it will automatically determine the step from which to continue the analysis.
4) *Interpret the output results of RepeatMasker*: The output of RepeatMasker is written into five different files in the same directory. Only three of them, those with the ".out", ".masked", and ".tbl" extensions, contain results; others only store processing log information and are therefore not described in detail here. The ".out" file is the annotation file that contains the repeats match summary results. The file is basically self-explanatory. The columns of the ".out" file are described in detail within the protocols of RepeatMasker (http://www.repeatmasker.org/webrepeatmaskerhelp.html). The ".masked" file is the same as the query sequence, except that the repetitive elements are masked using Ns, Xs, or lowercase letters. The ".tbl" file summarizes the annotation statistics shown in the ".out" file.

References

Ahmed, M. and Liang, P. (2012). Transposable elements are a significant contributor to tandem repeats in the human genome. *Comparative and Functional Genomics*, 2012, 947089.

Aishwarya, V., Grover, A. and Sharma, P.C. (2007) EuMicroSatdb: a database for microsatellites in the sequenced genomes of eukaryotes. *BMC Genomics*, **8**, 225.

Archak, S., Meduri, E., Kumar, P.S. and Nagaraju, J. (2007) InSatDb: a microsatellite database of fully sequenced insect genomes. *Nucleic Acids Research*, **35**, D36–D39.

Bailey, J.A., Liu, G. and Eichler, E.E. (2003) An Alu transposition model for the origin and expansion of human segmental duplications. *The American Journal of Human Genetics*, **73**, 823–834.

Bao, W., Kojima, K.K. and Kohany, O. (2015) Repbase Update, a database of repetitive elements in eukaryotic genomes. *Mobile DNA*, **6**, 11.

Bao, Z. and Eddy, S.R. (2002) Automated de novo identification of repeat sequence families in sequenced genomes. *Genome Research*, **12**, 1269–1276.

Benson, G. (1999) Tandem repeats finder: a program to analyze DNA sequences. *Nucleic Acids Research*, **27**, 573–580.

Bizzaro, J.W. and Marx, K.A. (2003) *Poly: a quantitative analysis tool for simple sequence repeat (SSR) tracts in DNA*, *BMC Bioinformatics*, p. 4.

Boby, T., Patch, A.M. and Aves, S.J. (2005) TRbase: a database relating tandem repeats to disease genes for the human genome. *Bioinformatics*, **21**, 811–816.

Boeva, V., Regnier, M., Papatsenko, D. and Makeev, V. (2006) Short fuzzy tandem repeats in genomic sequences, identification, and possible role in regulation of gene expression. *Bioinformatics*, **22**, 676–684.

Brosius, J. (2003) The contribution of RNAs and retroposition to evolutionary novelties. *Genetica*, **118**, 99–116.

Castelo, A.T., Martins, W. and Gao, G.R. (2002) TROLL—Tandem Repeat Occurrence Locator. *Bioinformatics*, **18**, 634–636.

Churbanov, A., Ryan, R., Hasan, N. *et al.* (2012) HighSSR: high-throughput SSR characterization and locus development from next-gen sequencing data. *Bioinformatics*, **28**, 2797–2803.

Delcher, A.L., Phillippy, A., Carlton, J. and Salzberg, S.L. (2002) Fast algorithms for large-scale genome alignment and comparison. *Nucleic Acids Research*, **30**, 2478–2483.

Delgrange, O. and Rivals, E. (2004) STAR: an algorithm to search for tandem approximate repeats. *Bioinformatics*, **20**, 2812–2820.

Du, L., Zhou, H. and Yan, H. (2007) OMWSA: detection of DNA repeats using moving window spectral analysis. *Bioinformatics*, **23**, 631–633.

Edgar, R.C. (2007) PILER-CR: fast and accurate identification of CRISPR repeats. *BMC Bioinformatics*, **8**, 18.

Edgar, R.C. and Myers, E.W. (2005) PILER: identification and classification of genomic repeats. *Bioinformatics*, **21** (1), i152–158.

Gelfand, Y., Rodriguez, A. and Benson, G. (2007) TRDB—the Tandem Repeats Database. *Nucleic Acids Research*, **35**, D80–87.

Gorni, C., Stella, A., Panzitta, F. and Mariani, R. (2005) Evaluation of Mammalian Interspersed Repeats to investigate the goat genome. *Italian Journal of Animal Science*, **4**, 52–54.

Gymrek, M., Golan, D., Rosset, S. and Erlich, Y. (2012) lobSTR: a short tandem repeat profiler for personal genomes. *Genome Research*, **22**, 1154–1162.

Hou, M., Berman, P., Hsu, C.H. and Harris, R.S. (2007) HomologMiner: looking for homologous genomic groups in whole genomes. *Bioinformatics*, **23**, 917–925.

Jjingo, D., Conley, A.B., Wang, J.R. *et al.* (2014) *Mammalian-wide interspersed repeat (MIR)-derived enhancers and the regulation of human gene expression, Mobile DNA*, p. 5.

Johnson, M., Zaretskaya, I., Raytselis, Y. *et al.* (2008) NCBI BLAST: a better web interface. *Nucleic Acids Research*, **36**, W5–9.

Jurka, J., Kapitonov, V.V., Pavlicek, A. *et al.* (2005) Repbase Update, a database of eukaryotic repetitive elements. *Cytogenetic and Genome Research*, **110**, 462–467.

Jurka, J., Klonowski, P., Dagman, V. and Pelton, P. (1996) CENSOR–a program for identification and elimination of repetitive elements from DNA sequences. *Computers & Chemistry*, **20**, 119–121.

Kazazian, H.H. Jr., (2004) Mobile elements: drivers of genome evolution. *Science*, **303**, 1626–1632.

Kofler, R., Schlotterer, C. and Lelley, T. (2007) SciRoKo: a new tool for whole genome microsatellite search and investigation. *Bioinformatics*, **23**, 1683–1685.

Kolpakov, R., Bana, G. and Kucherov, G. (2003) mreps: efficient and flexible detection of tandem repeats in DNA. *Nucleic Acids Research*, **31**, 3672–3678.

La Rota, M., Kantety, R.V., Yu, J.K. and Sorrells, M.E. (2005) Nonrandom distribution and frequencies of genomic and EST-derived microsatellite markers in rice, wheat, and barley. *BMC Genomics*, **6**, 23.

Lefebvre, A., Lecroq, T., Dauchel, H. and Alexandre, J. (2003) FORRepeats: detects repeats on entire chromosomes and between genomes. *Bioinformatics*, **19**, 319–326.

Levitsky, V.G. (2004) RECON: a program for prediction of nucleosome formation potential. *Nucleic Acids Research*, **32**, W346–349.

Li, R., Ye, J., Li, S., Wang, J., Han, Y., Ye, C. *et al.* (2005). ReAS: recovery of ancestral sequences for transposable elements from the unassembled reads of a whole genome shotgun. *PLoS Computational Biology*, **1**, e43.

Lim, K.G., Kwoh, C.K., Hsu, L.Y. and Wirawan, A. (2013) Review of tandem repeat search tools: a systematic approach to evaluating algorithmic performance. *Briefings in Bioinformatics*, **14**, 67–81.

Liu, Z. (2007) *Aquaculture genome technologies*, 1st edn, Blackwell Pub, Ames, Iowa.

Liu, Z., Li, P. and Dunham, R.A. (1998) Characterization of an A/T-rich family of sequences from channel catfish (Ictalurus punctatus). *Molecular Marine Biology and Biotechnology*, **7**, 232–239.

Ma, B., Tromp, J. and Li, M. (2002) PatternHunter: faster and more sensitive homology search. *Bioinformatics*, **18**, 440–445.

McCarthy, E.M. and McDonald, J.F. (2003) LTR_STRUC: a novel search and identification program for LTR retrotransposons. *Bioinformatics*, **19**, 362–367.

Merkel, A. and Gemmell, N. (2008) Detecting short tandem repeats from genome data: opening the software black box. *Briefings in Bioinformatics*, **9**, 355–366.

Morgante, M., Hanafey, M. and Powell, W. (2002) Microsatellites are preferentially associated with nonrepetitive DNA in plant genomes. *Nature Genetics*, **30**, 194–200.

Morgulis, A., Gertz, E.M., Schaffer, A.A. and Agarwala, R. (2006) A fast and symmetric DUST implementation to mask low-complexity DNA sequences. *Journal of*

Computational Biology : A Journal of Computational Molecular Cell Biology, **13**, 1028–1040.

Mudunuri, S.B. and Nagarajaram, H.A. (2007) IMEx: imperfect microsatellite extractor. *Bioinformatics*, **23**, 1181–1187.

Novak, P., Neumann, P., Pech, J. *et al.* (2013) RepeatExplorer: a Galaxy-based web server for genome-wide characterization of eukaryotic repetitive elements from next-generation sequence reads. *Bioinformatics*, **29**, 792–793.

O'Dushlaine, C.T. and Shields, D.C. (2006) Tools for the identification of variable and potentially variable tandem repeats. *BMC Genomics*, **7**, 290.

Parisi, V., De Fonzo, V. and Aluffi-Pentini, F. (2003) STRING: finding tandem repeats in DNA sequences. *Bioinformatics*, **19**, 1733–1738.

Pevzner, P.A., Tang, H. and Tesler, G. (2004) *De novo* repeat classification and fragment assembly. *Genome Research*, **14**, 1786–1796.

Price, A.L., Jones, N.C. and Pevzner, P.A. (2005) De novo identification of repeat families in large genomes. *Bioinformatics*, **21** (1), i351–358.

Richard, G.F., Kerrest, A. and Dujon, B. (2008) Comparative genomics and molecular dynamics of DNA repeats in eukaryotes. *Microbiology and Molecular Biology Reviews : MMBR*, **72**, 686–727.

Saha, S., Bridges, S., Magbanua, Z.V. and Peterson, D.G. (2008a) Computational approaches and tools used in identification of dispersed repetitive DNA sequences. *Tropical Plant Biology*, **1**, 11.

Saha, S., Bridges, S., Magbanua, Z.V. and Peterson, D.G. (2008b) Empirical comparison of ab initio repeat finding programs. *Nucleic Acids Research*, **36**, 2284–2294.

Singh, A., Keswani, U., Levine, D. *et al.* (2010) An algorithm for the reconstruction of consensus sequences of ancient segmental duplications and transposon copies in eukaryotic genomes. *International Journal of Bioinformatics Research and Applications*, **6**, 147–162.

Smit, A., Hubley, R. and Green, P. (1996). RepeatMasker Open-4.0. 2013–2015. Available at: http://www.repeatmasker.org.

Smit, A., Hubley, R. and Green, P. (2008). RepeatModeler Open-1.0. 2008–2015. Available at: http://www.repeatmasker.org.

Smith, C.D., Edgar, R.C., Yandell, M.D. *et al.* (2007) Improved repeat identification and masking in Dipterans. *Gene*, **389**, 1–9.

Sonnhammer, E.L. and Durbin, R. (1995) A dot-matrix program with dynamic threshold control suited for genomic DNA and protein sequence analysis. *Gene*, **167**, GC1–10.

Stojanovic, N. (2007) *Computational genomics: current methods*, Horizon Bioscience, Wymondham.

Temnykh, S., DeClerck, G., Lukashova, A. *et al.* (2001) Computational and experimental analysis of microsatellites in rice (Oryza sativa L.): frequency, length variation, transposon associations, and genetic marker potential. *Genome Research*, **11**, 1441–1452.

Tempel, S. (2012) Using and understanding RepeatMasker. *Methods in Molecular Biology*, **859**, 29–51.

Thiel, T., Michalek, W., Varshney, R.K. and Graner, A. (2003) Exploiting EST databases for the development and characterization of gene-derived SSR-markers in barley (*Hordeum vulgare* L.). *Theoretical and Applied Genetics*, **106**, 411–422.

Tu, Z. (2001) Eight novel families of miniature inverted repeat transposable elements in the African malaria mosquito. *Anopheles gambiae. Proceedings of the National Academy of Sciences of the United States of America*, **98**, 1699–1704.
Volfovsky, N., Haas, B.J. and Salzberg, S.L. (2001). A clustering method for repeat analysis in DNA sequences. *Genome Biology*, **2**, RESEARCH0027.1–RESEARCH0027.11.
Warburton, P.E., Giordano, J., Cheung, F. *et al.* (2004) Inverted repeat structure of the human genome: the X-chromosome contains a preponderance of large, highly homologous inverted repeats that contain testes genes. *Genome Research*, **14**, 1861–1869.
Wexler, Y., Yakhini, Z., Kashi, Y. and Geiger, D. (2005) Finding approximate tandem repeats in genomic sequences. *Journal of Computational Biology*, **12**, 928–942.
Yandell, M. and Ence, D. (2012) A beginner's guide to eukaryotic genome annotation. *Nature Reviews Genetics*, **13**, 329–342.
Yang, G. and Hall, T.C. (2003) MAK, a computational tool kit for automated MITE analysis. *Nucleic Acids Research*, **31**, 3659–3665.
Zhi, D., Raphael, B.J., Price, A.L. *et al.* (2006) Identifying repeat domains in large genomes. *Genome Biology*, **7**, R7.

6

Analysis of Duplicated Genes and Multi-Gene Families

Ruijia Wang and Zhanjiang Liu

Introduction

Gene duplication is an important source of genomic renovation and evolution, which is usually associated with lineage-specific adaptive traits during evolution, such as genes related to immunity, development, and reproduction. Although accurate identification of gene duplications is difficult, different computational approaches and pipelines have nevertheless been developed to identify duplicated/multi-member gene families. In this chapter, we mainly review the usage of these methods and pipelines on duplicated gene analysis, including MCL-edge, MCscan, MCscanX, OrthoMCL, and ParaAT.

Gene duplication is an important genomic behavior through which new genetic and functional material is generated during the evolution process of a species (Sankoff, 2001). Its appearance can be directly observed as any duplication of a region of DNA that contains a gene, which also usually leads to the generation of multi-member gene families and eventually functionally diversify the genome of a species (J. Zhang, 2003). As one of the sources of genomic evolution, the genes involved in the duplication are usually relevant to lineage-specific adaptive traits during evolution, such as those involved in immune responses, development, and reproduction (Bailey *et al.*, 2004; Liu & Bickhart, 2012).

Gene duplication can range from affecting single genes, large chromosomal fragments, or the whole genome (polyploidy) (Sankoff, 2001). Three major mechanisms are responsible for the generation of gene duplications: (1) unequal crossing-over, (2) retroposition, and (3) chromosomal (or genome) duplication (J. Zhang, 2003). The outcomes of gene duplication through these mechanisms are quite different. Unequal crossing-over usually generates tandem gene duplication; that is, duplicated genes are linked within a chromosome region next to each other (Long, 2001; Long *et al.*, 2003). Retroposition is a process in which an mRNA is retrotranscribed to cDNA and then inserted into the genome (Long, 2001; Long *et al.*, 2003). Unequal crossing-over and retroposition usually generate small-scale duplication. Large-scale duplications, such as chromosomal or genome duplications, happen possibly due to the lack of disjunction among offspring chromosomes after DNA replication (Long, 2001; Long *et al.*, 2003). Whole-genome duplications have a major impact on the generation of genetic diversity because all the genes in the genome are duplicated simultaneously.

Bioinformatics in Aquaculture: Principles and Methods, First Edition. Edited by Zhanjiang (John) Liu.

Tandem duplication and genome duplication are considered as two most important mechanisms contributing to vertebrate genome evolution (Long, 2001; Long *et al.*, 2003). Analysis of the human genome revealed yet another type of large-scale duplication, segmental duplications (SDs), which involve the duplication of blocks of genomic DNA, typically ranging in size from 1 to 200 kb (Bridges, 1936; Lynch & Conery, 2003; Maere *et al.*, 2005). However, the exact mechanism of segmental duplication is still unclear.

Accurate identification of gene duplication is difficult due to the complex duplication categories, including small (tandem duplication and retrotransposition) and large duplications (segmental duplication and whole-genome duplication) caused by genome rearrangement, gene loss, further duplication, and mutation after creation (Liu *et al.*, 2009; Meyer & Van de Peer, 2005). Different computational approaches have been developed to identify duplicated genes, which can be roughly summarized into two categories: (1) map-based approaches, and (2) sequence-based approaches (Durand & Hoberman, 2006). Map-based approaches are used to detect homologies between different chromosomal segments using the corresponding gene names. In this method, the genes are first retrieved from each chromosome and sorted according to their genome position. Using the gene list, genes that have same gene names in other chromosomal segments are identified, and then the local similarities in gene order and content are detected by comparing the map with itself (Durand & Hoberman, 2006; Van de Peer, 2004). This method is applied to detect recent tandem duplications and other large-scale duplications, which usually exhibit collinearity between genomic segments with roughly stable gene content and order (Durand & Hoberman, 2006; Van de Peer, 2004). Sequence-based approaches are the most currently applied strategies utilizing sequence similarity generated by self-blast to capture the duplicated genes in the target species. Such approaches have been wildly applied in aquaculture species, especially aquacultured teleost fish, for example, catfish (Lu *et al.*, 2012), half-smooth tongue sole (Chen *et al.*, 2014), carp (Wang, Lu *et al.*, 2015), and *Tetraodon* (Jaillon *et al.*, 2004).

Almost all the pipelines for the identification of gene duplication were development based on Markov Cluster Algorithm (MCL algorithm), which is a speedy unsupervised clustering algorithm according to the simulation of (stochastic) flow in graphs. This algorithm was first reported in 2001 (Van Dongen, 2001), and its corresponding analysis procedure was included in the classic MCL package, MCL-edge (Enright, Van Dongen & Ouzounis, 2002; Van Dongen, 2001), which is suitable for the analysis of small–large-scale duplicated/multi-member gene families. In the last decade, due to the demand in comparative genomics, two MCL-derived pipelines, MCscan (and its updated version MCscanX) (Tang *et al.*, 2008; Wang, Tang *et al.*, 2012) and OrthoMCL (Li, Stoeckert & Roos, 2003), were developed and used on the target species and its close species with genome reference and annotation. Integrated with the duplicated genes or multi-member families identified from MCL-based pipelines and the downstream Ka/Ks (performed by ParaAT) (Z. Zhang *et al.*, 2012), expression, and GO analysis (including GO term annotation and GO enrichment analysis, covered in great detail in Chapter 10), the analyst can explore the sequence feature, specific expression pattern, and relevant functions of duplicated genes. Here, we will provide descriptions of using MCscan, MCscanX, OrthoMCL, and ParaAT for the analysis of duplicated genes.

Pipeline Installations

MCL-edge

To get MCL-edge using the command line, type:

```
$ wget http://micans.org/mcl/src/mcl-14-137.tar.gz
```

To install, use the command line:

```
$ cd mcl-14-137
$ ./configure --prefix=the folder where to compile the
  software
$ make install
```

MCscan

To get MCscan, use the command line:

```
$ wget http://github.com/tanghaibao/mcscan/tarball/master/
  tanghaibao-mcscan-4cfe0f5.tar.gz
```

To install (Requirement pipeline: MCL-edge), use the command line:

```
$ tar zxf tanghaibao-mcscan-4cfe0f5.tar.gz
$ cd tanghaibao-mcscan-4cfe0f5/ && make
```

MCscanX

To get MCscanX, use the command line:

```
$ wget http://chibba.pgml.uga.edu/mcscan2/MCScanX.zip
```

To install (Requirement pipeline: MCL-edge), use the command line:

```
$ unzip MCscanX.zip
$ cd MCscanX
$ make
```

OrthoMCL

To get OrthoMCL, use the command line:

```
$ wget http://orthomcl.org/common/downloads/software/v2.0/
  orthomclSoftware-v2.0.9.tar.gz
```

Note: Due to the heavy computation intensity, OrthoMCL should be run with the computer clusters. Under this circumstance, it is recommended to contact the cluster administrator for the installation of MySQL, which will be used during the configuration and installation of OrthoMCL.

To configure and install (Requirement pipeline: MCL-edge and MySQL), use the command line:

```
$ tar zxf orthomclSoftware-v2.0.9.tar.gz
$ cd orthomclSoftware-v2.0.9
```

```
$ mkdir working
$ cd working
$cp ../doc/OrthoMCLEngine/Main/orthomcl.config.template.
```

Open this file ("orthomcl.config.template") and edit it as following:

```
dbVendor=mysql
```

No need to change this unless you are using other database packages.

```
dbConnectString=dbi:mysql:orthomcl:node69:3310
```

Setting your database name ("orthomcl"), your node ID ("node69"), and host ("3310") after communicating with your cluster administrator.

```
dbLogin=username
```

Your username in the MySQL of your clusters; get it from your cluster administrator.

```
dbPassword=password
```

Your password; get it from your cluster administrator.

```
similarSequencesTable=SimilarSequences
```

No need to change this.

```
orthologTable=Ortholog
```

No need to change this.

```
inParalogTable=InParalog
```

No need to change this.

```
coOrthologTable=CoOrtholog
```

No need to change this.

```
interTaxonMatchView=InterTaxonMatch
```

No need to change this.

```
percentMatchCutoff=50
```

Blast coverage% cutoff; change it based on your case.

```
evalueExponentCutoff=-5
```

Blast e-value cutoff; change it based on your case.

```
oracleIndexTblSpc=NONE
```

No need to change this.

After the editing of this configuration file, you need to access in MySQL to finish the installation of OrthoMCL:

```
$ mysql -u username -p password
$ mysql> create database orthomcl;
$ cp ~/orthomclSoftware-v2.0.9/working/orthomcl.config.
  template.
$ orthomclInstallSchema orthomcl.config.template
```

Identification of Duplicated Genes and Multi-Member Gene Family

MCL-edge

Prepare the Input Data

The first thing is to perform the self-blast on the target species:

```
$ blastall -p blastp -i protein_seq.fa -d protein_seq.fa
   -o self_blast.out -e [set the E-value based on your own
   case] -m 8
```

Then, covert the blast result to the format supported by MCL:

```
$ cut -f 1,2,11 self_blast.out > self_blast.abc
```

Identification of Duplication by MCL

Construct the similarity network with the reformatted self-blast result:

```
$ mcxload -abc self_blast.abc --stream-mirror
   --stream-neg-log10 -o self_blast.mci -write-tab
   self_blast.tab
```

Note that the options "--stream-mirror" and "--stream-neg-log10" are required to transfer the e-value into a readable similarity score.

Then the clustering can be performed:

```
$ mcl self_blast.mci -I 3
```

The preceding command generates a ".mci.I30" cluster file that can be used to identify the duplication genes; the option "-I" is used to set the clustering weight score for the clustering process. It usually can be set from 2 to 10 for different species.

The last step is to identify duplicated genes from the cluster file:

```
$ mcxdump -icl self_blast.mci.I30 -o dump.self_blast.mci.I30
```

Output Results

Open the ".mci.I30" file; the duplicated genes are listed as in Figure 6.1. Each line represents a duplicate genes cluster, which can be used for the downstream analysis.

MCscan and MCscanX

Due to the same core and command used in both MCscan and MCscanX (MSscan tools), their usage will be explained in the same section.

Prepare the Input Data

Two files are needed by MCscan tools—a self-blast output file that can be generated as mentioned earlier, and a "gff" file, which contains the genome annotation information of the species under analysis. The "gff" file contains gene positions in a tab-delimited format:

```
sp# geneID starting_position ending_position
```

Figure 6.1 An example output of MCL-edge.

```
g37649.t1   g57479.t1   g45423.t1   g37652.t1
g37855.t1   g8024.t1    g56820.t1   g8023.t1
g38360.t1   g45421.t1   g56157.t1   g56158.t1
g39471.t1   g56858.t1   g8547.t1    g4412.t1
g39598.t1   g54098.t1   g51675.t1   g39991.t1
g40566.t1   g48746.t1   g56578.t1   g56577.t1
g41229.t1   g50304.t1   g41230.t1   g53826.t1
g42143.t1   g42144.t1   g42145.t1   g42146.t1
g44769.t1   g45410.t1   g44770.t1   g44773.t1
g3948.t1    g28141.t1   g8091.t1
g54088.t1   g34341.t1   g48575.t1
g48795.t1   g49272.t1   g8388.t1
g26616.t1   g12931.t1   g56602.t1
g40746.t1   g40740.t1   g40741.t1
g40623.t1   g40627.t1   g43409.t1
g37444.t1   g11379.t1   g25450.t1
g52045.t1   g24052.t1   g40161.t1
g57484.t1   g35271.t1   g8022.t1
g33170.t1   g31671.t1   g46042.t1
g45952.t1   g44606.t1   g52499.t1
g27799.t1   g40933.t1   g43252.t1
g21293.t1   g12985.t1   g47908.t1
g52579.t1   g50813.t1   g53697.t1
g28649.t1   g44018.t1   g29920.t1
g55539.t1   g43201.t1   g13849.t1
```

sp is the two-letter short name for the species; # is the chromosome number (e.g., cc1 stands for the channel catfish chromosome 1).

The "blast" and "gff" files should be named with the same short name and put into the same folder for convenient reorganization by MCscan tools (e.g., use "cc.blast" and "cc.gff" as names for the input files and put them in the same folder).

Duplication Identification by MCscan Tools

When the input files are ready, the pairwise collinear block file (".collinearity") can be generated by:

```
$ ./MCscanX fish/cc
```

Then the groups, or clustered duplicated genes, can be retrieved by:

```
$ perl group_collinear_genes.pl -i cc. collinearity
   -o cc_duplication.out
```

Output Results

The format of output from MCscan tools is same as that from MCL-edge. Due to the usage of the genome annotation file, MCscan tools can also analyze the duplication category by:

```
$ ./duplicate_gene_classifier fish/cc
```

```
gene01010    1
gene01020    2
gene01030    4
gene01040    1
gene01050    4
gene01060    3
gene01070    1
gene01073    0
gene01080    1
```

Figure 6.2 An example output of MCscan and MCscanX. Note: 0, 1, 2, 3, 4 stand for singleton, dispersed, proximal, tandem, WGD/segmental duplications, respectively.

The output is a text file in the same directory as the input files, named as ".gene_type". It contains origin information for all the genes in a "cc.gff" file, in a tab-delimited format (Figure 6.2). Each gene will be assigned into only one category, which apparently is not realistic (e.g., gene A can be in tandem duplication with gene B and be in WGD/segmental duplication with gene C at the same time). Therefore, this module can be used as a tool for preliminary analysis, but not for final analysis.

OrthoMCL

Prepare the Input Data

The required input file of OrthoMCL is a protein sequence (".fasta") file, with specific formatted sequence header of each sequence as:

```
>taxoncode|unique_protein_id
```

"taxoncode" is the short name of the species, represented in three–four characters. For instance, one can use ">ccat|gene1" to represent the protein sequence of channel catfish gene1. The analysts can prepare their own ".fasta" file through:

```
$ orthomclAdjustFasta ccat sequence/protein.fasta 1
```

This command will generate a "ccat.fasta" file using "ccat" as the species short name and the first column of the sequence header in "protein.fasta" as the protein sequence ID.

Trim the Input Data

Before the self-blast, the input sequences needed to be trimmed to remove the sequences with length < 10aa and restrict codons proportion to >20%:

```
$ orthomclFilterFasta sequence/protein.fasta 10 20
```

This command will generate two files: "goodProteins.fasta" and "poorProteins.fasta"; "goodProteins.fasta" will be used in the downstream self-blast.

Self-Blast

```
$ makeblastdb -in goodProteins.fasta -dbtype prot
   -title selfblast -parse_seqids -out selfblast -logfile
   selfblast.log
$ blastp -db selfblast -query goodProteins.fasta -seg yes
   -out selfblast.out -evalue 1e-5 -outfmt 7 -num_threads 24
```

Transfer the Blast Result Format for OrthoMCL

```
$ grep -P "^[^#]" selfblast.out > blastresult
$ orthomclBlastParser blastresult sequence
   > similarSequences.txt
$ perl -p -i -e 's/\t(\w+)(\|.*)selfblast/\t$1$2$1/'
   similarSequences.txt
$ perl -p -i -e 's/0\t0/1\t-181/' similarSequences.txt
```

Load the "SimilarSequences.txt" into the Database

To shorten the running time of OrthoMCL, it is recommended to edit the configure file according to the size of the "similarSequences.txt" file; for example, if "similarSequences.txt" is 83 M, type:

```
myisam_max_sort_file_size = 424960
```

in /etc/my.cnf. And then, load the blast result through:

```
$ orthomclLoadBlast orthomcl.config.template
   similarSequences.txt
```

Identify Similar Sequence Pairs from the Blast Result

```
$ orthomclPairs orthomcl.config.template orthomcl_pairs.log
   cleanup=no
```

Retrieve Similar Sequence Pairs from Database

```
$ orthomclDumpPairsFiles orthomcl.config.template
```

This command will generate a folder named as "pairs" containing a "mclInput" file. This file will be used for the MCL clustering in next step.

MCL-edge Clustering

```
$ mcl mclInput --abc -I 3 -o mclOutput
```

Format the Clustering Results (Name the Clusters as "ccat1", "ccat2"...)

```
$ orthomclMclToGroups ccat 1 < mclOutput > groups.txt
```

Note: The process mentioned in the preceding text is for single species. OrthoMCL can also support the analysis on multi-species. For this purpose, the analyst can simply prepare the sequence as required, concatenate them in to one ".fasta" file, and run the same analysis process step by step.

Output Results

The results of OrthoMCL are very similar to other MCL-based pipelines, with the extra header and cluster name defined in the previously discussed commands (Figure 6.3).

```
ccat2269: ccat|ccat_gene18096 ccat|ccat_gene18097 ccat|ccat_gene18100 ccat|ccat_gene18099 ccat|ccat_gene18098
ccat2270: ccat|ccat_gene18112 ccat|ccat_gene30803 ccat|ccat_gene22540 ccat|ccat_gene22539 ccat|ccat_gene20074
ccat2271: ccat|ccat_gene18425 ccat|ccat_gene26194 ccat|ccat_gene26193 ccat|ccat_gene26195 ccat|ccat_gene26196
ccat2272: ccat|ccat_gene20988 ccat|ccat_gene28299 ccat|ccat_gene28298 ccat|ccat_gene28297 ccat|ccat_gene28300
ccat2273: ccat|ccat_gene19335 ccat|ccat_gene19336 ccat|ccat_gene19338 ccat|ccat_gene19337 ccat|ccat_gene19339
ccat2274: ccat|ccat_gene19353 ccat|ccat_gene25885 ccat|ccat_gene9468 ccat|ccat_gene20565 ccat|ccat_gene20566
ccat2275: ccat|ccat_gene19428 ccat|ccat_gene19432 ccat|ccat_gene19431 ccat|ccat_gene19430 ccat|ccat_gene19429
ccat2276: ccat|ccat_gene19937 ccat|ccat_gene19941 ccat|ccat_gene19940 ccat|ccat_gene19939 ccat|ccat_gene19938
ccat2277: ccat|ccat_gene20138 ccat|ccat_gene20139 ccat|ccat_gene21652 ccat|ccat_gene22318 ccat|ccat_gene22319
ccat2278: ccat|ccat_gene20377 ccat|ccat_gene20387 ccat|ccat_gene9299 ccat|ccat_gene20389 ccat|ccat_gene20388
ccat2279: ccat|ccat_gene20408 ccat|ccat_gene20410 ccat|ccat_gene20409 ccat|ccat_gene20411 ccat|ccat_gene20412
ccat2280: ccat|ccat_gene20665 ccat|ccat_gene20666 ccat|ccat_gene20668 ccat|ccat_gene20667 ccat|ccat_gene20669
ccat2281: ccat|ccat_gene20967 ccat|ccat_gene20968 ccat|ccat_gene20969 ccat|ccat_gene20970 ccat|ccat_gene20971
ccat2282: ccat|ccat_gene21074 ccat|ccat_gene21075 ccat|ccat_gene21078 ccat|ccat_gene21077 ccat|ccat_gene21076
ccat2283: ccat|ccat_gene21114 ccat|ccat_gene21115 ccat|ccat_gene21116 ccat|ccat_gene21118 ccat|ccat_gene21117
ccat2284: ccat|ccat_gene21319 ccat|ccat_gene21321 ccat|ccat_gene21320 ccat|ccat_gene21323 ccat|ccat_gene21322
ccat2285: ccat|ccat_gene21594 ccat|ccat_gene21595 ccat|ccat_gene21598 ccat|ccat_gene21597 ccat|ccat_gene21596
ccat2286: ccat|ccat_gene21599 ccat|ccat_gene21600 ccat|ccat_gene21603 ccat|ccat_gene21602 ccat|ccat_gene21601
ccat2287: ccat|ccat_gene21799 ccat|ccat_gene21800 ccat|ccat_gene27551 ccat|ccat_gene27552 ccat|ccat_gene27553
ccat2288: ccat|ccat_gene21951 ccat|ccat_gene21954 ccat|ccat_gene21955 ccat|ccat_gene21952 ccat|ccat_gene21953
ccat2289: ccat|ccat_gene22383 ccat|ccat_gene22384 ccat|ccat_gene22387 ccat|ccat_gene22386 ccat|ccat_gene22385
```

Figure 6.3 An example output of OrthoMCL.

```
g31636.t1	g48566.t1
g31636.t1	g29549.t1
g18403.t1	g22660.t1
g18403.t1	g17544.t1
g27793.t1	g24687.t1
g27793.t1	g5020.t1
g27793.t1	g14113.t1
g27793.t1	g9176.t1
g27793.t1	g27794.t1
g17534.t1	g34725.t1
g17534.t1	g48681.t1
g33467.t1	g16103.t1
g33467.t1	g16101.t1
g56239.t1	g56241.t1
g56239.t1	g56236.t1
g51771.t1	g23448.t1
g51771.t1	g32610.t1
```

Figure 6.4 An example input of ParaAT.

Calculate Ka/Ks by ParaAT

For temporal analysis, the fraction of synonymous substitutions per synonymous site (Ks) is a broadly used approach to estimate the time of duplication for individual gene pairs. The WGD event can be identified by a major peak in the histogram plot that plotted by the number of duplicated genes against estimated duplication times (Van de Peer, 2004). In this section, we demonstrate the usage of ParaAT as the pipeline to calculate the Ka/Ks for the identified duplicated/multi-member gene family. ParaAT has the advantage of less running time and a user-friendly output format.

Prepare the Input Data

The input files for ParaAT are:

1) A file that describes the relationships of multiple homologous groups (to ease the input of multiple homologous groups, ParaAT accepts a tab-delimited text file with each line representing a homologous group; see Figure 6.4)
2) A config file to set the number of processors
3) fasta-formatted nucleotide and amino acid sequences

```
Sequence         Ka           Ks          Ka/Ks
gene1-gene87     0.627636     1.64322     0.381954
gene4-gene1733   0.808        1.69054     0.477955
gene7-gene1037   0.675111     2.42607     0.278273
```

Figure 6.5 An example output of ParaAT.

Calculate Ka/Ks

```
$ ParaAT.pl -hcc.mcl.pair -n cc.cds.fa -a cc.aa.fa -p proc
   -m clustalw2 -o cc_out -f axt -kaks
```

In the preceding command, "-h" can be retrieved from the MCL result; "-n" and "–a" represent the fasta-formatted nucleotide and amino acid sequences; "-p" represents the number of processes; "-m" stands for the multiple sequence aligner ("clustalw2" in this case) used for sequence alignment; "-o" is the name index of the output folder; "-f" is to specify the alignment output format; and "-kaks" is to enable the Ka/Ks calculation for each cluster/pair list in the input file.

Summary of the Ka/Ks Results

```
$ cat /cc_out/*.kaks > ccout.kaks
$ head -n 1 ccout.kaks > header
$ grep -v 'Ks' ccout.kaks > results
$ cat header results | cut -f 1,3,4,5 > cc.kaks.out
```

Results

The formatted result of ParaAT is quite straightforward, as shown in Figure 6.5.

Downstream Analysis

Although more and more tools are available for the identification of gene duplications, there are no "standard" procedures for downstream analysis. Among various downstream analysis procedures, classification of duplications is almost the *sine qua non* for most research. According to the location of duplicated genes, the analyst can classify the duplicated genes into three major general categories:

1) *Tandem duplications*: if the duplicated gene copies are located on the same scaffold or chromosome within 10 KB of each other
2) *Intra-chromosomal (non-tandem) duplications*: if the duplicated gene copies are located on a different scaffold but on the same chromosome, or if they are located in the same scaffold but are separated by other genes with a distance greater than 10 KB
3) *Inter-chromosomal duplications*: if the duplicated gene copies are located on different chromosomes

In most cases, there are also mixed types of duplications that include combinations of the earlier-defined categories of duplications. For aquaculture species, due to the limitation of the linkage map and genome sequence assembly, some duplicated genes cannot yet be classified, and they should be placed into the "non-tandem unassigned duplication" category. In addition to classification, functional analysis is also a frequently demanded downstream analysis. Combined with the GO terms, KO Ids, and functional domains, the analyst can also explore the relevant biological functions, pathways, and second/third structures on the duplicated genes with specific Ka/Ks, deferential expression, and distinct genomic location pattern.

Perspectives

For aquaculture species, especially for the aquacultured fish species, sequence-based analyses (protein sequences self-blast followed by single-linkage clustering or Markov clustering) have been successfully performed for the identification of duplicated genes (Chen *et al.*, 2014; Jaillon *et al.*, 2004; Lu *et al.*, 2012). However, setting the proper sequence similarity and cluster weight cutoff is still a big challenge. When too low/high similarity cutoffs are set, the clustering bias is difficult to avoid; that is, under low similarity cutoff, members in the different gene family could be clustered into the same group due to their medium sequence similarity. Therefore, in most cases, gradient analysis is necessary. Using the different similarity and clustering cutoff combination, the analyst can check the clustering result in the target species and other closely related species to determine the optimal cutoffs. However, even if the optimal cutoffs are adopted, the clustering is still not perfect. There are always some genes that are clustered into the wrong groups due to the limitations of the MCL algorithm. Manual checking and correction are still indispensable for the analysis of gene duplication to maintain the confidence level of the analysis.

References

Bailey, J.A., Church, D.M., Ventura, M. *et al.* (2004) Analysis of segmental duplications and genome assembly in the mouse. *Genome research,* **14** (5), 789–801.

Bridges, C.B. (1936) The Bar "gene" a duplication. *Science (New York, NY),* **83** (2148), 210–211.

Chen, S., Zhang, G., Shao, C. *et al.* (2014) Whole-genome sequence of a flatfish provides insights into ZW sex chromosome evolution and adaptation to a benthic lifestyle. *Nature genetics,* **46** (3), 253–260.

Durand, D. and Hoberman, R. (2006) Diagnosing duplications – can it be done? *Trends in Genetics,* **22** (3), 156–164.

Enright, A.J., Van Dongen, S. and Ouzounis, C.A. (2002) An efficient algorithm for large-scale detection of protein families. *Nucleic Acids Research,* **30** (7), 1575–1584.

Jaillon, O., Aury, J.-M., Brunet, F. *et al.* (2004) Genome duplication in the teleost fish *Tetraodon nigroviridis* reveals the early vertebrate proto-karyotype. *Nature,* **431** (7011), 946–957.

Li, L., Stoeckert, C.J. and Roos, D.S. (2003) OrthoMCL: identification of ortholog groups for eukaryotic genomes. *Genome Research,* **13** (9), 2178–2189.

Liu, G.E. and Bickhart, D.M. (2012) Copy number variation in the cattle genome. *Functional & Integrative Genomics,* **12** (4), 609–624.

Liu, G.E., Ventura, M., Cellamare, A. *et al.* (2009) Analysis of recent segmental duplications in the bovine genome. *BMC Genomics*, **10** (1), 571.

Long, M. (2001) Evolution of novel genes. *Current Opinion in Genetics & Development*, **11** (6), 673–680.

Long, M., Betrán, E., Thornton, K. and Wang, W. (2003) The origin of new genes: glimpses from the young and old. *Nature Reviews Genetics*, **4** (11), 865–875.

Lu, J., Peatman, E., Tang, H. *et al.* (2012) Profiling of gene duplication patterns of sequenced teleost genomes: evidence for rapid lineage-specific genome expansion mediated by recent tandem duplications. *BMC Genomics*, **13** (1), 246.

Lynch, M. and Conery, J.S. (2003) The origins of genome complexity. *Science*, **302** (5649), 1401–1404.

Maere, S., De Bodt, S., Raes, J. *et al.* (2005) Modeling gene and genome duplications in eukaryotes. *Proceedings of the National Academy of Sciences of the United States of America*, **102** (15), 5454–5459.

Meyer, A. and Van de Peer, Y. (2005) From 2R to 3R: evidence for a fish-specific genome duplication (FSGD). *Bioessays*, **27** (9), 937–945.

Sankoff, D. (2001) Gene and genome duplication. *Current Opinion in Genetics & Development*, **11** (6), 681–684.

Tang, H., Bowers, J.E., Wang, X. *et al.* (2008) Synteny and collinearity in plant genomes. *Science*, **320** (5875), 486–488.

Van de Peer, Y. (2004) Computational approaches to unveiling ancient genome duplications. *Nature Reviews Genetics*, **5** (10), 752–763.

Van Dongen, S.M. (2001). Graph clustering by flow simulation. *PhD thesis, University of Utrecht*, pp. 15–67. (http://dspace.library.uu.nl/bitstream/handle/1874/848/full.pdf?sequence=1)

Wang, Y., Lu, Y., Zhang, Y. *et al.* (2015) The draft genome of the grass carp (*Ctenopharyngodon idellus*) provides insights into its evolution and vegetarian adaptation. *Nature genetics*, **47** (6), 625–631.

Wang, Y., Tang, H., DeBarry, J.D. *et al.* (2012) MCscanX: a toolkit for detection and evolutionary analysis of gene synteny and collinearity. *Nucleic Acids Research*, **40** (7), e49–e49.

Zhang, J. (2003) Evolution by gene duplication: an update. *Trends in Ecology & Evolution*, **18** (6), 292–298.

Zhang, Z., Xiao, J., Wu, J. *et al.* (2012) ParaAT: a parallel tool for constructing multiple protein-coding DNA alignments. *Biochemical and Biophysical Research Communications*, **419** (4), 779–781.

7

Dealing with Complex Polyploidy Genomes: Considerations for Assembly and Annotation of the Common Carp Genome

Peng Xu, Jiongtang Li and Xiaowen Sun

Introduction

In his seminal book *Evolution by gene duplication*, Ohno (1970) proposed that at least two rounds of whole-genome duplication (2R WGD) had occurred during the evolution of early vertebrates based on genome size (nuclear DNA content), chromosome number, meiotic chromosome behavior, and isozyme patterns. Later, it was hypothesized that an additional round (3R) of WGD had occurred in ray-finned fishes, and it was then proved with findings of abundant fish-specific gene duplications at the gene, gene family, and genome scales (Larhammar & Risinger, 1994; Postlethwait *et al.*, 2000; Van de Peer *et al.*, 2001; David *et al.*, 2003). The 3R WGD, also known as teleost-specific WGD, was estimated to happen around 320 million years ago (Vandepoele *et al.*, 2004). The duplication of entire genomes plays a significant role in evolution. Multiple rounds of WGD produced redundant genes, providing important genetic material basis for phenotypic complexity, which would potentially benefit an organism on its adaptation to environmental changes (Ohno, 1970).

Fishes show the most extensive polyploidy among vertebrates. Most of the well-characterized polyploidy fishes are included in Salmonidae (salmons and trouts) (Allendorf & Thorgaard, 1984; Phillips & Rab, 2001) and Cyprinidae (common carp *Cyprinus carpio*, crucian carp, etc.) (Ohno *et al.*, 1967; Wolf *et al.*, 1969). The polyploid salmonid genome may have resulted from an autopolyploidization event (genome doubling), while tetraploidized common carp and crucian carp genomes may have come from an allopolyploidization event (species hybridization). These polyploidization events are also called the fourth round of WGD, which bring more complexity in the genomes. Both families include important aquaculture species, such as Atlantic salmon, rainbow trout, and common carp.

Much effort has been made to develop genetic and genome resources for genetic breeding and selection in the past decades, including genome-sequencing projects on several polyploidy species. Unlike diploid genomes, polyploidy genomes involved multiple rounds of WGD and segmental duplications, which harbor much more complex structures and gene contents than diploid, posing significant challenges for producing high-quality genome assemblies.

Cyprinus carpio is cultured in over 100 countries worldwide, with an annual production of over 3.4 million metric tons (FAO Fisheries & Aquaculture Department,

Bioinformatics in Aquaculture: Principles and Methods, First Edition. Edited by Zhanjiang (John) Liu.

2007; Bostock *et al.*, 2010). In addition to its value as a food source, *C. carpio* is also an important ornamental fish species. One of its variants, koi, is the most popular ornamental fish because of its distinctive color and scale patterns. Cytogenetic evidence of *C. carpio*'s allotetraploidization has suggested that 50 bivalents rather than 25 quadrivalents are formed during meiosis (Ohno *et al.*, 1967), which implies that it retains two sets of ancestral genomes from the parental species. Molecular evidences collected in the past two decades suggest that the polyploidization event could have happened from 58 MYA to 12 MYA (Larhammar & Risinger, 1994; David *et al.*, 2003), which is considered a very recent event. Owing to its importance in aquaculture and evolution biology, we initiated the common carp genome project (CCGP) in 2009 with support from Chinese Academy of Fishery Sciences. The whole genome of *C. carpio* was completely sequenced with hybrid sequencing and assembly strategies by the end of 2012, and then annotated for over 52000 functional genes. The finished genome provides a valuable resource to common carp research community for the molecular-guided breeding and genetic improvement, and for genome evolution study on the polyploidies.

Although next-generation sequencing technologies and new assembly algorithms have overcome many major obstacles (cost, sequencing speed, and data output) for whole-genome sequencing and assembly, the allotetraploid nature and high levels of heterozygosity of the *C. carpio* genome bring us more challenges than the genome-sequencing projects on diploids. Thus, we have extensive technical concerns and considerations for the complex polyploidy genome-sequencing project. In this chapter, we share our experience and considerations for working with a polyploidy fish genome.

Properties of the Common Carp Genome

A variety of *C. carpio* genome resources and tools have been developed over the past decade before the genome project, including a large number of genetic markers (Ji *et al.*, 2012; Xu *et al.*, 2012), several low-density genetic maps (Sun & Liang, 2004; Cheng *et al.*, 2010; Zhang *et al.*, 2013), over 30000 expressed sequence tags (ESTs) (Christoffels *et al.*, 2006), and cDNA microarrays (Williams *et al.*, 2008). However, we were still short of large-scale genome data to assess the complexity of the *C. carpio* genome.

To provide the first insight into common carp genome, we first focused on developing the first batch of genome sequence data, the large-scale Bacterial Artificial Chromosome (BAC) end sequences with traditional Sanger's method. The common carp BAC library, containing a total of 92160 BAC clones with an average insert size of 141 KB, was constructed into the restriction site of *Hin*d III on BAC vector CopyControl pCC1BAC, covering 7.7 X haploid genome equivalents (Li *et al.*, 2011). A total of 40224 BAC clones were sequenced on both ends, generating 65720 clean BAC-end sequences (BESs) with an average read length of 647 bp after sequence processing, representing 42522168 bp or 2.5% of common carp genome (Xu *et al.*, 2011). Therefore, we conducted the first survey of the common carp genome based on the dataset.

The proportion of the repetitive elements in the common carp genome was assessed by using RepeatMasker against Vertebrates Repeat Database. RepeatMasking of the 42522168 bp of the carp BESs detected 7357899 (17.3%) base pairs of repeated sequences. The most abundant type of repetitive element in the common carp genome was DNA transposons (6.67%), mostly hobo-Activator (2.25%), followed by retroelements (4.52%) including LINEs (2.33%), LTR elements (1.98%), and SINEs (0.2%). The repeats divergence rate of DNA transposons (percentage of substitutions in the matching region compared with consensus repeats in constructed libraries) showed a nearly normal distribution with a peak at 24%. A fraction of LTR retrotransposons, LINEs, and SINEs had nearly the same divergence rates as DNA transposons (peaks at 30%, 28%, and 22%, respectively), indicating relatively old origin. To identify novel repetitive elements, repeat libraries were constructed using multiple *de novo* methods, and then combined into a non-redundant repeat library containing 1940 sequences. The repeat library was then used for repeat annotation of the common carp BES. An additional total of 4499836 bp were identified, representing approximately 10.6% of the BES, as *de novo* repeats. Therefore, we estimated that repetitive elements occupy 27.9% of the common carp genome, which is much higher than that in the genomes of Takifugu (7.1%), Tetroadon (5.7%), and stickleback (13.48%), but lower than that in the Atlantic salmon genome (30–35%) (Davidson *et al.*, 2010) and zebrafish genome (52.2%) (Howe *et al.*, 2013).

Zebrafish is the most closely related species to common carp among teleost fishes, with a high-quality whole genome assembly. They both belong to the same family of Cyprinidae. Zebrafish has a diploid genome with a genome size of 1.4 GB and a chromosome number of $n = 25$. To evaluate the similarity to the genome of closely related zebrafish, we aligned BES against the zebrafish genome. A total of 39335 BES of common carp have conserved homologs on the zebrafish genome, which demonstrated the high similarity between zebrafish and common carp genomes, indicating the feasibility of using zebrafish genome as a reference to assess architecture of common carp genome.

We then used the comparative mapping approach to assess and characterize the tetraploidized common carp genome. We genotyped a large number of BAC-anchored microsatellite and SNP markers and constructed a high-density genetic map with 1209 genetic markers on 50 linkage groups (Zhao *et al.*, 2013). Then, the BAC-based physical map and BAC sequences were anchored and integrated onto the genetic map. This was the first chromosome-scale integration map for the common carp genome before we initiated the genome-sequencing project. We mapped the anchored BESs onto the zebrafish reference genome, and constructed the first comparative map between the two Cyprinids genomes. A perfect "two versus one" homologous relationship was displayed between 50 common carp linkage groups and 25 zebrafish chromosomes, which gave us the first genome-scale evidence for the tetraploidization of the common carp genome (Zhao *et al.*, 2013).

The limited genome sequence data and genetic maps allowed us to evaluate the guanine-cytosine (CG) content and repetitive elements, and assess the manner of tetraploidization of the common carp genome. These results not only provided us guidance for the whole-genome-sequencing project, but also provided us valuable genome resources, including a large set of mate-paired BESs and genetic maps.

Genome Assembly: Strategies for Reducing Problems Caused by Allotetraploidy

Reduce Genome Complexity: Gynogen as Sequencing Template

Genome complexity and heterozygosity are the major obstacles for whole-genome sequencing and genome assembly, especially for polyploid species with relatively large genomes, such as common carp. It is one of the top priorities to reduce genome complexity before starting the whole-genome-sequencing project. Two approaches allowed us to collect fish samples with lower genome complexity. First, we collected mature female common carp from an inbred strain, Songpu carp, which has an inbreeding history of over 10 generations at Heilongjiang Fishery Research Institute. Second, gynogen carps were prepared by using gynogenesis technology in order to further reduce the genome complexity and heterozygosity of the sequencing template. Briefly, the eggs obtained from a normal female common carp were inseminated with UV-irradiated sperm from crucian carp. The eggs were then hatched at 22–25°C water temperature to the single cell stage, and then heat-shocked at 39.5°C for 105 seconds to double the chromosomes from haploid phase to diploid phase (they are also called "doubled haploids"). Thus, the embryos and newly hatched fry contained only maternal chromosomes, and heterozygosity was reduced significantly. Microsatellite markers were used to investigate and validate the homozygosity of the genogens.

Sequencing Strategies

The emerging of next-generation sequencing technologies has presented significant improvements in the speed, accuracy, and cost of genome sequencing, which has made it feasible to conduct whole-genome sequencing of non-model species. When CCGP was initiated in 2010, there were many different sequencing technologies available that had been developed by various biotechnology companies, each of which produced different sequencing reads in terms of accuracy and read length. These technologies included Roche 454, Illumina, and SOLiD. The Ion Torrent was also available during the CCGP progress. We made deep comparison of these available sequencing platforms. The Illumina and SOLiD platforms were much cheaper and more accurate than the pyrosequencing-based Roche 454 FLX platform. However, the read lengths of Illumina GA and SOLiD were only 50 bp or so at that time, which were much shorter than the read length of Roche 454 (~400 bp). In addition, the mate-paired library construction with large jumping size was also very challenging for the available sequencing facilities. According to our survey on the common carp genome, repetitive elements occupied about 30% of the allotetraploidized genome. This assembly may be computationally difficult and challenging when we dealt with such a large and complex genome with abundant repeats and duplications. These repeats can be so long that second-generation sequencing reads were not long enough to bridge the repeat. In order to resolve potential challenges during our genome assembly, we decided to adapt hybrid genome-sequencing and assembly strategies. First, we sequenced the genome with the Roche 454 FLX platform with shotgun strategies, and collected ~12 GB genome sequences with an average read length of 352 bp, equivalent to ~7 haploid genome of common carp. The longer reads could efficiently jump those short simple repeats in the common carp genome. The duplicated genome sequences that came from each ancestral genome (allotetraploid ancestors) may retain short haplotype diversity, which could be easily discriminated when sequence reads are relatively longer. The 454 data were primarily used for contig assembly. Secondly, various pair-end and mate-pair libraries were constructed for the Illumina, SOLiD, and Roche 454 platforms from 250 bp to 8 KB. These data were mainly used for contig gap filling and scaffolding to jump most repetitive elements in the common carp genome. Third, as mentioned in the preceding text, we sequenced a large number of BAC clones from both ends, providing us abundant mate-paired BESs with jumping length of ~141 kb. These BESs were finally used for long-region scaffolding to increase the connectivity of the common carp genome. Fourth, while the CCGP was in progress, the Ion Torrent platform became available, which has higher flexibility and efficiency for genome sequencing. Therefore, we decided to use it on BAC clone sequencing to validate the genome assembly.

Genome-sequencing technologies are developing very fast. A third-generation platform, the PacBio system, was soon available when we had almost completed the genome assembly. The PacBio system can produce long reads (maximum of 23 kb), but have a relatively low accuracy. It will be ideal in sequencing data for genome gap filling and contig connecting during assembly.

There is no perfect "all-in-one" genome-sequencing technology available so far. The hybrid sequencing strategy allowed us to take advantage of the strengths of various technologies and overcome impediments during the whole-genome sequencing and assembly of a large, complex, and polyploidized genome.

Genome Assembler Comparison and Selection

Currently, genome assemblers are classified into two categories: overlap–layout–consensus (OLC) and de bruijn graph strategy (Li *et al.*, 2012). Using OLC assemblers, sequence reads are compared to each other and combined together when a significant overlap is found, forming larger sequences called *contigs*. Many programs use this strategy, including Celera Assembler (Denisov *et al.*, 2008), Newbler (http://www.454.com/products/analysis-software/), and SGA Assembler (Simpson & Durbin, 2012). De bruijn graph assemblers use a different strategy: working by first chopping reads into much shorter k-mers and then using all the k-mers to form a de bruijin graph, and finally inferring the genome sequence. The strategy condenses the data from massive numbers of sequence reads into a single data structure. The size of the resulting graph scales with the size of the genome, whereas the memory required by an OLC approach scales with the number of reads. Therefore, it is very suitable for massive sets of reads. Many programs including ABySS (Simpson *et al.*, 2009), SOAPdenovo (Luo *et al.*, 2012), Velvet (Zerbino, 2010), and AllPath-LG (Utturkar *et al.*, 2014) use this strategy.

Although the assemblers based on the de bruijn graph strategy were widely applied in genome assembly, they were rarely used in the assembly of the polyploid genome. However, OLC assemblers have been applied to polyploid genomes. For instance, the genome of the rainbow trout, a tetraploid species (Berthelot *et al.*, 2014), was assembled with Newbler, an OLC assembler. To assemble the genome of common carp, we assembled the Roche 454 read data set and the Sanger BAC-end sequences into contigs using Celera Assembler. The pre-assembled contigs were used for further scaffolding.

Considering the high identities between two duplicated genes, an appropriate identity threshold is essential for assembly. If the identity was set too low, two duplicated genes might be merged into one gene, leading to gene loss. In contrast, if the identity was set too high, the identity threshold might not eliminate the sequence errors, resulting in chimeric gene redundancy. Therefore, to set the identity threshold, we analyzed the identity distribution of 20 pairs of known duplicated genes.

After preassembly of the contigs, a hierarchical scaffolding strategy was performed to increase genome sequences with reads from the Illumina libraries, the SOLiD libraries, and the Roche 454 mate-pair libraries. Reads from libraries of small insert size were first aligned to the genomic sequences for scaffolding. Then reads from libraries of larger insert size were used to construct larger scaffolds. Finally, BAC-end sequences were mapped to the scaffolds and used for further scaffolding.

Quality Control of Assembly

At least four indices were applied to the quality measurement of assembly. First, the mapping ratio of sequencing reads is widely applied in examining the assembly integrity. Short reads, including Illumina data and SOLiD reads, could be aligned again to the final assembly using BWA (Li & Durbin, 2010) or Bowtie (Langmead *et al.*, 2009). Long reads, for instance, 454 reads and BAC end sequences, could be aligned with BLAT (Kent, 2002) or the latest BWA (Li & Durbin, 2010). A total of 83% of raw reads can be mapped. The high mapping ratio indicated the assembly integrity.

The alignment ratio of full-length BAC sequences to the final assembly is the second widely adopted index of assembly integrity. The ultra-long read alignment was carried out using BLAT, MUSCLE (Edgar, 2004), or blastz (Schwartz *et al.*, 2003). In CCGP, we sequenced the full-length of one BAC sequence and then aligned it to the final assembly. The coverage reached 90%.

The third index is gene coverage. It was calculated as the ratio of aligned transcriptome reads to all reads. Generally, long reads, including Sanger EST sequences, 454 transcriptome reads, and assembled Illumina transcripts, were preferentially applied into gene coverage assessment. In CCGP, 454 transcriptome reads and Sanger ESTs were aligned to the assembly sequences using BLAT with default parameters and an identity cutoff at 90%. The mapping ratio of ESTs was about 95%.

Parra *et al.* define a set of conserved protein families that occur in a wide range of eukaryotes (Parra *et al.*, 2009). On the basis of sequence homologs, the core eukaryotic genes (CEGs) identified by them were also mapped to the assembled genome by BLAT to calculate the gene region coverage. This strategy has been applied to the quality assessment of multiple species (Parra, Bradnam & Korf, 2007). For the common carp genome, the coverage of CEGs is between 82% and 95%.

Especially considering that the common carp genome had a recent WGD event compared to other teleost, and that genome assembly may merge two duplicated genes into one copy, it is important to assess whether our assembly could differentiate highly similar genes. We estimated the missing rates of known duplicated genes. A total of 38 published paralogous mRNAs (19 pairs of paralogs) were used in the assessment (Wang *et al.*, 2012). The 38 mRNAs were aligned with the assembly using BLAT, and a region that aligned best with each of the mRNAs was selected. Only six genes in three homologous pairs were merged into three single genes on the assembly, while 32 genes in 16 homologous pairs were mapped to distinct genomic locations, indicating that 92% of homologous genes could be differentiated, and that our assembly was of high quality.

Annotation of Tetraploidy Genome

De novo gene prediction, sequence-homology-based predictions, and RNA-seq data were widely used for gene prediction.

De Novo Gene Prediction

Multiple *de novo* gene prediction software has been developed for gene prediction, including Fgenesh + (Salamov & Solovyev, 2000), AUGUSTUS (Stanke *et al.*, 2004), GeneMark (Besemer & Borodovsky, 2005), GENSCAN (Burge & Karlin, 1997), and Glimmer (Kelley *et al.*, 2012). There are mainly two steps in most de novo gene prediction software.

1) Training species-specific parameters with a high-quality set of genes for the gene prediction that follows. The training gene set could be public genes of this species. Gene prediction accuracy always depends on the quality of the training gene set.
2) Gene prediction with the species-specific parameters.

Among gene prediction programs, AUGUSTUS is usually the most accurate. In the independent gene finder assessment (EGASP) on the human ENCODE regions (Guigo *et al.*, 2006) and the more recent nGASP (worm) (Coghlan *et al.*, 2008), AUGUSTUS was the most accurate gene finder among all the tested *ab initio* programs. Although the *ab initio* programs were easily run, all have low prediction specificity. For instance, it is estimated that the specificities of AUGUSTUS, Fgenesh+, and GeneMark are 38.4%, 35.4% and 24.5%, respectively (Coghlan *et al.*, 2008).

Sequence-homology-based Prediction

Sequence-homology-based gene prediction included both raw and precise alignments. First, protein sequences from public databases, for instance, NCBI non-redundant (NR) database and Ensembl database (Flicek *et al.*, 2014), were collected for homolog alignment. To perform raw alignments, proteins were aligned to the assembled genome sequence in the database by BlastX. Adjacent and overlap matches were merged together, building the longest protein for each genomic sequence region. Second, the protein sequences were then aligned to these genome fragments by Genewise (Birney, Clamp & Durbin, 2004) to identify the accurate splicing sites. The alignment using Genewise allows for introns and frameshifting errors. The Genewise algorithm was developed from a principled combination of hidden Markov models (HMMs), is highly accurate, and can provide both accurate and complete gene structures when used with the correct evidence. It is heavily used by the Ensembl annotation system.

There were at least two disadvantages for sequence-homology-based prediction. First, although sequence-homology-based prediction is more accurate than de novo gene prediction, this method cannot predict species-specific genes. Second, the relationship between sequenced species and model species limited the predicted gene number. Therefore, for those species without closely related model species, the prediction might not cover all homologous genes.

Transcriptome Sequencing

Next-generation sequencing technologies have presented significant improvements in the speed, accuracy, and cost of transcriptome sequencing, which made the whole-genome sequencing of non-model species feasible. With species for which the whole-genome sequence is not yet available, transcriptome analysis is an alternative method that has been used to discover new genes and to investigate gene expression. A large set of common carp ESTs produced using Sanger sequencing has been developed and used to study traits in common carp. More recently, second-generation sequencing platforms have been applied to transcriptome sequencing, making the transcriptome more readily accessible.

To build RNA-seq gene models, transcriptome sequencing data were mapped to the genome using TopHat (Trapnell, Pachter & Salzberg, 2009), and gene models were predicted using Cufflinks (Trapnell *et al.*, 2012). Transcriptome-based gene prediction can be improved in several ways. First, because genes are spatially and temporally expressed, increased sequencing breadth, including multiple tissues and developmental stages, will cover more genes. Second, sufficient transcriptome sequencing coverage improves the completeness of gene models. To investigate whether the transcriptome sequencing coverage is sufficient or not, saturation analysis could be applied to ascertain whether sequencing coverage was sufficient to draw a comprehensive picture of the transcriptome. Rarefied libraries were constructed by randomly sampling from 10% to 100% of the transcriptome data. Then, new assemblies at each of the defined levels were produced to illustrate possible differences in gene discovery rates. The saturated curve indicated that a large part of the genes were detected.

Since all three prediction methods had their own advantages and disadvantages, integration of the results from these three methods would improve the accuracy of gene prediction. Multiple gene prediction integration programs were developed, for instance, Glean (Elsik *et al.*, 2007), EVidenceModeler (EVM) (Haas *et al.*, 2008), and Gaze (Howe, Chothia & Durbin, 2002).

In CCGP, we combined de novo gene prediction, sequence-homology-based predictions, and RNA-seq data with EVM. Two de novo prediction software programs, AUGUSTUS (Stanke *et al.*, 2004) and Fgenesh + (Salamov & Solovyev, 2000), were used to predict genes on repeat-masked genome sequences, and we did not mask low-complexity regions and simple repeats since they could be parts of coding sequences. Sequence-homology-based gene prediction included both raw and precise alignments. Adjacent and overlap matches were merged using Perl scripts, building the longest protein for each genomic sequence region. And then, each target region in the genome was extended by 10 KB from both ends of the aligned region, to cover potential untranslated regions. The protein sequences were then aligned to these genome fragments by Genewise. RNA-seq reads were mapped to genomic sequences by TopHat (Trapnell *et al.*, 2009), and then Cufflinks (Trapnell *et al.*, 2012) was used to get assemblies of transcripts. For a gene locus with several alternative splicing transcripts generated from Cufflinks, the transcript with the longest exon length was chosen. All the evidences were merged to form a consensus gene set by EVM. The integrated annotation pipeline generated a total of 52610 gene models. About 91.4% of these genes were detected in assembled transcripts from RNA-seq reads.

Assessment and Evaluation

Sequence structures, including exon number, exon length, and protein length, were conserved across species. Comparative analysis of the gene structure among zebrafish, medaka, stickleback, fugu, and tetraodon revealed that the average exon number and exon length were similar across the species (7 ~ 10 and 140 ~ 230 bp, respectively)

(Xu *et al.*, 2014). Comparing the gene structure of sequenced species with closely related species would provide a hint to the quality of gene prediction. The low quality of genome assembly or gene prediction decreases the gene length. The average gene and coding sequence lengths were 12145 bp and 1487 bp, respectively, and common carp genes had an average of 7.48 exons per gene (Xu *et al.*, 2014).

Sequence homolog is another index to assess the quality of the predicted genes. Protein homolog is much conserved in vertebrates. Even hundreds of genes are conserved across eukaryotes (Parra *et al.*, 2007). Howe *et al.* reported that 70% of human genes have at least one obvious zebrafish ortholog (Howe *et al.*, 2013). Sequence alignment with predicted proteins against public protein databases using blastp would indicate the quality of the predicted genes. However, erroneous gene prediction resulting from frameshifting or mis-assembly decreases the proportion of homologous genes. About 91% of common carp proteins were homologous to other species.

Taken together, the similar gene structures to other closely related species and the high proportion of homologous genes indicated the high quality of common carp gene annotations.

Conclusions

Fish species show the most extensive polyploidy among vertebrates, which brings significant difficulty and complexity for genome sequencing, sequence assembly, and annotation. Common carp is one of the typical instances. In order to unveil common carp genome information comprehensively and accurately, we employed hybrid sequencing strategies to (1) generate genome data from multiple sequencing platforms, (2) assemble the whole genome with multiple algorithms, and (3) annotate repeat elements and functional genes. A high-density genetic map served as the backbone for the scaffold integration. The well-assembled and annotated common carp genome provides a valuable resource for genetic, genomic, and biological studies of common carp, and for improving important aquaculture traits of farmed common carp. Our experience with common carp sequencing should be useful for those genome research projects with similar situations on genome duplication.

References

Allendorf, F.W. and Thorgaard, G.H. (1984) Tetraploidy and the evolution of salmonid fishes, in *Evolutionary Genetics of Fishes* (ed. B. Turner), Plenum Press, New York, pp. 1–53.

Berthelot, C., Brunet, F., Chalopin, D. *et al.* (2014) The rainbow trout genome provides novel insights into evolution after whole-genome duplication in vertebrates. *Nature Communications*, **5**, 3657.

Besemer, J. and Borodovsky, M. (2005) GeneMark: web software for gene finding in prokaryotes, eukaryotes and viruses. *Nucleic Acids Research*, **33**, W451–454.

Birney, E., Clamp, M. and Durbin, R. (2004) GeneWise and Genomewise. *Genome Research*, **14**, 988–995.

Bostock, J., McAndrew, B., Richards, R. *et al.* (2010) Aquaculture: global status and trends. *Philosophical Transactions of the Royal Society B: Biological Sciences*, **365**, 2897–2912.

Burge, C. and Karlin, S. (1997) Prediction of complete gene structures in human genomic DNA. *Journal of Molecular Biology*, **268**, 78–94.
Cheng, L., Liu, L., Yu, X. *et al.* (2010) A linkage map of common carp (*Cyprinus carpio*) based on AFLP and microsatellite markers. *Animal Genetics*, **41**, 191–198.
Christoffels, A., Bartfai, R., Srinivasan, H. *et al.* (2006) Comparative genomics in cyprinids: common carp ESTs help the annotation of the zebrafish genome. *BMC Bioinformatics*, **7**, S2.
Coghlan, A., Fiedler, T.J., McKay, S.J. *et al.* (2008) nGASP – the nematode genome annotation assessment project. *BMC Bioinformatics*, **9**, 549.
David, L., Blum, S., Feldman, M.W. *et al.* (2003) Recent duplication of the common carp (*Cyprinus carpio* L.) genome as revealed by analyses of microsatellite loci. *Molecular Biology and Evolution*, **20**, 1425–1434.
Davidson, W.S., Koop, B.F., Jones, S.J. *et al.* (2010) Sequencing the genome of the Atlantic salmon (*Salmo salar*). *Genome Biology*, **11**, 403.
Denisov, G., Walenz, B., Halpern, A.L. *et al.* (2008) Consensus generation and variant detection by Celera Assembler. *Bioinformatics*, **24**, 1035–1040.
Edgar, R.C. (2004) MUSCLE: multiple sequence alignment with high accuracy and high throughput. *Nucleic Acids Research*, **32**, 1792–1797.
Elsik, C.G., Mackey, A.J., Reese, J.T. *et al.* (2007) Creating a honey bee consensus gene set. *Genome Biology*, **8**, R13.
FAO Fisheries and Aquaculture Department (2007) *The State of World Fisheries and Aquaculture 2006*, Food and Agriculture Organization of the United Nations, Rome.
Flicek, P., Amode, M.R., Barrell, D. *et al.* (2014) Ensembl 2014. *Nucleic Acids Research*, **42**, D749–755.
Guigo, R., Flicek, P., Abril, J.F. *et al.* (2006) EGASP: the human ENCODE Genome Annotation Assessment Project. *Genome Biology*, **7** (1), S2.1–31.
Haas, B.J., Salzberg, S.L., Zhu, W. *et al.* (2008) Automated eukaryotic gene structure annotation using EVidenceModeler and the Program to Assemble Spliced Alignments. *Genome Biology*, **9**, R7.
Howe, K., Clark, M.D., Torroja, C.F. *et al.* (2013) The zebrafish reference genome sequence and its relationship to the human genome. *Nature*, **496**, 498–503.
Howe, K.L., Chothia, T. and Durbin, R. (2002) GAZE: a generic framework for the integration of gene-prediction data by dynamic programming. *Genome Research*, **12**, 1418–1427.
Ji, P., Zhang, Y., Li, C. *et al.* (2012) High throughput mining and characterization of microsatellites from common carp genome. *International Journal of Molecular Sciences*, **13**, 9798–9807.
Kelley, D.R., Liu, B., Delcher, A.L., Pop, M. and Salzberg, S.L. (2012). Gene prediction with Glimmer for metagenomic sequences augmented by classification and clustering. *Nucleic Acids Research*, **40**, e9.
Kent, W.J. (2002) BLAT – the BLAST-like alignment tool. *Genome Research*, **12**, 656–664.
Langmead, B., Trapnell, C., Pop, M. and Salzberg, S.L. (2009) Ultrafast and memory-efficient alignment of short DNA sequences to the human genome. *Genome Biology*, **10**, R25.
Larhammar, D. and Risinger, C. (1994) Molecular genetic aspects of tetraploidy in the common carp *Cyprinus carpio. Molecular Phylogenetics and Evolution*, **3**, 59–68.

Li, H. and Durbin, R. (2010) Fast and accurate long-read alignment with Burrows-Wheeler transform. *Bioinformatics*, **26**, 589–595.

Li, Y., Xu, P., Zhao, Z. *et al.* (2011) Construction and characterization of the BAC library for common carp *Cyprinus carpio* L. and establishment of microsynteny with zebrafish *Danio rerio*. *Marine Biotechnology (NY)*, **13**, 706–712.

Li, Z., Chen, Y., Mu, D. *et al.* (2012) Comparison of the two major classes of assembly algorithms: overlap-layout-consensus and de-bruijn-graph. *Briefings in Functional Genomics*, **11**, 25–37.

Luo, R., Liu, B., Xie, Y. *et al.* (2012) SOAPdenovo2: an empirically improved memory-efficient short-read de novo assembler. *Gigascience*, **1**, 18.

Ohno, S. (1970) *Evolution by gene duplication*, Springer-Verlag, Berlin, New York.

Ohno, S., Muramoto, J., Christian, L. and Atkin, N.B. (1967) Diploid–tetraploid relationship among old-world members of the fish family Cyprinidae. *Chromosoma*, **23**, 1–9.

Parra, G., Bradnam, K. and Korf, I. (2007) CEGMA: a pipeline to accurately annotate core genes in eukaryotic genomes. *Bioinformatics*, **23**, 1061–1067.

Parra, G., Bradnam, K., Ning, Z. *et al.* (2009) Assessing the gene space in draft genomes. *Nucleic Acids Research*, **37**, 289–297.

Phillips, R. and Rab, P. (2001) Chromosome evolution in the Salmonidae (Pisces): an update. *Biological Reviews of the Cambridge Philosophical Society*, **76**, 1–25.

Postlethwait, J.H., Woods, I.G., Ngo-Hazelett, P. *et al.* (2000) Zebrafish comparative genomics and the origins of vertebrate chromosomes. *Genome Research*, **10**, 1890–1902.

Salamov, A.A. and Solovyev, V.V. (2000) Ab initio gene finding in Drosophila genomic DNA. *Genome Research*, **10**, 516–522.

Schwartz, S., Kent, W.J., Smit, A. *et al.* (2003) Human–mouse alignments with BLASTZ. *Genome Research*, **13**, 103–107.

Simpson, J.T. and Durbin, R. (2012) Efficient de novo assembly of large genomes using compressed data structures. *Genome Research*, **22**, 549–556.

Simpson, J.T., Wong, K., Jackman, S.D. *et al.* (2009) ABySS: a parallel assembler for short read sequence data. *Genome Research*, **19**, 1117–1123.

Stanke, M., Steinkamp, R., Waack, S. and Morgenstern, B. (2004) AUGUSTUS: a web server for gene finding in eukaryotes. *Nucleic Acids Research*, **32**, W309–312.

Sun, X. and Liang, L. (2004) A genetic linkage map of common carp (*Cyprinus carpio* L.) And mapping of a locus associated with cold tolerance. *Aquaculture*, **238**, 165–172.

Trapnell, C., Pachter, L. and Salzberg, S.L. (2009) TopHat: discovering splice junctions with RNA-Seq. *Bioinformatics*, **25**, 1105–1111.

Trapnell, C., Roberts, A., Goff, L. *et al.* (2012) Differential gene and transcript expression analysis of RNA-seq experiments with TopHat and Cufflinks. *Nature Protocols*, **7**, 562–578.

Utturkar, S.M., Klingeman, D.M., Land, M.L. *et al.* (2014) Evaluation and validation of de novo and hybrid assembly techniques to derive high-quality genome sequences. *Bioinformatics*, **30**, 2709–2716.

Van de Peer, Y., Taylor, J.S., Braasch, I. and Meyer, A. (2001) The ghost of selection past: rates of evolution and functional divergence of anciently duplicated genes. *Journal of Molecular Evolution*, **53**, 436–446.

Vandepoele, K., De Vos, W., Taylor, J.S. *et al.* (2004) Major events in the genome evolution of vertebrates: paranome age and size differ considerably between ray-finned fishes and

land vertebrates. *Proceedings of the National Academy of Sciences of the United States of America*, **101**, 1638–1643.
Wang, J.T., Li, J.T., Zhang, X.F. and Sun, X.W. (2012) Transcriptome analysis reveals the time of the fourth round of genome duplication in common carp (*Cyprinus carpio*). *BMC Genomics*, **13**, 96.
Williams, D., Li, W., Hughes, M. *et al.* (2008) Genomic resources and microarrays for the common carp *Cyprinus carpio* L. *Journal of Fish Biology*, **72**, 2095–2117.
Wolf, U., Ritter, H., Atkin, N. and Ohno, S. (1969) Polyploidization in the fish family Cyprinidae, order Cypriniformes. *Humangenetik*, **7**, 240–244.
Xu, J., Ji, P., Zhao, Z., Zhang, Y., Feng, J. *et al.* (2012). Genome-wide SNP discovery from transcriptome of four common carp strains. *PLoS One*, **7**, e48140.
Xu, P., Li, J., Li, Y. *et al.* (2011) Genomic insight into the common carp (*Cyprinus carpio*) genome by sequencing analysis of BAC-end sequences. *BMC Genomics*, **12**, 188.
Xu, P., Zhang, X., Wang, X. *et al.* (2014) Genome sequence and genetic diversity of the common carp, *Cyprinus carpio*. *Nature Genetics*, **46** (11), 1212–1219.
Zerbino, D.R. (2010). Using the Velvet *de novo* assembler for short-read sequencing technologies. *Current Protocols in Bioinformatics*, pp. 11-5.
Zhang, X., Zhang, Y., Zheng, X. *et al.* (2013) A consensus linkage map provides insights on genome character and evolution in common carp (*Cyprinus carpio* L.). *Marine Biotechnology (NY)*, **15**, 275–312.
Zhao, L., Zhang, Y., Ji, P., Zhang, X., Zhao, Z. *et al.* (2013). A dense genetic linkage map for common carp and its integration with a BAC-based physical map. *PLoS One*, **8**, e63928.

Part II

Bioinformatics Analysis of Transcriptomic Sequences

8

Assembly of RNA-Seq Short Reads into Transcriptome Sequences

Jun Yao, Chen Jiang, Chao Li, Qifan Zeng and Zhanjiang Liu

Introduction

RNA-Seq is a technology to sequence transcriptomes using next-generation sequencing technologies. It has been widely used for analyses such as gene expression profiling and identification of differentially expressed genes (DEG). RNA-Seq can be done with a number of sequencing platforms including the Illumina sequencing platform, ABI Solid Sequencing, and less efficiently the Life Science's 454 sequencing. Of these, the Illumina HiSeq sequencers are the most popular because of their very high throughput and accuracy of sequencing reads.

RNA-Seq analysis starts with RNA samples. Before RNA-Seq, considerations need to be made to allow statistical analysis of the results with proper biological and technical replica. Since the analysis of DEGs is the subject of Chapter 9, we focus in this chapter on the design of RNA-Seq experiments and on the bioinformatics issues related to the assembly of RNA-Seq short reads into reference transcriptomes. We present procedures and command lines for both *de novo* assembly approaches and reference-sequence-guided assembly approaches.

RNA-Seq Procedures

RNA-Seq analysis starts with RNA samples. Such samples can be just one RNA sample if the researcher is interested in just the expression profiles of the sample under study. However, most often, RNA-Seq involves two or more samples in order to compare genome expression under a pair of or more conditions. For instance, a control sample can be compared with the infected samples at various times after infection. In order to allow statistical analysis of the results, a minimum of three biological replications are required for each condition (treatment). However, this does not necessarily mean that multiple lanes must be used. Instead, to reduce cost, each biological replica can be tagged with a molecular tag through the addition of a specific adaptor sequence. After the sequences are generated, bioinformatics processing would allow the identification of reads for each of the samples based on the molecular tags. RNA-Seq can be conducted in a single or more lanes depending on the level of desired output of the sequence

Bioinformatics in Aquaculture: Principles and Methods, First Edition. Edited by Zhanjiang (John) Liu.

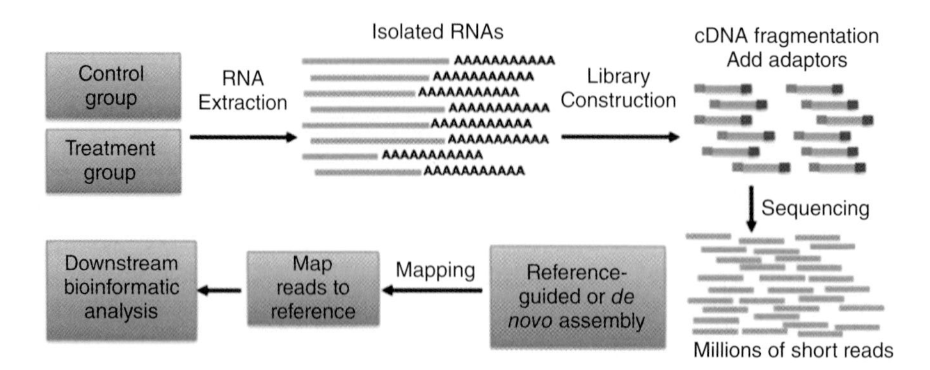

Figure 8.1 The procedures of RNA-Seq.

reads. For instance, we conducted an RNA-Seq to determine changes in transcriptional regulation induced in mucosal tissues, skin, and gill due to short-term (7 days) fasting (Liu, Li *et al.*, 2013). In this experiment, we had two treatments: fast and fed fish, and three biological replications in each treatment—therefore, a total of six samples. Based on the estimated number of transcripts of the channel catfish transcriptome, we decided to sequence these six samples in half a lane, which would provide deep enough sequencing coverage while being economical. In this particular example, a total of 209 million reads of 100 bp length were generated for the fasted and fed samples, that is, an average of over 34 million reads per sample, with greater than 26 million reads for any of the six libraries.

The procedures of RNA-Seq are schematically presented in Figure 8.1. The RNA is converted into cDNAs that are fractured into small segments such that overlapping sequences can be generated, allowing assembly of the transcriptome. Linkers are added to the cDNA fragments, allowing sequencing primers to have access to all the cDNA fragments. Such fragments are then subjected to sequencing. After sequencing, the fastq files are generated, ready for bioinformatics analysis to extract biological information out of the huge pool of short sequence tags.

The biological issues of RNA-Seq are actually extremely simple. The researchers are interested in which genes are expressed, how much they are expressed, and how the samples in different conditions compare. Before answering these questions, the immediate task is to assemble the short reads into a reference transcriptome. In general, one of the two types of assembly methods can be used for the assembly of RNA-Seq sequences, depending on the existing genome resources. If a reference genome sequence is available, reference-guided assembly methods can be used. In contrast, *de novo* RNA-Seq assembly methods must be used in the absence of a reference genome sequence.

Reference-Guided Transcriptome Assembly

For well-characterized species such as many of the model species, reference genome sequences are likely available (e.g., the known transcriptome or the reference genome). In those cases, one can simply map the short reads to the reference transcriptome or reference genome using reference-guided assembly approaches. At present, over a dozen

Table 8.1 Examples of fish species whose reference genome sequences are available.

Species	Total assembled size (MB)	References
Stickleback	463	Jones *et al.*, 2012
Catfish	783	Liu *et al.*, 2016
Grass carp	901	Wang *et al.*, 2015
Sea bass	675	Tine *et al.*, 2014
Shark	937	Venkatesh *et al.*, 2014
Cavefish	964	McGaugh *et al.*, 2014
Zebrafish	1410	Howe *et al.*, 2013
Medaka	700	Kasahara *et al.*, 2007
Platyfish	669	Schartl *et al.*, 2013
Common carp	1690	Xu *et al.*, 2014
Tetraodon	342	Jaillon *et al.*, 2004
Coelacanth	2860	Amemiya *et al.*, 2013
Sole	477	Chen *et al.*, 2014
Cod	753	Star *et al.*, 2011
Yellow croaker	644	Wu *et al.*, 2014
Rainbow trout	1900	Berthelot *et al.*, 2014
Lamprey	816	Smith *et al.*, 2012
Fugu	333	Aparicio *et al.*, 2002

fish and aquaculture species have reference genome sequences (Table 8.1). For species with a high-quality reference genome sequence, reference-guided transcriptome assembly is recommended. However, *de novo* assembly can also be conducted if desired. *De novo* assembly has the advantage of reconstructing transcripts from regions missing in the genome assembly.

The most widely used reference-guided assembly approach is the TopHat–Cufflinks assembly. TopHat is a fast splice junction mapper for RNA-Seq reads, which allows spliced alignment (Trapnell, Pachter & Salzberg, 2009). The reads are first mapped to the genome, and then the unmapped reads are split into shorter segments and aligned independently. Cufflinks suite has the ability to assemble transcripts, estimate their abundances, and test for differential expression and regulation in RNA-Seq samples. It accepts aligned RNA-Seq reads and assembles the alignments into a parsimonious set of transcripts. The assembly can be carried out mainly in four steps.

1) Before mapping reads to the reference genome, a reference index needs to be built using Bowtie2 (Langmead & Salzberg, 2013):

   ```
   $ bowtie2-build <reference_in> <bt2_base>
   ```

   Here, "`reference_in`" specifies the reference sequences to be aligned to in the form of a comma-separated list of FASTA files. "`bt2_base`" specifies the base name of the generated index files. By default, bowtie2-build generates files named "bt2_base.1.bt2", "bt2_base.2.bt2", "bt2_base.3.bt2", "bt2_base.4.bt2", "bt2_base.rev.1.bt2", and "bt2_base.rev.2.bt2".

2) Mapping reads to the reference genome using TopHat:

```
$ tophat -o . -p 8 -N 2 --read-gap-length 2 --GTF
   genes.gtf <genome_index_base> PE_reads_1.fq,SE_reads.fq
   PE_reads_2.fq
```

Here, "-o" specifies the name of the output directory. The default is "./tophat_out". "-p" specifies the number of threads used for alignment. The default is "1". "-N" specifies the number of mismatches in the final alignment. Alignments having more than the specified mismatches are discarded. The default is "2". "--read-gap-length" specifies the total length of indels (alignment gaps) in the final read alignments. Alignments having more than the specified total length of gaps are discarded. The default is "2". "--GTF:" specifies a set of gene model annotations and/or known transcripts to TopHat, in the format of a GTF 2.2 or GFF3 file. If GTF is provided, TopHat will first use Bowtie to align reads to the extracted transcript sequences. Reads that cannot fully map to this virtual transcriptome will then be mapped on the genome. The reads that are mapped on the transcriptome will be converted to genomic mappings (spliced as needed) and merged with the novel mappings and junctions in the final TopHat output. The name in the first column of the provided GTF/GFF file (column that indicates the chromosome or contig on which the feature is located) should match the name of the reference sequence built in the Bowtie2 index in step (1). "genome_index_base" is the "bt2_base" name of the genome index built in step (1).

In addition, TopHat allows the use of both paired and unpaired reads; the unpaired reads can be provided after the paired reads, which is shown in the example command (PE_reads_1.fq,SE_reads.fq PE_reads_2.fq).

3) After mapping, Cufflinks assembly is utilized for reference-guided assembly:

```
$ cufflinks -p 8 -o cufflinks_assembly -u -g genes.gtf
   aligned_reads.sam
```

Here, the option "-p" specifies the number of threads used during assembly. The default is "1". "-o" specifies the name of the output directory. The default is current directory. "-u" is used for correcting multiple aligned reads. If this option is used, Cufflinks will do an initial estimation procedure for more accurately judgment of reads mapping to multiple locations in the genome. The default is "FALSE". "-g" specifies the reference annotation GFF file supplied to Cufflinks for RABT (Reference Annotation Based Transcript) assembly. Reference annotation will be tiled with faux-reads to provide additional information for assembly. Note that the output will contain all reference transcripts along with any novel genes and isoforms that are assembled. "aligned_reads.sam" specifies the RNA-Seq read alignments file in the SAM format. The SAM file allows aligners to attach custom tags to individual alignments, which is required in Cufflinks assembly. Cufflinks produces three output files:

1) Transcriptome assembly: "transcripts.gtf"
2) Transcript-level expression: "isoforms.fpkm_tracking"
3) Gene-level expression: "genes.fpkm_tracking"

The output assembled GTF file contains Cufflinks' assembled isoforms. The first seven columns are standard GTF, and the last column contains attributes. The GTF file contains one GTF record per row, where each record represents an exon within a transcript.

The transcript-level expression file showed the estimated isoform-level expression values in the format of FPKM, and the gene-level expression file showed the estimated gene-level expression values in the format of FPKM as well.

4) When the RNA-Seq study consists of multiple libraries, assembly can be conducted first on each of the libraries, and then all of the cufflinks-assembled transcriptome can be merged into a master transcriptome using Cuffmerge. Cuffmerge also handles running Cuffcompare, and filters a number of transfrags that are potential artifacts. Providing a reference GTF file in the analysis could merge novel and known isoforms and maximize overall assembly quality.

   Before running Cuffmerge, create a file called "assemblies.txt" that lists the assembled files for each library:

```
./lib1/transcripts.gtf
./lib2/transcripts.gtf
./lib3/transcripts.gtf
...
```

   Then run Cuffmerge on all the listed assemblies to create a single merged transcriptome annotation:

```
$ cuffmerge -p 8 -s genome.fa -o cuffmerge -g genes.gtf
   assemblies.txt
```

   Here, "`-p`" specifies the threads used for merging assemblies. The default is "1". "`-s`" is the genomic DNA sequences of the reference sequences. If the reference sequences are provided in a directory, it needs to contain one fasta file per contig. If a multifasta file is provided, all contigs should be present. Cuffmerge will pass this option to the cuffcompare function, which will use the sequences to assist in classifying transfrags and excluding artifacts. "`-o`" specifies the output file instead of stdout. The default is "./merged_asm". "`-g`" specifies the provided reference annotation GTF file. The input assemblies are merged together with the reference annotation GTF included in the final output.

   Cuffmerge produces a GTF file named "merged.gtf" that merges together the input assemblies.

De novo Transcriptome Assembly

In spite of the rapid progress in genome sequencing with aquaculture species, the reference genome sequences or reference transcriptomes are not yet available for most aquaculture species. Therefore, in this case, one would need to perform *de novo* assembly with RNA-Seq datasets. Such *de novo* assembled transcriptome will be regarded as the reference for subsequent analyses. *De novo* transcriptome assembly is more challenging, especially for higher eukaryotes, because of the large number of genes, great variations in the expression levels, and large numbers of alternatively spliced transcript variants.

For *de novo* assembly, a number of software packages are available, such as ABySS (Birol *et al.*, 2009), Trans-ABySS (Robertson *et al.*, 2010), SOAPdenovo (Li *et al.*, 2009), Velvet (Zerbino & Birney, 2008), Oases (Schulz *et al.*, 2012), and Trinity (Grabherr *et al.*, 2011). Each package has its own advantages and disadvantages. However, Trinity (https://github.com/trinityrnaseq/trinityrnaseq/wiki) is the most popular package for *de novo* transcriptome assembly. It is designed specifically for transcriptome assembly, with superior performance when compared to the other *de novo* assemblers currently available, and thus has been widely used for *de novo* assembly of many transcriptomes (e.g., Chauhan *et al.*, 2014; Haas *et al.*, 2013; Li, Beck & Peatman, 2014; Wu *et al.*, 2015; Xu, Evensen & Munang'andu, 2015).

In order to reduce the memory requirement and improve the efficiency of assembly for large RNA-Seq datasets, data normalization can be performed after trimming (Haas *et al.*, 2013). The command for data normalization is:

```
$ TRINITY_HOME/util/insilico_read_normalization.pl
   --seqType fq --JM 100G --max_cov 30 --left input_left.fq
   --right input_right.fq --pairs_together --PARALLEL_STATS
   --output output_file --CPU 10
```

Here, "`--seqType`" is the type of input reads (fa, or fq); "`--JM`" is the amount of system memory (GB) to use k-mer counting by jellyfish; and "`--max_cov 100`" is the maximum coverage for reads. If paired reads are used for normalization, "`--left`" is left reads; "--right" is right reads; "`--pairs_together`" states the normalization averaging stats between paired reads; "`--PARALLEL_STATS`" stands for the paired reads statistics generated in this step; "`--output`" writes all output files to this directory; and "`--CPU`" is the number of CPUs to use.

After normalization, *de novo* assembly is performed using Trinity (Grabherr *et al.*, 2011).

```
$ Trinity --seqType fq --max_memory 400G --CPU 10
   --left input_forward_paired.normalized.fq
   --right input_reverse_paired.normalized.fq
   --output output_file
```

Here, "`--seqType`" is the type of input reads; "`--max_memory`" is the suggested max memory (GB) to use; and "`--CPU`" is the number of CPUs to use. If paired reads are used for Trinity assembly, "`--left`" is the left reads after normalization; "`--right`" is the right reads after normalization; and "`--output`" writes all output files to this directory.

Assessment of RNA-Seq Assembly

One question that many researchers ask is: "Is my assembly correct?". This is very difficult to answer. For reference-guided assemblies, the issue is twofold: (1) if the assembly of the reference genome is correct; and (2) if the transcriptome is complete. If all or most of the annotated genes are covered by the transcriptome, the RNA-Seq will be deep enough, and the assembly should be of high quality. For *de novo* assemblies, the situation is more complex in the absence of a reference genome sequence. However, the following will give you some basic assessment of the RNA-Seq assembly:

1) *Number of contigs in relation to the complexities of the transcriptome*: Obviously, the numbers of genes vary greatly depending on the species. For instance, the transcriptome of a bacterium may include several thousands of protein-encoding transcripts, while the transcriptome of a teleost fish may contain 20,000–50,000 protein-encoding transcripts. If the *de novo* assembly generates a similar or slightly larger number of contigs as expected from the number of protein-encoding genes in the organism, the assembly parameters are probably set at relatively reasonable levels. It should be noted that it is not the more the contigs are, the better the assembly is vice versa, a better assembly does not mean it contains a large number of contigs. Usually, the transcriptome for most diploid fish species should contain 20,000–30,000 genes. If too many contigs are generated in your reference, it may suggest problems of the assembly—for example, several contigs represent a single gene. On the other hand, if too few contigs are generated, it may indicate problems in the sequencing reactions, the assembly, or both. For instance, this may mean that the coverage of RNA-Seq is not deep enough, that reverse transcription reaction failed, or that many transcripts expressed at low levels were not captured. With various RNA-Seq projects of channel catfish, we usually had around 20,000–80,000 contigs, which has approximately 25,000 genes (Li *et al.*, 2012; Liu *et al.*, 2012; Liu, Wang *et al.*, 2013).
   Another parameter that needs to be considered is the average length of the contigs. In general, longer average length of RNA-Seq assembly is preferable, but this is really an issue of balance between a better assembly and the completeness of the assembly. Inclusion of shorter contigs would reduce the average contig length. We generally keep all contigs that have a minimum length of 200 bp. A good test is to determine if a set of full-length cDNAs are each assembled into a single contig. Obviously, it is acceptable to have more than one contig per transcript, but it is wrong when exons of different genes are assembled into a single transcript.
2) *Mapping percentage*: When mapping the short reads back to the reference, the crucial parameter you need to consider is the number of reads that can be mapped to the reference. In the perfect situation, 100% of the reads map to the reference, although that barely ever happens. In the various RNA-Seq projects we conducted, the mapping percentage was usually around 85%. If the mapping percentage is very low, it indicates that your assembly did not capture the whole transcriptome, or your sequencing did not have deep enough coverage, or your RNA-Seq may have contaminations. With pair-end RNA-Seq, one should determine how many reads can be mapped in pairs. If this number is low, it means that you have too many gaps with the transcriptome assembly. In our RNA-Seq projects, the percentage of reads that can be mapped in pairs is approximately 80%.

Conclusions

RNA-Seq is now the approach of choice for gene expression analysis among different treatments. Before various analyses, RNA-Seq short reads must first be assembled. The methods of choice for RNA-Seq assembly depend on existing genomic resources, especially the reference genome sequences. For species with a reference genome sequence, reference-guided transcriptome assembly can be conducted. However, for most aquaculture species without a reference genome sequence, *de novo* assembly methods must be used. TopHat–Cufflinks is the most popular reference-guided assembly method, while Trinity is the most popular *de novo* assembly method. For various analyses after transcriptome assembly, particularly the analysis of DEGs and coordinated expression, readers are referred to Chapter 9.

Acknowledgments

Research is supported by grants from USDA AFRI programs.

References

Amemiya, C.T., Alföldi, J., Lee, A.P. *et al.* (2013) The African coelacanth genome provides insights into tetrapod evolution. *Nature*, **496**, 311–316.

Aparicio, S., Chapman, J., Stupka, E. *et al.* (2002) Whole-genome shotgun assembly and analysis of the genome of *Fugu rubripes*. *Science*, **297**, 1301–1310.

Berthelot, C., Brunet, F., Chalopin, D. *et al.* (2014) The rainbow trout genome provides novel insights into evolution after whole-genome duplication in vertebrates. *Nature Communications*, **5**, 3657.

Birol, I., Jackman, S.D., Nielsen, C.B. *et al.* (2009) *De novo* transcriptome assembly with ABySS. *Bioinformatics*, **25**, 2872–2877.

Chauhan, P., Hansson, B., Kraaijeveld, K. *et al.* (2014) *De novo* transcriptome of *Ischnura elegans* provides insights into sensory biology, colour and vision genes. *BMC Genomics*, **15**, 808.

Chen, S., Zhang, G., Shao, C. *et al.* (2014) Whole-genome sequence of a flatfish provides insights into ZW sex chromosome evolution and adaptation to a benthic lifestyle. *Nature Genetics*, **46**, 253–260.

Grabherr, M.G., Haas, B.J., Yassour, M. *et al.* (2011) Full-length transcriptome assembly from RNA-Seq data without a reference genome. *Nature Biotechnology*, **29**, 644–652.

Haas, B.J., Papanicolaou, A., Yassour, M. *et al.* (2013) *De novo* transcript sequence reconstruction from RNA-seq using the Trinity platform for reference generation and analysis. *Nature Protocols*, **8**, 1494–1512.

Howe, K., Clark, M.D., Torroja, C.F. *et al.* (2013) The zebrafish reference genome sequence and its relationship to the human genome. *Nature*, **496**, 498–503.

Jaillon, O., Aury, J., Brunet, F. *et al.* (2004) Genome duplication in the teleost fish *Tetraodon nigroviridis* reveals the early vertebrate proto-karyotype. *Nature*, **431**, 946–957.

Jones, F.C., Grabherr, M.G., Chan, Y.F. *et al.* (2012) The genomic basis of adaptive evolution in threespine sticklebacks. *Nature*, **484**, 55–61.

Kasahara, M., Naruse, K., Sasaki, S. *et al.* (2007) The medaka draft genome and insights into vertebrate genome evolution. *Nature*, **447**, 714–719.

Langmead, B. and Salzberg, S.L. (2013) Fast gapped-read alignment with Bowtie2. *Nature Methods*, **9**, 357–359.

Li, C., Beck, B.H. and Peatman, E. (2014) Nutritional impacts on gene expression in the surface mucosa of blue catfish (*Ictalurus furcatus*). *Developmental & Comparative Immunology*, **44**, 226–234.

Li, C., Zhang, Y., Wang, R. *et al.* (2012) RNA-seq analysis of mucosal immune responses reveals signatures of intestinal barrier disruption and pathogen entry following *Edwardsiella ictaluri* infection in channel catfish. *Ictalurus punctatus. Fish and Shellfish Immunology*, **32**, 816–827.

Li, R., Yu, C., Li, Y. *et al.* (2009) SOAP2: an improved ultrafast tool for short read alignment. *Bioinformatics*, **25**, 1966–1967.

Liu, L., Li, C., Su, B. *et al.* (2013) Short-term feed deprivation alters immune status of surface mucosa in channel catfish (*Ictalurus punctatus*). *PloS One*, **8**, e74581.

Liu, S., Wang, X., Sun, F. *et al.* (2013) RNA-Seq reveals expression signatures of genes involved in oxygen transport, protein synthesis, folding, and degradation in response to heat stress in catfish. *Physiological Genomics*, **45**, 462–476.

Liu, S., Zhang, Y., Zhou, Z. *et al.* (2012) Efficient assembly and annotation of the transcriptome of catfish by RNA-Seq analysis of a doubled haploid homozygote. *BMC Genomics*, **13**, 595.

Liu, Z., Liu, S., Yao, J. *et al.* (2016) The channel catfish genome sequence and insights into evolution of scale formation in teleosts. *Nature Communications*, **7**, 11757.

McGaugh, S.E., Gross, J.B., Aken, B. *et al.* (2014) The cavefish genome reveals candidate genes for eye loss. *Nature Communications*, **5**, 5307.

Robertson, G., Schein, J., Chiu, R. *et al.* (2010) *De novo* assembly and analysis of RNA-seq data. *Nature Methods*, **7**, 909–912.

Schartl, M., Walter, R.B., Shen, Y. *et al.* (2013) The genome of the platyfish, *Xiphophorus maculatus*, provides insights into evolutionary adaptation and several complex traits. *Nature Genetics*, **45**, 567–572.

Schulz, M.H., Zerbino, D.R., Vingron, M. and Birney, E. (2012) Oases: robust de novo RNA-seq assembly across the dynamic range of expression levels. *Bioinformatics*, **28**, 1086–1092.

Smith, J.J., Kuraku, S., Holt, C. *et al.* (2012) Sequencing of the sea lamprey (*Petromyzon marinus*) genome provides insights into vertebrate evolution. *Nature Genetics*, **45**, 415–421.

Star, B., Nederbragt, A.J., Jentoft, S. *et al.* (2011) The genome sequence of Atlantic cod reveals a unique immune system. *Nature*, **477**, 207–210.

Tine, M., Kuhl, H., Gagnaire, P.A. *et al.* (2014) European sea bass genome and its variation provide insights into adaptation to euryhalinity and speciation. *Nature Communications*, **5**, 5770.

Trapnell, C., Pachter, L. and Salzberg, S. (2009) TopHat: discovering splice junctions with RNA-Seq. *Bioinformatics*, **25**, 1105–1111.

Venkatesh, B., Lee, A.P., Ravi, V. *et al.* (2014) Elephant shark genome provides unique insights into gnathostome evolution. *Nature*, **505**, 174–179.

Wang, Y., Lu, Y., Zhang, Y. *et al.* (2015) The draft genome of the grass carp (*Ctenopharyngodon idellus*) provides insights into its evolution and vegetarian adaptation. *Nature Genetics*, **47**, 625–631.

Wu, C., Zhang, D., Kan, M. *et al.* (2014) The draft genome of the large yellow croaker reveals well-developed innate immunity. *Nature Communications*, **5**, 5227.

Wu, J., Xiong, S., Jing, J., Chen, X., Wang, W., Gui, J.F. *et al.* (2015). Comparative Transcriptome Analysis of Differentially Expressed Genes and Signaling Pathways between XY and YY Testis in Yellow Catfish. *PloS One*, **10**, e0134626.

Xu, C., Evensen, O. and Munang'andu, H.M. (2015) *De novo* assembly and transcriptome analysis of Atlantic salmon macrophage/dendritic-like TO cells following type I IFN treatment and Salmonid alphavirus subtype-3 infection. *BMC Genomics*, **16**, 96.
Xu, P., Zhang, X., Wang, X. *et al.* (2014) Genome sequence and genetic diversity of the common carp, *Cyprinus carpio*. *Nature Genetics*, **46**, 1212–1219.
Zerbino, D.R. and Birney, E. (2008) Velvet: algorithms for *de novo* short read assembly using *de Bruijn* graphs. *Genome Research*, **18**, 821–829.

9

Analysis of Differentially Expressed Genes and Co-expressed Genes Using RNA-Seq Datasets

Chao Li, Qifan Zeng, Chen Jiang, Jun Yao and Zhanjiang Liu

Introduction

RNA-Seq has many applications. However, the two most popular analyses using RNA-Seq are the identification of differentially expressed genes (DEGs) and the analysis for coordinated gene expression. As RNA-Seq assembly was described in Chapter 8, we focus in this chapter on the identification of DEGs and the analysis of co-expressed genes. Demonstration of DEG identification using CLC Genomics Workbench and Trinity EdgeR is presented. Similarly, demonstration of using the weighted gene co-expression network analysis (WGCNA) package for the analysis of co-expressed genes is presented.

Transcriptomic studies allow the detection of dynamic changes resulting from changes in physiology, development, and environment conditions. For instance, one may be interested in DEGs after infection with pathogens, under elevated temperature, or under hypoxia treatment. Transitional study methods such as EST analysis and microarray analysis are limited by their low sensitivity. Under natural conditions, genes are not expressed at equal levels. For instance, some genes are expressed at very high levels, accounting for 1–2% of the entire cellular mRNA mass, while many other genes are expressed at very low levels, with only a few copies per cell, as demonstrated by EST analysis (Karsi *et al.*, 2002). Because of such polarized expression, traditional methods can only readily capture transcripts with relatively high levels of expression. The low-throughput sequencing, when coupled to the focus of protein-encoding genes using poly-adenylated RNA (poly $(A)^+$ RNA), led to the historical concept that only 1–5% of the genome is transcribed. Another limitation of the traditional method is the poor accuracy. Taking microarray analysis as an example, microarray technology adopts the hybridization reaction to measure gene expression levels. In a typical hybridization reaction, there is an optimal range for quantification. Therefore, genes cannot be detected when their amount falls out of the optimal range. Even within the window for optimal quantitation, the hybridization signal may never be “linear.”

RNA-Seq refers to the analysis of the entire transcriptome using high-throughput next-generation sequencing technologies.

With advances in the next-generation high-throughput sequencing, RNA-Seq has been extensively used as an approach for the analysis of gene expression profiles. It combines gene discovery and gene expression analysis in a single procedure. With the

Bioinformatics in Aquaculture: Principles and Methods, First Edition. Edited by Zhanjiang (John) Liu.

Table 9.1 Summary of the comparisons between microarray and RNA-Seq.

	Capacity	Samples	Replications	Cost (US$)
Microarray one color	Known genes	48	Yes	9000
Microarray two color	Known genes	48	Yes	7000
RNA-Seq without MIDs	Whole transcriptome	1	No	3000
RNA-Seq with MIDs	Whole transcriptome	12	Yes	8000

large volume of datasets, it allows discoveries of genes more efficiently than traditional EST sequencing. The sequencing reads of any specific transcripts in the RNA pool used for RNA-Seq are proportional to the levels of their presence in the RNA pool, thereby providing quantitative analysis on gene expression levels.

RNA-Seq also refreshed our understanding of transcription mechanisms. It is now known that the vast proportion of the genome is transcribed into RNA, including tens of thousands of protein-encoding transcripts and many non-protein-encoding transcripts that include the long non-coding RNAs, microRNA, antisense and sense RNAs, and the historically well-characterized rRNAs and tRNAs.

However, as with any other technology, RNA-Seq is by no means perfect. One of the major challenges is its cost. Currently, one full lane of RNA-Seq costs approximately US$4000–7000. This is much more than the cost of a parallel microarray experiment. For instance, the analysis of 48 samples using microarrays costs US$7000, while the same amount allows the analysis of only 12 samples using RNA-Seq with molecular barcoding tags (Li *et al.*, 2012). Table 9.1 compares the cost in various scenarios, which should give some general guidance for those readers who are not familiar with these technologies. Many commercial vendors of RNA-Seq provide much cheaper prices, but the throughput may not be a full lane. In most cases, commercial providers may co-run your samples with someone else's samples. In that case, you are actually being provided "a fraction of a lane" at a reduced cost.

RNA-Seq is more time consuming than microarray experiments. For instance, RNA-Seq requires over 10 days using traditional chemistry after the "libraries" are made. Even with the most recent chemistry and the most recent technology of sequencers, the sequencing process can take several days. Assembly of RNA-Seq without extremely powerful supercomputers may require another 2 weeks. Mapping reads to genome is challenging, and mapping RNAs consisting of small exons that may be separated by large introns can cause accuracy problem. Further analysis of read counts and identification of DEGs require additional statistical analysis and specialized software packages. In comparison, microarray experiments, upon availability of the array, require much less time and bioinformatics skills. Bioinformatics analysis poses the largest challenge for the aquaculture community because work associated with data generation, processing, and interpretation requires dedicated computer skills.

RNA-Seq can be used to address various biological questions, but it is often used to compare expression under various conditions or among treatments. In other words, it is most frequently used to determine DEGs. As bioinformatics analysis for the assembly of RNA-Seq short reads was described in Chapter 8, we will focus in this chapter on bioinformatics analysis for the identification of DEGs and of coordinated expression, that is, co-expressed genes.

Analysis of Differentially Expressed Genes Using CLC Genomics Workbench

CLC Genomics Workbench is developed by CLC bio, a bioinformatics software company based in Aarhus, Denmark. CLC Genomics Workbench is a comprehensive and user-friendly package to analyze, compare, and visualize next-generation sequencing data. It is commercially available.

Data Import

CLC Genomics Workbench supports all the next-generation sequencing platforms such as Roche 454, Illumina, SOLID, and Ion Torrent. The read files must be imported according to their sequencing platforms. The reference data need to be imported in Fasta format. CLC Genomics Workbench can support both paired reads and single reads. However, there are many advantages in using paired reads because of the greater level of accuracy in expression values with the paired data.

To determine expression levels, the sequencing reads need to be mapped to the reference. The mapping step generates the count matrix for the expression analysis. There are two options: use reference with annotations, or use reference without annotations, depending on if a reference genome is available or not. If you are working on a model species, and the annotated genome is available, you can select the first option. Otherwise, you need to select "use reference without annotations." In this later case, your de novo assembled transcriptome is used as the reference.

Mapping Reads to the Reference

To map the short reads to the reference sequence, several parameters need to be set properly:

1) *Maximum number of mismatches*: Only available when you use the short reads (<56 bp). This is the maximum number of mismatches allowed. Maximum value allowed is "3".
2) *Minimum length fraction*: This is the required minimal fraction of the short reads to be mapped to the reference. The default is 0.9, which means that at least 90% of the read (length) needs to be aligned to the reference in order to be included.
3) *Minimum similarity fraction*: This is how identical the matching region of the read should be. The default setting is at 0.8. When this default setting is used, it means that 90% of the read should align with 80% similarity in order to have the reads included.
4) *Maximum number of hits for a read*: A read may be able to align to multiple contigs. This option sets the threshold for the maximum number of contigs that one read can map to. Reads with high numbers of alignments will be discarded. Otherwise, it will be assigned according to the mapping quality. In case a read has identical mapping quality to several contigs, it will be assigned to the contig with the highest unique matches.
5) *Result handling*: You can use the output option of "create list of un-mapped sequences," which will export the unmapped reads in a single file. The "Create report" option allows the user to generate the statistical summary of the mapping. By default, the expression values are expressed as "RPKM".

Quantification of Gene Expression Value

The second task of DEG analysis is to quantify the expression value of each gene. The more times a transcript is detected, the more abundantly it is transcribed.

Normalization must be conducted for reliable reflection of the expression level of each gene. One issue is that the length of the gene can affect the reads count. The longer the gene sequence is, the more segments are present in the sequencing library. Therefore, the read counts need to be normalized according to the gene length. Reads per KB per million (RPKM) has been widely used as a reasonable normalizer. RPKM is the number of reads per KB length of a transcript from per 1000000 reads of sequences.

$$\text{RPKM} = 10^9 \times C/NL$$

Here, C is the total number of reads mapped onto the gene; N is total number of mapped reads; and L is the sum of the genes in base pairs. For instance, the RPKM of a 2 KB transcript with 3000 alignments in a sample of 10 million mapped reads is calculated as:

$$\text{RPKM} = 10^9 \times 3000/(2000 \times 10000000) = 150$$

In CLC Genomics Workbench, the standard output is a table that includes: "gene length," the length of each contig in your reference; "unique gene reads," the number of reads that are uniquely mapped to this contig; "total gene reads," including both unique mapped reads and multiple mapped reads that are assigned to this contig; and "RPKM." Additionally, CLC Genomics Workbench will create a report that includes the summary of sequencing reads, reference sequences, and mapping statistics. Of these mapping statistics, the mapping percentage is a very important index; it tells you how good your reference is. By calculating the percentage of reads mapped in pairs (reads mapped in pairs/total reads), one can determine whether there are too many gaps in your reference.

Set Up an Experiment

The next step is to identify the DEGs in the study. Once you have set up an experiment, you can select the samples you want to compare—"samples" here means the mapping results generated by the last step, which contain the expression values. These "samples" need to be assigned into different groups according to the "treatment"—specifically, if the samples are from the same individual under different conditions, such as before and after infection, or at different time points after treatment. You can use the "Paired" option; in this case, the statistical analysis allows the effects of the individuals to be taken into account, and comparisons are carried out by considering these corrected effects in each individual, instead of by group means.

Statistical Analysis for the Identification of DEGs

With CLC Genomics Workbench, proportions-based tests are applicable for the RNA-Seq datasets, in which the expression values are corrected by the sample size. Two proportions-based tests in CLC Genomics Workbench are available for RNA-Seq data, the test of Kal *et al.* (1999) and the test of Baggerly *et al.* (2003). Kal *et al.*'s Z-test is suitable for single sample comparisons, and Baggerly *et al.*'s method is suitable for group comparisons. In either method, you need to specify the control sample or group to be compared (the default is to use the first imported sample/group when you set up the experiment). CLC Genomics Workbench will generate the fold change and the "*p*-value" for each transcript. Here, the *p*-value is the two-sided *p*-value for the test. Additionally, there are options to either calculate the Bonferroni- or FDR-corrected *p*-values (Dudoit, Shaffer & Boldrick, 2003). Generally, transcripts with absolute fold change values larger than "2" and FDR values smaller than "0.05" are considered as differentially expressed.

Analysis of Differentially Expressed Genes Using Trinity

Trinity is developed by the Broad Institute (in Cambridge, Massachusetts, United States) and the Hebrew University of Jerusalem (Israel), and is now the most popular package for the analysis of RNA-Seq datasets (see also Chapter 8). In addition to the three modules (e.g., Inchworm, Chrysalis, and Butterfly) for the de novo reconstruction of transcriptomes, it is also compatible with other software for subsequent expression analysis. For instance, it employs Bowtie to align reads, either RSEM (Li & Dewey, 2011) or eXpress (Cappé & Moulines, 2009) to estimate the expression values after read alignments, and edgeR Bioconductor (Robinson, McCarthy & Smyth, 2010) for the identification of differentially expressed transcripts. Here, we will describe the detailed protocol and scripts for the identification of DEGs using Trinity.

Read Alignment and Abundance Estimation

The first step of DEG identification is to align the reads to the reference to estimate the abundance of each transcript. Bowtie is utilized for the alignment in this step. Then, RSEM (Li & Dewey, 2011) or eXpress (Roberts & Pachter, 2013) is employed to calculate the expression values for each transcript. These software packages need to be installed in your computer prior to mapping the reads to the reference.

In Trinity, you can format the reference for alignment and run the abundance estimation in one script, using the command line:

```
./align_and_estimate_abundance.pl --transcripts
   Trinity.fasta --seqType fq --left
   reads_1.fq--right reads_2.fq --est_method
   RSEM --aln_method bowtie --trinity_mode --prep_reference
```

Here "`--transcripts`" specifies your reference transcriptome file "`Trinity.fasta`", and "--seqType" is the data type of your clean reads in fastq or fasta format. If using paired-end reads, forward and reverse reads need to be specified with the "`--left`" and "`--right`" commands. Otherwise, "`--single <string>`" should be used to specify single reads. You can also choose tools for short read alignment and read counts estimation with "`--est_method`" and "`--aln_method`", respectively. It is noteworthy that RSEM is only compatible with bowtie for abundance calculation. The generated alignment file "bowtie.bam" will be imported in either RSEM or eXpress for estimation of abundance.

Using RSEM to Estimate Expression Value

The RSEM package supports both single-end and paired-end read data. Expression levels of both genes and isoforms can be estimated from RNA-Seq datasets with RSEM. Accordingly, the two output files are "RSEM.isoforms.results" and "RSEM.genes.results". The output files contain a few main columns: "transcript_id" is the transcript name of this transcript; "gene_id" is the gene name of this transcript; if no gene information is provided, "gene_id" and "transcript_id" will be the same; "length" is the sequence length of the transcript, with poly(A) tail not counted; "effective_length" only counts the positions that can generate a valid fragment; and "expected_count" stands for the sum of the posterior probability of each read from this transcript over all reads. "FPKM" is fragments per kilobase of transcript per million mapped reads.

Using eXpress to Estimate Expression Value

If the "--est_method" is set to use the eXpress software package, eXpress will also be used for the estimation of the abundances of each transcript from reference transcripts. eXpress is based on an online expectation–maximization algorithm (Cappé & Moulines, 2009) that is efficient in memory usage and can be used for huge RNA-Seq datasets. The files generated will include "results.xprs", which contains expression values for each transcript, and "results.xprs.genes", which stores expression values for each gene.

Generating Expression Value Matrices

To join the RSEM abundance estimation results for each group:

```
abundance_estimates_to_matrix.pl --est_method <method>
   RSEM sample1.results sample2.results ...>
   transcripts.counts.matrix
```

Two result files will be generated: "Trinity_trans.counts.matrix" contains the matrix of fragment raw counts, and "Trinity_trans.TMM.fpkm.matrix" contains the Trimmed Mean of M-values (TMM)-normalized FPKM expression values. The TMM normalization provides a scaling parameter to generate an effective library size for each sample (Robinson *et al.*, 2010). This effective library size is then used to calculate the FPKM value.

Identifying Differentially Expressed Transcripts

The next step is to identify the differentially expressed transcripts among samples. Trinity employs edgeR and DEseq packages to identify differentially expressed transcripts. To install the required R package:

```
source("http://bioconductor.org/biocLite.R")
biocLite("edgeR")
biocLite("DESeq")
biocLite("ctc")
biocLite("Biobase")
install.packages("gplots")
install.packages("ape")
```

A pairwise comparison among the samples will be performed to identify the DEGs. The "transcripts.counts.matrix" file is used for transcript-level analysis, while the "genes.counts.matrix" file is used for gene-level analysis. Before analysis, you need to prepare a "samples_described.txt" file, which should contain the meta-information of your samples. It contains two columns: the first column lists your group names, and the second lists your sample names. These names should be the same as all column names in the "counts.matrix" file.

```
DE_analysis.pl --matrix rnaseq.counts.matrix --method edgeR
   --samples_file samples_described.txt
```

This script will generate a folder named "edgeR" in your current folder. It contains all the comparison results of your samples. The next step is to analyze differentially expressed transcripts.

Extracting Differentially Expressed Transcripts

To set the cutoff values (p-values and fold-changes) of the significantly differentially expressed transcripts, you can run the following script:

```
analyze_diff_expr.pl --matrix matrix.TMM_normalized.FPKM
   -P 1e-3 -C 2
```

Here, the option "-P" stands for the FDR p-value cutoff, and "-C" is the threshold for the minimum absolute value for the log-transformed fold changes. In the command line shown here, the p-value is set at "0.001", and the fold change is set at "2". All the differentially expressed transcripts that passed the current cutoffs are exported as a result.

Analysis of Co-Expressed Genes

In addition to the identification of DEGs, the analysis of correlated or coordinated expression can be insightful. For instance, a single treatment such as high-temperature treatment may induce a common set of genes. The correlation of expression patterns among various genes, referred to as "co-expression," can be revealed by network analysis. An increasing number of studies have demonstrated associated behaviors of genes with related biological functions (Carter *et al.*, 2004; Rocke & Durbin, 2001). Given that most biological processes cannot be carried out by a single gene, analysis of co-expressed genes from RNA-Seq datasets may be quite informative.

Here, we provide a brief introduction to the usage of the weighted gene co-expression network analysis (WGCNA) package, a systems biology method for describing the correlated patterns among genes across samples (Langfelder & Horvath, 2008). WGCNA has been used in multiple studies, such as cancer (Horvath *et al.*, 2006), yeast cell cycle (Horvath *et al.*, 2006), tissue development (Oldham, Horvath & Geschwind, 2006), and plant stress response (Weston *et al.*, 2008).

WGCNA is provided as an R package that contains a comprehensive set of functions for weighted correlation network analysis. Generally, the package starts from the construction of a network based on pairwise correlations of gene expression profiles. Then, it clusters the whole dataset into several modules according to the topological overlap of the network. For each module, a weighted average expression profile is calculated and represented as the module eigengene. Each gene within a module is assigned with a "module membership", depending on its correlation with the eigengene. In the subsequent analysis, candidate modules are identified based on the statistical significance between the eigengene and sample traits. Genes with high module membership from significant modules are regarded as candidate genes for further functional analysis and validation.

Here, we start from the data import step to illustrate the usage of WGCNA:

```
workingDir="."
setwd(workingDir)
library(WGCNA)
expData= read.csv("expression_data.csv")
traitData=read.table("trait",header=T)
Samples= rownames(expData)
traitRows=match(Samples,traitData$sample)
datTraits=traitData[traitRows,-1]
```

The expression data "expression_data.csv" and sample trait data "trait" are imported accordingly. Next, we can check the data by clustering the samples to identify outlier samples (Figure 9.1):

```
sampleTree = hclust(dist(expData), method = "average")
sizeGrWindow(12,9)
par(cex = 0.6)
par(mar = c(0,4,2,0))
plot(sampleTree, main = "Sample clustering to detect out-
liers", sub="", xlab="", cex.lab = 1.5, cex.axis = 1.5,
    cex.main = 2)
```

Here, "`hclust()`" is the function for hierarchical cluster analysis on a set of dissimilarities. "`Par()`" is used to set or query graphical parameters. An example of sample distribution is given in Figure 9.1. It appears that "sample 3" falls apart from the other

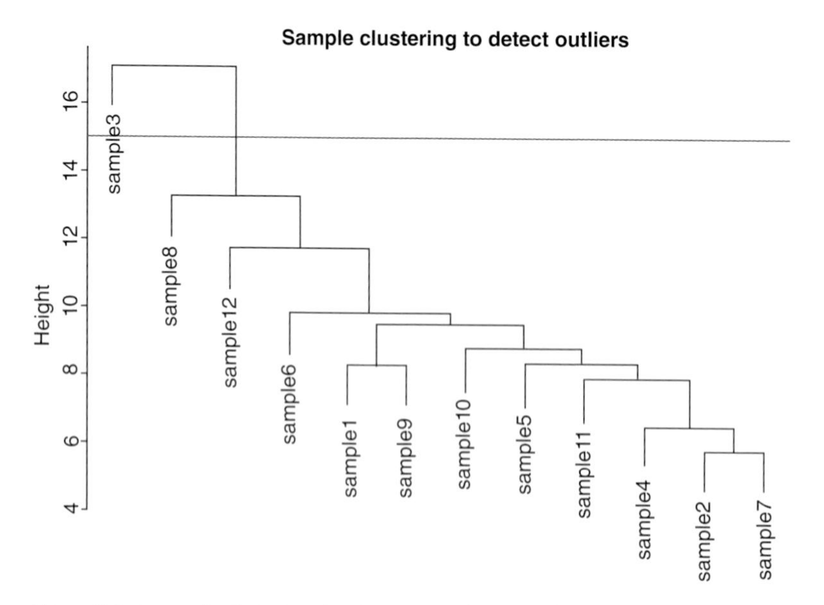

Figure 9.1 A sample clustering dendrogram to detect outliers generated by WGCNA.

samples. As samples from the example datasets are under the same conditions, the difference is obviously not because of biological reasons. It can be removed prior to subsequent analysis.

Network Construction

To construct a weighted co-expression network, WGCNA uses an adjacency matrix to represent the correlation patterns. The adjacency a_{ij} is defined by raising the co-expression similarity to a power β:

$$a_{ij} = s_{ij}^{\beta}$$

s_{ij} is the co-expression similarity that can be defined as the absolute value of the correlation coefficient between gene i and gene j:

$$s_{ij} = |cor(x_i x_j)|$$

It can also be defined as a biweight midcorrelation or the Spearman correlation. β represents the soft-thresholding power, which is selected based on the criterion of approximate scale-free topology (Zhang & Horvath, 2005). In practice, we can use the function "`pickSoftThreshold`" to analyze the network topology and choose a proper soft-thresholding power. First, a set of candidate powers is defined; then, the function will return the corresponding network indices (Figure 9.2):

```
powers=c(seq(from=1,to=20,by=1))
sft=pickSoftThreshold(expData,powerVector = powers)
```

The results can be plotted:

```
par(mfrow = c(1,2))
cex1 = 0.9
plot(sft$fitIndices[,1], -sign(sft$fitIndices[,3])
  *sft$fitIndices[,2], xlab="Soft Threshold (power)",
  ylab="Scale Free Topology Model Fit,signed R^2",type="n",
  main = paste("Scale independence"))
text(sft$fitIndices[,1], -sign(sft$fitIndices[,3])
  *sft$fitIndices[,2],  labels=powers,cex=cex1,col="red")
abline(h=0.9,col="red")
```

The mean connectivity result as a function of soft-thresholding power can be plotted:

```
plot(sft$fitIndices[,1], sft$fitIndices[,5],
  xlab="Soft Threshold (power)", ylab="Mean Connectivity",
   type="n", main = paste("Mean connectivity"))
text(sft$fitIndices[,1], sft$fitIndices[,5], labels=powers,
   cex=cex1,col="red")
softPower=18
```

By checking the model fit index in Figure 9.2, the smallest power value can be chosen that satisfies the scale-free topology.

With the soft thresholding power, the adjacencies can be calculated:

```
adjacency=adjacency(expData, power=softPower)
```

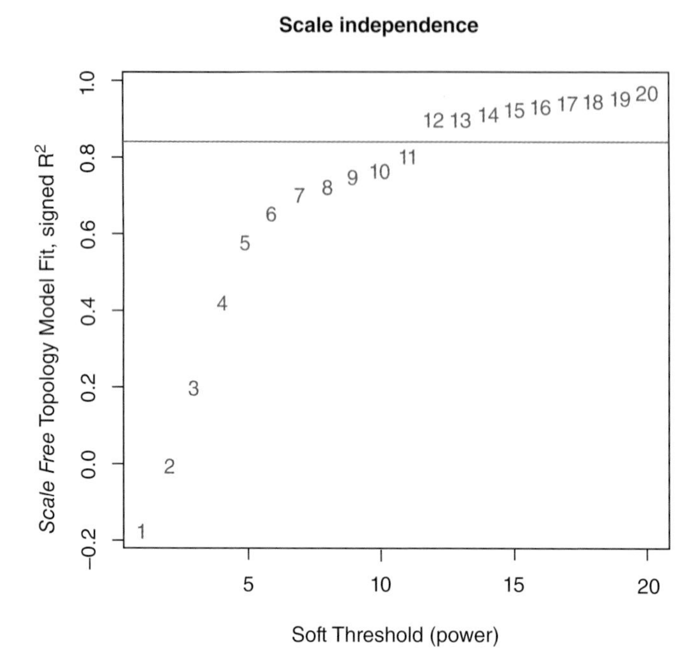

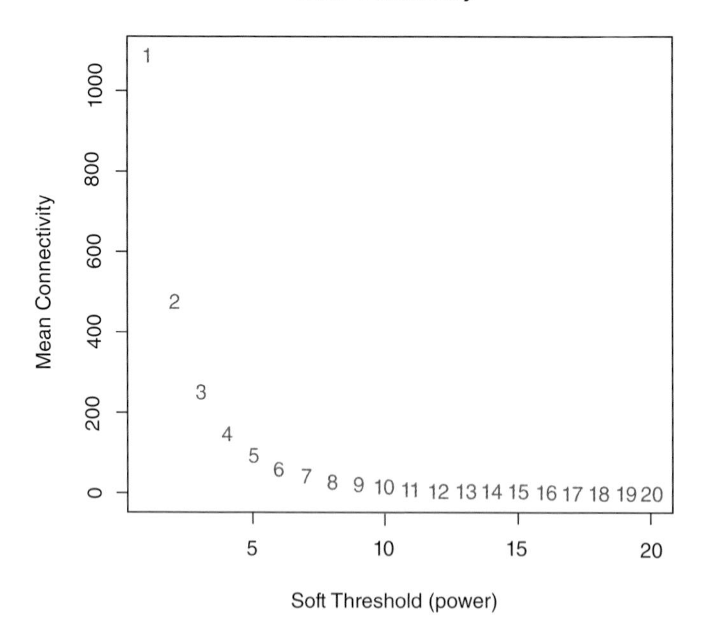

Figure 9.2 Analysis of signed network topology for various soft-thresholding powers. The left panel indicates that the network satisfies scale-free topology when the soft-threshold power increases above 12. The right panel shows the mean connectivity corresponding to each soft-threshold power.

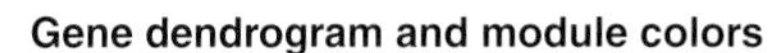

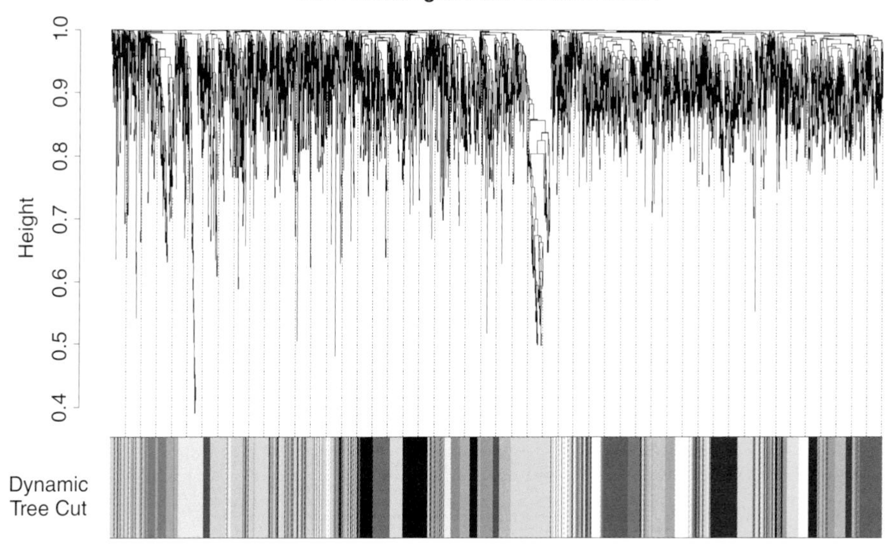

Figure 9.3 A dendrogram for the genes and detected modules. (*See color plate section for the color representation of this figure.*)

Module Detection

With the constructed network, we can detect the modules within the network. *Modules* are defined as the clusters of densely interconnected genes. By default, the network interconnectedness is measured by topological overlaps:

```
TOM=TOMsimilarity(adjacency)
dissTOM=1-TOM
```

The function "`hclust()`" can then be used for hierarchical clustering:

```
geneTree = hclust(as.dist(dissTOM), method = "average")
```

The clustering dendrogram of genes based on topological overlap can be plotted (Figure 9.3):

```
sizeGrWindow(12,9)
plot(geneTree, xlab="", sub="", main = "Gene clustering on
   TOM-based dissimilarity",labels = FALSE, hang = 0.04)
```

Branches of the hierarchical clustering dendrogram are identified as modules by using the Dynamic Branch Cut method (Langfelder, Zhang & Horvath, 2008).

```
minModuleSize = 30
dynamicMods = cutreeDynamic(dendro = geneTree,
   distM = dissTOM,deepSplit = 2, pamRespectsDendro = FALSE,
   minClusterSize = minModuleSize)
```

We can plot the module under the gene dendrogram.

```
dynamicColors = labels2colors(dynamicMods)
table(dynamicColors)
```

```
sizeGrWindow(8,6)
plotDendroAndColors(geneTree, dynamicColors,
   "Dynamic Tree Cut",dendroLabels = FALSE, hang = 0.03,
   addGuide = TRUE, guideHang = 0.05,
   main = "Gene dendrogram and module colors")
```

Relating the Modules to External Information

Depending on the research question, one of the sample traits can be used to calculate the absolute correlation between the trait and the expression profiles. Since the expression profile of a module can be represented by a weighted average value (eigengene), modules with the most significant associations can be determined by correlate eigengenes with external traits:

```
nGenes = ncol(datExpr)
nSamples = nrow(datExpr)
MEs0 = moduleEigengenes(datExpr, moduleColors)$eigengenes
MEs = orderMEs(MEs0)
moduleTraitCor = cor(MEs, datTraits, use = "p")
moduleTraitPvalue = corPvalueStudent(moduleTraitCor,
   nSamples)
```

A graphical view can be generated to present the results:

```
sizeGrWindow(10,6)
textMatrix = paste(signif(moduleTraitCor, 2),
   "\n(", signif(moduleTraitPvalue, 1), ")", sep = "");
dim(textMatrix) = dim(moduleTraitCor)
par(mar = c(6, 8.5, 3, 3))
labeledHeatmap(Matrix = moduleTraitCor,
   xLabels = names(datTraits), yLabels = names(MEs),
   ySymbols = names(MEs), colorLabels = FALSE,
   colors = greenWhiteRed(50), textMatrix = textMatrix,
   setStdMargins = FALSE, cex.text = 0.5,zlim =
   c(-1,1), main = paste("Module-trait relationships"))
```

Once the module–trait associations are quantified, the gene–trait association can be defined as the correlation between the gene expression level and the trait.

```
Total_fat = as.data.frame(datTraits)
names(Total_fat) = "Total_fat"
modNames = substring(names(MEs), 3)
geneModuleMembership=as.data.frame(cor(datExpr, MEs,
   use= "p"))
MMPvalue=as.data.frame(corPvalueStudent
  (as.matrix(geneModuleMembership),nSamples))
names(geneModuleMembership)=paste("MM", modNames, sep="")
names(MMPvalue) = paste("p.MM", modNames, sep="")
geneTraitSignificance = as.data.frame(cor(datExpr,
   Total_fat, use = "p"))
```

```
GSPvalue= as.data.frame(corPvalueStudent
  (as.matrix(geneTraitSignificance), nSamples))
names(geneTraitSignificance) = paste("GS.",
   names(Total_fat), sep="");
names(GSPvalue) = paste("p.GS.", names(Total_fat),sep="")
```

To identify genes that have high module membership as well as significant association with traits, the gene–trait association against the module membership can be plotted for each module:

```
module = "purple"
column = match(module, modNames)
moduleGenes = moduleColors==module
sizeGrWindow(7, 7)
par(mfrow = c(1,1))
verboseScatterplot(abs(geneModuleMembership
  [moduleGenes, column]), abs(geneTraitSignificance
  [moduleGenes, 1]), xlab = paste("Module Membership in",
   module, "module"), ylab = "Gene significance for sex
   differentiation", main = paste("Module membership vs.
   gene significance\n"), cex.main = 1.2, cex.lab = 1.2,
   cex.axis = 1.2, col = module)
```

Finally, this statistical information can be merged for the output of the results:

```
geneInfo0= data.frame(substanceBXH = names(datExpr),
   moduleColor=moduleColors, geneTraitSignificance, GSP-
value)
modOrder = order(-abs(cor(MEs, weight, use = "p")))
for(mod in 1:ncol(geneModuleMembership))
{oldNames=names(geneInfo0)
geneInfo0=data.frame(geneInfo0, geneModuleMembership
  [,modOrder[mod]], MMPvalue[,modOrder[mod]])
names(geneInfo0) = c(oldNames, paste("MM.",
   modNames[modOrder[mod]], sep=""),paste("p.MM.",
   modNames[modOrder[mod]], sep=""))
}
geneOrder = order(geneInfo0$moduleColor,
  -abs(geneInfo0$GS.weight))
geneInfo = geneInfo0[geneOrder, ]
write.csv(geneInfo, file = "geneInfo.csv")
```

Computational Challenges

Next-generation sequencing is deep sequencing with very high throughput. In just one lane, hundreds of millions of short sequence reads can be generated. This brings up computational challenges—storing, transferring, and analyzing large volumes of sequencing data. With the most popular sequencing platform Illumina, it can generate 400 million reads (100 bp pair-end reads) in one lane; this means about 40 GB data. During data analysis, it requires several hundred GBs of disk space. For example, ABySS needs at least 150 GB disk space and 24 CPUs in the assembly process for sequence data acquired from the sequencing of just one lane, while at least 200 GB of memory is needed for Trinity to do the assembly for one lane data, and much more disk space is required when mapping the reads back to the reference assembly. In order to do RNA-Seq data analysis, access to a computer server that has high levels of memory, CPUs, and disk space is required. There are some government-funded supercomputer providing free access to academic users. Use of such computational resources should allow savings to be made for the researchers.

Acknowledgments

Research in our laboratory is supported by grants from USDA AFRI programs.

References

Baggerly, K.A., Deng, L., Morris, J.S. and Aldaz, C.M. (2003) Differential expression in SAGE: accounting for normal between-library variation. *Bioinformatics*, **19**, 1477–1483.

Cappé, O. and Moulines, E. (2009) On-line expectation–maximization algorithm for latent data models. *Journal of the Royal Statistical Society: Series B (Statistical Methodology)*, **71**, 593–613.

Carter, S.L., Brechbuhler, C.M., Griffin, M. and Bond, A.T. (2004) Gene co-expression network topology provides a framework for molecular characterization of cellular state. *Bioinformatics*, **20**, 2242–2250.

Dudoit, S., Shaffer, J.P. and Boldrick, J.C. (2003) Multiple hypothesis testing in microarray experiments. *Statistical Science*, **18**, 71–103.

Horvath, S., Zhang, B., Carlson, M. *et al.* (2006) Analysis of oncogenic signaling networks in glioblastoma identifies ASPM as a molecular target. *Proceedings of the National Academy of Sciences*, **103**, 17402–17407.

Kal, A.J., van Zonneveld, A.J., Benes, V. *et al.* (1999) Dynamics of gene expression revealed by comparison of serial analysis of gene expression transcript profiles from yeast grown on two different carbon sources. *Molecular Biology of the Cell*, **10**, 1859–1872.

Karsi, A., Cao, D., Li, P. *et al.* (2002) Transcriptome analysis of channel catfish (*Ictalurus punctatus*): initial analysis of gene expression and micro satellite-containing cDNAs in the skin. *Gene*, **285**, 157–168.

Langfelder, P. and Horvath, S. (2008) WGCNA: an R package for weighted correlation network analysis. *BMC Bioinformatics*, **9**, 559.

Langfelder, P., Zhang, B. and Horvath, S. (2008) Defining clusters from a hierarchical cluster tree: the Dynamic Tree Cut package for R. *Bioinformatics*, **24**, 719–720.

Li, B. and Dewey, C.N. (2011) RSEM: accurate transcript quantification from RNA-Seq data with or without a reference genome. *BMC Bioinformatics*, **12**, 323.

Li, C., Zhang, Y., Wang, R. *et al.* (2012) RNA-Seq analysis of mucosal immune responses reveals signatures of intestinal barrier disruption and pathogen entry following *Edwardsiella ictaluri* infection in channel catfish, *Ictalurus punctatus*. *Fish & Shellfish Immunology*, **32**, 816–827.

Oldham, M.C., Horvath, S. and Geschwind, D.H. (2006) Conservation and evolution of gene coexpression networks in human and chimpanzee brains. *Proceedings of the National Academy of Sciences*, **103**, 17973–17978.

Roberts, A. and Pachter, L. (2013) Streaming fragment assignment for real-time analysis of sequencing experiments. *Nature Methods*, **10**, 71–73.

Robinson, M.D., McCarthy, D.J. and Smyth, G.K. (2010) edgeR: a Bioconductor package for differential expression analysis of digital gene expression data. *Bioinformatics*, **26**, 139–140.

Rocke, D.M. and Durbin, B. (2001) A model for measurement error for gene expression arrays. *Journal of Computational Biology*, **8**, 557–569.

Weston, D.J., Gunter, L.E., Rogers, A. and Wullschleger, S.D. (2008) Connecting genes, coexpression modules, and molecular signatures to environmental stress phenotypes in plants. *BMC Systems Biology*, **2**, 16.

Zhang, B. and Horvath, S. (2005) A general framework for weighted gene co-expression network analysis. *Statistical Applications in Genetics and Molecular Biology*, **4**, 1128.

10

Gene Ontology, Enrichment Analysis, and Pathway Analysis

Tao Zhou, Jun Yao and Zhanjiang Liu

Introduction

Genomic sequencing has revealed that a large fraction of the genes specifying the core biological functions are conserved across a broad spectrum of eukaryotes. Knowledge of the biological role of the shared genes in one organism can often be transferred to other organisms. Shared vocabularies are an important step toward unifying biological databases and making biological knowledge transferable. Gene ontology (GO) provides a set of structured, controlled vocabularies for community use in gene annotation. With various sequence datasets including peptide sequences, genes, ESTs, microarray datasets, RNA-Seq datasets, or whole genome sequences, GO analysis provides defined GO terms to genes. At the highest level, GO terms cover cellular components, molecular functions, and biological processes. Lower levels of GO terms can be applied to genes when relevant.

While GO analysis is useful, GO terms are generally too general to provide specific insights into the mechanisms of expression changes under a specific condition. Enrichment analysis is designed to help the researchers to interpret the expression data such that, from a list of differentially expressed genes, one can determine which sets of genes are over- or under-represented, providing insights into the potential molecular basis of the biological process under study. Further, pathway analysis has the ability to determine what pathways the enriched genes are involved with, reducing the complexity of analysis while increasing explanatory power.

GO and the GO Project

Gene ontology is a controlled vocabulary term to describe gene characteristics in terms of their localization and function. It is a major bioinformatics initiative to unify the representation of gene and gene product attributes across all species. The GO project (The Gene Ontology Consortium *et al.*, 2000) (http://www.geneontology.org/) is considered to be the most successful example of a systematic description of biological attributes of genes in order to allow the integration, retrieval, and computation of data (Yon Rhee *et al.*, 2008). GO was initially developed by researchers studying the genome of three

Bioinformatics in Aquaculture: Principles and Methods, First Edition. Edited by Zhanjiang (John) Liu.

model organisms: *Drosophila melanogaster* (fruit fly), *Mus musculus* (mouse), and *Saccharomyces cerevisiae* (yeast) in 1998 (The Gene Ontology Consortium *et al.*, 2000). Now, databases for many other model organisms have joined the GO Consortium, and made contributions to this project (The Gene Ontology Consortium, 2015). The GO Consortium was aimed at producing a dynamic, controlled vocabulary that could be applied to all eukaryotes. The GO project provides three structured ontologies that describe gene products in terms of their biological processes, cellular components, and molecular functions in a species-independent manner.

GO Terms

Each GO term within GO has a term name, which may be a word or string of words; a unique alphanumeric identifier; a definition with cited sources; or a namespace indicating the domain to which it belongs. Terms may also have synonyms, which are classified as being exactly equivalent, broader, narrower, or related to the term name. The GO vocabulary is designed to be species-neutral, and includes terms applicable to prokaryotes and eukaryotes, and single-cell and multicellular organisms.

Ontology

Ontology is a formal representation of a body of knowledge within a given domain. Ontologies usually consist of a set of terms with relations that operate between them. The domains that GO represents are biological processes, molecular functions, and cellular components.

Biological Process

Biological process refers to a biological objective to which the gene or gene product contributes. A biological process term describes a series of events accomplished by one or more organized assemblies of molecular functions. Examples of broad biological process terms are "GO:0044699 single organism process" or "GO:0032502 developmental process". Examples of more specific terms are "GO:0042246 tissue regeneration" or "GO:0031101 fin regeneration". The general rule to distinguish between a biological process and a molecular function is that a process must have more than one distinct step. We should also be aware that a biological process is not equivalent to a pathway.

Molecular Function

Molecular function is defined as the biochemical activity of a gene product, such as binding to specific ligands. Molecular function terms describe activities that occur at the molecular level, such as "catalytic activity" or "binding activity". GO molecular function terms represent activities rather than the entities (molecules or complexes) that perform the actions, and do not specify where, when, or in what context the action takes place. For instance, "GO:0042813 Wnt-activated receptor activity" is a molecular function term.

Cellular Component

Cellular component refers to the place in the cell where a gene product is active. Cellular component terms describe a component of a cell that is part of a larger object, such as an anatomical structure (e.g., rough endoplasmic reticulum or nucleus) or a

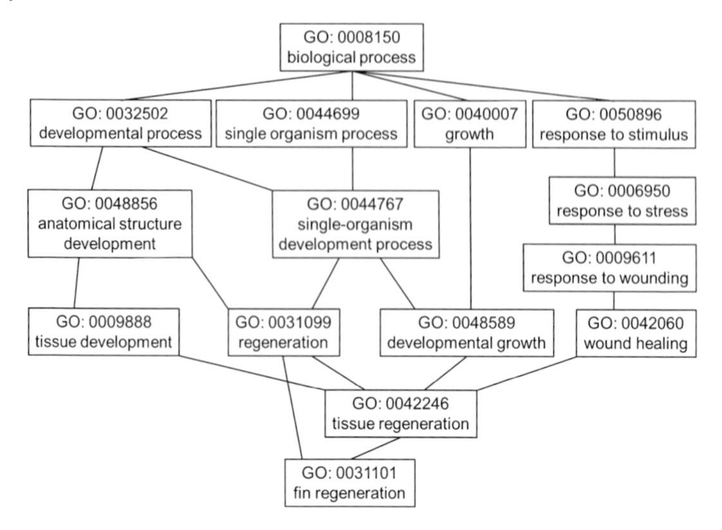

Figure 10.1 Schematic presentation of an example of the ontology structure.

gene product group (e.g., ribosome, proteasome, or a protein dimer). For instance, "GO:0009898 cytoplasmic side of plasma membrane" is a cellular component term.

Ontology Structure

The GO ontology is structured as a directed acyclic graph, and each term has defined relationships to one or more other terms in the same domain, and sometimes to other domains. For instance, ribosomal protein S2 is a component of the cellular component ribosome, it is involved in the translation process, and it binds to RNA. Each GO term is a node, and the relationships between the terms are edges between the nodes. Ontology structure is loosely hierarchical, with "child" terms being more specialized than their "parent" term. A GO term may have more than one parent term. Figure 10.1 shows a set of terms from the ontology. GO terms do not occupy strict fixed levels in the hierarchy. GO is structured as a graph, and terms would appear at different "levels" if different paths were followed through the graph.

GO Slim

GO slims are subsets of terms in the ontology. GO slims give a broad overview of the ontology content without the details of the specific fine-grained terms. GO slims are very useful when providing a summary of the results of the GO annotation of a genome, microarray, or cDNA collection where broad classification of gene product functions is required. GO slims are created according to the user's needs, and may be specific to species or to particular areas of the ontologies. GO also provides a generic GO slim that, as with GO itself, is not species-specific, and should be suitable for most purposes. The GO Consortium website provides GO slims related to different organisms or usages in Open Biomedical Ontologies (OBO) format for download (http://geneontology.org/page/go-slim-and-subset-guide).

Annotation

GO annotation is the process of assigning GO terms to gene products. Genome annotation is the practice of capturing data about a gene product. Although genome annotation provides annotation of genes, it is usually less detailed than GO annotations. The detailed information of a gene product can be found by GO annotations. Several computational methods can be used for mapping GO terms to gene products. The GO Consortium recommends three annotation processes: electronic annotation, literature annotation, and sequence-based annotation.

Electronic Annotation

The vast majority of GO annotations have been made using electronic annotation methods, without curators' oversight. Electronic annotation is very quick and produces large amounts of less detailed annotation. Electronic annotation is likely to tell you which of your genes are transcription factors, but unlikely to tell you in great detail what process the gene controls. Electronic annotation is especially useful for the annotation of new genomes or microarrays with thousands of sequences. However, it has been reported that electronic annotation is more reliable than generally believed, and that it can even compete with annotation inferred by curators when they use evidence other than experiments from primary literature (Škunca, Altenhoff & Dessimoz, 2012).

The primary method of generating electronic annotation is to manually map GO terms to corresponding concepts in the controlled vocabularies. Electronic annotations can be inferred from Enzyme Commission, InterPro, UniProt, Ensembl, HAMAP, BLAST, etc. All of these methods involve the use of mapping files, which can be downloaded from the GO Consortium website (http://geneontology.org/page/download-mappings). These mapping files are updated regularly, and can be downloaded in the ".txt" file format or viewed in a web browser. The line syntax for the mapping file is: "external database:term identifier (id/name) > GO:GO term name ; GO:id". Each line represents one mapping item. One external database term may direct to many GO terms.

InterPro provides the functional analysis of proteins by classifying them into families and predicting functional domains and important sites (Mitchell *et al.*, 2015). InterPro provides mapping of InterPro entries to GO terms, and can be downloaded from the InterPro website (www.ebi.ac.uk/interpro/download.html). For example: "InterPro:IPR000003 Retinoid X receptor/HNF4 > GO:DNA binding ; GO:0003677".

The UniProt Knowledgebase (UniProtKB) is a collection of functional information on proteins, with accurate, consistent, and rich annotations. UniProtkb captures the core data and as much annotation information as possible for each UniProtkb entry. UniProt keyword mapping, generated by the UniProtKB and UniProtKB-GOA teams, is aimed at assigning GO terms to UniprotKB keywords (Barrell *et al.*, 2009). For example: "UniProtKB-KW:KW-0067 ATP-binding > GO:ATP binding ; GO:0005524".

Kyoto Encyclopedia of Genes and Genomes (KEGG) is a database resource (www.genome.jp/kegg/) for understanding high-level functions and utilities of the biological system, such as cells, organisms, and ecosystems, from molecular-level information, especially large-scale molecular datasets generated by genome sequencing and other high-throughput experimental technologies. The detailed information for each KEGG entry, such as definition, enzyme, pathway, and ortholog, can be retrieved from the KEGG website. KEGG mapping is useful for subsequent pathway analysis. KEGG pathways and reactions mapping can map GO terms to KEGG entries. For example: "KEGG:R00004 > GO:inorganic diphosphatase activity ; GO:0004427".

Electronic annotations can also be made by aligning your own sequences to manually annotated sequences, then transferring the GO annotations across to your sequences. This is quite useful for huge amounts of sequences. The threshold of similarity or E-value in this process can be controlled. Many mapping files for different databases can be found on the GO Consortium website. All of the mapping files have the same format and usage. You may download the required mapping file and import it into Microsoft Access. By querying your data with the mapping file in Microsoft Access, it is easy to annotate large amounts of data.

Literature Annotation

Literature annotation in GO means capturing published information about the exact function of gene products as GO annotations. Literature annotation usually starts with a list of genes and is conducted through the following processes.

1) Read the publication about the gene and collect useful information such as function, biological process, and cellular location.
2) Search the NCBI taxonomy database (www.ncbi.nlm.nih.gov/taxonomy) and find the taxon id of the organism that the gene product is derived from (Format-Taxonomy ID: 7955).
3) Find the accession number of the gene product, for example, from UniProt, DDBJ, or GeneBank.
4) Find the GO terms that describe the function, biological process, or cellular location from the GO database (http://amigo.geneontology.org/amigo).
5) Choose the evidence code—for example, Inferred from Experiment (EXP)—that describes the experiment. The guide to GO evidence codes can be found in the Go Consortium website (http://geneontology.org/page/guide-go-evidence-codes).
6) The information can be submitted to the GO Consortium in the Gene Association File (GAF) format or the Gene Product Association Data (GPAD) format. GO annotation file formats can be found from the GO Consortium website (http://geneontology.org/page/go-annotation-file-formats).

Literature annotation is time-consuming, but produces high-quality annotation, and is worthwhile in the long term.

Sequence-Based Annotation

Researchers may start with some sequences, and want to assign GO terms to them. In this scenario, sequence-based annotation can be performed. This process consists of the following steps:

1) *Blast search.* Once a sequence has been selected for annotation, Blast searches can be conducted against UniProtKB to capture genes related to it and to identify homologs. The sequence and homologs can be merged into a single entry. The difference about the homologs, such as alternative splicing, natural variations, and frameshifts, can be documented for further comparison.
2) *Sequence analysis.* Sequence annotation prediction processes can be found in Chapter 4. The relevant gene, domain, protein topology, and protein family classification can be predicted after sequence analysis.

3) *Literature review.* The publications related to the gene can be identified by searching literature databases. Read the publications and collect information about the gene, such as function, biological process, or cellular location. The information collected through literature review and sequence analysis tools can be compared and verified. Annotation captured from literature review includes protein and gene name, function, catalytic activity, subcellular location, protein–protein interactions, patterns of expression, disease associated with deficiencies in a protein, RNA editing, etc. Relevant GO terms are assigned based on the experimental data documented in the literature.
4) *Family-based analysis.* Reciprocal Blast searches and phylogenetic resources (discussed in Chapter 2) can be used to identify putative homologs. Annotation is standardized and propagated across homologous proteins to ensure data consistency.
5) *Evidence attribution.* All information added to an entry during the manual annotation process should be linked to the original source, so that researchers can trace back to the origin of each piece of information and evaluate it. Evidence code can be assigned following the GO Consortium guidelines (http://geneontology.org/page/guide-go-evidence-codes).
6) *Quality assurance, integration, and update.* Each completed entry can be submitted to the GO Consortium in the GAF or GPAD format. After quality assurance, the complete entry will be updated and becomes available.

GO Tools

The GO Consortium develops and supports AmiGO 2, OBO-Edit, and GOOSE. There are a large number of third-party tools available that use the data provided by the GO project.

AmiGO 2

AmiGO 2 (http://amigo.geneontology.org/amigo) is a web-based application that allows users to query, browse, and view ontologies and gene product annotation data. AmiGO 2 can be used online at the GO website to access the data provided by the GO Consortium, or can be downloaded and installed for local use (only supports the Ubuntu platform). The AmiGO 2 manual can be found on the GO Consortium website (http://wiki.geneontology.org/index.php/Category:AmiGO_2_Manual). In the following text, we will introduce the usage of AmiGO 2.

Quick Search AmiGO 2 quick search is a powerful method that rapidly searches across all of the pre-computed indexes. Quick search supports Boolean operator, wildcard, and fuzzy searches. As you type in the search box, the autocomplete search will find matching terms and gene products. Once a term or gene product has been selected from the dropdown list, it will jump directly to that selection's detail page. Otherwise, by pressing return or clicking the search button, the user can jump to an annotation search for his or her text. Boolean logic and nesting may be used in the search. The following are some examples for quick search.

1) If you want to search the records that contain both angiogenesis and neurogenesis, you would enter: "angiogenesis and neurogenesis".

2) If you want to exclude neurogenesis from angiogenesis results, you would enter: "angiogenesis and – neurogenesis".
3) If you want to get all the results for angiogenesis, neurogenesis, or both, you would enter: "angiogenesis or neurogenesis". In this case, "angiogenesis or neurogenesis" and "angiogenesis neurogenesis" are functionally equivalent. Space between words are considered to be an implicit "or".
4) You need to use parentheses to club together words that you want to appear together. For example, if you want to search for records containing both "angiogenesis" and "wound healing", you would enter: "angiogenesis and (wound healing)".
5) If you want to search for "angiogenesis" in conjunction with either "neurogenesis" or "organogenesis", you would enter: "angiogenesis and (neurogenesis or organogenesis).

Advanced Search Advanced search can be used to search the GO database interactively. To start, click the search button, and select "annotations", "ontology", or "gene and gene products" from the drop-down list.

The following text lists the steps involved in searching for annotations. In this example, we will try to find the tgf-beta family proteins involved in angiogenesis:

1) Click "annotations" from the search button drop-down list.
2) Type "angiogenesis" into the free-text filtering box.
3) Click "PANTHER family" from the filters list.
4) Select "tgf-beta family".
5) From the found entities, you may browse the results.
6) More filters such as "source", "assigned by", "ontology", "evidence type", etc., can be used to narrow the results.

The following text lists the steps involved in searching for ontology. In this example, we will try to explore blood vessel development:

1) Click "ontology" from the search button drop-down list.
2) Type "blood vessel development" into the free-text filtering box.
3) Click "ontology source" from the filters list, and select "biological process".
4) Click "ancestor" from the filters list, and select "tissue development".
5) From the found entities, you may browse the results.

The following text lists the steps involved in searching for genes and gene products. In this example, we will try to find the genes and gene products in wnt family involved in fin regeneration:

1) Click "genes and gene products" from the search button drop-down list.
2) Type "fin regeneration" into the free-text filtering box.
3) Click "PANTHER family" from the filter list, and select "wnt related".
4) From the found entities, you may browse the results.

Grebe Grebe is designed for users who are unfamiliar with "quick search" and "advanced search", described in the preceding text. The Grebe search wizard can be used to quickly answer common questions using a fill-in-the-blanks approach. The use of the Grebe search wizard is straightforward: fill in the necessary blank columns with IDs, terms, or gene products; select from the autocomplete results; and clink the "GO" button at the end of the question.

GOOSE

GO Online SQL Environment (GOOSE) is a web environment that allows users to freely enter SQL queries to the GO database. Users familiar with SQL can enter queries into the box. After selecting one of the available mirrors, click the "Query" button. For users who are new to SQL, the GO Consortium provides many sample queries. Users can select the sample query that meets their needs, and substitute the keywords.

In this example, we will try to find all genes directly annotated to "angiogenesis" (excluding child terms):

1) Click the button "Go" under GOOSE in the main page of AmiGO 2, or visit the website (http://amigo.geneontology.org/goose) to enter the GO online SQL environment.
2) From the drop-down list of "use an example query", select "All genes directly annotated to 'nucleus' (excluding child terms)". You may notice that queries are automatically filled in the form field under "Directly query GO data using SQL".
3) Substitute the value of "term.name" from "nucleus" to "angiogenesis" at the end of the query.
4) Select one mirror that you believe will perform well for your needs from the available mirrors. Mirrors may have different load settings and frequencies.
5) Select the number of results you want to be returned.
6) Check the box "Download results directly in a text format" if you want to download the results.
7) Click the query button, and you will find the results.

Although many sample queries have been listed in GOOSE, users may have special or complex queries. The GO Consortium has listed various kinds of possible queries on the GO LEAD database in the GO wiki main page (http://wiki.geneontology.org/index.php/Example_LEAD_Queries). Users may develop their own queries by modifying these sample queries.

Blast2GO

Blast2GO (Conesa *et al.*, 2005) is a bioinformatics platform for the functional annotation and analysis of genomic datasets. Blast2GO provides a user-friendly interface for GO annotation. Users may design custom annotation styles through the many configurable parameters. Blast2GO does not only generate functional annotations, it also supports InterPro domains, RFAM IDs, enzyme codes (ECs), and KEGG maps. Additionally, Blast2GO provides a wide array of graphical and analytical tools for annotation manipulation and data mining. There are two editions of Blast2GO: basic and professional. Blast2GO basic is a free and simplified edition, with a limited number of databases and basic functions. Blast2GO PRO is the commercial edition, which provides more databases, parameters, and functions.

A typical analysis process of Blast2GO consists of five steps: BLAST, mapping, annotation, statistical analysis, and visualization. Basically, Blast2GO uses BLAST searches to find sequences similar to the input sequences. Mapping is the process of extracting the GO terms and ECs associated with each of the obtained hits. Annotation is an evaluation of the extracted GO terms. GO annotation can be viewed by reconstructing the structure of the GO relationships and ECs highlighted on the KEGG maps. Here, we will provide a general description for the usage of Blast2GO:

Blast2GO installation: Blast2GO can be downloaded from its website, (https://www.blast2go.com/blast2go-pro/download-b2g) and installed on Microsoft Windows, Mac, and Linux platforms. At least 1 GB of RAM and a working Internet connection are required for the application. Users may register for a free Blast2GO basic account or subscribe to Blast2GO PRO to activate the software.

Blast2GO user interface: There are four basic sections in the Blast2GO main user interface—the menu bar, the main analysis icons, the main sequence table, and the application tab.

General usage: Here, we provide an introduction to the general usage of Blast2GO. Users may find more detailed information from the Blast2GO manual (https://www.blast2go.com/images/b2g_pdfs/b2g_user_manual.pdf).

1) *Load sequence*. Click "File" - > "Load" - > "Load Sequence" and select the ".fasta" file containing the set of sequences. Blast results in xml format can also be loaded.
2) *Blast*. Clicking on the "Blast" icon will initiate the processes. At the Blast configuration dialog, select the Blast mode that is appropriate for your sequence type. You may select "blastx" for nucleotide and "blastp" for protein data. Click "Next" for advanced settings and choose the location to save the Blast results. Click "Run" to start the Blast search against an NCBI non-redundant (nr) database. Once the Blast analysis is completed, you may view the results by clicking the "Chart" icon -> "Blast statistics". On the main sequence table, right-click on a sequence and select "Show sequence result" to open the Blast results for the single sequence. The sequences with Blast hits will turn into orange, and those without Blast hits will turn into red.
3) *InterProScan*. Click on the "Interpro" icon, and the corresponding wizard will be shown. You may need to fill in your email address and choose applications from the wizard's list. Click "Run" after you have selected where to save the results. InterProScan can be run in parallel with Blast.
4) *Mapping*. Click on the "Mapping" icon and click "Run" to start mapping process. The mapping process will associate the Blast hits of each sequence with GO terms. Successfully mapped sequences will turn green. Mapping results can be viewed by clicking the "Charts" icon -> "Mapping Statistics".
5) *Annotation*. Clicking on the "Annot" icon will start the annotation configuration wizard. By clicking "Next", you will start the evidence code weights wizard. Change the parameters if needed and click "Run" to start the annotation. Successfully annotated sequences will turn blue. Mapping results can be viewed by clicking the "Charts" icon -> "Annotation Statistics".
6) *Enrichment analysis*. Blast2GO provides enrichment analysis for the statistical analysis of GO term frequency differences between two sets of sequences. Clicking "Analysis" - > "Enrichment Analysis (Fisher's exact test)" will open Fisher's exact test configuration page. Select a ".txt" file containing a sequence ID list for a subset of sequences. You may set all the loaded sequences as references, or upload a second set of files containing the reference sequences. Click on "Run" button to start the analysis. A table containing the results of the enrichment analysis will be displayed. Click on the "Make Enriched Graph" icon to view the results of the Fisher's exact

test on the GO DAG. Click on "Show Bar Chart" to obtain a bar chart representation of the GO term frequencies.

7) *Combined graph.* Blast2GO provides visualization of the combined annotation for a group of sequences. Select a group of sequences to generate their combined graph. The "Select by color" option under the "Select" icon may help in selecting the featured sequences quickly. Users can also select the sequences manually by clicking the checkbox near each sequence. After the sequences are selected, click "Graphs" -> "Make combined Graph", select the functional annotation to be viewed, and click "Run" to generate the graph.
8) *Save results.*
   Clicking "File" -> "Save" saves the current Blast2GO project as a ".b2g" file.
   Clicking "File" -> "Export" allows users to export various results in different formats. The Fisher's exact test results can be saved by clicking "Save as text".
   The graphs can be saved by clicking "Save as" in the toolbar located on the right side of the graphs.

Enrichment Analysis

Researchers usually get sets of differentially expressed genes after performing high-throughput experiments such as RNA-Seq and microarrays. Although such information is useful, just generating a list of genes ("listomics") is not insightful for biology. One may wish to retrieve a functional profile of the gene sets for a better understanding of the underlying biological processes. The comparison of differentially expressed gene sets with the gene sets involved in most biological processes may provide at least some insights into the involved biological mechanisms. This can be achieved by enrichment analysis, which compares the input gene set with the reference to determine if it is enriched. Gene set enrichment or functional enrichment analysis is a method to identify classes of genes or proteins that are over-represented in a set of genes or proteins (Subramanian *et al.*, 2005). The method uses statistical approaches to identify significantly enriched or depleted groups of genes. The principal foundation of enrichment analysis is that a gene set should have a higher chance to be selected if its underlying biological process is abnormal in a given study (Huang, Sherman & Lempicki, 2009).

Main Types of Enrichment Tools

There has been an explosion in the number of software packages available for annotation enrichment analysis. About 68 tools were uniquely categorized into three major classes according to their underlying enrichment algorithms: singular enrichment analysis (SEA), gene set enrichment analysis (GSEA), and modular enrichment analysis (MEA) (Huang *et al.*, 2009). Some tools with different capabilities belong to more than one class.

SEA is the most traditional enrichment approach that iteratively tests annotation terms one at a time against a list of users' pre-selected genes for enrichment. Enriched annotation terms that pass the enrichment p-value threshold are reported. SEA uses a simple strategy, which is very efficient in extracting the major biological meaning behind large gene lists. Most of the earlier tools used this strategy, and proved to be significantly successful in many genomic studies. However, a common weakness for this class of tools is that the linear output of terms can be very large and overwhelming.

GSEA adopts a "no-cutoff" strategy that takes all genes from an experiment without selecting significant genes. GSEA reduces the arbitrary factors in the gene selection step and uses all information obtained from the experiments. GSEA methods need biological values such as fold changes for each of the genes as input. It is sometimes difficult to summarize many biological aspects of a gene into one value when the experiment design or genomic platform is complex.

MEA considers the relationships existing between different annotation terms during enrichment. The GO structure is loosely hierarchical, in which some joint terms may contain unique biological meanings. Researchers can take advantage of term-to-term relationships by using the MEA method, which can also reduce redundancy and prevent the dilution of potentially important biological concepts. However, the disadvantage of MEA is that some terms or genes with weak relationships to neighboring terms or genes could be left out from the analysis.

Gene Set and Background

The input files for the three types of enrichment tools are different. SEA is a list-based method that needs a subset of all genes chosen by some relevant method and a list of annotations linked to genes. GSEA is a rank-based method. The inputs for GSEA are a set of all genes ranked by some metrics such as fold change, and a list of annotations linked to genes. The MEA method is also list-based, but needs term-to-term relationships. The MEA method requires a subset of all genes, and a list of annotations linked to genes that are organized in some relationship.

Defining the background is very important in enrichment methods. To get the real meaning of enrichment with respect to the experiment, researchers may need to upload the reference. For example, if you are looking at a set of genes from a particular tissue, a reference for that tissue would allow generation of more meaningful results than a reference for the whole genome.

Annotation Sources

Many different annotation categories can be used by enrichment analysis, including biological function (e.g., GO terms), physical position (e.g., chromosomal location), regulation (e.g., co-expression), protein domains, pathways, or other attributes for which prior knowledge is available. GO database is very suitable for enrichment analysis on gene sets. GO has become one of the most popular annotation sources.

Statistical Methods

The most popular statistical methods used in enrichment calculation are Fisher's exact test, chi-square test, hypergeometric distribution, and binomial distribution (Huang *et al.*, 2009). Binomial probability is believed to be suitable for analysis with a large population background on a principal level. Fisher's exact test, chi-square test, and hypergeometric distribution are better for analysis with a smaller population background (Khatri & Drăghici, 2005). Each statistical method has its own weakness and limitations. It is not realistic to choose enrichment analysis tools simply according to statistical methods. Different tests may produce very different ranges of p-values. Users may try different statistical methods on the same dataset and compare the results whenever possible.

Recommended Tools

It may be difficult for users to select suitable tools from the long list of available tools. It is better to choose tools that include the species that you are researching. Make sure the tools accept your input identifiers and have the most updated annotations. Try several tools to determine the one that best fits your study purpose. Database for annotation, visualization, and integrated discovery (DAVID; Huang, Sherman & Lempicki, 2008), protein analysis through evolutionary relationships (PANTHER; Mi, Muruganujan & Thomas, 2013), and GSEA (Subramanian *et al.*, 2005) are all very good tools to start with. In the following text, we will introduce DAVID.

Enrichment Analysis by Using DAVID

DAVID (Huang *et al.*, 2008) is a web-accessible program (https://david.ncifcrf.gov/) that provides a comprehensive set of functional annotation tools. It highlights the most relevant GO terms associated with a given gene list, and belongs to the MEA method. DAVID has over 40 annotation categories, including GO terms, protein–protein interactions, protein functional domains, disease associations, bio-pathways, sequence general features, homologies, gene functional summaries, gene tissue expressions, and literature. The extended annotation coverage provides researchers with much power to analyze their genes for many different biological aspects. DAVID accepts customized gene background, which more specifically meets user requirements for the best analytical results.

Universal Gene List Manager Panel The gene list manager panel is centralized and universal for all DAVID tools. Users can upload gene lists or backgrounds through this panel, so that different DAVID tools can access them. Users do not need to re-submit the same gene list for different DAVID tools.

The following are the specifics of the gene list manager panel's "Upload" tab:

1) *Enter gene list*. Users may paste the gene list into the box or upload it from a file. The gene list should be listed in the format of one gene per row, without a header row. DAVID is case-insensitive for all the accessions or IDs.
2) *Select identifier*. From the drop-down box, select the corresponding gene identifier type for the genes in the pasted/uploaded gene list. For users who are not sure about the identifier type, the "not sure" option can be selected. After submitting, users will be led to the gene ID conversion tool, where the possible source of gene IDs can be analyzed. Users may go back to select the identifier, or convert the gene ID, and then submit to DAVID as a gene list or background.
3) *Select list type*. If users are uploading gene lists for annotation analysis—for example, selected differentially expressed genes from the RNA-Seq experiment—the "Gene list" option can be selected. "Background" means the submitted genes as a customized gene background are for enrichment background calculation purpose—for example, the entire genes in an array. The customized background is useful when all the pre-built backgrounds in DAVID do not satisfy the user's particular purpose. The gene list submitted as a background will show up in the "Background" tab.

4) *Submit*. Click "Submit list" to submit the list to DAVID. You should see the corresponding gene lists listed in the "List" tab or "Background" tab if the submission is successful. Moreover, an expected gene number should also be associated with the gene lists.

The following are the specifics of the gene list manager panel's "List" tab:

1) The species information is listed in the "List" tab. Users may choose to limit annotations by one or more species to analyze together or separately from the top box if the gene list contains multiple species.
2) Click the "Select" button to switch species.
3) From the bottom box (list manager), users may highlight the gene list to be analyzed.
4) Click the "Use" button to switch gene list. Users may also rename, remove, combine, or show gene list by clicking each of the corresponding buttons.

The following are the specifics of the gene list manager panel's "Background" tab:

The DAVID default backgrounds and user-uploaded backgrounds are listed in the top box. Users may highlight a background, and click "Use" to switch backgrounds. There are many pre-built array backgrounds listed at the bottom part of the population manager. Users can click the checkbox near an array background to switch. The successfully switched background will be shown in "Current background" under "Annotation summary results".

The selection of a population background will affect results significantly. However, there is no background to fit all the situations of various studies. The DAVID default population background in enrichment calculation is the corresponding genome-wide genes with at least one annotation in the analyzing categories. For studies with genome-wide scope or close to genome-wide scope, the default background is a good choice. For a gene list derived from Affymetrix microarray or Illumina studies, the Affymetrix chip and Illumina chip backgrounds will be better choices, respectively. For studies far below genome-wide scope, a customized background may be a better choice.

Functional Annotation Tool The functional annotation tool mainly provides batch annotation and enrichment analysis. Functional annotation clustering uses a novel algorithm to measure relationships among annotation terms that reduce the burden of associating similar redundant terms and make the biological interpretation more focused at a group level. The functional annotation tool can display genes from a user's list on pathway maps to facilitate biological interpretation in a network context.

1) *Load gene list*. Users can load gene lists and references using to the universal gene list manager.
2) *View annotation summary results*. The selected gene list and background will be shown at the top of the annotation summary results. Users may view and select annotation categories from the middle part of the summary. The combined view for selected annotations will be presented by clicking the corresponding toolbar.
3) *Explore details using the functional annotation clustering report, chart report, and table report*. The functional annotation clustering report groups similar annotations together. The grouping algorithm is based on the hypothesis that similar annotations would have similar gene members. The functional annotation chart report lists annotation terms and their associated genes. Fisher's exact test statistics is calculated

based on the corresponding DAVID gene IDs. Options can be set to ensure the display of only statistically significant genes. In comparison, the functional annotation table report lists the genes and their associated annotation terms. There is no statistics applied in the annotation table report.

4) *Export and save results.* The functional annotation clustering report, chart report, and table report can be downloaded in ".txt" file format by clicking the "Download file" option in each of the reports.

Gene Functional Classification The gene functional classification tool generates a gene-to-gene similarity matrix based on shared functional annotations. The gene similarity matrix can systematically enhance biological interpretations of large lists of genes derived from high-throughput experiments. The functional classification tool provides a rapid means to organize large lists of genes into functionally related groups to help unravel the biological content captured by high-throughput technologies.

Users can click "Gene functional classification" in "Shortcut to DAVID tools menu" to initiate the tool. Then, the gene list needs to be submitted to the DAVID system. Users can then view the results in text mode or in heat map view.

Gene Accession Conversion Tool The gene accession conversion tool can be used to convert gene lists to identifiers of the most popular resources. The given gene accession can be quickly mapped to another, based on the user's choice. Suggestions of possible choices for ambiguous gene accessions in gene lists can also be automatically provided by this tool.

1) Select genes in the gene list and the corresponding species to be converted.
2) Select the final gene identifier type to be converted to.
3) Click submit.

After submission, the result page will be shown. The genes that have been successfully converted to the desired identifiers will be displayed in the right panel. The statistics summary of gene accession conversion will be displayed in the left panel. Users may download the list or submit the converted list to DAVID directly.

Gene Pathway Analysis

Genes do not function individually independently. A gene may be one of the many genes involved in a specific gene pathway. Given a list of genes involved in biological processes, researchers may wish to map them to known pathways and determine which pathways are over-represented in a given set of genes. Pathway analysis allows reduction of complexities and increases in explanatory power. It has become a good choice for gaining insight into the underlying biology for a given gene or protein list. In the following text, we discuss the different generations of pathway analysis tools and list some pathway analysis databases.

Definition of Pathway

Different biological definitions of *pathways* are used in different pathway databases. The Kyoto Encyclopedia of Genes and Genomes (KEGG) pathway combines multiple biological processes from different organisms to produce a substrate-centered reaction mosaic. The BioCyc ontology defines a pathway as a conserved, atomic module of the metabolic network of a single organism (Green & Karp, 2006). Different pathway concepts can lead to different outcomes from a computational study that relies on pathway databases.

The knowledge-driven pathway analysis refers to the methods that exploit pathway knowledge in public repositories such as GO or KEGG, rather than to methods that infer pathways from molecular measurements (Khatri, Sirota & Butte, 2012).

Pathway Analysis Approaches

The evolution of knowledge-driven pathway analysis has been distinctly divided into three generations, each with its advantages and disadvantages. For a comparison of pathway analysis tools for each generation, readers are referred to Khatri *et al.* (2012).

Over-representation Analysis (ORA) Approaches

ORA statistically evaluates the fraction of genes in a particular pathway found among the set of genes showing changes in expression. ORA uses the following strategy (Khatri *et al.*, 2012): First, create an input gene list using a certain threshold, such as differentially expressed genes. Then, count the genes involved in each pathway. Next, every pathway is tested based on the hypergeometric, chi-square, or binomial distribution for over- or under-representation in the list of input genes.

ORA has some limitations, despite that it is a widely used tool. ORA treats each gene equally, ignoring the correlation structure between genes and useful information about the extent of regulation, such as fold change. ORA may result in information loss by using only the most significant genes. Also, ORA assumes that each pathway is independent from others, which is erroneous.

Functional Class Scoring (FCS) Approaches

FCS hypothesizes that weaker but coordinated changes in sets of functionally related genes can have significant effects on pathways. FCS consists the following steps (Khatri *et al.*, 2012): First, gene-level statistics for the differential expression of individual genes is generated using the molecular measurements from an experiment. Second, the gene-level statistics for all genes in a pathway are aggregated into a single pathway-level statistic. Finally, the statistical significance of the pathway-level statistic is assessed. FCS overcomes some limitations of ORA, but it also has some limitations. Similar to ORA, FCS analyzes each pathway independently, which ignores the cross and overlap among pathways. FCS methods use changes in gene expression to rank genes in a given pathway, and discard the changes from further analysis.

Pathway Topology (PT)–based Approaches

PT-based methods have been developed to utilize additional information about gene products, including interaction with each other in a given pathway. PT-based methods are essentially the same as FCS methods, except for the use of pathway topology to compute gene-level statistics. They are improved from FCS methods, but also have several common limitations. True pathway topology is rarely available and is fragmented in the knowledge base. PT-based methods are unable to model the dynamic states of a system, and cannot consider interactions between pathways (Khatri *et al.*, 2012).

Pathway Databases

Pathway information is available through a large number of databases. These databases can display pathway diagrams, which combine metabolic, genetic, and signal networks. Most databases are created by extracting pathway information from journal articles and then organizing the information with pathway diagrams. It is necessary for users to select proper pathway databases for their research. In the following text, we introduce some of the major pathway databases.

KEGG

KEGG (Kanehisa & Goto, 2000) (http://www.genome.jp/kegg/) is an encyclopedia that collects all knowledge relevant to biological systems. KEGG pathway is a collection of manually drawn pathway maps representing knowledge on the molecular interaction and reaction networks for metabolism, genetic information processing, environmental information processing, cellular processes, organismal systems, human disease, and drug development. Users can map molecular datasets, especially large-scale datasets in genomics, transcriptomics, proteomics, and metabolomics, to the KEGG pathway maps for biological interpretation of higher-level systemic functions.

Reactome

Reactome (Joshi-Tope *et al.*, 2005) (http://www.reactome.org/) is a curated database of pathways and reactions (pathway steps). The goal of Reactome is to provide intuitive bioinformatics tools for the visualization, interpretation, and analysis of pathway knowledge to support basic research, genome analysis, modeling, systems biology, and education. Information in the database is authored by expert biologist researchers, maintained by Reactome editorial staff, and extensively cross-referenced to other resources—for example, NCBI, Ensembl, UniProt, UCSC Genome Browser, HapMap, KEGG, ChEBI, PubMed, and GO.

PANTHER

PANTHER pathway (http://www.pantherdb.org/pathway/) is a part of the PANTHER protein classification system (Mi *et al.*, 2005). The PANTHER pathway database consists of over 177 signaling and pathways. These primary signaling and pathways have subfamilies and protein sequences mapped to individual pathway components. A pathway component is usually a single protein in a given organism, but multiple proteins can sometimes play the same role. PANTHER pathways capture molecular-level events in both signaling and metabolic pathways. Pathway diagrams are interactive and include tools for viewing gene expression data in the context of the diagrams. The PANTHER pathway database aims to comprehensively represent biological knowledge concerning both metabolic and signaling pathways, and to provide structured visualization of pathways for biological and informatic analysis.

Pathway Commons

Pathway Commons (Cerami *et al.*, 2011) (http://www.pathwaycommons.org) is a collection of publicly available pathway information from multiple organisms. Pathway Commons provides a comprehensive collection of biological pathways from multiple sources represented in a common language for gene and metabolic pathway analysis.

Pathways in Pathway Commons include biochemical reactions, complex assembly, transport and catalysis events, physical interactions involving proteins, DNA, RNA, small molecules and complexes, gene regulation events, and genetic interactions involving genes. Researchers can search, view, and download information from Pathway Commons.

BioCyc

BioCyc (Caspi *et al.*, 2014) (http://biocyc.org/) databases provide reference on the genomes and metabolic pathways of sequenced organisms. BioCyc databases are generated by predicting the metabolic pathways of completely sequenced organisms, predicting the genes codes for missing enzymes in metabolic pathways, and predicting operons. BioCyc also integrates information from other bioinformatics databases, such as protein feature and GO information from UniProt. The BioCyc website provides a suite of software tools for database searching and visualization, for omics data analysis, and for comparative genomics and comparative pathway analysis.

Pathway Analysis Tools

Many pathway analysis tools are available. We have discussed the three generations of knowledge-driven pathway analysis that may be useful for users to select proper pathway analysis tools. In the following text, we introduce several popular pathway analysis tools.

Ingenuity Pathway Analysis (IPA)

IPA (http://www.ingenuity.com/products/ipa) is the software used to display pathway data from Ingenuity Knowledge Base by QIAGEN. IPA has been broadly adopted by the life science research community, and IPA licenses require a fee. For a given gene set, IPA automatically generates the pathways that are related to those genes.

KEGG Pathway Mapping

KEGG pathway mapping (http://www.genome.jp/kegg/pathway.html) is the process to map molecular datasets, especially large-scale datasets in genomics, transcriptomics, proteomics, and metabolomics, to the KEGG pathway maps for the biological interpretation of high-level systemic functions. Pathways can be searched from the KEGG pathway main page by using the KEGG pathway ID or pathway name. The pathway ID may be either a reference pathway ID (beginning with "map") or a species-specific pathway ID (beginning with the three-letter organism code). For example, the MAPK signaling pathway ID is "map04010", whereas the human MAPK signaling pathway ID is "hsa04010".

To search for a specific gene, the gene ID prefixed by the species code can be searched through the KEGG table of contents. Users may also search for a specific gene in the KEGG genes database (http://www.genome.jp/kegg/genes.html). For example, typing "fgf11" in the "Search genes" textbox and clicking "Go" will search for "fgf11" in the KEGG genes database. Clicking the result that fits your requirement will display the specific gene item. For example, if we click on "has:2256", information related to the human fgf11 gene, such as definition, pathway, and sequence, etc., will be presented. There are seven pathways that fgf11 is involved in. The pathway map can be viewed by clicking the pathway ID.

PANTHER

The PANTHER gene list analysis tool (http://www.pantherdb.org/) allows users to input a list of genes (and, optionally, quantitative data) for analysis. The PANTHER batch ID searches supported IDs, including Ensembl gene identifier, Ensembl protein identifier, Ensembl transcript identifier, EntrezGene ID, Gene symbol, NCBI GI number, HGNC, IPI, UniGene, and UniProt ID. The identifiers in the user's list are automatically mapped to the primary IDs in the PANTHER database. After selecting an organism, the gene list can be analyzed in three different ways. The functional classification tool provides the classification results of the uploaded list and displays them in a gene list page or pie chart. The statistical over-representation test compares a test gene list uploaded by the user to a reference gene list and determines whether a particular class (e.g., a GO biological process or the PANTHER pathway) of genes is over-represented or under-represented. The statistical enrichment test determines whether any ontology class or pathway has numeric values that are non-randomly distributed with respect to the entire list of values (Mi *et al.*, 2013). PANTHER is currently part of GO, and it integrates more updated GO curation data with the tools.

Reactome Pathway Analysis

The Reactome pathways can be viewed by the Reactome pathway browser (http://www.reactome.org/PathwayBrowser/). The "Pathways Overview" screen provides an intuitive visual overview of the Reactome hierarchical pathway structure. Most Reactome pathway topics are divided into smaller sub-pathways. Pathway diagrams show the steps of a pathway as a series of interconnected molecular events, known in Reactome as "reactions".

The Reactome pathway analysis tool (http://www.reactome.org/PathwayBrowser/#TOOL=AT) combines a number of analysis and visualization tools to permit the interpretation of user-supplied experimental datasets. Users can select the type of analysis to be performed and paste in or browse to a file containing their data. The ideal identifiers to use are UniProt IDs for proteins, ChEBI IDs for small molecules, and either HGNC gene symbols or Ensembl IDs for DNA/RNA molecules. Many other identifiers are also recognized and mapped to the appropriate Reactome molecules. After inputting the data, click "GO" to start the analysis. Analysis results are shown in the "Analysis" tab, within the "Details Panel". When you click on the name of a pathway in the "Analysis" tab, the "Pathway Hierarchy" will expand to show it, and its name will be highlighted in blue.

References

Barrell, D., Dimmer, E., Huntley, R.P. *et al.* (2009) The GOA database in 2009 – an integrated Gene Ontology Annotation resource. *Nucleic Acids Research*, **37**, D396–D403.

Caspi, R., Altman, T., Billington, R. *et al.* (2014) The MetaCyc database of metabolic pathways and enzymes and the BioCyc collection of pathway/genome databases. *Nucleic Acids Research*, **42**, D459–D471.

Cerami, E.G., Gross, B.E., Demir, E. *et al.* (2011) Pathway Commons, a web resource for biological pathway data. *Nucleic Acids Research*, **39**, D685–D690.

Conesa, A., Götz, S., García-Gómez, J.M. *et al.* (2005) Blast2GO: a universal tool for annotation, visualization and analysis in functional genomics research. *Bioinformatics*, **21**, 3674–3676.

Green, M.L. and Karp, P.D. (2006) The outcomes of pathway database computations depend on pathway ontology. *Nucleic Acids Research*, **34**, 3687–3697.

Huang, D.W., Sherman, B.T. and Lempicki, R.A. (2008) Systematic and integrative analysis of large gene lists using DAVID bioinformatics resources. *Nature Protocols*, **4**, 44–57.

Huang, D.W., Sherman, B.T. and Lempicki, R.A. (2009) Bioinformatics enrichment tools: paths toward the comprehensive functional analysis of large gene lists. *Nucleic Acids Research*, **37**, 1–13.

Joshi-Tope, G., Gillespie, M., Vastrik, I. *et al.* (2005) Reactome: a knowledgebase of biological pathways. *Nucleic Acids Research*, **33**, D428–D432.

Kanehisa, M. and Goto, S. (2000) KEGG: Kyoto Encyclopedia of Genes and Genomes. *Nucleic Acids Research*, **28**, 27–30.

Khatri, P. and Drăghici, S. (2005) Ontological analysis of gene expression data: current tools, limitations, and open problems. *Bioinformatics*, **21**, 3587–3595.

Khatri, P., Sirota, M. and Butte, A.J. (2012). Ten years of pathway analysis: current approaches and outstanding challenges. *PLoS Computational Biology*, **8**, e1002375. doi: 10.1371/journal.pcbi.1002375

Mi, H., Lazareva-Ulitsky, B., Loo, R. *et al.* (2005) The PANTHER database of protein families, subfamilies, functions and pathways. *Nucleic Acids Research*, **33**, D284–D288.

Mi, H., Muruganujan, A., Casagrande, J.T. and Thomas, P.D. (2013) Large-scale gene function analysis with the PANTHER classification system. *Nature Protocols*, **8**, 1551–1566.

Mi, H., Muruganujan, A. and Thomas, P.D. (2013) PANTHER in 2013: modeling the evolution of gene function, and other gene attributes, in the context of phylogenetic trees. *Nucleic Acids Research*, **41**, D377–D386.

Mitchell, A., Chang, H.-Y., Daugherty, L. *et al.* (2015) The InterPro protein families database: the classification resource after 15 years. *Nucleic Acids Research*, **43**, D213–D221.

Škunca, N., Altenhoff, A. and Dessimoz, C. (2012). Quality of computationally inferred gene ontology annotations. *PLoS Computational Biology*, **8**, e1002533. doi: 10.1371/journal.pcbi.1002533

Subramanian, A., Tamayo, P., Mootha, V.K. *et al.* (2005) Gene set enrichment analysis: a knowledge-based approach for interpreting genome-wide expression profiles. *Proceedings of the National Academy of Sciences*, **102**, 15545–15550.

The Gene Ontology Consortium (2015) Gene ontology consortium: going forward. *Nucleic Acids Research*, **43**, D1049–D1056.

The Gene Ontology Consortium, Ashburner, M., Ball, C.A. *et al.* (2000) Gene ontology: tool for the unification of biology. *Nature Genetics*, **25**, 25–29.

Yon Rhee, S., Wood, V., Dolinski, K. and Draghici, S. (2008) Use and misuse of the gene ontology annotations. *Nature Reviews Genetics*, **9**, 509–515.

11

Genetic Analysis Using RNA-Seq: Bulk Segregant RNA-Seq

Jun Yao, Ruijia Wang and Zhanjiang Liu

Introduction

RNA-Seq is an extremely efficient strategy for revealing global expression profiles under specific physiological conditions, development stages, or various environmental stimulations. As such, it is widely used to determine differentially expressed genes under a specific "treatment." Its application was previously limited to gene expression profiling. However, through the utilization of specific families combined with elaborate experimental designs and single-nucleotide polymorphism (SNP) analysis, it is possible to analyze genetic segregation between treatments. One of these experimental designs is called *bulk segregant RNA-Seq* (BSR-Seq). BSR-Seq is an integrated solution of bulk segregant analysis (BSA) and RNA-Seq. RNA samples are collected from pooled individuals that fall into phenotype extremes, for example, the best and the worst performers. Such pooled RNA samples are then subjected to RNA-Seq. The obtained RNA-Seq datasets can be analyzed not only for differentially expressed genes between the bulks (phenotypic extreme groups), but also for differences in allele usage. The strong association of differential usage of alleles in these phenotypic extreme bulks would suggest genetic linkage of the associated genes with the phenotype, whereby the positional candidate genes as well as the expression candidate genes can be identified. In this chapter, we provide an introduction to the experimental design, principles, and analysis of BSR-Seq datasets.

In 1991, Richard Michelson from University of California, Davis developed an experimental design for genetic analysis, referred to as *bulk segregant analysis* (BSA, Michelmore, Paran & Kesseli, 1991). In their original design, BSA was used to investigate the differential segregation patterns of plants with extreme phenotypes in disease resistance. The basic idea of BSA is that phenotypic extremes (in this case, the resistant plants versus the susceptible plants) should have drastic differences in their genotypes. When samples are selected from phenotypic extremes, say the best and the worst performers and their corresponding genotypes are analyzed in each bulk, a correlation of genotypes with phenotypes should be expected. In other words, the variations among individuals may be quite subtle and difficult to detect; however, the pooled samples (bulk) of the phenotypic extremes should pose a strong contrast in their genotypes at the genomic location linked to the trait. In the last 25 years, BSA has been widely used to study various traits, and it has been most popular for studies in plants (Asnaghi *et al.*, 2004; Dussle *et al.*, 2003; Hormaza, Dollo & Polito, 1994; Hyten *et al.*, 2009; Michelmore *et al.*, 1991;

Bioinformatics in Aquaculture: Principles and Methods, First Edition. Edited by Zhanjiang (John) Liu.

Quarrie *et al.*, 1999). However, BSA has also been applied in genetic studies involving animals (Lee, Penman & Kocher, 2003; Ruyter-Spira *et al.*, 1997).

The power of BSA analysis has been well correlated with the involved use of molecular markers. BSA has been evolving along with various types of molecular markers including restriction fragment length polymorphisms (RFLPs) (Lister & Dean, 1993), random amplified polymorphic DNAs (RAPDs) (Hormaza *et al.*, 1994; Suo *et al.*, 2000), simple sequence repeats (SSRs, or microsatellites) (Cheema *et al.*, 2008; Dussle *et al.*, 2003), amplified fragment length polymorphisms (AFLPs) (Asnaghi *et al.*, 2004; Dussle *et al.*, 2003), and SNPs (Becker *et al.*, 2011). The power of BSA was enhanced by its coupling use with next-generation sequencing (NGS) technologies and sequence-based markers such as restriction site associated DNA (RAD) markers (Baird *et al.*, 2008) and genome sequencing (Haase *et al.*, 2015; Wenger, Schwartz & Sherlock, 2010). It is apparent from the report that the power of BSA is correlated with the power of the marker systems deployed.

The advances in NGS allowed the rapid understanding of genome expression patterns through RNA-Seq analysis. In recent years, the application of RNA-Seq (Ayers *et al.*, 2013; Cloonan *et al.*, 2008; Mortazavi *et al.*, 2008; Oshlack, Robinson & Young, 2010; Pepke, Wold & Mortazavi, 2009; Wang, Gerstein & Snyder, 2009) has allowed rapid and comprehensive understanding variations at the transcriptome level. The application of RNA-Seq has accelerated gene expression profiling and the identification of gene-associated SNPs in many species. However, the integrated studies of gene expression along with SNP mapping have been lacking. Coupling of RNA-Seq with BSA has led to the development of BSR-Seq, which was first reported in 2012 (Liu *et al.*, 2012; Trick *et al.*, 2012). BSR-Seq is well suited to simultaneous studies of both expression patterns and associated SNPs with the phenotypes. It has been successfully applied in plants (Liu *et al.*, 2012; Trick *et al.*, 2012) and aquaculture species (Wang *et al.*, 2013) for the analysis of performance traits. As pooling of samples are involved, BSR-Seq is well suited to species with high fecundities because it is easier to collect samples from phenotypic extremes with such species. As such, it should be extremely useful for aquaculture species.

BSR-Seq: Basic Considerations

Performance is ultimately related to variations in the qualitative and quantitative differences in gene expression among individuals (Oleksiak, Roach & Crawford, 2005). As we previously described (Wang *et al.*, 2013), in well-defined families, the first level of variations comes from genetic segregation and recombination of chromosomes. As a result, each individual has a different genetic makeup. For a given genetic background, hereditary potential carried on DNA can only be realized when the genes are expressed. At the whole genome level, expression of each gene is affected by its genetic regulatory element as well as trans-acting factors activated or impacted by the environment. A composite of genes, transcriptional regulation, post-transcriptional modification and regulation, translational regulation, and post-translational modification and regulation, along with environmental impact and genotype–environment interactions eventually determines the phenotypes of individuals. When considered at the whole genome level, expressions of tens of thousands of genes and combinations of these genes make the variation of performance traits extremely complex with huge variability. The task of modern agricultural genomics is to gain understanding of such variations and their relationships for determining production and performance traits.

Genetic and molecular biological studies are conducted to dissect these variables at different levels. For instance, the effect of alleles can be dissected through genetic and QTL mapping analysis (Bevan & Uauy, 2013; Hallman *et al.*, 1991; He *et al.*, 2013; Young, 1996). Gene expression can be analyzed using high-throughput methodologies such as microarrays and RNA-Seq analysis (Bunney *et al.*, 2003; Oshlack *et al.*, 2010; Wang *et al.*, 2009). Various epigenetic regulations have also been studied to understand the differences in gene expressions with similar genetic backgrounds. Such analyses have been very powerful in determining genetic and epigenetic factors affecting performance and production traits (He, Hua & Yan, 2011; Molinie & Georgel, 2009; Ng *et al.*, 2008). However, performance and production traits are often highly complex, and the outcome of agricultural operations is affected by variations at all levels. For example, genetic background is very important because disease resistance genes allow the organism to survive serious infections (Gururani *et al.*, 2012; Zhao, Zhu & Chen, 2012). In most cases, disease resistance genes have been studied through genetic linkage and QTL analyses that allow the identification of genomic regions containing disease resistance genes. However, due to the redundancy of the genome on the DNA level (e.g., large segments of repeat sequences, introns, and the intergenic regions), even with the most powerful molecular approaches, analysis of complex traits such as disease resistance can be extremely challenging.

BSR-Seq combines BSA and RNA-Seq and, therefore, has the ability to simultaneously determine expression levels of all genes in the pooled bulk samples, and to determine the proportion of allele usage at SNP sites. In addition, it also has the ability to differentiate allele usage differences caused by segregation in different bulks from those caused by allele-specific expressions.

BSR-Seq Procedures

The analysis of expression levels of each gene within bulk samples and the determination of differentially expressed genes are the same as the traditional RNA-Seq analysis, which is described in Chapter 9 and hence not repeated here. Instead, we will focus on the analysis of allele usage in the bulked samples.

Identification of SNPs

Identification of SNPs from sequences depends on the alignment of multiple reads of sequences obtained from the same genomic location but from different chromosome sets. Most aquaculture species are diploid organisms, and, therefore, within a single individual, there are two sets of chromosomes. Thus, identification of SNPs is possible with just one fish. However, use of multiple individuals should increase the likelihood for SNP detection.

RNA-Seq generates sequences from RNA converted to cDNA. Therefore, identification of SNPs from RNA-Seq datasets depends on the sequencing of the same transcripts transcribed from different chromosome sets from a diploid individual or multiple individuals. RNA-Seq reads first need to be assembled into contigs (see Chapter 8). It does not matter if the assembly is done using de novo assembly or reference-guided assembly. However, the reads depth of RNA-Seq depends on the level of expression of genes. This is different from using a DNA template where the chances to sequence either chromosome with a single individual is the same, and sequencing is random.

For NGS datasets, a minimal number of reads and the minimal reads of the minor allele should be set. There is no absolute requirement for this, but a region that is sequenced with a high number of reads and the minor allele detected repeatedly usually increase SNP conversion rates. We suggest that, for reliable SNP identification, the minimal reads required should be 6–10 on each locus, and the minor allele should be detected at least two–three times to avoid the noise caused by sequencing errors.

SNPs are identified by the alignment of short reads to the assembly of the RNA-Seq. In order to be qualified for SNPs, once again, a reasonable cutoff of total mapped reads and minor allele reads count should be set. Sequence mapping for SNP identification analysis can be performed utilizing multiple aligners, for example, CLC Genomics Workbench, TopHat, and STAR (Dobin *et al.*, 2013; Trapnell, Pachter & Salzberg, 2009). Trimmed sequence reads are first aligned against the transcriptome assembly contigs. To keep the confidence level, the following cutoffs are recommended: mismatch cost of 2, deletion cost of 3, and insertion cost of 3; the highest scoring matches that shared ≥ 95% similarity with the transcriptome assembly contigs across ≥ 90% of their length. To keep the confidence level, the following cutoff non-unique mappings are removed. Finally, all mapping results are converted to BAM format for the downstream analysis. The BAM files from each group and replications are then combined accordingly through the mpileup function of SAMtools version 0.1.18 (Li *et al.*, 2009):

```
$ samtools mpileup -B group1.bam group2.bam >
   all_groups.mpileup
```

As the sample-pooling feature of BSR-seq, only few SNP identification pipelines are available for the circumstances. Among these pipelines, GATK (McKenna *et al.*, 2010) and PoPoolation2 (Kofler, Pandey & Schlötterer, 2011) are widely used in most cases. However, GATK requires high-quality genome assembly (usually only achievable in model species) and heavy computation intensity, which is obviously not suitable for most aquaculture species. Therefore, it seems PoPoolation2 is currently the optimal choice. The raw SNPs can be identified through:

```
$ java -ea -Xmx7g -jar mpileup2sync.jar
   --input all_groups.mpileup --output all_groups.sync
   --fastq-type illumina --min-qual 20 --threads 24
```

Please note that analysts can trim their mapping results though "`--min-qual`" and increase the running speed by adding more CPUs in "`---threads`".

As mentioned earlier, to increase the confidence level of the identified SNPs, the analysts can trim the raw results through:

```
$ perl snp-frequency-diff.pl --input all_groups.sync
   --output-allgroups --min-count 3 --min-coverage 6
   --max-coverage 20000
```

In this command, the cutoff was set as: (1) minimum reads in each group ≥ 6; (2) total minor allele reads count ≥ 3; and (3) maximum reads in each group ≤ 20,000. SNPs that passed all the cutoffs are written into an "_rc" file.

Identification of Significant SNPs

After the initial identification of SNPs, significant SNPs need to be identified between the treatment groups or between the bulks. For instance, SNPs displayed significantly genotypic differences between the resistant catfish group and susceptible catfish group. SNPs that displayed heterozygous genotype (e.g., allele variants = 2, and minor allele reads ≥ 2 in each group) can be retrieved to test the differences in the levels of allele frequencies between the resistant group and the susceptible group using two-tailed Fisher's Exact test (Fisher, 1922), with an FDR p-value cutoff ≤ 0.05.

Bulk Frequency Ratio

Although significant SNPs identified through Fisher's exact test and FDR reflect the final ratios of different alleles at some level, the p-value is actually affected by the total sequencing coverage at each loci, which can easily lead to the identification of false positives (Wang *et al.*, 2013). In order to compare the SNP allele frequencies more directly and accurately, the term "bulk frequency ratio" (BFR) was proposed. Calculation of BFR is necessary between the two bulks, for example, the resistant fish and the susceptible fish (Wang *et al.*, 2013). The frequency of the major allele should be first calculated in both bulks, and then BFR should be calculated by generating the ratio of these frequencies between the bulks (Figure 11.1). Using catfish as an example, in the resistant bulk, the A allele at the locus was detected nine times, while the G allele was detected just one time. The major allele frequency = 9/10; in contrast, in the susceptible bulk, the frequency of A allele is 1/10. BFR then equals 0.9/0.1 = 9.

The threshold to be set for the significant level for BFR really depends on the experiment design. If a single known family is used, the situation is much less complex, because the expected allele segregation ratio is known. For instance, if the parental genotype is AG × AG, then the expected ratio of segregation is 1 AA, 2 AG, and 1 GG. Statistical analysis can be used to test if the observed ratio is significantly deviated from the expected ratio. If multiple families are used and the parental genotypes are unknown, the situation can be more complex. Nonetheless, the greater the BFR, the more likely the locus has something to do with the traits. A BFR value of 4–6 is suggested as a starting point for one–two families (Ramirez to two families, 2015; Wang *et al.*, 2013).

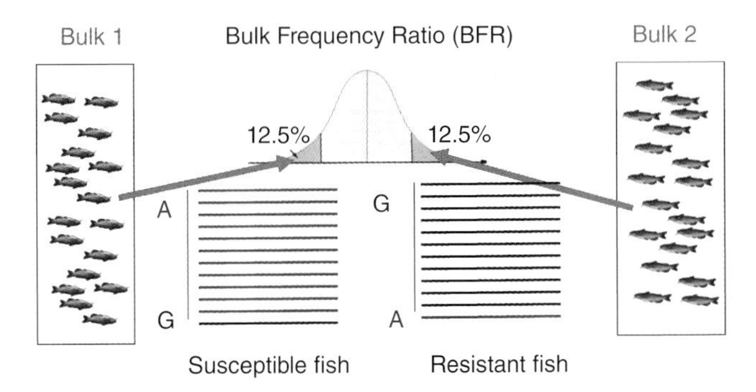

Figure 11.1 Schematic presentation of bulk frequency ratio (BFR). Phenotypic extremes are pooled into bulks (bulk 1 for susceptible fish and bulk 2 for resistant fish). A allele is indicated by blue lines, and G allele is indicated by black lines. In this example, BFR = (11/12)/(1/12) = 11. (*See color plate section for the color representation of this figure.*)

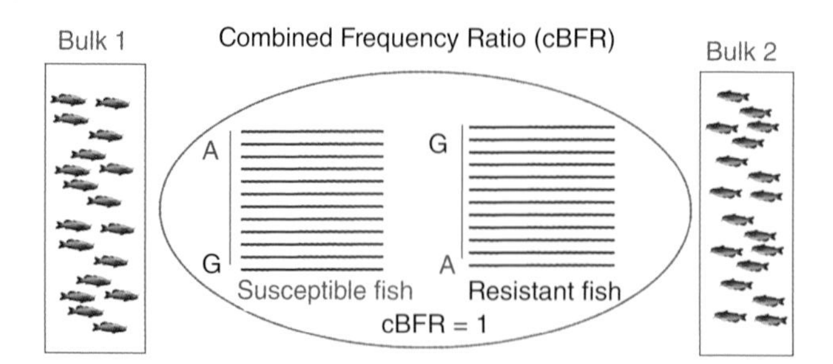

Figure 11.2 Schematic presentation of combined frequency ratio (cBFR). Allele frequency of A in bulk 1 is 11/12, and in bulk 2 is 1/12. However, when the bulks are combined, the overall allele frequency ratio is 12A/12G = 1. (*See color plate section for the color representation of this figure.*)

Combined Bulk Frequency Ratio

With DNA samples, allele ratio can be expected if there is no linkage. However, RNA-Seq data is complex because the sequencing frequency of a specific transcript heavily depends on the expression level of that gene, and two alleles of the gene can be regulated differentially, leading to allele-specific expression. As RNA-Seq data is analyzed in terms of RPKM at the RNA level, the allele ratios obtained by RNA-Seq are compounded by two factors: the genotype allele frequencies at the DNA level, and the relative expression levels of the two alleles at the RNA level. For instance, the two alleles may have very different genotype allele frequencies in the two bulked samples, and, in these cases, even if the expression is not regulated at the transcriptional level, the final ratio of the two alleles between the bulked samples are expected to be different. However, if one of the two alleles is differentially regulated, the final allele ratio at the RNA level would be different from the allele ratio at the DNA level.

In order to qualify the locus to be linked to the trait, lack of allele-specific expression is assumed. In other words, the differences in RNA-Seq should be caused by segregation of the alleles in different bulks rather than by allele-specific regulation of gene expression. Therefore, the genes with large BFRs must be also tested to determine if allele-specific expression is involved. Combined BFR is the parameter designed to detect if allele-specific expression is involved.

Combined BFR is the allele frequency of all the samples, combining both bulks. If indeed the large BFR is caused by segregation as shown in Figure 11.2, when all the alleles are combined, the combined BFR should equal to a value close to 1 (in the example in Figure 11.2, it is 1). For instance, the total number of A alleles now in the combined bulks in 10, and the total number of G alleles in the combined bulks is also 10. Therefore, the combined BFR is 1.

Through the analysis of BFR and combined BFR (cBFR), the genes containing significant SNPs can be classified into several categories:

1) *Genes containing significant SNPs with large BFR and small cBFR*. This category is highly interesting for the traits because segregation of different alleles into different bulks is the major reason of allele frequency difference.
2) *Genes containing significant SNPs with large BFR and large cBFR*. This category of genes may or may not be genetically linked to the traits. They are interesting in that

allele-specific expression may be involved, but they are also interesting as genetic segregation may also be involved.

3) *Genes containing significant SNPs with relatively small BFR but large cBFR.* This category of genes likely represents genes with allele-specific regulation. They are interesting for the analysis of allele-specific expression, but they may not be genetically linked to the trait.

If the large BFR is caused by different allele frequencies at the DNA level in the two bulks but not by allele-specific expression, the ratio of the two alleles in the combined bulk (e.g., resistant bulk plus susceptible bulk) should be relatively small and predictable with the family structure. For instance, at an AA × AG SNP site in a single family, the progenies should have a 3A:1G allele frequency at the genomic level. However, the situation is more complex if more than one family is used. Nevertheless, the largest allele ratio at the DNA level can still be predicted. For instance, if two families are used at an SNP site, the largest possible allele ratio at the DNA level is 7:1 (i.e., AA × AG in one family, and AA × AA [not polymorphic at this site] in the second family), where the largest possible allele ratio at the DNA level is 7A:1G. Any other combinations would result in a smaller combined allele ratio at the DNA level. Therefore, it is possible to differentiate the large BFR caused by genetic segregation and those caused by allele-specific expression based on the combined allele ratio of the bulks. Whereas the large BFRs with very large combined allele ratios are likely caused, at least in part, by allele-specific expression, and the large BFRs with small combined allele ratios are likely caused by genetic segregation, the BFRs in the transitional zone could be caused by both genetic segregation and allele-specific expression.

Location of Genes with High BFR

To determine the genomic location of SNPs with high levels of BFR, genes containing SNPs with high BFR (e.g., BFR ≥ 6) can be used initially as query for BLAST searches against the genome sequence of the species under study. Genes harboring significant SNPs with high BFR and low combined allele ratio (e.g., ≤ 4) can be used as query to map to the whole genome sequence by BLASTN with an e-value of 1e-20/1e-10. The linkage groups or chromosomes contain a large number of genes with significant SNPs, which indicates their harboring of QTL regions controlling the trait.

In summary, BSR-Seq is a rapid, relatively inexpensive approach for the analysis of traits-associated genes. It has full capacity for the identification of differentially expressed genes, the capacity to identify significant SNPs between phenotypic bulks, the capacity to potentially identify the positional candidate genes, and the ability to identify allelic expressed genes. Even in the absence of a whole genome sequence, BSR-Seq can still be useful for the analysis of genes controlling performance traits.

Acknowledgments

Our research involving this line of studies has been supported by Agriculture and Food Research Initiative Competitive Grants from the USDA National Institute of Food and Agriculture (NIFA).

References

Asnaghi, C., Roques, D., Ruffel, S. *et al.* (2004) Targeted mapping of a sugarcane rust resistance gene (*Bru1*) using bulked segregant analysis and AFLP markers. *Theoretical and Applied Genetics*, **108**, 759–764.

Ayers, K.L., Davidson, N.M., Demiyah, D. *et al.* (2013) RNA sequencing reveals sexually dimorphic gene expression before gonadal differentiation in chicken and allows comprehensive annotation of the W-chromosome. *Genome Biology*, **14**, R26.

Baird, N.A., Etter, P.D., Atwood, T.S. *et al.* (2008) Rapid SNP discovery and genetic mapping using sequenced RAD markers. *PloS One*, **3**, e3376.

Becker, A., Chao, D.-Y., Zhang, X. *et al.* (2011) Bulk segregant analysis using single nucleotide polymorphism microarrays. *PLoS One*, **6**, e15993.

Bevan, M.W. and Uauy, C. (2013) Genomics reveals new landscapes for crop improvement. *Genome Biology*, **14**, 206.

Bunney, W.E., Bunney, B.G., Vawter, M.P. *et al.* (2003) Microarray technology: a review of new strategies to discover candidate vulnerability genes in psychiatric disorders. *The American Journal of Psychiatry*, **160**, 657–666.

Cheema, K.K., Grewal, N.K., Vikal, Y. *et al.* (2008) A novel bacterial blight resistance gene from *Oryza nivara* mapped to 38 kb region on chromosome 4 L and transferred to *Oryza sativa* L. *Genetics Research (Camb)*, **90**, 397–407.

Cloonan, N., Forrest, A.R., Kolle, G. *et al.* (2008) Stem cell transcriptome profiling via massive-scale mRNA sequencing. *Nature Methods*, **5**, 613–619.

Dobin, A., Davis, C.A., Schlesinger, F. *et al.* (2013) STAR: ultrafast universal RNA-seq aligner. *Bioinformatics*, **29**, 15–21.

Dussle, C., Quint, M., Melchinger, A. *et al.* (2003) Saturation of two chromosome regions conferring resistance to SCMV with SSR and AFLP markers by targeted BSA. *Theoretical and Applied Genetics*, **106**, 485–493.

Fisher, R.A. (1922) On the interpretation of $\chi 2$ from contingency tables, and the calculation of P. *Journal of the Royal Statistical Society*, **106**, 87–94.

Gururani, M.A., Venkatesh, J., Upadhyaya, C.P. *et al.* (2012) Plant disease resistance genes: current status and future directions. *Physiological and Molecular Plant Pathology*, **78**, 51–65.

Haase, N.J., Beissinger, T., Hirsch, C.N. *et al.* (2015) Shared genomic regions between derivatives of a large segregating population of maize identified using bulked segregant analysis sequencing and traditional linkage analysis. *G3: Genes| Genomes| Genetics*, **5**, 1593–1602.

Hallman, D.M., Boerwinkle, E., Saha, N. *et al.* (1991) The apolipoprotein E polymorphism: a comparison of allele frequencies and effects in nine populations. *The American Journal of Human Genetics*, **49**, 338–349.

He, G., Chen, B., Wang, X. *et al.* (2013) Conservation and divergence of transcriptomic and epigenomic variation in maize hybrids. *Genome Biology*, **14**, R57.

He, H., Hua, X. and Yan, J. (2011) Epigenetic regulations in hematopoietic Hox code. *Oncogene*, **30**, 379–388.

Hormaza, J., Dollo, L. and Polito, V. (1994) Identification of a RAPD marker linked to sex determination in *Pistacia vera* using bulked segregant analysis. *Theoretical and Applied Genetics*, **89**, 9–13.

Hyten, D.L., Smith, J.R., Frederick, R.D. *et al.* (2009) Bulked segregant analysis using the GoldenGate assay to locate the locus that confers resistance to soybean rust in soybean. *Crop Science*, **49**, 265–271.

Kofler, R., Pandey, R.V. and Schlötterer, C. (2011) PoPoolation2: identifying differentiation between populations using sequencing of pooled DNA samples (Pool-Seq). *Bioinformatics*, **27**, 3435–3436.

Lee, B.Y., Penman, D. and Kocher, T. (2003) Identification of a sex-determining region in Nile tilapia (*Oreochromis niloticus*) using bulked segregant analysis. *Animal Genetics*, **34**, 379–383.

Li, H., Handsaker, B., Wysoker, A. *et al.* (2009) The sequence alignment/map format and SAM tools. *Bioinformatics*, **25**, 2078–2079.

Lister, C. and Dean, C. (1993) Recombinant inbred lines for mapping RFLP and phenotypic markers in Arabidopsis thaliana. *The Plant Journal*, **4**, 745–750.

Liu, S., Yeh, C.T., Tang, H.M., Nettleton, D. and Schnable, P.S. (2012) Gene mapping via bulked segregant RNA-Seq (BSR-Seq). *PLoS One*, **7**, e36406.

McKenna, A., Hanna, M., Banks, E. *et al.* (2010) The Genome Analysis Toolkit: a MapReduce framework for analyzing next-generation DNA sequencing data. *Genome Research*, **20**, 1297–1303.

Michelmore, R.W., Paran, I. and Kesseli, R. (1991) Identification of markers linked to disease-resistance genes by bulked segregant analysis: a rapid method to detect markers in specific genomic regions by using segregating populations. *Proceedings of the National Academy of Sciences*, **88**, 9828–9832.

Molinie, B. and Georgel, P. (2009) Genetic and epigenetic regulations of prostate cancer by genistein. *Drug News & Perspectives*, **22**, 247–254.

Mortazavi, A., Williams, B.A., McCue, K. *et al.* (2008) Mapping and quantifying mammalian transcriptomes by RNA-Seq. *Nature Methods*, **5**, 621–628.

Ng, J.H., Heng, J.C., Loh, Y.H. and Ng, H.H. (2008) Transcriptional and epigenetic regulations of embryonic stem cells. *Mutation Research*, **647**, 52–58.

Oleksiak, M.F., Roach, J.L. and Crawford, D.L. (2005) Natural variation in cardiac metabolism and gene expression in *Fundulus heteroclitus*. *Nature Genetics*, **37**, 67–72.

Oshlack, A., Robinson, M.D. and Young, M.D. (2010) From RNA-seq reads to differential expression results. *Genome Biology*, **11**, 220.

Pepke, S., Wold, B. and Mortazavi, A. (2009) Computation for ChIP-seq and RNA-seq studies. *Nature Methods*, **6**, S22–32.

Quarrie, S.A., Lazić-Jančić, V., Kovačević, D. *et al.* (1999) Bulk segregant analysis with molecular markers and its use for improving drought resistance in maize. *Journal of Experimental Botany*, **50**, 1299–1306.

Ramirez-Gonzalez, R.H., Segovia, V., Bird, N. *et al.* (2015) RNA-Seq bulked segregant analysis enables the identification of high-resolution genetic markers for breeding in hexaploid wheat. *Plant Biotechnology Journal*, **13**, 613–624.

Ruyter-Spira, C.P., Gu, Z., Van der Poel, J. and Groenen, M. (1997) Bulked segregant analysis using microsatellites: mapping of the dominant white locus in the chicken. *Poultry Science*, **76**, 386–391.

Suo, G., Huang, Z., He, C. *et al.* (2000) Identification of the molecular markers linked to the salt-resistance locus in the wheat using RAPD-BSA technique. *Acta Botanica Sinica*, **43**, 598–602.

Trapnell, C., Pachter, L. and Salzberg, S.L. (2009) TopHat: discovering splice junctions with RNA-Seq. *Bioinformatics*, **25**, 1105–1111.

Trick, M., Adamski, N.M., Mugford, S.G. *et al.* (2012) Combining SNP discovery from next-generation sequencing data with bulked segregant analysis (BSA) to fine-map genes in polyploid wheat. *BMC Plant Biology*, **12**, 14.

Wang, R., Sun, L., Bao, L. *et al.* (2013) Bulk segregant RNA-seq reveals expression and positional candidate genes and allele-specific expression for disease resistance against enteric septicemia of catfish. *BMC Genomics*, **14**, 929.

Wang, Z., Gerstein, M. and Snyder, M. (2009) RNA-Seq: a revolutionary tool for transcriptomics. *Nature Reviews Genetics*, **10**, 57–63.

Wenger, J.W., Schwartz, K. and Sherlock, G. (2010) Bulk segregant analysis by high-throughput sequencing reveals a novel xylose utilization gene from *Saccharomyces cerevisiae. PLoS Genetics*, **6**, e1000942.

Young, N.D. (1996) QTL mapping and quantitative disease resistance in plants. *Annual Review of Phytopathology*, **34**, 479–501.

Zhao, S., Zhu, M. and Chen, H. (2012) Immunogenomics for identification of disease resistance genes in pigs: a review focusing on Gram-negative bacilli. *Journal of Animal Science and Biotechnology*, **3**, 34.

12

Analysis of Long Non-coding RNAs

Ruijia Wang, Lisui Bao, Shikai Liu and Zhanjiang Liu

Introduction

Long non-coding RNAs (lncRNAs) are a class of non-coding RNAs with length greater than 200 bases. They are involved in many biological processes. In recent years, along with the development of high-throughput sequencing technologies, systematic identification and characterization of lncRNAs has been reported in many model species. Analyses of lncRNAs have been limited in non-model species, especially in aquaculture species. In this chapter, we provide the bioinformatics analysis of lncRNAs. We will describe the general considerations and bioinformatics methods for the identification and analysis of lncRNAs with RNA-Seq datasets.

During the first two decades of molecular biology research in the 1970s and 1980s, it was widely believed that only a small fraction (1–5%) of the genome was transcribed. Part of the reason for this was the technological limitations for the detection of transcripts expressed at low levels. In addition, most research was focused on the so-called central dogma, according to which, from DNA to RNA to protein, genes are first transcribed into RNAs and then be translated into proteins. However, this notion was challenged by the discovery of new classes of regulatory non-coding RNAs (ncRNAs). As such, the term "transcript" is now used in a more broad sense, from the original usage to mostly refer to the protein-encoding mRNA, now to various transcriptional products that cover almost the entire genome. Thus, the concept of pervasive transcription evolved to include various types of RNAs, in addition to the traditional mRNA, rRNA, and tRNA, now to also include various non-coding RNAs (Mercer, Dinger & Mattick, 2009). The proportion of such non-coding RNAs may vary among species, but are becoming more and more represented. In humans, for instance, current research reveals that only one-fifth of the transcription across the human genome is associated with protein-coding genes (Kapranov *et al.*, 2007), suggesting at least four times more non-coding than coding RNA sequences.

Various types of non-coding RNAs have been identified, including lncRNAs, microRNAs (miRNAs), short interfering RNAs (siRNAs), Piwi-interacting RNAs (piRNAs), small nucleolar RNAs (snoRNAs), etc. (Kapranov *et al.*, 2007). Among these, lncRNAs are non-coding RNAs whose sizes are greater than 200 bases. Obviously, such classification is arbitrary, but is based on practical considerations, including the separation of RNAs in common experimental protocols.

Bioinformatics in Aquaculture: Principles and Methods, First Edition. Edited by Zhanjiang (John) Liu.

Based on their genomic location, context, and origin of transcription relative to protein-encoding genes, lncRNAs can be classified into intergenic and intragenic lncRNAs (Kung, Colognori & Lee, 2013). Intergenic lncRNAs (lincRNAs) are standalone lncRNAs positioned in intergenic regions that may be transcribed by polymerase II independent of protein-encoding genes (Cabili *et al.*, 2011; Guttman & Rinn, 2012; Ulitsky *et al.*, 2011). Intragenic lncRNAs are located within the protein-encoding gene sequences. Several types of lncRNAs are intragenic in nature:

1) Those that are transcribed from pseudogenes that harbor functional transcriptional regulatory elements, but lost full translational potential due to nonsense, frameshift, or other mutations (Balakirev & Ayala, 2003; Pink *et al.*, 2011)
2) Those that are transcribed as antisense transcripts (Katayama *et al.*, 2005; He *et al.*, 2008; Faghihi & Wahlestedt, 2009), most often located around the 5' or 3' ends of the sense transcripts (Kung & Lee, 2013)
3) Those that are transcribed from introns of the protein-encoding genes as intronic lncRNAs.

Some intragenic lncRNAs can harbor sequences beyond the protein-encoding genes. They are most often located in the vicinity of transcription start sites in both sense and antisense orientations, corresponding to peaks of RNA polymerase II occupancy due to pausing (Core, Waterfall & Lis, 2008; He *et al.*, 2008; Preker *et al.*, 2008; Seila *et al.*, 2008). These lncRNAs include transcription start site-associated RNAs (TSSa-RNAs), upstream antisense RNAs (uaRNAs), promoter upstream transcripts (PROMPTs), promoter associate short RNAs (pasRNAs), and enhancer RNAs (eRNAs). It is unclear at present that these are just transcriptional byproducts (Kung *et al.*, 2013), but a hinted concept is that these are being transcribed with the associated transcription power of the protein-encoding genes. Transcription events tend to ripple away from "legitimate" transcripts, leading to leaky expression of neighboring regions (Ebisuya *et al.*, 2008). Theoretically, lncRNAs can be transcribed using the promoters of protein-encoding genes or their own promoters within or outside of protein-encoding genes. Most lncRNAs resemble those of protein-coding genes in terms of the CpG islands, multi-exonic structures, and poly(A)-signals, but they have no more than chance potential to code for proteins and are translated poorly from relatively short reading frames, if at all (Ingolia, Lareau & Weissman, 2011; Numata *et al.*, 2003).

The functions of non-coding RNAs are being unraveled, and the discoveries are continuing. Among the non-coding RNAs, the functions for the small regulatory non-coding RNAs such as miRNA are probably the best studied (Bartel, 2009). The functions for lncRNAs are not well understood, but recent research indicates that lncRNAs could be involved in a number of functions (Huarte & Rinn, 2010; Pauli *et al.*, 2012; Wang & Chang, 2011), including the following:

1) lncRNAs can interact with and modulate the activity of the chromatin-modifying machinery (Huarte *et al.*, 2010; Nagano *et al.*, 2008; Rinn *et al.*, 2007; Tian, Sun & Lee, 2010)
2) lncRNAs serve as decoys in the sequestration of miRNAs (Poliseno *et al.*, 2010), transcription factors (Hung *et al.*, 2011), or other proteins (Tripathi *et al.*, 2010)
3) lncRNA may serve as precursors for the generation of sRNAs (Fejes-Toth *et al.*, 2009; Kapranov *et al.*, 2007; Wilusz, Sunwoo & Spector, 2009)

In addition, with the co-expression of lncRNAs and mRNAs, the co-localized expression of lncRNA and protein-coding genes were also observed (Ponjavic *et al.*, 2009), suggesting their cooperative actions and/or sharing of cis-regulatory elements in the transcription process.

In many instances, the act of lncRNA transcription alone is sufficient to regulate the expression of nearby genes (e.g., Martens, Laprade & Winston, 2004; Petruk *et al.*, 2006; Wilusz & Sharp, 2013) or distant genes through the modification of chromatin complexes (e.g., Tsai *et al.*, 2010) or by binding to transcription elongation factors (Yang, Froberg & Lee, 2014). Based on the effects exerted on DNA sequences, they can be classified as cis-lncRNA and trans-lncRNA (Rinn & Chang, 2012). In addition, a number of mechanisms have been proposed for functions of lncRNAs.

In mammals, most genes express antisense transcripts, which might constitute a class of ncRNA that is particularly adept at regulating mRNA dynamics (He *et al.*, 2008). In this case, antisense ncRNAs can modify key cis-elements in mRNA by the formation of RNA duplexes, such as Zeb2, an antisense RNA that complements the 5' splice site of an intron in the 5' UTR of the zinc finger Hox mRNA Zeb2 (Beltran *et al.*, 2008), and expression of the lncRNA prevents the splicing of an intron that contains an internal ribosome entry site required for efficient translation.

The non-coding RNA, *XIST*, first reported in 1991 (Borsani *et al.*, 1991), is located on the X chromosome of the placental mammals that acts as a major effector of the X inactivation process. It is a component of the Xic-X-chromosome inactivation center, along with two other RNA genes (*Jpx* and *Ftx*) and two proteins (Tsx and Cnbp2) (Brockdorff *et al.*, 1992). It is only expressed on the inactive chromosome and not on the active chromosome, and is processed in a similar way as mRNAs, through splicing and polyadenylation, and remains untranslated. X chromosomes lacking *XIST* will not be inactivated, while duplication of the *XIST* gene on another chromosome causes inactivation of that chromosome (Brown *et al.*, 1992). Since the discovery of *XIST*, progress on lncRNA research has been relatively slow. During the following decade, less than a dozen lncRNAs were identified and catalogued in all eukaryotes (Erdmann *et al.*, 2001). The slow progress was mostly due to the lack of high-throughput sequencing and the lack of databases with satisfactory classifications of such transcripts. It is only in the recent years that genome-wide identification of lncRNAs has become possible, because of the advances made in high-throughput sequencing technologies of cDNAs (RNA-Seq).

LncRNAs were first reported as a class of transcripts during the large-scale sequencing of full-length cDNA libraries in mouse (Okazaki *et al.*, 2002). Since then, thousands of other lncRNAs have been reported in mammals and other vertebrates (Carninci *et al.*, 2005; Gerstein *et al.*, 2010; Grabherr *et al.*, 2011; Guttman *et al.*, 2009; Kim *et al.*, 2010; Numata *et al.*, 2003; Ørom *et al.*, 2010; Ulitsky *et al.*, 2011; Wang & Chang, 2011). Although lncRNA has been a hot spot of research in mammals, studies of lncRNAs with fish are still rare. The first study of lncRNA identification and annotation with fish was published in zebrafish (Pauli *et al.*, 2012). Through a series of RNA-Seq experiments, a set of 1133 non-coding multi-exonic transcripts were identified (Pauli *et al.*, 2012).

Although systematic identification and characterization of lncRNAs has been reported in humans (Wapinski & Chang, 2011) and some model species including mouse (Guttman *et al.*, 2009), chicken (Li *et al.*, 2012), zebrafish (Pauli *et al.*, 2012), and *Caenorhabditis elegans* (Nam & Bartel, 2012), such analyses are lacking in non-model species, especially in aquaculture species, only rare studies were reported (Mu *et al.*, 2016; Wang *et al.*, 2016), which hampers a thorough understanding of the evolutionary perspective of pervasive transcription, and the origin and extent of the transcription of lncRNAs. Hence, we believe that research in the area of lncRNA with aquaculture species will drastically increase, and this chapter will attempt to to demonstrate the processes involved, particularly the bioinformatics analysis of lncRNAs. We will describe the general considerations and bioinformatics methods for the identification and analysis of lncRNAs with RNA-Seq datasets.

Data Required for the Analysis of lncRNAs

To identify lncRNAs from a transcriptome, a considerable amount of transcriptome datasets is required. Of course, the researcher can create such datasets by conducting RNA-Seq, or it can be downloaded from the Sequence Read Archive (SRA; http://www.ncbi.nlm.nih.gov/sra). If RNA-Seq is planned to capture lncRNAs, it is advised that the researchers consider the various biological questions that can be asked—for instance, how the expression of lncRNAs is affected by various treatments such as different tissues, developmental stages, bacterial infection, stress treatments, or various other treatments as compared to their controls. In addition, for lncRNAs, strand-specific information is often needed to properly annotate the lncRNAs, such as if they are transcribed from the sense strand or antisense strand. Therefore, strand-specific RNA-Seq is preferred over traditional RNA-Seq.

Many RNA-Seq datasets are available for bioinformatics data mining or additional analysis in the SRA database. From SRA, RNA-Seq datasets can be selected based on the study, the species, or various other key words of interest. For instance, in December 2014, there were 46392 sets of short read datasets in the SRA database that were generated by the next-generation sequencing technologies. Of these, many are genomic sequences, but many are RNA-Seq datasets. By searching for the key word "RNA-Seq", over 6500 records are exhibited. Combinations of key words can be used to search the short reads database. For example, the use of "RNA-Seq and zebrafish" would allow the identification of the 108 datasets in the SRA database, as of December 2014. Apparently, this database is the most dynamic and rapidly increasing database among all DNA and RNA databases, and, therefore, the number of datasets available is expected to grow rapidly. Table 12.1 summarizes the currently available RNA-Seq datasets from several major aquaculture species. It must be understood that this list is incomplete, and is used here only for illustration purposes.

Assembly of RNA-Seq Sequences

For the most part, the assembly of the RNA-Seq datasets for the analysis of lncRNAs is the same as that for the assembly of transcriptomes, as described in great detail in Chapter 8; therefore, it will not be repeated in this chapter. However, with lncRNAs, the focus is now not on the protein-encoding genes, but on non-coding RNAs. In addition, as reflected by its definition, lncRNAs include all transcripts that are non-coding with

Table 12.1 The availability of RNA-Seq datasets from fish and aquaculture species in the SRA database as of December 2014.

Species	Number of RNA-Seq datasets	Accession numbers
Zebrafish	108	Not listed
Fugu	6	SRP015822; SRP015823; SRP015849; SRP030658; SRP032428; SRP032431
Medaka	6	SRP017400; SRP021892; SRP029233; SRP032993; SRP041650; SRP043653
Catfish	11	SRP003906; SRP008839; SRP009069; SRP010406; SRP012586; SRP017689; SRP018265; SRP020252; SRP028159; SRP028517; SRP041359
Common carp	7	SRP010352; SRP010735; SRP011159; SRP011162; SRP011277; SRP023984; SRP026407
Grass carp	6	DRP002413; SRP033018; SRP033615; SRP040125; SRP040126; SRP049081
Rohu carp	1	SRP004499
Shrimp	7	DRP000406; SRP014749; SRP015348; SRP018120; SRP022057; SRP033328; SRP034965
Oysters	15	SRP007959; SRP013236; SRP014559; SRP017345; SRP019967; SRP019969; SRP021170; SRP028617; SRP029304; SRP029305; SRP029373; SRP032997; SRP039435; SRP042090; SRP042159
Rainbow trout	9	DRP000322; ERP000696; ERP003742; SRP005674; SRP009644; SRP022881; SRP028233; SRP032774; SRP033406
Atlantic salmon	2	SRP043420; SRP035898
Tilapia	6	SRP008027; SRP009911; SRP014017; SRP019938; SRP026706; SRP028106
Scallops	5	SRP018281; SRP018710; SRP019933; SRP028882; SRP037741
Sea bass	3	DRP000610; SRP028235; SRP033113
Striped bass	1	SRP039910
Seabream	1	SRP007197
Cod	4	SRP013269; SRP015940; SRP029145; SRP029148
Flounder	2	SRP043651; DRP001292
Sole	3	SRP019995; SRP022228; SRP031456
Turbot	1	SRP008289

a minimum length of 200 bases. As such, a number of specific bioinformatics analysis steps are required. In the following text, we will provide a brief description for the assembly of the transcriptome, but focus more on the specific processes for the analysis of lncRNAs.

With the RNA-Seq datasets, the first step is the assembly of the transcriptome into contigs by transcriptome assemblers, such as Trinity, Cufflink, and TransABySS (see Chapter 8).

Identification of lncRNAs

Length Trimming

As lncRNAs are defined as non-coding transcripts with a minimum length of 200 bases, the transcriptome contigs must be longer than 200 bp in order to be qualified for further analysis. Therefore, the input dataset must be trimmed by length before analysis for lncRNAs. This step can be readily accomplished with most transcriptome assemblers, such as Trinity, Cufflink, and TransABySS.

Coding Potential Analysis

To identify lncRNAs, the first step is to determine the coding potential of the transcriptome contigs. If the contig sequences have similarities to known protein-encoding genes, such sequences are excluded from lncRNAs. Here, the phrase is "coding potential", not "coding sequences." In other words, if the sequences have some hint of being similar to any coding sequences, they will not be considered as lncRNAs.

Many software pipelines are available for the analysis of coding potential. A few popular examples are Coding Potential Calculator (CPC) (http://cpc.cbi.pku.edu.cn/), RNAcode (http://wash.github.io/rnacode/), PhyloCSF (https://github.com/mlin/PhyloCSF/wiki), and CPAT (http://sourceforge.net/projects/rna-cpat/). The software packages for the analysis of coding potential can generally be grouped into two categories based on their principles: those based on sequence alignments, and those based on searches of linguistic features.

The first category of software pipelines relies on sequence alignments, either pairwise to search for evidence similar to existing proteins (e.g., CPC, Coding Potential Calculator) or multiple alignments to calculate phylogenetic conservation score (e.g., RNAcode and PhyloCSF). The use of alignment-based pipelines for the analysis of coding potential is straightforward. Their advantages are user-friendliness and high accuracy. The disadvantage of these pipelines is their dependence on existing information in the databases related to the datasets under study. If the sequences of your species are highly divergent from those in the databases, the coding potential would be significantly underestimated, leading to the overestimation of the lncRNAs. For instance, genomic information, particularly the transcriptome or protein-encoding genes of sea cucumber, have not been well characterized, and, therefore, many of its coding sequences may not generate any significant hits, simply because similar sequences are lacking in the GenBank or other databases.

The second category of software pipelines relies on using a logistic regression model based on linguistic features. One most popular packages of this category is CPAT, the Coding-Potential Assessment Tool (Wang *et al.*, 2013). The CPAT package relies on four pure sequence linguistic features: (1) open reading frame (ORF) size, (2) ORF coverage, (3) Fickett TESTCODE, and (4) Hexamer usage bias. The ORF size feature assumes that, if the ORF of a certain size can be identified from the transcript sequences, this may suggest its coding potential because true non-coding sequences should not carry long ORFs. ORF length is one of the most fundamental features used to distinguish ncRNA from messenger RNA because a long putative ORF is unlikely to be observed by random chance in non-coding sequences (Wang *et al.*, 2013). ORF coverage is the ratio of ORF size to the transcript size. Apparently, the larger the ORF coverage is, the more likely the transcript has a coding potential. Large ncRNAs such as lncRNAs usually have much lower ORF coverage than protein-coding RNAs (Wang *et al.*, 2013). Fickett test-code (Fickett, 1982) is a statistic that measures computational bias between coding and non-coding sequences based on a periodicity of three (every three nucleotides encode one amino acid). Similarly, hexamer usage bias (hexamer score) between coding and non-coding sequences was used to differentiate genes and non-coding DNA sequences (Fickett & Tung, 1992). This second category of pipelines using linguistic feature-based methods does not require genome or protein databases to perform alignments, and they are more robust. Because of being alignment-free, these pipelines run much faster and are also easier to use. Compared to alignment-based approaches, they perform better with improved sensitivity and specificity. For instance, CPAT was able to achieve 96.6% accuracy when used for the identification of human lncRNAs (Wang *et al.*, 2013).

Coding Potential Calculator (CPC)

Pre-requisite

1) NCBI BLAST package
2) A relatively comprehensive protein database, such as UniRef100, UniRef90, UniRef50, or the NCBI nr, is highly recommended. The UniRef databases contain reference protein sequences that hide the redundant sequences. With UniRef100, all the protein sequences are clustered with 100% identity. Because the identity requirement is 100% (identical), the total entries of clusters are the highest, at over 49 million at its current release. UniRef90 takes the UniRef clusters and reclusters at 90% sequence identity and 80% sequence overlap with the longest sequence, and also excludes sequences that are shorter than 11 amino acids. Similarly, UniRef50 reclusters the sequences within UniRef90 by using 50% sequence identity and 80% sequence length overlap with the longest sequence in the cluster. As a result, the current release of UniRef90 has over 28 million clusters, and the UniRef50 has over 12 million clusters (http://www.uniprot.org/uniref/).

Getting CPC Software

```
$ wget http://cpc.cbi.pku.edu.cn/download/cpc-0.9.tar.gz
```

Compiling and Installing

```
$ gzip -dc cpc-0.9.tar.gz | tar xf -
$ cd cpc-0.9
$ export CPC_HOME="$PWD"
$ cd lib/libsvm
$ gzip -dc libsvm-2.8.1.tar.gz | tar xf -
$ cd libsvm-2.8.1
$ make clean && make
$ cd ../..
$ gzip -dc estate.tar.gz | tar xf -
$ make clean && make
```

Database Construction Format your database using “formatdb” (see the “NCBI” section for detailed usage), name as "prot_db", and put it under the “cpc/data/” directory.

```
AF282387          528    coding    3.32462
Tsix_mus          4300   noncoding           -1.30047
Evf1_Rat          2704   noncoding           -0.991937
ENST00000361290   7834   coding    17.7115
```

Figure 12.1 An example of output from Coding Potential Calculator (CPC).

Run CPC

```
$ bin/run_predict.sh (input_seq) (result_in_table)
   (working_dir) (result_evidence)
```

Output An example of CPC output is provided in Figure 12.1, where the first column is the input sequence ID; the second column is the input sequence length; the third column is coding status; and the fourth column is the coding potential score.

The coding potential score can then be parsed and ranked. A cutoff coding potential score is determined. Generally, to be qualified for lncRNA, the coding potential of the sequence should be no more than 0.4 (Wang *et al.*, 2013).

RNAcode

Pre-requisite ClustalW to build your multiple sequence alignment files

Getting RNAcode Download from https://github.com/wash/rnacode.

Compiling and Installing

```
$ ./configure
$ make
$ make install (as root)
```

Input Alignment The input alignment needs to be formatted in the ClustalW or MAF formats (see details of usage in the "sequence alignment" section).

Note: RNAcode uses the first sequence as a reference sequence, that is, all results and reported coding regions apply to this reference sequence.

Running RNAcode

```
$ RNAcode [OPTIONS] alignment.aln
```

Here, "alignment.aln" is the alignment file, and "OPTIONS" is one of the following command-line options either given in one-letter form with a single dash or as long option with double dash:

```
--outfile -o
```

File to which the output is written (default: "stdout"). Defaults to standard output.

```
--cutoff -p
```

Show only regions that have a p-value below the given number (default: 1.0). By default, all hits are shown.

```
--num-samples -n
```

Numbers of random alignments (default: 100) that are sampled to calculate the p-value. RNAcode estimates the significance of a coding prediction by sampling a given number of random alignments. Default is "100", which gives reasonably stable p-values that are useful for assessing the relevance of a prediction.

`--stop-early -s`

Setting this option stops the sampling process as soon as it is clear that the best hit will not fall below the given p-value cutoff. For example, assume a p-value cutoff of 0.05 (see `--cutoff`), and a sample size of 1000 is given (see `--num-samples`). As soon as 50 random samples score better than the original alignment, the process is stopped, and all hits in the original alignment are reported as $p > 0.05$ (or by convention as "1.0" in gtf and tabular output).

`--best-region -r`

Show only best non-overlapping hits. By default, all positive scoring segments are shown in the output if they fall below the given p-value cutoff. If two hits overlap (different frame or different strand) and the `--best-region` option is given, only the hit with the highest score is shown. Strong coding regions often lead to statistically significant signals in other frames also. These hits are suppressed by this option, and only the correct reading frame is reported.

`--best-only -b`

Show only best hit. This option shows only the best hit for each alignment.

`--pars -c`

Scoring parameters as comma-separated strings.

```
--gtf -g
--tabular -t
```

Changes the default output to two different machine-readable formats.

Output In the default output, each prediction is reported on one line by 10 fields.

1) *HSS id*: Unique running number for each high-scoring segment predicted in one RNAcode call.
2) *Frame*: The reading frame phasing relative to the starting nucleotide position in the reference sequence. "+1" means that the first nucleotide in the reference sequence is in the same frame as the predicted coding region. Negative frames indicate that the predicted regions are on the reverse complement strand.
3) *Length*: The length of the predicted region in amino acids.
4) *From*: The position of the first/last amino acid in the translated protein.
5) *To*: Nucleotide sequence of the reference sequence starting with 1.
6) *Name*: The name of the reference sequence as given in the input alignment.
7) *Start*: The nucleotide start position in the reference sequence of the predicted coding region. If no genomic coordinates are given (if you provide a ClustalW alignment file as input), the first nucleotide position in the references sequence is set to "1"; otherwise, the positions are the 1-based genomic coordinates as given in the input MAF file.

8) *End*: The nucleotide end position in the reference sequence of the predicted coding region.
9) *Score*: The coding potential score. High scores indicate high coding potential.
10) *P*: The p-value associated with the score. This is the probability that a random alignment with same properties contains an equally good or better hit.

Note: If the `--tabular` option is given, the output is printed as a tab-delimited list without header or any other output. With the `--gtf` option, the output is formatted as a GTF genome annotation file.

PhyloCSF

1) PhyloCSF requires a MAF-format whole genome alignment file as the input, which needs both whole genome assembly sequences (.fa) and the corresponding annotation files (.gff or .gtf) to generate. Therefore, if the analyst does not have an annotated genome for his or her species, this part may be skipped.
2) PhyloCSF has been reported to lose accuracy when the lncRNAs are poorly conserved (Wang *et al.*, 2013). Analysts who have similar situation of the datasets need be cautious when using it.

Pre-requisite: OCaml Package Manager (OPAM)

```
$ wget https://github.com/ocaml/opam/releases/download
  /1.1.1/opam-full-1.1.1.tar.gz
$ ./configure && make && make Install
```

Getting PhyloCSF

```
$ git clone git://github.com/mlin/PhyloCSF.git
```

Compiling and Installing

```
$ cd PhyloCSF
$ make
```

Preparing Whole Genome Alignments See details in the section on comparative genomics.

Run PhyloCSF

```
$ ./PhyloCSF 29fish /path_to_your_target_seqence_alignment
   /seq.fa --frames=6
```

Here, “29fish” is the whole genome alignment file, and “seq.fa” is the file containing your target sequence and the corresponding alignments.

Output In the output of PhyloCSF, each line stands for the result of one sequence of your input. The first column is the name and path of the target sequence. The second column is a score, positive if the alignment is likely to represent a conserved coding region, and negative otherwise. This score has a precise theoretical interpretation: it quantifies how much more probable the alignment is under PhyloCSF’s model of protein-coding sequence evolution than under the non-coding model. As indicated in the output, this likelihood ratio is expressed in units of decibans. A score of 10 decibans means the coding model is 10:1 more likely than the non-coding model; 20 decibans, 100:1; 30 decibans, 1000:1; and so on. A score of −20 means the non-coding model is 100:1 more likely, and a score of 0 indicates the two models are equally likely. The third column stands for the start position of the possible orf in this sequence. The fouth column stands for the end position of the possible orf in this sequence. And the fifth column stands for the strands for this orf.

CPAT

Pre-requisite gcc; python2.7+; numpy; cython; R

Getting CPAT

```
$ wget http://cpat.googlecode.com/files/CPAT-1.2.1.tar.gz
```

Compiling and Installing

```
$ tar zxf CPAT-VERSION.tar.gz
$ cd CPAT-VERSION
$ python setup.py install #will install CPAT in system
   level. require root previledge
$ python setup.py install --root=/home/user/CPAT #will
   install CPAT at user specified location
$ export PYTHONPATH=/home/user/CPAT/usr/local/lib/python2.7
  /site-packages:$PYTHONPATH
$ export PATH=/home/user/CPAT/usr/local/bin:$PATH
   #setup PATH
```

Input File

a) *Sequence input file*: CPAT takes both BED and FASTA files as input. When using a BED file (standard 12-column format), the user also needs to provide the reference genome sequence (in FASTA format).
b) *Hexamer frequency table file*: CPAT needs a frame hexamer (6mer) frequency table file from mRNA sequences in at least one close species to calculate the hexamer usage score. It can be generated by "`make_hexamer_tab.py`" integrated in the CPAT package. Its usage is shown as follows:

   ```
   $ python ../bin/make_hexamer_tab.py
      -c CODING_sequence_FILE -n NONCODING_sequence_FILE
   ```

   Both coding sequences (must be CDS without UTR, i.e., from start codon to stop codon) and noncoding sequences must be in FASTA format. Analysts can get the CDS sequence of a bed file using the UCSC table browser.
c) *Logistic regression model*: A logistic regression model ("prefix.logit.RData") required by CPAT needs to be built before running CPAT. It can be generated by "make_logitModel.py". Its usage is shown as follows:

   ```
   $ python ../bin/make_logitModel.py
      -c CODING_sequence_FILE -n NONCODING_sequence_FILE
      -o OUT_FILE -x HEXAMER_table -s START_CODONS
      (default=ATG) -t STOP_CODONS (default=TAG,TAA,TGA)
   ```

ID	RNA_size	ORF_size	Fickett_score	Hexamer_score	coding_prob
TCONS_00000129	1944	480	0.6027	−0.0044	0.2385
TCONS_00001909	1106	510	0.5087	0.0174	0.2815
TCONS_00000779	575	171	0.6882	0.0566	0.0230
TCONS_00000796	578	114	0.4917	−0.1089	0.0023
TCONS_00000445	1389	387	0.7759	0.1170	0.3129
TCONS_00000446	630	180	0.5092	−0.1551	0.0036
TCONS_00002277	596	288	0.5682	−0.0496	0.0276
TCONS_00000073	436	195	0.7074	0.1108	0.0448
TCONS_00000799	558	381	0.5961	0.0822	0.1685
TCONS_00000135	268	90	0.5495	−0.1511	0.0016

Figure 12.2 An example of output from CPAT.

Run CPAT Run CPAT using BED file:

```
$ python ../bin/cpat.py -d prefix.logit.RData -r ref.fa
   -x prefix_Hexamer.tab -g sequence.bed
-o out
```

Run CPAT using FASTA file

```
python ../bin/cpat.py -d prefix.logit.RData
   -x prefix_Hexamer.tab -g seq.fa -o out
```

Output The output of CPAT is quite straightforward (Figure 12.2). The only parameter that needs to be treated carefully is the coding probability (CP). The threshold of CP depends on the species. Some optimal CP cutoff is recommended by the authors of CPAT:

Human: 0.364 ($CP \geq 0.364$ indicates coding sequence, $CP < 0.364$ indicates non-coding sequence).
Mouse: 0.44
Fly: 0.39
Teleost fish: 0.38

Warning Please do not claim that anything “is” or “is not” a non-coding RNA based solely on coding potential prediction pipelines. These pipelines only provide preliminary evidences based entirely on bioinformatics analysis either for or against the hypothesis that the input sequence represents a coding region. If strong statements need to be issued about whether coding or non-coding regions are found, additional analysis is necessary, such as homology searches of known protein sequence databases (e.g., BLASTX on nr or HMMER on Pfam), which will be described in the following text.

Homology Search

As mentioned earlier, even if the sequences obtained remarkably low coding probability (CP) scores, it is still necessary to further confirm them through homology searching. In general, all the transcripts with low CP will be employed as queries to search against known protein and domain databases. Among the different types of databases,

NCBI "non-redundant" database (nr) and Pfam domain database are usually utilized in this step.

Homology Protein Search

Transcripts with low CP will be used as queries against NCBI nr database through blastx in this step. The recommended cutoff E-value is 1E − 4. Transcripts that have hits to known proteins in nr will be removed from the lncRNA datasets.

Homology Domain Search

The Pfam database is a large collection of protein families (http://pfam.xfam.org/). Proteins generally have one or more functional domains. The homology to known protein domains can therefore provide insights into their functions.

After sequence similarity searches, the remaining transcripts need to be analyzed for the presence of known domains because the presence of protein domains is a sign of coding potential. The transcripts are used as queries against the Pfam domain database through HMMER in this step. The recommended E-value cutoff is also 1E − 4. Any transcripts with significant hits to known domains in nr will be removed from the candidate lncRNA dataset. The usage of HMMER is described in the following text.

Getting HMMER

```
$ wget ftp://selab.janelia.org/pub/software/hmmer3/3.1b1
  /hmmer-3.1b1-linux-intel-x86_64.tar.gz
```

Compiling and Installing

```
$ tar zxf hmmer-3.1b1.tar.gz
$ cd hmmer-3.1b1
$ ./configure
$ make
```

Preparing Pfam Databases for hmmscan

```
$ wget ftp://ftp.sanger.ac.uk/pub/databases/Pfam
  /current_release/Pfam-A.hmm.gz
$ wget ftp://ftp.sanger.ac.uk/pub/databases/Pfam
  /current_release/Pfam-B.hmm.gz
$ cp *.hmm.gz hmmer-3.1b1/db/
$ cd hmmer-3.1b1/db/
$ hmmpress Pfam-A.hmm
$ hmmpress Pfam-B.hmm
```

Running HMMER

```
$ hmmscan [options] <hmmdb> <seqfile>
```

For example,

```
$ hmmscan --tblout hitpfamA.out --qformat fasta -E 1e-4
   --noali Pfam-A.hmm seq.fa
$ hmmscan --tblout hitpfamB.out --qformat fasta -E 1e-4
   --noali Pfam-B.hmm seq.fa
```

Options for Controlling Output

"`-o <f>`": Direct the main human-readable output to a file "<f>" instead of the default "stdout".

"`--tblout <f>`": Save a simple tabular (space-delimited) file summarizing the per-target output, with one data line per homologous target model found.

"`--domtblout <f>`": Save a simple tabular (space-delimited) file summarizing the per-domain output, with one data line per homologous domain detected in a query sequence for each homologous model.

"`--acc`": Use accessions instead of names in the main output, where available, for profiles and/or sequences.

"`--noali`": Omit the alignment section from the main output. This can greatly reduce the output volume.

"`--notextw`": Unlimit the length of each line in the main output. The default is a limit of 120 characters per line, which helps in displaying the output cleanly on terminals and in editors, but can truncate target profile description lines.

"`--textw <n>`": Set the main output's line length limit to "<n>" characters per line. The default is 120.

Options for Reporting Thresholds Reporting thresholds control the hits that are reported in the output files (the main output, `--tblout`, and `--domtblout`).rvi

"`-E <x>`": Report the target profiles with an E-value of ≤ < x > in the per-target output. The default is 10.0, meaning that, on average, about 10 false positives will be reported per query; the analyst can determine the top of the noise and decide if it is really noise.

"`-T <x>`": Report target profiles with a bit score ≥ <x>, instead of thresholding the per-profile output on the E-value.

"`--domE <x>`": Report individual domains with a conditional E-value ≤ <x > in the per-domain output, for target profiles that have already satisfied the per-profile reporting threshold. The default is 10.0. A conditional E-value stands for the expected number of additional false positive domains in the smaller search space of those comparisons that have already satisfied the per-profile reporting threshold (and thus must have at least one homologous domain already).

"`--domT <x>`": Report domains with a bit score ≥ <x>, instead of thresholding the per-domain output on the E-value.

Other Options

"`--nonull2`": Turn off the null2 score corrections for biased composition.

"`-Z <x>`": Assert that the total number of targets in your searches is "<x>", for the purposes of per-sequence E-value calculations, rather than the actual number of targets seen.

"--domZ < x>": Assert that the total number of targets in your searches is < x>, for the purposes of per-domain conditional E-value calculations, rather than the number of targets that passed the reporting thresholds.

"--seed < n>": Set the random number seed to "<n>". Some steps in post-processing require the Monte Carlo simulation. The default is to use a fixed seed (42), so that results are exactly reproducible. Any other positive integer will give different (but also reproducible) results. A choice of "0" ensures the use of an arbitrarily chosen seed.

"--qformat <s>": Assert that the query sequence file is in the format "<s>". Accepted formats include FASTA, embl, genbank, ddbj, uniprot, stockholm, Pfam, a2m, and afa.
"--cpu <n>": Set the number of parallel worker threads to "<n>".
"--mpi": Run in MPI master/worker mode, using mpirun (only available if optional MPI support was enabled at compile-time).

Output *(`--tblout`)* An example output of HMMER is provided in Figure 12.3. Detailed information for some of the columns are described as in the following text:

1) *target name*: The name of the target sequence or profile
2) *accession*: The accession of the target sequence or profile, or "-" if none
3) *query name*: The name of the query sequence or profile
4) *accession*: The accession of the query sequence or profile, or "-" if none
5) *E-value (full sequence)*: The expectation value (statistical significance) of the target. This is a per-query E-value; that is, calculated as the expected number of false positives achieving this comparison's score for a single query against the Z sequences in the target dataset. If you search with multiple queries and want to control the overall false positive rate of that search rather than the false positive rate per query, you will want to multiply this per-query E-value by the number of queries in the search.
6) *score (full sequence)*: The score (in bits) for this target/query comparison. It includes the biased composition correction (the "null2" model).
7) *Bias (full sequence)*: The biased-composition correction—the bit score difference contributed by the null2 model. High bias scores may be a red flag for a false positive, especially when the bias score is as large as, or larger than, the overall bit score. It is difficult to correct for all possible ways in which nonrandom but nonhomologous biological sequences can appear to be similar, such as short-period tandem repeats, so there are cases where the bias correction is not strong enough (creating false positives).
8) *E-value (best 1 domain)*: The E-value if only the single best-scoring domain envelope is found in the sequence, and none of the others. If this E-value is not good, but the full-sequence E-value is good, this is a potential red flag. Weak hits, none of which are good enough on their own, are summing up to lift the sequence up to a high score. Whether this is good or bad is not clear; the sequence may contain several weak homologous domains, or it might contain a repetitive sequence that is hit by chance (i.e., once one repeat hits, all the repeats hit).
9) *score (best 1 domain)*: The bit score if only the single best-scoring domain envelope is found in the sequence, and none of the others (inclusive of the null2 bias correction).
10) *bias (best 1 domain)*: The null2 bias correction that was applied to the bit score of the single best-scoring domain.
11) *exp*: Expected number of domains, as calculated by posterior decoding on the mean number of begin states used in the alignment ensemble.

ORF Length Trimming (Optional)

If the analysts are handling the data from a non-model species without a decent annotated genome, this step is recommended, because novel proteins may exist in

```
# target name        accession  query name           accession    E-value  score  bias    E-value  score  bias   exp reg clu  ov env dom rep inc description of target
#------------------- ---------- -------------------- ---------- --------- ------ ----- --------- ------ -----   --- --- --- --- --- --- --- --- ---------------------
Trypan_PARP          PF05887.6  g10141.t1            -            5.5e-12   45.5  35.9   1.6e-09   37.5  13.3   2.1   2   0   0   2   2   2   2 Procyclic acidic repetitive protein (PARP)
Collagen             PF01391.13 g10144.t1            -            1.6e-16   59.4  66.3   5.8e-09   35.2  23.3   3.2   3   0   0   3   3   3   3 Collagen triple helix repeat (20 copies)
DUF2051              PF09738.4  g10153.t1            -            1.9e-08   33.8   2.8     2e-08   33.7   1.9   1.1   1   0   0   1   1   1   1 Double stranded RNA binding protein (DUF2051)
Pentapeptide_4       PF13599.1  g1017.t1             -            2.4e-19   68.8   0.0   5.1e-08   32.5   0.0   3.1   1   1   2   3   3   3   3 Pentapeptide repeats (9 copies)
Pentapeptide_3       PF13576.1  g1017.t2             -            1.2e-06   28.2   2.3      0.45   10.3   0.1   5.0   1   1   1   2   2   2   2 Pentapeptide repeats (9 copies)
gag-asp_proteas      PF13975.1  g10558.t1            -            2.5e-07   30.3   0.0   5.8e-07   29.2   0.0   1.5   1   0   0   1   1   1   1 gag-polyprotein putative aspartyl protease
Ten_N                PF06484.7  g10634.t1            -            2.6e-12   46.4   4.2   2.7e-12   46.3   2.9   1.0   1   0   0   1   1   1   1 Teneurin Intracellular Region
```

Figure 12.3 An example of output from HMMER.

your species. To avoid the inclusion of novel genes in lncRNAs, the ORF region of the remaining transcripts will need to be predicted, and a maximal ORF cutoff less than 100 aa will be imposed to get candidate lncRNA datasets. The prediction of ORF can be finished by NCBI Orf finder (http://www.ncbi.nlm.nih.gov/gorf/gorf.html) or EMBOSS getorf (http://emboss.bioinformatics.nl/cgi-bin/emboss/getorf).

UTR Region Trimming

The final step of lncRNA identification is to eliminate the transcripts representing only the UTR regions of coding RNAs due to the assembly of contigs located in the UTR regions. Analysts have two options based on their species. For species with a transcriptome database containing full-length cDNAs, these "lncRNAs" located within the 5'- and 3'-UTR regions can be easily removed through Blastn searches. For species without a transcriptome database of full-length cDNAs, the UTRdb database (http://utrdb.ba.itb.cnr.it/home/download) can be searched. However, as the sequences in the UTR regions may or may not be conserved, the effectiveness depends on the database contents per se. Currently, the UTRdb database contains UTR sequences from 79 species (http://utrdb.ba.itb.cnr.it/).

Analysis of lncRNA Expression

The expression profiling of lncRNAs is similar to expression analysis in RNA-Seq. The expression level of lncRNAs can be readily achieved using the Trinity RESM module or CLC Genomics Workbench (both described in Chapter 9 under the RNA-Seq section). The RPKM/FPKM reads counts and relative expression fold changes may both be needed for the downstream analysis. Several parameters are important considerations for the selection of differentially (induced) expressed lncRNA: (1) false discovery rate (e.g., FDR-adjusted p-value <0.05); (2) minimal mapped reads (e.g., >5), and (3) fold change (e.g., weighted proportions of fold change $\geq |2|$).

Analysis and Prediction of lncRNA Functions

With the mapping of thousands of lncRNA loci, the biggest challenge is to determine the functions of lncRNAs. As with the functions of genes, but perhaps more difficult, a first step in hypothesis generation is to use the expression patterns of lncRNAs to speculate their functions. For instance, spatial and temporal expression patterns, induced expression under specific conditions, and positively or negatively correlated expression with a set of genes, etc., can be used as clues to predict their functions. For instance, lncRNAs may be expressed within a specific cell type or biological process associated with certain functions. For example, in situ hybridization analysis allowed revealing expression patterns of certain types of lncRNAs in the mouse brain (Mercer *et al.*, 2008). A similar study by the same group identified numerous lncRNAs in the mouse brain that were tightly correlated with pluripotency transcription factors, suggesting that many lncRNAs may function in stem cell pluripotency transcriptional networks (Dinger *et al.*, 2008).

More recently, a bioinformatics method termed "Guilt by Association" allowed a global understanding of lncRNAs and protein-coding genes that are tightly co-expressed and thus presumably co-regulated (Guttman *et al.*, 2009). This method identifies protein-coding genes and pathways significantly correlated with a given lncRNA using gene-expression analyses. Thus, based on known functions of the co-expressed protein-coding genes, hypotheses are generated for the functions and potential regulators of the candidate lncRNA. Moreover, this analysis revealed "families" of lncRNAs based on the pathways with which they do and do not associate. This approach has allowed the prediction of diverse roles for lncRNAs, ranging from stem cell pluripotency to cancer (Guttman *et al.*, 2009). For example, numerous lncRNAs that were tightly correlated with p53 were induced in a p53-dependent manner—many more than would be expected by chance (Guttman *et al.*, 2009; Huarte *et al.*, 2010; Khalil *et al.*, 2009). These lncRNAs also were enriched for the p53-binding motif in their promoters. Moreover, one of these lncRNAs, termed lincRNA-p21, predicted to be associated with the p53, was found to be directly regulated by p53 and subsequently forming lncRNA-RNP with a nuclear factor to serve as a global transcriptional repressor facilitating p53- mediated apoptosis (Huarte *et al.*, 2010). Similarly, several lncRNAs predicted to be associated with adipogenesis and pluripotency have recently been identified to be required for maintaining these cellular states (Loewer *et al.*, 2010).

Other expression correlation analyses have revealed additional functional roles of lncRNAs. For example, a recent study profiled lncRNAs across over 130 breast cancers comprised of varying grades of tumor and clinical information (Gupta *et al.*, 2010). This study identified numerous lncRNAs that are specifically up- or down-regulated in tumor subtypes. For example, it was identified that an lncRNA named HOTAIR encoded in the HOXC cluster was a strong predictor of breast cancer metastasis. In fact, enforced expression of HOTAIR was sufficient to drive breast cancer metastasis. More global expression studies of lncRNAs overlapping promoter regions of protein-coding genes identified numerous lncRNAs associated with cell-cycle regulation (Hung *et al.*, 2011). This lead to the functional characterization of an lncRNA named PANDA that plays a critical role inhibiting p53-mediated apoptosis. The Guilt by Association method is universally applicable to any biological system. For example, a family of telomere-encoded lncRNAs in the malaria parasite (*P. falciprum*) was identified by their stage-specific co-expression with PfsiP2 as a key virulence transcription factor (Broadbent *et al.*, 2011).

These and other correlation studies have allowed the identification of specific roles of lncRNAs in global transcriptional regulation. Such analyses allowed the generation of hypotheses that can be experimentally tested; yet, the full scope of lncRNA transcriptional regulation and function is far from being understood. To understand the more global regulatory roles of lncRNAs, comprehensive knock down/out experiments need to be performed. It is possible that many lncRNAs may have multiple roles or may have no role in biological regulation.

Future Perspectives

Although more lncRNAs will undoubtedly be found, the identification of lncRNAs in aquaculture species, with initial characterization of their evolution, genomics, and expression, is still necessary at present. It will provide a starting point for the study of lncRNA biology in the non-model species along the evolutionary spectrum. The expression or sequence features of lncRNAs associated with functional genes may make them valuable resources and biomarkers for exploring the regulation of coding–non-coding gene networks on disease resistance, stress response, and various other phenotypes, even sex determination, in aquaculture species.

References

Balakirev, E.S. and Ayala, F.J. (2003) Pseudogenes: are they "junk" or functional DNA? *Annual Review of Genetics*, **37** (1), 123–151.

Bartel, D.P. (2009) MicroRNAs: target recognition and regulatory functions. *Cell*, **136**, 215–233.

Beltran, M., Puig, I., Peña, C. *et al.* (2008) A natural antisense transcript regulates Zeb2/Sip1 gene expression during Snail1-induced epithelial–mesenchymal transition. *Genes & Development*, **22**, 756–769.

Borsani, G., Tonlorenzi, R., Simmler, M.C. *et al.* (1991) Characterization of a murine gene expressed from the inactive X chromosome. *Nature*, **351**, 325–329.

Broadbent, K.M., Park, D., Wolf, A.R. *et al.* (2011) A global transcriptional analysis of Plasmodium falciparum malaria reveals a novel family of telomere-associated lncRNAs. *Genome Biology*, **12**, R56.

Brockdorff, N., Ashworth, A., Kay, G.F. *et al.* (1992) The product of the mouse *XIST* gene is a 15 kb inactive X-specific transcript containing no conserved ORF and located in the nucleus. *Cell*, **71**, 515–526.

Brown, C.J., Hendrich, B.D., Rupert, J.L. *et al.* (1992) The human *XIST* gene: Analysis of a 17 kb inactive X-specific RNA that contains conserved repeats and is highly localized within the nucleus. *Cell*, **71**, 527–542.

Cabili, M.N., Trapnell, C., Goff, L. *et al.* (2011) Integrative annotation of human large intergenic noncoding RNAs reveals global properties and specific subclasses. *Genes & Development*, **25** (18), 1915–1927.

Carninci, P., Kasukawa, T., Katayama, S. *et al.* (2005) The transcriptional landscape of the mammalian genome. *Science*, **309**, 1559–1563.

Core, L.J., Waterfall, J.J. and Lis, J.T. (2008) Nascent RNA sequencing reveals widespread pausing and divergent initiation at human promoters. *Science*, **322** (5909), 1845–1848.

Dinger, M.E., Amaral, P.P., Mercer, T.R. *et al.* (2008) Long noncoding RNAs in mouse embryonic stem cell pluripotency and differentiation. *Genome Research*, **18**, 1433–1445.

Ebisuya, M., Yamamoto, T., Nakajima, M. and Nishida, E. (2008) Ripples from neighbouring transcription. *Nature Cell Biology*, **10** (9), 1106–1113.

Erdmann, V., Barciszewska, M., Hochberg, A. *et al.* (2001) Regulatory RNAs. *Cellular and Molecular Life Sciences (CMLS)*, **58**, 960–977.

Faghihi, M.A. and Wahlestedt, C. (2009) Regulatory roles of natural antisense transcripts. *Nature Reviews Molecular Cell Biology*, **10** (9), 637–643.

Fejes-Toth, K., Sotirova, V., Sachidanandam, R. *et al.* (2009) Post-transcriptional processing generates a diversity of 5'-modified long and short RNAs. *Nature*, **457**, 1028–1032.
Fickett, J.W. (1982) Recognition of protein coding regions in DNA sequences. *Nucleic Acids Research*, **10** (17), 5303–5318.
Fickett, J.W. and Tung, C.-S. (1992) Assessment of protein coding measures. *Nucleic Acids Research*, **20** (24), 6441–6450.
Gerstein, M.B., Lu, Z.J., Van Nostrand, E.L. *et al.* (2010) Integrative analysis of the *Caenorhabditis elegans* genome by the modENCODE project. *Science*, **330**, 1775–1787.
Grabherr, M.G., Haas, B.J., Yassour, M. *et al.* (2011) Full-length transcriptome assembly from RNA-Seq data without a reference genome. *Nature Biotechnology*, **29**, 644–652.
Gupta, R.A., Shah, N., Wang, K.C. *et al.* (2010) Long non-coding RNA HOTAIR reprograms chromatin state to promote cancer metastasis. *Nature*, **464**, 1071–1076.
Guttman, M. and Rinn, J.L. (2012) Modular regulatory principles of large non-coding RNAs. *Nature*, **482** (7385), 339–346.
Guttman, M., Amit, I., Garber, M. *et al.* (2009) Chromatin signature reveals over a thousand highly conserved large non-coding RNAs in mammals. *Nature*, **458**, 223–227.
He, S., Liu, C., Skogerbø, G. *et al.* (2008) NONCODE v2. 0: decoding the non-coding. *Nucleic Acids Research*, **36**, D170–D172.
Huarte, M. and Rinn, J.L. (2010) Large non-coding RNAs: missing links in cancer? *Human Molecular Genetics*, **19**, R152–R161.
Huarte, M., Guttman, M., Feldser, D. *et al.* (2010) A large intergenic noncoding RNA induced by p53 mediates global gene repression in the p53 response. *Cell*, **142**, 409–419.
Hung, T., Wang, Y., Lin, M.F. *et al.* (2011) Extensive and coordinated transcription of noncoding RNAs within cell-cycle promoters. *Nature Genetics*, **43**, 621–629.
Ingolia, N.T., Lareau, L.F. and Weissman, J.S. (2011) Ribosome profiling of mouse embryonic stem cells reveals the complexity and dynamics of mammalian proteomes. *Cell*, **147**, 789–802.
Kapranov, P., Cheng, J., Dike, S. *et al.* (2007) RNA maps reveal new RNA classes and a possible function for pervasive transcription. *Science*, **316**, 1484–1488.
Katayama, S., Tomaru, Y., Kasukawa, T. *et al.* (2005) Antisense transcription in the mammalian transcriptome. *Science*, **309** (5740), 1564–1566.
Khalil, A.M., Guttman, M., Huarte, M. *et al.* (2009) Many human large intergenic noncoding RNAs associate with chromatin-modifying complexes and affect gene expression. *Proceedings of the National Academy of Sciences*, **106**, 11667–11672.
Kim, T.-K., Hemberg, M., Gray, J.M. *et al.* (2010) Widespread transcription at neuronal activity-regulated enhancers. *Nature*, **465**, 182–187.
Kung, J.T., Colognori, D. and Lee, J.T. (2013) Long noncoding RNAs: past, present, and future. *Genetics*, **193** (3), 651–669.
Kung, J.T. and Lee, J.T. (2013) RNA in the Loop. *Developmental Cell*, **24** (6), 565–567.
Li, T., Wang, S., Wu, R. *et al.* (2012) Identification of long non-protein coding RNAs in chicken skeletal muscle using next generation sequencing. *Genomics*, **99** (5), 292–298.
Loewer, S., Cabili, M.N., Guttman, M. *et al.* (2010) Large intergenic non-coding RNA-RoR modulates reprogramming of human induced pluripotent stem cells. *Nature Genetics*, **42**, 1113–1117.
Martens, J.A., Laprade, L. and Winston, F. (2004) Intergenic transcription is required to repress the *Saccharomyces cerevisiae SER3* gene. *Nature*, **429** (6991), 571–574.

Mercer, T.R., Dinger, M.E. and Mattick, J.S. (2009) Long non-coding RNAs: insights into functions. *Nature Reviews Genetics*, **10**, 155–159.

Mercer, T.R., Dinger, M.E., Sunkin, S.M. *et al.* (2008) Specific expression of long noncoding RNAs in the mouse brain. *Proceedings of the National Academy of Sciences*, **105**, 716–721.

Mu, C., Wang, R., Li, T. *et al.* (2016) Long non-coding RNAs (lncRNAs) of sea cucumber: large-scale prediction, expression profiling, non-coding network construction, and lncRNA-microRNA-gene interaction analysis of lncRNAs in *Apostichopus japonicus* and *Holothuria glaberrima* during LPS challenge and radial organ complex regeneration. *Marine Biotechnology*, **18** (4), 485–499.

Nagano, T., Mitchell, J.A., Sanz, L.A. *et al.* (2008) The air noncoding RNA epigenetically silences transcription by targeting G9a to chromatin. *Science*, **322**, 1717–1720.

Nam, J.-W. and Bartel, D.P. (2012) Long noncoding RNAs in *C. elegans*. *Genome Research*, **22** (12), 2529–2540.

Numata, K., Kanai, A., Saito, R. *et al.* (2003) Identification of putative noncoding RNAs among the RIKEN mouse full-length cDNA collection. *Genome Research*, **13**, 1301–1306.

Okazaki, Y., Furuno, M., Kasukawa, T. *et al.* (2002) Analysis of the mouse transcriptome based on functional annotation of 60,770 full-length cDNAs. *Nature*, **420**, 563–573.

Ørom, U.A., Derrien, T., Beringer, M. *et al.* (2010) Long noncoding RNAs with enhancer-like function in human cells. *Cell*, **143**, 46–58.

Pauli, A., Valen, E., Lin, M.F. *et al.* (2012) Systematic identification of long noncoding RNAs expressed during zebrafish embryogenesis. *Genome Research*, **22**, 577–591.

Petruk, S., Sedkov, Y., Riley, K.M. *et al.* (2006) Transcription of bxd noncoding RNAs promoted by trithorax represses Ubx in cis by transcriptional interference. *Cell*, **127** (6), 1209–1221.

Pink, R.C., Wicks, K., Caley, D.P. *et al.* (2011) Pseudogenes: pseudo-functional or key regulators in health and disease? *RNA*, **17** (5), 792–798.

Poliseno, L., Salmena, L., Zhang, J. *et al.* (2010) A coding-independent function of gene and pseudogene mRNAs regulates tumour biology. *Nature*, **465**, 1033–1038.

Ponjavic, J., Oliver, P.L., Lunter, G. and Ponting, C.P. (2009) Genomic and transcriptional co-localization of protein-coding and long non-coding RNA pairs in the developing brain. *PLoS Genetics*, **5** (8), e1000617.

Preker, P., Nielsen, J., Kammler, S. *et al.* (2008) RNA exosome depletion reveals transcription upstream of active human promoters. *Science*, **322** (5909), 1851–1854.

Rinn, J.L. and Chang, H.Y. (2012) Genome regulation by long noncoding RNAs. *Annual Review of Biochemistry*, **81**, 145–166.

Rinn, J.L., Kertesz, M., Wang, J.K. *et al.* (2007) Functional demarcation of active and silent chromatin domains in human *HOX* loci by noncoding RNAs. *Cell*, **129**, 1311–1323.

Seila, A.C., Calabrese, J.M., Levine, S.S. *et al.* (2008) Divergent transcription from active promoters. *Science*, **322** (5909), 1849–1851.

Tian, D., Sun, S. and Lee, J.T. (2010) The long noncoding RNA, *Jpx*, is a molecular switch for X chromosome inactivation. *Cell*, **143**, 390–403.

Tripathi, V., Ellis, J.D., Shen, Z. *et al.* (2010) The nuclear-retained noncoding RNA MALAT1 regulates alternative splicing by modulating SR splicing factor phosphorylation. *Molecular Cell*, **39**, 925–938.

Tsai, M.-C., Manor, O., Wan, Y. *et al.* (2010) Long noncoding RNA as modular scaffold of histone modification complexes. *Science*, **329** (5992), 689–693.

Ulitsky, I., Shkumatava, A., Jan, C.H. *et al.* (2011) Conserved function of lincRNAs in vertebrate embryonic development despite rapid sequence evolution. *Cell*, **147**, 1537–1550.

Wang, J., Fu, L., Koganti, P.P. *et al.* (2016) Identification and functional prediction of large intergenic noncoding RNAs (lincRNAs) in rainbow trout (*Oncorhynchus mykiss*). *Marine Biotechnology*, **18** (2), 271–282.

Wang, K.C. and Chang, H.Y. (2011) Molecular mechanisms of long noncoding RNAs. *Molecular Cell*, **43**, 904–914.

Wang, L., Park, H.J., Dasari, S., Wang, S., Kocher, J.-P. and Li, W. (2013). CPAT: Coding-Potential Assessment Tool using an alignment-free logistic regression model. *Nucleic Acids Research*, **41** (6), e74, 1–7.

Wapinski, O. and Chang, H.Y. (2011) Long noncoding RNAs and human disease. *Trends in Cell Biology*, **21** (6), 354–361.

Wilusz, J.E. and Sharp, P.A. (2013) A circuitous route to noncoding RNA. *Science*, **340** (6131), 440–441.

Wilusz, J.E., Sunwoo, H. and Spector, D.L. (2009) Long noncoding RNAs: functional surprises from the RNA world. *Genes & Development*, **23**, 1494–1504.

Yang, L., Froberg, J.E. and Lee, J.T. (2014) Long noncoding RNAs: fresh perspectives into the RNA world. *Trends in Biochemical Sciences*, **39**, 35–43.

13

Analysis of MicroRNAs and Their Target Genes

Shikai Liu and Zhanjiang Liu

Introduction

Small non-coding RNAs are being rapidly recognized as significant effectors of gene regulation in various organisms. Several distinct classes of small non-coding RNAs, including microRNAs (miRNAs), short interfering RNAs (siRNAs), Piwi-interacting RNAs (piRNAs), and repeat associated siRNAs (rasiRNAs), have been identified. These molecules are typically ~18–40 nucleotides in length, and play profound roles in many cellular processes. miRNAs, with a length of ~22 nucleotides, play critical roles in post-transcriptional regulation of gene expression. With the advances in next-generation sequencing technologies, miRNAs have been extensively studied in numerous animal and plant species. Along with the explosive accumulation of miRNA data, programs and web servers have been developed for miRNA data analysis. In this chapter, we provide an overview of tools for analyzing miRNA and its target genes, with emphasis on analyzing data generated from next-generation sequencing platforms.

A central dogma of biology holds that genetic information is transferred from DNA to RNA and to protein. Therefore, it has been generally recognized that genes encoding proteins are responsible for not only most structural and catalytic functions but also regulatory functions in cells (Maniatis & Tasic, 2002). This is essentially true in prokaryotes whose genomes are almost entirely composed of protein-coding genes. However, in complex organisms, protein-coding genes occupy only a small fraction of the genome, whereas non-coding DNAs constitute the vast majority (97–98%) of the genome (Mattick, 2001, 2003).

Recent studies have shown that the vast majority of the genome is transcribed including non-protein-coding portions of the genome (Clark *et al.*, 2011). The dominance of non-protein-coding RNAs (ncRNAs) in the genomic output suggests that these ncRNAs might not be simply transcription byproducts; rather, they may constitute an unrecognized, but extensive regulatory network. It is now widely recognized that various ncRNAs are functional and play critical roles in a wide range of biological processes (Morris & Mattick, 2014; Pauli, Rinn & Schier, 2011). The pervasive transcription of the genomes in higher organisms suggests the presence of a second tier of genetic output and a network of parallel RNA-mediated interactions, which may enable the coordination of sophisticated suites of gene expression required for differentiation and

Bioinformatics in Aquaculture: Principles and Methods, First Edition. Edited by Zhanjiang (John) Liu.

development (Morris & Mattick, 2014; Pauli *et al.*, 2011). The expansion of ncRNAs in higher organisms also suggests that the evolution of complexity may not have been simply dependent on an expanded repertoire of proteins and protein isoforms, but also on a much larger set of genomic units with regulatory roles (Mattick, 2005).

Non-coding RNAs consist of long non-coding RNAs (lncRNAs) with lengths greater than 200 bp, and small non-coding RNAs. As lncRNAs are separately discussed in Chapter 12, we will not repeat it here. Several distinct classes of small non-coding RNAs, including miRNA, siRNA, piRNA, and rasiRNA, have been identified and characterized. These molecules are typically ~18–40 nucleotides in length, and have profound effects on cellular processes. Small ncRNAs are rapidly being recognized as significant effectors of gene regulation in various organisms. Small RNAs function by binding to their targets and negatively affecting gene expression via diverse mechanisms including translational repression, mRNA decay, heterochromatic modification, and even nascent peptide turnover (Filipowicz, Bhattacharyya & Sonenberg, 2008; He & Hannon, 2004; Lippman & Martienssen, 2004; Petersen *et al.*, 2006). Small ncRNAs have been reported to be involved in many cellular processes such as developmental timing, cell fate, tumor progression, neurogenesis, transposon silencing, and viral defense (Aalto & Pasquinelli, 2012; Caron, Lafontaine & Massé, 2010; Filipowicz *et al.*, 2008; Grosshans & Filipowicz, 2008; Houwing *et al.*, 2007; Kim, Han & Siomi, 2009; Lindsay, 2008; Xiao & Rajewsky, 2009).

miRNAs are a class of small ncRNAs with lengths of ~22 nucleotides, which play critical roles in post-transcriptional regulation of gene expression in animals and plants (Ambros, 2004; Carrington & Ambros, 2003; He & Hannon, 2004; Voinnet, 2009). miRNAs regulate gene expression in a variety of ways, including translation repression, mRNA cleavage, and deadenylation (Bartel, 2004; Chekulaeva & Filipowicz, 2009; Filipowicz *et al.*, 2008).

The first miRNA was identified by traditional cloning and Sanger sequencing in early 1990s (Lee, Feinbaum & Ambros, 1993). Discovery and analysis of miRNAs have been limited by laborious sample preparation and targeted detection technology. Due to these limitations, miRNA studies have been restricted to humans and several other model organisms. In recent years, advances in high-throughput sequencing technologies have allowed comprehensive identification and characterization of miRNAs in any organism. However, the abundance of miRNAs is often as low as degraded products of annotated or un-annotated transcripts, which adds complexity in classification. Therefore, advanced post-filtering steps are required for high-throughput data mining for miRNA identification (Friedländer *et al.*, 2012).

Great efforts have been made to predict and identify novel miRNAs within various eukaryotic genomes. A number of tools have been developed to predict miRNAs in genomes, either using a comparative phylogenetic approach (Altuvia *et al.*, 2005; Hertel & Stadler, 2006; Lim *et al.*, 2003) or a non-comparative, *ab initio* prediction approach (Batuwita & Palade, 2009; Sewer *et al.*, 2005). With the emergence of high-throughput sequencing techniques, a great resource of small RNA species has been characterized, which greatly enhances the power of prediction algorithms to efficiently identify novel miRNAs (Friedländer *et al.*, 2012; Hackenberg, Rodríguez-Ezpeleta & Aransay, 2011; Hansen *et al.*, 2014; Huang *et al.*, 2010). Currently, miRDeep2 is among the most reliable prediction tools available (Friedländer *et al.*, 2012; Hansen *et al.*, 2014; Williamson *et al.*, 2012).

In this chapter, we provide an overview of the tools used for miRNA data analysis, including identification, expression profiling, prediction of target genes, and co-expression networks with functional protein genes. Moreover, we demonstrate the analysis with miRDeep2, one of the most popularly used programs for the miRNA data analysis of small RNA deep sequencing data generated from next-generation sequencing platforms.

miRNA Biogenesis and Function

The canonical biogenesis of miRNAs is involved in a complex pathway that starts in the nucleus with an array of proteins, including Drosha and DGCR8, which are able to recognize and excise the small RNA loops formed during transcription (Bartel, 2004; Han *et al.*, 2004, 2006; Lee *et al.*, 2004). These excised loops (i.e., pre-miRNAs) are exported to the cytoplasm, mediated by Exportin-5. In the cytoplasm, the pre-miRNAs are recognized by a type III ribonuclease, Dicer, which cuts the loop to generate a small dsRNA fragment of 19–23 bp (Bartel, 2004). The small dsRNA remains bound to the Dicer enzyme to form a complex, which recruits several proteins of the argonaute family to form the RNA-induced silencing complex (RISC). RISC selects one of the chains of the dsRNA to generate a mature miRNA molecule. An alternative pathway for the generation of miRNAs from intronic RNA transcripts was also reported (Okamura *et al.*, 2007; Ruby, Jan & Bartel, 2007). In this pathway, the pre-miRNA is generated from small introns that are excised exclusively by the splicing machinery, whereas nuclear processing of RNA hairpins by Drosha/DGCR8 is skipped. These miRNAs are called *mirtrons* (Okamura *et al.*, 2007; Ruby *et al.*, 2007). Moreover, a Dicer-independent miRNA biogenesis was also reported (Cifuentes *et al.*, 2010; Yang *et al.*, 2010), which relies on the Argonaute2 slicer catalytic activity.

miRNAs function by targeting the mature RISC complexes to complementary sequences in mRNA transcripts to interfere with the translational processes and, consequently, reduce the protein production from mRNA transcripts (Bartel, 2004; Krützfeldt, Poy & Stoffel, 2006). The RISC complex can also induce mRNA-targeted degradation with catalysis of the Argonaute2 protein (Ambros, 2004; Bartel, 2004; Jones-Rhoades, Bartel & Bartel, 2006). One miRNA can be complementary to one or more target mRNAs. Animal miRNA is usually complementary to the 3′-UTRs, while plant miRNA is usually complementary to coding regions. In animal cells, the miRNA is more often partly complementary to its targets, whereas plant miRNA is highly complementary to target mRNAs (Jones-Rhoades *et al.*, 2006).

Tools for miRNA Data Analysis

Along with the explosive increase in the number of miRNA studies and the rapid accumulation of sequencing data, large numbers of packages, web servers, and databases have been developed for miRNA data analysis. The algorithms and programs are evolving in accordance with the generation of sequencing data of various scales. Some useful websites are available to provide resources related to the analysis of miRNAs (such as https://sites.google.com/site/mirnatools and http://mirnablog.com). Here,

Table 13.1 A list of tools for *in-silico* miRNA prediction.

Tool	Description	Application	URL
MiRscan	A classic miRNA prediction server	Web server	http://genes.mit.edu/mirscan/
microPred	A tool to classify pre-miRNAs from pseudo hairpins and other ncRNAs	Package	http://www.cs.ox.ac.uk/people/manohara.rukshan.batuwita/microPred.htm
miRAbela	A tool to predict if a sequence contains miRNA-like structures	Web server	http://www.mirz.unibas.ch/cgi/pred_miRNA_genes.cgi
miRNAFold	*Ab initio* miRNA prediction in genomes	Web server	http://evryrna.ibisc.univ-evry.fr/miRNAFold/
MirEval	A tool for evaluating sequences for possible miRNA-like properties	Web server	http://mimirna.centenary.org.au/mireval/
CID-miRNA	Identification of miRNAs from a single sequence or complete genome	Web server and package	https://github.com/alito/CID-miRNA

we provide an overview of the currently available packages that are developed for miRNA identification, expression analysis, target gene prediction, and miRNA–mRNA integrated analysis. We do not intend to catalog all the packages, and only use some popular packages as examples. Readers interested in the use of a specific package can skip to the relevant section.

miRNA Identification

miRNAs can be computationally predicted from genomic sequences, based on their canonical characteristics and secondary structures. Numerous tools have been developed (Table 13.1), including, but not limited to, the tools briefly introduced herein. *MiRscan*, developed by Bartel Lab, is a classical miRNA prediction web server (Lim *et al.*, 2003). *microPred* is an effective tool for classifying human pre-miRNAs from both genome pseudo hairpins and other non-coding RNAs (Batuwita & Palade, 2009). *miRAbela* is a tool from Swiss Biozentrum in Basel that predicts for miRNA-like structures, and produces output with a score and putative pre-miRNA sequences. *miRNAFold* is a fast *ab-initio* algorithm to search for pre-miRNA precursors in genomes. *MirEval* is a tool for the evaluation of a sequence in regards to its possible miRNA-like properties including miRNA precursor-like structure, sequence similarity with a known miRNA, and sequence conservation. *CID-miRNA* is a web server developed for the identification of miRNA precursors in DNA sequences, utilizing secondary-structure-based filtering systems and an algorithm based on stochastic context-free grammar trained on human miRNAs. CID-miRNA scans a given sequence for the presence of putative miRNA precursors, and the generated output lists all the potential regions that can form miRNA-like structures. It can also scan large genomic sequences for the presence of potential miRNA precursors in its stand-alone form.

Besides the *ab-initio* prediction tools, some homology-based packages are also developed. One widely used example is **MapMi** (Guerra-Assunção & Enright, 2010),

Table 13.2 A list of tools for miRNA analysis of deep sequencing data.

Tools	Description	Application	URL
miRDeep/ miRDeep2	Adapter removal, filter, alignment, identification and quantification of expression	Package	https://www.mdc-berlin.de/ 8551903/en/
miRanalyzer	Detects all known miRNA sequences and predicts new miRNAs	Web server and package	http://bioinfo5.ugr.es/ miRanalyzer/miRanalyzer.php
mirTools/ mirTools 2.0	Classification of the large-scale short reads into known categories. Identification of novel miRNAs and differential expression	Web server and package	http://centre.bioinformatics.zj .cn/mirtools/start.php
UEA sRNA workbench: miRCat	Adapter removal, filter, alignment, expression, target, visualization	Package	http://srna-workbench.cmp .uea.ac.uk/
MIREAP	Identification of genuine miRNAs from deeply sequenced small RNA libraries	Package	http://sourceforge.net/ projicts/mireap

which is developed by European Bioinformatics Institute. It is a tool designed to locate miRNA precursor sequences in existing genomic sequences using known mature miRNA sequences as input. After searching the genome, sequences with hits are extended into the flanking regions, and the miRNA is classified based on the structural properties of known miRNA precursors. MapMi uses other third-party tools including Bowtie (Langmead & Salzberg, 2012) and RNAfold (Hofacker, 2009).

In recent years, with the advance of next-generation sequencing technologies (Table 13.2), many tools for miRNA analysis using deep sequencing data have been developed. Some of the widely used packages and web servers are briefly introduced here.

The *miRDeep* package is widely used to discover known or novel miRNAs from deep sequencing data (Illumina, 454, etc.). This package consists of the core miRDeep algorithm in addition to a handful of scripts to preprocess the mapped data. *miRDeep2* is a completely overhauled tool related to miRDeep. This tool has been demonstrated in several species representing the animal clades to identify known and hundreds of novel miRNAs with high accuracy. The low demand on time and memory, combined with user-friendly interactive graphic output, makes miRDeep2 one of the most popular packages used in various studies.

miRanalyzer is a web server tool that requires a simple input file containing a list of unique reads and its copy numbers (expression levels). Using these data, miRanalyzer can: (i) detect all known miRNA sequences annotated in miRBase, (ii) find all perfect matches against other libraries of transcribed sequences, and (iii) predict new miRNAs. The prediction of new miRNAs is an especially important point, as there are many species with very few known miRNAs. The stand-alone version of miRanalyzer is also available, which is accessible from http://bioinfo5.ugr.es/miRanalyzer/standalone.html.

mirTools is a tool to classify the large-scale short reads into known categories, such as known miRNAs, other non-coding RNAs, genomic repeats, or coding sequences. It can be used for the discovery of novel miRNAs from the high-throughput sequencing data, and the identification of differentially expressed miRNAs according to read tag counts. *mirTools 2.0* is an updated version of mirTools, which includes many new features (Zhu *et al.*, 2010). These include: (i) detecting and profiling various types of ncRNAs, such as miRNA, tRNA, snRNA, snoRNA, rRNA, and piRNA; (ii) identifying miRNA-targeted genes and performing detailed functional annotation of miRNA targets, including gene ontology, KEGG pathway, and protein–protein interaction; (iii) detecting differentially expressed ncRNAs between two experimental groups or among multiple samples; and (iv) detecting novel miRNAs and piRNAs. A stand-alone version of mirTools 2.0 is also available at http://122.228.158.106/mr2_dev/.

miRCat is a tool from the UEA sRNA workbench that identifies miRNAs in high-throughput small RNA sequencing data. miRCat takes a FASTA file of small RNA reads as input and maps them to a reference genome. The tool then looks at genomic hit distribution patterns and the secondary structure of genomic regions corresponding to sRNA hits to predict miRNAs and their precursor structures.

MIREAP (http://sourceforge.net/projicts/mireap) is designed specifically to identify genuine miRNAs from deeply sequenced small RNA libraries. It considers miRNA biogenesis, sequencing depth, and structural features to improve the sensitivity and specificity of miRNA identification.

miRNA Hairpin Structure

The computational prediction of miRNAs from nucleotide sequences depends heavily on assessment of the secondary structures. Therefore, a large number of packages and web servers are developed for calculation and analysis of RNA secondary structures (Table 13.3). The *RNAfold* web server is a collection of web utilities for the calculation and analysis of RNA secondary structures. *MFold* is developed for prediction of melting profiles for nucleic acids, including DNA and RNA. *iFoldRNA* is an interactive RNA folding simulation by discrete molecular dynamics calculations. *RNAbor* is developed to compute the number and Boltzmann probability of a set of secondary structures of a given RNA sequence at base pair distance δ from a given structure. *RNAshapes* is a tool for minimum free energy (MFE) RNA structure prediction based on abstract shapes. Shape abstraction retains the adjacency and nesting of structural features, but disregards helix lengths, thus reducing the number of suboptimal solutions without losing significant information. Furthermore, shapes represent classes of structures for which probabilities based on Boltzmann-weighted energies can be computed. *Sfold* is a tool for the statistical analysis of RNA folding, and a rational design tool for nucleic acids. *CentroidFold* predicts an RNA secondary structure from an RNA sequence. FASTA and the one-sequence-in-a-line format are accepted for predicting a secondary structure per sequence. It also predicts a consensus secondary structure when a multiple alignment (CLUSTALW format) is given.

miRNA Expression Profiling

Similarly, as in the situation for the expression analysis of protein-coding genes, hybridization-based microarray technology has been used for miRNA expression

Table 13.3 List of tools for prediction of RNA secondary structures.

Tools	Description	Application	URL
RNAfold	A collection of web utilities for the analysis of RNA secondary structures	Web server/ package	http://www.tbi.univie.ac.at/RNA/
MFold	Prediction of melting profiles for nucleic acids	Web server/ package	http://mfold.rna.albany.edu/?q=mfold
iFoldRNA	Interactive RNA folding simulations by discrete molecular dynamics calculations	Web server	http://troll.med.unc.edu/ifoldrna.v2/index.php
RNAbor	Compute the number and Boltzmann probability of the set of secondary structures of a given RNA sequence	Web server	http://bioinformatics.bc.edu/clotelab/RNAbor
RNAshapes	MFE RNA structure prediction based on abstract shapes	Package	http://bibiserv.techfak.uni-bielefeld.de/rnashapes/
Sfold	Statistical analysis of RNA folding	Web server	http://sfold.wadsworth.org/cgi-bin/index.pl
CentroidFold	Prediction of RNA secondary structures	Web server	http://www.ncrna.org/centroidfold/

profiling. However, the limitations pertinent to hybridization-based methods, such as narrow detection range, more sensitivity to technical variations, and the lack of ability to characterize novel miRNAs and sequence variations, hinder its wide application in miRNA expression profiling. The miRNA profiling through next-generation sequencing overcomes these limitations, providing a new avenue for expression analysis. However, the analysis of miRNA sequencing data is challenging due to the requirements of extensive computational resources and bioinformatics expertise. Several analytical tools have been developed over the past few years, most of which are web servers.

omiRas is a web server for the differential expression analysis of miRNAs derived from small RNA-Seq data (Müller *et al.*, 2013). The web server is designed for the annotation, comparison, and visualization of interaction networks of ncRNAs derived from next-generation sequencing experiments under two different conditions. The web tool allows the user to submit raw sequencing data, and the analysis results are presented as: (i) static annotation results including length distribution, mapping statistics, alignments, and quantification tables for each library, as well as lists of differentially expressed ncRNAs between the conditions; and (ii) an interactive network visualization of user-selected miRNAs and their target genes based on the combination of several miRNA–mRNA interaction databases. The omiRas web server is implemented in Python, PostgreSQL, n R, and can be accessed at http://tools.genxpro.net/omiras/. However, only a limited number of organisms are currently supported, including human (hg19), mouse (mm10), swine (Sscrofa10.2), *Arabidopsis thaliana* (TAIR9), *Medicago truncatula* (Mt4.0), and *Solanum lycopersicum* (SL2.40).

Web-server-based analysis is limited due to the poor reliability and lack of flexibility of web services, such as unknown parameters, server crashes, and slow performance. In addition, although some web tools provide differential miRNA analysis, they are either limited to a pair of samples or use a model not suitable to a study design. Therefore, differential miRNA expression between samples is often determined by using a combination of packages. The expression values are first determined, and are then used as input for differential expression analysis through packages such as EdgeR or DESeq (see Chapter 9).

miRExpress is one of the packages used to extract miRNA expression profiles from sequencing reads generated by next-generation sequencing platforms (Wang *et al.*, 2009). miRExpress contains miRNA information from miRBase, and efficiently reveals miRNA expression profiles by aligning sequencing reads against the sequences of known miRNAs. Some miRNA analysis pipelines integrate the expression analysis modules—for instance, *CAP-miRSeq*, which is a comprehensive analysis pipeline for deep miRNA sequencing data (Sun *et al.*, 2014).

miRNA Target Prediction

Many packages have been developed to predict the miRNA target (binding) sites in mRNAs (Table 13.4). The packages differ based on the algorithms used and on filtering factors such as consideration of the secondary structure of the target mRNAs. Here, a brief introduction is provided for the miRNA target prediction packages.

miRanda is a widely used algorithm for finding genomic targets for miRNAs. miRanda was developed in the Computational Biology Center at Memorial Sloan Kettering Cancer Center (New York, United States). It can be run locally, and is also available as a web server. The limitation of this program as a web server is that it includes only a few model organisms, including human, mouse, rat, fruit fly, and *C. elegans*. *TargetScan* is another widely used package, which predicts miRNA targets for mammals, fish, fruit flies, and *C. elegans*. TargetScan is frequently updated and can be searched by gene symbol or by miRNA ID. It gives information about the conservation of different miRNA families in all the scanned genomes. *TargetScanS* is similar to TargetScan but it also focuses on the search of miRNA targets within ORFs of vertebrate genomes. *PicTar* is a classical algorithm for the prediction of miRNA targets in several organisms. PicTar site predictions can be searched from the online database, and are also available for downloading for local use. *Diana microT (v.4.0)* is a web server for miRNA target prediction, which is based on artificial neural networks. It can be used to search for target genes of annotated or user-defined miRNA sequences. *PITA* is a target prediction tool that takes into consideration the secondary structure of the target mRNA. The interface is easily customizable and can be used to search for miRNA, genes, and also for UTR sequences. *RNA22* is an algorithm for miRNA target prediction that uses input sequences for miRNA and UTR. Whole genomic predictions are available for downloading and browsing.

A number of platforms are also developed to simultaneously use several individual target prediction packages and present the results in a comparative way. This allows the users to determine if a particular miRNA target is predicted by one or more packages. *miRecords* is a resource for the analysis of miRNA–target interactions (MTIs) in animals. miRecords consists of two components: a database of validated targets and a combined

Table 13.4 List of tools for prediction of miRNA targets.

Tools	Description	Application	URL
miRanda	An algorithm for finding genomic targets	Web server/ package	http://www.microrna.org/microrna/getMirnaForm.do
TargetScan	Prediction of targets for mammals, fish, fruit fly, and *Caenorhabditis elegans*	Web server/ package	http://www.targetscan.org/
PicTar	A classical algorithm for target prediction in several organisms	Web server	http://pictar.mdc-berlin.de/
Diana microT	Target prediction based on artificial neural networks	Web server	http://diana.imis.athena-innovation.gr/DianaTools/index.php?r=site/page&view=software
PITA	A target prediction tool that takes into consideration the secondary structure of the target mRNA.	Package	http://genie.weizmann.ac.il/pubs/mir07/mir07_exe.html
RNA22	An algorithm for miRNA target prediction that uses input sequences for miRNA and UTR.	Web server/ package	https://cm.jefferson.edu/rna22/
miRecords	A resource for the analysis of miRNA–target interactions in animals	Web server	http://c1.accurascience.com/miRecords/
mirDIP	A tool for integrating multiple databases to obtain robust predictions	Web server	http://ophid.utoronto.ca/mirDIP/

resource that simultaneously uses 11 applications for miRNA target prediction. *mirDIP* integrates 12 miRNA prediction datasets from six miRNA prediction databases, allowing users to customize their miRNA target searches. Combining miRNA predictions allows users to obtain more robust target predictions.

miRNA–mRNA Integrated Analysis

Tools for the integrative analysis of miRNA and mRNAs are also developed. These packages are mainly designed to infer the effects of miRNAs on the transcriptomic or proteomic outputs. Typically, the expression levels of miRNAs were compared and associated with the mRNA levels of their predicted or validated targets in a particular experiment. To run the applications, the users must supply expression data for miRNAs and genes or proteins, or a list of miRNAs and/or targets. Some examples of this kind of tools are introduced in Table 13.5, even though these tools currently only support a few model organisms. It is expected that similar packages that are applicable to many more other organisms will be developed.

MMIA, a web server for *mi*RNA and *m*RNA *i*ntegrated *a*nalysis, is involved in several sequential steps: (i) miRNA data analysis, (ii) miRNA target search, (iii) mRNA

Table 13.5 A list of tools for miRNA–mRNA integrated analysis.

Tools	Description	Application	URL
MMIA	Array-based miRNA–mRNA integrated analysis system	Web server	http://epigenomics.snu.ac.kr/biovlab_mmia_ngs/
MAGIA	MiRNA and Genes Integrated Analysis web tool	Web server	http://gencomp.bio.unipd.it/magia/start/
TaLasso	Determination of miRNA–mRNA relationships	Web server and package	http://talasso.cnb.csic.es/
miRConnX	Inferring, displaying, and parsing mRNA and miRNA gene regulatory networks	Web server	http://www.benoslab.pitt.edu/mirconnx.
miRTrail	Analysis of potential relationships between a set of miRNAs and a set of mRNAs	Web server	http://mirtrail.bioinf.uni-sb.de
miRror	Analysis of the cooperative regulation by miRNAs on gene sets and pathways	Web server	http://www.proto.cs.huji.ac.il/mirror/
Mirvestigator	Identifying miRNAs responsible for co-regulated gene expression patterns	Web server	http://mirvestigator.systemsbiology.net/
DIANA-miRPath	Investigating the combinatorial effect of miRNAs in pathways	Web server	http://diana.imis.athena-innovation.gr/DianaTools/index.php?r=mirpath/index
miRTar	Identifying the biological functions and regulatory relationships between miRNAs and coding genes.	Web server	http://mirtar.mbc.nctu.edu.tw/human

data analysis, and (iv) miRNA and mRNA combined analysis. *BioVLAB-MMIA-NGS* is a cloud-based system for miRNA–mRNA integrated analysis using next-generation sequencing data. This system allows identification of differentially expressed miRNAs (DEmiRNAs) and differentially expressed genes (mRNAs, DEGs), and identification of DEGs targeted by DEmiRNAs and having negative correlation between them. The use of MMIA/BioVLAB-MMIA-NGS is limited to four organisms—human, mouse, rhesus monkey, and rice.

MAGIA, *m*iRNA *a*nd *g*enes *i*ntegrated *a*nalysis, is designed to cope with the low sensitivity of target prediction algorithms by exploiting the integration of target predictions with miRNA and gene expression profiles to improve the detection of functional miRNA–mRNA relationships. The use of MAGIA is limited to four species—human, mouse, rat, and fruit fly.

Talasso is an integrative analysis tool for the determination of miRNA–mRNA relationships. It uses several predictors that can be easily customized by the user. *miRConnX* is a user-friendly web interface for inferring, displaying, and parsing mRNA and miRNA gene regulatory networks. mirConnX combines sequence information with gene expression data analysis to create a disease-specific, genome-wide regulatory network. *miRTrail* allows the analysis of potential relationships between a set of miRNAs and a set of mRNAs. This enables the assessing of possible important implications of the miRNAs on the given disease. *miRror* (2.0) reports a ranked list of gene-targets according to their likelihood to be targeted by the provided miRNA set. It supports the analysis for human, mouse, rat, fruit fly, *C. elegans*, and zebrafish. *Mirvestigator* is a portal to analyze a list of co-expressed genes with output of the most likely miRNAs regulating these genes. The web server computes the over-represented sequence motif in the 3′-UTR of the input list of genes and compares it with a list of miRNA seeding sequences using a hidden Markov model.

The regulatory effect of miRNAs within biological processes is normally exerted by groups of these non-coding RNAs that act in a coordinated manner. In order to systematically determine the overall effect of miRNA regulatory control, clustering and grouping of the target genes has been recently explored. This section includes several server platforms to identify the cellular pathways potentially affected by a group of miRNAs. *miRPath 2.0* is a web-based computational tool that identifies potentially altered molecular pathways by the expression of a single or multiple miRNAs. *miRTar* is a comprehensive tool for identifying the biological function and regulatory relationships between a group of known or putative miRNA and their targets. It also provides integrative pathway analysis.

miRNA Analysis Pipelines

Several pipelines have been developed for miRNA analysis. For instance, *CAP-miRSeq* is a comprehensive analysis pipeline for deep miRNA sequencing (Sun *et al.*, 2014), which integrates sequence read preprocessing, alignment, mature/precursor/novel miRNA qualification, variant detection in miRNA coding region, and flexible differential expression between experimental conditions. *DSAP* is an automated multi-task web server designed to analyze deep-sequencing small RNA datasets generated by next-generation sequencing (Huang *et al.*, 2010). DSAP uses the input format of tab-delimited files that hold the unique sequence reads (tags) and their corresponding number of copies generated by the sequencing platform. The input data will go through four analysis steps in DSAP: (i) cleanup to remove adaptors and poly-A/T/C/G/N nucleotides; (ii) clustering to group cleaned sequence tags into unique sequence clusters; (iii) non-coding RNA (ncRNA) matching to conduct sequence homology mapping against a transcribed sequence library from the ncRNA database Rfam (http://rfam.sanger.ac.uk/); and (iv) known miRNA matching to detect known miRNAs in miRBase (http://www.mirbase.org/) based on sequence homology. The expression levels corresponding to matched ncRNAs and miRNAs are summarized in multi-color clickable bar charts linked to external databases. DSAP is also capable of displaying miRNA expression levels from different jobs using a log2-scaled color matrix. Furthermore, a cross-species comparative function is also provided to show the distribution of identified miRNAs in different species as deposited in miRBase. DSAP is available at http://dsap.cgu.edu.tw.

CPSS is a computational platform for the analysis of small RNA deep sequencing data (Zhang *et al.*, 2012). It integrates most functions of currently available bioinformatics tools, and provides all the information wanted by the majority of users from small RNA deep sequencing datasets. The *UEA sRNA toolkit* provides a set of software tools for the analysis of high-throughput small RNA data, including sequence reprocessing, filtering, miRNA identification, expression profiling, and graphical visualization. It can be accessed at http://srna-tools.cmp.uea.ac.uk/.

Oasis is a web application that allows for the fast and flexible online analysis of small-RNA-Seq (sRNA-Seq) data (Capece *et al.*, 2015). It was designed for the end user in the lab, providing user-friendly web front end on the analysis of sRNA-Seq data. The exclusive selling points of Oasis are a differential expression module that allows for the multivariate analysis of samples, a classification module for robust biomarker detection, and an application program interface (API) that supports the batch submission of jobs. Both modules include the analysis of novel miRNAs, miRNA targets, and functional analyses including GO and pathway enrichment analysis. Oasis generates downloadable interactive web reports for easy visualization, exploration, and analysis of data on a local system.

miRNA and Target Databases

A large number of databases was developed as repositories of miRNA sequences and their targets. *miRBase* is hosted and maintained at the University of Manchester, and was previously hosted and supported by the Wellcome Trust Sanger Institute. It includes a complete repository of miRNA sequences and targets and contains the official rules for miRNA nomenclature. The *MirOrtho* database provides a catalog of miRNA orthologs in several species. It contains predictions of precursor miRNA genes covering several animal genomes combining orthology and a support vector machine, providing homology-extended alignments of already known miRBase families and putative miRNA families. *IntmiR* is a manually curated database of published intronic miRNAs in human and mouse. *miRGen* is an integrated database of miRNA gene transcripts, transcription factor binding sites, miRNA expression profiles, and single nucleotide polymorphisms associated with miRNAs. *mESAdb* is a database for the multivariate analysis of sequences and expression of miRNAs from multiple taxa. *miRFocus* is a database for in-depth analysis of miRNA–target gene pathways and the related miRNA annotations. *miRNEST* is a database of miRNAs predicted from expressed sequence tags (ESTs) of animals and plants. *deepBase* is a novel database, developed to facilitate the comprehensive annotation and discovery of small RNAs from transcriptomic data. The current release of deepBase contains deep sequencing data from 185 small RNA libraries from diverse tissues and cell lines of seven organisms: human, mouse, chicken, *Ciona intestinalis*, *Drosophila melanogaster*, *C. elegans*, and *Arabidopsis thaliana*. For the purpose of comparative analysis, deepBase provides an integrative, interactive, and versatile display. A convenient search option, related publications, and other useful information are also provided for further investigation.

The databases for miRNA targets are developed by scanning the published references for experimental evidence related to the regulatory effect of a particular miRNA over its cognate targets. Many such databases have been developed. For instance, *miRTarbase* contains more than 3000 MTIs that are collected by manually surveying pertinent literature after data mining of the text systematically to filter research articles related to functional studies of miRNAs. It contains a customizable search engine that allows personalized data mining using several parameters, including the experimental technique used to validate miRNA targets.

Tarbase is a database hosting detailed information for each miRNA–gene interaction, ranging from miRNA- and gene-related facts to information specific to their interaction, the experimental validation methodologies, and their outcomes. *miRecords* is a high-quality database of experimentally validated miRNA targets resulting from meticulous manual literature curation. It is a resource for analysis of animal MTIs. miRecords consists of two components. The "validated targets" component is a large, high-quality database of experimentally validated miRNA targets resulting from meticulous literature curation. The "predicted targets" component of miRecords is an integration of predicted miRNA targets produced by 11 established miRNA target prediction programs.

miRNA Analysis: Computational Identification from Genome Sequences

miRNAs are generated from long precursors, suggesting that it is possible to identity these precursors by searching publicly available genomic sequences. The principle is to identify sequences containing mature miRNAs and to determine if expected hairpin secondary structures can be predicted from the identified sequences. Two types of computational approaches are generally used: (i) *ab initio* prediction, which is only based on sequence and structural features, and (ii) comparative genomics approach, which is based on evolutionary conservation.

The *ab initio* approach uses *in-silico* prediction packages, while the comparative genomics approach uses both pre-miRNA and mature miRNA sequence conservation across species coupled with the predictable hairpin structures to reveal homologs of known miRNAs. The *ab initio* prediction is able to identify novel miRNAs, but the prediction is interfered by other types of small ncRNAs, such as rRNA, snoRNA, piRNA, tRNA, and fragmented mRNA, because these molecules can also be folded into hairpin structures. Although the mature miRNAs are evolutionarily conserved, ranging from nematodes to humans (Altuvia *et al.*, 2005; Pasquinelli *et al.*, 2000), their precursor sequences are less conserved. Therefore, comparative genomic approach is not effective to detect non-conserved miRNAs across genomes. The integrative approach of comparative genomics and *ab initio* prediction could not only identify highly conserved miRNAs, but also predict less conserved and novel miRNAs.

Here, we demonstrate the use of a homology-based approach, *MapMi* program (version 1.5.9), to identify the channel catfish homologous miRNAs by mapping all the vertebrate miRNAs in miRBase against the channel catfish genome. In brief, the vertebrate mature miRNAs in the miRBase are aligned to the channel catfish genome by Bowtie. The upstream and downstream regions of each genome alignments are retrieved to produce a pair of potential miRNA stem-loops through an extension of 110 bp. The hairpin structures of each potential miRNA stem-loop are predicted by the RNAfold program from the ViennaRNA package. The putative miRNAs were selected based on the MapMi score, which was calculated by considering both the quality of the genome alignment and hairpin structure of the stem-loops (Guerra-Assunção & Enright, 2010, 2012).

Installing MapMi

A stand-alone package can be downloaded at http://www.ebi.ac.uk/enright-srv/MapMi/download.html. Both binary distribution and source distribution are available, and installation of the binary distribution is recommended. Several accessory scripts are provided with the main package, which are the helper applications for running MapMi. In addition, MapMi requires three Perl modules, including Config::General (available from cpan), Bio::DB::Fasta (available from cpan integrated in the BioPerl distribution), and RNA (part of the Vienna RNA package available from http://www.tbi.univie.ac.at/~ivo/RNA/index.html). The step-by-step procedures are provided as the following:

- To download the latest version of MapMi (version 1.5.9-build 32),

```
$ wget http://www.ebi.ac.uk/enright-srv/MapMi/SiteData
  /MapMi-SourceRelease-1.5.9-b32.zip
```

- To decompress the main package,

```
$ unzip MapMi-SourceRelease-1.5.9-b32.zip
```

- To decompress the correct binaries for the helper applications,

```
$ tar xvf HelperPrograms-Linux.tar.bz2
```

Using MapMi

The working directory of MapMi generally contains two subdirectories and two input files. One of the subdirectories, "HelperPrograms", contains the programs to help execute MapMi, while the other subdirectory, "RawData", contains the genome sequence data for analysis, which are either retrieved from Ensembl database and put into the "Ensembl" subdirectory or prepared locally and put into the "Others" subdirectory. One of the input files is the configuration file that contains all the settings to run the program. The other file contains query sequences that are known mature miRNA sequences from other species.

MapMi can be run by executing the main script:

```
$ ./MapMi-MainPipeline-v159-b32.pl ConfigFile
  QuerySequences OutputPrefix
```

Here, `ConfigFile` is a required option (Table 13.6), which by default is a file named "PipelineConfig.conf". `QuerySequences` is an optional parameter to override the query file specified in the configuration file. To use this parameter, `ConfigFile` must be specified as well. `OutputPrefix` is an optional argument that enables the user to specify the name of the result file. To use this parameter, the two previous parameters must be specified.

Using Genomic Sequences from Ensembl

To use MapMi for miRNA identification in genomes deposited in the Ensembl database, simply set the "Ensembl_Species" and "Ensembl_Metazoa_Species" parameters in the config file to designate the name of species used for analysis.

Using Custom Genomic Sequences

MapMi supports using custom genomic sequences. This feature is more useful for non-model species such as aquaculture species because most of their genome sequences

Table 13.6 Options of MapMi (version 1.5.9-Build32)

Option	Description
[Required: Option 1]	
`--configFile=`	MapMi configuration file. *Note*: Specifying a configuration file will override every other option given as input.
[Required: Option 2]	
`--queryFile=`	FASTA file containing query mature sequences/sequencing reads
`--ensembl_species=`	Comma-separated list of Ensembl species
`--metazoa_species=`	Comma-separated list of Ensembl Metazoa species
`--ensembl_version=`	Specify release of Ensembl
`--metazoa_version=`	Specify release of Ensembl Metazoa
[Optional/Advanced Options]	
`--run_id=`	Specifies the ID for the current job
`--outputPrefix=`	Specifies the prefix for the output files produced
`--min_score=`	Specifies the minimum score for a candidate hairpin [recommended = 35]
`--min_prec=`	Minimum precursor length
`--min_ratio=`	Minimum match/mismatch ratio
`--min_len=`	Minimum "match" length
`--exclude_loop=`	Binary toggle to exclude mature
`--loop_overlap=`	Maximum allowed mature base pairs overlapping loop (depends on `--exclude_loop`)
`--penalty=`	Mapping mismatch penalty
`--max_mismatch=`	Maximum number of mapping mismatches
`--bowtie_m=`	Bowtie "m" parameter
`--short_ext=`	Short side extension (nt)
`--long_ext=`	Long side extension (nt)
[General]	
`-h, --help`	Prints this help page

are not available in the Ensembl database. To use this feature, put genome sequences in FASTA format into the directory "RawData/Others/". The FASTA file is required to have the extension of ".fa". A bowtie index will be automatically created for the sequences when running the program for the first time. If run on genomes locally, the "Ensembl_Species" and "Ensembl_Metazoa_Species" parameters need to be left empty in the config file. If you want to deactivate the use of your own genomic sequences, you just need to rename the FASTA file without filename extension of ".fa".

Output File

Each run of MapMi should produce a file under the "Results/ScoredMapFile/" directory, with a file extension of "dust." The description of each column of the result file is described in Table 13.7. The MapMi score, calculated by considering both the quality of the genome alignment and the hairpin structure of the stem-loops, is important to filter the putative miRNAs. A MapMi score threshold of 35 is often used according to an empirical analysis of authentic and permutated miRNAs (Guerra-Assunção & Enright, 2010, 2012). Moreover, typical stem-loop hairpins are often verified by fulfilling three criteria: (i) mature miRNAs are present in one arm of the hairpin precursors, which do not contain large internal loops or mismatches; (ii) the secondary structures of the hairpins are stable with minimum free energy less than −20 kcal/mol; and (iii) the hairpins are located in intergenic regions or introns. The putative miRNAs mapped to protein-coding sequences, repetitive elements, or other classes of non-coding RNAs are excluded.

miRNA Analysis: Empirical Identification by Small RNA-Seq

The advent of next-generation sequencing techniques allow for high-throughput small RNA sequencing (RNA-Seq) for discovery and expression profiling. Small RNA-Seq is a powerful method to identify and quantify miRNA expression, and discover novel miRNAs and various miRNA isoforms. Small RNA-Seq starts with total RNA extraction, followed by fractioning and size selection for 17–25 nucleotides. The fragments are then ligated to adapters, followed by reverse transcription and PCR amplification. The adapter-ligated PCR products are sequenced using next-generation sequencers such as Illumina HiSeq platforms.

Procedures of Small RNA Deep Sequencing

Total RNA is extracted from tissues using the RNeasy Plus Kit (Qiagen), following manufacturer's instructions and treated with RNase-free DNase I (Qiagen) to remove genomic DNA. RNA concentration and integrity of each sample is measured on an Agilent 2100 Bioanalyzer using a RNA Nano Bioanalysis chip. The small RNA libraries can be constructed using NEBNext® Multiplex Small RNA Library Prep Set for Illumina® (NEB, USA.), following manufacturer's protocol. The index codes are usually added to attribute sequences to each sample. Briefly, the small RNA was first ligated with 3' adaptors followed by hybridization of reverse transcription primer to the excess of 3' adaptors. The 5' adaptors are then ligated to the 5' ends of small RNAs. Then, first-strand cDNA is synthesized, and PCR amplification is performed. Size selection of the small RNA library can be conducted on Pippin Prep instrument using the 3% Agarose, dye-free gel with internal standards. Before performing size selection, the quality of library was assessed on the Agilent Bioanalyzer 2100 system using DNA 1000 chip according to the manufacturer's instructions. The miRNA library should appear as a peak at 147 bp, which corresponds to 21 nucleotide inserts. DNA fragments with the sizes ranging from 140 bp to 160 bp are re-isolated and dissolved in 8 μL elution buffer. The clustering of the index-coded samples is performed on a cBot Cluster Generation System using TruSeq SR Cluster Kit v3-cBot-HS (Illumia) according to the manufacturer's instructions. After cluster generation, sequencing is performed on an Illumina HiSeq 2000 platform for 50 bp singe-end reads.

Table 13.7 Example of MapMi output. The example is one miRNA identification extracted from this output of miRNA identification from the channel catfish genome (Ipg1.fa) using all vertebrate mature miRNAs as queries.

Column name	Example output
Name of query sequence	dre-let-7a
Query sequence	TGAGGTAGTAGGTTGTATAGTT
Species name	Ipg1.fa
Extension side	Right
Number of mismatches in mature sequences when mapping	0
Chromosome name	lg7
Strand	+
Mature start	22327944
Mature end	22327965
Mature length	22
Precursor start	22327929
Precursor end	22328031
Precursor length	102
Number of structural matches	27
Minimal free energy (RNAfold delta G)	−47.09999847
MapMi score	49.87558063
Small extension (parameter)	40
Long extension (parameter)	70
Mismatch penalty (parameter)	10
Maximum number of mismatches (parameter)	1
Precursor sequence	TGTGTTTCCACAAGAtgaggtagtaggttgtatagtt TGTGGGAAACATCACATCCTATTGAGGTG AGAACTATACAGTCTATTGCCTTCCTTG TGAGACACA
Fraction of precursor that is flagged by dust	0

Workflow of Data Analysis

A workflow of small RNA-Seq data analysis is illustrated in Figure 13.1. The raw sequence data is obtained in FASTQ format, and is subject to quality control and trimming using FastQC and Trimmomatic (Bolger, Lohse & Usadel, 2014) to remove sequencing adaptors, low-quality reads ($Q < 30$), and reads less than 15 bp in length. The clean reads are annotated by searching against the Rfam database (http://rfam.sanger.ac.uk/) using BLAST. The reads that are annotated as tRNA, rRNA, snoRNA, snRNA, and other ncRNAs are excluded for further analysis.

The remaining reads are then annotated with all known miRNAs in miRBase (Release 21) using miRDeep2 (Friedländer *et al.*, 2012). Sequences that are identical (no more

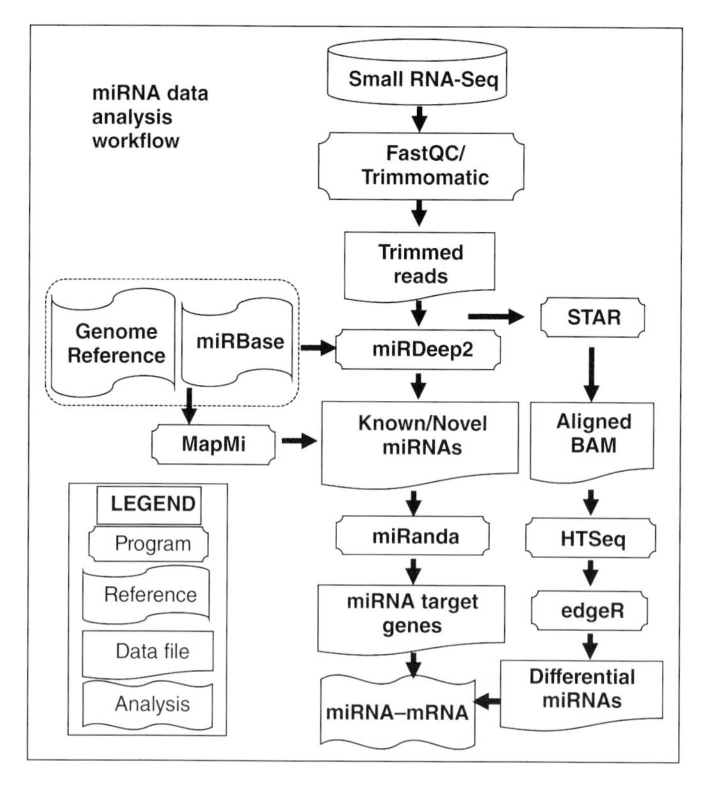

Figure 13.1 A workflow of miRNA data analysis.

than one mismatch) to the reference mature miRNAs are identified as conserved miRNAs. The sequences that are not identical to the conserved miRNAs are used to BLAST against the genome assembly to identify novel miRNAs. miRNA expression profiling is conducted using the STAR aligner for mapping, HTSeq for determining read counts, and edgeR for differential expression analysis. Target prediction is achieved using the program miRanda. Integrative analysis of miRNA–mRNA interactions can be achieved by integrating the results of expression profiling and target prediction (Figure 13.1).

In the following sections, we demonstrate the installation and use of miRDeep2 for miRNA identification and expression quantification, and of the target prediction programs miRanda and TargetScan. The analysis of expression profiling using the pipeline of STAR–HTSeq–edgeR is the same as the regular RNA-Seq expression profiling, which is referred to Chapter 9.

miRNA Analysis Using miRDeep2

Package Installation

1) Download the installation package using the `wget` command:

```
$ wget https://www.mdc-berlin.de/43969303/en/research
  /research_teams/systems_biology_of_gene_regulatory_
  elements/projects/miRDeep/mirdeep2_0_0_7.zip
```

Decompress zipped package:

```
$ unzip mirdeep2_0_0_7.zip
```

After decompressing, a folder is obtained with the name "mirdeep2_0_0_7", which contains all scripts and documentations.

2) Installation with the provided "install.pl" script:

```
$ perl install.pl, or ./install.pl
```

If it works, the script will run a set of commands to automatically download and install all prerequisite packages, including bowtie short read aligner, Vienna package with RNAfold, and Perl package PDF::API2. It will then add all installed programs into the environmental path. Everything that is downloaded by the installer will be in a directory called "the_path_to_mirdeep2/essentials". Following installation with "install.pl", the programs can be used after restarting the shell.

Sometimes, the "install.pl" script cannot install all the required packages, which need to be downloaded and installed manually. The instructions for manual installation can be found in the "README" file in the package folder, "mirdeep2_0_0_7".

Using miRDeep2

The algorithm of miRDeep2 is based on the miRNA biogenesis model, which aligns sequencing reads to potential hairpin structures in a manner consistent with Dicer processing, and assigns log-odds scores to measure the probability that hairpins are true miRNA precursors. The output of the analysis is a scored list of known and novel miRNAs with their expression levels.

The miRDeep2 analyses are mainly executed using three scripts—"mapper.pl", "quantifier.pl", and "miRDeep2.pl" (Table 13.8). Many other scripts are also provided in the miRDeep2 package, which are useful to preprocess input files and post-process the outputs. A brief tutorial on the analysis of deep sequencing data can be found in the TUTORIAL file, which was provided in the package folder. In general, five input files are required to start analysis: (1) "genome.fa", a FASTA file with the reference genome, (2) "mature_ref_this.fa", a FASTA file with the reference miRBase mature miRNAs for the species under study, (3) "mature_ref_other.fa", a FASTA file with the reference miRBase mature miRNAs for the related species, (4) "precursor_ref_this.fa", a FASTA file with the reference miRBase precursor miRNAs for the species under study, and (5) "reads.fa", a fast file with the deep sequencing reads.

Five steps are involved in the analysis using miRDeep2, described in the following subsections.

Step 1: Prepare miRNA References and Build Genome Index miRDeep2 analysis requires the known miRNAs as references, which can be downloaded from miRbase (http://www.mirbase.org/ftp.shtml). The file "mature.fa" contains FASTA format sequences of all mature miRNA sequences, and the file "hairpin.fa" contains FASTA format sequences of all miRNA hairpins. Both the mature miRNAs and hairpins are in RNA format. miRDeep2 requires the reference miRNAs only from a single species, and the RNA sequences to be converted to DNA. A couple of Perl scripts in the miRDeep2 packages can be used to convert these files.

Table 13.8 A list of major scripts and options in the miRDeep2 package.

Script	Options
mapper.pl	-a: input file is "seq.txt" format
	-b: input file is "qseq.txt" format
	-c: input file is FASTA format
	-e: input file is FASTQ format
	-d: input file is a config file; -a, -b, or -c must be given with option -d.
	-g: three-letter prefix for reads
	-h: parse to FASTA format
	-i: convert RNA to DNA alphabet
	-j: remove all entries that have a sequence that contains ambiguous letters
	-k seq: clip 3′ adapter sequence
	-l int: discard reads shorter than int nucleotides
	-m: collapse reads
	-p genome: map to genome
	-q: map with one mismatch in the seed
	-r int: cutoff of multiple mapping in the genome
	-s file: print processed reads to this file
	-t file: print read mappings to this file
	-u: do not remove directory with temporary files
	-v: outputs progress report
	-n: overwrite existing files
quantifier.pl	-u: list all values allowed for the species parameter
	-p precursor.fa: miRNA precursor sequences
	-m mature.fa: miRNA sequences
	-r reads.fa: your read sequences
	-c [file]: config.txt file with different samples
	-s [star.fa]: optional star sequences from miRBase
	-t [species]: species being analyzed
	-y [time]: optional, otherwise generating a new one
miRDeep2.pl	-a int: minimum read stack height
	-b int: minimum score cutoff for predicted novel miRNAs
	-c: disable randfold analysis
	-g int: maximum number of precursors to analyze
	-t species: species being analyzed
	-u: output list of UCSC browser species that are supported
	-v: remove directory with temporary files
	-s file: File with known miRBase star sequences

Using "extract_miRNAs.pl" to extract sequences for a species:

```
$ perl extract_miRNAs.pl hairpin.fa ipu >hairpin.ipu.fa
$ perl extract_miRNAs.pl mature.fa ipu >mature.ipu.fa
```

Using "rna2dna.pl" to substitutes "u"s and "U"s to "T"s:

```
$ perl rna2dna.pl hairpin.ipu.fa > hairpin.ipu.dna.fa
$ perl rna2dna.pl mature.ipu.fa > mature.ipu.dna.fa
```

The genome index file is produced using bowtie-build:

```
$ bowtie-build Ipg1.fa Ipg1
```

Before building the genome index, the genome FASTA file needs to be checked for correctness using the script "sanity_check_genome.pl". The identifier lines of the FASTA file are not allowed to contain whitespaces, and must be unique. Sequence lines are not allowed to contain characters others than [A|C|G|T|N|a|c|g|t|n].

Step 2: Process the Reads and Map Them to the Genome The "mapper.pl" script processes reads and/or maps them to the reference genome. The input is a file in FASTA, FASTQ, seq.txt, or qseq.txt format. More inputs can be given depending on the options used. The output depends on the options used, either a FASTA file with processed reads, or an arf file with mapped reads, or both.

```
$ mapper.pl esc0hr.fastq -e -h -i -j -l 18 -m -p Ipg1 -s
   esc0hr.reads_collapsed.fa -t esc0hr.reads_collapsed.arf
   -v -n
```

Input files:

"esc0hr.fastq": input file for analysis
"Ipg1": genome index file

Output files:

"esc0hr.reads_collapsed.fa": processed reads
"esc0hr.reads_collapsed.arf": alignment file

Options:

`-e`: input file is FASTQ format
`-h`: parse to FASTA format
`-i`: convert RNA to DNA letters to map against genome
`-j`: remove all entries that have a sequence that contains letters other than a, c, g, t, u, n, A, C, G, T, U, and N
`-l`: discard reads shorter than 18 bp
`-m`: collapse reads
`-p`: map to genome (must be indexed by bowtie-build)
`-s file`: print processed reads to this file
`-t file`: print read mappings to this file
`-v`: outputs progress report
`-n`: overwrite existing files

The sequencing data from different samples, for example, control and treatment, can be processed and mapped to the genome together, by providing a "contig.txt" file (using the `-d` option) in which each line designates file locations and a sample-specific three-letter code.

Step 3: Fast Quantitation of Reads Mapped to Known miRBase Precursors The "quantifier.pl" script maps the deep sequencing reads to predefined miRNA precursors and determines the expression levels of the corresponding miRNAs. First, the predefined mature miRNA sequences are mapped to the predefined precursors. Optionally, predefined star sequences can be mapped to the precursors as well. Based on the mapping, the mature and star sequences in the precursors are determined. Second, the deep sequencing reads are mapped to the precursors. The number of reads falling into an interval 2 nt upstream and 5 nt downstream of the mature/star sequence is determined.

```
$ quantifier.pl -p Ipu.hairpin.dna.fa -m Ipu.mature.dna.fa
    -r esc0hr.reads_collapsed.fa -y esc0hr
```

Input files:

"Ipu.hairpin.dna.fa": a FASTA file with precursors
"Ipu.mature.dna.fa": a FASTA file with mature miRNAs
"esc0hr.reads_collapsed.fa": a FASTA file with deep sequencing reads

Output files: a tab-separated file named "miRNAs_expressed_all_samples.csv" with miRNA identifiers and its read count, a signature file called "miRBase.mrd", a file called "expression.html" that gives an overview of all miRNAs, the input data, and a directory called "pdfs" that contains PDF files, each of which shows a miRNA's signature and structure.

Options:

`-p precursor.fa`: miRNA precursor sequences
`-m mature.fa`: mature miRNA sequences
`-r reads.fa`: small RNA-Seq sequences
`-y [time]`: optional, otherwise the script will generate a new one

Step 4: Identification of Known and Novel miRNAs in the Deep Sequencing Data The "miRDeep2.pl" script runs all the necessary scripts of the miRDeep2 package to perform a miRNA detection analysis with deep sequencing data.

```
$ miRDeep2.pl esc0hr.reads_collapsed.fa Ipg1.fa
   esc0hr.reads_collapsed_vs_genome.arf Ipu.mature.
   rna2dna.fa Dre.mature.rna2dna.fa Ipu.hairpin.rna2dna.fa
   -q expression_analyses/expression_analyses_now
  /miRBase.mrd 2>report.log
```

Input files:

"esc0hr.reads_collapsed.fa": a FASTA file with deep sequencing reads
"Ipg1.fa": a FASTA file of the corresponding genome
"esc0hr.reads_collapsed.arf": a file of mapped reads to the genome in miRDeep2 arf format
"Ipu.mature.dna.fa": a FASTA file with known miRNAs of the species under study
"Dre.mature.dna.fa": a FASTA file of known miRNAs of related species
"Ipu.hairpin.dna.fa": a FASTA file with known miRNA precursors of the species under study

Output files: a spreadsheet and an html file with an overview of all detected miRNAs in the deep sequencing input data.

Options:

-q <.mrd file>: use the output of "quantifier.pl" to get information on the miRNAs
2>: pipe all progress output to the "report.log" file

Step 5: Browse the Results Open the "results.html" file using an Internet browser to view the results.

Prediction of miRNA Targets

The mechanism of miRNA-mediated gene regulation requires perfect or near-perfect complementarity between the miRNAs and their targeted mRNAs for directly cleaving mRNAs or repressing protein translation (Rhoades *et al.*, 2002). Therefore, miRNA targets can be identified by BLASTN search based on the complementarity between miRNAs and their targets. To date, large numbers of miRNA targets have been predicted based on BLASTN searches followed by validation using one or several experimental approaches, such as Northern blotting, qRT-PCR, and 5' rapid amplification of cDNA ends (5'RACE). Two of the popularly used miRNA target prediction software are miRanda and TargetScan.

We demonstrate miRNA target prediction using miRanda, which is an algorithm for the detection of potential miRNA target sites in genomic sequences. miRanda reads RNA sequences (such as miRNAs) from one file ("file1") and genomic DNA/RNA sequences from another file ("file2"). Both of these files should be in FASTA format.

One or more miRNA sequences from "file1" are scanned against all sequences in "file2" to identify potential target sites using a two-step strategy. First, a local alignment is carried out between the query miRNA sequence and the reference sequence. This alignment procedure provides a score based on sequence complementarity (e.g., A:U and G:C matches), but not on sequence identity (e.g., A:A, G:G matches). The G:U wobble base pair is also permitted, but it is generally scored less than the perfect matches. Second, high-scoring alignments are determined with a score threshold (defined by -sc), and the thermodynamic stabilities of RNA duplexes are estimated based on these alignments. The second step of the method utilizes folding routines from the RNAlib library, which is a part of the ViennaRNA package. In this step, a constrained fictional single-stranded RNA composed of the query sequence, a linker, and the reference sequence (reversed) is generated, and the structure is then folded using RNAlib to calculate the minimum free energy (DG kcal/mol). Finally, detected targets with energies less than an energy threshold (defined by -en) are selected as potential target sites. Target site alignments passing both thresholds and other information are produced as output.

Package Installation

The source package of miRanda (version 3.3a) can be downloaded at: http://cbio.mskcc.org/microrna_data/miRanda-aug2010.tar.gz

```
$ wget http://cbio.mskcc.org/microrna_data
  /miRanda-aug2010.tar.gz
```

Decompress the package by:

```
$ tar -xvzf miRanda- aug2010.tar.gz
```

To install the package:

1) “cd” to the directory containing the package’s source code and type “./configure --prefix=/home/aubsxl/miRNA” to configure the package for your system. The option “--prefix=PATH” specifies the path to install the package instead of “/usr/local/bin”, “/usr/local/man”, etc., by default settings.
2) Type “make” to compile the package.
3) Type “make install” to install the programs and any data files and documentation.
4) Attach the miRanda executable path to the PATH:

```
$ echo 'export PATH=$PATH:/home/aubsxl/miRNA/bin'
    >> ~/.bashrc.local.dmc
```

Using miRanda

miRanda can be used in the following syntax:

```
miranda file1 file2 [-sc score] [-en energy] [-scale scale]
   [-strict] [-go X] [-ge Y] [-out fileout] [-quiet]
   [-trim T] [-noenergy] [-restrict file]
```

The files “file1” and “file2” are two input files containing miRNA sequences and genomic sequences, respectively. The remaining are options, of which `-sc` and `-en` are two important options to set the alignment score threshold and the energy threshold. The detailed information on miRanda options is given in Table 13.9. Default command line arguments are set to `-sc 140 -go -9 -ge -4 -en 1 -scale 4` (no energy filtering). The following sets of parameters are reported to increase the prediction precision (Marín & Vaníček, 2011), including score cutoff ≥ 140, energy cutoff ≤ -20 kcal/mol, gap opening $= -9.0$, and gap extension $= -4.0$.

The following is a sample demonstration:

```
$ miranda bantam_stRNA.fasta hid_UTR.fasta -sc 140 -en -20
   -go -9 -ge -4 -out bantam.target.pred
```

The “bantam_stRNA.fasta” file contains the miRNA sequences, the “hid_UTR.fasta” file contains UTR sequences, and the output file, “bantam.target.pred”, contains the target prediction results with current settings.

Another widely used target prediction program, TargetScan, is available as a Perl script (http://www.targetscan.org//vert_50//vert_50_data_download/targetscan_50.zip), which identifies miRNA targets and then determines whether a given target is conserved. The script takes two input files: a tab-delimited file that lists the miRNA seed sequences, and a tab-delimited multiple sequence alignment of the 3′ UTRs of genes from the desired species.

The format of the input files is important for the script to work correctly. Each line of the miRNA seed sequence file consists of three tab-separated entries: (1) name of the miRNA family, (2) the seven nucleotide long seed region sequence, and (3) species ID of this miRNA family (which should match the species ID in the UTR input file).

Table 13.9 A list of miRanda options for miRNA target prediction.

Options	Description
`-sc score`	Set score threshold for the alignment.
`-en energy`	Set the energy threshold for the alignment.
`-scale scale`	Set the scaling parameter to scale. This scaling is applied to match/mismatch scores in the critical 7 bp region near the 5′ end of the miRNA.
`-strict`	Require strict alignment in the seed region (offset positions 2–8).
`-go X`	Set the gap-opening penalty to X for alignments. This value must be negative.
`-ge Y`	Set the gap-extend penalty to Y for alignments. This value must be negative.
`-out fileout`	Print results to an output file: "fileout".
`-quiet`	Quiet mode, omit notices.
`-trim T`	Trim reference sequences to T nucleotides.
`-noenergy`	Turn off thermodynamic calculations from RNAlib; the `-en` setting will be ignored.
`-restrict file`	Restrict scans to those between specific miRNAs and UTRs. The file should contain lines of tab-separated pairs of sequence identifiers: "miRNA_id <tab> target_id".
`--help -h`	Display help, usage information, and command-line options.
`--version -v` `--license`	Display version and license information.

Each line of the alignment file consists of three tab-separated entries: (1) gene symbol or transcript ID, (2) species/taxonomy ID (which should match the species ID in the miRNA input file), and (3) sequence.

In the package installation directory, samples of both these files are provided: "UTR_sequences_sample.txt" and "miR_Family_info_sample.txt". The script can be executed as:

```
$ ./targetscan_50.pl miR_Family_info_sample.txt
   UTR_Sequences_sample.txt TargetScan_50_output.txt
```

The output file is "TargetScan_50_output.txt".

Conclusions

The tools for the analysis of miRNA and its target genes are actively being developed, such as various stand-alone programs, web servers, and databases. However, most tools are developed for only a few model organisms; web servers usually only support a limited number of species. With the rapid accumulation of miRNA deep sequencing data in non-model organisms including aquaculture species, programs and web tools that support customizable genome references will be more useful in these species.

In the near future, various web servers and databases are anticipated to be developed for investigating miRNA and its target genes in aquaculture fish and shellfish species. Selection of proper package for analysis largely depends on the familiarity of the user to the software packages, and the user-friendliness of the packages. At genomic levels, minor differences obtained by using different packages are normal, and packages will not provide "correct" or "wrong" results, but just different results. Validation through experiments is the ultimate test for the identified miRNAs and their targets.

References

Aalto, A.P. and Pasquinelli, A.E. (2012) Small non-coding RNAs mount a silent revolution in gene expression. *Current Opinion in Cell Biology*, **24**, 333–340.

Altuvia, Y., Landgraf, P., Lithwick, G. *et al.* (2005) Clustering and conservation patterns of human microRNAs. *Nucleic Acids Research*, **33**, 2697–2706.

Ambros, V. (2004) The functions of animal microRNAs. *Nature*, **431**, 350–355.

Bartel, D.P. (2004) MicroRNAs: genomics, biogenesis, mechanism, and function. *Cell*, **116**, 281–297.

Batuwita, R. and Palade, V. (2009) microPred: effective classification of pre-miRNAs for human miRNA gene prediction. *Bioinformatics*, **25**, 989–995.

Bolger, A.M., Lohse, M. and Usadel, B. (2014) Trimmomatic: a flexible trimmer for Illumina sequence data. *Bioinformatics*, **30**, 2114–2120.

Capece, V., Vizcaino, J.C.G., Vidal, R. *et al.* (2015) Oasis: online analysis of small RNA deep sequencing data. *Bioinformatics*, **31**, 2205–2207.

Caron, M.-P., Lafontaine, D.A. and Massé, E. (2010) Small RNA-mediated regulation at the level of transcript stability. *RNA Biology*, **7**, 140–144.

Carrington, J.C. and Ambros, V. (2003) Role of microRNAs in plant and animal development. *Science*, **301**, 336–338.

Chekulaeva, M. and Filipowicz, W. (2009) Mechanisms of miRNA-mediated post-transcriptional regulation in animal cells. *Current Opinion in Cell Biology*, **21**, 452–460.

Cifuentes, D., Xue, H., Taylor, D.W. *et al.* (2010) A novel miRNA processing pathway independent of Dicer requires Argonaute2 catalytic activity. *Science*, **328**, 1694–1698.

Clark, M.B., Amaral, P.P., Schlesinger, F.J. *et al.* (2011). The reality of pervasive transcription. *PLoS Biology*, **9**, e1000625.

Filipowicz, W., Bhattacharyya, S.N. and Sonenberg, N. (2008) Mechanisms of post-transcriptional regulation by microRNAs: are the answers in sight? *Nature Reviews Genetics*, **9**, 102–114.

Friedländer, M.R., Mackowiak, S.D., Li, N. *et al.* (2012) miRDeep2 accurately identifies known and hundreds of novel microRNA genes in seven animal clades. *Nucleic Acids Research*, **40**, 37–52.

Grosshans, H. and Filipowicz, W. (2008) Molecular biology: the expanding world of small RNAs. *Nature*, **451**, 414–416.

Guerra-Assunção, J.A. and Enright, A.J. (2010) MapMi: automated mapping of microRNA loci. *BMC Bioinformatics*, **11**, 133.

Guerra-Assunção, J.A. and Enright, A.J. (2012) Large-scale analysis of microRNA evolution. *BMC Genomics*, **13**, 218.

Hackenberg, M., Rodríguez-Ezpeleta, N. and Aransay, A.M. (2011) miRanalyzer: an update on the detection and analysis of microRNAs in high-throughput sequencing experiments. *Nucleic Acids Research*, **39**, W132–W138.

Han, J., Lee, Y., Yeom, K.-H. *et al.* (2004) The Drosha-DGCR8 complex in primary microRNA processing. *Genes & Development*, **18**, 3016–3027.

Han, J., Lee, Y., Yeom, K.-H. *et al.* (2006) Molecular basis for the recognition of primary microRNAs by the Drosha-DGCR8 complex. *Cell*, **125**, 887–901.

Hansen, T.B., Venø, M.T., Kjems, J. and Damgaard, C.K. (2014) miRdentify: high stringency miRNA predictor identifies several novel animal miRNAs. *Nucleic Acids Research*, **42**, e124.

He, L. and Hannon, G.J. (2004) MicroRNAs: small RNAs with a big role in gene regulation. *Nature Reviews Genetics*, **5**, 522–531.

Hertel, J. and Stadler, P.F. (2006) Hairpins in a Haystack: recognizing microRNA precursors in comparative genomics data. *Bioinformatics*, **22**, e197–e202.

Hofacker, I.L. (2009) RNA secondary structure analysis using the Vienna RNA package. *Current Protocols in Bioinformatics*, **12**, 12.11–12.12.

Houwing, S., Kamminga, L.M., Berezikov, E. *et al.* (2007) A role for Piwi and piRNAs in germ cell maintenance and transposon silencing in Zebrafish. *Cell*, **129**, 69–82.

Huang, P.-J., Liu, Y.-C., Lee, C.-C. *et al.* (2010). DSAP: deep-sequencing small RNA analysis pipeline. *Nucleic Acids Research*, **38**, W385–W391.

Jones-Rhoades, M.W., Bartel, D.P. and Bartel, B. (2006) MicroRNAs and their regulatory roles in plants. *Annual Review of Plant Biology*, **57**, 19–53.

Kim, V.N., Han, J. and Siomi, M.C. (2009) Biogenesis of small RNAs in animals. *Nature Reviews Molecular Cell Biology*, **10**, 126–139.

Krützfeldt, J., Poy, M.N. and Stoffel, M. (2006) Strategies to determine the biological function of microRNAs. *Nature Genetics*, **38**, S14–S19.

Langmead, B. and Salzberg, S.L. (2012) Fast gapped-read alignment with Bowtie 2. *Nature Methods*, **9**, 357–359.

Lee, R.C., Feinbaum, R.L. and Ambros, V. (1993) The *C. elegans* heterochronic gene lin-4 encodes small RNAs with antisense complementarity to lin-14. *Cell*, **75**, 843–854.

Lee, Y., Kim, M., Han, J. *et al.* (2004) MicroRNA genes are transcribed by RNA polymerase II. *The EMBO Journal*, **23**, 4051–4060.

Lim, L.P., Glasner, M.E., Yekta, S. *et al.* (2003) Vertebrate microRNA genes. *Science*, **299**, 1540.

Lindsay, M.A. (2008) microRNAs and the immune response. *Trends in Immunology*, **29**, 343–351.

Lippman, Z. and Martienssen, R. (2004) The role of RNA interference in heterochromatic silencing. *Nature*, **431**, 364–370.

Maniatis, T. and Tasic, B. (2002) Alternative pre-mRNA splicing and proteome expansion in metazoans. *Nature*, **418**, 236–243.

Marín, R.M. and Vaníček, J. (2011) Efficient use of accessibility in microRNA target prediction. *Nucleic Acids Research*, **39**, 19–29.

Mattick, J.S. (2001) Non-coding RNAs: the architects of eukaryotic complexity. *EMBO Reports*, **2**, 986–991.

Mattick, J.S. (2003) Challenging the dogma: the hidden layer of non-protein-coding RNAs in complex organisms. *Bioessays*, **25**, 930–939.

Mattick, J.S. (2005) The functional genomics of noncoding RNA. *Science*, **309**, 1527–1528.

Morris, K.V. and Mattick, J.S. (2014) The rise of regulatory RNA. *Nature Reviews Genetics*, **15**, 423–437.
Müller, S., Rycak, L., Winter, P. *et al.* (2013) omiRas: a Web server for differential expression analysis of miRNAs derived from small RNA-Seq data. *Bioinformatics*, **29**, 2651–2652.
Okamura, K., Hagen, J.W., Duan, H. *et al.* (2007) The mirtron pathway generates microRNA-class regulatory RNAs in Drosophila. *Cell*, **130**, 89–100.
Pasquinelli, A.E., Reinhart, B.J., Slack, F. *et al.* (2000) Conservation of the sequence and temporal expression of let-7 heterochronic regulatory RNA. *Nature*, **408**, 86–89.
Pauli, A., Rinn, J.L. and Schier, A.F. (2011) Non-coding RNAs as regulators of embryogenesis. *Nature Reviews Genetics*, **12**, 136–149.
Petersen, C.P., Bordeleau, M.-E., Pelletier, J. and Sharp, P.A. (2006) Short RNAs repress translation after initiation in mammalian cells. *Molecular Cell*, **21**, 533–542.
Rhoades, M.W., Reinhart, B.J., Lim, L.P. *et al.* (2002) Prediction of plant microRNA targets. *Cell*, **110**, 513–520.
Ruby, J.G., Jan, C.H. and Bartel, D.P. (2007) Intronic microRNA precursors that bypass Drosha processing. *Nature*, **448**, 83–86.
Sewer, A., Paul, N., Landgraf, P. *et al.* (2005) Identification of clustered microRNAs using an ab initio prediction method. *BMC Bioinformatics*, **6**, 267.
Sun, Z., Evans, J., Bhagwate, A. *et al.* (2014) CAP-miRSeq: a comprehensive analysis pipeline for microRNA sequencing data. *BMC Genomics*, **15**, 423.
Voinnet, O. (2009) Origin, biogenesis, and activity of plant microRNAs. *Cell*, **136**, 669–687.
Wang, W.-C., Lin, F.-M., Chang, W.-C. *et al.* (2009) miRExpress: analyzing high-throughput sequencing data for profiling microRNA expression. *BMC Bioinformatics*, **10**, 328.
Williamson, V., Kim, A., Xie, B. *et al.* (2012) Detecting miRNAs in deep-sequencing data: a software performance comparison and evaluation. *Briefings in Bioinformatics*, **14**, 36–45.
Xiao, C. and Rajewsky, K. (2009) MicroRNA control in the immune system: basic principles. *Cell*, **136**, 26–36.
Yang, J.-S., Maurin, T., Robine, N. *et al.* (2010) Conserved vertebrate mir-451 provides a platform for Dicer-independent, Ago2-mediated microRNA biogenesis. *Proceedings of the National Academy of Sciences USA*, **107**, 15163–15168.
Zhang, Y., Xu, B., Yang, Y. *et al.* (2012) CPSS: a computational platform for the analysis of small RNA deep sequencing data. *Bioinformatics*, **28**, 1925–1927.
Zhu, E., Zhao, F., Xu, G. *et al.* (2010) mirTools: microRNA profiling and discovery based on high-throughput sequencing. *Nucleic Acids Research*, **38**, W392–W397.

14

Analysis of Allele-Specific Expression

Yun Li, Ailu Chen and Zhanjiang Liu

Introduction

The analysis of allele-specific expression (ASE) of genes has been of great interest in the study of gene regulatory mechanisms. Such an interest has been fueled by the determination of phenotypes by allelic expression of genes, mainly through *cis*- and *trans*-regulations, as well as by the advances in sequencing technologies. ASE analysis is also used as an efficient tool to detect parent-of-origin effects, including classical imprinting and the allelic imbalance in expression favoring one parent. However, analyzing ASE on a genome-scale continues to be a technical challenge. In this chapter, we provide introductions to strategic approaches for genome-scale ASE identification, and demonstrate with specific examples for ASE and downstream analyses. We will focus on RNA-Seq-based ASE analysis since RNA-Seq is the only approach that can provide information both on the level of expression and on the ratio of allelic expression genome-wide.

Gene expression is influenced by *cis*- and *trans*- acting genetic variations, epigenetic variations, and environmental factors. Genetic variations can modulate gene expression transcriptionally or post-transcriptionally, thereby altering phenotypes (Wood *et al.*, 2015). A diploid organism has two sets of chromosomes and thereby two alleles at a given locus. ASE refers to the phenomenon that the two alleles are not equally expressed, up to the exclusive expression of only one of the two alleles. ASE is usually associated with a variety of mechanisms, including genomic imprinting, X-chromosome inactivation, and transcriptional and post-transcriptional regulation (Wood *et al.*, 2015). The importance of regulatory sequence has become increasingly apparent in recent studies (Crowley *et al.*, 2015; Gan *et al.*, 2011; Keane *et al.*, 2011). Changes in *cis*- regulatory elements and transcribed regions affect transcription and mRNA stability (Shi *et al.*, 2012). A number of recent studies have demonstrated that ASE is also common among autosomal, non-imprinted genes (Lo *et al.*, 2003; Yan *et al.*, 2002).

Furthermore, evidence showed that these allele-specific differences in expression are heritable (Yan *et al.*, 2002). Interest in the existence of ASE in non-imprinted autosomal genes has increased with awareness of the important role that variations in non-coding DNA sequences can play in determining phenotypic diversity (Knight, 2004).

ASE is an effective method for the identification of *cis*- acting genetic factors. For instance, gene expression is under strict negative-feedback control (i.e., transcription is

Bioinformatics in Aquaculture: Principles and Methods, First Edition. Edited by Zhanjiang (John) Liu.

repressed by an increasing amount of the transcriptional product), and thus, the total expression of genes varied little across samples with different *cis*- regulatory genotypes (Pastinen, 2010). However, in a heterozygous individual, both alleles are under the control of the same negative feedback and share other non-*cis*-acting factors (e.g., environment), which means, even in the presence of extreme feedback, measurement of ASE can readily detect differences between the alleles, which reflects the intrinsic activity of *cis*-regulatory genetic variations.

A number of approaches have been used for the detection of ASE. In early studies, ASE was detected by single-base extension of a primer adjacent to the variable single-nucleotide polymorphism (SNP) (Carrel & Willard, 2005; Cowles *et al.*, 2002). Several recent studies applied a variety of technologies to scale up the tested genes (Guo *et al.*, 2008; Jeong *et al.*, 2007), of which array-based approaches were the most widely used. A number of array-based ASE studies have been published in the past decade (Bjornsson *et al.*, 2008; Daelemans *et al.*, 2010; Lo *et al.*, 2003; Serre *et al.*, 2008; Zhang & Borevitz, 2009). Due to rapidly increasing throughput and decreasing costs, next-generation sequencing (NGS) is rapidly replacing array-based technology for functional genomic assays (Rozowsky *et al.*, 2011). In addition, the ability to resolve single base differences, digital quantification, and comprehensive genome-wide coverage provide information on the abundance and the allelic biases in transcripts or regulatory DNA, which otherwise could not be achieved using hybridization-based arrays (Wood *et al.*, 2015). NGS-based approaches such as RNA-Seq can be used for global functional genomic assays by interrogating allelic effects when reads overlap a site that provides a high-quality heterozygote call (Pastinen, 2010). RNA-Seq technology using high-throughput sequencing platforms allows relatively unbiased measurements of expression across the entire length of a transcript. This technology has several advantages, including the ability to detect transcription of unannotated exons, measure both overall and exon-specific expression levels, and assay ASE (Pickrell *et al.*, 2010). Notably, RNA-Seq is the only technology that provides concurrent allelic and total expression data, which is attractive for the study of *cis*-regulatory changes detected by total expression, and for further quantifying the relative contribution of *cis*- versus *trans*- regulatory changes (Pastinen, 2010).

Measuring ASE is vital to better understanding global mechanisms of genetic variations. ASE analysis has been widely employed in mammals, insects, and plant systems (Bell *et al.*, 2013; Combes *et al.*, 2015; Gregg *et al.*, 2010; Serre *et al.*, 2008; Shi *et al.*, 2012; Springer & Stupar, 2007; Wittkopp, Haerum & Clark, 2008). In aquaculture species, despite many reports regarding expression of specific genes, very little is known about ASE (Murata, Oda & Mitani, 2012; Shen *et al.*, 2012). In general, these observations have been made with small gene sets. In this chapter, we will provide detailed methodologies and bioinformatics procedures for genome-wide allele-specific analysis by using high-throughput RNA-Seq data.

Genome-wide Approaches for ASE Analysis

In early studies, ASE was detected by PCR-based methods, such as real-time quantitative PCR or discrimination of PCR products by differing primer extensions (Carrel & Willard, 2005; Cowles *et al.*, 2002). However, these methods can only identify and characterize variations for a small number of gene sets. Genome-wide ASE patterns are now accessible with recent advances in genomic technologies. With the availability of high-density arrays and NGS, genome-wide assessment of ASE has recently become feasible. The general approaches for genome-wide allele-specific analysis can be grouped into two categories: polymorphism-directed approach and the *NGS-based* approach (Pastinen, 2010).

Polymorphism-based Approach

The polymorphism-based approach relies on differential hybridization of mRNA (convert to cDNA) to arrays on which previously characterized SNP probes are displayed (Liu *et al.*, 2012). In the past decade, numerous array-based ASE studies have been reported (Bjornsson *et al.*, 2008; Daelemans *et al.*, 2010; Lo *et al.*, 2003; Serre *et al.*, 2008; Zhang & Borevitz, 2009). The advantage of this approach is its relatively low cost. However, its obvious limiting factor is the dependence upon preselected regions present on the arrays, which results in only a small fraction of the genome being evaluated (Bell & Beck, 2009). For example, the low coverage of ASE sites in regulatory elements is a concern for this method because current standard SNP arrays contain only a small subset of polymorphic regulatory elements (<5%) (Pastinen, 2010).

The use of padlock probes is an alternative way to assay allelic expression (Lee *et al.*, 2009; Zhang *et al.*, 2009). Padlock probes are oligonucleotides that become circularized by DNA ligation in the presence of an appropriate target sequence. The digital quantification of the alleles in captured sequences from a genomic DNA control and corresponding cDNA is carried out by NGS, followed by allele counting of the short reads (Pastinen, 2010).

Padlock probes exhibit very high specificity and allow targeted analysis of tens of thousands of sites in a genome (Antson *et al.*, 2000). This method has been used to detect genomic DNA or cDNA variations for gene expression analysis in response to disease- and drug-associated studies (Banér *et al.*, 2003), and was recently extended to understanding allelic variation in DNA methylation by the capture of CpG islands in bisulfite-treated DNA (Shoemaker *et al.*, 2010).

NGS-based Assays

The rapid development in NGS technologies allows unbiased views of allelic variations in the transcripts or regulatory DNA with a high resolution of single bases, which cannot be achieved by using hybridization-based arrays. As a result, for functional genomic assays, next-generation short read sequencing is rapidly taking the place of array-based technologies.

RNA-Seq has become the method of choice for in-depth analysis of the whole transcriptome of an organism. The RNA-Seq approach allows genome-wide measurements of ASE for both protein-coding genes and noncoding RNA, and also offers an improved dynamic range over microarrays and results in digital allele counts with precision limited only by depth of coverage (Skelly *et al.*, 2011). Moreover, RNA-Seq is considered as the only method that provides both current allelic and total expression data, which provides the possibility for investigating *cis-* and *trans-* regulatory effects.

The basic principles of ASE analysis rely on defining the allele-specific origin of RNA using the polymorphisms of a transcribed marker (in most cases, SNP) and quantifying

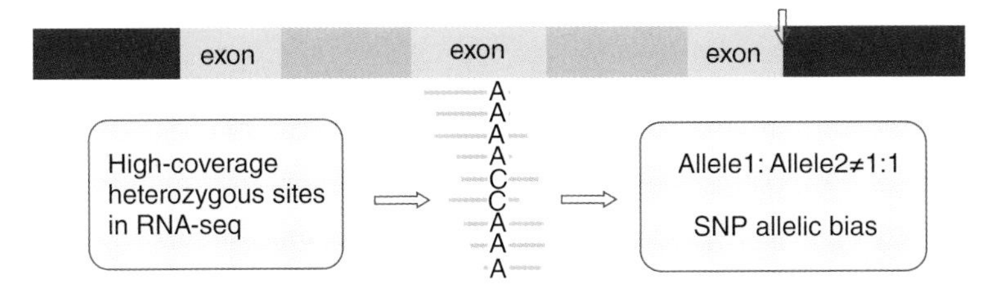

Figure 14.1 Allelic bias at high-coverage heterozygous sites is determined by RNA-Seq assay. Allelic bias at such sites is determined if the ratio of reads from each allele (Allele 1:Allele 2) deviates from 1:1. Here in the example, 7A:2C.

the relative abundance of the two alleles within transcripts (Singer-Sam *et al.*, 1992). For RNA-Seq, short sequence reads are collected relatively uniformly across expressed transcripts, and these reads correlate with the abundance of the tested samples (Pastinen, 2010). Sequencing reads at the polymorphic sites provides quantitative information about allelic abundance (Figure 14.1). In that case, genes with abundant genetic variation in their mRNA and are robustly expressed in the tested sample can yield high coverage for the expressed polymorphic sites. The allelic bias can be detected by simple statistical approaches (e.g., binomial tests, Fisher's exact test).

Applications of ASE Analysis

Detection of *Cis-* and *Trans-* Regulatory Effects

Differences in gene expression are important sources of phenotypic variations, which can arise from *cis*-regulatory changes or *trans*-regulatory changes. *Cis*-regulatory variants affect transcription initiation, transcription rate, and transcript stability in an allele-specific manner. *Trans*-regulatory changes modify the activity or expression of factors that interact with *cis*-regulatory sequences (Wittkopp, Haerum & Clark, 2004). As such, distinguishing the relative contribution of *cis-* and *trans-* effects to the divergence of gene expression is an essential step in elucidating its genetic basis (Tirosh *et al.*, 2009). The contribution of *cis-* versus *trans-* changes can be determined by comparing the allelic expression ratios between two parental species/strains to the F1 hybrids (Wittkopp *et al.*, 2008). Within the hybrid, both alleles of each gene are exposed to the same nuclear environment; thus, the *trans-* expression difference between two parental genes would be eliminated. Therefore, differential expression of two alleles in the F1 hybrid would reflect the *cis-* effect. *Trans-* effect can then be inferred from the total expression differences that are not explained by *cis-*. Consequently, *trans*-regulatory divergence could be inferred for genes with significant differences in the ratio of two alleles between F1 hybrids and the parental species (Shi *et al.*, 2012; Wittkopp *et al.*, 2004).

Based on the preceding principles, the mode of gene regulation can be classified in the four patterns shown in Figure 14.2.

Cis- effect only: the same allelic expression bias in both parents and the hybrid (case 1 in Figure 14.2); *trans-* effect only: biased expression in the parents, but equal expression in the hybrid (case 2 in Figure 14.2); *cis-* plus *trans-* regulation: biased expression in the

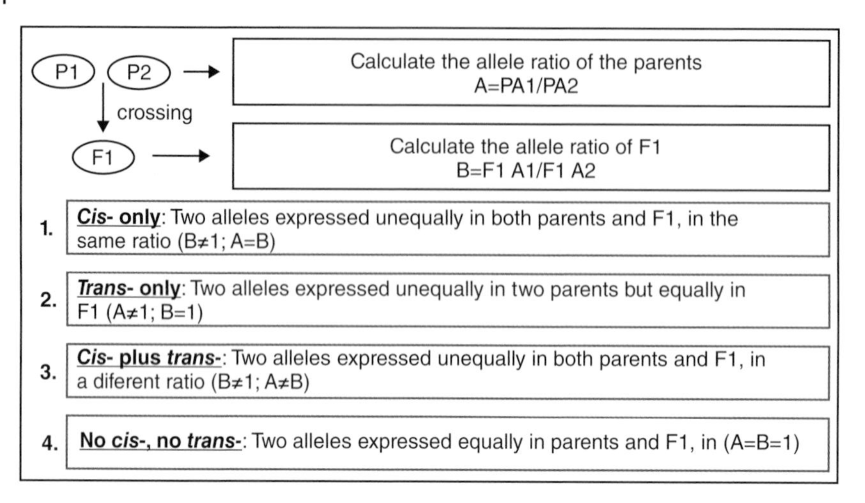

Figure 14.2 Quantification of *cis-* and *trans-* effects and classification of the genes with *cis-* and *trans-* effects. First step is to determine the expression ratio (A) at the same locus in two parents (may have different *trans-* and *cis-* environments). The second step is to determine the allele expression ratio of this gene in F1 hybrid (now the two alleles have the same *trans-* environment, but may have different *cis-* regulatory elements). The regulation can be classified as (1) *cis-* only when the two alleles are expressed unequally but in the same ratio in parents and in F1; (2) *trans-* only when the two alleles are expressed unequally in the two parents but equally in F1; (3) *cis-* plus *trans-* when the two alleles are expressed unequally in the parents and in F1, and the ratios are different. If the two alleles are expressed all equally, then there is no *cis-* or *trans-*regulation (4). For situation (3) (*cis-* plus *trans-*), the regulation is referred to as *enhancing* if the ratio of the parents and of the F1 are in the same direction, or as *compensating* when the ratio of the parents and of the F1 are in different directions.

hybrid that is different from the expression bias in the parents (case 3 in Figure 14.2). Under this latest situation, the regulation is referred to as *enhancing* if the allele ratio of the parent and the ratio of the F1 are in the same direction, or as *compensating* when the ratios are in opposite directions. Of course, if the two alleles are expressed equally, there is no *cis-* or *trans-* regulation.

Cis- and *trans-* effects have been investigated in intra- and inter-specific hybrids of Drosophila (Meiklejohn *et al.*, 2014; Wittkopp *et al.*, 2004, 2008), yeast (Schaefke *et al.*, 2013; Tirosh *et al.*, 2009), mouse (Goncalves *et al.*, 2012; Wilson *et al.*, 2008), and various species of plants (Bell *et al.*, 2013; Combes *et al.*, 2015). These studies indicated that the *cis*-regulatory divergences predominated and explained more of the expression differences between species than within species, whereas *trans*-regulation is often associated with *cis*-regulation and sensory responses to the environment (Shi *et al.*, 2012). Further, these studies also indicated that most of the genes exhibiting both *cis-* and *trans-* effects were subjected to compensating interaction, implying that compensatory regulation to stabilize gene expression levels is widespread (Bell *et al.*, 2013; Goncalves *et al.*, 2012; Shi *et al.*, 2012). Only a few studies were conducted with ASE analysis in aquaculture species. In stickleback, ASE analysis provided evidence that *cis*-regulation contributes to the differential gene expression, but those studies were only restricted to a couple of genes (Kitano *et al.*, 2010; Miller *et al.*, 2007). There are several other reports on the analysis of ASE (Garcia *et al.*, 2014; Pala, Coelho & Schartl, 2008; Shen *et al.*, 2012; Wang, Sun *et al.*, 2013). However, for the most part, the molecular basis of gene expression divergence of aquaculture species is largely unknown.

Detection of Parent-of-Origin Effects

Another important application of ASE analysis is to detect parent-of-origin effects. Parent-of-origin effects are epigenetic phenomena that appear as phenotypic differences between heterozygotes depending on the allelic origin from the parents. Several phenomena including genomic imprinting, gene-specific trinucleotide expansions

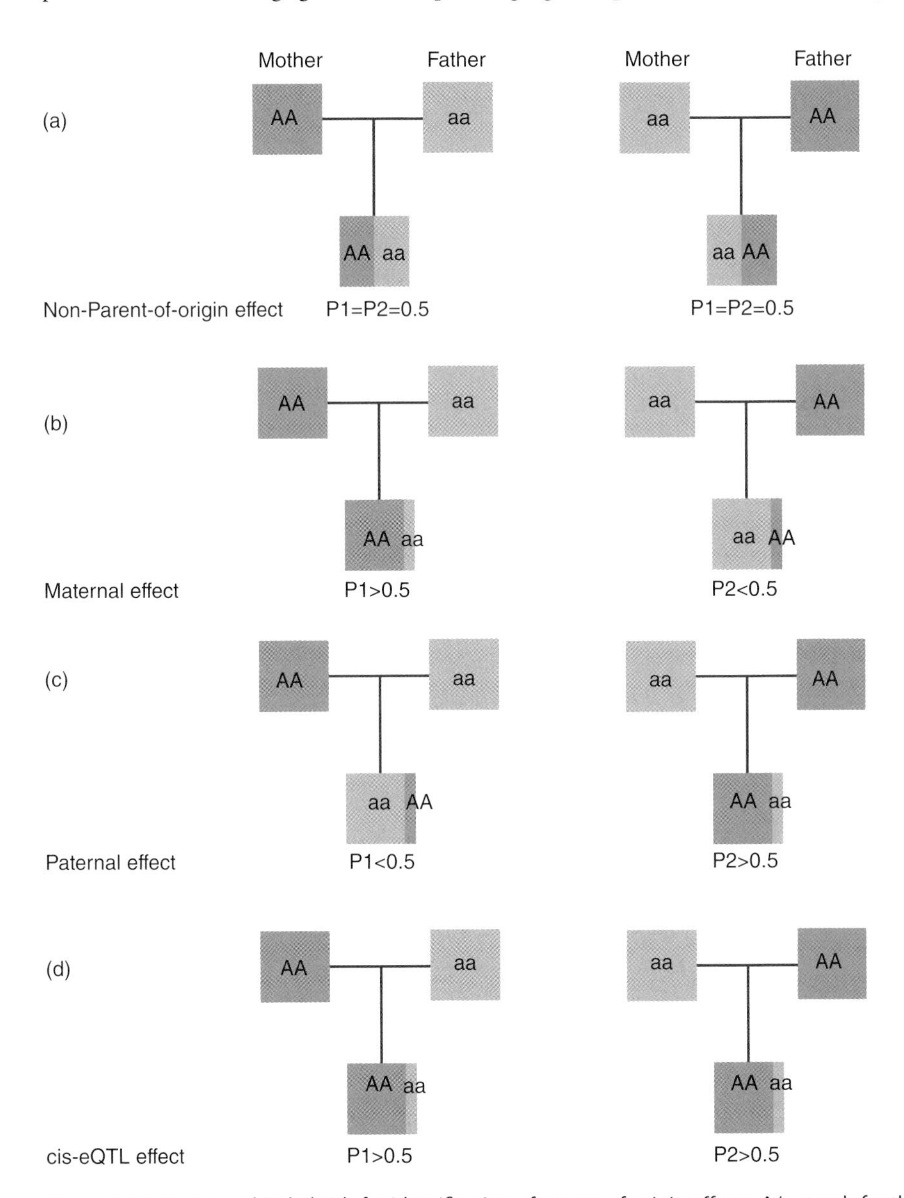

Figure 14.3 Reciprocal F1 hybrids for identification of parent-of-origin effects. A/a stands for the two alleles of informative SNP site in polymorphic parents; (a) alleles are expressed equally, exhibiting with a pattern of non-parent-of-origin effect; (b) maternal alleles are expressed more abundantly, exhibiting a pattern of maternal effect; (c) paternal alleles are expressed more abundantly, exhibiting a pattern of paternal effect; (d) a specific allele is preferentially expressed, exhibiting a pattern of *cis*-eQTL effect. (*See color plate section for the color representation of this figure.*)

(Pearson, 2003; Tome *et al.*, 2011), and parental genetic effects (Hager, Cheverud & Wolf, 2008) can cause parent-of-origin effects. Genomic imprinting leads to functional non-equivalency of the two alleles at a locus. It is considered to be the primary epigenetic phenomenon that can lead to the manifestation of parent-of-origin effects. Increasing genomic data has demonstrated that imprinted genes can be important contributors to complex trait variations (see review by Lawson, Cheverud & Wolf, 2013). Therefore, interpreting of these parent-of-origin effects is essential to understand the genetic architecture and evolution of complex traits.

Genomic imprinting has been observed in both animals and plants (Kohler & Weinhofer-Molisch, 2010), but has not been reported for fish. In animals, genomic imprinting is only known in therian mammals, and most imprinted genes are expressed and imprinted in the brain and placenta (Pask, 2012; Renfree, Suzuki & Kaneko-Ishino, 2013; Wang, Soloway & Clark, 2011). However, the taxonomic distribution and the breadth of imprinting remain uncertain. For identifying genes with parent-of-origin effects, the direct way is to score the differential ASE in the offspring of reciprocal crosses, where one can identify which parent transmitted each allele (Figure 14.3).

To detect significant parent-of-origin expression, allele ratios can be determined from reciprocal F1 crosses. When alleles are expressed equally, there is no parent-of-origin effect (Figure 14.3a). When the maternal allele is preferentially expressed, there is maternal effect (Figure 14.3b). When the paternal allele is preferentially expressed, there is paternal effect (Figure 14.3c), and, finally, when a specific allele is preferentially expressed regardless of the parent of origin, there is *cis*-eQTL effect (Figure 14.3d), which means the regulatory effect is caused by eQTL (expression Quantitative Trait Loci) close to the associated transcripts or genes.

Considerations of ASE Analysis by RNA-Seq

Although ASE analysis using RNA-Seq is simple, considerable caution should be exercised at several key steps to avoid false positives. Here, we discuss challenges that should be considered before ASE analysis using RNA-Seq.

Biological Materials

Studies on *cis*- and *trans*-regulation are often performed in inter- or intra-specific F1 hybrids. RNA samples from reciprocal hybrid F1 should be used for the detection of parent-of-origin effects. The use of crosses of two inbred stains/lines is most efficient because every single SNP detected in the transcripts of F1 is informative with trackable transmission direction. However, the inbred samples are not always easy to obtain for many animal species. It could be challenging to perform a genome-wide survey for ASE genes as SNP positions are not always heterozygous in all genes of the assayed individuals. In that case, a larger sample size is needed to capture representative individuals expressing an identified paternal/maternal allele in F1 hybrids and to achieve genome-wide coverage (Wang & Clark, 2014). Moreover, independent biological replication for each cross and technical replicates should be used instead of pooling the samples in order to identify ASE more accurately.

RNA-Seq Library Complexity

Adequate library complexity is critical for quantitative ASE analysis using RNA-Seq. It has been reported that the allele expression ratios could not be accurately quantified for lowly and moderately expressed genes. The errors can occur at all stages when the number of molecules during the library preparation is small. To improve the library complexity and ensure proper data analysis, researchers can follow the suggestions made by Wang and Clark (2014), listed in the following text:

1) *Starting with more input RNA*: It is suggested that at least 2 μg input total RNA should be used.
2) *Planning for biological and technical replicates.* As deeper sequencing and pooling samples before library preparation cannot solve library complexity problem, independent biological replication in each of the two reciprocal crossed and/or different technique replicates for a single sample at library preparation level are suggested.
3) *Determining library complexity*: The high value of the over-dispersion parameter for beta-binomial distribution can be used as an indicator of poor library complexity. Moreover, the empirical Bayesian method implemented in pre-Seq software can also be used for characterizing the molecular complexity (Daley & Smith, 2013).
4) *Validating the candidate genes*: Because of the cost limit, most of the earlier RNA-Seq studies lack biological or technical replicates. As a result, it is essential to validate the candidate genes with ASE across multiple individual samples by allele-specific pyrosequencing or other methods.
5) *Selecting candidate genes with multiple independent SNP support*: Generally, the candidate gene with multiple independent SNP support has a higher rate of successful verification. Here, the independent SNPs stand for those informative SNPs separated by a distance greater than the read length of RNA-Seq (most often ~100 bp).

Genome Mapping Bias

When aligning the RNA-Seq reads from F1 hybrids, in most cases, only one reference genome is available for an organism, resulting in a bias toward the reference allele. As a result, reads from the parental strain or species that is more closely related to the reference genome will be over-represented by using the same cutoff values of mismatches. The allele expression ratios will shift toward the reference allele. This alignment bias can have a significant effect on the estimation of allele expression ratios, and the effect is more severe in regions with high SNP density (Wang & Clark, 2014).

Two methods have been applied to solve this problem. The first strategy is to apply different mismatch cutoff values for the reference and alternative allele-containing reads (Wang *et al.*, 2008). However, deciding the exact cutoff values is difficult, and this method is not applicable when a single read contains both the reference allele and the alternative allele for different SNP sites. The best way to remove mapping bias is to align the sequencing reads to both the reference genome and a pseudo-genome constructed by substituting the reference alleles with the alternative alleles in all transcribed regions, and then take the average counts from the reference and pseudo-genome (Wang, Miller *et al.*, 2013).

Removing Problematic SNPs

Problematic SNPs need to be removed to reduce the false positive rate. These problematic SNPs are caused due to the following reasons: (1) genotyping errors; (2) RNA editing; (3) SNPs being located in repetitive/paralogous regions; (4) mis-alignment over INDELs; and (5) wrongly aligned junction SNPs nearing the exon-intron boundaries.

The effect of problematic SNPs can be eliminated by using RNA-Seq and gDNA-Seq data from parental cross as follows: (1) correct positions with sequence inconsistence in the reference genome according to gDNA-Seq data from parental strain; (2) exclude the edited positions from the analysis by searching for different base positions between RNA-Seq and gDNA-Seq data in parental strains; (3) perform local realignment over INDEL positions using software such as GATK to remove problematic SNPs; and (4) remove mis-aligned SNPs near the exon–intron boundaries.

Step-by-Step Illustration of ASE Analysis by RNA-Seq

Trimming of Sequencing Reads

Before ASE analysis, it is suggested to remove the poor quality sequences or technical sequences such as adapters in NGS data. The step can be done by a software named Trimmomatic, which can meet the requirements in terms of flexibility, correct handling of paired-end data, and high performance (Bolger, Lohse & Usadel, 2014).

Trimmomatic is licensed under GPL V3. It is a cross-platform (Java 1.5+ required) software, and is available at http://www.usadellab.org/cms/index.php?page=trimmomatic.

Trimmomatic includes a variety of processing steps for read trimming and filtering, but the main algorithm is related to the identification of adapter sequences and quality filtering. For trimming of paired end sequences, the basic command line is:

```
java -jar <path to trimmomatic.jar> PE [-threads <threads]
   [-phred33 | -phred64] <input 1> <input 2> <paired
   output 1> <unpaired output 1> <paired output 2> <unpaired
   output 2> ILLUMINACLIP: <fastaWithAdaptersEtc>:<seed
   mismatches>: <palindrome clip threshold>:<simple clip
   threshold> LEADING:<quality> TRAILING:<quality>
   SLIDINGWINDOW: <window Size>:<requiredQuality> HEADCROP:
   <length> MINLEN: <length>
```

Aligning the RNA-Seq Reads to the Reference

A number of mapping software packages are available for short-read sequence alignment. Popular spliced read aligners include Tophat (Trapnell, Pachter & Salzberg, 2009), Subread (Liao, Smyth & Shi, 2013), and STAR (*S*pliced *T*ranscripts *A*lignment to a *R*eference) (Dobin *et al.*, 2013). Here, we recommend STAR, which is an ultrafast universal RNA-Seq aligner. The advantages of STAR also include the accurate alignment of contiguous and spliced reads; capability to detect polyA-tails, non-canonical splices, and fusion junction; and the ability to efficiently map reads of any length generated by current or emerging sequencing platforms. The detailed list of other software for read alignment can be found in Wikipedia (https://en.wikipedia.org/wiki/List_of_sequence_alignment_software).

Obtaining STAR

STAR is a free, open-source software package distributed under the GPL V3 license, and can be downloaded from http://code.google.com/p/rna-star/. STAR is implemented as C++ code. The only pre-requisite with STAR is computer RAM—STAR requires at least 30 GB RAM to align to the human or mouse genomes. The pre-compiled STAR executables are located in the directory "bin/subdirectory". For the Linux environment, run "make" in the source directory to compile STAR from the source files.

Build a Genome Index

As with all aligners, users need to build the genome index first. The reference genome sequences (FASTA files) and annotation (GTF file) should be supplied. To build the index, the command is:

```
$ STAR --runMode genomeGenerate --runThreadN <Number of
   CPUs> --genomeDir <genome output directory>
   --genomeFastaFiles <Genome FASTA file> --sjdbGTFfile
   <Annotation file> --sjdbOverhang <ReadLength-1>
```

Mapping Reads to the Reference Genome

For the mapping step, users should supply the genome files generated in the first step, as well as the trimmed RNA-Seq reads in the form of FASTQ files. To align RNA-Seq reads with STAR, run the following command:

```
STAR -runThreadN < Number of CPUs > --genomeDir <Directory
   with the Genome Index> --readFilesIn <FASTQ file>
   --outFileNamePrefix <Output Prefix directory> --<option1>
   --<option2> ...
```

In addition to the basic options mentioned in the preceding text, there are many advanced options that control the mapping behavior in STAR. The most commonly used options include: "`--outFilterMismatchNoverLmax`", which sets the maximum number of mismatches per pair relative to read length (e.g., for a paired read with length 100 bp, the maximum number of mismatches is $0.04 \times 200 = 8$ if we set 0.04 for this option); "`--outFilterMismatchNmax`", which means alignment will be output only if it has fewer mismatches than this value; and "`--outFilterScoreMin`", which indicates that the alignment will be removed if the mapping quality score (MAPQ) is lower than this value. For STAR mapping, MAPQ = 255 stands for unique mapping as the default setting. For ASE analysis, usually only the unique mapped results should be used.

The Output Files

STAR creates several output files. The summary mapping statistics will be listed in the "Log.final.out" file, which includes some important mapping information such as number of uniquely mapped reads, percentage of uniquely mapped reads, number of splices, and number of multi-mapped reads.

The most important output file is "*Aligned.out.sam", which is the alignments in standard SAM format. Alternatively, users can convert this to a BAM file and sort it by using other programs (e.g., SAMtools), or obtain *Aligned.out.bam" directly by setting the "`--outSAMtype`" option. These alignment files will be used for downstream analysis.

SNP Calling

After getting the mapping results, the next step is the identification of SNPs that are used for allelic analysis. Here, we show the detailed procedure of SNP calling by using SAMtools (Li *et al.*, 2009), followed by VarScan (Koboldt *et al.*, 2009).

SAMtools

SAMtools is a universal tool for processing read alignments. To use the variant calling features of VarScan, the inputs are needed in mpileup format, which should be generated by SAMtools. It does merging and indexing, and allows retrieving reads in any regions swiftly. SAMtools is free software, available at http://sourceforge.net/projects/samtools/files/.

Installing SAMtools For compilation, type "cd" to the directory containing the package's source, and type "make" to compile SAMtools. Then type "make install" to install the SAMtools executable, various scripts and executables from "misc/", and a manual page to "/usr/local".

Quality Control and Sort To remove the ambiguously mapped reads and generate sorted BAM files simultaneously, the following command can be used:

```
samtools view -q value -bu XXX.bam | samtools sort
```

The option "`-q value`" means only sequences with a quality score higher than this value are to be kept; and "`-bu`" means the output file is in unzipped BAM format, which is the input format for the mpileup procedure.

Building the mpileup File The next step is to identify the transcriptomic variants. The mpileup command in SAMtools is used to calculate the genotype likelihoods supported by the aligned reads in tested samples. The mpileup command automatically scans every position supported by an aligned read, computes all the possible genotypes supported by these reads, and then computes the probability that each of these genotypes is truly present in the sample. The sorted BAM files are used to generate the "mpileup" file with the following command:

```
samtools mpileup -f reference.fasta sample1.bam sample2.bam
   sample3.bam ... > XXX.mpileup
```

The output "mpileup" file is a merged file with all the information of these variants in tested groups, and both groups-specific and shared variants are included. The "mpileup" file is used to call variants in VarScan.

VarScan

VarScan is a central, platform-independent tool for variant detection. VarScan employs a robust heuristic/statistic approach to call variants that meet the desired thresholds for read depth, base quality, variant allele frequency, and statistical significance (Koboldt *et al.*, 2009).

Installing VarScan VarScan is a free software tool written in Java that runs on any operating system (Linux, UNIX, macOS, and Windows). To install it, users should download the VarScan JAR file from SourceForge (http://sourceforge.net/projects/varscan/files/). Then, run the following command line to get the usage information:

```
Java -jar VarScan.jar
```

Variants Calling The following command is used to call SNPs from the "mpileup" file:

```
java -jar VarScan.jar mpileup2snp XXX.mpileup option1
   option2 ...
```

Four options are important for excluding false SNPs caused by sequencing errors: (1) "`--min-coverage minimum`" means minimum read depth at a position to make a call; (2) "`--min-reads2`" stands for minimum variant allele frequency threshold; (3) "`--min-var-freq`" is used for setting the minimum supporting reads at a position to call variant; and (4) "`--p-value`" sets the *p*-value threshold for calling variants. An optimal combination of these factors is used for screening qualified SNPs.

The output file generated by VarScan shows SNP calls in a tab-delimited format.

Filtering and Normalization

In order to identify reliable SNPs, further quality controls are applied as follows: (1) the variant alleles detected at each SNP site must only contain "A", "T", "G", or "C"; (2) each SNP must consist of only two alleles (one for reference allele, another for variant allele); (3) the read number of minor alleles at each SNP site should be ≥2; and (4) the total read number of alleles at each SNP site should be ≥20.

When comparing functional data between samples, it is necessary to consider normalization before any comparisons are made. To prevent making misleading inferences due to technical differences in sequencing depth between libraries, a total-count scaling method of normalization is usually applied to standardize the total number of reads between samples (Bell *et al.*, 2013; Marioni *et al.*, 2008; Robinson & Smyth, 2007). Then, normalized count data should be rounded to the nearest integer to satisfy the requirements of downstream statistical tests (e.g., Fisher's exact test, binomial test). This step can also be performed by using the DESeq package (Anders & Huber, 2010).

Quantification of ASE

Constructing a Pseudo-genome

As we mentioned earlier, RNA-Seq reads are often mapped only to one reference genome, and there will be mapping bias toward the reference allele if the same cut-off value is used. To remove mapping bias, a pseudo-genome should be constructed by replacing the reference allele in the genome with the alternative allele at the SNP sites (Figure 14.4).

Realigning the RNA-Seq Reads to the Pseudo-Genome

After constructing the pseudo-genome, the reads should be realigned to the pseudo-genome with the same cut-off value as when aligning to the original reference genome.

Ref	GTGTGTGTGTGTGTGTGTGTGTTTTTTTTACCAATATGTTGTCGAAGGCATA
Pseudo	GTGTGTGTGTGTGTGTGTGTGTGTGTTTTACCGATATGTTGTCGAAGGCATA

Figure 14.4 Example of constructing a psudo-genome. Compared with the reference genome, the pseudo-genome is constructed by replacing the base at the SNP site with the variant allele (indicated by shade).

Counting the Allele ratios

The averaged counts from the original reference and pseudo-genome are used as the final SNP count. For identification of genes with ASE pattern, only SNP sites at both parents that are homozygous with different genotypes are selected for analysis. Fixed polymorphisms between parents must be required to assign a parental origin unambiguously to hybrid reads. Allelic expression ratios should be calculated on a per-gene basis by summarizing all informative SNP positions. With the informative SNPs and the SNP counts, the ASE ratios can be determined by the relative counts from the reference and alternative alleles.

Downstream Analysis

Evaluating *Cis*- and *Trans*-regulatory Changes

As described in the preceding text, gene expression is regulated by *cis*- and/or *trans*-factors. Comparison of ASE in the hybrids with allelic expression in their parents is done to determine the relative contributions of *cis*- or *trans*-acting regulators. Binomial exact test with a null hypothesis of equal expression of the two parental alleles in the hybrids is used to identify the *cis*- effect (H_0: F1A1/F1A2 (hybrid) = 1:1). To test the *trans*-regulatory divergence, Fisher's exact test is used to test for the unequal allelic abundance between parental and F1 hybrids (H_0: PA1 : PA2 [parent] = F1A1 : F1A2 [hybrid])]. An FDR of 5% is applied to determine statistical significance. Practically, log2-transformed ratios are usually used to estimate the extent of allelic expression divergence. The *cis*-regulatory difference is calculated as *cis* = log2 (F1A1/F1A2). The *trans*-regulatory difference is calculated as the difference between the total expression and *cis*-regulatory differences: *trans* = log2 (PA1/PA2) − log2 (F1A1/F1A2). Based on these tests, genes are categorized as conserved, *cis*- only, *trans*- only, compensating, and enhancing regulatory effects (Figure 14.2). An example of *cis- and trans*-regulatory analysis is shown in Table 14.1.

Detecting of Parental of Origin Effects

As described in the preceding text, imprinting is required of a paternal or maternal bias greater than 50% in both reciprocal crosses. If we define P1 as the expression percentage from the maternal allele in F1 hybrid cross and P2 as the paternal allele percentage in its reciprocal cross, for non-imprinted genes, the expression value should be P1 = P2 = 0.5. To quantify the degree of genomic imprinting, P1 − P2 is used as a measurement of the parent-of-origin effect, ranging from −1 to +1 (where −1 indicates 100% maternally expressed imprinted gene; 0 indicates non-imprinted genes, and +1 indicates 100%

Table 14.1 Example of *cis*- and *trans*-regulatory analysis. ID is gene ID; PA1 is the allele count in parent 1; PA2 is the allele count in parent 2; F1A1 is the count of the first allele in F1; F1A2 is the count of the second allele in F1.

ID	PA1	PA2	F1A1	F1A2	log2 (PA1/PA2)	log2 (F1A1/F1A2)	log2 (PA1/PA2) – log2 (F1A1/F1A2)	FDRp (PA1/PA2 = F1A1/F1A2)	FDRp (F1A1 = F1A2)	Regulatory effect
g1	203	35	130	32	2.54	2.02	0.51	0.409804	6.33E – 14	Only *cis*
g3	182	207	205	149	−0.19	0.46	−0.65	0.014638	0.009809	Compensating
g4	189	18	91	7	3.39	3.7	−0.31	0.930481	2.82E – 18	Only *cis*
g6	133	141	96	79	−0.08	0.28	−0.37	0.397454	0.340036	Conserved
g10	161	135	98	42	0.25	1.22	−0.97	0.013815	1.41E – 05	Compensating
g15	122	106	67	41	0.2	0.71	−0.51	0.329935	0.037278	Only *cis*
g16	111	126	55	42	−0.18	0.39	−0.57	0.267139	0.33575	Conserved
g19	113	93	79	42	0.28	0.91	−0.63	0.203734	0.003233	Only *cis*
g23	163	237	150	115	−0.54	0.38	−0.92	0.000771	0.075503	Only *trans*
g25	221	127	131	68	0.8	0.95	−0.15	0.812724	4.8E – 05	Only *cis*
g29	103	82	49	35	0.33	0.49	−0.16	0.849161	0.250064	Conserved
g33	120	83	43	51	0.53	−0.25	0.78	0.109387	0.601401	Conserved
g36	128	66	79	43	0.96	0.88	0.08	0.985198	0.004523	Only *cis*
g39	102	46	67	41	1.15	0.71	0.44	0.484893	0.037278	Only *cis*
g50	178	141	152	91	0.34	0.74	−0.4	0.270907	0.000451	Only *cis*
g55	202	166	122	79	0.28	0.63	−0.34	0.367096	0.008625	Only *cis*
g57	184	108	81	75	0.77	0.11	0.66	0.091231	0.798462	Conserved
g61	231	375	132	152	−0.7	−0.2	−0.5	0.071598	0.379707	Conserved
g66	112	108	49	55	0.05	−0.17	0.22	0.738772	0.745157	Conserved
g69	142	156	74	34	−0.14	1.12	−1.26	0.001799	0.000596	Compensating
g72	247	447	144	161	−0.86	−0.16	−0.69	0.005216	0.489941	Enhancing

Table 14.2 Example of genes showing imprinting patterns. ID is gene ID; Hyb_reads1: reads count for allele 1 in F1 hybrid; Hyb_reads2: reads count for allele 2 in F1 hybrid; Rec_Reads1: reads count of allele 1 in reciprocal hybrid; Rec_Reads2: reads count of allele 2 in reciprocal hybrid; P1 is the expression percentage from the maternal allele in F1 hybrid; and P2 is the expression percentage from the paternal allele in reciprocal hybrid.

ID	Hyb_ Reads1	Hyb_ Reads2	Rec_ Reads1	Rec_ Reads2	P1	1 − P1	1 − P2	P2	Storer–Kim test *p*-value	Parent-of-origin
g1	374	8	18	58	0.02	0.98	0.24	0.76	0	Paternal
g103	247	39	37	85	0.14	0.86	0.30	0.70	4E − 31	Paternal
g130	170	11	28	59	0.06	0.94	0.32	0.68	7.94E − 27	Paternal
g135	37	258	88	39	0.87	0.13	0.69	0.31	2.01E − 34	Maternal
g15	220	13	12	78	0.06	0.94	0.13	0.87	0	Paternal
g155	141	16	16	58	0.10	0.90	0.22	0.78	3.85E − 26	Paternal
g165	172	17	24	55	0.09	0.91	0.30	0.70	1.53E − 23	Paternal
g168	193	36	30	74	0.16	0.84	0.29	0.71	2.25E − 23	Paternal
g175	36	189	67	11	0.84	0.16	0.86	0.14	6.63E − 30	Maternal
g180	180	25	25	61	0.12	0.88	0.29	0.71	3.91E − 23	Paternal
g189	176	25	25	61	0.12	0.88	0.29	0.71	7.57E − 23	Paternal
g19	18	170	58	8	0.90	0.10	0.88	0.12	0	Maternal

paternally expressed imprinted gene). Chi-squared test or Fisher's exact test can be used, but the Storer–Kim test has more power when the four counts are smaller (Storer & Kim, 1990; Wang & Clark, 2014). An arbitrary cut-off of $P1 > 0.65$ and $P2 < 0.35$ for maternally expressed candidates and $P1 < 0.35$ and $P2 > 0.65$ for paternally expressed ones are used to include the significant partially imprinted candidates. An example of genes exhibiting imprinting patterns and their calculations are shown in Table 14.2.

Validating of the ASE Ratio

To validate the ASE genes identified from RNA-Seq, an independent validation experiment is needed. Allele-specific quantitative reverse transcriptase-PCR (Singer-Sam & Gao, 2001), Sanger sequencing, or allele-specific pyrosequencing (Wang & Elbein, 2007) are common assays for validation of the ASE ratios. Although Sanger sequencing of the F1 cDNA can be used for validation of all-or-none or nearly fully silenced alleles, it cannot provide accurate allelic expression ratios because only a limited number of clones can be sequenced. Allele-specific pyrosequencing is an accurate method for quantifying allele ratios with only 2–5% error rate, which has better resolution than quantitative reverse transcriptase-PCR method. The validation assay should be performed in a panel of independent individuals to exclude the possibility of random monoallelic expression (Wang & Clark, 2014).

References

Anders, S. and Huber, W. (2010) Differential expression analysis for sequence count data. *Genome Biology*, **11**, R106.

Antson, D.O., Isaksson, A., Landegren, U. and Nilsson, M. (2000) PCR-generated padlock probes detect single nucleotide variation in genomic DNA. *Nucleic Acids Research*, **28**, E58.

Ban.r, J., Isaksson, A., Waldenström, E. *et al.* (2003) Parallel gene analysis with allele-specific padlock probes and tag microarrays. *Nucleic Acids Research*, **31**, e103.

Bell, C.G. and Beck, S. (2009) Advances in the identification and analysis of allele-specific expression. *Genome Medicine*, **1**, 56.

Bell, G.D., Kane, N.C., Rieseberg, L.H. and Adams, K.L. (2013) RNA-seq analysis of allele-specific expression, hybrid effects, and regulatory divergence in hybrids compared with their parents from natural populations. *Genome Biology and Evolution*, **5**, 1309–1323.

Bjornsson, H.T., Albert, T.J., Ladd-Acosta, C.M. *et al.* (2008) SNP-specific array-based allele-specific expression analysis. *Genome Research*, **18**, 771–779.

Bolger, A.M., Lohse, M. and Usadel, B. (2014) Trimmomatic: a flexible trimmer for Illumina sequence data. *Bioinformatics*, btu170.

Carrel, L. and Willard, H.F. (2005) X-inactivation profile reveals extensive variability in X-linked gene expression in females. *Nature*, **434**, 400–404.

Combes, M.-C., Hueber, Y., Dereeper, A., *et al.* (2015) Regulatory divergence between parental alleles determines gene expression patterns in hybrids. *Genome Biology and Evolution*, **7**, 1110–1121.

Cowles, C.R., Hirschhorn, J.N., Altshuler, D. and Lander, E.S. (2002) Detection of regulatory variation in mouse genes. *Nature Genetics*, **32**, 432–437.

Crowley, J.J., Zhabotynsky, V., Sun, W., *et al.* (2015) Analyses of allele-specific gene expression in highly divergent mouse crosses identifies pervasive allelic imbalance. *Nature Genetics*, **47**, 353–360.

Daelemans, C., Ritchie, M.E., Smits, G., *et al.* (2010) High-throughput analysis of candidate imprinted genes and allele-specific gene expression in the human term placenta. *BMC Genetics*, **11**, 25.

Daley, T. and Smith, A.D. (2013) Predicting the molecular complexity of sequencing libraries. *Nature Methods*, **10**, 325–327.

Dobin, A., Davis, C.A., Schlesinger, F., *et al.* (2013) STAR: ultrafast universal RNA-seq aligner. *Bioinformatics*, **29**, 15–21.

Gan, X., Stegle, O., Behr, J., *et al.* (2011) Multiple reference genomes and transcriptomes for Arabidopsis thaliana. *Nature*, **477**, 419–423.

Garcia, T.I., Matos, I., Shen, Y., *et al.* (2014) Novel method for analysis of allele specific expression in triploid *Oryzias latipes* reveals consistent pattern of allele exclusion. *PLoS One*, **9**, e100250.

Goncalves, A., Leigh-Brown, S., Thybert, D., *et al.* (2012) Extensive compensatory *cis–trans* regulation in the evolution of mouse gene expression. *Genome Research*, **22**, 2376–2384.

Gregg, C., Zhang, J., Weissbourd, B., *et al.* (2010) High-resolution analysis of parent-of-origin allelic expression in the mouse brain. *Science*, **329**, 643–648.

Guo, M., Yang, S., Rupe, M., *et al.* (2008) Genome-wide allele-specific expression analysis using massively parallel signature sequencing (MPSS™) reveals *cis-* and *trans-*effects on gene expression in maize hybrid meristem tissue. *Plant Molecular Biology*, **66**, 551–563.

Hager, R., Cheverud, J.M. and Wolf, J.B. (2008) Maternal effects as the cause of parent-of-origin effects that mimic genomic imprinting. *Genetics*, **178**, 1755–1762.

Jeong, S., Hahn, Y., Rong, Q. and Pfeifer, K. (2007) Accurate quantitation of allele-specific expression patterns by analysis of DNA melting. *Genome Research*, **17**, 1093–1100.

Keane, T.M., Goodstadt, L., Danecek, P., *et al.* (2011) Mouse genomic variation and its effect on phenotypes and gene regulation. *Nature*, **477**, 289–294.

Kitano, J., Lema, S.C., Luckenbach, J.A., *et al.* (2010) Adaptive divergence in the thyroid hormone signaling pathway in the stickleback radiation. *Current Biology*, **20**, 2124–2130.

Knight, J.C. (2004) Allele-specific gene expression uncovered. *Trends in Genetics*, **20**, 113–116.

Koboldt, D.C., Chen, K., Wylie, T., *et al.* (2009) VarScan: variant detection in massively parallel sequencing of individual and pooled samples. *Bioinformatics*, **25**, 2283–2285.

Kohler, C. and Weinhofer-Molisch, I. (2010) Mechanisms and evolution of genomic imprinting in plants. *Heredity*, **105**, 57–63.

Lawson, H.A., Cheverud, J.M. and Wolf, J.B. (2013) Genomic imprinting and parent-of-origin effects on complex traits. *Nature Reviews Genetics*, **14**, 609–617.

Lee, J.-H., Park, I.-H., Gao, Y., *et al.* (2009) A robust approach to identifying tissue-specific gene expression regulatory variants using personalized human induced pluripotent stem cells. *PLOS Genetics*, **5**, e1000718.

Li, H., Handsaker, B., Wysoker, A., *et al.* (2009) The sequence alignment/map format and SAMtools. *Bioinformatics*, **25**, 2078–2079.

Liao, Y., Smyth, G.K. and Shi, W. (2013) The Subread aligner: fast, accurate and scalable read mapping by seed-and-vote. *Nucleic Acids Research*, **41**, e108.

Liu, R., Maia, A.T., Russell, R., *et al.* (2012) Allele-specific expression analysis methods for high-density SNP microarray data. *Bioinformatics*, **28**, 1102–1108.

Lo, H.S., Wang, Z., Hu, Y., *et al.* (2003) Allelic variation in gene expression is common in the human genome. *Genome Research*, **13**, 1855–1862.

Marioni, J.C., Mason, C.E., Mane, S.M., *et al.* (2008) RNA-seq: an assessment of technical reproducibility and comparison with gene expression arrays. *Genome Research*, **18**, 1509–1517.

Meiklejohn, C.D., Coolon, J.D., Hartl, D.L. and Wittkopp, P.J. (2014) The roles of *cis-* and *trans-*regulation in the evolution of regulatory incompatibilities and sexually dimorphic gene expression. *Genome Research*, **24**, 84–95.

Miller, C.T., Beleza, S., Pollen, A.A., *et al.* (2007) *cis-* Regulatory changes in Kit ligand expression and parallel evolution of pigmentation in sticklebacks and humans. *Cell*, **131**, 1179–1189.

Murata, Y., Oda, S. and Mitani, H. (2012) Allelic expression changes in Medaka (*Oryzias latipes*) hybrids between inbred strains derived from genetically distant populations. *PLoS One*, **7**, e36875.

Pala, I., Coelho, M.M. and Schartl, M. (2008) Dosage compensation by gene-copy silencing in a triploid hybrid fish. *Current Biology (CB)*, **18**, 1344–1348.

Pask, A. (2012) Insights on imprinting from beyond mice and men. *Methods in Molecular Biology*, **925**, 263–275.

Pastinen, T. (2010) Genome-wide allele-specific analysis: insights into regulatory variation. *Nature Reviews Genetics*, **11**, 533–538.
Pearson, C.E. (2003) Slipping while sleeping? Trinucleotide repeat expansions in germ cells. *Trends in Molecular Medicine*, **9**, 490–495.
Pickrell, J.K., Marioni, J.C., Pai, A.A., *et al.* (2010) Understanding mechanisms underlying human gene expression variation with RNA sequencing. *Nature*, **464**, 768–772.
Renfree, M.B., Suzuki, S. and Kaneko-Ishino, T. (2013) The origin and evolution of genomic imprinting and viviparity in mammals. *Philosophical Transactions of the Royal Society of London Series B, Biological Sciences*, **368**, 20120151.
Robinson, M.D. and Smyth, G.K. (2007) Moderated statistical tests for assessing differences in tag abundance. *Bioinformatics*, **23**, 2881–2887.
Rozowsky, J., Abyzov, A., Wang, J., *et al.* (2011) AlleleSeq: analysis of allele-specific expression and binding in a network framework. *Molecular Systems Biology*, **7**, 522.
Schaefke, B., Emerson, J., Wang, T.-Y., *et al.* (2013) Inheritance of gene expression level and selective constraints on *trans-* and *cis-* regulatory changes in yeast. *Molecular Biology and Evolution*, mst114.
Serre, D., Gurd, S., Ge, B., *et al.* (2008) Differential allelic expression in the human genome: a robust approach to identify genetic and epigenetic *cis-* acting mechanisms regulating gene expression. *PLoS Genetics*, **4**, e1000006.
Shen, Y., Catchen, J., Garcia, T., *et al.* (2012) Identification of transcriptome SNPs between Xiphophorus lines and species for assessing allele specific gene expression within F1 interspecies hybrids. *Comparative Biochemistry and Physiology Part C: Toxicology & Pharmacology*, **155**, 102–108.
Shi, X., Ng, D.W., Zhang, C., *et al.* (2012) *Cis-* and *trans-* regulatory divergence between progenitor species determines gene-expression novelty in Arabidopsis allopolyploids. *Nature Communications*, **3**, 950.
Shoemaker, R., Deng, J., Wang, W. and Zhang, K. (2010) Allele-specific methylation is prevalent and is contributed by CpG-SNPs in the human genome. *Genome Research*, **20**, 883–889.
Singer-Sam, J. and Gao, C. (2001) Quantitative RT-PCR-based analysis of allele-specific gene expression. *Methods in Molecular Biology*, **181**, 145–152.
Singer-Sam, J., LeBon, J.M., Dai, A. and Riggs, A.D. (1992) A sensitive, quantitative assay for measurement of allele-specific transcripts differing by a single nucleotide. *Genome Research*, **1**, 160–163.
Skelly, D.A., Johansson, M., Madeoy, J., *et al.* (2011) A powerful and flexible statistical framework for testing hypotheses of allele-specific gene expression from RNA-seq data. *Genome Research*, **21**, 1728-1737.
Springer, N.M. and Stupar, R.M. (2007) Allele-specific expression patterns reveal biases and embryo-specific parent-of-origin effects in hybrid maize. *Plant Cell*, **19**, 2391–2402.
Storer, B.E. and Kim, C. (1990) Exact properties of some exact test statistics for comparing two binomial proportions. *Journal of the American Statistical Association*, **85**, 146–155.
Tirosh, I., Reikhav, S., Levy, A.A. and Barkai, N. (2009) A yeast hybrid provides insight into the evolution of gene expression regulation. *Science*, **324**, 659–662.
Tome, S., Panigrahi, G.B., Lopez Castel, A., *et al.* (2011) Maternal germline-specific effect of DNA ligase I on CTG/CAG instability. *Human Molecular Genetics*, **20**, 2131–2143.
Trapnell, C., Pachter, L. and Salzberg, S.L. (2009) TopHat: discovering splice junctions with RNA-Seq. *Bioinformatics*, **25**, 1105–1111.

Wang, H. and Elbein, S.C. (2007) Detection of allelic imbalance in gene expression using pyrosequencing. *Methods in Molecular Biology*, **373**, 157–176.

Wang, R., Sun, L., Bao, L., *et al.* (2013) Bulk segregant RNA-seq reveals expression and positional candidate genes and allele-specific expression for disease resistance against enteric septicemia of catfish. *BMC Genomics*, **14**, 929.

Wang, X. and Clark, A. (2014) Using next-generation RNA sequencing to identify imprinted genes. *Heredity*, **113**, 156–166.

Wang, X., Miller, D.C., Harman, R., *et al.* (2013) Paternally expressed genes predominate in the placenta. *Proceedings of the National Academy of Sciences*, **110**, 10705–10710.

Wang, X., Soloway, P.D. and Clark, A.G. (2011) A survey for novel imprinted genes in the mouse placenta by mRNA-seq. *Genetics*, **189**, 109–122.

Wang, X., Sun, Q., McGrath, S.D., *et al.* (2008) Transcriptome-wide identification of novel imprinted genes in neonatal mouse brain. *PloS One*, **3**, e3839.

Wilson, M.D., Barbosa-Morais, N.L., Schmidt, D., *et al.* (2008) Species-specific transcription in mice carrying human chromosome 21. *Science*, **322**, 434–438.

Wittkopp, P.J., Haerum, B.K. and Clark, A.G. (2004) Evolutionary changes in *cis*- and *trans*-gene regulation. *Nature*, **430**, 85–88.

Wittkopp, P.J., Haerum, B.K. and Clark, A.G. (2008) Regulatory changes underlying expression differences within and between Drosophila species. *Nature Genetics*, **40**, 346–350.

Wood, D.L., Nones, K., Steptoe, A., *et al.* (2015) Recommendations for Accurate Resolution of Gene and Isoform Allele-Specific Expression in RNA-Seq Data. *PloS One*, **10**, e0126911.

Yan, H., Yuan, W., Velculescu, V.E., *et al.* (2002) Allelic variation in human gene expression. *Science*, **297**, 1143.

Zhang, K., Li, J.B., Gao, Y., *et al.* (2009) Digital RNA allelotyping reveals tissue-specific and allele-specific gene expression in human. *Nature Methods*, **6**, 613–618.

Zhang, X. and Borevitz, J.O. (2009) Global analysis of allele-specific expression in Arabidopsis thaliana. *Genetics*, **182**, 943–954.

15

Bioinformatics Analysis of Epigenetics

Yanghua He and Jiuzhou Song

Introduction

In eukaryotes, the epigenetic modifications of DNA and histones are essential for the interactions between environment and genetic information that regulates gene expression, which in turn determines the phenotypes. The epigenetic modification is found to be associated with various diseases, such as developmental abnormalities, cancers, neurological diseases, and autoimmune diseases, which supports the idea that epigenetic signatures may play an important role in disease susceptibility. The rapid progress of experimental methods, especially the development of high-throughput sequencing technologies, has driven the field of epigenetics. The exponential increases in the speed of data generation in turn require advanced computational methods for data processing and quality control, prompting understanding of epigenetic information. In this chapter, we describe the concepts and methods used for the analysis of high-throughput data, and summarize commonly used tools for epigenetic data analysis as well as available resources for epigenetic studies.

In eukaryotes, DNA is coined into the chromatin. The fundamental subunit of chromatin is nucleosome, which is composed of DNA (about 147 bp) wrapping on a histone octamer. The regulation of the chromatin structure and the epigenetic modification on the DNA and histone tails were found to regulate gene expression, which determines the phenotype of a cell, and thereby the ultimate phenotype of the organism (Jaenisch & Bird, 2003). The study of DNA methylation and histone modifications is termed *epigenetics*, coined by Conrad Hal Waddington from the Greek prefix "epi-", meaning *over* or *above*, and "genetics" to describe the different phenotypes of a zygote during development (Waddington, 1942). Epigenetics has been defined as a functional modification to the DNA that does not involve an alteration of the DNA sequence itself (Meaney, 2010). However, this definition has recently been revised, so that the essential features of epigenetic mechanisms are: (a) structural modifications to chromatin either at the level of the histone proteins or DNA; (b) regulation of the structure and function of chromatin; (c) regulation of gene expression; and (d) that these modifications occur in the absence of any changes in the nucleotide sequences. The functional byproduct of epigenetic modifications is the change in gene transcription (Bird, 2007; Hake & Allis, 2006).

DNA methylation and histone modification patterns are very important in various biological processes such as X-chromosome inactivation (Kaslow & Migeon, 1987),

Bioinformatics in Aquaculture: Principles and Methods, First Edition. Edited by Zhanjiang (John) Liu.

imprinting (Li, Beard & Jaenisch, 1993), lineage-specific and tissue-specific gene expression (Eden & Cedar, 1994), transcription, signal transduction, and development and differentiation (Kouzarides, 2007; Zhang & Reinberg, 2001). It was reported that the behavior of the mother can directly alter cellular signals during embryonic development, which then actively sculpt the epigenetic landscape of the offspring, influencing the activity of specific regions of the genome and the phenotype of the offspring (Cameron *et al.*, 2008; Weaver, 2007). Dysregulation of DNA methylation and histone modifications have been found to be related to various disease conditions, including developmental abnormalities (Kota & Feil, 2010), cancers (Dawson & Kouzarides, 2012), neurological diseases (Jakovcevski & Akbarian, 2012), and autoimmune diseases (Meda *et al.*, 2011). Although the mechanism of epigenetic modifications in the etiology of diseases is still unclear, some unique epigenetic modification patterns identified in diseases are used as markers for diagnostics and new therapeutic strategies (Arai *et al.*, 2009; Arai & Kanai, 2010; Suzuki *et al.*, 2011; Tong *et al.*, 2010).

Recently, the development of next-generation sequencing (NGS) makes it possible to detect genome-wide DNA methylation and histone modification profiles, which drive the concept of epigenome—that is, the entire DNA methylation and histone modifications of a cell (Bernstein *et al.*, 2010; Milosavljevic, 2011). However, because of unclear distribution of epigenetic data, proper informatics analysis and interpretation of epigenetic information have become major challenges. In this chapter, we introduce the basic concepts of epigenetics and the available methods used for epigenetic research. Furthermore, we summarize relevant computational and bioinformatics pipelines used for epigenetic data analysis. Finally, we provide future perspectives and comment on some challenges in the emerging field of computational epigenetics.

Mastering Epigenetic Data

DNA Methylation

The methylation on 5 base of cytosine (5-methylcytosine; 5mC) was thought for a long time to be the only modification found on DNA sequence (Hotchkiss, 1948). However, recent studies in embryonic stem (ES) cells and brain development revealed other types of DNA methylation variants, including 5-hydroxymethylcytosine (5hmC) (Kriaucionis & Heintz, 2009; Tahiliani *et al.*, 2009; Wyatt & Cohen, 1952), 5-formylcytosine (5fC) (He *et al.*, 2011; Ito *et al.*, 2010, 2011), and 5-carboxylcytosine (5aC) (He *et al.*, 2011; Ito *et al.*, 2010, 2011). In vertebrates, 5mC was mainly found in the context of CG dinucleotide (Goll & Bestor, 2005; Zemach *et al.*, 2010). However, other non-CG methylations, including methylation on CT, CA, CNG, and CNN, were also found in invertebrates, plants, and fungi (Suzuki & Bird, 2008). In mammals, at least three enzymes were found to be related with DNA methylation, including DNA methyltransferase 3A and 3B (*DNMT3A* and *DNMT3B*) catalyzing the *de novo* DNA methylation, and *DNMT1* maintaining the status of DNA methylation (Figure 15.1) (Branco, Ficz & Reik, 2012; Wu & Zhang, 2010). Other DNMTs, such as DNMT3L, are regulators of DNMT3A and DNMT3B in maternal imprinting establishments (Bourc'his *et al.*, 2001; Hata *et al.*, 2002). DNA methyltransferase were also found in plants but with different names, including DNA methyltransferase 1 (MET1), domains rearranged

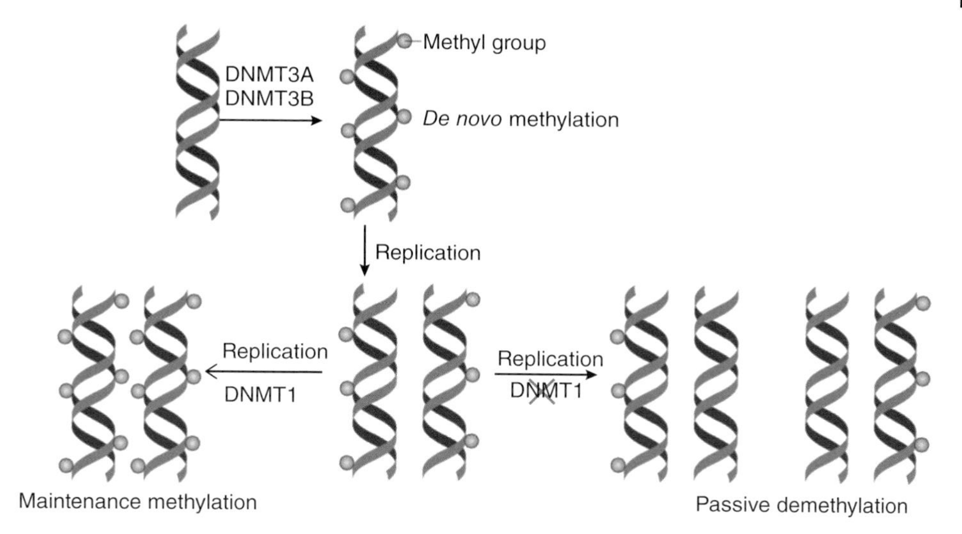

Figure 15.1 Mechanisms of DNA methylation and demethylation. During early development, methylation patterns are initially established by the *de novo* DNA methyltransferases DNMT3A and DNMT3B. When DNA replication and cell division occur, these methyl marks are maintained by the maintenance methyltransferase, DNMT1, which has a preference for hemi-methylated DNA. If DNMT1 is inhibited or absent when the cell divides, the newly synthesized strand of DNA will not be methylated, and successive rounds of cell division will result in passive demethylation (Branco *et al.*, 2012).

methyltransferase 2 (DRM2), and chromomethylase 3 (CMT3) (Cao & Jacobsen, 2002; Jones, Ratcliff & Baulcombe, 2001).

The DNA demethylation process is more complicated, including DNA demethylation by TET (ten eleven translocation) protein and passive DNA demethylation through deletion of the functions of DNMTs (He *et al.*, 2011; Ito *et al.*, 2011; Morgan *et al.*, 2004). It is reported that there are two pathways in the DNA demethylation process (Figure 15.2). One pathway, containing TET protein–induced DNA demethylation, constitutes iterative oxidations of 5mC to generate 5hmC, 5fC, and 5aC, respectively (He *et al.*, 2011; Ito *et al.*, 2011). The decarboxylase then removes the carboxyl group of 5aC to generate the unmethylated C (Wu & Zhang, 2010). The other pathway is through DNA glycosylase activity, including the deamination of 5hmC by activation-induced deaminase (AID) and apolipoprotein B mRNA-editing enzyme complex (APOBEC) to produce 5-hydroxymethyluracil (5hmU). Then, the 5hmU is converted to C by DNA glycosylases and base excision repair (BER) (Cortellino *et al.*, 2011; Guo *et al.*, 2011).

In order to understand the role of DNA methylation in development and diseases, demonstrating the detailed distribution of methylation in the genome is essential. Fortunately, the availability of reference genome assemblies and massively parallel sequencing has led to many high-throughput methods, which makes it possible to provide high-resolution, genome-wide profiles of 5-methylcytosine (Lister *et al.*, 2008, 2009). In contrast to the microarray method, sequencing-based methods interrogate DNA methylation in repetitive sequences, and more readily allow epigenetic states to be assigned to specific alleles. Currently, the four most frequently used sequencing-based technologies are the bisulfite-based methods MethylC-seq and reduced representation bisulfite sequencing (RRBS), and the enrichment-based techniques methylated DNA

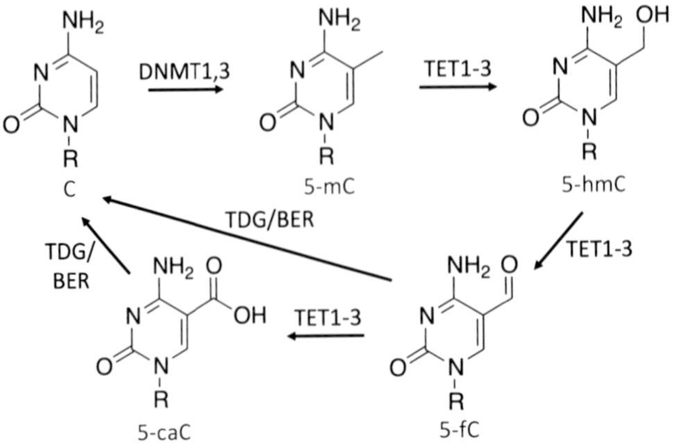

Figure 15.2 The pathway of DNA demethylation process. Genomic 5-methylcytosine (5mC) can be removed passively during replication, but several pathways for active demethylation have also been proposed, including those in which 5-hydroxymethylcytosine (5hmC) is an intermediate. 5hmC may be further oxidized to 5-formylcytosine (5fC) and 5-carboxylcytosine (5caC) by TET enzymes. Deformylation of 5fC and decarboxylation of 5caC probably convert these intermediates directly back to cytosine; however, thymine DNA glycosylase (TDG) was shown to cleave 5fC and 5caC during this conversion, implicating the BER pathway in DNA demethylation.

immunoprecipitation sequencing (MeDIP-seq) and methylated DNA binding domain sequencing (MBD-seq).

Bisulfite-based Methods

Bisulfite genomic sequencing is regarded as a gold-standard technology for the detection of DNA methylation because it is a quantitative and efficient approach to identify 5-methylcytosine at single base pair resolution (Li & Tollefsbol, 2011). This method was first introduced by Frommer *et al.* (1992), and it is mainly based on the finding that the amination reactions of cytosine and 5-methylcytosine (5mC) proceeding with very different consequences after the treatment of sodium bisulfite (Frommer *et al.*, 1992). In this regard, cytosines in single-stranded DNA is converted into uracil residues and recognized as thymine in subsequent PCR amplification and sequencing. However, 5mCs are immune to this conversion and remain as cytosines, allowing 5mCs to be distinguished from unmethylated cytosines. Treatment of DNA with bisulfite converts cytosine residues to uracil, but leaves 5-methylcytosine residues unaffected. Thus, bisulfite treatment introduces specific changes in the DNA sequences that depend on the methylation status of individual cytosine residues, yielding single-nucleotide resolution information about the methylation status of a segment of DNA.

The second bisulfite-based method, RRBS (Gu *et al.*, 2011; Meissner *et al.*, 2005), reduces the portion of the genome analyzed through Msp I digestion and fragment size selection. Notably, although CpGs are not distributed uniformly in the genome, every Msp I RRBS sequence read includes at least one informative CpG position, which makes the approach highly efficient (Meissner *et al.*, 2008).

These two bisulfite methods reached a concordance of 82% for CpG methylation levels and 99% for non-CpG cytosine methylation levels (Figure 15.3) (Harris *et al.*, 2010). For bisulfite-based methods, reads that mapped to the positive and negative strands

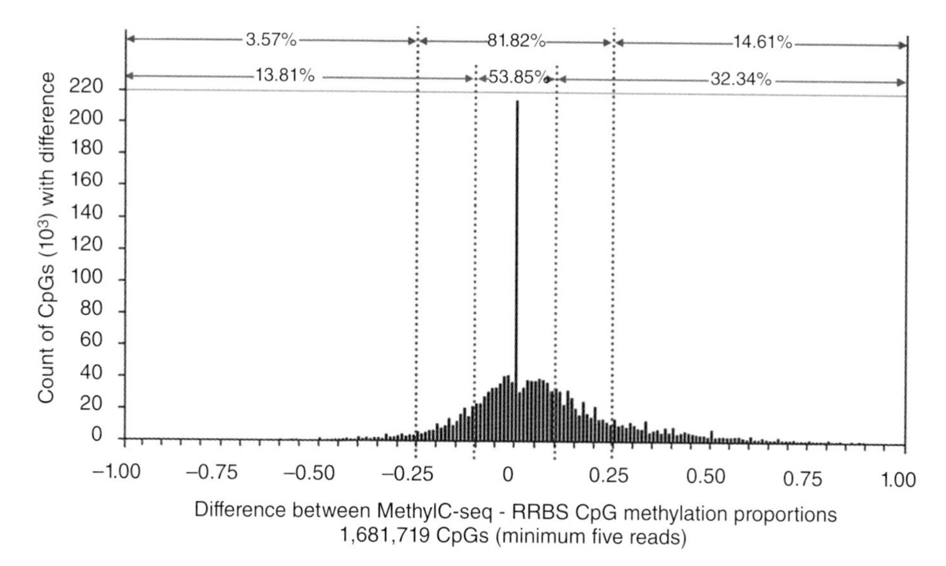

Figure 15.3 The difference in methylation proportions between MethylC-seq and RRBS at a minimum read depth of 5 was calculated for individual CpGs. Of the CpGs compared between MethylC-seq and RRBS, only 12.75% displayed identical methylation levels with a difference threshold of zero. If the difference threshold is relaxed to 0.1 (green dashed lines) or 0.25 (red dashed lines), the concordance increased to 53.85% or 81.82%, respectively (Meissner *et al.*, 2008). (*See color plate section for the color representation of this figure.*)

were combined for CpG methylation calculations, but not for CHG and CHH methylation calculations due to the strand asymmetry of non-CpG methylation (Frommer *et al.*, 1992; Meissner *et al.*, 2008). The methylated proportion was calculated for each CpG or non-CpG as $\frac{\text{(methylated reads)}}{\text{(methylated reads + unmethylated reads)}}$.

Enrichment-based Methods

MeDIP-seq (methylated DNA immunoprecipitation sequencing) (Down *et al.*, 2008; Jacinto, Ballestar & Esteller, 2008) and MBD-seq (methyl-CpG binding domain protein 2 sequencing) (Serre, Lee & Ting, 2010) involve proteins to enrich methylated DNA regions by immunoprecipitation followed by sequencing. In MeDIP-seq, a highly specific antibody that recognizes 5-methylcytosine is used to immunoprecipitate sonicated genomic DNA, resulting in the enrichment of genomic regions that are methylated.

MBD-seq utilizes the MBD2 protein methyl-CpG binding domain to enrich for methylated double-stranded DNA fragments. As a complementary approach for use in conjunction with methylated fragment enrichment methods, unmethylated CpGs are identified by the methyl-sensitive restriction enzymes (MREs) from parallel DNA digestions fragments, a 5-methylcytosine antibody, methylated DNA binding proteins, or proteins that primarily bind unmethylated DNA. We found that, by integrating two enrichment-based methods that are complementary in nature, that is, MeDIP-seq (or MBD-seq) and MRE-seq, the efficiency of whole DNA methylome profiling can be significantly increased (Li *et al.*, 2015).

The method involving Methyl-MAPS (methylation mapping analysis by paired-end sequencing) (Edwards *et al.*, 2010) uses both methyl-sensitive enzymes (AciI, HhaI,

HpaII, HpyCH4V, and BstUI) and a methyl-dependent enzyme (McrBC) to generate two mate-paired libraries for sequencing on the Illumina high-throughput sequencing platform. Sequence tags from the two libraries are used to estimate the DNA methylation level for a single CpG site. By using this method, the methylation of up to 82.4% of the CpG sites in humans was determined.

Histone Modifications

Histone modifications, including acetylation, methylation, phosphorylation, ubiquitylation, SUMOylation, ADP ribosylation, deimination, and proline isomerization, were identified in more than 60 amino acid residues on the N-terminal tails of the histones (Kouzarides, 2007). The widely studied histone modifications in recent years are acetylation and methylation on arginines (R) and lysines (K). Histone acetylation is catalyzed by histone acetyltransferases (HATs) that have more than 10 members, including general control non-depressible 5 (Gcn5)–related N-acetyltransferases (GNATs), E1A-binding protein p300, CREB-binding protein (p300/CBP), MYST histone acetyltransferase family, elongator protein 3 (Elp3), establishment of cohesion 1 (Eco1), and chromodomain on Y chromosome (CDY) (Yang & Seto, 2007). The acetyl groups on histones can be removed by histone deacetylases (HDACs) that were divided into four groups of proteins according to their sequence similarities and cofactor dependency (de Ruijter *et al.*, 2003). Correspondingly, the methylation on histones can be divided into three categories, mono-, di-, and tri-, depending on the number of methyl groups added on them by histone methyltransferases (HMTs) (Dillon *et al.*, 2005; Qian & Zhou, 2006). Usually, HMTs catalyze methylation in a site-specific manner, which allows the histone H3 methylation on arginine on 2, 8, 17, 26 positions and lysine on 4, 27, 36, 79 positions (Bedford & Clarke, 2009; Dillon *et al.*, 2005; Gary & Clarke, 1998; Qian & Zhou, 2006). Two different types of histone demethylases have been identified: lysine-specific demethylase (LSD1) domain (Shi *et al.*, 2004) and Jumonji C (JmjC) domain (Tsukada *et al.*, 2006). Different catalytic reactions were found in the LSD1 and JmjC domains. The LSD1 domain catalyzes the demethylation of histone H3 lysine 4 di-/mono-methylation (H3K4me2/me1); and the JmjC domain catalyzes the demethylation at tri-methylated histone and other sites of histone.

Different histone modifications have different functions on gene expression regulation, depending on the type of histones, residues, and modifications. Histone acetylations activate gene expression in the beginning of genes (Wang *et al.*, 2008). Histone methylations at different amino acid residues or having different methyl groups have different effects on transcription activity. While histone H3 lysine 4 trimethylation (H3K4me3) is correlated with gene activation around the TSS region, H3K27me3 is correlated with gene repression in the gene body region (Barski *et al.*, 2007). In addition, H3K9me1 is correlated with gene activation, but H3K9me2 and H3K9me3 are correlated with gene repression (Barski *et al.*, 2007).

The chromatin immunoprecipitation (ChIP) assay allows investigators to characterize DNA–protein interactions *in vivo*. ChIP followed by high-throughput sequencing (ChIP-Seq) is a powerful tool for identifying genome-wide profiles of transcription factors (TFs), histone modifications, and nucleosome positioning (Xu & Sung, 2012).

Genomic Data Manipulation

Working with genomic data demands some knowledge of computer languages (see Chapter 1). Various computer languages such as C++, S, R, Perl, Python, and many others were used to write genomic packages. Some of these programs require specific formats to function properly. Specific tools for data conversion and handling have been developed—for example, BEDtools (Quinlan & Hall, 2010) and SAMtools (Li & Durbin, 2009). Some of these languages, such as Perl and Python, are good options and possess the capacity to interact with other languages as well, facilitating the integration of different processes in one set called "pipelines." Genomic studies demand the comparison of results; they should be obtained in the most possible similar conditions in order to distinguish the real signals from the noises. Pipelines are good alternatives for establishing work methodologies, and specifically for situations that require the same processes.

R Language

R is a freely accessible scripting software environment that is used for statistical computing and graphics (Mittal, 2011). It is a GNU project that runs on different platforms such as Microsoft Windows, macOS, and UNIX, and is administered by various CRAN mirrors. R is an implementation of the S language, which was developed at Bell Laboratories by John Chambers (Matloff, 2011). Robert Gentleman and Ross Ihaka at the University of Auckland created R in 1991, but it was not released to the public until 1993. The name R refers to the first letter of both founders' names. Currently, the R Development Core Team develops R. R usually employs a command line interface; however, some graphical user interfaces (GUIs) were created to make it user-friendlier for non-informaticians. RStudio, StatET, and ESS are some of the GUI examples.

R is a very lean and functional language that offers the possibility of combining various commands, each of which uses the output of the previous one (pipes) (Quinlan & Hall, 2010). It also gives the possibility to divide complex processes into modules (packages) that can be tailored depending on the user requisites (Paradis, Claude & Strimmer, 2004). Importantly, it also offers sophisticated graphic capabilities to visualize intricate data, a characteristic that is often absent in most other statistical packages (Guy, Roat Kultima & Andersson, 2010). For more details about R, please visit the web page http://cran.r-project.org

Bioconductor

Bioconductor is an R-based open-source software that provides tools for the manipulation, analysis, and assessment of high-throughput genomic data. The project started in 2001 and is presently under the supervision of the Bioconductor core team. It is mainly located at the Fred Hutchinson Cancer Research Center, although other national and international institutions participate as active members. It can be installed from the website http://www.bioconductor.org/install/. Currently, most of the genomic data analyses are performed using add-on modules (packages) that can be accessed through the web page www.bioconductor.org (Dai *et al.*, 2012; Delhomme *et al.*, 2012). This site also displays the situations in which Bioconductor could be employed: microarrays, variants, sequence data, annotation, and high-throughput assays. After accessing any of these categories, explanations, scripts, and samples for different procedures are provided. The instructions are simple, but in order to use them proficiently, knowledge of molecular genomics and R is highly recommended, although the last one is not determinative since R and Bioconductor can be learned on-the-job, owing to its interactivity. Almost all the procedures, descriptions, and comments from other users can be obtained on the Internet, making the use and implementation of the packages easier than closed programs. Often, adjustment to personal requirements should be performed to obtain an adequate analysis, and R has the flexibility to attain it because the package scripts are accessible and can be modified. Although R presents many advantages, it has to be used carefully because it does not display any warning if used incorrectly, a problem that many sealed programs have. This situation demands that users know the statistical and programming concepts according to the analysis to avoid erroneous results.

USCS Genome Bioinformatics Site

The UCSC Genome Browser was developed and is maintained by the Genome Bioinformatics Group, which is a team formed by members from the Center for Biomolecular Science and Engineering (CBSE) at the University of California (Fujita *et al.*, 2011; Kent *et al.*, 2002). The website can be accessed at http://genome.ucsc.edu/index.html. The site provides reference sequences and draft assemblies for various organism genomes, visualization of the chromosomes with zoom capabilities for a close observation of the annotated features, portal access for other genomic projects as ENCODE and Neandertal, and other helpful web-based tools for genomic analyses (Dreszer *et al.*, 2012; Green *et al.*, 2010; Maher, 2012).

Here is a list of the currently available tools:

BLAT: Used to identify 95% or more similarity between sequences of 25 nucleotides or more (Kent, 2002).
Table Browser: Permits retrieving data associated with a track in text format, to calculate intersections between tracks, and to retrieve DNA sequence covered by a track (Karolchik *et al.*, 2004).
Gene Sorter: Could be used to sort related genes based on different criteria such as similar gene expression profiles or genomic proximity (Kent *et al.*, 2005).
Genome Graphs: Provides different graphical capabilities for visualizing genome-wide data sets (Durinck *et al.*, 2009).
In-Silico PCR: Employed for searching a sequence database with a couple of PCR primers, applying an indexing strategy for faster performance (Schuler, 1997).
Lift Genome Annotation: Converts genomic positions and genome annotation files between different assemblies.
VisiGene: Allows the use of a virtual microscope in order to visualize *in situ* images, showing where the gene is employed in the respective organism (Kuhn *et al.*, 2009).
Other Utilities: Gives a list of links and instructions for other tools and utilities developed by the UCSC Genome Bioinformatics Group, but are more specific for determined genomic studies.

Various web tools are available for the investigator to choose from. The choice of tool will depend on the species, cost, availability of antibodies, and many other factors that could be critical for adopting a decision. On the other hand, it is important to develop a methodology while analyzing the same type of data in order to reduce errors due to the use of different strategies to make similar studies comparable. This will allow the transformation of older methods into more fine-tuned procedures with traceable records of the improvements.

Galaxy

Galaxy is an open, scientific workflow, data integration (Blankenberg *et al.*, 2011; Blankenberg, Gordon *et al.*, 2010; Blankenberg, Von Kuster *et al.*, 2010; Goecks *et al.*, 2010; Taylor *et al.*, 2007), and analysis persistence and publishing platform (Blankenberg, Von Kuster *et al.*, 2010; Goecks *et al.*, 2010; Taylor *et al.*, 2007). It aims to make computational biology accessible to research scientists who do not have computer programming experience. Galaxy is now used as a general bioinformatics workflow management system, providing a means to build multi-step computational analyses akin to a recipe. Galaxy supports a range of widely used biological data formats, and translation between those formats. It typically provides a GUI (Schatz, 2010) for specifying what data to operate on, what steps to take, and what order to operate. Besides, the set of available tools has been greatly expanded over the years, and Galaxy is now also used for gene expression, genome assembly, proteomics, epigenomics, transcriptomics, and a host of other disciplines in the life sciences. As a free public web server supported by the Galaxy Project, Galaxy can be accessed at https://galaxyproject.org. Researchers can create logins, and also save histories, workflows, and datasets on the server. These saved items can also be shared with others.

DNA Methylation and Bioinformatics

DNA methylation, as an important epigenetic mark involved in a diverse range of biological processes, including gene silencing, X inactivation, genomic imprinting, plus roles in maintaining cellular function and development of autoimmunity (Richardson, 2002; Su, Shao *et al.*, 2012). Aberrant DNA methylation may be associated with the disorder of gene expression in carcinogenesis. In comparison with the relative stability of genomic DNA sequences, DNA methylomes dynamically change among different cells and even vary along with the change of conditions in a single cell (Suzuki & Bird, 2008). Therefore, it is of great significance to study the biological function of DNA methylation and its underlying mechanisms in various organisms. At present, the development of sequencing technology makes it easier to measure the genome-wide DNA methylation profiling, which is a premise of understanding the role of methylation in development and disease. We will focus on bioinformatics tools for the processing and analysis of DNA methylation data generated by NGS technologies (Figure 15.4).

Demo of Analysis for DNA Methylation

Pre-processing

Modern sequencing technologies can generate a massive number of sequence reads in a single experiment. However, no sequencing technology is perfect, and each instrument will generate different types and amounts of errors. Therefore, it is necessary to understand, identify, and exclude error types that may impact the interpretation of the

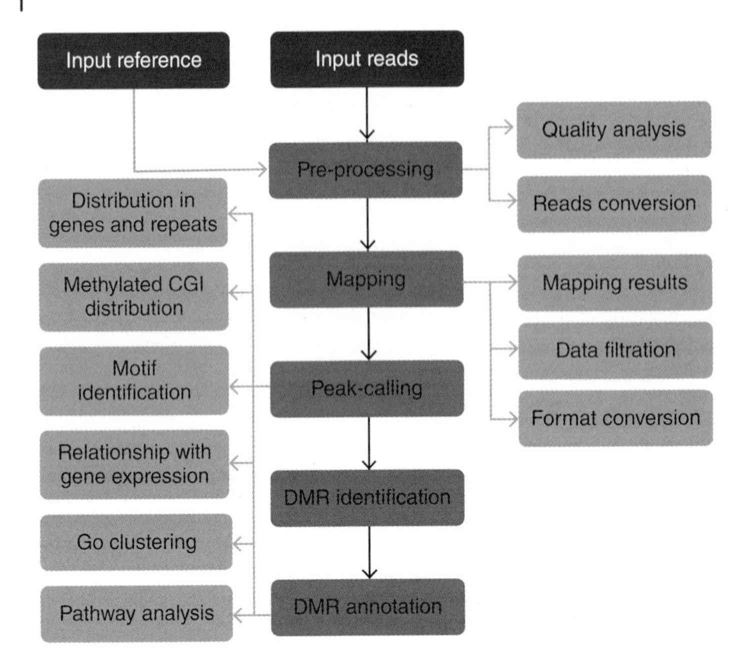

Figure 15.4 The workflow of DNA methylation analysis. The workflow shows the basic process of high-throughput DNA methylation analysis. The key steps in the pipeline are in red boxes; sub-steps for each key step are shown in blue; input data are in dark green boxes.

data. The objective of pre-processing is to understand some relevant properties of raw sequence data. We will use methylation data from enrichment-based methods as an example, and focus on properties such as length, quality scores, and base contents to assess the quality of the data and discard low-quality or uninformative reads.

Basic UNIX commands, FastQC, and Trimmomatic can be applied to sequencing datasets, but FastQC is recommended primarily because of its simplicity and accuracy. The software can be downloaded—with versions available for Microsoft Windows, Linux, or macOS, depending on your preference—from the website http://www.bioinformatics.babraham.ac.uk/projects/fastqc. When the operating environment is ready, you can import your data with BAM, SAM, or FastQ files, and then the result interface will show up when it is done. The output results include basic statistics, per-base or per-sequence quality scores, and per-base GC content, etc. If the quality scores for more than 70% of bases are higher than 30 (which means the error probability is less than 0.1%), the sequence quality is good (Edgar & Flyvbjerg, 2015). More attention needs to be paid to per-base sequence content; the content percentage of bases G and C across the whole read should be higher than the content of bases A and T for DNA methylation datasets. After an adequate quality confirmation, quality filtering and/or adaptor trimming need to be conducted. When FastQC is used to detect adaptors and primers, these sequences can be removed with Cutadapt. Other trimming toolkits are also useful, such as AlienTrimmer (Criscuolo & Brisse, 2013), which is a tool to quickly and accurately trim off multiple short contaminant sequences from high-throughput sequencing reads; and Fastx toolkit, which allows you to handle paired end data as well. For detailed command lines, readers are referred to Chapter 3 and Chapter 8 of this book.

Mapping

A ubiquitous and fundamental step in high-throughput sequencing analysis is the alignment (mapping) of the generated reads to a reference sequence. To accomplish this task, numerous software tools have been developed. Alignment is the process where a sequenced segment (read) is compared to a reference based on its nucleotide sequence similarity. Pairwise sequence alignment is employed to find similar regions that can explain structural, functional, or evolutionary relationships between samples.

There are many available software packages for alignment such as BOWTIE (Langmead *et al.*, 2009), SOAP2 (Li *et al.*, 2008), and TOPHAT (Trapnell, Pachter & Salzberg, 2009). Table 15.1 shows the most commonly used tools for mapping high-throughput sequencing data from DNA, RNA, miRNA, or bisulfite sequences. Determining the mappers that are most suitable for a specific application is not trivial. For MBD-seq or MeDIP-seq data, Bowtie (ultrafast, memory-efficient short read aligner) is often recommended for the alignment process. The trimmed Fastq files are aligned to the reference genome obtained from the UCSC genome browser (http://genome.ucsc.edu). For data manipulation, filtration, and format conversion, a combination of procedures available in SAMtools and BEDtools is commonly applied (Li, Handsaker *et al.*, 2009; Quinlan & Hall, 2010). An important step that should be considered is the removal of duplicated reads, which can be achieved using the "`samtools rmdup -sS`" option included in the SAMtools suite. This action will affect subsequent procedures. Besides, unique mapped files can be converted to normalized BedGraph format by using the *genomecov* tool, included in the BEDTools suite for visualized viewing or comparing the data in the UCSC genome browser or Integrative Genomics Viewer (see Figure 15.5 as an example).

Peak-calling

Peak areas represent *in vivo* locations where proteins of interest (e.g., modified histones or TFs) are associated with DNA. When the protein is a TF, the enriched area may be its TF binding site (TFBS); when the protein is a DNase enzyme, the enriched area is the DNase I hypersensitive site (DHS); when the protein is an antibody for histone modifications, the enriched area is a histone modification peak; and for DNA methylation experiments, the enriched area is the methylated enriched region. Currently, popular software programs to call peaks include F-Seq (Boyle *et al.*, 2008), MACS (Zhang *et al.*, 2008), BALM (Lan *et al.*, 2011), PeakSeq (Rozowsky *et al.*, 2009), and WaveSeq (Mitra & Song, 2012). From the distribution patterns of peaks, MACS and PeakSeq are appropriate for capturing point/punctate peaks (e.g., TFBS); F-Seq is useful for obtaining broad peaks (e.g., open chromatin, DHS); and WaveSeq can detect both narrow and broad peaks with a high degree of accuracy even in low signal-to-noise ratio datasets. For DNA methylation data, most people use the MACS package. However, BALM could detect the methylation level of CpG dinucleotides with high efficiency and resolution for MBD-seq data due to its low cost, high specificity, efficiency, and resolution. Many peak callers have multiple parameters that can affect the number of peaks called, and understanding these parameters takes time. This is perhaps why MACS is still the dominantly used software.

DMR Identification

After the identification of methylated enriched peaks, the differentially methylated peaks/regions (DMRs) need to be identified for case/control design. DMR identification

Table 15.1 Tools for mapping high-throughput sequencing data. The "Sequencing platform" column indicates whether the mapper natively supports reads from a specific sequencing platform (I, Illumina; So, ABI Solid; 454, Roche 454; Sa, ABI Sanger; H, Helicos; Ion, Ion torrent; and P, PacBio), or not (N). The "Input" and "Output" columns indicate, respectively, the file formats accepted and produced by the mappers. *Input formats*: FASTA, FASTQ, CFASTA, CFASTQ, and Illumina sequence and probability files' format. *Output formats*: SAM, tab-separated values (TSV), BED file format, different versions of GFF, and number of reads mapped to genes/exons (Counts).

Mapper	Data	Sequencing platform	Input	Output	References
MAQ	DNA	I,So	(C)FAST(A/Q)	TSV	Li, Ruan & Durbin (2008)
SOAP	DNA	I	FASTA/Q	TSV	Li *et al.* (2008)
BWA	DNA	I,So,454,Sa,P	FASTA/Q	SAM	Li & Durbin (2009)
GenomeMapper	DNA	I	FASTA/Q	BED TSV	Schneeberger *et al.* (2009)
Bowtie	DNA	I,So,454,Sa,P	(C)FAST(A/Q)	SAM TSV	Langmead *et al.* (2009)
SOAP2	DNA	I	FASTA/Q	SAM TSV	Li, Yu *et al.* (2009)
BWA-SW	DNA	I,So,454,Sa,P	FASTA/Q	SAM	Li & Durbin (2010)
BFAST	DNA	I,So,454, Hel	(C)FAST(A/Q)	SAM TSV	Homer, Merriman & Nelson (2009)
Bowtie2	DNA	I,454,Ion	FASTA/Q	SAM TSV	Langmead & Salzberg (2012)
TopHat	RNA	I	FASTA/Q, GFF	BAM	Trapnell *et al.* (2009)
MapSplice	RNA	I	FASTA/Q	SAM BED	Wang *et al.* (2010)
SpliceMap	RNA	I	FASTA/Q	SAM BED	Au *et al.* (2010)
Supersplat	RNA	N	FASTA	TSV	Bryant Jr *et al.* (2010)
SOAPSplice	RNA	I,454	FASTA/Q	TSV	Huang *et al.* (2011)
mrFAST	miRNA	I	FASTA/Q	SAM	Alkan *et al.* (2009)
MicroRazerS	miRNA	N	FASTA	SAM TSV	Emde *et al.* (2010)
mrsFAST	miRNA	I,So	FASTA/Q	SAM	Hach *et al.* (2010)
BSMAP	Bisulfite	I	FASTA/Q	SAM TSV	Xi & Li (2009)
BS Seeker	Bisulfite	I	FASTA/Q	SAM	Chen, Cokus & Pellegrini (2010)
Bismark	Bisulfite	I	FASTA/Q	SAM	Krueger & Andrews (2011)
RRBSMAP	RRBS data	I	FASTA/Q	SAM	Xi *et al.* (2012)

can be accomplished by using the DiffBind R package (Ross-Innes *et al.*, 2012). It computes differentially bound sites using affinity data. The set of peaks identified by the peaks caller (e.g., MACS) and the BAM files containing aligned reads for each sample are the inputs for DiffBind. The program creates a matrix with the consensus peaks that are determined by the number of replications in the experiment. After creating a

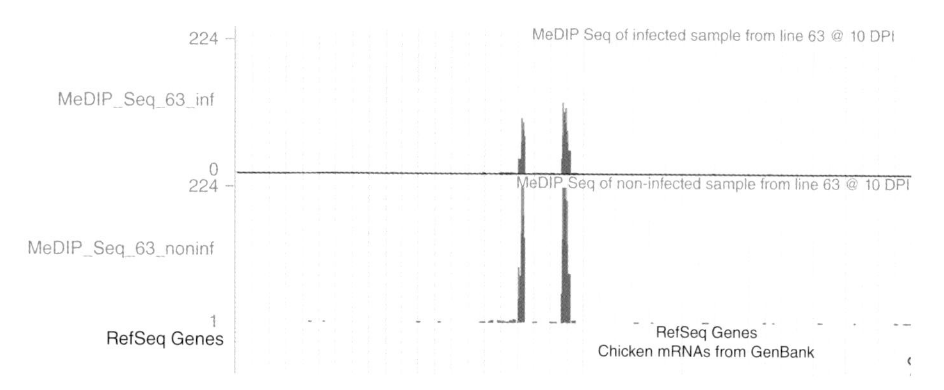

Figure 15.5 The methylation map spanning 50 KB of length on chromosome 1 is shown by the UCSC genome browser. The upper and bottom panels are samples from Line 63 infected and non-infected chickens, respectively, at 10 days post-infection (dpi).

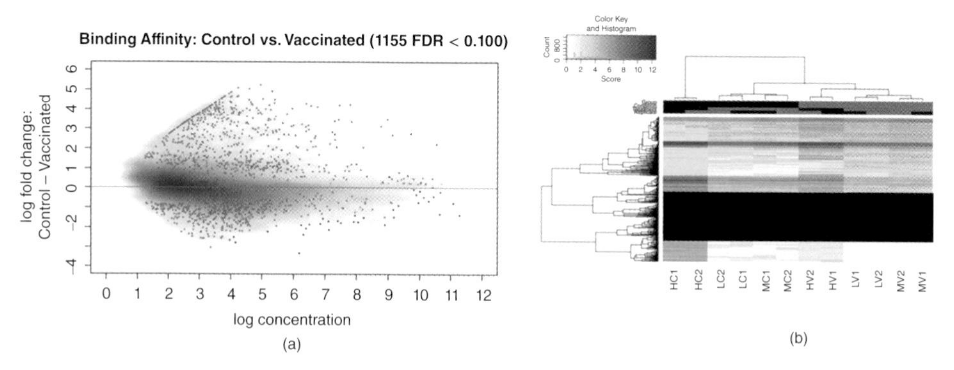

Figure 15.6 Visualization and cluster of differentially methylated region (DMR) samples considering their methylation levels (Eck *et al.*, 2009): (a) the MA plot shows in red the DMRs between chickens immunized with infectious laryngotracheitis (ILT) vaccine and control individuals with FDR < 0.1; (b) heat-map of samples demonstrates a perfect classification of the condition of the individuals based on methylation levels in the DMRs. Each condition (control and vaccinated) has three elute concentrations (high, medium, and low), with two replications for MBD sequencing. (*See color plate section for the color representation of this figure.*)

contrast between case and control as a block effect, DiffBind runs an edgeR analysis (readers are also referred to Chapter 9 for details on edgeR), which is an empirical Bayes method (Robinson, McCarthy & Smyth, 2010). For normalization, the default method can be applied—trimmed mean of M-values (TMM), which subtracts the control's reads and considers the effective library size (reads in peaks). A set of DMRs between two conditions will be included in the output with a particular false discovery rate (FDR). Then, these DMRs can be visualized by MA plot, and samples can be clustered according to their methylation levels (see Figure 15.6 as an example).

In some DNA methylation projects, we would expect to capture not only the differentially methylated regions (DMRs) between different conditions, but also conservative methylated regions through different groups. For example, in the study of the heritability of methylation states through generations, obtaining conservative and stable methylated regions is important. The MEDIPS R package was developed for analyzing data derived from methylated DNA immunoprecipitation (MeDIP) experiments followed by sequencing (MeDIP-seq) (Down *et al.*, 2008). However, MEDIPS provides several functionalities for the analysis of other kinds of quantitative sequencing data (e.g., ChIP-Seq, MBD-seq, CMS-seq, and others), including calculation of differential coverage between groups of samples as well as saturation and correlation analyses. MEDIPS can calculate genome-wide sequence pattern densities (e.g., CpGs) at a user-specified resolution, and export the results in the wiggle track format. Besides, MEDIPS starts where the mapping tools stop (BAM or BED files), and can be used for any genome of interest. Therefore, we can ignore the peak-calling step and jump the analysis of DNA methylation from the mapping step to DMR identification directly. In addition, genome-wide DNA methylation densities through all samples are equalized at a specified resolution, which is convenient for comparing between different groups and obtaining differentially and conservative methylated regions.

In order to reveal the functional roles of DNA methylation in specific conditions, gene expression data has to be introduced for integrate analysis. WIMSi is a tool for discovering patterns of methylation that correlate with differential expression using whole-genome bisulfite sequencing (WGBS) or Methyl-MAPS data (single-base resolution), with minimal prior assumptions about what patterns should exist in the data (Vanderkraats *et al.*, 2013). It can also be used for DNA methylation data from enrichment-based methods, for example, MBD-seq and MeDIP-seq. Differentially expressed genes between comparison pairs need to be first prepared from microarray or RNA-Seq technologies. Using a shape-similarity metric known as the *coupling distance*, groups of genes with similar methylation signatures that also have corresponding expression changes can be identified. Finally, WIMSi generates a list of genes for which methylation changes putatively correlate with expression changes indicated in the pipeline (Figure 15.7). It can successfully detect a variety of introduced DNA methylation regulation patterns for exploring the relationships between changes in epigenetic patterns and transcription, both in normal cellular function and in human diseases.

Gene Expression Analysis

Here, we focus on gene expression data by RNA-Seq technology. Figure 15.8 indicates the pipeline of RNA-Seq analysis for detecting differential expression (Haas & Zody, 2010). Readers who are interested in identification of differentially expressed genes are also referred to Chapter 9 of this book. First, millions of short reads from RNA-Seq are mapped to the reference genome or transcriptome that uses junction libraries to map reads across exon boundaries; and then mapped reads are assembled into expression summaries by coding region, exon, gene, or junction by using some tools (e.g., HTSeq; Anders, Pyl & Huber, 2015). The assembled fragments are aggregated based on the overlapping reads falling into the given feather region and then normalized in order to diminish the noise and make the real signals more detectable. For normalizing the data, various methods have been proposed; total count and upper-quartile normalizations are the most commonly used. Regarding models, Poisson and negative binomial distributions are suggested for straightforward read counts (Bullard *et al.*, 2010; Robinson *et al.*, 2010). Some R packages for statistical testing of differential expression (DE) are performed to produce a list of genes with associated p-values and fold changes (e.g., edgeR [Robinson *et al.*, 2010] and DESeq [Anders & Huber, 2010]). Systems biology approaches can then be used to gain biological insights from these lists.

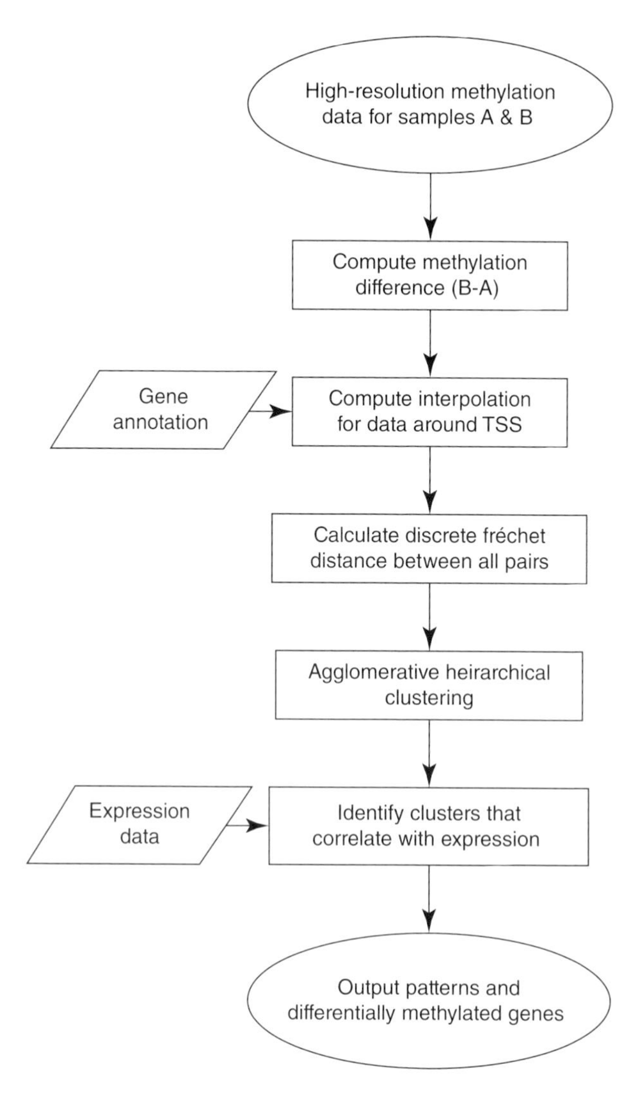

Figure 15.7 WIMSi method overview (Vanderkraats *et al.*, 2013). Overview of the approach for associating spatially similar DNA methylation changes with corresponding changes in transcription.

DMR Annotation and Functional Prediction

For the genomic annotation of the DMRs, the software ChIPpeakAnno can be used (Zhu *et al.*, 2010). ChIPpeakAnno provides information about the overlaps, relative position, and distances for the inquired feature. The annotation information is obtained from the Biomart database in the archive site. The CpG island annotation is retrieved from the UCSC genome browser corresponding to your studied species. For the enrichment analysis, genes that are annotated based on the nearest TSS or overlaps with the peaks are considered. In order to study epigenetic variability on tissue or cell-type specific, Haystack pipeline (Pinello *et al.*, 2014) is recommended to study epigenetic variability, specially crossing tissue-type or cross-cell-type plasticity of chromatin states and TFs, motifs providing mechanistic insights into chromatin structure, cellular identity, and

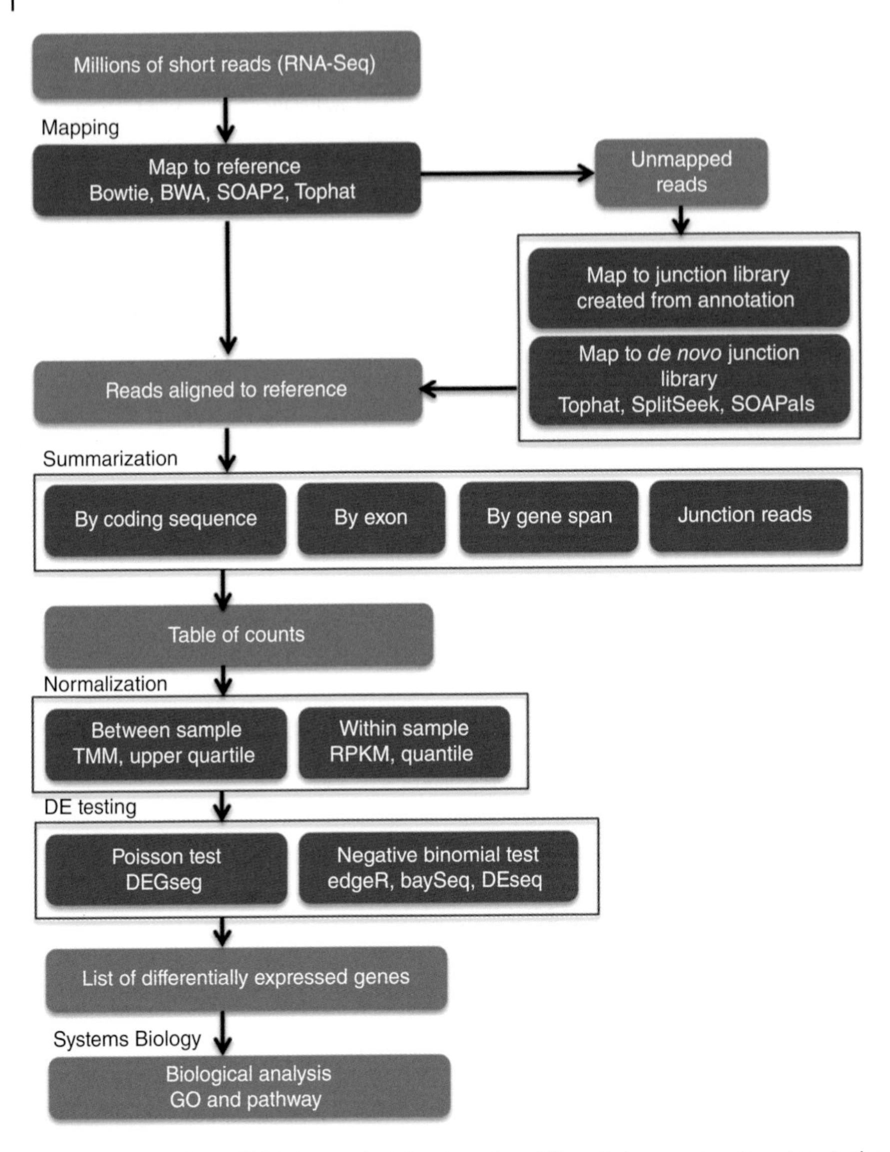

Figure 15.8 Pipeline of RNA-Seq analysis for detecting differential expression. Key steps in the pipeline are in red boxes; methodological components of the pipeline are shown in blue boxes and bold text; software examples and methods for each step (a non-exhaustive list) are shown by regular text in blue boxes.

gene regulation. Haystack can be used with epigenetic data. In addition, it is also possible to integrate gene expression data obtained from array-based or RNA-Seq approaches to quantify TF motifs activity and specificity on nearby genes. You can find this pipeline on the website https://github.com/lucapinello/haystack. Besides, DAVID (Huang, Sherman & Lempicki, 2008), IPA (http://www.ingenuity.com), and PANTHER (Mi *et al.*, 2010) can be performed. GO and pathway analysis can be conducted for genes and proteins of interest and visualization. STRING (Szklarczyk *et al.*, 2011) can be used to obtain

Table 15.2 Web-based tools for biological functional prediction.

Name	Main advantages	Availability	Website
DAVID Bioinformatics Resources 6.7	Identify enriched biological terms Discover enriched functionally related gene groups Cluster redundant annotation terms Visualize genes on BioCarta and KEGG pathway maps List interacting proteins Link gene–disease associations Convert gene identifiers from one type to another	Free	https://david.ncifcrf.gov
Interactive Pathway Analysis (IPA)	Variant analysis Pathway analysis	Commercial	http://www.ingenuity.com
Protein ANalysis THrough Evolutionary Relationships (PANTHER)	Classify proteins (and their genes) according to family and subfamily, molecular function, biological process, and pathway.	Free	http://www.pantherdb.org
Animal Transcription Factor DataBase (AnimalTFDB)	Identify and classify all the genome-wide TFs, transcription co-factors, and chromatin remodeling factors in 50 sequenced animal genomes.	Free	http://www.bioguo.org/AnimalTFDB/index.php
STRING 10	Known and predicted protein–protein interactions	Free	http://string-db.org

known and predicted protein interactions for mining of molecular and cellular functions (Table 15.2).

Other Ways of Analysis for Methylation

DNA methylation status can be interpreted by comparing the sequencing results and the original DNA sequence. However, for bisulfite treatment sequencing, such as WGBS or RRBS, one important step in calling methylation of a genome is to map bisulfite reads. Mapping of bisulfite reads is different from that of ChIP-Seq and RNA-Seq data since the non-methylated Cs are converted to Ts by bisulfite treatment and subsequent PCR. The bisulfite reads are difficult to map to the reference genome due to the high number of mismatches between the converted Ts and the original Cs. There are some mappers developed for bisulfite sequencing data, for example, *BatMeth* (Lim *et al.*, 2012). Additional pipelines have been developed for genome-wide bisulfite sequencing data analysis. For example, *WBSA* (Liang *et al.*, 2014) is a simple data processing software, and it is freeware. It can generate mapping result report, sequence annotation, analysis of methylation level and mC distribution, and even correlation analysis between methylation levels and gene functional regions. Researchers only need to upload the NGS bisulfite sequencing data and fill in some necessary parameters, and then an auto-email will be received as to when the task will be finished. Besides, *BSmooth* is also a pipeline for analyzing WGBS data. It includes tools for aligning

Table 15.3 Web-based tools for bisulfite treatment sequencing.

Name	Application	Website
WBSA	Web service for bisulfite sequencing data analysis (for high-throughput sequencing data)	http://wbsa.big.ac.cn/index.jsp
BSmooth	Pipeline for analyzing WGBS data. Includes tools for aligning data, quality control, and identifying DMRs	http://rafalab.jhsph.edu/bsmooth
methylKit	Comprehensive R package for analysis of genome-wide DNA methylation profiles	https://github.com/al2na/methylKit
methtools	Toolbox to visualize and analyze DNA methylation data	https://github.com/bgruening/methtools
MethylSeq v1.0 app	Analyzes DNA samples prepared using TruSeq DNA Methylation Kit	http://blog.basespace.illumina.com/2015/05/08/methylseq-v1-0-app-for-dna-methylation-calling
QUMA	Quantification tool for methylation analysis (for bisulfite-cloning sequencing)	http://quma.cdb.riken.jp/
BISMA	Bisulfite sequencing DNA methylation analysis (for bisulfite-cloning sequencing)	http://services.ibc.uni-stuttgart.de/BDPC/BISMA/
MethPrimer	Design PCR primers for bisulfite-treated sequence	http://www.urogene.org/cgi-bin/methprimer/methprimer.cgi

the data, quality control, and identifying DMRs (Hansen, Langmead & Irizarry, 2012; Lan *et al.*, 2011). *methylKit* and *methtools* are toolboxes to visualize and analyze high-throughput bisulfite sequencing data (Akalin *et al.*, 2012; Grunau *et al.*, 2000). In addition to genome-wide bisulfite sequencing, bisulfite-cloning sequencing can be used to validate or confirm DNA methylation features for short sequences of interest. Table 15.3 indicates some web-based tools for bisulfite treatment sequencing, including data analysis, result visualization, and primer design.

Histone Modifications and Bioinformatics

Histone Modifications

Chromatin states are the key to gene regulation and cell identities. ChIP is a typical method widely used to detect interaction of DNA and proteins. Readout from ChIP in combination with deep sequencing, termed *ChIP-Seq*, has distinct advantages in functional genomics for mapping genome-wide chromatin modification profiles and TF binding sites. ChIP-Seq was first used to map the genome-wide sites of 20 histone methylation marks, along with CTCF (CCCTC-binding factor), the histone variant H2A.Z, and RNA polymerase II in human cells (Barski *et al.*, 2007). The genome-wide epigenetic information at a finer resolution delivered by NGS platforms is of great value to biomedical research. In particular, histone modifications have been found to play a key role in transcription—for example, histone acetylation and H3K4me3 (histone H3 trimethyl Lys4) have been found to be associated with active genes, while H3K27me3 is associated with silenced genes (Cuddapah *et al.*, 2009). It is now widely recognized that epigenetic regulation is essential to understanding how animals develop and respond to their environmental changes and challenges. Epigenetic signatures are very useful in identifying biomarkers of resistance to diseases, and can potentially lead to improved diagnosis, etiology, and risk prediction.

Data Analysis

ChIP-Seq has been increasingly used to map *in vivo* TF binding sites and chromatin states of the genome. Whereas binding of TFs is mainly governed by their sequence specificity and therefore is typically associated with much localized ChIP-Seq signals, the signals for histone modifications, histone variants, and histone-modifying enzymes are usually diffuse, spanning from tens of nucleosomes to large domains that are multi-million bases long (Mitra & Song, 2012). Therefore, the key step of sequencing data analysis for histone modifications is to capture enriched signals of ChIP-Seq. As with DNA methylation data analysis, ChIP-Seq data also goes through pre-processing, mapping, peak calling, and (or) identification of differentially expressed peaks (Figure 15.4). Most mapping tools used in sequencing data of enrichment-based DNA methylation experiments can also be used for ChIP-Seq data. For example, F-Seq can be used for broad chromatin marks, for example, H3K36me3 and H3K27me3, while MACS can analyze a sharp mark, for example, H3K4me3 (Boyle *et al.*, 2008; Zhang *et al.*, 2008). Many peak callers have multiple parameters that can affect the number of peaks called. However, WaveSeq is a highly sensitive, data-driven method capable of detecting significantly enriched regions in data having diverse characteristics, especially in histone modification data (Mitra & Song, 2012).

Perspectives

With more and more research focused on epigenetic studies in recent years, especially the genome-wide epigenetic research, a large amount of epigenetic information related to human and animal health is being generated and stored in public databases (Table 15.4). This growth presents several challenges, especially regarding the epigenetic data produced by novel molecular technologies. Different types of studies demand tailored analysis methodologies, fueling the invention of new approaches in order to analyze them according to the experimental requirements. Data capturing, manipulation, processing, analysis, results storage, and utilization are examples of common issues for all high-throughput assays. New advances allow faster and inexpensive sequencing, accruing the problem with extremely large data sets (Pevzner & Shamir, 2011). This scenario is a fertile field for ingenious bioinformatics people who can develop new strategies for solving the custom requirements in particular cases.

Some programs are being launched to better understand the epigenetic mechanisms in diseases and health (Table 15.4). Some of these allow the identification of genetic markers for the genetic selection of resistant lines for different conditions in animal genetic studies (Eck *et al.*, 2009; Su, Brondum *et al.*, 2012). Epigenetic studies also enable

Table 15.4 Public databases for epigenetic information.

Description	Website
Epigenetic databases	
The Epigenomics NCBI browser	http://www.ncbi.nlm.nih.gov/epigenomics
The UCSC Epigenome browser	http://www.epigenomebrowser.org/
Epigenetic programs	
NIH Roadmap Epigenomics Program	http://nihroadmap.nih.gov/epigenomics/
The ENCODE Project	http://www.genome.gov/10005107
Introduction to epigenetics and epigenomes	
EpiGeneSys Science	http://www.epigenesys.eu/
EpiGenie	http://epigenie.com/
Learn.Genetics	http://learn.genetics.utah.edu/content/epigenetics/

the design of drugs for disease therapy. The local aberrant epigenetic modifications identified in diseases will provide a better and precise way to cure and prevent diseases.

In spite of these advances, epigenetic studies in aquaculture species have been lacking. However, it is anticipated that epigenetic analysis in aquaculture species will increase. Many of the methodologies used in mammals are adaptable to studies in aquaculture species.

References

Alkan, C., Kidd, J.M., Marques-Bonet, T., *et al.* (2009) Personalized copy number and segmental duplication maps using next-generation sequencing. *Nature Genetics*, **41**, 1061–1067.

Akalin, A., Kormaksson, M., Li, S., *et al.* (2012) methylKit: a comprehensive R package for the analysis of genome-wide DNA methylation profiles. *Genome Biology*, **13**, R87.

Anders, S. and Huber, W. (2010) Differential expression analysis for sequence count data. *Genome Biology*, **11**, R106.

Anders, S., Pyl, P.T. and Huber, W. (2015) HTSeq—a Python framework to work with high-throughput sequencing data. *Bioinformatics*, **31**, 166–169.

Arai, E. and Kanai, Y. (2010) DNA methylation profiles in precancerous tissue and cancers: carcinogenetic risk estimation and prognostication based on DNA methylation status. *Epigenomics*, **2**, 467–481.

Arai, E., Ushijima, S., Fujimoto, H., *et al.* (2009) Genome-wide DNA methylation profiles in both precancerous conditions and clear cell renal cell carcinomas are correlated with malignant potential and patient outcome. *Carcinogenesis*, **30**, 214–221.

Au, K.F., Jiang, H., Lin, L., *et al.* (2010) Detection of splice junctions from paired-end RNA-seq data by SpliceMap. *Nucleic Acids Research*, **38**, 4570–4578.

Barski, A., Cuddapah, S., Cui, K., *et al.* (2007) High-resolution profiling of histone methylations in the human genome. *Cell*, **129**, 823–837.

Bedford, M.T. and Clarke, S.G. (2009) Protein arginine methylation in mammals: who, what, and why. *Molecular Cell*, **33**, 1–13.

Bernstein, B.E., Stamatoyannopoulos, J.A., Costello, J.F., *et al.* (2010) The NIH roadmap epigenomics mapping consortium. *Nature Biotechnology*, **28**, 1045–1048.
Bird, A. (2007) Perceptions of epigenetics. *Nature*, **447**, 396–398.
Blankenberg, D., Coraor, N., Von Kuster, G., *et al.* (2011) Integrating diverse databases into an unified analysis framework: a Galaxy approach. *Database (Oxford)*, **2011**, bar011.
Blankenberg, D., Gordon, A., Von Kuster, G., *et al.* (2010) Manipulation of FASTQ data with Galaxy. *Bioinformatics*, **26**, 1783–1785.
Blankenberg, D., Von Kuster, G., Coraor, N., *et al.* (2010) Galaxy: a web-based genome analysis tool for experimentalists. *Current Protocols in Molecular Biology*, Chapter 19, Unit **19**.10.11–21.
Bourc'his, D., Xu, G.L., Lin, C.S., *et al.* (2001) Dnmt3L and the establishment of maternal genomic imprints. *Science*, **294**, 2536–2539.
Boyle, A.P., Guinney, J., Crawford, G.E. and Furey, T.S. (2008) F-Seq: a feature density estimator for high-throughput sequence tags. *Bioinformatics*, **24**, 2537–2538.
Branco, M.R., Ficz, G. and Reik, W. (2012) Uncovering the role of 5-hydroxymethylcytosine in the epigenome. *Nature Reviews Genetics*, **13**, 7–13.
Bryant, D.W. Jr, Shen, R., Priest, H.D., *et al.* (2010) Supersplat-spliced RNA-seq alignment. *Bioinformatics*, **26**, 1500–1505.
Bullard, J., Purdom, E., Hansen, K. and Dudoit, S. (2010) Evaluation of statistical methods for normalization and differential expression in mRNA-Seq experiments. *BMC Bioinformatics*, **11**, 94.
Cameron, N.M., Shahrokh, D., Del Corpo, A., *et al.* (2008) Epigenetic programming of phenotypic variations in reproductive strategies in the rat through maternal care. *Journal of Neuroendocrinology*, **20**, 795–801.
Cao, X. and Jacobsen, S.E. (2002) Locus-specific control of asymmetric and CpNpG methylation by the DRM and CMT3 methyltransferase genes. *Proceedings of the National Academy of Sciences (USA)*, **99**(4), 16491–16498.
Chen, P.Y., Cokus, S.J. and Pellegrini, M. (2010) BS Seeker: precise mapping for bisulfite sequencing. *BMC Bioinformatics*, **11**, 203.
Cortellino, S., Xu, J., Sannai, M., *et al.* (2011) Thymine DNA glycosylase is essential for active DNA demethylation by linked deamination-base excision repair. *Cell*, **146**, 67–79.
Criscuolo, A. and Brisse, S. (2013) AlienTrimmer: a tool to quickly and accurately trim off multiple short contaminant sequences from high-throughput sequencing reads. *Genomics*, **102**, 500–506.
Cuddapah, S., Jothi, R., Schones, D.E., *et al.* (2009) Global analysis of the insulator binding protein CTCF in chromatin barrier regions reveals demarcation of active and repressive domains. *Genome Research*, **19**, 24–32.
Dai, Y., Guo, L., Li, M. and Chen, Y.B. (2012) Microarray R US: a user-friendly graphical interface to bioconductor tools that enables accurate microarray data analysis and expedites comprehensive functional analysis of microarray results. In: *BMC Research Notes*, p. 282.
Dawson, M.A. and Kouzarides, T. (2012) Cancer epigenetics: from mechanism to therapy. *Cell*, **150**, 12–27.
de Ruijter, A.J., van Gennip, A.H., Caron, H.N., *et al.* (2003) Histone deacetylases (HDACs): characterization of the classical HDAC family. *Biochemical Journal*, **370**, 737–749.
Delhomme, N., Padioleau, I., Furlong, E.E. and Steinmetz, L. (2012) easyRNASeq: a bioconductor package for processing RNA-Seq data. *In: Bioinformatics*, **28**(19), 2532–2533.

Dillon, S.C., Zhang, X., Trievel, R.C. and Cheng, X. (2005) The SET-domain protein superfamily: protein lysine methyltransferases. *Genome Biology*, **6**, 227.

Down, T.A., Rakyan, V.K., Turner, D.J., *et al.* (2008) A Bayesian deconvolution strategy for immunoprecipitation-based DNA methylome analysis. *Nature Biotechnology*, **26**, 779–785.

Dreszer, T.R., Karolchik, D., Zweig, A.S., *et al.* (2012) The UCSC Genome Browser database: extensions and updates 2011. *Nucleic Acids Research*, **40**(Database issue), D918–923.

Durinck, S., Bullard, J., Spellman, P.T. and Dudoit, S. (2009) GenomeGraphs: integrated genomic data visualization with R. In: *BMC Bioinformatics (BioMed Central Ltd)*, p. 2.

Eck, S.H., Benet-Pages, A., Flisikowski, K., *et al.* (2009) Whole genome sequencing of a single *Bos taurus* animal for single nucleotide polymorphism discovery. *Genome Biology*, **10**, 6.

Eden, S., and Cedar, H. (1994) Role of DNA methylation in the regulation of transcription. *Current Opinion in Genetics & Development*, **4**, 255–259.

Edgar, R.C. and Flyvbjerg, H. (2015) Error filtering, pair assembly and error correction for next-generation sequencing reads. *Bioinformatics*, **31**, 3476–3482.

Edwards, J.R., O'Donnell, A.H., Rollins, R.A., *et al.* (2010) Chromatin and sequence features that define the fine and gross structure of genomic methylation patterns. *Genome Research*, **20**, 972–980.

Emde, A.K., Grunert, M., Weese, D., *et al.* (2010) MicroRazerS: rapid alignment of small RNA reads. *Bioinformatics*, **26**, 123–124.

Frommer, M., McDonald, L.E., Millar, D.S., *et al.* (1992) A genomic sequencing protocol that yields a positive display of 5-methylcytosine residues in individual DNA strands. *Proceedings of the National Academy of Sciences (USA)*, **89**, 1827–1831.

Fujita, P.A. Rhead, B., Zweig, A.S., *et al.* (2011) The UCSC Genome Browser database: update 2011. *Nucleic Acids Research*, **39**.

Gary, J.D. and Clarke, S. (1998) RNA and protein interactions modulated by protein arginine methylation. *Progress in Nucleic Acid Research and Molecular Biology*, **61**, 65–131.

Goecks, J., Nekrutenko, A., Taylor, J. and Team, G. (2010) Galaxy: a comprehensive approach for supporting accessible, reproducible, and transparent computational research in the life sciences. *Genome Biology*, **11**, R86.

Goll, M.G. and Bestor, T.H. (2005) Eukaryotic cytosine methyltransferases. *Annual Review of Biochemistry*, **74**, 481–514.

Green, R.E., Krause, J., Briggs, A.W., *et al.* (2010) A draft sequence of the Neandertal genome. *Science*, **328**, 710–722.

Grunau, C., Schattevoy, R., Mache, N. and Rosenthal, A. (2000) MethTools—a toolbox to visualize and analyze DNA methylation data. *Nucleic Acids Research*, **28**, 1053–1058.

Gu, H., Smith, Z.D., Bock, C., *et al.* (2011) Preparation of reduced representation bisulfite sequencing libraries for genome-scale DNA methylation profiling. *Nature Protocols*, **6**, 468–481.

Guo, J.U., Su, Y., Zhong, C., *et al.* (2011) Emerging roles of TET proteins and 5-hydroxymethylcytosines in active DNA demethylation and beyond. *Cell Cycle*, **10**, 2662–2668.

Guy, L., Roat Kultima, J. and Andersson, S.G.E. (2010) genoPlotR: comparative gene and genome visualization in R. *Bioinformatics*, **26**, 2334–2335.

Haas, B.J. and Zody, M.C. (2010) Advancing RNA-Seq analysis. *Nature Biotechnology*, **28**, 421–423.

Hach, F., Hormozdiari, F., Alkan, C., *et al.* (2010) mrsFAST: a cache-oblivious algorithm for short-read mapping. *Nature Methods*, **7**, 576–577.

Hake, S.B. and Allis, C.D. (2006) Histone H3 variants and their potential role in indexing mammalian genomes: the "H3 barcode hypothesis." *Proceedings of the National Academy of Sciences (USA)*, **103**, 6428–6435.

Hansen, K.D., Langmead, B. and Irizarry, R.A. (2012) BSmooth: from whole genome bisulfite sequencing reads to differentially methylated regions. *Genome Biology*, **13**, R83.

Harris, R.A., Wang, T., Coarfa, C., *et al.* (2010) Comparison of sequencing-based methods to profile DNA methylation and identification of monoallelic epigenetic modifications. *Nature Biotechnology*, **28**, 1097–1105.

Hata, K., Okano, M., Lei, H. and Li, E. (2002) Dnmt3L cooperates with the Dnmt3 family of de novo DNA methyltransferases to establish maternal imprints in mice. *Development*, **129**, 1983–1993.

He, Y.F., Li, B.Z., Li, Z., *et al.* (2011) TET-mediated formation of 5-carboxylcytosine and its excision by TDG in mammalian DNA. *Science*, **333**, 1303–1307.

Homer, N., Merriman, B. and Nelson, S.F. (2009) BFAST: an alignment tool for large scale genome resequencing. *PLoS One*, **4**, e7767.

Hotchkiss, R.D. (1948) The quantitative separation of purines, pyrimidines, and nucleosides by paper chromatography. *The Journal of Biological Chemistry*, **175**, 315–332.

Huang, D.W., Sherman, B.T. and Lempicki, R.A. (2008) Systematic and integrative analysis of large gene lists using DAVID bioinformatics resources. *Nature Protocols*, **4**, 44–57.

Huang, S., Zhang, J., Li, R., *et al.* (2011) SOAPsplice: genome-wide ab initio detection of splice junctions from RNA-Seq data. *Frontiers in Genetics*, **2**, 46.

Ito, S., D'Alessio, A.C., Taranova, O.V., *et al.* (2010) Role of TET proteins in 5mC to 5hmC conversion, ES-cell self-renewal and inner cell mass specification. *Nature*, **466**, 1129–1133.

Ito, S., Shen, L., Dai, Q., *et al.* (2011) TET proteins can convert 5-methylcytosine to 5-formylcytosine and 5-carboxylcytosine. *Science*, **333**, 1300–1303.

Jacinto, F.V., Ballestar, E. and Esteller, M. (2008) Methyl-DNA immunoprecipitation (MeDIP): hunting down the DNA methylome. *Biotechniques*, **44**, 35, 37, 39 passim.

Jaenisch, R. and Bird, A. (2003) Epigenetic regulation of gene expression: how the genome integrates intrinsic and environmental signals. *Nat Genet*, **33** Suppl, 245–254.

Jakovcevski, M. and Akbarian, S. (2012) Epigenetic mechanisms in neurological disease. *Nat Med*, **18**, 1194–1204.

Jones, L., Ratcliff, F. and Baulcombe, D.C. (2001) RNA-directed transcriptional gene silencing in plants can be inherited independently of the RNA trigger and requires Met1 for maintenance. *Curr Biol*, **11**, 747–757.

Karolchik, D., Hinrichs, A.S., Furey, T.S., *et al.* (2004) The UCSC Table Browser data retrieval tool. *Nucleic Acids Research*, **32**(Database issue), D493–496.

Kaslow, D.C. and Migeon, B.R. (1987) DNA methylation stabilizes X chromosome inactivation in eutherians but not in marsupials: evidence for multistep maintenance of mammalian X dosage compensation. *Proceedings of the National Academy of Sciences (USA)*, **84**, 6210–6214.

Kent, W.J. (2002) BLAT—The BLAST-Like Alignment Tool. *Genome Research*, **12**, 656–664.

Kent, W.J., Hsu, F., Karolchik, D., *et al.* (2005) Exploring relationships and mining data with the UCSC Gene Sorter. *Genome Research*, **15**, 737–741.

Kent, W.J., Sugnet, C.W., Furey, T.S., *et al.* (2002) The human genome browser at UCSC. *Genome Research*, **12**, 996–1006.

Kota, S.K. and Feil, R. (2010) Epigenetic transitions in germ cell development and meiosis. *Developmental Cell*, **19**, 675–686.

Kouzarides, T. (2007) Chromatin modifications and their function. *Cell*, **128**, 693–705.

Kriaucionis, S. and Heintz, N. (2009) The nuclear DNA base 5-hydroxymethylcytosine is present in Purkinje neurons and the brain. *Science*, **324**, 929–930.

Krueger, F. and Andrews, S. (2011) Bismark: a flexible aligner and methylation caller for Bisulfite-Seq applications. *Bioinformatics*, **27**, 1571–1572.

Kuhn, R.M., Karolchik, D., Zweig, A.S., *et al.* (2009) The UCSC Genome Browser Database: update 2009. *Nucleic Acids Research (England)*, **37**(Database issue), D755–761.

Lan, X., Adams, C., Landers, M., *et al.* (2011) High resolution detection and analysis of CpG dinucleotides methylation using MBD-Seq technology. *PLoS One*, **6**, e22226.

Langmead, B. and Salzberg, S. (2012) Fast gapped-read alignment with Bowtie 2. *Nature Methods*, **9**, 357–359.

Langmead, B., Trapnell, C., Pop, M. and Salzberg, S.L. (2009) Ultrafast and memory-efficient alignment of short DNA sequences to the human genome. *Genome Biology*, **10**, R25.

Li, D., Zhang, B., Xing, X. and Wang, T. (2015) Combining MeDIP-seq and MRE-seq to investigate genome-wide CpG methylation. *Methods*, **72**, 29–40.

Li, E., Beard, C. and Jaenisch, R. (1993) Role for DNA methylation in genomic imprinting. *Nature*, **366**, 362–365.

Li, H. and Durbin, R. (2009) Fast and accurate short read alignment with Burrows-Wheeler transform. *Bioinformatics*, **25**, 1754–1760.

Li, H. and Durbin, R. (2010) Fast and accurate long-read alignment with Burrows–Wheeler transform. *Bioinformatics*, **26**, 589–595.

Li, H., Handsaker, B., Wysoker, A., *et al.* (2009) The Sequence Alignment/Map format and SAMtools. *Bioinformatics*, **25**, 2078–2079.

Li, H., Ruan, J. and Durbin, R.(2008) Mapping short DNA sequencing reads and calling variants using Fast and accurate short read alignment with mapping quality scores. *Genome Research*, **18**, 1851–1858.

Li, R., Li, Y., Kristiansen, K. and Wang, J. (2008) SOAP: short oligonucleotide alignment program. *Bioinformatics*, **24**, 713–714.

Li, R., Yu, C., Li, Y., *et al.* (2009) SOAP2: an improved ultrafast tool for short read alignment. *Bioinformatics*, **25**, 1966–1967.

Li, Y. and Tollefsbol, T.O. (2011) DNA methylation detection: bisulfite genomic sequencing analysis. *Methods in Molecular Biology*, **791**, 11–21.

Liang, F., Tang, B., Wang, Y., *et al.* (2014) WBSA: web service for bisulfite sequencing data analysis. *PLoS One*, **9**, e86707.

Lim, J.Q., Tennakoon, C., Li, G., *et al.* (2012) BatMeth: improved mapper for bisulfite sequencing reads on DNA methylation. *Genome Biology*, **13**, R82.

Lister, R., O'Malley, R.C., Tonti-Filippini, J., *et al.* (2008) Highly integrated single-base resolution maps of the epigenome in Arabidopsis. *Cell*, **133**, 523–536.

Lister, R., Pelizzola, M., Dowen, R.H., *et al.* (2009) Human DNA methylomes at base resolution show widespread epigenomic differences. *Nature*, **462**, 315–322.

Maher, B. (2012) ENCODE: the human encyclopaedia. *Nature News*, **489**, 46.

Matloff, N. (2011) *The art of R programming: a tour of statistical software design*. San Francisco, CA, USA: No Starch Press.

Meaney, M.J. (2010) Epigenetics and the biological definition of gene x environment interactions. *Child Development*, **81**, 41–79.

Meda, F., Folci, M., Baccarelli, A. and Selmi, C. (2011) The epigenetics of autoimmunity. *Cell Mol Immunol*, **8**, 226–236.

Meissner, A., Gnirke, A., Bell, G.W., *et al.* (2005) Reduced representation bisulfite sequencing for comparative high-resolution DNA methylation analysis. *Nucleic Acids Research*, **33**, 5868–5877.

Meissner, A., Mikkelsen, T.S., Gu, H., *et al.* (2008) Genome-scale DNA methylation maps of pluripotent and differentiated cells. *Nature*, **454**, 766–770.

Mi, H., Dong, Q., Muruganujan, A., *et al.* (2010) PANTHER version 7: improved phylogenetic trees, orthologs and collaboration with the Gene Ontology Consortium. *Nucleic Acids Research*, **38**, D204–210.

Milosavljevic, A. (2011) Emerging patterns of epigenomic variation. *Trends in Genetics*, **27**, 242–250.

Mitra, A. and Song, J. (2012) WaveSeq: a novel data-driven method of detecting histone modification enrichments using wavelets. *PLoS One*, **7**, e45486.

Mittal, H. (2011) *R graph cookbook*. Birmingham, UK: Packt Publishing.

Morgan, H.D., Dean, W., Coker, H.A., *et al.* (2004) Activation-induced cytidine deaminase deaminates 5-methylcytosine in DNA and is expressed in pluripotent tissues: implications for epigenetic reprogramming. *The Journal of Biological Chemistry*, **279**, 52353–52360.

Paradis, E., Claude, J. and Strimmer, K. (2004) APE: analyses of phylogenetics and evolution in R language. *Bioinformatics*, **20**, 289–290.

Pevzner, P. and Shamir, R. (2011) *Bioinformatics for biologists*. Cambridge, New York: Cambridge University Press.

Pinello, L., Xu, J., Orkin, S.H. and Yuan, G.C. (2014) Analysis of chromatin-state plasticity identifies cell-type-specific regulators of H3K27me3 patterns. *Proceedings of the National Academy of Sciences (USA)*, **111**, E344–353.

Qian, C. and Zhou, M.M. (2006) SET domain protein lysine methyltransferases: Structure, specificity and catalysis. *Cellular and Molecular Life Sciences*, **63**, 2755–2763.

Quinlan, A.R. and Hall, I.M. (2010) BEDTools: a flexible suite of utilities for comparing genomic features. *Bioinformatics*, **26**, 841–842.

Richardson, B.C. (2002) Role of DNA methylation in the regulation of cell function: autoimmunity, aging and cancer. *Journal of Nutrition*, **132**, 2401S–2405S.

Robinson, M.D., McCarthy, D.J. and Smyth, G.K. (2010) edgeR: a bioconductor package for differential expression analysis of digital gene expression data. *Bioinformatics*, **26**, 139–140.

Ross-Innes, C.S., Stark, R., Teschendorff, A.E., *et al.* (2012) Differential oestrogen receptor binding is associated with clinical outcome in breast cancer. *Nature*, **481**, 389–393.

Rozowsky, J., Euskirchen, G., Auerbach, R.K., *et al.* (2009) PeakSeq enables systematic scoring of ChIP-seq experiments relative to controls. *Nature Biotechnology*, **27**, 66–75.

Schatz, M.C. (2010) The missing graphical user interface for genomics. *Genome Biology*, **11**, 128.

Schneeberger, K., Hagmann, J., Ossowski, S., *et al.* (2009) Simultaneous alignment of short reads against multiple genomes. *Genome Biology*, **10**, R98.

Schuler, G.D. (1997) Sequence mapping by electronic PCR. *Genome Research*, **7**, 541–550.

Serre, D., Lee, B.H. and Ting, A.H. (2010) MBD-isolated genome sequencing provides a high-throughput and comprehensive survey of DNA methylation in the human genome. *Nucleic Acids Research*, **38**, 391–399.

Shi, Y., Lan, F., Matson, C., *et al.* (2004) Histone demethylation mediated by the nuclear amine oxidase homolog LSD1. *Cell*, **119**, 941–953.

Su, G., Brondum, R.F., Ma, P., *et al.* (2012) Comparison of genomic predictions using medium-density (approximately 54,000) and high-density (approximately 777,000) single nucleotide polymorphism marker panels in Nordic Holstein and Red Dairy Cattle populations. *Journal of Dairy Science*, **95**, 4657–4665.

Su, J., Shao, X., Liu, H., *et al.* (2012) Genome-wide dynamic changes of DNA methylation of repetitive elements in human embryonic stem cells and fetal fibroblasts. *Genomics*, **99**, 10–17.

Suzuki, H., Takatsuka, S., Akashi, H., *et al.* (2011) Genome-wide profiling of chromatin signatures reveals epigenetic regulation of MicroRNA genes in colorectal cancer. *Cancer Research*, **71**, 5646–5658.

Suzuki, M.M. and Bird, A. (2008) DNA methylation landscapes: provocative insights from epigenomics. *Nature Reviews Genetics*, **9**, 465–476.

Szklarczyk, D., Franceschini, A., Kuhn, M., *et al.* (2011) The STRING database in 2011: functional interaction networks of proteins, globally integrated and scored. *Nucleic Acids Research*, **39**, D561–568.

Tahiliani, M., Koh, K.P., Shen, Y., *et al.* (2009) Conversion of 5-methylcytosine to 5-hydroxymethylcytosine in mammalian DNA by MLL partner TET1. *Science*, **324**, 930–935.

Taylor, J., Schenck, I., Blankenberg, D. and Nekrutenko, A. (2007) Using galaxy to perform large-scale interactive data analyses. *Current Protocols in Bioinformatics*, Chapter 10, Unit 10.15.

Tong, W.G., Wierda, W.G., Lin, E., *et al.* (2010) Genome-wide DNA methylation profiling of chronic lymphocytic leukemia allows identification of epigenetically repressed molecular pathways with clinical impact. *Epigenetics*, **5**, 499–508.

Trapnell, C., Pachter, L. and Salzberg, S.L. (2009) TopHat: discovering splice junctions with RNA-Seq. *Bioinformatics*, **25**, 1105–1111.

Tsukada, Y., Fang, J., Erdjument-Bromage, H., *et al.* (2006) Histone demethylation by a family of JmjC domain-containing proteins. *Nature*, **439**, 811–816.

Vanderkraats, N.D., Hiken, J.F., Decker, K.F. and Edwards, J.R. (2013) Discovering high-resolution patterns of differential DNA methylation that correlate with gene expression changes. *Nucleic Acids Research*, **41**, 6816–6827.

Waddington, C.H. (1942) The epigenotype. *Endeavour*, **1**, 18–20.

Wang, K., Singh, D., Zeng, Z., *et al.* (2010) MapSplice: accurate mapping of RNA-seq reads for splice junction discovery. *Nucleic Acids Research*, **38**, e178.

Wang, Z., Zang, C., Rosenfeld, J.A., *et al.* (2008) Combinatorial patterns of histone acetylations and methylations in the human genome. *Nature Genetics*, **40**, 897–903.

Weaver, I.C. (2007) Epigenetic programming by maternal behavior and pharmacological intervention. *Nature versus nurture: let's call the whole thing off. Epigenetics*, **2**, 22–28.
Wu, S.C. and Zhang, Y. (2010) Active DNA demethylation: many roads lead to Rome. *Nature Reviews Molecular Cell Biology*, **11**, 607–620.
Wyatt, G.R. and Cohen, S.S. (1952) A new pyrimidine base from bacteriophage nucleic acids. *Nature*, **170**, 1072–1073.
Xi, Y., Bock, C., Müller, F., *et al.* (2012) RRBSMAP: a fast, accurate and user-friendly alignment tool for reduced representation bisulfite sequencing. *Bioinformatics*, **28**(3), 430–432.
Xi, Y. and Li, W. (2009) BSMAP: whole genome bisulfite sequence MAPping program. *BMC Bioinformatics*, **10**, 232.
Xu, H. and Sung, W.K. (2012) Identifying differential histone modification sites from ChIP-seq data. *Methods in Molecular Biology*, **802**, 293–303.
Yang, X.J. and Seto, E. (2007) HATs and HDACs: from structure, function and regulation to novel strategies for therapy and prevention. *Oncogene*, **26**, 5310–5318.
Zemach, A., McDaniel, I.E., Silva, P. and Zilberman, D. (2010) Genome-wide evolutionary analysis of eukaryotic DNA methylation. *Science*, **328**, 916–919.
Zhang, Y., Liu, T., Meyer, C.A., *et al.* (2008) Model-based analysis of ChIP-Seq (MACS). *Genome Biology*, **9**, R137.
Zhang, Y. and Reinberg, D. (2001) Transcription regulation by histone methylation: interplay between different covalent modifications of the core histone tails. *Genes & Development*, **15**, 2343–2360.
Zhu, L.J., Gazin, C., Lawson, N.D., *et al.* (2010) ChIPpeakAnno: a Bioconductor package to annotate ChIP-seq and ChIP-chip data. *BMC Bioinformatics*, **11**, 237.

Part III

Bioinformatics Mining and Genetic Analysis of DNA Markers

16

Bioinformatics Mining of Microsatellite Markers from Genomic and Transcriptomic Sequences

Shaolin Wang, Yanliang Jiang and Zhanjiang Liu

Introduction

In recent years, single-nucleotide polymorphisms (SNPs) have become the markers of choice. However, microsatellites are still useful for a number of applications for aquaculture because of their very high levels of polymorphism. For instance, microsatellites are perhaps still the most powerful marker type among all markers for stock identification. Recent advances in next-generation sequencing made it unnecessary to develop microsatellites from the scratch. Instead, microsatellites can be identified by bioinformatics mining from existing sequences. In this chapter, we provide an overview of the informatics processes for the identification of microsatellites, and provide detailed methodologies for microsatellite identification and primer design, using several popular software packages as examples.

Microsatellites, also known as *simple sequence repeats* (SSRs) or *short tandem repeats* (STRs), are repeated sequences of 1–6 base pairs, for example, (AT)n, (CAG)n, and (ATGT)n (Tautz, 1989). They have been used as popular molecular markers because of the following reasons:

1) Microsatellites are highly abundant in various eukaryotic genomes. For instance, in most aquaculture fish species, there is at least one microsatellite every 10 KB on average.
2) They are highly polymorphic. Although microsatellites can be relatively stably inherited, they have high levels of mutation rates because of polymerase slippage during DNA replication. As a result, the number of repeat units in each microsatellite locus varies among alleles, making microsatellites highly variable.
3) They are co-dominant markers.
4) Flanking sequences are well conserved in most cases, allowing locus-specific genotyping by PCR using primers designed from the flanking sequences.

Microsatellites have been widely used for various genetic analyses, including linkage analysis, marker-assisted selection, analysis of kinship, population structure analysis, and various other studies. The molecular basis and applications of microsatellite markers have been detailed in the book titled *Aquaculture Genome Technologies* (Liu, 2007), and also in several other books. Here, we will only cover procedures for bioinformatics mining of microsatellites from existing genomic resources of aquaculture species.

Bioinformatics in Aquaculture: Principles and Methods, First Edition. Edited by Zhanjiang (John) Liu.

Bioinformatics Mining of Microsatellite Markers

For the development of microsatellite markers, microsatellites need to be identified along with sufficient flanking sequences for primer design. Therefore, sequence information is required prior to microsatellite identification. Historically, microsatellite markers were developed through sequencing of microsatellite-enriched libraries. Of course, that can still be done. However, in recent years, the application of next-generation sequencing technologies has changed the way research is conducted. Next-generation sequencing technologies can be directly used for the development of microsatellite markers from whole genome shotgun sequencing or transcriptome sequencing (Liu *et al.*, 2011).

Sequence Resources for Microsatellite Markers

About 10–15 years back, expressed sequence tags (ESTs) were the most abundant sequence resources for aquaculture species. ESTs were used as a rich resource for molecular marker identification, such as microsatellites. EST-derived microsatellite markers are gene-associated markers, which can provide information for linkage analysis and help in locating complex traits associated genes (Chistiakov *et al.*, 2005; Coimbra *et al.*, 2003; Gilbey *et al.*, 2004; Kocher *et al.*, 1998; Kucuktas *et al.*, 2009; Sakamoto *et al.*, 2000; Waldbieser *et al.*, 2001). However, the importance of these historically useful resources is diminishing with the huge datasets from next-generation sequencing.

BAC end sequences (BESs) are another resource for microsatellite marker development. BES-derived microsatellite markers can be used to integrate the physical map and linkage map, which could facilitate draft genome assembly and comparative genome analysis (Liu *et al.*, 2009; Wang *et al.*, 2007; Xu *et al.*, 2007). However, BESs are less important now, with the increasing volume of next-generation sequences.

Next-generation sequencing platforms have been widely used in the last decade. With these platforms, terabytes of sequencing data can be produced in a very short period, which can be used for microsatellite marker development. However, the sequences from these platforms are usually short, such as those produced by Illumina sequencers. In some cases such as PacBio sequencing, long reads are generated, but the accuracy of the sequences is relatively low. Therefore, assembly is required with the high-throughput short sequences to generate sequences that are both accurate and long enough to provide sufficient flanking sequences for the design of PCR primers. The assembly of short reads into contigs is covered in Chapter 4, and hence we will not discuss it here.

Microsatellite Mining Tools

With genomic sequences, microsatellites can be easily identified through informatics analysis. Numerous algorithms and related software packages have been developed to identify microsatellites. Generally speaking, online web-based programs may not be always available. Standalone programs are a better choice for long-term use, especially for the analysis of large datasets. Most of the microsatellite mining programs were developed in the early 2000s, and many websites are no longer available. With time, other programs too may become inactive. Part of the reason is that demands on microsatellite markers are declining. Here, we will only introduce a few programs that are still available online and/or can be downloaded for local computers. Table 16.1 provides a comparison

Table 16.1 Comparisons of selected programs for microsatellite discovery. Plus symbols indicate degrees of user-friendliness.

Programs	Standalone	Online interface	Input sequence format	Primer design	Easy to use
Msatfinder	Perl Script	—	EMBL, FASTA GBK, Swissprot	Yes	++
MISA	Perl Script	—	FASTA	No	+
IMEx	GUI	Yes	FASTA	Yes	+
Msatcommander	GUI	—	FASTA	Yes	+
QDD	Perl Script	Yes	FASTA/FASTQ	Yes	+++

of the programs and their features. Several programs can only be run as commands, such as Msatfinder, MISA, and QDD (Meglecz *et al.*, 2010; Thiel *et al.*, 2003; Thurston & Field, 2005); some others have a general user interface (GUI), such as Msatcommander and IMEx (Faircloth, 2008; Mudunuri & Nagarajaram, 2007). Most of these programs have been designed to utilize Primer3 for primer design.

Msatfinder

Msatfinder is very powerful and easy to use; however, it requires EMBOSS and Primer3 to be installed first (Rice, Longden & Bleasby, 2000; Rozen & Skaletsky, 2000). The program can be downloaded from http://web.archive.org/web/20071026090642/http://www.genomics.ceh.ac.uk/msatfinder. Msatfinder can handle a variety of sequence formats, such as EMBL, FASTA, GENBANK, SWISSPROT, and even ASCII raw files. Under the terminal, the command to run Msatfinder is "`msatfinder.pl <FASTA file>`", and "FASTA file" is the sequence file. A single command can finish both microsatellites identification and primer design. Here, we introduce microsatellites identification; primer design will be discussed in a later section. The following is an example of standard output information after searching microsatellites from the sequences in FASTA format.

```
This looks like a nucleic acid sequence.
Inspecting sequence: 31879966...
Please wait: <---
#This part is the annotation.
Found (gt)8 msat at 202:
#Miscrosatellite (gtgtgtgtgtgtgtgt) was
FT   repeat_region   202..217
#identified between 202 and 217bp
FT                    /label=(gt)8
FT                    /note="(gt)8"
FT                    /note="Coding:  0"
FT                    /note="Gene: N/A"
FT                    /note="Product: N/A"
FT                    /note="Protein: N/A"
Opening output file 31879966.msat_tab
```

```
#Microsatellite information was stored
Writing MINE file 31879966.202.gt.8.db
#in db and fasta format
Opening fasta FILE Fasta/31879966.202.gt.8.fasta
# fasta format will be used to design primers
Determining PCR primers.
# Primer3 was evoked to design primers
Outfile: Primers/31879966.202.gt.8.primers.txt
#Primer information was stored in text file
Running command: eprimer3 -sequence Primers/31879966.202.gt
  .8.temp -outfile Primers/31879966.202.gt.8.primers.txt
   -target 200,218 -primer # Primer3 was evoked to design
   primers
Unlinking file: Primers/31879966.202.gt.8.temp
```

The Msatfinder program has a configuration file, "msatfinder.rc", and microsatellite repeat types of any length can be determined in this file. The default setting is "motif_threshold" = "1,12|2,8|3,5|4,5|5,5|6,5", which indicates that the mono-nucleotide repeat requires a length of at least 12 repeats, eight repeats for di-nucleotide, and at least five repeats for tri-, tetra-, penta-, and hexo-nucleotide repeats. The default values are recommended for most sequences, but if microsatellites cannot be identified with the default settings, you may consider lowering the requirement threshold to detect repeats of shorter lengths.

Msatfinder creates several directories to store the output files—this is convenient, as there are often large numbers of output files. An html file ("results.html") is created in the same directory in which Msatfinder is running—open this in a browser to view all of Msatfinder's output. The "results.html" file contains links to the contents of the following seven subdirectories:

1) "Repeats": This directory contains various data files summarizing sequence and microsatellite information—for example, taxonomy, number of microsatellites, etc. (see the following text).
2) "Counts": A summary of all microsatellites found in a sequence by length and type of exact motif. Unlike the index and count files in the "Repeats" directory, the files here preserve the exact motifs rather than just their content.
3) "Msat_tabs" and "Flank_tabs": These two directories contain feature tables for use with "artemis". This command can show sequence information of each microsatellite. The "msat_tabs" files show only the microsatellite itself, whereas the "flank_tabs" files also include the flanking regions.
4) "Fasta": The sequence of the microsatellite plus flanking regions, in FASTA format. These would be suitable for use in BLAST searches and primer design.
5) "MINE": These files contain the same information as shown in the "repeats" file in the "Repeats" directory (a summary of the details of each microsatellite), but one file per repeat is produced. These contain HTML formatting, and are designed to allow the creation of simple databases using our related MINE software, and are turned off by default.
6) "Primers": Primer files are produced for each microsatellite, if possible. By default, it produces information for possible PCR primers.

The following files will be found in the directory named "Repeats", and all the information can be accessed from the summary results file ("results.html"):

1) "Sequence": This file contains the information on each sequence with microsatellites, including number of microsatellites found, GC content, etc.
2) "Repeats": This file contains the details of each individual microsatellite, plus similar genomic information as the sequence file.
3) "Type.count": This is a table showing the number of microsatellites found, categorized by motif type (mono, di, tri, etc.) and number of repeat units.
4) "Motif.count": This file is similar to the "type.count" file, and it shows the number of microsatellites found, categorized by the bases/amino acids in the motif. These are ordered by the total base content only.
5) "Index": This is an overall summary of the microsatellite results.
6) "Primers.csv": This is a CSV format file that contains the primer information in all the primer files in the "Primers" directory.

Although Msatfinder does not have a GUI, it is easy to use on the Linux and Microsoft Windows (Cygwin needs to be installed) platforms without any other installations necessary. EMBOSS and Primer3 are required for the primer design. The version of Primer3 is very important, as only version 1.1.4 is compatible with the program Msatfinder without any modification (it can be downloaded from http://sourceforge.net/projects/primer3/files/primer3/1.1.4/).

MIcroSAtellite Identification Tool (MISA)

MISA is just a simple Perl script, which does not require any other package or software. The command syntax is "`misa.pl <FASTAfile>`". "FASTAfile" is the name of the file containing DNA sequences in FASTA format. An additional file containing the search parameters is required, named "misa.ini"—an example of "misa.ini" parameters is shown here:

```
definition (unit_size, min_repeats): 1-10 2-6 3-5 4-5 5-5
    6-5
```

In a single line beginning with "definition", a sequence of number pairs is expected, where the first number defines the microsatellite repeat sizes and the second number defines the minimum length of repeats for that specific microsatellite.

```
interruptions (max_difference_between_2_SSRs): 50
```

In a single line beginning with "interruptions", a single number is expected, defining the maximum number of bases between two adjacent microsatellites to be recognized as being a compound microsatellite. The minimum distance is 50 bp, which gives sufficient sequence for primer design. The results of the microsatellites search are stored in two files—in "<FASTAfile>.misa", where the location and type of the identified microsatellites are stored in a tabulated format; and in a file named "<FASTAfile>.statistics", which summarizes different statistics as the frequency of a specific SSR type according to unit size or individual motifs. MISA result output includes: sequence ID, SSR numeric order in the sequence, SSR type (p2 = dinucleotide), SSR, SSR size, SSR start, and SSR end.

Msatcommander

Msatcommander is a Python program written to search microsatellites within FASTA-formatted sequence or consensus files. This GUI-based program can be run on Microsoft Windows/Linux/macOS systems without any other programs or extra installations. It can be downloaded from https://code.google.com/p/msatcommander/. Msatcommander can design and tag primers using Primer3 as its primer design engine. The settings for Msatcommander is very straightforward, the length of repeats can be determined by the user, and the program interface is self-explanatory. For instance, under "Choose Array Type", mark di-, tri-, and tetra-nucleotides as appropriate; and specify the repeat length for each type of microsatellites as appropriate (e.g., eight repeats for di-, six for tri-, and five for tetra-nucleotide repeats). Msatcommander will generate one file in CSV format to store all the microsatellite results, and one directory to store all the microsatellite primers ("Primer3"). This program can automatically pick up the best pair of primers based on its algorithm, and it will also give a few extra pairs of primers for your choice.

Imperfect Microsatellite Extractor (IMEx)

IMEx is a tool for extracting perfect, imperfect, and compound microsatellites from sequences. It is an efficient, fast, and user-friendly program. Both online and GUI interfaces have the exact same design, and the parameters can be determined in the same fashion. IMEx has a unique design, which can search both perfect and imperfect microsatellites. IMEx has three modes: basic, intermediate (standalone program only), and advanced modes. The basic and intermediate modes have limited features with default values, which are recommended for most situations, such as only extracting perfect microsatellites. The advanced mode includes all possible options of imperfect microsatellite mismatch, such as the level of mismatch allowed in the microsatellite sequence, which can be determined by the user. For the online version, there is another mode, Genome-wide mode, which is designed to search for microsatellites in bacteria and virus genomes, and this mode is only available for the online version.

The output of microsatellite searching includes: "Consensus"—meaning the microsatellite repeat type; "Iterations"—meaning the repeat length of microsatellites; "Tract-size"— meaning the total length of microsatellite repeats; "Start" and "End"—meaning the start and stop positions of microsatellites in the sequence; and "Imperfection"—meaning how much mismatch was identified in the microsatellite.

If you want to design primers for the microsatellites, just click on the button "Design Primer" at the right end of the table; you will be redirected to the Primer3 website, hosted on the IMEx server, and the IMEx server will automatically fill the microsatellite sequence and parameters for primer design; then, you need to click on the "Pickup Primer" button.

QDD

QDD is the latest program developed for microsatellite mining, and it has two versions: (1) one version of QDD is written in Perl language and runs as a standalone application on Microsoft Windows or Linux systems from a terminal (command line version), which is similar to Msatfinder and other standalone programs; and (2) the other version of QDD is integrated into the Galaxy server (https://usegalaxy.org/), which can provide an online interface similar to IMEx. However, the Galaxy server is a complicated bioinformatics platform that has integrated many handy bioinformatics software and requires more computing power on the High-performance Cluster System and extra knowledge for installation. This program can be downloaded from http://net.imbe.fr/~emeglecz/qdd_download.html, and detailed installation instruction is also available online (http://net.imbe.fr/~emeglecz/qdd_installation.html). The best features of this software are: (1) it is very powerful on a large, high-performance server, and the online version can provide access for many users simultaneously; (2) this program can analyze the traditional Sanger sequences (BESs and ESTs), and also the raw 454 sequences and the low-coverage Illumina sequences (less than 500 MB) in FASTA or FASTQ formats without adapter filtering and pre-assembly, which is very useful for microsatellite mining from small transcriptome sequencing data. Here, will only introduce the standalone program and its usage.

The QDD program has four pipelines, and you can run all the four pipelines either together or separately. The following are the four pipelines running separately:

1) Sequence preparation and microsatellite detection: the microsatellite markers will be extracted from either pre-assembled sequences (contigs, scaffold, or chromosomes) or non-assembled sequences based on the flanking sequencing length defined by the user.

   ```
   perl pipe1.pl -input_file qdd_data\example.fasta
   ```

   Microsatellites are extracted with 80 bp flanking regions on both sides, and are found in the file "example1_pipe1_for_pipe2.fas" under "qdd_output" directory, and the flanking region size can be defined by the user with the "-flank_length" setting.
2) The sequences with microsatellites will be compared with each other to remove duplicated sequences. The unique sequences are found in the file "example_pipe2_for_pipe3.fas" in the "qdd_output" directory.

   ```
   perl pipe2.pl -input_file qdd_output\example_pipe1_for_
     pipe2.fas
   ```

3) The unique microsatellite sequences are loaded into Primer3 for primer design. The primer information is stored in the file named "example_pipe3_primers.tabluar" in the "qdd_output" directory.

   ```
   perl pipe3.pl -input_file qdd_output\example_pipe2_for_
     pipe3.fas
   ```

4) The last step is to check the sequence with primers against the NCBI database to remove any potential contamination or transposable elements.

   ```
   perl pipe4.pl -input_file qdd_output\example_pipe3_
     primers.tabular -check_contamination 1
   perl pipe4.pl -input_file qdd_output\example_pipe3_
     primers.tabular -rm 1 -rm_lib insecta
   ```

This program can also be done all at once by running "QDD.pl" with the following command:

```
perl QDD.pl -input_folder c:\data_example -check_
  contamination 1
```

The detailed results will not be shown here, as both the input and output file examples can be found on the website (http://net.imbe.fr/~emeglecz/qdd_run.html#ex1).

Primer Design for Microsatellite Markers

Primer Design

Primer design is the most crucial step for success in using microsatellite markers. The quality of microsatellite marker primers highly depends on the length and quality of microsatellite flanking sequences. Here, we will talk about how to design the primers using the software and how to choose the best pair of primers.

Primer3 is one of the most popular free primer design software for microsatellite primer design. Primer3 has both a standalone program and an online interface. For a handful of microsatellites, the online interface is easy to use. If large numbers of microsatellites are used for primer design, the standalone Primer3 program will be more efficient.

Selection of Primer Pairs

Primer3 usually generates at least five pairs of primers, and ranks the primers 1–5 based on its algorithm. However, the primers chosen by the program may not be the best primers; it is very essential to choose the proper primers for application of microsatellite markers. Here are some tips:

1) Before designing the primers from the flanking sequences, at least 50 bp of flanking sequences are required on both sides of the microsatellite; longer flanking sequences may sometimes be needed to design the primers, as the sequences content close to the microsatellites are not ideal for primer design.
2) If the microsatellites have sufficient flanking sequences, the flanking sequences should be checked to see if the sequences contain secondary structure and/or unbalanced GC-rich and AT-rich sequences. Some online tools are very handy for this checkup, such as GC content calculator (http://www.endmemo.com/bio/gc.php).
3) In general, the length of primers between 18 and 25 would be good. GC content should be between 40% and 60%, with the 3' of a primer ending in GC. Both forward and reverse primers should have similar lengths and melting temperatures (Tm; between 55 °C and 70 °C). The difference of Tm between two primers should be less than 5 °C—the smaller the difference, the better. If the melting temperature is less than 55 °C, higher GC content or longer primer length would be necessary. If multiplex reactions will be conducted in the same run, all the primers should have similar Tm values. Usually, the annealing temperature should be set 5 °C lower than Tm for the amplification.
4) Intra-primer homology (more than three bases that complement within the primer) or inter-primer homology (forward and reverse primers have complementary sequences) should be avoided, as these can lead to hairpins or primer-dimers instead of annealing to the desired DNA templates. The primer secondary structure and primer-dimer can be checked using online tools such as OligoAnalyzer

(http://www.idtdna.com/calc/analyzer) and Multiple primer analyzer (https://www.thermofisher.com/in/en/home/brands/thermo-scientific/molecular-biology/molecular-biology-learning-center/molecular-biology-resource-library/thermo-scientific-web-tools/multiple-primer-analyzer.html).

5) A product length ranging from 100 to 300 bp is desirable, which is ideal for visualization through either electrophoresis gels or automated genotyping instruments. The product size should be optimized to allow ideal separation of different microsatellite markers when multiplexing primers are used in the amplification.

Conclusions

In spite of being a traditional type of molecular marker, microsatellites continue to be used for various studies. In some cases, they will continue to be advantageous as compared to any other types of molecular markers. In the aquaculture field, microsatellite markers are useful for the construction of linkage maps, characterization of genetic stocks, sex determination, QTL mapping, and marker-assisted breeding (Chistiakov, Hellemans & Volckaert, 2006).

Although microsatellites can be identified through traditional approaches, the current availability of genomic sequences from various aquaculture species and the application of next-generation sequencing make it unnecessary to develop microsatellites through experimentation. Rather, microsatellite markers can be readily identified through bioinformatics mining of genomic or transcriptomic sequences using various software packages, among which Msatfinder and QDD are efficient for searching microsatellite markers from hundreds of, even thousands, of sequences. Both programs can automatically invoke the Primer3 program, and design PCR primers immediately after microsatellites are identified. However, these two programs can only be used in the command line mode, which requires some extra knowledge of Microsoft Windows and the Linux shell. Msatcommander and IMEx are intuitive, as they have a GUI that is very user-friendly, without demanding knowledge about command line operations. However, they are not very efficient in searching for microsatellite markers and design primers from large numbers of sequences. Most of the programs have been designed to find only the perfect repeat type of microsatellites, but IMEx is capable of searching for imperfect microsatellite repeats. Although microsatellite markers can now be readily identified, designing PCR primers requires a set of considerations including the length of primers, GC contents, hairpin structures, primer-dimer formation, and 3' base composition.

References

Chistiakov, D.A., Hellemans, B., Haley, C.S., *et al.* (2005) A microsatellite linkage map of the European sea bass *Dicentrarchus labrax* L. *Genetics*, **170**, 1821–1826.

Chistiakov, D.A., Hellemans, B. and Volckaert, F.A.M. (2006) Microsatellites and their genomic distribution, evolution, function and applications: A review with special reference to fish genetics. *Aquaculture*, **255**, 1–29. doi: 10.1016/j.aquaculture.2005.11.031

Coimbra, M.R.M., Kobayashi, K., Koretsugu, S., *et al.* (2003) A genetic linkage map of the Japanese flounder, *Paralichthys olivaceus*. *Aquaculture*, **220**, 203–218. doi: 10.1016/S0044-8486(02)00353-8

Faircloth, B.C. (2008) msatcommander: detection of microsatellite repeat arrays and automated, locus-specific primer design. *Molecular Ecology Resources*, **8**, 92–94. doi: 10.1111/j.1471-8286.2007.01884.x

Gilbey, J., Verspoor, E., McLay, A. and Houlihan, D. (2004) A microsatellite linkage map for Atlantic salmon (*Salmo salar*). *Animal Genetics*, **35**, 98–105. doi: 10.1111/j.1365-2052.2004.01091.x

Kocher, T.D., Lee, W.J., Sobolewska, H., *et al.* (1998) A genetic linkage map of a cichlid fish, the tilapia (*Oreochromis niloticus*). *Genetics*, **148**, 1225–1232.

Kucuktas, H., Wang, S., Li, P., *et al.* (2009) Construction of genetic linkage maps and comparative genome analysis of catfish using gene-associated markers. *Genetics*, **181**, 1649–1660. doi: 10.1534/genetics.108.098855

Liu, H., Jiang, Y., Wang, S., *et al.* (2009) Comparative analysis of catfish BAC end sequences with the zebrafish genome. *BMC Genomics*, **10**, 592. doi: 10.1186/1471-2164-10-592

Liu, S., Zhou, Z., Lu, J., *et al.* (2011) Generation of genome-scale gene-associated SNPs in catfish for the construction of a high-density SNP array. *BMC Genomics*, **12**, 53. doi: 10.1186/1471-2164-12-53

Liu, Z. (2007) *Aquaculture genome technologies*, John Wiley & Sons Publishing, Ames, IA USA.

Meglecz, E., Costedoat, C., Dubut, V., *et al.* (2010) QDD: a user-friendly program to select microsatellite markers and design primers from large sequencing projects. *Bioinformatics*, **26**, 403–404. doi: 10.1093/bioinformatics/btp670

Mudunuri, S.B. and Nagarajaram, H.A. (2007) IMEx: Imperfect Microsatellite Extractor. *Bioinformatics*, **23**, 1181–1187. doi: 10.1093/bioinformatics/btm097

Rice, P., Longden, I. and Bleasby, A. (2000) EMBOSS: the European Molecular Biology Open Software Suite. *Trends in Genetics*, **16**, 276–277.

Rozen, S. and Skaletsky, H. (2000) Primer3 on the WWW for general users and for biologist programmers. *Methods in Molecular Biology*, **132**, 365–386.

Sakamoto, T., Danzmann, R.G., Gharbi, K., *et al.* (2000) A microsatellite linkage map of rainbow trout (*Oncorhynchus mykiss*) characterized by large sex-specific differences in recombination rates. *Genetics*, **155**, 1331–1345.

Tautz, D. (1989) Hypervariability of simple sequences as a general source for polymorphic DNA markers. *Nucleic Acids Research*, **17**(16), 6463–6471.

Thiel, T., Michalek, W., Varshney, R.K. and Graner, A. (2003) Exploiting EST databases for the development and characterization of gene-derived SSR-markers in barley (*Hordeum vulgare* L.). *Theoretical and Applied Genetics*, **106**, 411–422. doi: 10.1007/s00122-002-1031-0

Thurston, M.I. and Field, D. (2005) *Msatfinder: detection and characterisation of microsatellites*.

Waldbieser, G.C., Bosworth, B.G., Nonneman, D.J. and Wolters, W.R. (2001) A microsatellite-based genetic linkage map for channel catfish, *Ictalurus punctatus*. *Genetics*, **158**, 727–734.

Wang, S., Xu, P., Thorsen, J., *et al.* (2007) Characterization of a BAC library from channel catfish *Ictalurus punctatus*: indications of high levels of chromosomal reshuffling among teleost genomes. *Marine Biotechnology (NY)*, **9**, 701–711. doi: 10.1007/s10126-007-9021-5

Xu, P., Wang, S., Liu, L., *et al.* (2007) A BAC-based physical map of the channel catfish genome. *Genomics*, **90**, 380–388. doi: 10.1016/j.ygeno.2007.05.008

17

SNP Identification from Next-Generation Sequencing Datasets

Qifan Zeng, Luyang Sun, Qiang Fu, Shikai Liu and Zhanjiang Liu

Introduction

The advancement of next-generation sequencing (NGS) and computational abilities on data analysis has made single-nucleotide polymorphisms (SNPs) the most commonly used markers in a wide range of genetic analyses. Thus, the demands to obtain large numbers of high-quality SNPs from large volumes of NGS datasets are becoming increasingly high. In this chapter, we will introduce specific statistical models to separate true variations from artifacts, and detailed protocols (including command lines) for SNP identification from NGS datasets. Selection of proper analysis parameters will be discussed when possible. A number of optional steps, such as local realignment, base quality recalibration, and variant quality score recalibration, can greatly improve the SNP calling quality.

Genomic variations include SNPs, insertions and deletions (INDELs), inversions, translocations, segmental duplications, and copy-number variations (CNVs). The polymorphisms caused by these genomic variations form the basis of phenotypic differences among individuals in one species. Therefore, identification of genetic variations conferring biological diversity is becoming increasingly important for genomic studies. It is technically also increasingly feasible because of the advances in NGS technologies (Kidd *et al.*, 2004).

Among various genomic variations, SNPs are the most common type of genetic variations; about 90% of sequence variants are single-base substitutions in humans (Collins, Brooks & Chakravarti, 1998). SNPs have become the molecular markers of choice for many reasons, including their abundance in the genome, their relatively even distribution across the genome, and their adaptation to automation for genotyping (Helyar *et al.*, 2011). Many automated genotyping platforms are now available for mid–high-throughput genotyping using SNPs. These include Sequenom MassArray (Gabriel, Ziaugra & Tabbaa, 2009), Illumina BeadArray (Oliphant *et al.*, 2002), and Affymetrix Axiom Array (Rabbee & Speed, 2006) technologies. With any of these platforms, a large number of SNPs for multiple samples can be genotyped simultaneously. Because of these advantages, SNPs have taken the place of microsatellites as the primary choice of genetic markers (Mardis, 2008).

Large numbers of SNPs have been identified from various species. A comprehensive database of SNPs in humans have been constructed to provide a valuable resource for

Bioinformatics in Aquaculture: Principles and Methods, First Edition. Edited by Zhanjiang (John) Liu.

disease research, forensics, and diagnostics (Riva & Kohane, 2004; Stenson *et al.*, 2003). Large collections of SNPs have been identified from various livestock as well, including cattle (Bovine HapMap *et al.*, 2009; Matukumalli *et al.*, 2009), sheep (Kijas *et al.*, 2009), chicken (Marklund & Carlborg, 2010), swine (Ramos *et al.*, 2009; Wiedmann, Smith & Nonneman, 2008), and turkey (Aslam *et al.*, 2012; Kerstens *et al.*, 2009). With aquaculture species, large numbers of SNPs have been identified from a number of species, including Atlantic Herring (Helyar *et al.*, 2012), common carp (Xu *et al.*, 2012), plateau fish (Wang *et al.*, 2015), tiger puffer (Cui *et al.*, 2014), and blunt snout bream (Gao *et al.*, 2012). High-density SNP arrays have been developed in common carp (Xu *et al.*, 2014), rainbow trout (Palti *et al.*, 2015), Atlantic salmon (Houston *et al.*, 2014), and catfish (Liu *et al.*, 2014).

In this chapter, we provide an introduction on the mining of SNPs from NGS datasets, focusing on the bioinformatics aspects of the processes.

SNP Identification and Analysis

SNPs can be rapidly identified with NGS datasets. Data fragments containing SNPs can be isolated efficiently from almost all types of sequencing data, including whole genome sequencing, exome sequencing, RNA sequencing, and chromatin immunoprecipitation sequencing (ChIP-seq). With whole genome sequencing datasets, SNPs can be identified from the genome. SNPs covering the whole genome are more useful for genetic and genomic analysis. With RNA-Seq datasets, SNPs can be identified from the sequenced transcripts, which are correlated with the expression patterns in the tissue and under the specific environment. In addition, SNPs identified from transcriptome cover only the transcribed regions of the genome.

Identification of SNPs from raw sequencing data consists of multiple steps involving utilization of a diverse set of software. In the following sections, we provide a general introduction on the essential building blocks for an SNP identification pipeline, including quality control of sequencing reads, alignment of reads to the reference, processing of the alignment, filtering of SNP candidates, and SNP annotation.

Quality Control of Sequencing Data

Despite advances in NGS technologies, base calling errors, poor-quality reads, and adaptor contamination are still inevitable. Sequencing data quality assessment and control can dramatically reduce artifacts and improve SNP quality. Assessment of raw data includes several key statistics, including distribution of base calling quality scores, adapter content, *k*-mer frequency spectrum, and average GC content. Several open-source software packages can be used for quality assessment, such as FastQC (http://www.bioinformatics.bbsrc.ac.uk/projects/fastqc/), kPAL (Anvar *et al.*, 2014), SolexaQA (Cox, Peterson & Biggs, 2010), and NGSQC Toolkit (Patel & Jain, 2012).

Base calling qualities are measured by the Phred quality score of identification of each nucleobase generated during sequencing. Quality scores of higher than 20 ($p < 0.01$) are generally regarded as acceptable (Figure 17.1). The higher the scores are, the better the base calling qualities. Adapter contamination is a major issue with NGS datasets. Removal of adapter sequences is essential before the identification of SNPs.

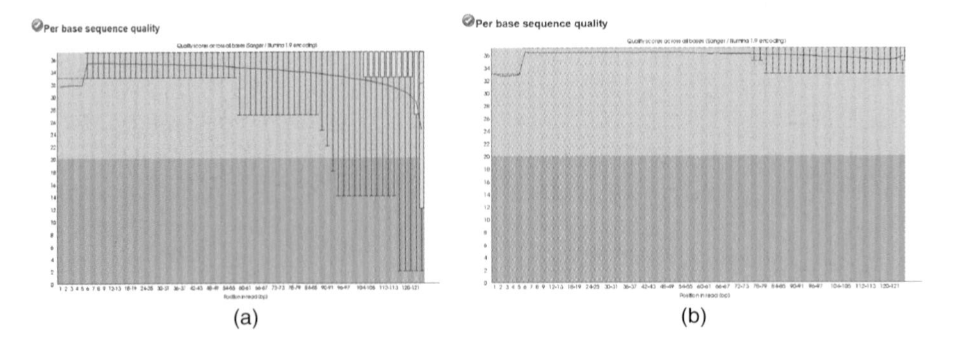

Figure 17.1 Summary of reads quality generated by FastQC: an overview of the range of base quality values at each position before (a) and after (b) trimming. (*See color plate section for the color representation of this figure.*)

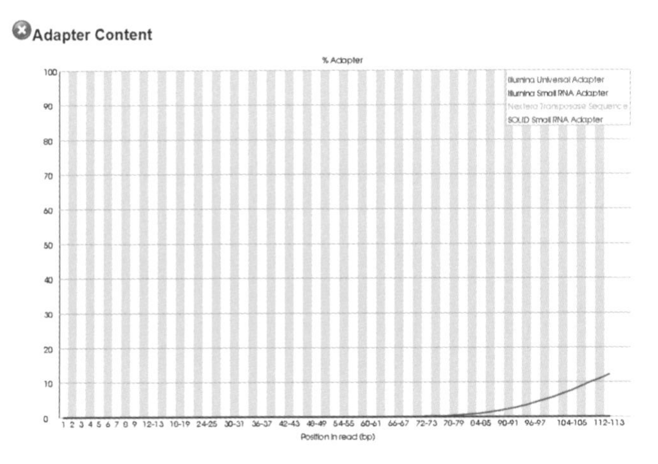

Figure 17.2 Summary of reads quality generated by FastQC: detection of adapter contamination. X-axis indicates base positions, and Y-axis is the percentage of adapter sequences. Note the presence of adapter sequences starting at the base 79 position.

For instance, with the analysis of the specific dataset shown in Figure 17.2, adapter sequences were evident after base position 79. Before subsequent analysis is conducted, adapter sequences must be removed (see the following text). The k-mer content measures the enrichment of all the subsequences of a specific length k. Theoretically, overrepresented k-mers from true repeat regions should be distributed equally within a sequence. The k-mers with positionally biased enrichment may indicate problems. For instance, in Figure 17.3, 7-mers "GATCTCG", "ATCTCGG", "TCGGTGG", and "TGGTCGC" are truncated from "GATCTCGGTGGTCGC", which is a part of the reverse complementary sequence of TruSeq universal adapter. Therefore, they have similar distribution along the sequencing cycle. Average GC content reflects the A/T vs G/C ratio of the genome. A single peak in roughly normal distribution is expected. For instance, the catfish genome has approximately 60.7% A/T and 39.3% G/C (Xu, 2007), and the distribution of the G/C content curve should peak at a position of 40% G/C (Figure 17.4). The better the actual read curve coincides with the theoretical curve, the higher the quality of the dataset is.

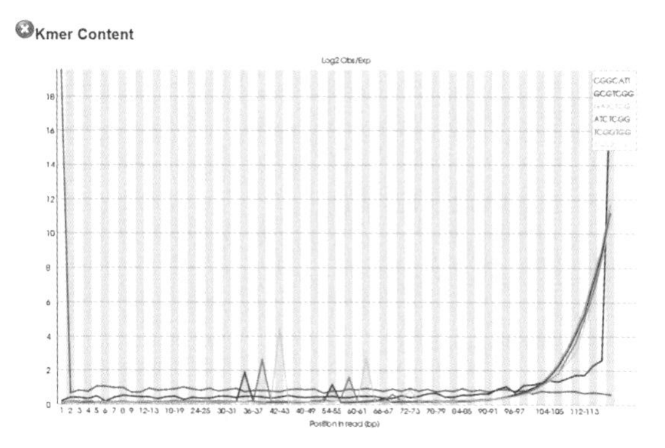

Figure 17.3 Summary of reads quality generated by FastQC. Analysis of overrepresented sequences. The underlined part of the TruSeq universal adapter sequence "AATGATACGGCGACCACCGAGATCTACACTCTTTCCCTACACGACGCTCTTCCGATCT" was identified with positional biases enrichment. (*See color plate section for the color representation of this figure.*)

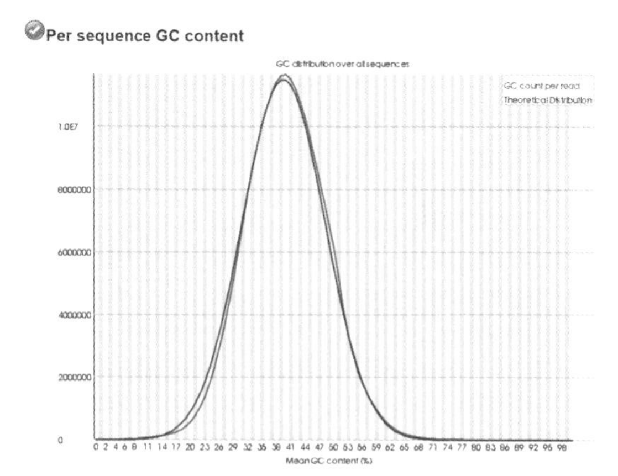

Figure 17.4 An example of the output file for GC content analysis: well-aligned distribution between the theoretical and observed G/C content indicate good quality of the data. (*See color plate section for the color representation of this figure.*)

Many packages and scripts are freely available for trimming the low-quality bases and adaptors. These include Trimmomatic (Bolger, Lohse & Usadel, 2014), UrQt (Modolo & Lerat, 2015), and Btrim (Kong, 2011). Analysis of base calling quality after trimming the adapters should reveal the retention of only high-quality sequences (Figure 17.1B).

Alignment of Short Reads to the Reference Sequence

The next step is to align the clean reads to reference sequences. Alignment of millions of short reads with variations requires relatively large amounts of computing resources. To make tradeoffs intelligently among speed, accuracy, and memory utilization, a number of efficient algorithms have been developed.

Two of the most common ones are Burrows–Wheeler transform (BWT)–based algorithms, which is adopted by Bowtie (Langmead *et al.*, 2009), Bowtie2 (Langmead & Salzberg, 2012), and BWA (Li & Durbin, 2009); and hash-based algorithms, which is adopted by Mosaik (Quinlan *et al.*, 2008), MAQ (Li, Ruan & Durbin, 2008), and stampy (Lunter & Goodson, 2011). BWT-based algorithms are efficient in data compression and allow short reads to simultaneously align to similar substrings of the reference. It is faster than hash-based algorithms, especially in mapping reads to genome references with many repetitive regions. However, hash-based algorithms are more sensitive and efficient in approximate matching. Introduction of multiple seeds also improve the performance of hash-based algorithms (Lunter & Goodson, 2011). The increasing length of NGS reads drive the alignment strategy away from the former algorithm, which is speedy but not error-tolerant. Adoption of affine-gap alignment instead of traditional end-to-end alignment can detect multiple non-overlapping local matches caused by structure variations or mis-assembly in the reference genome (Li, 2013). In our experience, BWA-MEM should be a good choice for its banded dynamic programming, which promises linear time complexity in the length of query reads. It also possesses an automatic selection strategy of affine-gap and end-to-end alignment, which could boost its sensitivity and reduce reference bias (Li, 2013).

Proper settings of parameters are also of importance to get a fair proportion of aligned reads with controlled false alignments and mismatches. SNP rates vary significantly among teleosts (Table 17.1). The selection of criteria for allowed mismatches typically depends on the species under study.

Mapping results can also be visually inspected through visualization tools. Currently, multiple genome browsers are available to display reference sequence, aligned reads, mapping quality, and identified mutations. One of the advantages of genome browsers is the graphical user interface (GUI), which makes the tools intuitive and user-friendly. Another advantage is the support for public databases of genomic annotations, such as the UCSC (University of California, Santa Cruz) Genome Browser Annotation Database. Many genome browsers with GUI are implemented in Java and, therefore, are available on multiple systems. Integrative Genomics Viewer (IGV) is one of the high-performance tools for the interpretation of large and integrated genomic datasets (Robinson *et al.*, 2011). It supports multiple data types such as NGS sequencing data, array-based data, and genomic annotations.

In addition to stand-alone software, genome browsers also have web-based editions supported by dedicated web servers (Wang *et al.*, 2013). Common web-based applications include Ensembl Genome Browser (Birney *et al.*, 2004), Vertebrate Genome Annotation (Vega) Genome Browser (Ashurst *et al.*, 2005), and UCSC Genome Browser (Kent *et al.*, 2002). Computationally intensive procedures are carried out on the server; therefore, users do not need to download and install additional annotation files and applications. However, uploading large volumes of datasets to the server is time-consuming and poses problems for data security.

Table 17.1 Identified SNP number and rates in several teleosts with genome references.

Species	SNP number	Genome reference length	SNP rates	Data source	References
Zebrafish	7 million	$1.41e^9$	201	—	Howe *et al.* (2013)
Stickleback	1190483	$3.84e^8$	322	DRR014002	(Ensembl:BROADS1)
Medaka	1281882	$6.6e^8$	514	DRR014054	(Ensembl:MEDAKA1)
Channel catfish	8.3 million	$7.83e^8$	93	—	Liu *et al.* (2016)
Common carp	962621	$1.67e^9$	1734	SRR924318	(ENA:PRJEB7241)
Tilapia	775424	$7.4e^8$	954	SRR071614	(Ensembl:Orenill.0)
Atlantic cod	1047875	$7.53e^8$	718	—	Star *et al.* (2011)
Torafugu	59701	$2.8e^8$	4690	—	Zhang *et al.* (2013)

Processing of the Post-alignment File

The results of the alignment are commonly stored as Sequence Alignment/Map (SAM) format (Li, Handsaker *et al.*, 2009). It is a standardized text file that contains a header section and an alignment section. The header section records generic information, including the format version, the sorting order of the short reads, the reference sequence dictionary, the read group information, and the program used for alignment. The alignment section describes details of each query reads on how and where they align to the reference genome. For more information on the SAM file format, please refer to Sequence Alignment/Map Format Specification (https://samtools.github.io/hts-specs/SAMv1.pdf). The aligned reads should be sorted according to their position on the reference. The "`sort`" command in SAMtools (Li, Handsaker *et al.*, 2009) and the "`SortSam`" module in Picard (http://picard.sourceforge.net) can both be used. Once the SAM file is sorted and indexed, one can check the mapping statistic of the SAM file with the "`CollectAlignmentSummaryMetrics`" command in Picard.

A few post-alignment processing steps can improve the accuracy in variant calling. The first step is to mark and remove duplicated reads. Duplicated reads are sets of reads with identical unclipped alignments. As for Illumina sequencing technology, duplicate reads could result from PCR amplification during library preparation or optical duplicates that occur during base calling. Sequencing errors in duplicates could affect the allele frequency or result in false-positive SNPs. In practice, the "`MarkDuplicatesWithMateCigar`" tool in Picard can be used to identify and remove redundant copies of the duplicates, which would mitigate the effects of sequencing errors.

Realignment of regions near INDELs is also helpful for elimination of the potential false-positive SNPs. Due to defects in mapping software, INDELs near the ends of reads are usually misaligned as mismatches, which would affect identification of true SNPs. By using the "`RealignerTargetCreator`" tool of GATK (McKenna *et al.*, 2010), one can perform local realignments to accommodate INDELs. Known INDELs can also be imported to reduce computational time. However, this software is not compatible with a few data types, such as Roche 454. In that case, one can consider filtering the regions near INDELs by using the "`identify-genomic-indel-regions`" and "`filter-pileup-by-gtf`" commands of PoPoolation (Kofler *et al.*, 2011).

Another helpful step is recalibration of base quality score of query reads. For multiple types of sequencing platforms, Phred-like quality scores assigned to each base may deviate from the true error rate (DePristo *et al.*, 2011; Li, Li *et al.*, 2009). For Illumina sequences, the reported quality scores are calculated by a quality model that uses a set of empirically determined quality values such as optical signal intensities and signal-to-noise ratios. Even for reads with high reported qualities, base mismatches with the reference are more frequently at the 3' ends than at the 5' ends. Also, dinucleotide AC is usually with lower reported qualities than TG. These biases indicate that the pre-calibrated scores are usually correlated with sequencing features. The "`BaseRecalibrator`" command in the GATK package could be applied to recalibrate the quality scores on the sequencing data of various platforms. In practice, bases from the same read group are categorized with respect to several features (e.g., raw quality, dinucleotide content, position with the read) in order to analyze the co-variations. The calculated co-variates will then be applied through a piecewise tabular recalibration to correct raw quality scores. For a detailed introduction on how exactly the base quality is recalibrated, readers are referred to Mark Depristo's work (DePristo *et al.*, 2011). The system also allows users to add new co-variates by creating a Java class implemented with the required interface. "ContextCovariate.java" and other covariate Java classes in the package could be referred to as examples.

SNP and Genotype Calling

Variant and genotype calling for individual sequencing data can be performed simply by counting the number of mapped reads of each allele and applying hard filters. For instance, an SNP could be called if the proportion of the minor allele is greater than 20%. It works well for data with high sequencing depth, as the probability of a heterozygous site with allele frequencies falling outside the range of 20–80% is small. However, using a fixed cutoff will increase the false-negative rate in case of moderate or low sequencing depth. Moreover, variant calling by hard filter does not quantify the uncertainty about the genotypes. Consequently, approaches with probabilistic frameworks have been developed to provide a posterior probability for each genotype (Li, Gao *et al.*, 2009; Li, Li *et al.*, 2009). Taking the algorithm of Haplotypecaller (DePristo *et al.*, 2011) as an example, the program first identifies genome regions of samples with significant evidence of variation related to the reference. For these active regions, a De Brujin–like graph is built to reassemble plausible haplotypes. Then each read is aligned to these haplotypes in turn and assigned a per-read allele likelihood $P(D|H)$, which is used for computing the posterior probabilities for each possible genotype by applying Bayes' theorem. The most likely genotypes on each variant site and their associate metrics are then annotated in a VCF file. The basic formula used for computing genotype posterior probability is:

$$P(G|D) = \frac{P(G)P(D|G)}{\sum_i P(G_i)P(D|G_i)}$$

Where G refers to a particular diploid genotype, and D refers to the evidence from observed data. $P(G|D)$ is the conditional probability of genotype G, given the observed data D. The denominator on the right side of the equation represents the prior probability for the evidence from observed data. It remains constant for all the genotypes, and therefore could be omitted. The term $P(G)$ refers to the prior probability of genotype G. The SNP frequency found from previous experiments based on allele frequencies can be used, such as dbSNP (Sherry *et al.*, 2001). Alternatively, flat prior can be used in the absence of external information. The term $P(D|G)$ refers to the probability of observing the particular sequencing data from a specific genotype. For diploid organisms, the formula is:

$$P(D|G) = \prod j \left(\frac{P(D_j|H_1)}{2} + \frac{P(D_j|H_2)}{2} \right)$$

H refers to a particular haplotype. The term $P(D_j|H_n)$ is the per-read allele likelihood calculated in the previous step.

The contemporary algorithms with probabilistic frameworks could also incorporate linkage disequilibrium information for the imputation of missing genotypes (Browning & Browning, 2007). One can improve the genotype calling result by using the BEAGLE and BEAGLECALL software packages (Browning & Yu, 2009).

Alternatively, sequencing pooled DNA samples is a cost-effective choice. Despite the missing information on haplotypes, the extraordinary coverage of resequencing data enables imputation of genome-wide polymorphism and detection of rare variants in a population (Brockman *et al.*, 2008). PoPoolation is a useful toolbox to retrieve polymorphism from genome-wide data, and link them with functional information. It allows calculating an unbiased estimator of population dynamics ($\theta_{\text{Watterson}}$ and θ_{π}) for pooled samples with a sliding window approach:

$$\theta_{W_{b,pool}} = \frac{\theta_{W_b} c_C}{c_n \sum_{m=b}^{C-b} \sum_{r=1}^{n-1} P(X_C = m|Y_n = r) P(Y_n = r)}$$

$$\theta_{\pi_{b,pool}} = \frac{\theta_{\pi_b}}{c_n \sum_{m=b}^{C-b} \theta_{\pi}(m) \sum_{r=1}^{n-1} P(X_C = m|Y_n = r) P(Y_n = r)}$$

These two estimators explain the truncated allele frequency spectrum. θ_{W_b} and θ_{π_b} are modified classical θ_W and θ_π that only consider SNPs with a minimum allele count of b. Here, n is the pool size, C is the observed coverage, and $c_n = \sum_{k=1}^{n-1} \frac{1}{k}$. The term $P(X_C = m|Y_n = r)$ is the conditional possibility of observing the allele frequency m, given an allele frequency of r in the pool. The term $P(Y_n = r)$ refers to the prior probability that an allele has a frequency of r in the pool.

A corrected Tajima's D test used for pooled data is implemented in the package to evaluate the null hypothesis of neutral evolution:

$$D_{b,pool} = \frac{d_{b,pool}}{\sqrt{Var(d_{b,pool})}}$$

Here, $d_{b,pool} = \theta_{\pi_{b,pool}} - \theta_{W_{b,pool}}$. The term $Var(d_{b,pool})$ is the variance of $d_{b,pool}$, and is calculated as:

$$Var(d_{b,pool}) = \theta c_n \sum_{m=b}^{C-b} (d_{b,pool}(m))^2 \sum_{r=1}^{n-1} P(X_C = m|Y_n = r) P(Y_n = r)$$

θ can be estimated with $\theta_{\pi_{b,pool}}$ on the same window of the estimated $D_{b,pool}$. When $D_{b,pool}$ is statistically indistinguishable from zero, the population is supposed to have no size change and no pattern of selection at the locus. If $D_{b,pool}$ is negative, the population may be increasing in size or having a purifying selection at the locus. If $D_{b,pool}$ is positive, the population may suffer a bottleneck or have overdominance at the locus.

Generally, the called SNP candidates are stored in Variant Call Format (VCF) files (Danecek *et al.*, 2011). The VCF file is a standard text file that contains verbose annotation about the SNPs, INDELs, and other structural variations. A VCF file contains two major parts: the header and the variant calling data. The header contains data lines start with a "##" string. This part describes meta-information about the dataset and reference sources, including file format version, applied filters, and information about reference assembly. The variant calling data is composed of eight mandatory columns:

CHROM: chromosome name
POS: the left-most position of the variant
ID: unique variant identifier (could be "NULL")
REF: the reference allele
ALT: the alternate allele(s), separated by commas
QUAL: variant/reference quality
FILTER: filters applied
INFO: information related to the variant

In addition to general information on the position and quality, detailed annotations on variation callings are stored in the "INFO" field. For more detailed information of the VCF format, please visit http://samtools.github.io/hts-specs/.

By using BCFtools (Li, 2011), one can easily manipulate VCF files, such as merging multiple files and extracting SNPs from defined regions.

Filtering SNP Candidates

It is important to select appropriate filtering strategies to keep as many true SNPs as possible while removing artifacts. As SNP calling from pooled samples is less sensitive to sequencing errors and erroneous mapping of short reads, applying more stringent filters could minimize SNP calling artifacts caused by stochastic errors. Typical filters include checks for deviations of the Hardy–Weinberg equilibrium (HWE), minor allele counts, strand bias, adjacency to INDELs, and read depth. The filtering steps can be fulfilled by using SAMtools (Li, Handsaker *et al.*, 2009), VCFtools (Danecek *et al.*, 2011), and other packages. One can also parse VCF files directly with customized scripts.

Depending on the goals of the project, researchers need to trade off sensitivity and specificity. Adjustment of hard filters is time consuming. One practical method is variant quality score recalibration (VQSR). The idea of this process is to evaluate what a real variant looks like, and to select the real ones from the candidates. In practice, this method builds a model for the real genetic variations and ranks all the variant callings according to their likelihood of being real. As discussed in the last section, diverse sets of annotations are reported along with each SNP calling in the VCF file. Variants with similar statistics will be grouped into clusters. By linking to high-confidence known variant databases, one can distinguish the real variants from pseudo-variants. Moreover, one can quantify the likelihood by assigning probability scores to all variants with respect to their distance to the true cluster. Taking the algorithm of the "`VariantRecalibrator`" tool in the GATK package as an example, the tool first builds a Gaussian mixture model using the annotation values over high-confidence variants, and then each variant is assigned a log odds ratio of being a true variant versus being false under the trained model. By using the "`ApplyRecalibration`" tool, variants are filtered according to the well-calibrated probability score cutoffs.

SNP Annotation

Context annotations (e.g., sequence quality, variant frequency, etc.) help refine the estimate of the possibility of a variant being real, which are mostly generated together with variant calling. Functional annotations help interpret the links between variants and traits.

Variant calling from whole genome sequencing data allows identification of millions of SNPs. Manual analysis for the potential functional consequences on each SNP is extremely time consuming. Therefore, adoption of automated annotation is essential to predict potential causal mutations of traits. One of the packages that can be used for variant functional annotation is Oncotator (Ramos *et al.*, 2015). The tool aggregates data relevant to cancer studies. By detecting the effects of mutations on protein sequences, it is also adaptable for other research. Another powerful software is ANNOVAR (Wang, Li & Hakonarson, 2010). It is written in Perl, and therefore can be used on diverse systems with standard Perl modules preinstalled. By incorporation with UCSC Genome Browser Annotation Database and multiple third-party datasets, the tool can perform gene-based, region-based, and filter-based annotations. Gene-based annotations can identify amino acids and protein structure changes caused by structural variants. Region-based annotations allow detection of variants within or near specific genomic regions (e.g., conserved regions among species, transcription factor binding sites, duplication regions, and etc.). Filter-based annotations allow the identification of variants with external databases, such as dbSNP, and calculate SIFT (Ng & Henikoff, 2003), PolyPhen (Ramensky, Bork & Sunyaev, 2002), MetaLR (Dong *et al.*, 2015), and other scores.

Detailed Protocols of SNP Identification

A number of software can be used for SNP identification. Here, we introduce several popular ones—including GATK, SAMtools, and VarScan (all of which are suitable for individual sequencing); and PoPoolation2, which is designed for population genetic analysis and can cope with Pool-Seq data.

Quality Control

After getting the NGS sequencing data, a convenient tool for quality assessment is FastQC. Download the appropriate version according to the operating system from http://www.bioinformatics.babraham.ac.uk/projects/fastqc/. FastQC is a Java application, which requires the installation of the Java Runtime Environment (JRE, v1.6–v1.8). Under the Linux system, the wrapper script "`fastqc`" can be used. First, to make this script executable:

```
chmod 755 fastqc
```

To specify the files to be processed:

```
fastqc data.fq
```

By default, the generated files will be stored in the current directory. The output files can be specified with the "`-o`" flag, and the "`-t`" flag can be used to specify the number of files to be processed simultaneously. Note that each thread will be allocated 250 MB of memory, so you need to check your available memory before setting this option.

To check other available options you can run:

```
fastqc --help
```

Prior to short read alignment, sequences need to be trimmed off low-quality bases and adaptors. Trimmomatic is a fast and multithreaded tool that can trim Illumina FASTQ data and remove adapters. The tool can be downloaded from http://www.usadellab.org/cms/?page=trimmomatic. According to the types of sequencing data, choose paired-end mode or single-end mode. The paired-end mode maintains correspondence of the read pairs and tries to find adapter or PCR primer fragments. An example of a command is:

```
java -jar trimmomatic-0.33.jar PE input1.fq input2.fq
   paired_output1.fq unpaired_output1.fq paired_output2.fq
   unpaired_output2.fq ILLUMINACLIP:TruSeq3-PE-2.fa:2:30:10
   LEADING:3 TRAILING:3 SLIDINGWINDOW:4:15 MINLEN:25
```

Here, "`PE`" indicates the paired-end mode; "`input1.fq`" and "`input2.fq`" are your paired-end data; "`paired_output1.fq`" and "`paired_output2.fq`" are output which both reads of a pair pass the filters; and "`unpaired_output1.fq`" and "`unpaired_output2.fq`" are output which only one read of a pair pass the filters. "`ILLUMINACLIP`", "`LEADING`", "`TRAILING`", and "`SLIDINGWINDOW`" are optional processing steps that take one or more settings, delimited by a colon. The "`ILLUMINACLIP`" option is to cut adapter and other illumine-specific sequences from the read. Here, we specify that the path to the adapter file is "`TruSeq3-PE-2.fa`", and the maximum mismatch that allows a full match is "2". The inputs "`30`" and "`10`" means "palindromeClipThreshold" and "simpleClipThreshold", respectively. They represent the match score between adapters and reads under different mapping modes. "`LEADING`" and "`TRAILING`" represent the minimum quality required to keep a base at the beginning and at the end, respectively. "`SLIDINGWINDOW`" indicates that the tool performs a sliding window trimming, and that bases within the window will be cut once the average quality is less than the threshold. Here, we set the window size as "4", and the required average quality as "15". "`MINLEN`" means that reads falling below a minimal length will be removed. Here, we set the value as "25". By assessing the reads quality after trimming, the threshold can be tuned to get good results.

Short Reads Alignment

Before SNP identification, raw reads should be mapped to reference assembly using mapping software such as BWA, Bowtie2, or Tophat2. Generated SAM or BAM files are used in the following SNP identification steps. We take BWA as an example. Download the latest version of BWA from http://sourceforge.net/projects/bio-bwa/files/. BWA consists of three mapping algorithms: BWA-backtrack, BWA-SW, and BWA-MEM. BWA-backtrack is designed for Illumina 100 bp or shorter sequence reads, while the other two are designed for long reads of up to 1 Mbp. We recommend BWA-MEM, which is faster and more accurate. It works by seeding alignments with maximal exact matches (MEMs), and then expanding the seed with the affine-gap Smith–Waterman (SW) algorithm. For all the algorithms, you need to construct the index file for the reference genome first:

```
bwa index reference.fa
```

Here, "`reference.fa`" is the reference assembly in FASTA format, and, by default, "reference" will be the prefix for the generated BWT index files.

To align short reads to the reference sequence:

```
bwa mem -t 4 -R '@RG\tID:catfish1\tDS:channel\tSM:hatchery'
   -M reference.fa input1.fq input2.fq > output.sam
```

With this command, BWA aligns a set of paired reads "input1.fq" and "input2.fq" to the reference genome "reference.fa". The resulting SAM file is stored as "output.sam". The option "-t" indicates the number of threads to use; "-R" instructs to add a read group line for each read; and "-M" instructs to mark short split hits as a secondary alignment, which makes the output file compatible with downstream applications. In the paired-end mode, the tool will infer the read orientation and insert size.

Processing of the Post-alignment File

To convert SAM files to BAM files and sort the reads by the leftmost coordinates, SAMtools need to be installed. Download SAMtools from http://www.htslib.org/. An instruction on SAMtools installation can be seen in the file named "INSTALL" that is present in the unzipped SAMtools folder.

To convert a SAM file to a BAM file and sort it, run:

```
samtools view -b input.sam | samtools sort - output
```

The "view" command prints the input alignment file "input.sam" to a standard output; and "-b" outputs the file in the BAM format. A sorted BAM file named "output.bam" will be generated.

To identify and remove redundant copies of duplicates, you can use the "MarkDuplicatesWithMateCigar" command in Picard. Download the latest version of Picard from https://github.com/broadinstitute/picard/releases/tag/1.141. An example is:

```
java -jar picard.jar MarkDuplicatesWithMateCigar
   INPUT=input.bam OUTPUT=output.bam
```

Due to defects in the mapping software, regions near INDELs are usually misaligned. To perform a local realignment near the INDELs, you can use the "RealignerTargetCreator" and "IndelRealigner" commands in GATK. Download the latest version of the GATK package from https://www.broadinstitute.org/gatk/download/. First, find the intervals that need realignment:

```
java -jar GenomeAnalysisTK.jar -T RealignerTargetCreator -R
   reference.fa -I input.bam -known INDELs.vcf -o realigner
  .intervals
```

Here, "-T" indicates the tool option; "-R" is the reference sequence; "-I" is the input BAM file (which is not necessary if processing only at known INDELs); "-known" specifies the known record for the INDELs (which can be downloaded from external databases such as dbSNP); and "-o" specifies the output intervals file. Once we get the intervals file, we can perform the realignment at the target intervals by using the "IndelRealigner" command:

```
java -jar GenomeAnalysisTK.jar -T IndelRealigner -R ref-
erence.fa -I input.bam -known INDELs.vcf -targetIntervals
realigner.intervals -o realigned.bam
```

Here, we import the intervals with the "`-targetIntervals`" option, and the output file is stored as "`realigned.bam`".

The next step is to recalibrate the quality scores on the sequencing data. Here, we demonstrate the usage of the "`BaseRecalibrator`" and "`PrintReads`" commands in the GATK package. First, to build the recalibration model with "`BaseRecalibrator`":

```
java -jar GenomeAnalysisTK.jar -T BaseRecalibrator -R
   reference.fa -I realigned.bam -knownSites SNP.vcf -o
   recal.table
```

Here, "`-I`" is used to input the generated BAM file after local realignment near the INDELs, and "`-knownSites`" specifies the known SNP sites that can be obtained from external databases or previous studies. The recalibration table is stored as "`recal.table`". Then, we create a new BAM file with the recalibrated quality:

```
java -jar GenomeAnalysisTK.jar -T PrintReads -R reference.fa
   -I realigned.bam -BQSR recal.table -o recal.bam
```

Here, we use the "`-BQSR`" option to import the generated recalibration table. The output file is stored as "`recal.bam`", which is the analysis-ready reads.

SNP Identification

GATK

GATK is a genome analysis toolkit developed by Broad Institute to analyze NGS data. It is widely used for genome variant discovery and genotyping. To call SNP with GATK, additional processes need to be done on both the reference assembly file and the BAM file. First, generate the index file for reference assembly:

```
samtools faidx reference.fa
```

Then, generate a dictionary file for reference assembly using the "`CreateSequenceDictionary`" command in Picard:

```
java -jar picard.jar CreateSequenceDictionary R=reference.fa
   O=reference.dict
```

Index your analysis-ready reads:

```
samtools index recal.bam
```

SNP can be identified by using the "`UnifiedGenotyper`" command in GATK, a variant caller with Bayesian genotype likelihood method.

```
java -jar GenomeAnalysisTK.jar -R reference.fa -T Uni-
fiedGenotyper -I recal.bam -o snps.raw.vcf
```

This step will generate a raw, unfiltered, highly sensitive SNP list in VCF format. To recalibrate the variant quality score and filter the low-quality variant callings, the "`VariantRecalibrator`" and "`ApplyRecalibration`" tools can be used. To build the Gaussian mixture model:

```
java -jar GenomeAnalysisTK.jar -T VariantRecalibrator -R
   reference.fa -input snps.raw.vcf -resource:SNPdataset1,
   known=false,training=true,truth=true,prior=15.0
   SNPdataset1.vcf -resource:SNPdataset2,known=false,
   training=true,truth=false,prior=12.0 SNPdataset2.vcf
   -an QD -an MQ -an ReadPosRankSum -mode SNP -recalFile
   raw.recal -tranchesFile raw.tranches
```

Here, the "`resource`" option is used to import external datasets "`SNPdataset1`" and "`SNPdataset2`" to construct the model. The two datasets are for different usages, and "`SNPdataset1`" contains only SNPs with a very high degree of confidence. The tool will consider all the SNPs in the calling set as true variants, and use them to train the recalibration model. Hence, we set the option for "`SNPdataset1`" as "`training=true`" and "`truth=true`". The option "`known`" is used to stratify output metrics depending on whether variants are present in dbSNP or not; it is not required to train the model. Thus, a dataset labeled with "`known=truth`" is optional for this step. "`SNPdataset2`" contains high-confidence SNPs and false positive SNPs, and the program uses the truth sites within the dataset to determine the cutoff in VQSLOD sensitivity. Therefore, we set the option for "`SNPdataset2`" as "`training=true`" and "`truth=false`". The "`prior`" option is used to specify the prior likelihood of how reliable the true sites are, with greater values for reliable datasets. The "`-an`" option is used to specify the annotation field to use for model construction. Here, we use "`QD`", "`MQ`", and "`ReadPosRankSumTest`"—"`QD`" is the quality score normalized by allele depth for a variant; "`MQ`" is the RMS mapping quality; and "`ReadPosRankSumTest`" is the u-based z-approximation from the Mann–Whitney–Wilcoxon rank sum test for site position within reads. In this step, all input variants are assigned with a VQSLOD score according to the model.

The next step is to apply the filters to the callset:

```
java -jar GenomeAnalysisTK.jar -T ApplyRecalibration -R
   reference.fa -input snps.raw.vcf -mode SNP -recalFile
   raw.recal -tranchesFile raw.tranches -o recal.SNPs.vcf
   -ts_filter_level 99.0
```

With the tranche file "raw.trances" and recalibration table "raw.recal" generated in the previous step, the tool checks on each variant and decides which tranche it belongs to. Then, the tranche level of each variant is record in the FILTER field. We set the "ts_filter_level" to 99, which means that variants with a VQSLOD score of above 99% of the variants in the training callset are reported as high-confidence SNPs.

SAMtools/BCFtools

SAMtools is a collection of tools to manipulate alignment files. It can also be used for SNP calling in combination with BCFtools, which is a set of tools for variant calling files in VCF and BCF. BCFtools can be downloaded from https://github.com/samtools/bcftools. First, we need to generate the pileup file:

```
samtools mpileup -ugf reference.fa input.bam > data.mpileup
```

Here, the option "`-u`" is used to export files in an uncompressed format; "`-g`" is used to compute genotype likelihoods and export them in the BCF format; and "`-f`" is used to add reference assembly information into the output.

For convenience, multiple BAM files can be piled up together simultaneously:

```
samtools mpileup -ugf ref.fa input1.bam input2.bam input3
  .bam > data.mpileup
```

The next step is to call putative SNPs and estimate allele frequencies:

```
bcftools call -vMmO z -o snp.raw.vcf.gz data.mpileup
```

The pileup files generated in the last step are taken as input files by "bcftools" for SNP calling. The option "-v" means only variant sites are generated. Here, "-M" is used to keep the variants when the reference allele is "N". It is useful when the quality of the reference assembly is not good. The option "-m" is used to adopt the new multi-allelic calling model, which is recommended for most tasks; and the "-O" option is used to specify the format of the output file—we set it as "z", which means compressed VCF.

To generate the statistics of the SNP results:

```
bcftools index snp.raw.vcf.gz
bcftools stats -F reference.fa -s - snp.raw.vcf.gz > snp.raw
  .vcf.stats
```

The report is written in file "snp.raw.vcf.stats".

Varscan

Varscan is a software tool designed for variant detection in data from multiple sequencing platforms. It employs a robust heuristic/statistic approach to call variants instead of probabilistic frameworks, which have better performance on data with extreme read depth, pooled samples, and contaminated samples. Varscan can be downloaded from http://sourceforge.net/projects/varscan/files/. It is written in Java and can be used on any system with JRE preinstalled. Varscan can only process files in a pileup format. Hence, we need to generate the pileup file by using SAMtools:

```
samtools mpileup -B -f reference.fa input.bam > data.mpileup
```

Then, call SNP with Varscan:

```
Java -jar VarScan.jar mpileup2snp data.mpileup --p-value
   0.05 --min-coverage 10 --min-avg-qual 20 --min-var-freq
   0.05 > snp.varscan
```

Here, the option "--p-value" means the significance of Fisher's exact test on variant read count versus expected baseline error; "--min-coverage" sets the threshold of minimum read depth at a position to make a call; "--min-avg-qual" sets the threshold for base quality to count a read; and "--min-var-freq" sets the minor allele frequency to call a variant. For other available filters, you can refer to the documentation (http://varscan.sourceforge.net/using-varscan.html#v2.3_mpileup2snp). The SNP callings are stored in the file "snp.varscan", which is a tab-delimited file with variant position and other annotation information.

PoPoolation2

PoPoolation2 is an SNP calling toolkit that is suitable for identifying SNPs from pooled genomic DNA sequencing data (Kofler, Pandey & Schlotterer, 2011). The software also allows comparing allele frequencies of SNPs between multiple populations. PoPoolation2 is written in Perl, and it can be downloaded from https://code.google.com/p/popoolation2/downloads/list. To speed up analysis, PoPoolation2 expects its input files to be sync files, which can be derived from the pileup files for SNP calling and other downstream analyses:

```
java -jar mpileup2sync.jar --input data.mpileup --output
   data.sync --fastq-type sanger --min-qual 20 --threads 8
```

Note that there is a quality control step here, which can be set by the researcher. Here, we have set it as 20. Now, we can call the SNPs and calculate allele frequency differences among populations:

```
perl snp-frequency-diff.pl --input data.sync --output-prefix
   data --min-count 10 --min-coverage 50 --max-coverage 300
```

Several options are invoked in the command: "`--min-count`" indicates the threshold for the minimum count of the minor allele; "`--min-coverage`" is the threshold for the minimum coverage; and "`--max-coverage`" is a mandatory option that sets the threshold for the maximum coverage of each population. These parameters are used as filters for SNPs identification.

The scripts generate two output files: "data_rc" and "data_pwc". Both of them are tab-delimited. The file "data_rc" contains the SNPs identified from each population and the variant context annotations, such as SNP position, genotype, allele count, etc. The file "data_pwc" records the allele frequency differences among populations.

The PoPoolation2 toolkit can also perform other analyses, such as the measurement of differentiation between populations (e.g., Fixation index calculation), estimation of the significance of allele frequency differences (e.g., Fisher's exact test), and detection of consistent allele frequency differences among independent measurements (e.g., CMH test). For additional details, readers are referred to the user manual (https://code.google.com/p/popoolation2/wiki/Manual).

Which Software Should I Choose?

As many software packages can be used for SNP identification, choosing the appropriate software and the associated efforts for choosing the proper parameters for analysis is a substantial challenge. Most software tools are implanted with specific algorithms that serve a particular purpose. For instance, GATK is mainly used for mining high-confidence SNPs, and thus it fits individual sequencing data well and can make use of external variant databases. In contrast, PoPoolation is designed for population genetic analysis; therefore, it can cope with pool-seq data. It is therefore difficult to say which software is better. It all depends on the purpose of the study and the situations involved. The truth is that most software packages will deliver the intended purpose for SNP identification, but the output may slightly differ. Similarly, various parameters need to be tried out for best performance. A recommendation is to try different software and compare the results. Candidate callings obtained with more than one software tool should be more reliable (Howe *et al.*, 2013; Star *et al.*, 2011; Zhang *et al.*, 2013).

References

Anvar, S.Y., Khachatryan, L., Vermaat, M. *et al.* (2014) Determining the quality and complexity of next-generation sequencing data without a reference genome. *Genome Biology*, **15**, 555.

Ashurst, J.L., Chen, C.K., Gilbert, J.G. *et al.* (2005) The Vertebrate Genome Annotation (Vega) database. *Nucleic Acids Research*, **33**, D459–465.

Aslam, M.L., Bastiaansen, J.W., Elferink, M.G. *et al.* (2012) Whole genome SNP discovery and analysis of genetic diversity in Turkey (*Meleagris gallopavo*). *BMC Genomics*, **13**, 391.

Birney, E., Andrews, T.D., Bevan, P. *et al.* (2004) An overview of Ensembl. *Genome Research*, **14**, 925–928.

Bolger, A.M., Lohse, M. and Usadel, B. (2014) Trimmomatic: a flexible trimmer for Illumina sequence data. *Bioinformatics*, **30**, 2114–2120.

Bovine HapMap Consortium, Gibbs, R.A., Taylor, J.F. *et al.* (2009). Genome-wide survey of SNP variation uncovers the genetic structure of cattle breeds. *Science*, **324**, pp. 528–532.

Brockman, W., Alvarez, P., Young, S. *et al.* (2008) Quality scores and SNP detection in sequencing-by-synthesis systems. *Genome Research*, **18**, 763–770.

Browning, B.L. and Yu, Z.X. (2009) Simultaneous genotype calling and haplotype phasing improves genotype accuracy and reduces false-positive associations for genome-wide association studies. *The American Journal of Human Genetics*, **85**, 847–861.

Browning, S.R. and Browning, B.L. (2007) Rapid and accurate haplotype phasing and missing-data inference for whole-genome association studies by use of localized haplotype clustering. *The American Journal of Human Genetics*, **81**, 1084–1097.

Collins, F.S., Brooks, L.D. and Chakravarti, A. (1998) A DNA polymorphism discovery resource for research on human genetic variation. *Genome Research*, **8**, 1229–1231.

Cox, M.P., Peterson, D.A. and Biggs, P.J. (2010) SolexaQA: At-a-glance quality assessment of Illumina second-generation sequencing data. *BMC Bioinformatics*, **11**, 485.

Cui, J., Wang, H.D., Liu, S.K. *et al.* (2014) SNP discovery from transcriptome of the swimbladder of *Takifugu rubripes*. *PloS One*, **9**, e92502.

Danecek, P., Auton, A., Abecasis, G. *et al.* (2011) The variant call format and VCFtools. *Bioinformatics*, **27**, 2156–2158.

DePristo, M.A., Banks, E., Poplin, R. *et al.* (2011) A framework for variation discovery and genotyping using next-generation DNA sequencing data. *Nature Genetics*, **43**, 491–498.

Dong, C.L., Wei, P., Jian, X.Q. *et al.* (2015) Comparison and integration of deleteriousness prediction methods for nonsynonymous SNVs in whole exome sequencing studies. *Human Molecular Genetics*, **24**, 2125–2137.

Gabriel, S., Ziaugra, L. and Tabbaa, D. (2009) SNP genotyping using the Sequenom MassARRAY iPLEX platform. In: *Current Protocols in Human Genetics* (eds Haines, J.L., Korf, B.R., Morton, C.C. *et al.*), John Wiley & Sons, Inc., New York, Chapter 2, Unit 2.12.

Gao, Z.X., Luo, W., Liu, H. *et al.* (2012) Transcriptome analysis and SSR/SNP markers information of the blunt snout bream (*Megalobrama amblycephala*). *PloS One*, **7**, e42637.

Helyar, S.J., Hemmer-Hansen, J., Bekkevold, D. *et al.* (2011) Application of SNPs for population genetics of nonmodel organisms: new opportunities and challenges. *Molecular Ecology Resources*, **11** (1), 123–136.

Helyar, S.J., Limborg, M.T., Bekkevold, D. *et al.* (2012) SNP discovery using next generation transcriptomic sequencing in Atlantic herring (*Clupea harengus*). *PloS One*, **7**, e42089.

Houston, R.D., Taggart, J.B., Cezard, T. *et al.* (2014) Development and validation of a high density SNP genotyping array for Atlantic salmon (*Salmo salar*). *BMC Genomics*, **15**, 90.
Howe, K., Clark, M.D., Torroja, C.F. *et al.* (2013) The zebrafish reference genome sequence and its relationship to the human genome. *Nature*, **496**, 498–503.
Kent, W.J., Sugnet, C.W., Furey, T.S. *et al.* (2002) The human genome browser at UCSC. *Genome Research*, **12**, 996–1006.
Kerstens, H.H.D., Crooijmans, R.P.M.A., Veenendaal, A. *et al.* (2009) Large scale single nucleotide polymorphism discovery in unsequenced genomes using second generation high throughput sequencing technology: applied to turkey. *BMC Genomics*, **10**, 479.
Kidd, K.K., Pakstis, A.J., Speed, W.C. *et al.* (2004) Understanding human DNA sequence variation. *Journal of Heredity*, **95**, 406–420.
Kijas, J.W., Townley, D., Dalrymple, B.P. *et al.* (2009) A genome wide survey of SNP variation reveals the genetic structure of sheep breeds. *PloS One*, **4**, e4668.
Kofler, R., Orozco-terWengel, P., De Maio, N. *et al.* (2011) PoPoolation: a toolbox for population genetic analysis of next generation sequencing data from pooled individuals. *PloS One*, **6**, e15925.
Kofler, R., Pandey, R.V. and Schlotterer, C. (2011) PoPoolation2: identifying differentiation between populations using sequencing of pooled DNA samples (Pool-Seq). *Bioinformatics*, **27**, 3435–3436.
Kong, Y. (2011) Btrim: a fast, lightweight adapter and quality trimming program for next-generation sequencing technologies. *Genomics*, **98**, 152–153.
Langmead, B. and Salzberg, S.L. (2012) Fast gapped-read alignment with Bowtie 2. *Nature Methods*, **9**, 357–359.
Langmead, B., Trapnell, C., Pop, M. *et al.* (2009) Ultrafast and memory-efficient alignment of short DNA sequences to the human genome. *Genome Biology*, **10**, R25.
Li, H. (2011) A statistical framework for SNP calling, mutation discovery, association mapping and population genetical parameter estimation from sequencing data. *Bioinformatics*, **27**, 2987–2993.
Li, H (2013) *Aligning sequence reads, clone sequences and assembly contigs with* BWA-MEM. https://arxiv.org/pdf/1303.3997v2.pdf.
Li, H. and Durbin, R. (2009) Fast and accurate short read alignment with Burrows–Wheeler transform. *Bioinformatics*, **25**, 1754–1760.
Li, H., Handsaker, B., Wysoker, A. *et al.* (2009) The sequence alignment/map format and SAMtools. *Bioinformatics*, **25**, 2078–2079.
Li, H., Ruan, J. and Durbin, R. (2008) Mapping short DNA sequencing reads and calling variants using mapping quality scores. *Genome Research*, **18**, 1851–1858.
Li, J.B., Gao, Y., Aach, J. *et al.* (2009) Multiplex padlock targeted sequencing reveals human hypermutable CpG variations. *Genome Research*, **19**, 1606–1615.
Li, R., Li, Y., Fang, X. *et al.* (2009) SNP detection for massively parallel whole-genome resequencing. *Genome Research*, **19**, 1124–1132.
Liu, S., Sun, L., Li, Y. *et al.* (2014) Development of the catfish 250K SNP array for genome-wide association studies. *BMC Research Notes*, **7**, 135.
Liu, Z., Liu, S., Yao, J. *et al.* (2016) The channel catfish genome sequence provides insights into the evolution of scale formation in teleosts. *Nature Communications*, **7**, 11757.
Lunter, G. and Goodson, M. (2011) Stampy: a statistical algorithm for sensitive and fast mapping of Illumina sequence reads. *Genome Research*, **21**, 936–939.

Mardis, E.R. (2008) The impact of next-generation sequencing technology on genetics. *Trends in Genetics (TIG)*, **24**, 133–141.
Marklund, S. and Carlborg, O. (2010) SNP detection and prediction of variability between chicken lines using genome resequencing of DNA pools. *BMC Genomics*, **11**, 665.
Matukumalli, L.K., Lawley, C.T., Schnabel, R.D. *et al.* (2009) Development and characterization of a high density SNP genotyping assay for cattle. *PloS One*, **4**, e5350.
McKenna, A., Hanna, M., Banks, E. *et al.* (2010) The genome analysis toolkit: a MapReduce framework for analyzing next-generation DNA sequencing data. *Genome Research*, **20**, 1297–1303.
Modolo, L. and Lerat, E. (2015) UrQt: an efficient software for the Unsupervised Quality trimming of NGS data. *BMC Bioinformatics*, **16**, 137.
Ng, P.C. and Henikoff, S. (2003) SIFT: predicting amino acid changes that affect protein function. *Nucleic Acids Research*, **31**, 3812–3814.
Oliphant, A., Barker, D.L., Stuelpnagel, J.R. *et al.* (2002). BeadArray technology: enabling an accurate, cost-effective approach to high-throughput genotyping. *BioTechniques*, Suppl, pp. **56–58**, 60–51.
Palti, Y., Gao, G., Liu, S. *et al.* (2015) The development and characterization of a 57K single nucleotide polymorphism array for rainbow trout. *Molecular Ecology Resources*, **15**, 662–672.
Patel, R.K. and Jain, M. (2012) NGS QC Toolkit: a toolkit for quality control of next generation sequencing data. *PloS One*, **7**, e30619.
Quinlan, A.R., Stewart, D.A., Stromberg, M.P. *et al.* (2008) Pyrobayes: an improved base caller for SNP discovery in pyrosequences. *Nature Methods*, **5**, 179–181.
Rabbee, N. and Speed, T.P. (2006) A genotype calling algorithm for affymetrix SNP arrays. *Bioinformatics*, **22**, 7–12.
Ramensky, V., Bork, P. and Sunyaev, S. (2002) Human non-synonymous SNPs: server and survey. *Nucleic Acids Research*, **30**, 3894–3900.
Ramos, A.H., Lichtenstein, L., Gupta, M. *et al.* (2015) Oncotator: cancer variant annotation tool. *Human Mutation*, **36**, E2423–2429.
Ramos, A.M., Crooijmans, R.P.M.A., Affara, N.A. *et al.* (2009) Design of a high density SNP genotyping assay in the pig using SNPs identified and characterized by next generation sequencing technology. *PloS One*, **4**, e6524.
Riva, A. and Kohane, I.S. (2004) A SNP-centric database for the investigation of the human genome. *BMC Bioinformatics*, **5**, 33.
Robinson, J.T., Thorvaldsdottir, H., Winckler, W. *et al.* (2011) Integrative genomics viewer. *Nature Biotechnology*, **29**, 24–26.
Sherry, S.T., Ward, M.H., Kholodov, M. *et al.* (2001) dbSNP: the NCBI database of genetic variation. *Nucleic Acids Research*, **29**, 308–311.
Star, B., Nederbragt, A.J., Jentoft, S. *et al.* (2011) The genome sequence of Atlantic cod reveals a unique immune system. *Nature*, **477**, 207–210.
Stenson, P.D., Ball, E.V., Mort, M. *et al.* (2003) Human gene mutation database (HGMD): 2003 update. *Human Mutation*, **21**, 577–581.
Wang, J., Kong, L., Gao, G. *et al.* (2013) A brief introduction to web-based genome browsers. *Brief Bioinform*, **14**, 131–143.
Wang, K., Li, M. and Hakonarson, H. (2010) ANNOVAR: functional annotation of genetic variants from high-throughput sequencing data. *Nucleic Acids Research*, **38**, e164.

Wang, Y., Yang, L.D., Wu, B. *et al.* (2015) Transcriptome analysis of the plateau fish (*Triplophysa dalaica*): Implications for adaptation to hypoxia in fishes. *Gene*, **565**, 211–220.
Wiedmann, R.T., Smith, T.P. and Nonneman, D.J. (2008) SNP discovery in swine by reduced representation and high throughput pyrosequencing. *BMC Genetics*, **9**, 81.
Xu, J., Ji, P., Zhao, Z. *et al.* (2012) Genome-wide SNP discovery from transcriptome of four common carp strains. *PloS One*, **7**, e48140.
Xu, J., Zhao, Z.X., Zhang, X.F. *et al.* (2014) Development and evaluation of the first high-throughput SNP array for common carp (*Cyprinus carpio*). *BMC Genomics*, **15**, 307.
Xu, P. (2007) Physical map construction and physical characterization of channel catfish genome. (Doctoral dissertation). http://www.ag.auburn.edu/fish/wp-content/uploads/formidable/XU_PENG_49.pdf.
Zhang, H., Hirose, Y., Yoshino, R. *et al.* (2013) Assessment of homozygosity levels in the mito-gynogenetic torafugu (*Takifugu rubripes*) by genome-wide SNP analyses. *Aquaculture*, **380**, 114–119.

18

SNP Array Development, Genotyping, Data Analysis, and Applications

Shikai Liu, Qifan Zeng, Xiaozhu Wang and Zhanjiang Liu

Introduction

The advances in sequencing technology have enabled efficient generation of large-scale genomic resources, including identification of genome-scale genetic variations used in genetic studies. Single-nucleotide variations/polymorphisms (SNPs) are well-recognized markers for genetic studies due to their amenability to be genotyped with high automation and high abundance in the genome. High-density SNP arrays are developed for high-throughput SNP genotyping. With the availability of high-density SNP arrays, the success of a genetic study such as determining the association of an allele with a trait is greatly dependent on proper study design and high-quality data. Owing to the nature of large-scale SNP data, false positive results are often observed, and can be caused by even small sources of systematic or random errors in the data, reinforcing the need for careful attention to study design and data quality control (DQC). Systematic variations can be derived from numerous sources, such as from the SNP array design, experiment design to SNP genotyping, and data analysis. In this chapter, we seek to provide an overview of the technical aspects of SNP array development and genotyping with the Affymetrix Axiom genotyping technology in order to pinpoint the critical considerations and procedures in data acquisition to reduce the systematic or random errors to increase the power for genetic analysis.

SNPs are now well recognized as markers of choice for genetic studies because they are the most abundant genomic variations in all the organisms; they are distributed in the genomes to provide broad and full coverage of the genomes; and they are generally bi-allelic, and therefore are amenable to automated genotyping (Kruglyak, 1997). Because of these advantages, SNPs now are the most popular and dominant molecular markers for genetic studies with various species, with the exception of species in which extensive genomic studies have not been conducted, and where other types of molecular markers such as microsatellites may still be used.

SNPs can be discovered by sequencing and bioinformatics analysis. Recent advances in sequencing technologies have enabled efficient identification of SNPs from genomes of various organisms (Davey *et al.*, 2011; Nielsen *et al.*, 2011). Assuming the sequences are correct, analysis of sequence variations between and among chromosomes allows the generation of SNPs. For the discussion in this chapter, we will focus on diploid organisms in which a set of two homologous chromosomes are harbored by the

Bioinformatics in Aquaculture: Principles and Methods, First Edition. Edited by Zhanjiang (John) Liu.

organism, with one set being derived from the maternal parent, and the other from the paternal parent. Therefore, direct sequencing of DNA from a single individual will generate sequences coming from both sets of chromosomes. Sequence differences from the two chromosomes at the same loci represent SNPs that can be revealed through high-throughput sequencing. However, caution must be exercised with the use of next-generation sequencing platforms because they generate short sequence tags of approximately 100–150 bp length. The generation of longer sequences depending on *de novo* assembly using bioinformatics programs is likely problematic because mistakes can be made in bringing paralogous sequences into single contigs. This problem would become paramount when polyploidy organisms are involved, thereby complicating analysis.

With the availability of a large number of SNPs, high-density SNP arrays can be developed for efficient and cost-effective genotyping of large numbers of individuals. A variety of SNP array platforms are available, including Illumina Infinium iSelect HD BeadChip (Illumina, San Diego, California, United States), Sequenom MassArray iPLEX (Sequenom, San Diego, California, United States), and Affymetrix MyGeneChip Custom Array (Affymetrix, Santa Clara, California, United States). These platforms differ in their requirements of SNP number, sample size, cost, and automation. In general, MassArray technology is quite efficient when a relatively small number of SNPs are analyzed across a large number of individuals; and Illumina technology is very efficient for genotyping a large number of individuals with a large number of SNPs, but becomes cost-inefficient when the number of SNPs is too high, for example, over 50000–100000 SNPs. More recently, Affymetrix adopted the Axiom genotyping technology, which allows the development of customized arrays containing 1500–2.6 million SNPs (Hoffmann *et al.*, 2011). The Affymetrix Axiom genotyping platform provides a flexible solution for designing customizable high-density SNP arrays in non-model organisms. There are no technology barriers for the development of high-density SNP arrays. With the development of genotyping technologies, the unit cost per genotype is currently declining, and the total cost for high-density SNP arrays with large numbers of SNPs becomes relatively more affordable for most research groups, even for those who work on species within small research communities.

SNP arrays have been developed in various species for genetic analysis (Table 18.1). As one would expect, those developed for humans or various model species lead the way, with the highest densities and the largest numbers of applications. For instance, various human SNP arrays have been developed for genetic research in humans, with the highest density of SNP arrays containing over 1000000 genetic markers, such as Affymetrix Genome-Wide Human SNP array 6.0, which features 1.8 million genetic markers, including 906600 SNPs and over 946000 probes for detection of copy number variations (CNVs; Korn *et al.*, 2008). Similarly, with zebrafish, a teleost model species for developmental biology, high-density SNP arrays have been developed with 201917 SNPs (Howe *et al.*, 2013). In agricultural and companion animal species, SNP arrays have been developed for cattle (Matukumalli *et al.*, 2009), horse (McCue *et al.*, 2012), pig (Ramos *et al.*, 2009), sheep (Miller *et al.*, 2011), dog (Mogensen *et al.*, 2011), and chicken (Groenen *et al.*, 2011). Due to its high flexibility and relatively low unit cost per genotype for extremely high-density arrays, Affymetrix Axiom genotyping technology is now being widely used. For instance, a high-density 600K chicken SNP array has been developed (Kranis *et al.*, 2013), and the Axiom Genome-Wide BOS 1 Array is designed

Table 18.1 Examples of high-density arrays developed for humans and various animal species.

Species	SNP Array	Chemistry	Array density	Selected references
Human	Affymetrix Human SNP Array 6.0	Axiom	906K[a)]	Korn *et al.* (2008)
Human	Illumina HumanOmniZhongHua-8 Beadchip	Infinium	890K	Chen *et al.* (2014)
Mouse	Affymetrix Mouse Diversity Genotyping Array	Axiom	623K	Yang *et al.* (2009)
Dog	Affymetrix Canine SNP Genotyping Array	Axiom	127K	Meurs *et al.* (2010)
Dog	Illumina CanineHD Beadchip	Infinium	170K	Mogensen *et al.* (2011)
Cattle	Illumina BovineSNP50 Beadchip	Infinium	54K	Matukumalli *et al.* (2009)
Cattle	Illumina BovineHD BeadChip	Infinium	770K	Rincon *et al.* (2011)
Cattle	Affymetrix Genome-Wide BOS 1 Bovine Array	Axiom	640K	Rincon *et al.* (2011)
Swine	Illumina Porcine60K BeadChip	Infinium	60K	Ramos *et al.* (2009)
Chicken	Affymetrix Genome-Wide Chicken Genotyping Array	Axiom	600K	Kranis *et al.* (2013)
Horse	Illumina EquineSNP50 BeadChip	Infinium	60K	McCue *et al.* (2012)
Sheep	Illumina OvineSNP50 BeadChip	Infinium	54K	Miller *et al.* (2011)
Zebrafish	Affymetrix Zebrafish Genotyping Array	Axiom	200K	Howe *et al.* (2013)
Catfish	Affymetrix Catfish Genotyping Array	Axiom	250K	Liu *et al.* (2014)
Common carp	Affymetrix Common Carp Genotyping Array	Axiom	250K	Xu *et al.* (2014)
Rainbow trout	Affymetrix Rainbow Trout Genotyping Array	Axiom	57K	Palti *et al.* (2014)
Atlantic salmon	Affymetrix Atlantic Salmon Genotyping Array	Axiom	130K	Houston *et al.* (2014)

a) Affymetrix Human SNP Array 6.0 contains 906600 SNPs, with over 946000 probes in addition for the detection of copy-number variations.

to cover more than 640000 SNP markers for bovine genetic studies (Taylor *et al.*, 2012). Much progress has also been made recently in the development of SNP arrays in aquaculture species, and a number of high-density SNP arrays have been developed, such as for catfish (Liu *et al.*, 2014), carp (Xu *et al.*, 2014), Atlantic salmon (Houston *et al.*, 2014), and rainbow trout (Palti *et al.*, 2014).

SNP arrays have been applied to diverse genetic studies and breeding programs, such as genome-wide association studies (GWAS), linkage mapping and quantitative trait loci (QTL) analysis, population and evolutionary genetics studies, and whole-genome-based prediction and selection. In spite of the various applications, GWAS analysis and QTL analysis are probably the most important uses of SNP arrays. With the availability of high-density SNP arrays, the success of GWAS to associate an allele with a trait is highly dependent on proper study design and data quality. For instance, large numbers of variants have been associated with various complex diseases and traits by using GWAS, most of which account for only a small proportion of phenotypic variations. While large numbers of false positives are being discovered, inabilities in detecting truly significant associations are also frequently encountered, thereby requiring more samples to enable statistical significance for the identification and validation of these findings. False positive results can be caused by even small sources of systematic or random errors, reinforcing the need for careful consideration of study design and data quality. In this chapter, we provide an overview of the technical aspects of SNP array development, genotyping, and data analysis with the Affymetrix Axiom genotyping technology. Readers are referred to other relevant chapters for individual genetic applications after accurate SNP genotypes are determined, for example, Chapter 23 for GWAS analysis using SNP genotypes.

Development of High-density SNP Array

Marker Selection

To design SNP arrays, the first requirement is the availability of a large number of SNPs. In the current stage of genomic research, sequence resources are available for many species, but may not always be available for the species of your interest. This is particularly true with aquaculture species, where existing genome sequence resources can often be lacking. In such cases, the first step would be genome sequencing for the discovery and identification of SNPs. Readers are referred to Chapter 17 for detailed descriptions of genome sequencing and identification of SNPs. Here, we will provide a detailed description of SNP selection using existing resources.

Various SNPs sources can be utilized for the selection of SNPs, including validated SNPs from public databases, novel SNPs discovered during whole-genome sequencing using a single individual, pooled sequencing of genomic DNA from multiple individuals, or SNPs identified from transcriptome sequencing using RNA-Seq. For SNP arrays that are focused on a specific genomic region, SNPs derived from targeted re-sequencing of candidate genomic regions of interest can be used. In general, the sources of SNPs can either be from analysis of existing sequence resources, or from sequencing projects of various kinds.

For humans, Affymetrix has generated SNP resources from existing human genome databases. Affymetrix's Axiom Genomic Database contains approximately 11 million genomic markers across the human genome from major populations, including over 8.8 million Axiom-validated markers, and the Axiom Design Center, which is an interactive web application enabling users to select markers from the Axiom Genomic Database based on the needs of their study to optimize an Axiom myDesign Array (https://www.affymetrix.com/analysis/netaffx/axiomtg/index.affx).

However, for non-model species, such as aquaculture species, the SNP resources for array design are likely unavailable in the databases. Researchers are required to generate such resources by high-throughput sequencing followed by bioinformatics data analysis. Therefore, the first task is to identify a large set of quality SNPs for inclusion in the SNP arrays.

In the case of catfish, we generated SNPs using several approaches, including whole genome sequencing of multiple individuals using Illumina sequencing and transcriptome sequencing using RNA-Seq for gene-associated SNPs (Liu *et al.*, 2011; Sun *et al.*, 2014). High levels of genome coverage in genome sequencing should allow the assembly of genomic sequences into contigs where genomic variations in the form of SNPs can be readily detected. However, parameters must be set up to assess SNP qualities to differentiate true SNPs from sequencing errors without large-scale validation.

Although many parameters can be used for the assessment of SNP quality without experimental validation, two parameters are the most important (Wang *et al.*, 2008; Sun *et al.*, 2014). These are *minimal sequencing coverage* and *minor allele frequency*. Those SNPs identified from regions with a minimum number of 30 sequence reads, with a minimum read of five for minor allele, appeared to provide a high level of success. However, even if the SNPs are real, a number of parameters affect their success toward generating accurate genotypes when they are included in the SNP arrays.

One of the important factors for SNP genotyping success is the neighboring sequences around the SNPs. For the development of the Affymetrix SNP array, SNPs are filtered following specific requirements. Flanking sequences of 70 bp for each SNP are required, 35 bp on each side of the SNP, where no other variations (SNPs and/or indels) are allowed within 30 bp of the SNPs. The balanced A/T/G/C ratio of flanking sequences are required with a GC content of 30–70%. No repetitive elements or simple sequences (e.g., a single base repeats or microsatellites) are allowed in the flanking sequences. For example, single base repeats of G or C greater than 4, and A or T greater than 6, are not allowed. The SNPs that fulfill these requirements are submitted to Affymetrix for design score assessment, where a *p*-convert value is assigned for each probe that flanks an SNP.

The *p*-convert value, which arises from a random forest model, is intended to predict the probability that the involved SNP will convert in the array. The model provides *p*-convert values by considering factors including probe sequence, binding energy, and the expected degree of non-specific binding and hybridization to multiple genomic regions. The highest *p*-convert value is 1.0, and the lowest value is 0. SNPs with probes of high *p*-convert values are more likely to be convertible. A *p*-convert value threshold is determined by excluding the tail of the lowest performing probes to facilitate selection of a final SNP list. For the SNPs with both probes passing the *p*-convert value threshold, the probe with the greater value can be selected. For the SNPs with both probes having low *p*-convert values, both probes are selected to ensure conversion of at least one probe at the SNP site. For catfish SNP arrays, the lowest *p*-convert value was 0.62 (Liu *et al.*, 2014).

One of the most important goals of SNP array development is to select markers that have a good coverage of the whole genome, that is., the SNPs are selected from all chromosomes, all chromosome regions, and with marker spacing not exceeding a desired distance. To achieve this, markers need to be selected from a fixed distance of the genome. For instance, if 250000 SNPs are used to cover a genome of 1000000000 bp size, one SNP needs to be selected from every 4 KB of genomic DNA. This is easy for species with a well-assembled genome because SNPs can simply be selected based on their genome coordinates. However, for species without sequenced genomes, such as the vast majority of aquaculture species, selection of well-spaced markers across the genome is relatively more difficult.

With catfish, we used various genome resources to facilitate marker selection to ensure a better coverage of the whole genome. We selected at least one (when quality is high) but no more than two SNPs per transcript contig for gene-associated SNPs. This would ensure at least one SNP from each protein-encoding gene in the genome. Assuming the distribution of genes is more or less even across the genome, selection of SNPs from all genes would practically place SNPs corresponding to the gene distributions in the genome. We also used the preliminary catfish genome assembly to facilitate selection of the anonymous genomic SNPs based on contig length. For instance, one SNP per contig was selected from the contigs with lengths less than 4 KB, and two or more SNPs per contig were selected from the contigs with lengths greater than 4 KB, aiming at the rough marker interval of 4 KB; as we developed the SNP array containing 250000 SNPs, the catfish genome was hence approximately 1 GB.

Axiom myDesign Array Plates

The Axiom myDesign Genotyping Array Plates are fully customizable with mid–high-density array plate selections to be optimized with the most relevant markers based on the study. The Axiom arrays can hold 1500–2.6 million markers for every sample to be analyzed, in various array configurations. The configurations available for customization include 1 × 96, 2 × 48, 3 × 32, and 4 × 24 (Table 18.2). Each plate contains 96 arrays, and it can be designed to have the same contents on all arrays on the plate, or divide the plate into two (2 × 48 configuration), three (3 × 32 configuration), or four (4 × 24 configuration) regions, with each region having different contents, enabling the interrogation of more markers per sample but fewer samples per plate. Each array within an array plate can accommodate 1500–675000 markers (Axiom myDesign Genotyping Arrays design guide).

SNP Array Design

Principles of Probe Design for the Differentiation of SNPs and Indels

With Affymetrix Axiom technology, SNPs are differentiated by the use of a pair of probes. One probe is synthesized and attached to the solid phase on the array, and the other is the "solution probe." The first reaction is the hybridization of the genomic

Table 18.2 Affymetrix Axiom SNP array configurations.

	Maximum SNPs on each plate	Number of DNA samples to be analyzed per plate
1 × 96	675000	96
2 × 48	1350000	48
3 × 32	2025000	32
4 × 24	2700000	24

DNA to the probe on the solid phase, and then the hybridization of the solution probe to the genomic DNA. One of the two solution probes would have a perfect match to the genomic template DNA when homozygous, and both solution probes will hybridize if heterozygous. As A and T in the solution are labeled with the same fluorescence color. and G and C are labeled with the same fluorescence color, two pairs of probes are required to differentiate A/T or G/C SNPs. The second reaction is the ligation reaction. When the solution probe base-pairs with the genomic DNA with no gap, the ligation reaction occurs. Once the ligation occurs, the fluorescence is retained when further processed. Otherwise, the free solution probes are not retained after further processing, leading to the non-detection of fluorescence, thereby indicating a mismatch at the SNP site, and now inference can be made that the base at the SNP site in the sample is an alternative allele.

As shown in Figure 18.1, Axiom SNP probes are typically 30mers that contain a 5'-phosphate group and are synthesized onto arrays. After the hybridization of fragmented whole-genome amplified (WGA) genomic DNA (indicated by the dash lines in Figure 18.1) to the array-bound probes, the polymorphic loci are interrogated using two differentially labeled sets of ligation probes (Label solution probes).

Different probe sets are required to interrogate different types of SNPs (or indels, not discussed here). The SNP probes for the interrogation of non A/T or G/C SNPs are 30-mers that end before the SNPs, while the probes for A/T or G/C SNPs include the SNP sites (i.e., 31-mers). For non G/C or A/T SNPs, each SNP requires only one probe set, while two probe sets must be designed for interrogating the G/C or A/T SNPs (Figure 18.1). Apparently, inclusion of the G/C or A/T SNPs will decrease the number of SNPs that can be interrogated, because the total number of probes synthesized onto arrays is fixed.

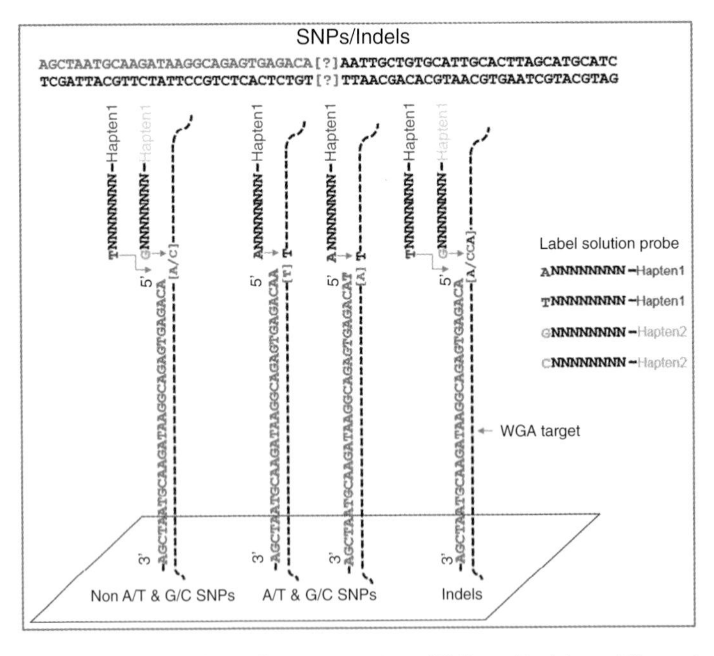

Figure 18.1 Probe design for interrogation of SNPs and indels on Affymetrix SNP array. (*See color plate section for the color representation of this figure.*)

Forward strand of reference genome

5' GGATGTCTGTCTCTTTGTTTTAATTTCTCT[G/T]TAAATTUTTTTATTTCCAGAGTTTGAAGG 3'

5' GGATGTCTGTCTCTTTGTTTTAATTTCTCT[A/T]TAAATTTTTTTTATTTCCAGAGTTTGAAGG 3'

5' GGATGTCTGTCTCTTTGTTTTAATTTCTCT[G/-]TAAATTTTTTTTATTTCCAGAGTTTGAAGG 3'

SNP type	SNP allele	A/B code
Non A/T & C/G	G T	B A
A/T & C/G	A T	A B
Insertion/ Deletion	G -	B A

Figure 18.2 Examples illustrating the allele naming convention for the Affymetrix Axiom genotyping platform.

The SNPs are interrogated using four solution probes, which are random 9-mers with a fixed 3' nucleotide. The solution probes with A and T 3' nucleotides are labeled with one dye, and the solution probes with G and C 3' nucleotides are labeled with another dye. The SNPs are determined based on which solution probes are ligated to the array. For non–G/C or A/T SNPs (i.e., A/C, A/G, T/C, or T/G SNPs), the alleles can be detected according to each label (A/T or C/G). For A/T and G/C SNPs, both alleles are represented on the array, and the genotype is determined by measuring the signal intensity from each of the two probes.

Allele Naming Convention for Affymetrix Axiom Genotyping Platform

With Affymetrix Axiom technology, all bases at SNP sites are coded into two letters, A or B. The coding rules are shown in Figure 18.2. For non A/T or C/G SNPs (A/G, A/C, C/T, G/T), bases A or T are defined as A, while bases C or G are defined as B. For A/T or C/G SNPs, A and C are coded as A, and G and T are coded as B. When the marker is an insertion/deletion, the deletion is defined as A, while the insertion based on the reference strand is defined as B.

DQC

DQC probes are designed from non-SNP sites. In other words, these sites are always homozygous with known sequences. DQC probes are required on the SNP array, serving as negative controls. They are non-polymorphic 31-mers that are randomly collected from the non-repetitive regions of the genome.

The specific requirements for designing DQC probes are: (1) avoid sequences with any SNPs; (2) avoid extreme GC content (aim for GC proportion between 0.3 and 0.7); (3) avoid sequences with high levels of 16-mer hits (select sequences with 16-mer counts < 50), or avoid sequences that are repetitive in the genome; (4) avoid sequences with "GGGG" or "CCCC"; (5) for each sequence, randomly pick forward or reverse strand; and (6) for each autosomal chromosome, randomly pick 100 probes with A or T as the ligation base (thirty-first base in sequence), and randomly pick 100 probes with C or G as the ligation base.

Case Study: the Catfish 250K SNP Array Design

A total of 250113 SNPs were selected for developing the 250K catfish SNP array, which included 103185 (41.3%) gene-associated SNPs and 146928 (58.7%) anonymous SNPs. Of the gene-associated SNPs, 32188 SNPs were identified from channel catfish, 31392

SNPs were identified from blue catfish, and 39605 inter-specific SNPs were identified between channel catfish and blue catfish (Liu *et al.*, 2014).

A total of 316706 SNP probe sets were synthesized for the interrogation of these 250113 SNPs, of which 66593 SNPs were tiled with two probes (designed from forward and reverse strands each) (Liu *et al.*, 2014). In addition to SNP probes, a total of 2000 DQC probes were also designed, including 1000 DQC probes with A or T at the thirty-first base, and the other 1000 DQC probes with G or C at the thirty-first base (Liu *et al.*, 2014).

SNP Genotyping: Biochemistry and Workflow

Axiom Genotyping Solution

Axiom Genotyping Solution allows for complete automation of DNA target preparation, DNA amplification, and enzymatic fragmentation of post-amplified products on the Biomek FX Target Prep Express platform. Following target preparation, arrays are processed using GeneTitan Multi-Channel (MC) Instrument, which can handle the Axiom SNP arrays with different layouts depending on the number of samples genotyped on one plate, including 24-, 32-, 48-, 96-, and 384-sample array layouts. The latest design of 384-sample arrays offers the capability of very high throughput, which enables genotyping of approximate 50000 variants, and processing of greater than 3000 samples per week.

The Axiom Genotyping Solution includes array plates, complete reagents, automatic and manual workflow for sample preparation, instrumentation, and software, including:

1) *Axiom Reagent Kit*: A complete reagent system that includes all consumables for processing 96 reactions per array plate
2) *GeneTitan MC Instrument*: A single instrument that combines a hybridization oven, fluidics processing, and imaging to provide automated array plate processing from target hybridization to data generation
3) *Affymetrix Genotyping Console (GTC)*: A software for automated allele calling and quality evaluation of called genotypes

Biochemistry and Workflow

Several steps are involved in the genotyping process, as shown in Figure 18.3:

1) Whole-genome amplification of total genomic DNA (100 ng).
2) The whole-genome amplified products were then randomly fragmented into 25–125 bp fragments.
3) Fragments are purified, re-suspended, and hybridized to customized Axiom myDesign Array plates.
4) After hybridization, the bound target is washed under stringent conditions to remove non-specific backgrounds caused by random ligation events. Each polymorphic locus is interrogated via multi-color ligation events carried out on the array surface.
5) Following ligation, the arrays are stained and imaged on GeneTitan MC Instrument.

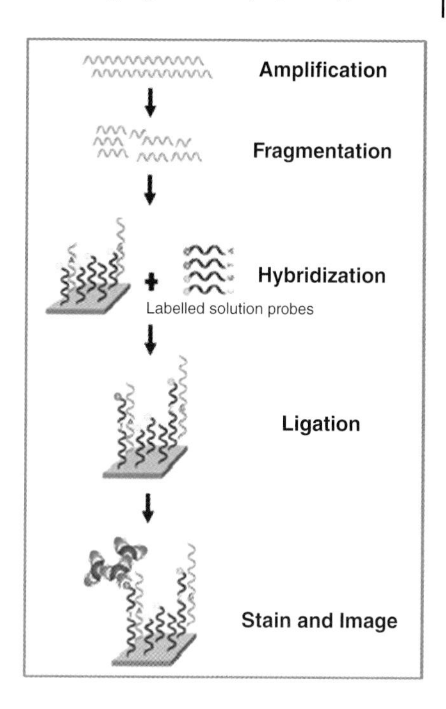

Figure 18.3 Overview of Affymetrix Axiom genotyping: chemistry and workflow. (*See color plate section for the color representation of this figure.*)

SNP Genotyping: Analysis of Axiom Genotyping Array Data

Genotyping Workflow

The analysis of Axiom genotyping array data can be conducted by using the Affymetrix Power Tools (APT) or Affymetrix GTC software. Both APT and GTC software packages can be used to generate quality control (QC) metrics and genotype calls. The genotyping analysis workflow and algorithm are identical for both APT and GTC. The genotyping algorithm, AxiomGT1, is a tuned version of the BRLMM-P algorithm that adapts pre-positioned clusters to the data using an expectation–maximization (EM) algorithm. Clustering is carried out in two dimensions—size: $\log_2(A + B)/2$, and contrast: $\log_2(A) - \log_2(B)$. Five steps are typically involved in the genotyping with both APT and GTC (Figure 18.4), even though the actual commands used to accomplish these steps differ between the two software tools.

Step 1: Group Samples into Batches

A genotyping batch consists of a set of samples that are processed together through the genotyping chemistry, hybridization, and scanning stages. For Axiom genotyping projects, a batch usually contains a 96-array plate. If the genotyping performance of some plates systematically differs from that of other plates in the study, these differences may manifest themselves as putative differences in the allele frequencies of SNPs, which may increase the chance to generate false positive results during data analysis.

To determine which CEL files (data files created by Affymetrix DNA microarray image analysis software) should be genotyped together as a batch—that is, the choice to analyze the data from each 96-array plate individually, or analyze some of the study samples together—no universally accepted approach is available. The choice of the

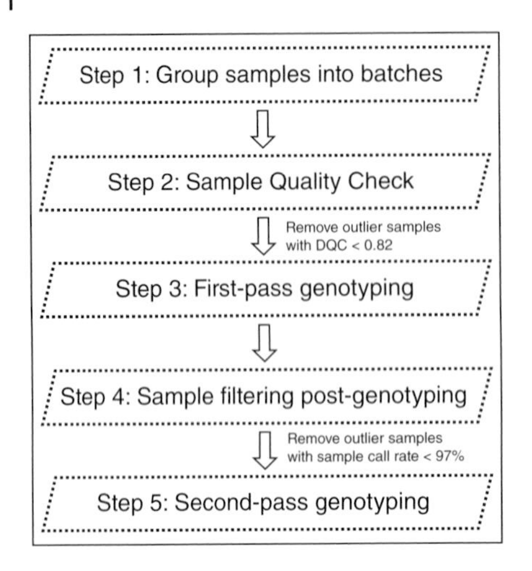

Figure 18.4 Steps for genotyping workflow.

approach is usually dependent on which is the most convenient for the laboratory and study workflow. However, the key consideration is to exclude plates (or batches) of samples that systematically differ from other plates in order to obtain optimal genotyping data, because these plates/batches may contribute to mis-clustering events, whether processed separately or with all other plates. If, in the initial genotyping workflow, all plates/batches were clustered together, then samples from outlier plates/batches should be removed, and the genotyping process should be repeated prior to further analysis.

Step 2: Perform a Sample QC

The QC of each individual sample should be first performed before conducting genotyping analysis. Sample QC is an important step for genotyping analysis, because inclusion of low-quality samples in a genotyping clustering run can negatively impact the genotyping quality of high-quality samples.

Axiom arrays use a two-color system, and the associated sample QC metrics are based on the principle that the two channels must be separable to achieve good genotyping results. The signal to background ratio (SBR) is the simplest form of QC; therefore, the indicator signals are obtained by including probes based on known non-polymorphic sequences during the array design. In addition to SBR, a novel metric DQC has been developed to take into account both inter-channel and intra-channel signal separation and spread. The standard practice to qualify samples is based on their DQC values.

DQC operates by measuring signals at a collection of non-polymorphic sites in the genome. DQC monitors non-polymorphic locations where it is known which of the two channels in the assay represents the allele present in the genome—and therefore, which channel should have a signal intensity above the background signal, and which channel's signal should be indistinguishable from the background are known. DQC is a measure of the extent to which the distribution of signal values for the two channels are separated from the signal values for the background, with 0 indicating no separation and 1 indicating perfect separation. Samples with a DQC value less than the default DQC threshold of 0.82 should be either reprocessed or dropped from the study.

Step 3: Perform First-pass Genotyping of the Samples

The DQC filtering is necessary but not sufficient for identifying all problematic samples. For instance, sample contamination will not be detected with DQC alone as contaminated samples can have good DQC but relatively low sample genotype call rates. Therefore, genotyping using a set of SNPs whose genotypes are representative of the expected performance needs to be performed for further QC based on sample genotype call rate. The SNPs used for QC call rate may be the same as the full SNP set or a subset, depending on the array design.

Step 4: Perform Sample Filtering Post-genotyping

To achieve the highest genotyping performance, outlier samples should be filtered post-genotyping so that these samples do not pull down the cluster quality of the non-outlier samples. The most basic post-genotyping filter is based on the sample call rate. Any samples that have overall sample call rates less than 97% should be removed and not included in the downstream analysis.

Step 5: Perform Second-Pass Genotyping

After removing the outlier samples identified in step 3 (earlier), the remaining samples should be re-genotyped. The results from the initial round of genotyping should be discarded or moved to another location to ensure they are not mistakenly included in downstream analysis.

Genotyping Analysis Software

The analysis of Axiom genotyping array data, particularly from Axiom myDesign Genotyping Array plates, is conducted by using APT or GTC software to perform QC analysis, and sample and/or SNP filtering prior to downstream analysis.

APT is a set of cross-platform command line programs that implement algorithms for working with Affymetrix arrays. Because APT programs are command-line based, its usage requires expertise in running programs under scripting environments. APT is available as a Microsoft Windows installer package, as pre-built binaries for Linux and macOS, and as source code. The latest version of APT is accessible at http://www.affymetrix.com/estore/partners_programs/programs/developer/tools/powertools.affx?hightlight=true&rootCategoryId=34002#1_2. The old versions of APT packages are archived on the same website for historical use. Multiple versions of APT can be installed in parallel on the same computer.

GTC is a software package that provides a graphical user interface to the algorithms contained in the APT, and is designed to streamline whole-genome genotyping analysis and QC. GTC includes QC and visualization tools to easily identify and segregate sample outliers, including a cluster graph visualization tool that enables a detailed look at the performance of SNPs of interest. GTC integrates SNP genotyping, indel detection, CNV identification, and cytogenetic analyses into one application. It generates genotyping calls, copy number calls for CNV and individual probe sets, loss of heterozygosity (LOH) data, cluster plots, and QC metrics. The latest version of GTC, version 4.2 (updated on 04/07/2014), is available for Microsoft Windows 7 (64-bit) and Microsoft Windows 8.1. GTC 4.2 can be downloaded from Affymetrix.com (http://www.affymetrix.com/products_services/software/download/genotyping_console/genotyping_console_download_terms.affx?v=4.2_64). For use on 32-bit and/or Microsoft Windows XP systems, GTC 4.1.4 (posted on 02/19/2013) is available (http://www.affymetrix.com/products_services/software/download/genotyping_console/genotyping_console_download_terms.affx?v=4.1.4_32).

Genotyping Analysis Reference Files

Both GTC and APT require a set of files for genotyping, which are collectively referred to as the *analysis reference files,* including library files and annotations that are generated during array development. As shown in Table 18.3, all the analysis files needed for genotyping with either GTC or APT are listed using the catfish 250K SNP array as an example (Liu *et al.*, 2014). The SNP array name is "IpCcSNP". The extra APT files are XML files that specify analysis parameters for QC and genotyping, while the extra GTC files are annotations that are used for displaying SNP annotations in SNP result tables, for cluster graph visualizations, and for some export functionality.

Table 18.3 Axiom Genotyping array reference files using the catfish 250K SNP array as an example.

Analysis Library Files	GTC	APT
Axiom_IpCcSNP.analysis	Required	N/A
Axiom_IpCcSNP.array_set	Required	N/A
Axiom_IpCcSNP.AxiomGT1.gc_analysis_parameters	Required	N/A
Axiom_IpCcSNP.gc_analysis_configuration	Required	N/A
Axiom_IpCcSNP.geno_intensity_report	Required	N/A
Axiom_IpCcSNP.gt_thresholds	Required	N/A
Axiom_IpCcSNP.qc_thresholds	Required	N/A
Axiom_IpCcSNP_CsvAnnotationFile.r1.Catfish.txt	Required	N/A
Axiom_IpCcSNP.r1._step1.ps	Required	Required
Axiom_IpCcSNP.r1.AxiomGT1.models	Required	Required
Axiom_IpCcSNP.r1.AxiomGT1.sketch	Required	Required
Axiom_IpCcSNP.r1.Catfish.ps	Required	Required
Axiom_IpCcSNP.r1.cdf	Required	Required
Axiom_IpCcSNP.r1.generic_prior.txt	Required	Required
Axiom_IpCcSNP.r1.psi	Required	Required
Axiom_IpCcSNP.r1.qca	Required	Required
Axiom_IpCcSNP.r1.qcc	Required	Required
Axiom_IpCcSNP.r1.spf	Required	Required
Axiom_IpCcSNP.20121213.annot.db	N/A	Required
Axiom_IpCcSNP.r1.apt-geno-qc.AxiomQC1.xml	N/A	Required
Axiom_IpCcSNP_96orMore_Step1.r1.apt-probeset-genotype.AxiomGT1.xml	N/A	Required
Axiom_IpCcSNP_96orMore_Step2.r1.apt-probeset-genotype.AxiomGT1.xml	N/A	Required

Genotyping Using GTC

Software Installation

1) Download the Microsoft Windows installer software from Affymetrix.com: http://www.affymetrix.com/catalog/131535/AFFY/Genotyping+Console+Software#1_1.
2) Unzip the downloaded software package, which includes the installation program, release notes, and software manual.
3) Review the release notes and manual for installation instructions before proceeding with the installation
4) Double-click the installation program ("GenotypingConsoleSetup.exe") to install the software.
5) Follow the directions provided by the installer to complete the installation.

Starting GTC

1) Double-click the GTC shortcut on the desktop, or from the Microsoft Windows start menu, select "Programs > Affymetrix > Genotyping Console". GTC opens with the "User Profile" window displayed.
2) Create a user profile for first-time use, or select an existing user profile.
3) After creating a user profile, the library path notice appears. Following the directions, select or create a location for the library files and click "OK".
4) After setting up the library folder, the temporary files folder needs to be selected or created, which stores temporary files during data analysis.
5) The locations of the library and temporary files folder can be changed in the "Options" dialog box: "Edit > Options". The new locations for library and temporary files can be specified by either entering the absolute paths in the appropriate box or by clicking the "Browse" button to select the new locations.
6) The analysis reference files need to be available in the library folder for genotyping data analysis.
7) The workspace can be created after setting up the library and temporary files folders: "File > New Workspace ..."
8) After creating a workspace, GTC prompts the user to create a "Data Set" in the workspace, which can be done by clicking the "Create Data Set" shortcut on the main toolbar, or by selecting "Data Sets > Create Data Set" from the "Workspace" menu.
9) Select the array type for the data set from the "Array" drop-down list, and enter a name for the data set.
10) To add data to a data set, either right-click the data set created earlier and select "Add Data ..." on the shortcut menu, or click on the "Add Data" shortcut button on the main toolbar, or select "Add Data" from the menu ("Workspace > Data Sets > Add data ...").
11) Select data to add to the data set, usually by checking "Intensity and QC files" to import the raw files in CEL format. Click "OK" to import CEL files.

Genotyping with GTC

Step 1: Sample QC The QC analysis provides an assessment of the overall quality for a sample based on the QC algorithm prior to performing a full clustering analysis. After importing the CEL files into the data set, a sample QC can be performed in GTC. GTC can calculate the QC metrics at the time of importing, or by doing one of the following:

1) Clicking the "QC" button on the toolbar menu
2) Selecting the "Intensity Data" option from the "Workspace" pull-down menu
3) Right-clicking on a data group displayed in the data tree and selecting the "Perform QC" option from the drop-down menu.

After the completion of the QC process, the resulting DQC values and other quality metrics are displayed in an Intensity QC table. Any samples with a DQC value of less than 0.82 (by default, meaning that the signal–noise ratio is above a certain threshold; once again, 1 means no noise, 0 means all the noise) are highlighted and grouped into the "Out of Bounds" group, so that they can be easily excluded from the genotyping analysis. The QC thresholds and the other metrics can be changed as needed (see the GTC user manual for details).

Step 2: First-pass Genotyping After sample QC, the CEL files that passed the DQC threshold are ready to be genotyped, which can be performed by doing one of the following:

1) Clicking the "Genotyping" button on the toolbar menu
2) Selecting it from the "Workspace > Intensity Data > Perform Genotyping ..." drop-down menu
3) Right-clicking on the "In Bounds" data group in the data tree and selecting "Performing Genotyping..." from the pop-out menu

To proceed with genotyping, select the analysis configuration "96orMore_Step1_AxiomGT1"; the prior models file "Axiom_IpCcSNP.r1.generic_prior.txt"; and the SNP list file "Axiom_IpCcSNP.r1._step1.ps". Then, select the output root folder, which is the folder where the CEL files are located by default, and then click "OK" to continue with genotyping.

Step 3: Second-pass Genotyping After the first-pass genotyping, the outlier samples with a call rate of less than 97% can be identified, and should be excluded from further analysis. The second-pass genotyping is performed after creating the intensity data group that includes only the CEL files with sample call rates greater than or equal to 97% after first-pass genotyping. The second round of genotyping will cluster only the high-quality samples, and will yield improved clustering performance and more accurate genotype calls. The intensity data group can be created by first selecting all the CEL files with call rates of >97%, and then by right-clicking to select "Create Custom Intensity Data Group Using Selected Sample". Right-click the intensity data group that is created, and select "Perform genotyping" to conduct second-pass genotyping. In this round of genotyping, select the analysis configuration as "96orMore_Step2_AxiomGT1", and select the SNP list file "Axiom_IpCcSNP.r1.Catfish.ps"; others can be left as in the first-pass genotyping step.

Step 4: Export Genotype Calls for Downstream Analysis The genotype calls for passing samples and SNPs can be exported from GTC into text files by right-clicking the genotyping results batch in the data tree, and then selecting "Export Genotype Results..." from the pop-out menu. It can also be exported in a PLINK-compatible format and used for downstream analysis using PLINK software, which is a free, open-source whole genome association analysis toolset (Purcell *et al.*, 2007). Alternatively, the genotype calls for each sample can be found as CHP files in the output directory. These files can be located by right-clicking the dataset and choosing the "Show File Locations..." option.

Genotyping Using APT

Although APT is available as a Microsoft Windows installer package and as pre-built binaries for Linux and macOS, the command line codes used for genotyping are same among different operating systems. Here, we used APT version 1.16.1, running on Microsoft Windows, to illustrate the genotyping.

Software Installation

1) Download the Microsoft Windows installer software from Affymetrix.com: http://www.affymetrix.com/estore/partners_programs/programs/developer/tools/powertools.affx?hightlight=true&rootCategoryId=34002#1_2.
2) Unzip the downloaded software package, which includes the installation program.
3) Double-click the installation program ("Affymetrix Power Tools v1.16.1.msi") to install the software.
4) Follow the directions provided by the installer to complete the installation.

Genotyping with APT

1) To start APT, click on the Microsoft Windows start menu, select "Programs > Affymetrix Power Tools > APT-1.16.1 > APT Command Prompt".
2) To perform a sample QC, change the directory to the folder containing all binary executables, and execute the following command line codes:

```
apt-geno-qc --analysis-files-path E:\GTC_genotyping
  \GT_lib --xml-file E:\GTC_genotyping\GT_lib\
   Axiom_IpCcSNP.r1.apt-geno-qc.AxiomQC1.xml --cel-files
   E:\GTC_genotyping\ESC-P2\cel_file_list.ESC-P2.txt
   --out-file E:\GTC_genotyping\ESC-P2\qc-report.txt
```

   Here, "`apt-geno-qc`" is the program for performing QC in the APT package; "`--analysis-files-path`" provides the location of the library files; "`--xml-file`" provides the file with parameter settings for QC; "`--cel-files`" provides the file listing the CEL files for genotyping, which is a tab-delimited file containing one column with the header of "cel_files" to provide the absolutes path and names of the CEL files; and "`--out-file`" provides the file that receives the QC outputs.

3) To perform first-pass genotyping, first update the CEL file list "cel_file_list_passQC .ESC-P2.txt" to only include CEL files that have passed DQC threshold, and then perform genotyping for these CEL files with the following command line code to generate CHP files:

```
apt-probeset-genotype --analysis-files-path
    E:\GTC_genotyping\GT_lib --xml-file E:\GTC_genotyping
    \GT_lib\Axiom_IpCcSNP_96orMore_Step1.r1.apt-probeset-
    genotype.AxiomGT1.xml --cel-files E:\GTC_genotyping
    \ESC-P2\cel_file_list_passQC.ESC-P2.txt --out-dir
    E:\GTC_genotyping\ESC-P2\cel_file_list_passQC
    --cc-chp-output
```

Here, "apt-probeset-genotype" is the program for genotyping in the APT package; "--analysis-files-path" provides the location of the library files; "--xml-file" provides the file with parameter settings for first-pass genotyping; "--cel-files" provides the file listing the CEL files for genotyping, which includes only CEL files that have passed DQC threshold (DQC >0.82 by default); "--out-dir" provides the directory that receives the genotyping outputs; and "--cc-chp-output" is an option to generate CHP files in addition to text file output.

4) To perform second-pass genotyping, first update the CEL file list "cel_file_list_passQC_excludeOutlier.ESC-P2.txt" to include only CEL files that have passed the DQC threshold, and the additional sample filtering criteria such as sample call rate; then, perform genotyping for these CEL files with the following command line codes to generate CHP files:

```
apt-probeset-genotype --analysis-files-path
    E:\GTC_genotyping\GT_lib --xml-file E:\GTC_genotyping
    \GT_lib\Axiom_IpCcSNP_96orMore_Step2.r1.apt-probeset-
    genotype.AxiomGT1.xml --cel-files E:\GTC_genotyping
    \ESC-P2\cel_file_list_passQC_excludeOutlier.ESC-P2.txt
    --out-dir E:\GTC_genotyping\ESC-P2\cel_file_list_
    passQC_excludeOutlier --cc-chp-output --write-models
```

Here, "apt-probeset-genotype" is the program for genotyping in the APT package; "--analysis-files-path" provides the location of the library files; "--xml-file" provides the file with parameter settings for second-pass genotyping; "--cel-files" provides the file listing the CEL files for genotyping, which includes only CEL files that have passed the DQC threshold and sample call rate cut-off (DQC >0.82 and sample call rate ≥97% by default); "--out-dir" provides the directory that receives the genotyping outputs; "--cc-chp-output" is an option to generate CHP files in addition to text file output; and "--write-models" writes SNP-specific models in an output file for analysis

SNP Analysis After Genotype Calling

Following genotyping, the next step is to filter the SNPs to exclude those with low calling rate and low genotyping quality. Both GTC and APT provide some metrics for SNP filtering. In GTC, some SNP summary statistics can be viewed in the SNP summary table by creating an SNP list by using the "Create SNP list..." menu option in GTC. The cluster plots of SNPs can be visually inspected for QC. Alternatively, an R package, SNPolisher (Affymetrix), can be used to post-process the genotyping results with the SNP summary metrics for efficiently filtering SNPs after genotyping.

View SNP Cluster Plots

To inspect genotype clusters from the final SNP list, the "Cluster Graph" visualization tool in GTC needs to be used. The "Show SNP Cluster Graphs..." command can be selected as an option from a pop-out menu. An example of an SNP cluster graph is shown in Figure 18.5. The clustering pattern of a particular SNP can be color-coded with respect to the plate ID, which allows the identification of any plate effects within the clusters.

In a well-clustered SNP, the three genotype clusters are well resolved, and the location of the AB (het) cluster is higher than the homozygous clusters (Figure 18.5). In a mis-clustered SNP example, some of the samples in the actual BB and AA genotype clusters are split into an AB (het), and are separated visually as a collection of "No Call" data.

Metrics Used for SNP Post-processing

Several metrics can be used to identify and filter out SNPs that have the potential to produce false-positive associations due to genotype mis-clustering. With SNPolisher, the objective of SNP filtering is to initially filter out SNPs that fall below the threshold levels set for CallRate, FLD, HetSO, and HomRO (Table 18.4).

SNP call rate (CallRate) is the ratio of the samples with genotypes called by the total number of samples used for genotyping. SNP call rate is a measure of data completeness and genotype cluster quality. Low SNP call rates occur due to failures in resolving genotype clusters. Poor cluster resolution can produce incorrect genotypes, and, even if the

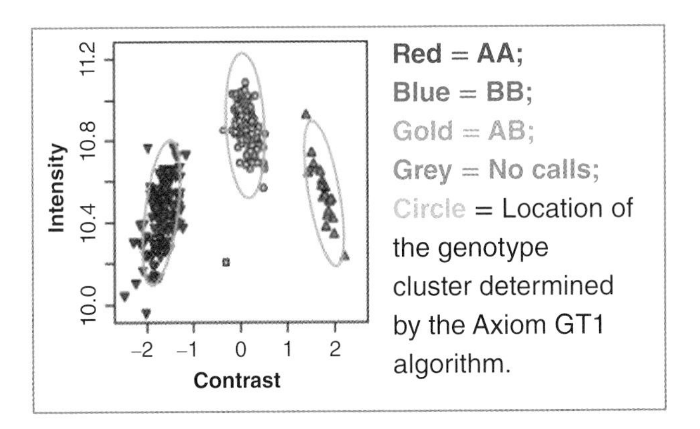

Figure 18.5 GTC cluster graph for an example SNP. In this figure, each cluster plot is for one SNP, and each point in the cluster plot is for one sample. For each sample, "A" indicates the summarized intensity of the A allele, and "B" indicates the summarized intensity of the B allele. The values are transformed into the cluster space of the genotyping algorithm, referred to as *contrast* and *size*. Contrast = $Log_2(A/B)$, and size = $[Log_2(A) + Log_2(B)]/2$. The samples are colored by the genotype called by the algorithm, where AA calls are red, BB calls are blue, AB calls are gold, and no calls are grey. (*See color plate section for the color representation of this figure.*)

Table 18.4 Metrics that are used for SNP post-filtering.

Metrics	Parameters	Threshold[a)]
CallRate	SNP call rate	<95–97%
FLD	Fisher's linear discriminant scores	<3.6
HetSO	Heterozygous cluster strength offset values	<−0.1
HomRO	Homozygote ratio offset	If the number of clusters (nClus) < 3, then HomRO <0.3; if the number of clusters (nClus) == 3, then HomRO < −0.9

a) Higher values correspond to higher quality of SNPs obtained.

called genotypes are correct, the missing information may lead to false positive results in association analysis if the no-calls are not randomly distributed across the genotype clusters. Therefore, it is important to filter out SNPs with low call rates before downstream analysis. Although SNP call rate is correlated with genotype quality, there is no perfect threshold for filtering out problematic SNPs. An SNP call rate of 95–97% is generally recommended as the filtering threshold, but visually examining the cluster plots for SNPs with call rates just above or below the threshold is required. The examination may result in the inclusion of some SNPs with SNP call rates below the threshold, as well as the removal of some SNPs with SNP call rates just above the threshold.

Fisher's linear discriminant (FLD) is a direct measure of the cluster quality of an SNP. High-quality SNP clusters have well-separated clusters with respect to other clusters, and small variance of the center of the cluster (e.g., Figure 18.6(i)). The posteriors (i.e., an ellipse that identifies the cluster location and variance of each genotype cluster that is produced by the genotyping algorithm) for the individual clusters should be narrow and have centers that are well separated from each other. It is recommended that SNPs with an FLD value of less than 3.6 should be removed from downstream analysis. SNP call rates and FLD values are normally correlated, but in some cases FLD will detect problems not detected by the SNP call rate metric.

Heterozygous Cluster Strength Offset (HetSO) is defined as the vertical distance (as measured by [A + B]/2 or signal size) from the center of the het cluster to the line connecting the centers of the hom clusters. Low HetSO values are produced by mis-clustering events and also by the inclusion of samples with high degrees of mismatches relative to the reference genome (i.e., samples with variations in the genome region against which the 30mer SNP probe sequence was designed). When there is a high degree of mismatch between the sample and reference genome, the A and B intensities produced by the samples are low, and fall into the location of the heterozygous cluster. The clustering algorithm would incorrectly call these cases as "AB" instead of "No Calls", if large numbers of samples fall into this heterozygous cluster location. Therefore, low HetSO values can be used to identify these samples. The average signal value for the heterozygous cluster (as measured along the y-axis)

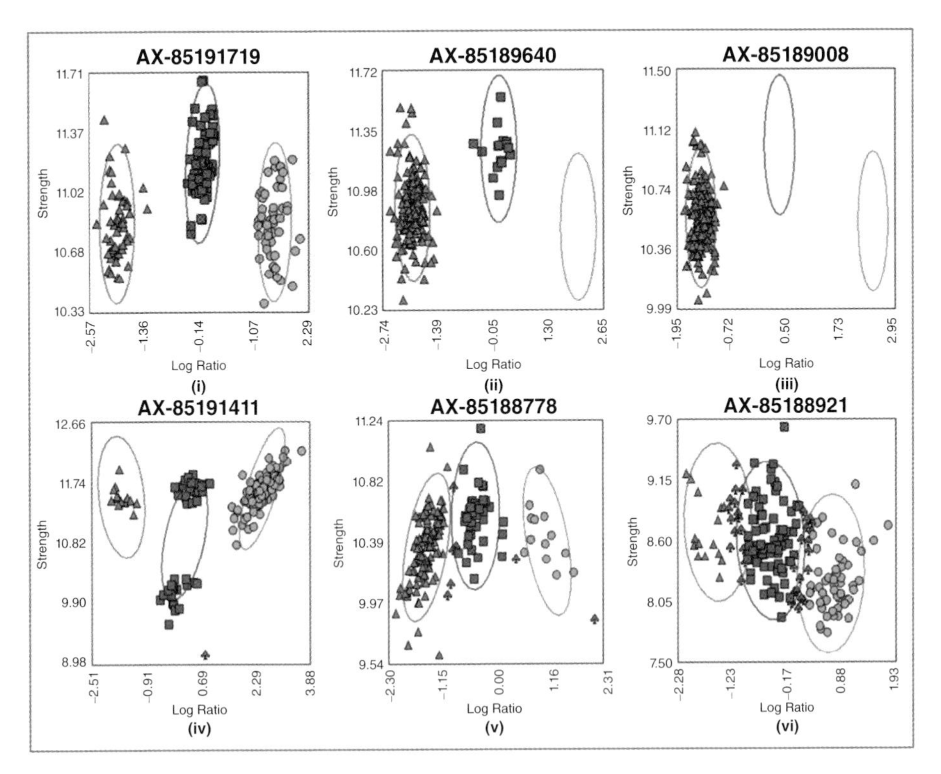

Figure 18.6 Examples of six SNP categories. SNPs were classified into six categories according to cluster properties: **(i)** "PolyHighResolution"; **(ii)** "NoMinorHom"; **(iii)** "MonoHighResolution"; **(iv)**, "OTV" off-target variants; **(v)** "CallRateBelowThreshold"; and **(vi)** "Other". The figure is adapted from Liu *et al.* (2014). (*See color plate section for the color representation of this figure.*)

is visually much lower than that for homozygous clusters. SNPs with HetSO less than −0.1 are recommended to be removed from the downstream genotyping analysis.

Homozygote Ratio Offset (HomRO) is the location in the contrast dimension of the homozygous genotype cluster center that is closest to zero (the heterozygote position), and/or most likely to be misplaced. If the homozygous cluster is on the expected side of zero (the het position), the value is positive, otherwise negative. A low or negative value tends to indicate that the genotyping algorithm has mislabeled the clusters, producing incorrect genotype calls (see details in Axiom Best Practice Supplement User Guide).

Perform SNP Filtering Using SNPolisher

SNPolisher calculates the QC metrics for each SNP, classifies SNPs into categories, and generates cluster plots for each SNP to evaluate quality. The input files are the standard output files from Affymetrix GTC or APT for Axiom arrays. The output files generated by the GTC genotyping process can be located by right-clicking the dataset and choosing the "Show File Locations…" option. The output of APT can be found in the location specified in the script.

Software Installation

1) Both R and Perl (64-bit) must be installed for SNPolisher to run.
2) To install SNPolisher (version 1.5.0), download the zipped package file from http://www.affymetrix.com/Auth/support/developer/downloads/Tools/SNPolisher_package.zip). This folder contains the SNPolisher R package file, the User's Guide, the Quick Reference Card, and several example data sets.
3) Unzip the folder, and install the R package ("SNPolisher_1.5.0.tar.gz"). Before the installation of the SNPolisher package, R and Perl must be installed in the Microsoft Windows environment, while R must be installed in macOS and Linux environments as Perl is usually preinstalled in these. It is important to install recent versions of R and Perl, else the SNPolisher functions will not run correctly. For version 1.5.0 of SNPolisher, the version of R installed must be 2.11.0 or later, and the Perl version installed must be 5.8.0 or later.
4) To install SNPolisher, type this command line in R:

```
Install.packages ("Path-to-package/SNPolisher_1.5.0
  .tar.gz", repos=NULL, type="source")
```

The "`Path-to-package`" option provides the absolute path for where the "SNPolisher_1.5.0.tar.gz" package file is located on the computer. Using the `type="source"` option makes R use the already downloaded package source file that is provided, instead of downloading the package from CRAN.

Using SNPolisher

Input files: Summary, calls, confidences, and posteriors files are produced from GTC and APT. In GTC, they are stored in the "Output" folder, while, in APT, they are located in the output folder selected using the "`--out-dir`" option with the "`apt-probeset-genotype`" tool. APT does not produce the posteriors and summary files by default. They can be produced by using the "`--write-models`" and "`--summaries`" options when running "`apt-probeset-genotype`". When analyzing results from Axiom arrays that include *de novo* SNP content or multiple probesets per SNP, the file "ps2snp.txt" is also required. The "ps2snp" file contains two columns named "probeset_id" and "snp_id" that lists all probesets on the array and the SNPs that they interrogate. This list must be used with the *Ps_Classification* function to select the best probeset for each SNP. If an array only has one probeset per SNP for the entire array, then the "ps2snp" file is not generated.

Functions: SNPolisher contains nine functions for performing post-processing analyses on the genotyping results produced by GTC or APT genotyping for Axiom arrays. It is not necessary to run all nine functions; the three functions *Ps_Metrics*, *Ps_Classication*, and *Ps_Visualization* are most often used.

The *Ps_Metrics* function generates SNP QC metrics that help identify problematic SNPs. This function requires two GTC/APT output files as inputs: the calls and posteriors files. The input files required for Ps_Metrics should either be located in the working directory, or the file path must be provided. If the file path is wrong or no file path is given, and the files are not in the working directory, R will return an error of "file not found". The Ps_Metrics function can be run by:

```
Ps_Metrics (posteriorFile="AxiomGT1.snp-posteriors.txt",
   callFile="AxiomGT1.calls.txt",
   output.metricsFile="metrics.txt")
```

Here, "`PosteriorFile`" is the name of the posteriors file produced by GTC/APT (usually "AxiomGT1.snp-posteriors.txt"); "`callFile`" is the name of the calls file produced by GTC/APT (usually "AxiomGT1.calls.txt"); and "`output.metricsFile`" is the name of the metrics file produced by Ps_Metrics. This file will be read into other SNPolisher functions such as Ps_Classification (see the following text). If no name is supplied, the file is automatically written to the output directory and named as "metrics.txt".

The output metrics file contains 12 QC metrics (see Table 18.5)—*Call Rate, FLD, HomFLD, Het Strength O_set, Hom Ratio O_set, Minor Allele Count, Number of Clusters, Number of AA Calls, Number of AB Calls, Number of BB Calls, Number of No Calls,* and *Hemizygous.* Of these, *Minor Allele Count, Number of Clusters, Number of AA Calls, Number of AB Calls, Number of BB Calls,* and *Number of No Calls* are self-explanatory. *Hemizygous* is a logical operator that indicates if an SNP appears to be hemizygous, such as the SNPs in the Y chromosome of human males. The remaining four QC metrics—*Call Rate, FLD, Het Strength Offset,* and *Hom Ratio Offset*—are used for filtering SNPs, and to aid in the classification step.

The *Ps_Classification* function classifies SNPs into categories based on the QC metrics from Ps_Metrics, including *PolyHighResolution, MonoHighResolution, Off-Target Variant* (OTV), *CallRateBelowThreshold, NoMinorHom,* and *Other.* One additional category of *Hemizygous* (chrY/chrW/Mitochondrial SNPs) will also be available, given the relevant SNP information is known on the array.

Ps_Classification takes one required input file ("metrics.txt") and one optional input file ("ps2snp.txt"). If there are any *de novo* SNPs, then they should have two or more probesets and be listed in the "ps2snp.txt" file. Ps_Classification has three required arguments.

```
Ps_Classification (metricsFile="metrics.txt",
   ps2snpFile="ps2snp.txt", output.dir=".",
   SpeciesType="Diploid")
```

Here, "`metricsFile`" is the name of the output metrics file from the Ps_Metrics function; "`output.dir`" is the path to the preferred output directory for results files—to use the current directory, type a period ("."), and if an output directory is not provided, a folder named "classification" will be created in the working directory, and all output will be written to it; and "`SpeciesType` takes one of the three options as its input—"Human", "Diploid (non-human)", and "Polyploid". The default value is "Human".

Ps_Classification outputs eight or nine files—one file for each of the seven categories, a summary file for all SNPs (named "Ps.performance.txt"), and an optional file with the list of converted SNPs ("converted.ps") if requested. The files are written to the specified output directory.

Examples of SNPs classified into six categories are shown in Figure 18.6: (i) "PolyHighResolution", where three clusters are formed with good resolution; (ii) "NoMinorHom", where two clusters are formed with no samples of the minor homozygous genotypes;

Table 18.5 Example of metrics file. The first 20 rows of "metrics.txt" opened in Microsoft Excel.

probeset_id	CR	FLD	HomFLD	HetSO	HomRO	nMinorAllele	Nclus	n_AA	n_AB	n_BB	n_NC
AX-85188750	100	6.174281	16.0583	0.312115	0.808468	123	3	47	29	103	0
AX-85188875	100	6.435538	NA	0.522114	1.57622	29	2	150	29	0	0
AX-85188894	100	NA	NA	NA	1.30013	0	1	0	0	179	0
AX-85189006	98.883	4.699605	9.529431	0.085061	0.731244	151	3	34	83	60	2
AX-85189011	100	6.887224	17.42846	0.285353	1.44422	174	3	45	94	40	0
AX-85189133	98.883	5.182815	10.58462	0.113191	0.42302	100	3	7	86	84	2
AX-85189270	100	NA	NA	NA	1.59783	0	1	0	0	179	0
AX-85189421	100	9.160119	NA	0.267335	1.67607	66	2	0	66	113	0
AX-85189424	97.765	4.823505	11.31037	0.107955	0.535493	173	3	87	3	85	4
AX-85189535	100	8.083286	NA	−0.00893	2.44174	91	2	88	91	0	0
AX-85189540	100	5.337368	NA	0.065053	0.903928	94	2	85	94	0	0
AX-85189563	99.441	9.058236	NA	0.372498	1.15545	95	2	83	95	0	1
AX-85189676	100	NA	NA	NA	2.36967	0	1	0	0	179	0
AX-85189684	100	NA	NA	NA	0.27246	0	1	0	0	179	0
AX-85189692	100	NA	NA	NA	0.64628	0	1	179	0	0	0
AX-85189924	100	10.09114	NA	0.091871	1.60609	86	2	93	86	0	0
AX-85189943	100	NA	NA	NA	0.999967	0	1	0	0	179	0
AX-85189947	98.324	4.179805	9.616739	0.42397	1.23303	146	3	71	64	41	3
AX-85190043	98.324	3.016386	12.34585	0.029954	0.390225	104	3	1	102	73	3
AX-85190081	100	7.797094	NA	0.248618	1.26047	89	2	0	89	90	0

(iii) "MonoHighResolution", in which only one cluster is formed; (iv) "OTV", off-target variants, where three good clusters are formed, but with one extra OTV cluster that is caused by sequence dissimilarity between probes and target genome regions; (v) "CallRateBelowThreshold", where SNP call rate is below threshold, but other cluster properties are above threshold; and (vi) "Other", where one or more cluster properties are below threshold.

Classification can be used to determine which SNPs should be convertible for further downstream analysis. For diploid species that are outbred, or for a mixture of outbred and inbred, SNPs of categories (i) to (iii) were considered as convertible SNPs, and SNPs of categories (i) to (ii) were considered as polymorphic SNPs (Figure 18.6). However, "MonoHighRes" SNPs are recommended with caution as convertible SNPs. An additional test is to check if both probesets (if available) for the SNP are classified as "MonoHighRes", and if both were called with identical genotypes.

SNPs that are classified as OTV (iv) may also be considered convertible after the "OTV_Caller" function has been used to re-label the genotype calls. Hemizygous SNPs should be visually inspected for any problems. Chromosomes Y and W, mitochondrial genomes, and other hemizygous genomes produce only two genotype clusters. These two clusters should be clearly resolved from one another. A visual check of the cluster plots as part of the overall SNP conversion analysis is needed for this category.

The recommended SNP list is created by combining the list of SNPs that are classified into the recommended categories. Ps_Classification produces an optional output file named "converted.ps" that contains the list of PolyHighRes, MonoHighRes, NoMinorHom, and Hemizygous SNPs. These SNPs are recommended for conversion when the species type is "Human" or "Diploid" that are not inbred. If the species is either polyploid or inbred diploid, only PolyHighRes SNPs are recommended for conversion, and MonoHighRes SNPs may be used with caution.

Ps_Visualization generates plots of the genotype cluster patterns for a set of selected SNPs. Plots include the posterior information (default) and prior information (optional). OTV genotypes and reference genotypes can also be included. The cluster plots can help QC SNPs and diagnose underlying genotyping problems. All plots are output as PDFs (.ps files), which can be read into GTC for visualization.

Ps_Visualization takes six required arguments:

```
Ps_Visualization (pidFile="probesetID.txt",
    output.pdfFile="output.pdf", summaryFile="AxiomGT1.
    summary.txt", callFile="AxiomGT1.calls.txt",
    confidenceFile="AxiomGT1.con_dences.txt",
    posteriorFile="AxiomGT1.snp-posteriors.txt")
```

Here, "`pidFile`" is the text file listing probeset IDs—the first line of `pidFile` should always be "probeset_id"; "`output.pdfFile`" is the name for the output PDF file; "`summaryFile`" is the GTC/APT output summary file (usually "AxiomGT1.summary.txt"); "`callFile`" is the GTC/APT output calls file (usually "AxiomGT1.calls.txt"); "`confidenceFile`" is the GTC/APT confidences file (usually "AxiomGT1.con_dences.txt"); and "`posteriorFile`" is the GTC/APT posteriors file (usually "AxiomGT1.snp-posteriors.txt").

Ps_Visualization outputs a PDF file that contains one or more pages of SNP genotype cluster plots. Each page has six columns (one SNP per column) and four rows (four plots per SNP). The SNP genotyping quality can be determined by visually inspecting these cluster plots.

Applications of SNP Arrays

The use of SNP arrays for genome-wide SNP genotyping has been mainly concentrated on the association studies for the identification of genomic regions associated with various complex traits, and for the assessment of livestock breeding values. The application of SNP arrays in genome-wide association studies has allowed the investigation of numerous performance and production traits in various livestock animals. The application of SNP arrays in the practice of animal breeding enables whole-genome-based prediction and selection, providing molecular estimates of breeding values with high accuracy (Taylor *et al.*, 2012). Besides GWAS and genomic selection, genome-wide SNP genotyping with SNP arrays are also useful for studies on different aspects of genome structure, variability, diversity, and evolution.

Genome-Wide Association Studies

SNP arrays have long been used for genome-wide genotyping to detect common and rare SNPs, and CNVs that can contribute to complex diseases in humans. For genome-wide association studies on human diseases with SNP arrays, readers are referred to extensive review papers such as those by Hirschhorn & Daly (2005), Hardy & Singleton (2009), and Manolio (2010). In recent years, along with the development of high-density SNP arrays, GWAS was extended to the field of agricultural and companion animals (Zhang *et al.*, 2012), such as cattle (Hayes *et al.*, 2009), pig (Becker *et al.*, 2013; Bolormaa *et al.*, 2010; Sahana *et al.*, 2013), horse (Brooks *et al.*, 2010), sheep (Zhao *et al.*, 2011), dog (Meurs *et al.*, 2010), and chicken (Wolc *et al.*, 2012; Xie *et al.*, 2012).

In spite of being very powerful, GWAS has barely been applied to aquaculture species, the primary reason being the lack of high-density SNP arrays. Recently, with the development of SNP arrays in aquaculture species—such as those for catfish (Liu *et al.*, 2014), carp (Xu *et al.*, 2014), Atlantic salmon (Houston *et al.*, 2014) and rainbow trout (Palti *et al.*, 2014)—the application of genome-wide SNP genotyping for GWAS is being made in aquaculture species too. For instance, a genome-wide association study conducted using the catfish 250K SNP array allowed identification of four genomic regions within QTLs associated with Columnaris disease resistance (Geng *et al.*, 2015).

In addition to GWAS, SNP arrays are useful for fine-scale linkage analysis and QTL mapping of economically important traits when applied to genotype samples from families (Li *et al.*, 2015; Liu *et al.*, 2016). By genotyping three large resource families using the catfish 250K SNP array, we constructed a high-density and high-resolution genetic map for channel catfish (Li *et al.*, 2015). In parallel, a high-density genetic linkage map was also constructed using the interspecific hybrid catfish resource families (Liu *et al.*, 2016). Comparison of these genetic maps with those constructed using channel catfish resource families allowed the identification of genomic regions underlying the incompatibility between channel catfish and blue catfish, possibly accounting, at least in part, for the reproductive isolation of the two species.

Analysis of Linkage Disequilibrium

Linkage disequilibrium (LD), that is, non-random association of alleles at two or more loci, is influenced by population history and its evolution. The knowledge of LD between markers is not only essential to determine the effective number of markers used for GWAS and fine-scale QTL mapping, but is also important for the investigation of population history, breeding systems, and patterns of geographic distribution (Gurgul *et al.*, 2014). Genome-wide SNP genotyping have allowed the investigation of LD within and across populations in agricultural and companion animals (McCue *et al.*, 2012; Miller *et al.*, 2011; Xu *et al.*, 2014). For instance, the analysis of genome-wide LD in domestic horse by genotyping a panel of samples from 14 domestic horse breeds and 18 evolutionarily related species revealed that the extent of LD was higher within a breed that across breeds (McCue *et al.*, 2012). In common carp, the LD analysis in five domesticated strains revealed that the extent of LD varied among strains, with three strains (Yellow River carp, Hebao carp, and Xingguo red carp) having longer haplotype blocks (Xu *et al.*, 2014).

Population Structure *and* Discrimination

Genetic markers are useful in identifying and verifying the origin of individuals when sufficient genetic heterogeneity is present among populations (Wilkinson *et al.*, 2011). Genome-wide SNP genotyping allows identifying the most informative genetic markers to discriminate among populations. The utility of the rainbow trout 57K SNP array for population traceability was evaluated for identifying markers that were homozygote to an allele in one population and an alternative allele in another population (Palti *et al.*, 2014). Similarly, population-specific SNPs among several aquaculture strains and wild populations were identified in catfish to be used for distinguishing fish from different origins and for differentiating individuals from farmed and wild populations (Sun *et al.*, 2014).

Genomic Signatures of Selection and Domestication

Animal breeding and domestication are closely related to artificial selection, which lead to changes in the frequency of the genetic variants associated with a trait under selection. The genomic regions underlying selection signatures can be detected by the analysis of differences in allele or haplotype frequencies of genome-wide SNPs between populations with different levels of selected traits (Gurgul *et al.*, 2014). Most studies aiming at detecting selection signatures have been performed in cattle, a species most widely subjected to genomic selection. For instance, the analysis of the allele frequency distribution between dairy and beef cattle breeds in Japan allowed the identification of 11 candidate regions associated with different types of production traits (Hosokawa *et al.*, 2012). Candidate genes were identified within the regions, including those genes previously associated with meat quality and milk yield traits (Hosokawa *et al.*, 2012). In a recent study, by analyzing allele frequency differences of genome-wide SNPs between catfish from domesticated strains and wild populations, we identified 23 genomic regions that were putatively under selection and domestication in the catfish genome (Sun *et al.*, 2014). Several genes with known functions related to aquaculture performance traits were included in the regions (Sun *et al.*, 2014).

Analysis of selection signatures can be an important step in identifying biological factors that are responsible for physiological and production characteristics in farmed animals. The selected genomic regions could harbor the functional elements involved in the development of desired traits.

Conclusion

In summary, the advances in high-throughput sequencing and SNP array genotyping technologies allow for the generation of large-scale SNP genotype data at the genome level for genetic analysis. In this stage, genome-wide association studies and whole genome mapping of quantitative traits are becoming routine in non-model species, including many aquaculture species. However, considering the flood of genotype data, the quality of the data is essential for drawing solid conclusions. Since even small sources of systematic or random errors can cause false positive results, QC of the data from the initial genotyping is essential to account for plate/batch effect, to exclude outlier samples, and to post-filter SNPs-based specific genotyping metrics. This chapter provides an overview of SNP array development and genotyping with the Affymetrix Axiom genotyping technology, covering the specific steps that are essential for DQC. This chapter can serve as a practical guide for working with the Affymetrix Axiom genotyping platform.

Further Readings

Axiom myDesign Genotyping Arrays Design Guide: http://media.affymetrix.com/support/downloads/manuals/axiom_genotyping_array_design_guide.pdf

Axiom Genotyping Solution Data Analysis Guide: http://www.affymetrix.com/support/downloads/manuals/axiom_genotyping_solution_analysis_guide.pdf

GTC 4.2 User Guide: http://www.affymetrix.com/support/downloads/manuals/gtc_4_2_user_manual.pdf

Affymetrix® Power Tools User Manual: http://media.affymetrix.com/support/developer/powertools/changelog/apt-probeset-genotype.html

References

Becker, D., Wimmers, K., Luther, H., *et al.* (2013) A genome-wide association study to detect QTL for commercially important traits in Swiss large white boars. *PLoS One*, **8**, e55951.

Bolormaa, S., Pryce, J., Hayes, B. and Goddard, M. (2010) Multivariate analysis of a genome-wide association study in dairy cattle. *Journal of Dairy Science*, **93**, 3818–3833.

Brooks, S.A., Gabreski, N., Miller, D., *et al.* (2010) Whole-genome SNP association in the horse: identification of a deletion in myosin Va responsible for Lavender foal syndrome. *PLoS Genetics*, **6**, e1000909.

Chen, K.X., Ma, H.X., Li, L., *et al.* (2014) Genome-wide association study identifies new susceptibility loci for epithelial ovarian cancer in Han Chinese women. *Nature Communications*, **5**, 4682.

Davey, J.W., Hohenlohe, P.A., Etter, P.D., *et al.* (2011) Genome-wide genetic marker discovery and genotyping using next-generation sequencing. *Nature Reviews Genetics*, **12**, 499–510.

Geng, X., Sha, J., Liu, S., *et al.* (2015) A genome-wide association study in catfish reveals the presence of functional hubs of related genes within QTLs for Columnaris disease resistance. *BMC Genomics*, **16**, 196.

Groenen, M.A.M., Megens, H.J., Zare, Y., *et al.* (2011) The development and characterization of a 60K SNP chip for chicken. *BMC Genomics*, **12**, 274.

Gurgul, A., Semik, E., Pawlina, K., *et al.* (2014) The application of genome-wide SNP genotyping methods in studies on livestock genomes. *Journal of Applied Genetics*, **55**, 197–208.

Hardy, J. and Singleton, A. (2009) CURRENT CONCEPTS genomewide association studies and human disease. *The New England Journal of Medicine*, **360**, 1759–1768.

Hayes, B.J., Bowman, P.J., Chamberlain, A.J., *et al.* (2009) A validated genome wide association study to breed cattle adapted to an environment altered by climate change. *PLoS One*, **4**, e6676.

Hirschhorn, J.N. and Daly, M.J. (2005) Genome-wide association studies for common diseases and complex traits. *Nature Reviews Genetics*, **6**, 95–108.

Hoffmann, T.J., Kvale, M.N., Hesselson, S.E., *et al.* (2011) Next generation genome-wide association tool: Design and coverage of a high-throughput European-optimized SNP array. *Genomics*, **98**, 79–89.

Hosokawa, D., Ishii, A., Yamaji, K., *et al.* (2012) Identification of divergently selected regions between Japanese Black and Holstein cattle using bovine 50K SNP array. *Animal Science Journal*, **83**, 7–13.

Houston, R.D., Taggart, J.B., Cezard, T., *et al.* (2014) Development and validation of a high density SNP genotyping array for Atlantic salmon (*Salmo salar*). *BMC Genomics*, **15**, 90.

Howe, K., Clark, M.D., Torroja, C.F., *et al.* (2013) The zebrafish reference genome sequence and its relationship to the human genome. *Nature*, **496**, 498–503.

Korn, J.M., Kuruvilla, F.G., McCarroll, S.A., *et al.* (2008) Integrated genotype calling and association analysis of SNPs, common copy number polymorphisms and rare CNVs. *Nature Genetics*, **40**, 1253–1260.

Kranis, A., Gheyas, A.A., Boschiero, C., *et al.* (2013) Development of a high density 600K SNP genotyping array for chicken. *BMC Genomics*, **14**, 59.

Kruglyak, L. (1997) The use of a genetic map of biallelic markers in linkage studies. *Nature Genetics*, **17**, 21–24.

Li, Y., Liu, S., Qin, Z., *et al.* (2015) Construction of a high-density, high-resolution genetic map and its integration with BAC-based physical map in channel catfish. *DNA Research*, **22**, 39–52.

Liu, S., Li, Y., Qin, Z., *et al.* (2016) High-density interspecific genetic linkage mapping provides insights into genomic incompatibility between channel catfish and blue catfish. *Animal Genetics*, **47**, 81–90.

Liu, S., Sun, L., Li, Y., *et al.* (2014) Development of the catfish 250K SNP array for genome-wide association studies. *BMC Research Notes*, **7**, 135.

Liu, S.K., Zhou, Z.C., Lu, J.G., *et al.* (2011) Generation of genome-scale gene-associated SNPs in catfish for the construction of a high-density SNP array. *BMC Genomics*, **12**, 53.

Manolio, T.A. (2010) Genomewide association studies and assessment of the risk of disease. *The New England Journal of Medicine*, **363**, 166–176.

Matukumalli, L.K., Lawley, C.T., Schnabel, R.D., *et al.* (2009) Development and characterization of a high density SNP genotyping assay for cattle. *PLoS One*, **4**, e5350.

McCue, M.E., Bannasch, D.L., Petersen, J.L., *et al.* (2012) A high density SNP array for the domestic horse and extant Perissodactyla: utility for association mapping, genetic diversity, and phylogeny studies. *PLOS Genetics*, **8**, e1002451.

Meurs, K.M., Mauceli, E., Lahmers, S., *et al.* (2010) Genome-wide association identifies a deletion in the 3′ untranslated region of striatin in a canine model of arrhythmogenic right ventricular cardiomyopathy. *Human Genetics*, **128**, 315–324.

Miller, J.M., Poissant, J., Kijas, J.W., *et al.* (2011) A genome-wide set of SNPs detects population substructure and long range linkage disequilibrium in wild sheep. *Molecular Ecology Resources*, **11**, 314–322.

Mogensen, M.S., Karlskov-Mortensen, P., Proschowsky, H.F., *et al.* (2011) Genome-wide association study in Dachshund: identification of a major locus affecting intervertebral disc calcification. *Journal of Heredity*, **102**, S81–S86.

Nielsen, R., Paul, J.S., Albrechtsen, A. and Song, Y.S. (2011) Genotype and SNP calling from next-generation sequencing data. *Nature Reviews Genetics*, **12**, 443–451.

Palti, Y., Gao, G., Liu, S., *et al.* (2014) The development and characterization of a 57K single nucleotide polymorphism array for rainbow trout. *Molecular Ecology Resources*, **15**(3), 662–672.

Purcell, S., Neale, B., Todd-Brown, K., *et al.* (2007) PLINK: a tool set for whole-genome association and population-based linkage analyses. *The American Journal of Human Genetics*, **81**, 559–575.

Ramos, A.M., Crooijmans, R.P.M.A., Affara, N.A., *et al.* (2009) Design of a high density SNP genotyping assay in the pig using SNPs identified and characterized by next generation sequencing technology. *PLoS One*, **4**, e6524.

Rincon, G., Weber, K.L., Van Eenennaam, A.L., *et al.* (2011) Hot topic: performance of bovine high-density genotyping platforms in Holsteins and Jerseys. *Journal of Dairy Science*, **94**, 6116–6121.

Sahana, G., Kadlecova, V., Hornshoj, H., *et al.* (2013) A genome-wide association scan in pig identifies novel regions associated with feed efficiency trait. *Journal of Animal Science*, **91**, 1041–1050.

Sun, L.Y., Liu, S.K., Wang, R.J., *et al.* (2014) Identification and analysis of genome-wide SNPs provide insight into signatures of selection and domestication in channel catfish (*Ictalurus punctatus*). *PLoS One*, **9**, e109666.

Taylor, J.F., McKay, S.D., Rolf, M.M., *et al.* (2012) Genomic selection in beef cattle, in *Bovine Genomics* (ed. J.E. Womack), John Wiley & Sons, Iowa, pp. 211–233.

Wang, S., Sha, Z., Sonstegard, T.S., *et al.* (2008) Quality assessment parameters for EST-derived SNPs from catfish. *BMC Genomics*, **9**, 450. doi:10.1186/1471-2164-9-450

Wilkinson, S., Wiener, P., Archibald, A.L., *et al.* (2011) Evaluation of approaches for identifying population informative markers from high density SNP Chips. *BMC Genetics*, **12**, 45.

Wolc, A., Arango, J., Settar, P., *et al.* (2012) Genome-wide association analysis and genetic architecture of egg weight and egg uniformity in layer chickens. *Animal Genetics*, **43**, 87–96.

Xie, L., Luo, C., Zhang, C., *et al.* (2012) Genome-wide association study identified a narrow chromosome 1 region associated with chicken growth traits. *PLoS One*, 7, e30910.

Xu, J., Zhao, Z., Zhang, X., *et al.* (2014) Development and evaluation of the first high-throughput SNP array for common carp (*Cyprinus carpio*). *BMC Genomics*, **15**, 307.

Yang, H., Ding, Y.M., Hutchins, L.N., *et al.* (2009) A customized and versatile high-density genotyping array for the mouse. *Nature Methods*, **6**, U663–U655.

Zhang, H., Wang, Z., Wang, S. and Li, H. (2012) Progress of genome wide association study in domestic animals. *Journal of Animal Science and Biotechnology*, **3**, 26.

Zhao, X., Dittmer, K.E., Blair, H.T., *et al.* (2011) A novel nonsense mutation in the DMP1 gene identified by a genome-wide association study is responsible for inherited rickets in Corriedale sheep. *PLoS One*, **6**, e21739.

19

Genotyping by Sequencing and Data Analysis: RAD and 2b-RAD Sequencing

Shi Wang, Jia Lv, Jinzhuang Dou, Qianyun Lu, Lingling Zhang and Zhenmin Bao

Introduction

Single-nucleotide polymorphisms (SNPs) have become the markers of choice, but development of their genotyping platforms involves very large upfront investments. In contrast, sequencing cost is decreasing steadily. The increased throughput and reduced costs of next-generation sequencing (NGS) make it possible to consider genotyping by sequencing (GBS). GBS has been widely used in model species and in humans, but its relatively high cost is still beyond the capacity of most laboratories working with non-model species such as aquaculture species. As a result, targeted whole genome sequencing methods have been developed, such as restriction-site-associated DNA (RAD), 2b-RAD. In this chapter, we will cover RAD and 2b-RAD principles, experimental procedures, and bioinformatics analysis methodologies.

Understanding the genetic basis of complex traits is a central task of modern genetics. Scanning the whole genome to pin down the causal loci or genes underlying complex traits represents an effective approach to achieving this goal, but a key prerequisite is the availability of a large number of genetic markers. Historically, genetic markers in aquatic organisms have predominantly been traditional marker systems such as RAPD, AFLP, and SSR (Liu, 2007). However, these marker systems suffer from low marker output, making it difficult to fulfill a high-resolution genome scan. Recently, SNPs have attracted significant attention, as they represent the most abundant class of genetic variations in eukaryotic genomes and are the major genetic source of phenotypic variability. In human and model organisms, SNPs have already become the markers of choice in large-scale genotyping applications, such as genome-wide association, high-resolution linkage mapping, quantitative trait locus (QTL) mapping, genomic selection, and comparative genome analysis. However, development of a large number of SNP markers in a high-throughput and cost-effective manner has been a challenging task in most aquatic organisms, mostly due to the lack of sufficient genome information for large-scale SNP discovery. Most recently, this difficulty has been overcome by the application of NGS technologies in aquatic species, making SNPs readily available. However, high-throughput genotyping using large numbers of SNPs of whole genome coverage involves huge investments in genotyping assays (e.g., SNP arrays).

Recent advances in NGS technologies (e.g., Roche's 454 and Illumina) now allow for rapid and affordable generation of extensive genomic resources, making it possible to

Bioinformatics in Aquaculture: Principles and Methods, First Edition. Edited by Zhanjiang (John) Liu.

efficiently discover and genotype a large number (e.g., thousands to tens of thousands) of SNPs in less-studied aquatic organisms. Currently, sequencing costs are not yet low enough to allow the use of genome re-sequencing for population studies with aquatic organisms. In addition, many types of studies do not benefit from the extremely high density of genotyped polymorphisms provided by complete genome re-sequencing, and would already attain maximal power if only a fraction of the genome was reproducibly re-sequenced across individuals. Based on this idea, a variety of genotyping techniques using sequencing methods have recently been developed, such as RAD, reduced representation library (RRL), and GBS, most of which utilize restriction enzymes to reduce genome complexity (Davey *et al.*, 2011). Among these methods, RAD has become particularly popular (Figure 19.1), because it allows for nearly every restriction site in the genome to be screened in parallel. RAD has already been widely applied in a variety of studies, such as fine-scale linkage mapping, population genetics, phylogenetics and phylogeography, and genome scaffolding. For example, using RAD markers, Baird *et al.* (2008) successfully fine-mapped the causal gene *Eda*, which is responsible for lateral plate armor loss in the three-spined stickleback. Hohenlohe *et al.* (2010) conducted genome-wide scans to identify genomic regions that are involved in the parallel adaptation of oceanic and freshwater stickleback populations. Emerson *et al.* (2010) used RAD markers to reveal previously unresolved genetic structure and direction of evolution in pitcher-plant mosquitoes. A RAD-based linkage map was also utilized to facilitate butterfly genome assembly, with 83% of the sequenced genome ordered onto the 21 chromosomes (Heliconius Genome Consortium, 2012). However, the traditional RAD method uses a relatively long and complex procedure for library preparation, which reduces technical reproducibility and may lead to several sources of bias (Davey *et al.*, 2013). Most recently, several simpler RAD methods have been developed with technical improvements that greatly enhance their applicability (Peterson *et al.*, 2012; Toonen *et al.*, 2013; Wang *et al.*, 2012). For example, 2b-RAD, recently developed by Wang *et al.* (2012), represents a streamlined RAD method that utilizes type IIB restriction enzymes to produce uniform fragments for sequencing. It features even and tunable genome coverage, therefore providing a flexible genotyping platform to meet different research needs.

Currently, the demand for the genotyping of aquatic organisms by sequencing methods is rapidly increasing. In this chapter, we will mainly focus on RAD and 2b-RAD by presenting descriptions of their methodology principles, experimental procedures, and bioinformatics analyses for genetic mapping and other purposes.

Methodology Principles

RAD

The preparation of a RAD library begins with the digestion of genomic DNA using a typical type II restriction enzyme, producing various sizes of fragments with definite sticky ends (Figure 19.2). These fragments are then ligated with the first adaptor (P1), which contains forward amplification and sequencing primer sites, as well as a sample-specific barcode (usually 4–5 bp) that enables the adaptor-ligated fragments from different samples to be pooled together. The pooled fragments are randomly sheared and size-selected

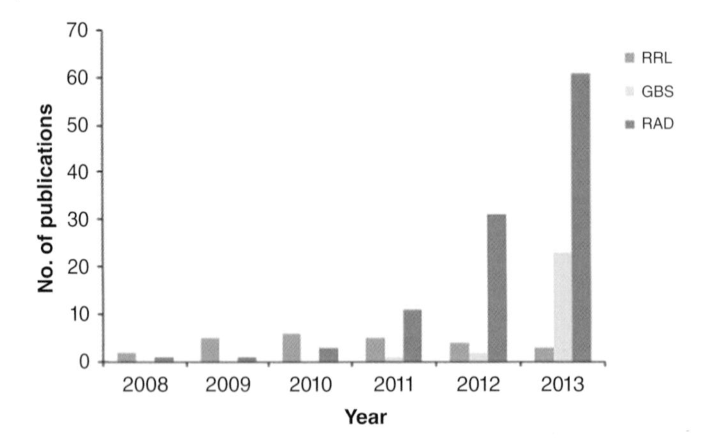

Figure 19.1 A tendency overview of RRL, GBS, and RAD publications (data were summarized based on the search records of Google Scholar without including review articles). From this chart, we can see that, since the invention of RAD-Seq in 2008, the number of RAD publications has grown very fast, especially in the past 3 years.

to 300–700 bp, and the selected fragments are subjected to fragment end repair (i.e., blunting the fragment ends, followed by the addition of 3'-A overhangs) to facilitate subsequent ligation with the second adaptor (P2). P2 is a divergent Y-shaped adaptor that is not able to bind to the P2 primer until the complementary sequence amplified by the P1 adaptor is completed; this design is to prevent amplification of fragments that lack a P1 adaptor. After a second ligation with the P2 adaptor, PCR amplification is conducted using P1 and P2 primers to generate the desired construct, which is composed of a P1 adaptor, a partial restriction site, a flanking sequence of interest, and a P2 adaptor. The amplicons are then size-selected (300–700 bp), and the resultant libraries are ready to be sequenced on NGS platforms.

2b-RAD

2b-RAD is a streamlined RAD method that is based on sequencing the uniform fragments produced by type IIB restriction enzymes. The special feature of these enzymes is that they can cleave genomic DNA both upstream and downstream of the target site, producing tags of uniform length (typically 30–37 base pairs) that are ideally suited for sequencing on existing next-generation platforms. A schematic overview of the 2b-RAD library preparation is shown in Figure 19.3. The 2b-RAD procedure begins with the digestion of genomic DNA by a type IIB restriction enzyme, which generates uniform fragments with arbitrary overhangs. Then, two adaptors with compatible overhangs (e.g., NNN for BsaXI) are added to the fragment ends. These adaptors can be modified to adjust marker density to meet different research needs. For example, a wide range of representations can be potentially achieved in *Arabidopsis*, ranging from one quarter of all sites (NNR overhang on both adaptors) to 1/256th of all sites (NGG overhangs on both adaptors). The adaptor-ligated fragments undergo subsequent amplifications to incorporate sample-specific barcodes, and the purified amplicons are then ready for high-throughput sequencing. The 2b-RAD protocol is substantially simpler than RAD protocols because it uses type IIB enzymes, thereby eliminating

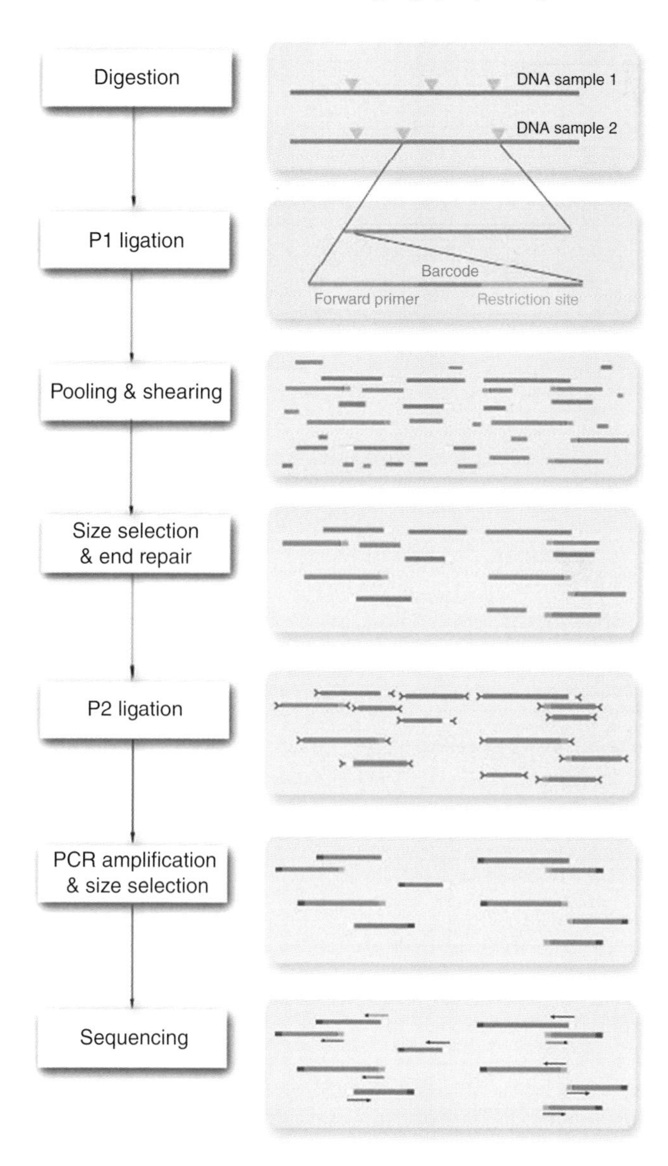

Figure 19.2 Schematic overview of RAD library preparation. (*See color plate section for the color representation of this figure.*)

the need for multiple steps such as mechanical fragmentation, fragment end repair, secondary ligation and size selection.

The Experimental Procedure of 2b-RAD

RAD experimentation has been documented in many publications. Most recently, Etter *et al.* (2011) provided a more detailed RAD protocol with step-by-step instructions, which is publicly available at http://www.ncbi.nlm.nih.gov/pmc/articles/PMC3658458/.

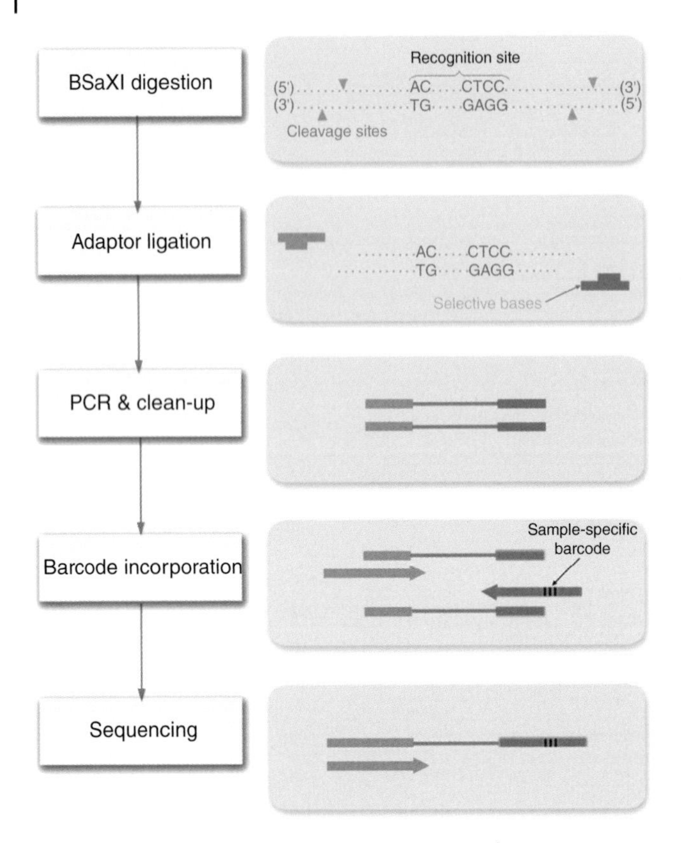

Figure 19.3 Schematic overview of 2b-RAD library preparation. (*See color plate section for the color representation of this figure.*)

Since RAD protocols have been sufficiently described elsewhere, here we will only focus on a detailed description of 2b-RAD library preparation for high-throughput Illumina sequencing.

DNA Input

Relatively pure, high-molecular-weight, RNA-free genomic DNA is required for efficient restriction enzyme digestion, and is important for the overall success of the protocol. At least 100–200 ng of genomic DNA is needed, with an optimal concentration of 40 ng/μl or greater. High-quality DNA can be obtained using most commercial DNA isolation kits, such as DNeasy Blood & Tissue kit (Qiagen, for animals) and DNeasy Plant Mini kit (Qiagen, for plants).

Note 1: We have found that lower-quality or slightly degraded DNA can be used in some cases, but the amount of DNA will likely need to be increased to maintain the appropriate molar ratio in the digest product and adaptors for efficient ligation.

Note 2: We recommend checking DNA integrity using a 1% agarose gel before digestion; the genomic DNA observed on the gel should be a fairly tight, high-molecular-weight band without any visible degradation products or smears. The presence of residual organics, metal ions, salts, or nucleases in the DNA sample could affect downstream enzymatic reactions, possibly causing the inhibition of enzymatic activity or DNA degradation. The purity of the DNA sample should be checked using a UV spectrometer to perform ratio absorbance measurements at A260/A280 and A260/A230. A260/280 and A260/230 values greater than 1.8 are typically suitable for the next digestion step.

Restriction Enzyme Digestion

Many types of IIB enzymes (e.g., BsaXI, AlfI, BcgI, etc.) can be utilized in 2b-RAD experiments. Here, we will use BsaXI as an example to demonstrate the procedure for 2b-RAD library preparation. BsaXI digestion is conducted using 100–200 ng DNA with an appropriate restriction enzyme in a 20 μl reaction volume. For BsaXI digestion, the following are combined in a microcentrifuge tube: 1.5 μl 10Xbuffer 4; 2 μl BsaXI (2 U/μl, NEB); 100–200 ng DNA; and sterile H_2O to 20 μl. The reaction is incubated at 37°C for 4 h to complete digestion; then, 4 μl of the digested product are electrophoresed on a 1% agarose gel to check the digestion results and make sure the genomic DNA has been completely digested.

Note 1: Depending on the genomic features of the species (e.g., GC content), the number of recognition sites across the genome may vary substantially for different enzymes. Therefore, it is possible to adjust marker density just by choosing different enzymes. For example, although the recognition sites of the two enzymes (AlfI and BsaXI) are equal in length (6 bp), the number of AlfI sites in the *Arabidopsis* genome is less than one-third that of BsaXI sites.

Note 2: We recommend preparing a "negative control reaction" consisting of all the digestion reaction components except for the enzyme to check whether the reaction condition performs well in terms of DNA sample stability. The incubation condition is the same as the enzyme reaction; 4 μl of control product is electrophoresed on a 1% agarose gel to compare DNA digests with and without the enzyme. The control product, consisting of untreated genomic DNA, should be an intact band without any visible degradation products or smears in the gel, indicating that the genomic DNA was stable in the digestion reaction. In the digestion reaction, many sites of genomic DNA are cut by the enzyme, generating a large number of fragments that would appear as a smear rather than as distinct bands on the gel. Therefore, complete digestion is indicated when a smear of DNA appears on the gel (Figure 19.4). It is normal not to see any visible smear of DNA on the gel if the starting amount of input DNA is small. If the band of intact genomic DNA is still visible on the gel after digestion, the genomic DNA may appear to be partially digested, and the reaction time or the number of units of enzyme may need to be increased.

Adaptor Ligation

Adaptor Preparation

The sense and antisense oligonucleotides of adaptor-1 and adaptor-2 are separately dissolved in sterile H_2O (10 μM). For each adaptor, 25 μl of each sense and antisense oligonucleotide solutions are mixed in one tube to a final concentration of 5 μM. A total of 50 μl of this mixture is denatured at 95°C for 5 min, and then cooled down to 20°C for oligonucleotide annealing. Then, the annealed adaptors are ready for use.

Adaptor sequences (5' to 3')
Adaptor-1 sense: ACACTCTTTCCCTACACGACGCTCTTCCGATCTNNN
Adaptor-1 antisense: AGATCGGAAGAGC

Figure 19.4 Agarose gel electrophoresis for the products from the treatment and control reactions. Lane 1 contains **5 μl** of the control product. The intact DNA band demonstrates that the DNA sample was stable in the digestion reaction. Lane 2 contains **5 μl** of the digestion product. A DNA smear appeared on the gel, indicating that the genomic DNA was completely digested.

Adaptor-2 sense: GTGACTGGAGTTCAGACGTGTGCTCTTCCGATCTNNN
Adaptor-2 antisense: AGATCGGAAGAGC

Ligation Reaction

For ligation of adaptors, a 22 μl reaction solution composed of 2.2 μl T4 ligase buffer, 0.8 μl adaptor-1 (5 μM), 0.8 μl adaptor-2 (5 μM), 2 μl ATP (10 mM), 2 μl T4 DNA ligase (400000 U/ml, NEB), and 14.2 μl sterile H_2O is added to the digestion product. The reaction mixture is then incubated at 4°C for 12 h or overnight.

Note: Reductions in marker density can be achieved in 2b-RAD using modified adaptors targeting a subset of all fragments. For example, reduced tag representation libraries constructed using adaptors with 5'-NNT-3' theoretically overhangs only 1/16th of all target BsaXI restriction sites in the scallop genome, and sequencing of such a library to 20X would require only 0.6 million reads per individual (in contrast to 6 million reads per individual for a standard BsaXI library) (Jiao *et al.*, 2014). The appropriate adaptors for a desired marker density can be determined through bioinformatics analysis of the species genome (or 2b-RAD sequencing data derived from standard adaptors). The ligation condition for modified adaptors is the same as that for standard adaptors.

PCR Amplification and Gel Purification

The following two steps are performed to ensure that the BsaXI fragments are amplified in a high-fidelity manner and that the fragments containing both adaptors are to be enriched.

PCR Amplification A test amplification should first be performed to determine the library quality. For the test amplification, a 20 μl amplification reaction, containing

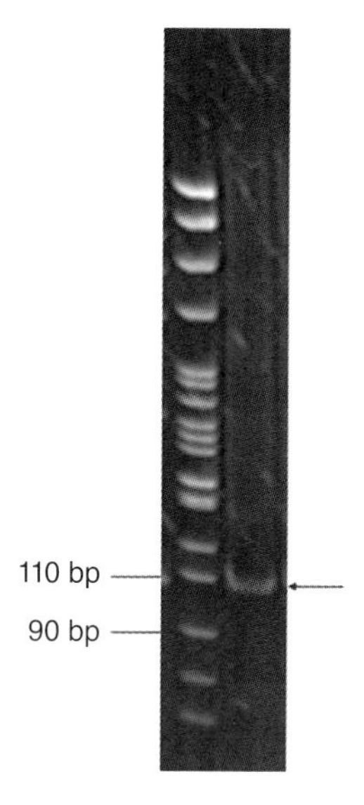

Figure 19.5 The picture of the gel shows the result of a 10% PAGE run of the first PCR product. The band of the expected PCR product (first PCR product; 100 bp) is indicated by an arrow. Only a band of this size is excised from the gel. Left lane: GeneRuler 100 bp DNA ladder.

7 μl ligated DNA, 0.2 μl Solexa primer 1 (10 μM), 0.2 μl Solexa primer 2 (10 μM), 0.6 μl dNTP (10 mM), 4 μl 5X Phusion HF buffer, 0.2 μl Phusion high-fidelity DNA polymerase (2 U/μl, NEB), and 7.8 μl sterile H_2O, is prepared. Then, 18 cycles of amplification are performed in a thermal cycler as follows: 98°C for 5 s, 60°C for 20 s, and 72°C for 10 s, with a final extension at 72°C for 5 min, and then held at 4°C. Then, 5 μl PCR product is mixed with 1 μl 6X loading buffer and loaded onto a 10% polyacrylamide gel at 300 V for 60 min, using GeneRuler 100 bp DNA Ladder Plus (Thermo Scientific) for a size reference. The size of the expected amplified fragments is 100 bp (Figure 19.5). After the PCR results are confirmed, PCR amplifications are repeated under the same conditions in two additional tubes.

Primer sequences (5' to 3')
Primer 1: ACACTCTTTCCCTACACGACGCT
Primer 2: GTGACTGGAGTTCAGACGTGTGCT

Gel Purification The PCR products from all tubes are pooled together, and 10 μl 6X loading buffer is added. A 10% polyacrylamide gel is prepared using a wide comb that can accommodate 70 μl sample loading. The PCR products are loaded onto the gel and run at 300 V for 60 min. The gel should be viewed on a UV transilluminator at low intensity; and the target 100 bp band should be rapidly excised from the gel and transferred to a 1.5 ml microtube. The handling time should be kept to a minimum to minimize possible DNA damage under UV light exposure. The gel is then pestled, and the tube is centrifuged at 13000 g for 1 min. Then, the gel pieces are suspended in 45 μl nuclease-free water, and the tube is kept at 4°C for 12 h. The eluent PCR product is transferred into a new tube and quantified using a fluorometer.

Barcode Incorporation

This step enables the incorporation of sample-specific barcodes into 2b-RAD libraries for multiplex Illumina sequencing. First, 50 μl PCR reactions, containing 60 ng

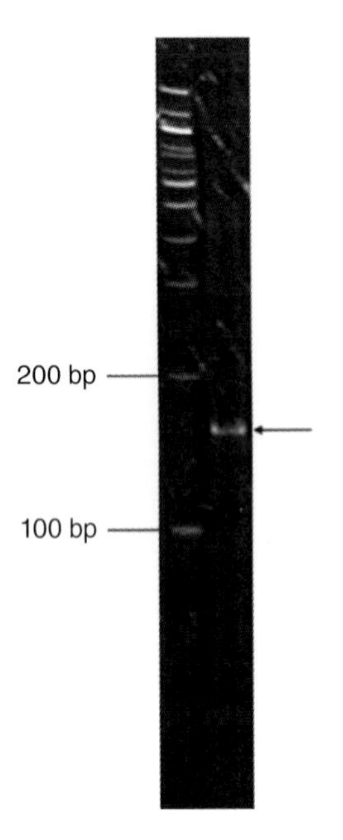

Figure 19.6 The gel picture shows the result of a 10% PAGE run of the second PCR product. The band of the expected PCR product (second PCR product; 155 bp) is indicated by an arrow. Left lane: pBR322 DNA-MspI Digest ladder.

gel-extracted PCR product, 0.5 μl Solexa primer 3 (10 μM), 0.5 μl Solexa index primer (10 μM), 1.5 μl dNTP (10 mM), 10 μl 5X Phusion HF buffer, 0.5 μl Phusion high-fidelity DNA polymerase (2 U/μl, NEB), and sterile H_2O, are prepared. Then, four to nine cycles of amplification are performed in a thermal cycler: 98°C for 5 s, 60°C for 20 s, 72°C for 10 s, and a final extension at 72°C for 5 min, then held at 4°C. PCR products are then subjected to column-based purification (e.g., QIAquick PCR purification kit, Qiagen) according to the manufacturer's instructions. Then, 5 μl of the purified PCR product, plus 1 μl loading buffer, is run on a 10% polyacrylamide gel at 300 V for 60 min. The expected size of the purified product should be a distinct band at 155 bp (Figure 19.6). We recommend measuring the concentration of each library using the Qubit Fluorometer (Life Technologies). Based on this quantification, the barcoded libraries can be mixed at equal or the desired ratio. The pooled libraries are then ready for Illumina sequencing.

Primer sequences (5' to 3')

Primer 3: AATGATACGGCGACCACCGAGATCTACACTCTTTCCCT ACACGACGCT

Index primer: CAAGCAGAAGACGGCATACGAGATXXXXXXGTGACTG GAGTTCAGACGTGT

Note: We recommend that you choose one or two libraries randomly for validation with clone-based Sanger sequencing. If TA cloning is desired, Taq DNA polymerase should be used for this step as Phusion DNA polymerase can only produce blunt-end products. For each library, 10 or so clones should be chosen for Sanger sequencing. It should be confirmed that the insert sequences contain correct barcodes and the restriction site, and are indeed derived from the genome of the species of interest.

Bioinformatics Analysis of RAD and 2b-RAD Data

Overview

RAD and 2b-RAD focus on sequencing the same subset of genomic regions across multiple individuals, allowing tens or hundreds of thousands of SNP markers that are spread evenly throughout the genome to be identified and genotyped. However, as with other types of NGS data, bioinformatics analysis of RAD and 2b-RAD data is not a straightforward process. A series of factors must be considered to achieve accurate genotyping, such as distinguishing SNPs from sequencing errors, inferring a heterozygous genotype in the face of various sequencing depths, and excluding false SNPs arising from paralogous or repetitive genomic regions. Several tools such as RADtools (Baxter *et al.*, 2011), RApiD (Willing *et al.*, 2011), Stacks (Catchen *et al.*, 2011, 2013), and RADtyping (Fu *et al.*, 2013) have recently been developed to analyze RAD or 2b-RAD data for linkage mapping and other purposes. The main features of these tools are summarized in Table 19.1. RADtools is the first pipeline that was developed to analyze RAD data *de novo* and has been demonstrated in linkage mapping of diamondback moth. RApiD can take advantage of paired-end data, allowing *de novo* assembly of the paired-end data to produce extended contigs flanking a restriction site for marker detection. Among the existing tools, Stacks has received the broadest application in linkage mapping and population genomic analyses, due to its robust read clustering approach, sophisticated genotyping algorithm, and user-friendly interface. However, Stacks and other tools were primarily developed for typing codominant markers; RADtyping features the incorporation of a novel statistical framework for *de novo* dominant genotyping in mapping populations. Next, we will focus on Stacks and RADtyping to introduce the analytical approaches behind these tools.

Reference-based and *De Novo* Analytical Approaches

There are two different approaches (i.e., reference-based and *de novo*) to analyze RAD sequencing data, depending on whether the species of interest has a reference genome. In the first approach, RAD reads are directly mapped to the reference genome, and SNPs

Table 19.1 Summary of available software for RAD and 2b-RAD data analysis.

Name	Strategy		Read Processing Mode		Genotyping Algorithm		Applicability	
	De Novo	Reference-Based	Read Clustering	Read Mapping	Codominant (AA/Aa/aa)	Dominant (presence/absence)	Preferred Method	Data Type
RADtools	Yes	No	RADtags	—	Threshold	—	RAD	Family/population
RApiD	Yes	No	Locas	Genome mapper	Threshold	—	RAD	Family/population
Stacks	Yes	Yes	Ustacks	Bowtie	ML	—	RAD	Family/population
RADtyping	Yes	Yes	Ustacks	Soap	iML	dom_calling	2b-RAD	Family/population

can be identified from the sequence alignment and then genotyped by choosing one of the SNP calling algorithms, such as the threshold or maximum likelihood method. Though the reference-based approach is very accurate and most desirable, it is difficult to apply to aquatic organisms, most of which still do not have a reference genome.

In the second approach, RAD data are analyzed *de novo*, beginning with assembling reads with respect to one another to establish reference sites (i.e., read-clustering process). For unique genomic regions, this assembly process works as well as aligning reads against the genome. However, short reads that are derived from paralogous or repetitive genomic regions are usually unavoidably clustered together (i.e., forming composite clusters) due to high sequence similarity. False SNPs could arise and be miscalled from such clusters. To address this problem, *ustacks*, the core component of Stacks, calculates the average depth of coverage, identifies clusters that are two standard deviations above the mean, and excludes these clusters along with all clusters that are one nucleotide apart from these extremely deep clusters. This strategy is effective in removing deep clusters, but identifying many low-depth composite clusters remains difficult. Instead, RADtyping adopts an alternative way to identify composite clusters. Theoretically, the read-depth distribution of composite clusters is expected to show a repeating pattern occurring at multiples of the average sequencing depth, which corresponds to the copy number variations of repetitive sites. RADtyping therefore fits the RAD data with a mixed Poisson/normal distribution model to identify and exclude composite clusters from genotyping. Once the reliable clusters are built, Stacks creates a catalog of reference loci, and then matches RAD data against the catalog, which defines alleles at each locus in each individual. Unlike Stacks, RADtyping combines all reads from selected individuals (e.g., mapping parents) and assembles them into exactly matching read clusters that represent individual alleles. Then, these "allele" clusters are further merged into "locus" clusters by allowing certain mismatches. A collection of consensus sequences from all "locus" clusters comprises the reference sites, which are further classified into two categories for codominant and dominant genotyping, respectively.

Codominant and Dominant Genotyping

Codominant Genotyping

The sequencing errors and read-sampling variation across RAD sites that are inherent in NGS platforms become significant sources of inferential confusion when millions of reads are considered simultaneously. An excellent approach to identify and genotype true SNPs from noisy RAD data is the maximum-likelihood (ML) statistical framework, proposed by Hohenlohe *et al.* (2010). More precisely, for a given site in an individual, let n be the total number of reads at that site, where $n = n_1 + n_2 + n_3 + n_4$, and n_i is the read count for each possible nucleotide at the site. For a diploid individual, there are ten possible genotypes (AA, AC, AT, AG, CC, CT, CG, TT, TG, and GG). A multinomial sampling distribution gives the probability of observing a set of read counts (n_1, n_2, n_3, n_4) for a given particular genotype, which translates into the likelihood for that genotype. For example, the likelihoods of a homozygote or a heterozygote can be calculated as follows:

$$L1 = \Pr(n_1, n_2, n_3, n_4 | homozygote) = \frac{n!}{n_1!n_2!n_3!n_4!}\left(1 - \frac{3\varepsilon}{4}\right)^{n_1}\left(\frac{3\varepsilon}{4}\right)^{n_2+n_3+n_4}$$

$$L2 = \Pr(n_1, n_2, n_3, n_4|heterozygote) = \frac{n!}{n_1!n_2!n_3!n_4!}\left(0.5 - \frac{\varepsilon}{4}\right)^{n_1+n_2}\left(\frac{3\varepsilon}{4}\right)^{n_3+n_4} \tag{19.1}$$

In this expression, ε is the sequencing error rate. At a significance level of $\alpha = 0.05$, the most likely genotype is assigned to that locus; otherwise, the genotype is uncalled. In RADtyping, we further developed an improved ML algorithm (iML) that incorporates the mixed Poisson/normal model to exclude repetitive loci from genotyping efficiently (Dou *et al.*, 2012). Unlike ML, iML calculates the posterior probabilities for each of the three possible categories (homozygote, heterozygote, and undetermined) as follows:

$$\Pr(n_1, n_2, n_3, n_4|homozygote) = \frac{n!}{n_1!n_2!n_3!n_4!}poisson(n|C)\left(1 - \frac{3\varepsilon}{4}\right)^{n_1}\left(\frac{3\varepsilon}{4}\right)^{n_2+n_3+n_4}$$

$$\Pr(n_1, n_2, n_3, n_4|heterozygote) = \frac{n!}{n_1!n_2!n_3!n_4!}poisson(n|C)\left(0.5 - \frac{\varepsilon}{4}\right)^{n_1+n_2}\left(\frac{3\varepsilon}{4}\right)^{n_3+n_4}$$

$$\Pr(n_1, n_2, n_3, n_4|undetermined) = \sum_{k\geq 2}\frac{a_k}{1 - a_1}poisson(n|kC) \tag{19.2}$$

In this expression, $\sum_{1\leq i\leq M} a_i = 1$ and M indicate the copy number of repetitive sites; all parameters, including C and $a_1 \sim a_M$, can be directly estimated from the sequencing dataset using the expectation–maximization algorithm. The iML algorithm can efficiently prevent incorrect SNP calls that result from repetitive genomic regions, thus outperforming the ML algorithm by achieving much higher genotyping accuracy. For example, iML generated lower false positive rates than ML with an approximately 17% reduction for an *Arabidopsis thaliana* 2b-RAD dataset, and a 4–7% reduction for a stickleback RAD dataset (Dou *et al.*, 2012).

Dominant Genotyping

Unlike codominant markers, dominant markers are scored as present or absent to reflect whether a recognition site is intact or disrupted. It has been shown that dominant markers can provide a large amount of additional genotypic information (e.g., in the three-spined stickleback, dominant markers account for ~ 40% of total markers; Baird *et al.*, 2008). The obvious benefit of using dominant markers is a significant increase of marker numbers at no additional sequencing cost. However, most RAD tools are being developed predominantly for scoring codominant markers. RADtyping is the first tool that implements a novel statistical framework for *de novo* dominant genotyping in mapping populations. The principle of dominant genotyping in RADtyping is outlined in Figure 19.7 and briefly described in the following text.

Parent-specific reference sites that are identified from the read-clustering step are subjected to dominant genotyping. First, reference sites that are not supported by parental reads in sufficient depth are filtered out by the requirements $d_{p1} > l_{p1}$ and $d_{p2} > l_{p2}$, where the thresholds l_{p1} and l_{p2} are determined by:

$$l_{pj} = \max\left\{d|\sum_{0\leq m\leq d} poisson(m|C_j) \leq 0.05\right\} \tag{19.3}$$

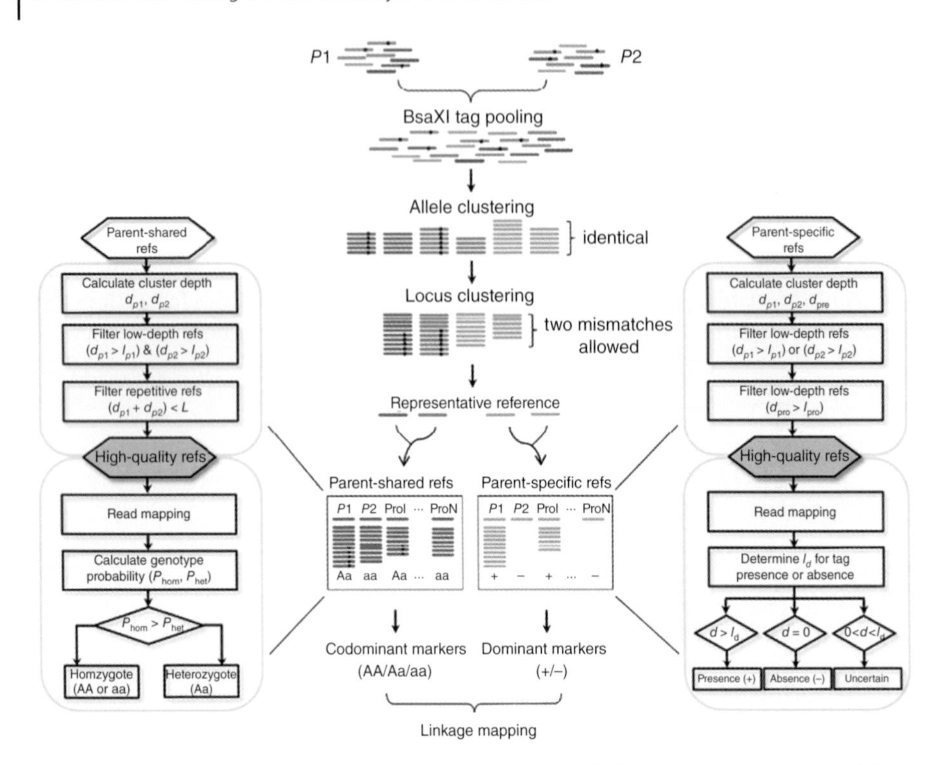

Figure 19.7 An overview of the RADtyping analytical approach for *de novo* codominant and dominant genotyping in a mapping population (adopted from Fu *et al.*, 2013). The main principles of codominant and dominant genotyping are shown in flowcharts. (*See color plate section for the color representation of this figure.*)

In this expression, C_j is the mean sequencing depth of the j -th parent ($j = 1, 2$). In addition, to avoid incorrect "absence" calls from low-coverage sites in progeny, the reference sites are further filtered to remove those with d_{pro} less than l_{pro}, where d_{pro} is calculated for each site by summarizing all progeny having reads derived from that site, and the threshold l_{pro} is determined for each site using equation (3), with C being the mean depth of the progeny. Sequencing reads of all progeny are then mapped to the high-quality reference sites obtained from the read-clustering step, and the absence or presence of each site is determined for each progeny. Supposing that the cluster depth of the i -th site for the j -th progeny is d_{ij}, this site is genotyped as "presence" if $d_{ij} > l_d$, "absence" if $d_{ij} = 0$, and "unknown" if $d_{ij} \in (0,\ l_d)$, where the threshold l_d is determined using equation (2), with C representing the mean sequencing depth of the i -th site.

In fact, when the average sequencing depth is low, dominant loci can be more reliably genotyped than codominant loci, as shown in a recent study (Fu *et al.*, 2013), where the genotyping consistency between 2b-RAD replicate datasets was 99% for dominant markers and 96% for codominant markers.

Usage Demonstration

To get started with these tools, we recommend readers to explore the websites of these tools for acquiring more detailed information about their usage—for example,

Stacks (http://catchenlab.life.illinois.edu/stacks/) and RADtyping (http://www2.ouc.edu.cn/mollusk/detailen.asp?Id=727). Here, we just provide a step-by-step example to demonstrate how to run a linkage mapping analysis using the RADtyping pipeline.

Example for Running a Linkage Mapping Analysis

If all your input files are stored in a directory called "rawdata" that contains Illumina sequencing data for two parents and 20 progeny:

```
mgb@M:~/djz/rawdata$ ls
Progeny_001.fastq   Progeny_007.fastq  Progeny_013.fastq
  Progeny_019.fastq
Progeny_002.fastq   Progeny_008.fastq  Progeny_014.fastq
  Progeny_020.fastq
Progeny_003.fastq   Progeny_009.fastq  Progeny_015.fastq
  Parent_001.fastq
Progeny_004.fastq   Progeny_010.fastq  Progeny_016.fastq
  Parent_002.fastq
Progeny_005.fastq   Progeny_011.fastq  Progeny_017.fastq
Progeny_006.fastq   Progeny_012.fastq  Progeny_018.fastq
```

The high-quality (HQ) reads can be obtained by running:

```
mgb@M:~/djz/$ perl reads_proc.pl -i rawdata -p1
   rawdata/Parent_001.fastq -p2
rawdata/Parent_002.fastq -b [ATGC]{9}AC[ATGC]{5}CTCC
  [ATGC]{7} -l 27
```

The HQ reads obtained are stored in a directory named "proc_data".

```
mgb@M:~/djz/proc_data$ ls
Progeny_001.fasta   Progeny_007.fasta  Progeny_013.fasta
  Progeny_019.fasta
Progeny_002.fasta   Progeny_008.fasta  Progeny_014.fasta
  Progeny_020.fasta
Progeny_003.fasta   Progeny_009.fasta  Progeny_015.fasta
  P1.fasta
Progeny_004.fasta   Progeny_010.fasta  Progeny_016.fasta
  P2.fasta
Progeny_005.fasta   Progeny_011.fasta  Progeny_017.fasta
Progeny_006.fasta   Progeny_012.fasta  Progeny_018.fasta
```

If your data have paired ends, it is necessary to put the data together in the directory "proc_data".

```
mgb@M:~/djz/proc_data$ ls
Progeny_001_p.fasta  Progeny_007_p.fasta Progeny_013_p.fasta
Progeny_019_p.fasta  Progeny_002_p.fasta Progeny_008_p.fasta
Progeny_014_p.fasta  Progeny_020_p.fasta Progeny_003_p.fasta
Progeny_009_p.fasta  Progeny_015_p.fasta Progeny_004_p.fasta
```

```
Progeny_010_p.fasta  Progeny_016_p.fasta Progeny_005_p.fasta
Progeny_011_p.fasta  Progeny_017_p.fasta Progeny_006_p.fasta
Progeny_012_p.fasta  Progeny_018_p.fasta  P1_p.fasta
  P2_p.fasta
Progeny_001.fasta   Progeny_007.fasta  Progeny_013.fasta
  Progeny_019.fasta
Progeny_002.fasta   Progeny_008.fasta  Progeny_014.fasta
  Progeny_020.fasta
Progeny_003.fasta   Progeny_009.fasta  Progeny_015.fasta
  P1.fasta
Progeny_004.fasta   Progeny_010.fasta  Progeny_016.fasta
  P2.fasta
Progeny_005.fasta   Progeny_011.fasta  Progeny_017.fasta
Progeny_006.fasta   Progeny_012.fasta  Progeny_018.fasta
```

Next, you can finish the genotyping procedure by either:

i) Executing the integrated pipeline "RADtyping.pl" (default parameters):

```
mgb@M:~/djz/$ perl RADtyping.pl -p1 proc_data/P1.fasta
   -p2 proc_data/P2.fasta -l 27
```

This will produce 10 output files that are created in two directories, "ref" and "genotype":

```
mgb@M:~/djz/ref$ ls
ref_codom  ref_dom  HQ_ref_codom  HQ_ref_dom
mgb@M:~/djz/genotype$ ls
all_codom  all_dom  codom_JM  dom_JM  poly_dodom
    poly_dom
```

or

ii) Going through a few executions step by step:
Step 1: Run "ref_build.pl" to reconstruct the representative reference sites.

```
mgb@M:~/djz/$ perl ref_build.pl -p1 proc_data/P1.fasta
   -p2 proc_data/P2.fasta
```

The reference sites are stored in two output files, "ref/ref_codom" and "ref/ref_dom".

Step 2: Run "ref_filter.pl" to obtain HQ reference sites.

```
mgb@M:~/djz/$ perl ref_filter.pl -m P
```

The HQ sites are stored in two output files, "ref/HQ_ref_codom" and "ref/HQ_ref_dom".

Step 3: Run "reads_map.pl" to map the HQ reads to the HQ reference sites.

```
mgb@M:~/djz/$ perl reads_map.pl -l 27
```

All mapping results are stored in a directory named "reads_mapping":

```
mgb@M:~/djz/reads_mapping$ ls
Progeny_001     Progeny_007    Progeny_013    Progeny_019
```

```
Progeny_002    Progeny_008    Progeny_014    Progeny_020
Progeny_003    Progeny_009    Progeny_015    P1
Progeny_004    Progeny_010    Progeny_016    P2
Progeny_005    Progeny_011    Progeny_017
Progeny_006    Progeny_012    Progeny_018
```

Step 4: Run "codom_calling.pl" for performing codominant genotyping.

```
mgb@M:~/djz/$ perl codom_calling.pl -a 0.05 - p 0.8
```

The codominant genotypes are stored in two output files, "genotype/all_codom" and "genotype/poly_codom".

Step 5: Run "dom_calling.pl" for performing dominant genotyping.

```
mgb@M:~/djz/$ perl dom_calling.pl - p 0.8
```

The dominant genotypes are stored in two output files, "genotype/all_dom" and "genotype/poly_dom".

Step 6: Run "joinmap_trans.pl" to transform the genotyping results in a "joinmap"-ready format.

```
mgb@M:~/djz/$ perl joinmap_trans.pl
```

The two "joinmap" format files are "genotype/codom_JM" and "genotype/dom_JM".

The Benefits and Pitfalls of RAD and 2b-RAD Applications

As with other genotyping-by-sequencing methods, RAD and 2b-RAD can achieve genome-wide genotyping cost-effectively by only sequencing a fraction of the whole genome, making it possible to genotype a common set of thousands to hundreds of thousands of markers in tens to hundreds of individuals. However, investigators should be aware of the advantages and disadvantages of each method in order to choose the right method for fulfilling the research aims in specific applications.

RAD is the most widely applied genotyping-by-sequencing method to date. There are also sophisticated tools such as Stacks for RAD data analysis. However, RAD has a complex library preparation procedure that is labor-intensive and time-consuming, making it difficult to parallelize for high-throughput sample processing. The complex nature of RAD library preparation could also lead to several sources of sequencing bias (reviewed by Davey *et al.*, 2013). One of the most significant biases is the restriction fragment bias—that is, longer fragments tend to be over-represented compared to shorter fragments because they can be sheared more easily during sonication. This problem renders the evenness of genome representation, and poses a challenge to genotyping tools for calling genotypes in shorter fragments.

2b-RAD features even and tunable genome coverage by taking advantage of the special features of type IIB restriction enzymes. It has a very simple library preparation procedure, and is therefore suitable for the parallel genotyping of large numbers of samples. It can fine-tune genome coverage, providing great flexibility for meeting different research needs. The major drawback of 2b-RAD is that tag lengths are constrained by the activity of type IIB enzymes; therefore, the increasing read lengths of NGS platforms cannot be exploited.

Nevertheless, both RAD and 2b-RAD interrogate a very limited range of sequences around the restriction site, making it difficult to reveal key genes adjacent to the markers of interest (e.g., QTL markers). The availability of a full genome is obviously key in resolving this problem. However, even when NGS platforms are used, the cost of whole-genome *de novo* sequencing remains prohibitively high, and is out of reach for most laboratories focusing on aquatic organisms. Indeed, only a few genome-sequencing plans have been initiated for aquatic organisms to date. Recently, we have made an attempt to improve the utility of short RAD or 2b-RAD tags by proposing that such tags can be effectively extended using the scaffolds obtained from genome survey sequencing. For example, through ~ 52X genome survey sequencing, we were able to generate a preliminary reference genome for the scallop *Chlamys farreri*, for which 57% of 2b-RAD markers on a linkage map can be anchored to scaffolds (Jiao *et al.*, 2014). Although a majority of scaffolds are not very long (scaffold N50 = 1.5 KB), they are already useful for identifying adjacent genes around 2b-RAD tags. Based on this preliminary reference genome, we were able to pin down a growth-hormone regulatory gene *PROP1* within a growth-related QTL region, suggesting that our proposed approach can effectively improve the utility of RAD or 2b-RAD tags.

References

Baird, N.A., Etter, P.D., Atwood, T.S., *et al.* (2008) Rapid SNP discovery and genetic mapping using sequenced RAD markers. *PLoS One*, **3**, e3376.

Baxter, S.W., Davey, J.W., Johnston, J.S., *et al.* (2011) Linkage mapping and comparative genomics using next-generation RAD sequencing of a non-model organism. *PLoS One*, **6**, e19315.

Catchen, J.M., Amores, A., Hohenlohe, P., *et al.* (2011) Stacks: building and genotyping loci *de novo* from short-read sequences. *G3 (Bethesda)*, **1**, 171–182.

Catchen, J., Hohenlohe, P.A., Bassham, S. and Amores, A. (2013) Stacks: an analysis tool set for population genomics. *Molecular Ecology*, **22**, 3124–3140.

Davey, J.W., Cezard, T., Fuentes-Utrilla, P., *et al.* (2013) Special features of RAD sequencing data: implications for genotyping. *Molecular Ecology*, **22**, 3151–3164.

Davey, J.W., Hohenlohe, P.A., Etter, P.D., *et al.* (2011) Genome-wide genetic marker discovery and genotyping using next-generation sequencing. *Nature Reviews Genetics*, **12**, 499–510.

Dou, J., Zhao, X., Fu, X., *et al.* (2012) Reference-free SNP calling: Improved accuracy by preventing incorrect calls from repetitive genomic regions. *Biology Direct*, **7**, 17.

Emerson, K.J., Merz, C.R., Catchen, J.M., *et al.* (2010) Resolving postglacial phylogeography using high-throughput sequencing. *Proceedings of the National Academy of Sciences USA*, **107**, 16196–16200.

Etter, P.D., Bassham, S., Hohenlohe, P.A., *et al.* (2011) SNP discovery and genotyping for evolutionary genetics using RAD sequencing. *Methods in Molecular Biology*, **772**, 157–178.

Fu, X., Dou, J., Mao, J., *et al.* (2013) RADtyping: an integrated package for accurate *de novo* codominant and dominant RAD genotyping in mapping populations. *PLoS One*, **8**, e79960.

Jiao, W., Fu, X., Dou, J., *et al.* (2014) High-resolution linkage and quantitative trait locus mapping aided by genome survey sequencing: building up an integrative genomic framework for a bivalve mollusc. *DNA Research*, **21**, 85–101.

Heliconius Genome Consortium. (2012) Butterfly genome reveals promiscuous exchange of mimicry adaptations among species. *Nature*, **487**, 94–98.

Hohenlohe, P.A., Bassham, S., Etter, P.D., *et al.* (2010) Population genomics of parallel adaptation in three spine stickleback using sequenced RAD tags. *PLoS Genetics*, **6**, e1000862.

Liu, Z. (2007) *Aquaculture genome technologies*, Blackwell Publishing Ltd., Oxford, UK.

Peterson, B.K., Weber, J.N., Kay, E.H., *et al.* (2012) Double digest RADseq: an inexpensive method for *de novo* SNP discovery and genotyping in model and non-model species. *PLoS One*, 7, e37135.

Toonen, R.J., Puritz, J.B., Forsman, Z.H., *et al.* (2013) ezRAD: a simplified method for genomic genotyping in non-model organisms. *Peer J*, **1**, e203.

Wang, S., Meyer, E., McKay, J.K. and Matz, M.V. (2012) 2b-RAD: a simple and flexible method for genome-wide genotyping. *Nature Methods*, **9**, 808–810.

Willing, E.M., Hoffmann, M., Klein, J.D., *et al.* (2011) Paired-end RAD-seq for *de novo* assembly and marker design without available reference. *Bioinformatics*, **27**, 2187–2193.

20

Bioinformatics Considerations and Approaches for High-Density Linkage Mapping in Aquaculture

Yun Li, Shikai Liu, Ruijia Wang, Zhenkui Qin and Zhanjiang Liu

Introduction

Construction of genetic linkage maps is essential for genetic and genomic studies. As such, linkage maps have been constructed for a number of aquaculture species. Recent advances in sequencing and genotyping technologies have made it possible to generate high-density and high-resolution genetic linkage maps. However, the use of large numbers of molecular markers have also brought greater challenges for linkage analysis. In this chapter, we provide a practical introduction to genetic linkage mapping, and describe map construction procedures using large numbers of markers. Our experience and lessons learned should be helpful for those who are facing similar challenges.

Genetics as a branch of science has a relative short history of less than 160 years. It started in the mid-1800s, when Gregor Johann Mendel conducted a series of experiments with the inheritance of a number of traits in peas. Through observations of phenotypes and analysis of the distribution of phenotypes among individuals in different generations, Mendel established his groundbreaking principles of inheritance, the so-called *Mendel's laws*, which included the law of segregation (the "First Law"), the law of independent assortment (the "Second Law"), and the law of dominance (the "Third Law"). The law of segregation states that, during gamete formation, the alleles for each gene segregate from each other, so that each gamete carries only one allele for each gene; the law of independent assortment states that genes for different traits segregate independently during the formation of gametes; and the law of dominance states that some alleles are dominant, while others are recessive, and that recessive alleles are masked by dominant alleles (Bateson & Mendel, 1909; Miko, 2008). These principles of inheritance set the foundation for genetics. However, considering what is known today as *genetic linkage*, it can be inferred that Mendel was only partially correct, since genes next to each other, because of their linkage, most often do not segregate independently.

In 1905, William Bateson, Edith Rebecca Saunders, and Reginald Punnett examined flower color and pollen shape in sweet pea plants by performing crosses similar to those carried out by Gregor Mendel, but they observed severe deviations from the predicted Mendelian independent assortment ratios (Bateson, Waunders & Punnett, 1909). They hypothesized that there was coupling, or connection, between the parental alleles for flower color and pollen shape, and this coupling resulted in the observed deviation from

Bioinformatics in Aquaculture: Principles and Methods, First Edition. Edited by Zhanjiang (John) Liu.

independent assortment. By 1910, geneticist Thomas Hunt Morgan made similar discoveries with the fruit fly (*Drosophila melanogaster*), in which he found that the trait of eye color was correlated or "connected" with the sex factor. By analysis of the phenotypic ratios of the traits (Morgan, 1910), he hypothesized that the two traits were "linked". Thus, the term *genetic linkage* was born—that is, when two genes are closely associated on the same chromosome, they do not assort independently as Mendel initially proposed in his law of independent assortment (Morgan, 1911). With this knowledge in place, Morgan's student, Alfred Sturtevant, developed the first linkage map.

Genetic linkage maps are developed by placing "genes" on chromosomes based on the distances among them. If two genes segregate absolutely independently, they are considered to be located on two separate chromosomes—that is, the original Mendel's law of independent assortment. However, if the two genes are on the same chromosome, they do not assort independently; rather, they are linked and, therefore, assort depending on the distances between them. The greater the distance between linked genes is, the greater the chance that non-sister chromatids would cross over in the region between the genes. Therefore, the recombination rate between genes is used as a measurement of genetic distance. This original concept for the measurement of genetic distance between genes is the basis for the construction of genetic linkage maps.

The entire history of genetic linkage analysis is a mirror of technologies for the detection of polymorphisms. In the earliest days, polymorphisms were detected at the level of phenotypes or traits. It was only possible to detect linkage among simple traits at the phenotypic level, due to the lack of molecular markers. The discovery of DNA in 1953 (Watson & Crick, 1953) allowed the start of genetic analysis at the DNA level. However, for over 20 years after DNA was discovered, the progress of linkage analysis was slow, because it was difficult to analyze genomic DNA with very high molecular weights. It was the discovery of restriction endonucleases in 1973 that made molecular analysis possible. With restriction endonucleases, high-molecular-weight DNA can be digested into smaller pieces that can be readily examined for their polymorphisms. Even faster advances were made possible by the invention of DNA sequencing technologies in 1977 (Maxam & Gilbert, 1977; Sanger, Nicklen & Coulson, 1977), which allowed greater abilities for the detection of polymorphism at the single nucleotide level. Thereafter, for the following decade, what was then considered rapid progress was made in the area of linkage mapping through the use of DNA markers, mostly in the form of restriction fragment length polymorphisms (RFLPs), the most popular marker system by the mid-1980s. Then, in 1986, the polymerase chain reaction (PCR) technology was invented. The coupling of DNA sequencing technologies with PCR technology allowed rapid progress to be made in DNA marker technologies. In particular, the application of microsatellites allowed much progress in linkage mapping. For almost another decade and half, linkage mapping was among the hottest research areas in genetics—using various polymorphic DNA markers such as microsatellites, RFLP, rapid amplification of polymorphic DNA (RAPD), and amplified fragment length polymorphism (AFLP), among many types of polymorphic markers. However, all these technologies were relatively slow, laborious, and costly when compared to the current marker technologies.

Further progress in linkage mapping was limited by the lack of large numbers of polymorphic markers and automated technologies. Although single-nucleotide polymorphism (SNP) was discovered as early as the 1970s, effective technologies were not available for efficient and cost-effective analysis of SNPs until the beginning of this century. Similarly, methodologies for the identification of large numbers of SNPs were not available. It was only with the most recent advances in next-generation sequencing (Ren *et al.*, 2000) that it became possible to identify SNPs at a genome scale in a cost-effective manner (Davey *et al.*, 2011). Along with the availability of highly efficient genotype array platforms, the development of high-density and high-resolution genetic linkage maps became possible.

While it is relatively easy to manage a linkage mapping project with just a few hundred markers, it is much more difficult to manage a linkage mapping project with hundreds of thousands or millions of markers. The use of large numbers of markers makes bioinformatics analysis much more complicated because of the computational challenges, but, more importantly, large numbers of markers make the differentiation of true linkage from pseudo-linkage much more difficult. In this chapter, we will provide a description of the considerations and detailed methodologies and bioinformatics analysis procedures for construction of high-density and high-resolution genetic linkage maps. We assume that readers are familiar with the concept and procedures for the development and application of various molecular markers. If not, readers are referred to discussions in Chapter 16 for microsatellite markers and Chapter 17 for SNP markers.

Basic Concepts

A genome is composed of a fixed number of chromosomes. On each chromosome, a certain number of genes are arranged in a linear fashion. Genetic linkage mapping aims to establish a genetic map that reflects the linear ordering and relative positions of genes, most often in the form of genetic markers on the chromosomes. In this section, we will provide the basic principles for genetic linkage analysis.

Principles of Genetic Mapping

Genetic mapping is the procedure for the construction of a genetic map (Van Ooijen & Jansen, 2013). A genetic map is made with markers or genes in a linear arrangement on chromosomes with relative distances between them. In contrast to physical distances where the distances between markers are measured by the physical length between genes or markers in base pairs, genetic distances are measured by frequencies of recombination between markers during crossovers of homologous chromosomes. The basic principles of linkage mapping lie in the fact that the farther apart the two markers on the same chromosome, the more likely that recombination can happen between the two markers, and conversely, the closer the two markers, the less likely that recombination can happen between them.

As stated in Mendel's law of independent assortment, chromosomes assort randomly into gametes, and thus the segregation of alleles of one gene is independent of alleles of another gene if they reside on different chromosomes. However, independent assortment does not apply to genes residing on the same chromosome, simply because they are physically linked together. Unless recombination occurs, alleles carried on the same chromatids go together as a "unit," and they are said to be linked (Semagn, Bjørnstad & Ndjiondjop, 2006).

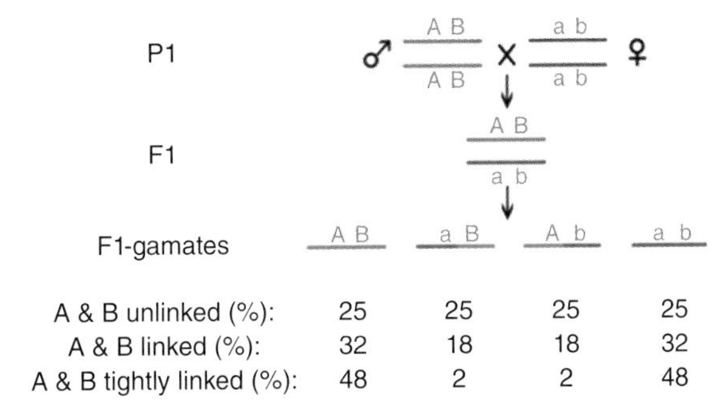

Figure 20.1 An example of the percentage of gametes generated by independent segregation and linkage. (*See color plate section for the color representation of this figure.*)

For two genes located at different chromosomes, they are unlinked. The chance that A/B or a/b co-inherit to the offspring is 0.5. Increases of this chance indicate linkage (Figure 20.1). Note that aB and Ab did not appear in the parental cells; they were called recombinants due to recombination. Recombination is the process by which, during meiosis, the chromosome often breaks and rejoins with the homologous chromosome (indicated as crossover), resulting in the production of the new alleles, the recombinant alleles aB and Ab (Whitehouse, 1982). The greater the frequency of recombination between two genetic markers or genes, the further apart they are on the same chromosome, and when the recombinant frequency increases to 0.5, the two genes are independently assorted, reflecting that they are indefinitely distant—they are on different chromosomes.

The *recombination frequency* (also called *recombination fraction*) between two loci is defined as the ratio of the number of recombinants to the total number of gametes produced (Xu, 2013). Therefore, the minimum recombination fraction is 0, indicating no recombinants (i.e., the two genes are absolutely linked together). The maximum recombination fraction is 0.5, indicating that the genes are on different chromosomes.

Linkage Phase

Two types of gametes are generated following genes or alleles on the same chromosome:

a) If crossing over does not occur, the products are parental gametes.
b) If crossing over occurs, the products are recombinant gametes.

The two possible arrangements of alleles on homologous chromosomes are referred to as *linkage phase*. The allelic composition of parental and recombinant gametes depends upon whether the original cross involved genes in coupling or repulsion phase (Semagn *et al.*, 2006). In diploid species, the most prevalent gametes in a coupling phase are two dominant alleles or two recessive alleles linked on one chromosome—for example, for genes A and B, AB and ab, respectively. For repulsion phase, gametes containing one dominant allele linked with one recessive allele will be the most abundant type (Figure 20.2).

The distribution of gamete types should allow the differentiation of parental types from the recombinant types. In general, the two most frequent types of gametes are parental types, while the two least frequent types of gametes are recombinants. This is because recombinant gametes can never account for more than 50% of the gametes. In addition, by examination of the frequency of gametes, one can determine if the original

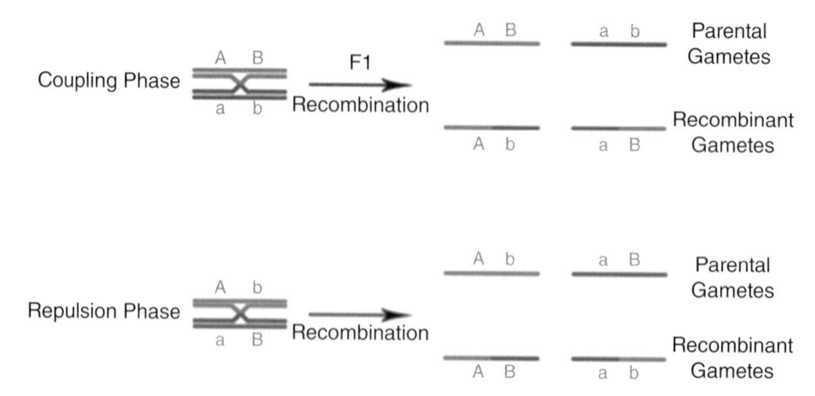

Figure 20.2 An example of the gamete composition for linked markers or genes for coupling and repulsion crosses. (*See color plate section for the color representation of this figure.*)

cross was a coupling or repulsion phase cross, which is important for estimating linkage distances.

The LOD Score

As described in the preceding text, linkage maps are made based on the genetic distances between markers, and genetic distances are estimated based on the chances for recombination. When chances are in question, statistical significance must be in place to ensure the confidence one would have on the map. In this context, one key concept in genetic linkage analysis is the LOD score [logarithm (base 10) of odds], which is the most commonly used statistic to test whether or not two loci are truly linked. This method was developed by Newton E. Morton (Morton, 1955), and is an iterative approach where a series of LOD scores are calculated from a number of proposed linkage distances. The LOD score compares the likelihood of obtaining the test data if the two loci are indeed linked, to the likelihood of observing the same data purely by chance. The LOD score is calculated from the following formula:

$$\mathrm{LOD} = \log_{10}\frac{\textit{likelihood if the loci are linked}}{\textit{likelihood if unlinked}} = \log 10\frac{(1-\theta)^{NR}X\theta^{R}}{0.5^{(NR+R)}}$$

Here, R denotes the number of recombinant offspring, NR denotes the number of non-recombinant offspring, and θ denotes the recombinant fraction. In linkage analysis, an LOD of more than 3 is considered as strong support for the linkage, as it indicates that the linkage is 1000 times more probable than no linkage. On the contrary, an LOD score of less than −2 is required for linkage exclusion. However, this criterion should be modified depending on the type of species, complexity of traits, and number of genetic markers.

Mapping Function

In linkage mapping, the distance between two genes is determined by their recombination faction. Map construction based on distance and order of genetic markers. However, it is difficult to use only recombination fraction as a linkage metrics, because recombination factions themselves are not additive. Consider the loci A, B, and C. If the recombination fraction between A and B is 0.2, and between B and C is 0.4, the recombination fraction between A and C cannot be 0.6, as the maximum would be 0.5. Therefore, the recombination fraction between A and C is not equal to the sum of the recombination fractions AB and BC. Furthermore, the distance between AC depends on the existence of interference, which is the effect in which the occurrence of a crossover in a certain region reduces the probability of a crossover in the adjacent region (Kinghorn & van der Werf, 2000). In that case, mathematical transformations were called mapping functions, with the most well-known ones being Haldane (Haldane, 1919) and Kosambi (Kosambi, 1943). The Haldane function assumes that there is no interference (all crossovers occur independently of one another):

$$D = -1/2 \ln(1 - 2\theta)$$

Whereas the Kosambi map function allows the occurrence of interference:

$$D = 1/4 \ln[(1 + 2\theta)(1 - 2\theta)]$$

Here, D represents the genetic map distance between two marker loci, and θ represents the recombination fraction. Mapping functions translate recombination fractions between two loci into a map distance in cM (centiMorgan). In theory, 1% of recombinants equal approximately 1 cM. Both mapping functions act over the range $0 < \theta < 0.5$. From a practical point of view, the Kosambi function has been more widely used because of its advantage over Haldane function for interference. Moreover, studies have reported that the measured distances are more accurate in Kosambi function than in Haldane function in a multi-point analysis of data (Huehn, 2010; Vinod, 2011).

Requirements for Genetic Mapping

Genetic mapping requires four elements: (1) polymorphic markers; (2) genotyping platforms; (3) reference families; and (4) mapping software.

Polymorphic Markers

For linkage map construction, one should first decide the type of molecular markers. SNPs are now the markers of choice for large-scale and high-throughput genomic analysis because of their abundance, genomic distribution, and easy adaptation to automation. Prior to the emergence of next-generation sequencing technologies, the identification of SNPs has been fuelled by mining large numbers of expressed sequence tags (ESTs) available in many species. However, it was severely limited by sequence coverage and depth (Liu *et al.*, 2011). Now, the advent of next-generation sequencing technologies has revolutionized genomic and transcriptomic approaches to SNP discovery. For the sake of this chapter, we will not provide details on SNP discovery using next-generation sequencing. Interested readers are referred to Chapter 17.

Theoretically, SNPs may be bi-, three-, and four-allelic; however, in practice, even three-allelic SNPs are very rare (less than 0.1% of all human SNPs; Lai, 2001). The overwhelming majority of SNPs are bi-allelic. There are four types of bi-allelic SNPs, including one transition C/T (G/A), and three types of transversions: C/A (G/T), C/G (G/C), and T/A (A/T). The most abundant are transitions originating from the deamination of 5'-methylcytosine to thymine (Holliday & Grigg, 1993). The bi-allelic nature of SNPs leads to a lower polymorphism information content (PIC) value as compared to most other marker types, which are often multi-allelic (Kruglyak, 1997), but this shortcoming is greatly compensated for by their high frequency.

Genotyping Platforms

With the availability of large numbers of SNPs, high-density SNP array platforms can be developed for high-throughput and efficient genotyping. Alternatively, next-generation-sequencing-based genotyping by sequencing (GBS) (Baird *et al.*, 2008; Davey *et al.*, 2011; Miller *et al.*, 2007) is also highly efficient. SNP arrays and GBS have both been used to genotype large-scale SNPs for genetic mapping and association analyses. Interested readers are referred to Chapters 18 and 19.

Reference Families

The design of reference families is the foundation of a genetic mapping program because reference families affect the quality and cost of the linkage maps (Da *et al.*, 1998). F1 families generated by crossing genetically diverse individuals, F2 populations derived from F1 hybrids, and backcross populations derived by crossing the F1 hybrid to one of the parents are the most commonly used types of mapping population for aquaculture species (Yue, 2014). Although some studies have attempted to determine the ideal population size and type for constructing accurate maps (Ferreira, Silva & Cruz, 2006; van der Beek, Groen & van Arendonk, 1993), there are no standard rules for creating reference families. Based on previous experience, some universal comments that apply to any type of mapping population can be made as follows:

- The amount of effective information available for mapping is based on the number of informative meiosis. In general, informative meiosis means we can identify whether or not the gamete is recombinant. In Figure 20.3, we assume that the father has a dominant condition, and that the child inherited allele A from him. Both (C) and (D) show informative meiosis conditions. In practice, full-sib families are more efficient than half-sib families, as full-sib families can yield more informative offspring. In that case, the requirement for the number of genotyping attempts is lower for the same number of informative meiosis.
- It is better to use more mapping families because one pair of parents may be homozygous at the locus, which will not produce informative meiosis.
- The greater the number of individuals, the higher the map resolution. In practice, although any number of individuals can be used, a large population size is needed for high-resolution mapping (van der Beek *et al.*, 1993).

Linkage Mapping Software

There are a number of software packages available for linkage mapping. The detailed list of software for linkage mapping, QTL analysis, and related studies can be found on the website http://www.jurgott.org/linkage/ListSoftware.html. The type of software used is mainly dependent on the experimental design, including the population structure and the number of genetic markers. In Table 20.1, we list the experimental population types supported by some popular software currently in use.

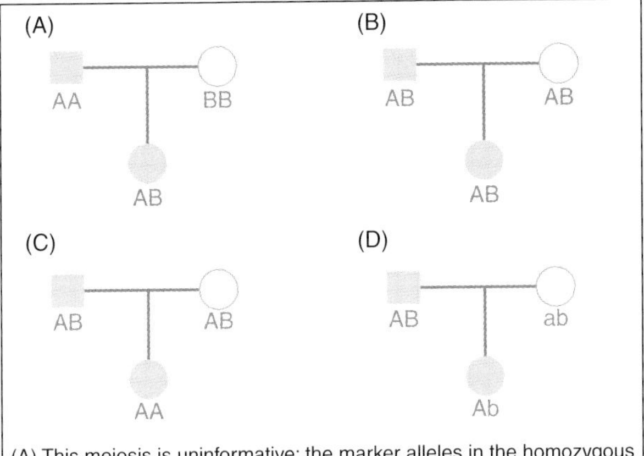

(A) This meiosis is uninformative: the marker alleles in the homozygous father cannot be distinguished.
(B) This meiosis is uninformative: the child could have inherited A from father and B from mother, or vice versa.
(C) This meiosis is informative: the child inherited A from the father.
(D) This meiosis is informative: the child inherited A from the father.

Figure 20.3 An example of informative and non-informative meiosis. (*See color plate section for the color representation of this figure.*)

Table 20.1 The experimental population types supported by popular packages for the construction of linkage maps in aquaculture species. Abbreviations: DH, double haploid; HAP, haploid population; BC, backcross; RIL, recombinant inbred line; and F2, F2 intercross.

Software	Population Type	License	Reference
JoinMap	DH, HAP, BC, RIL, F2, outcross	Commercial	Stam, 1993
OneMap	RIL, BC, F2, outcross	Free	Margarido, Souza & Garcia, 2007a
MSTMap	BC, DH, HAP, RIL	Free	Wu, Close & Lonardi, 2008
LepMap	F2, F1, diploid, gynogenetic diploid family, gynogenetic haploid family	Free	Rastas *et al.*, 2013
CRIMAP	Pedigree	Free	Green, Falls & Crooks, 1990
MapMaker	F2, BC, RIL, DH	Free	Lincoln, 1992
Map Manager QTX	F2, BC, RIL	Free	Manly, Cudmore Jr & Meer, 2001
LINKMFEX	Outcross	Free	Danzmann, 2001

JoinMap (Stam, 1993), CRI-MAP (Green, Falls & Crooks, 1990), MapMaker (Lincoln, 1992), LINKMFEX (Danzmann, 2001), and Map Manager QTX (Manly, Cudmore & Meer, 2001) for linkage map construction, and MapChart (Voorrips, 2002) for map drawing are the most widely used tools in aquaculture species. As marker density becomes very high, which leads to the exponential increase in computational intensity

Table 20.2 Some examples of linkage mapping studies in aquaculture species using various software packages.

Species	Software	Experimental Crosses	Genetic Markers	References
Common carp	JoinMap 4.0	Outbreeding full-sib family	Microsatellites SNPs	Zhao *et al.*, 2013
Common carp	JoinMap 4.0	Outbreeding full-sib family	Microsatellites SNPs	Zheng *et al.*, 2011
Atlantic salmon	CRIMAP 2.4	Outbreeding full-sib family	SNPs	Lien *et al.*, 2011
Atlantic salmon	CRIMAP 2.4 (modified version)	Outbreeding full-sib family	SNPs	Gonen *et al.*, 2014
Atlantic salmon	JoinMap 4.0	Outbreeding full-sib family	SNPs	Hubert *et al.*, 2010
Channel catfish	JoinMap 4.0	Backcross family	Microsatellites SNPs	Ninwichian *et al.*, 2012
Channel catfish	OneMap JoinMap 4.0	Outbreeding full-sib family	SNPs	Li *et al.*, 2015
Rainbow trout	CARTHAGENE	Doubled-haploid lines	Microsatellites SNPs	Guyomard *et al.*, 2012
Rainbow trout	MULTIMAP CRIMAP	Outbreeding full-sib family	Microsatellites SNPs	Palti *et al.*, 2011
Red drum	JoinMap 4.1	Outbreeding full-sib family	Microsatellites	Hollenbeck, Portnoy & Gold, 2015
Red drum	LINKMFEX 2.1	Outbreeding full-sib family	Microsatellites	Portnoy *et al.*, 2010
Asian sea bass	CRIMAP 2.4	Outbreeding full-sib family	Microsatellites SNPs	Wang *et al.*, 2011
Zhikong scallop	JoinMap 4.1	Outbreeding full-sib family	SNPs	Jiao *et al.*, 2013
Pearl oyster	CRIMAP2.4 (modified version)	Full-sib family Half-sib family	SNPs	Jones *et al.*, 2013
Pearl oyster	JoinMap 4.0	Outbreeding full-sib family	SNPs	Shi *et al.*, 2014
White shrimp	CRIMAP	F2 family	SNPs	Du *et al.*, 2010

(van Os *et al.*, 2005b), several new packages such as OneMap (Margarido, Souza & Garcia, 2007a), MSTmap (Wu, Bhat *et al.*, 2008), and HighMap (Liu *et al.*, 2014) have been developed for constructing high-density linkage maps. Some examples of the applications of these software packages in aquaculture species in recent years (since 2010) are listed in Table 20.2.

Linkage Mapping Process

Data Filtering

For SNP markers, only those that have high quality of genotype calls and are heterozygous in at least one parent can be retained for linkage analysis. Based on segregation patterns, SNPs are categorized into three types: 1:2:1 type (AB x AB, segregating in both parents); 1:1 type (AB x AA or AB x BB, segregating only in female); and 1:1 type (AA x AB or BB x AB, segregating only in male). The other types should be removed at the beginning because they are not informative for linkage mapping. It is also necessary to be aware that, just as with any estimation procedure, linkage map estimation is prone to errors. The errors may be genotyping errors, or may originate from missing data or segregation distortion. Genotyping errors may have a large impact on the accuracy of a linkage map. Even a low frequency of errors can inflate map lengths and support incorrect marker orders. This was observed during linkage map construction in zebrafish, where the genetic sizes of the initial map were over 1000 cM per chromosome. After the removal of genotyping errors, genetic sizes were reduced to around 100 cM (Howe *et al.*, 2013). Genotype errors can also increase the proportion of incorrectly ordered maps, and this problem will be more severe as the distances between loci decreases.

Missing data can also lead to incorrect marker orders, especially in map regions with dense markers. Furthermore, it was reported that missing values tend to reduce the map lengths for widely separated markers, particularly under the weighted least-squares criterion (Hackett & Broadfoot, 2003). In such case, it is suggested to delete the markers with genotype errors and those with too many missing values before further analysis.

Markers with segregation distortion can also lead to inaccuracy in linkage mapping, although the effects of distorted markers on the final linkage map may be dataset-specific. In order to remove markers with segregation distortion, segregation is tested against the normal Mendelian expectation ratios using the Chi-square test. Hackett and Broadfoot (2003) indicated that segregation distortion had little effect on the accuracy of map length or marker order, while Semagn *et al.* (2006) showed reduction in map length due to the presence of distorted markers. In many cases, markers that are distorted significantly with $p < 0.001$ should not be used in the linkage mapping, because they could create inaccurate distances and false linkages (Cervera *et al.*, 2001; Recknagel, Elmer & Meyer, 2013). For less serious distorted markers, it is better to study these loci after calculating the map and then decide which ones should be removed (Lu, Romero-Severson & Bernardo, 2002; Matsushita *et al.*, 2003).

The availability of a large number of SNP markers and efficient SNP genotyping technology allows the construction of high-density linkage maps. However a new problem arises. In many practical cases, the number of markers may exceed the resolution of recombination for the given population size, which will lead to marker stacking—that is, mapping of many markers to a single genetic location. For practical considerations of reducing the computational workload, stacked markers can be removed to keep only one marker with the most informative meiosis at the locus. After building the map with "representative markers", the stacked markers with "zero recombination clusters" that were excluded during the initial mapping steps should be relocated onto the linkage maps based on the positions of their representative anchor marker. Such stacked markers are still valuable for the chromosome-scale scaffolding of genomic sequences.

Assigning Markers into Linkage Groups

Determining which markers belong to the same linkage group is the first step in the linkage mapping process. Linkage groups are formed by sets of linked loci. The task of assigning markers to linkage groups is essentially a clustering problem (Wu, Bhat *et al.*, 2008). Several strategies have been applied in different software packages. One strategy employs the principle of *agglomerative hierarchical clustering*, such as nearest neighbor locus. It uses a two-point linkage analysis module to infer linkage groups, and clusters of markers are grown by sequentially adding that marker which shows the lowest recombination value to the current members of groups. Several popular software tools are built based on this theory, such as MapMaker (Lander *et al.*, 1987) and JoinMap (Stam, 1993). They consider linkage to be transitive, such that if marker A is significantly linked to marker B, and marker B is linked to marker C, then A, B, and C tend to belong to the same linkage group. Another strategy employs ideas from graph theory. As with MSTmap (Wu, Bhat *et al.*, 2008), this strategy involves creating a complete graph over the clustered markers, with connecting graph edges weighted by some two-point function of the data. All edges over a certain threshold value are chopped. The graph will be broken up into a number of distinct subunits, each of which corresponds to a linkage group (Cheema & Dicks, 2009).

Ordering Markers Within Each Linkage Group

Establishing marker orders within each linkage group is the main step and task for linkage map construction. Several types of algorithmic methods have been developed for marker ordering, such as *simulated annealing* (SA) (Thompson, 1984), *the greedy* or *nearest-neighbor algorithm* (Stam, 1993), *tabu search* (Schiex & Gaspin, 1997), *evolution strategy* (Mester *et al.*, 2003), *genetic algorithms* (Gaspin & Schiex, 1998), and *ant colony optimization* (Iwata & Ninomiya, 2006). These algorithms have been implemented in various linkage mapping software packages. For example, JoinMap has implemented *simulated annealing*, ANTMAP (Iwata & Ninomiya, 2006) uses the ant colony optimization heuristic, and CARTHAGENE (Schiex & Gaspin, 1997) employs a combination of *simulated annealing*, *tabu search*, and *genetic algorithms*.

In order to characterize the quality of orders, various scoring functions (objective functions) have been developed—for example, minimum sum of square errors (Stam, 1993), minimum sum of adjacent recombination fractions (Falk & Chakravarti, 1992), minimum number of recombination events (van Os *et al.*, 2005a), maximum likelihood (Jansen, De Jong & Van Ooijen, 2001), maximum sum of adjacent LOD scores (Weeks & Lange, 1987), and minimum product of adjacent recombination fractions (Wilson, 1988).

Take JoinMap as an example: one can choose between the regression mapping algorithms (Stam, 1993) and the *Monte Carlo* maximum likelihood mapping algorithms (Jansen *et al.*, 2001) as calculation options. The regression mapping algorithm applies an iterative process of marker ordering that is based on a weighted least squares procedure (linear regression). First, the most informative pair of markers is selected, followed by adding other loci one by one. The best positions of the newly added markers are searched by comparing the goodness-of-fit test (Chi-square) of the calculated map for each tested position. A big normalized difference in goodness-of-fit Chi-square before and after adding a locus (*jump*) indicates a poor fit of the added marker. When the newly added marker causes a large jump or negative distance estimated in the map, the marker is removed. This procedure is repeated until all loci are tested once (*first round*). Some markers that have been removed in the first round can be added to the map in the second round; however, the markers that produce too large a jump or negative distances should still not be used. In the third round, all markers previously removed are forced to add to the map without constraints of maximum allowed jump and negative distances. In that case, the results from the third round should not be considered as a map of high quality because some poor-fitting loci still exist. This approach works well with relatively small numbers of markers. However, it requires tremendous amount of computational powers with large numbers of markers. To cope with large numbers of markers, a maximum likelihood mapping algorithm was developed to construct high-density maps. It combines several techniques, including *Gibbs sampling*, *simulated annealing*, and *spatial sampling* to order markers and compute marker spacing. Gibbs sampling is used for the likelihood calculation based on multipoint recombination frequencies and for missing-data imputation. Simulated annealing searches for the best map order that has the maximum likelihood. An alternate use of simulated annealing and Gibbs sampling is employed to achieve an improved map order. In practice, three of four iterations are adequate. Spatial sampling is needed to escape from local optima, which is caused by missing information and genotyping errors. A new multipoint maximum likelihood mapping algorithm is introduced in the latest version 4.1 of JoinMap, which can deal with a linkage group of about 250 good-quality SNP markers of the outbreeding species full-sib family within 8 minutes. For OneMap (R package, version 2.0; Margarido, Souza & Garcia, 2007b; Wu *et al.*, 2002), marker ordering in each linkage group can be done by using the two-point-based algorithm, including seriation (Buetow & Chakravarti, 1987), the rapid chain delineation algorithm (Doerge, 1996), the recombination counting and ordering (van Os *et al.*, 2005a), and unidirectional growth (Tan & Fu, 2006). Then, the multipoint analysis may be used to check the local order of markers in a linkage group and to refine the map distance between adjacent markers (Margarido *et al.*, 2007b).

Step-by-Step Illustration of Linkage Mapping

A number of software packages are available for genetic linkage mapping. Here, we provide an example to illustrate the detailed processes of linkage map construction.

JoinMap

JoinMap is a Microsoft Windows–based, user-friendly program for linkage mapping using experimental populations of diploid species. During mapping processes, users can perform several diagnostic tests and simply remove erroneous loci and individuals from the map calculations. As stated earlier, two marker order strategies, the regression mapping algorithm and the maximum likelihood mapping algorithm, can be used for map calculation. However, the previous versions cannot be applied to the outbreeding full-sib family (CP), which is the most commonly used population type in aquaculture species. The most important enhancement of version 4.1 of JoinMap is the ability to use the multipoint maximum likelihood mapping algorithm on CP populations.

Obtaining JoinMap

JoinMap is a commercial package, and licenses for use of the software should be purchased from Kyazma. One can download the evaluation version of the software online (http://www.kyazma.nl/index.php/mc.JoinMap/sc.Evaluate/) to test whether the package provides the appropriate functions for the target dataset.

Input Data Files

JoinMap can use several types of data files, including the *locus genotype file* (*loc-file*), the *pairwise recombination frequencies data file* (*pwd-file*), and *map file*. For original SNP genotype files, AA/BB x AB must be converted to < nn x np>, the reciprocal configurations AB x AA/BB must be converted to < lm x ll>, and the fully heterozygous markers AB x AB should be converted to < hk x hk > according to the coding scheme. The < nn x np > and < lm x ll > markers are used to map for the male and female parents, respectively. The < hk x hk > marker can be used for both parents and, therefore, allow anchoring of the male and female maps.

Genotype data stored in Microsoft Excel spreadsheets can be directly copied into JoinMap's data sheet (Figure 20.4).

After the genotype data is successfully loaded, a population node can be created, which is the starting point for genetic mapping. Several tabsheets appear in the panel. You can see the summary of the data by clicking on the *Info* tabsheet. The empty tabsheets will be filled with results of the corresponding calculations. The most important tabsheet before grouping is *Locus Genot. Freq*, which displays the genotype frequencies for each locus, including the Chi-square test results for segregation according to the Mendelian ratio. Markers of poor quality and individuals with significant genotyping errors should be identified and removed from further analysis by checking their corresponding checkboxes in the *Exclude* column of *Loci* and *Individuals* tabsheets, based on the Chi-square values or the numbers of missing genotypes.

Then, the genotypes of the currently selected set of loci and individuals can be analyzed toward obtaining the linkage groups. JoinMap assigns grouping using four test

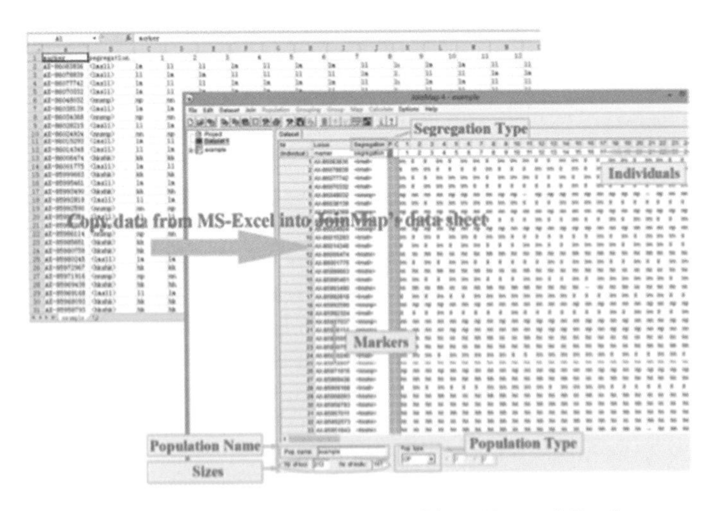

Figure 20.4 The input data matrix imported from Microsoft Excel.

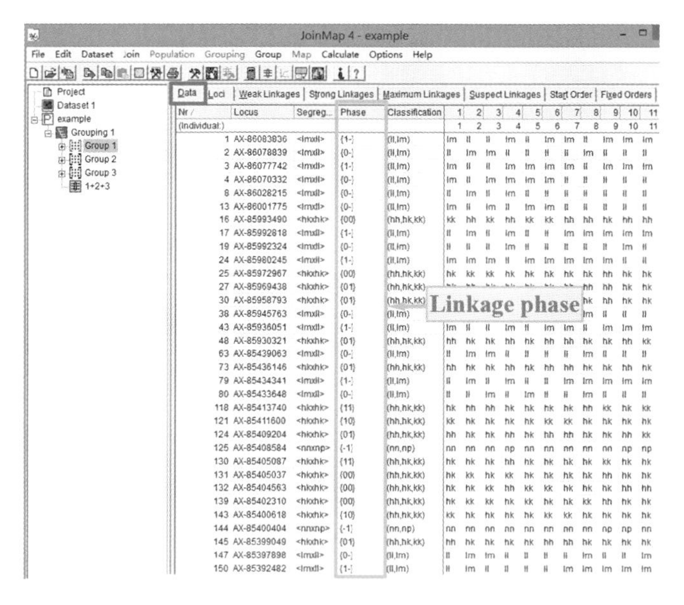

Figure 20.5 An example shows that the linkage phase of each marker can be automatically determined by JoinMap.

statistics (parameters), including LOD value of the test for independence, p-value of the test for independence, recombination frequency, and linkage LOD. The grouping will be done at several significance levels of increasing stringency. The *Groupings (tree)* tabsheet shows how the loci fall apart in groups at increasing stringency levels of a test for linkage. Each node in the tree represents a group of linked loci based on one of the four available test statistics for grouping. A LOD score of 3 is suggested as the lowest threshold to determine the initial linkage, and true linked markers tend to stay together even when the LOD score is increased.

In each group node, we can see that the linkage phases of population type CP can be determined automatically (Figure 20.5). It should be noted that, in the earlier versions of JoinMap, only the regression mapping algorithm can be used as a marker order search strategy for CP population. The two most generally used mapping functions, Haldane and Kosambi, can be selected from "Calculation Options". Then, map calculations will be performed using the *Group* menu function *Calculate Map*. After the map calculations are performed successfully, you can check the details of the procedure in the *Session Log* tabsheet and view the maps generated after each round.

JoinMap allows checking of the quality of each map through the information given in the map node. There are a number of functions in JoinMap that measures how well a marker fits in its current assigned position: (1) the *Mean Chi-square Contribs* tabsheet shows the contribution of each locus to the goodness-of-fit averaged over all pairs among which the locus was involved; (2) the nearest neighbor fit (*N.N.Fit*) measure indicates whether a locus fits well between its neighboring loci; and (3) the *Genotype Probabilities* tabsheet is for the detection of unlikely genotype scores. Large values of Mean Chisquare Contribs and N.N.fit suggest that the marker does not fit well. In the example, the marker AX-85404080 has the largest contribution (4.322) to the

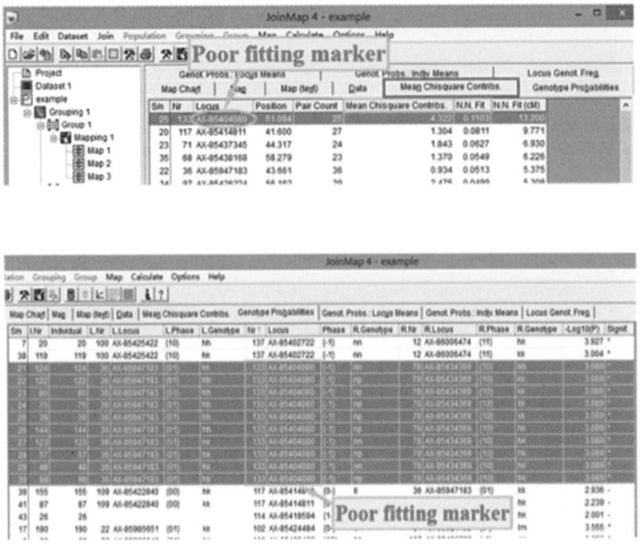

Figure 20.6 An example shows poor fitting markers that stand out after sorting the *Mean Chi-square Contribs* and *Genotype Probabilities* tabsheets.

Chi-square goodness-of-fit measure of the map, as well as the highest nearest neighbor fit (0.1103). In addition, it is involved many times in the *Genotype Probabilities* tabsheet (Figure 20.6), all suggestive of its poor fit in its current map position, and this marker should hence be removed from recalculation.

After two rounds of regression mapping, there are still some markers that cannot be located in the map. Although JoinMap allows all markers previously removed to be forcefully added into the map after the third round's calculation, we suggest exclusion of such markers.

OneMap

OneMap is a recently developed package that is freely available. It is capable of handling several experimental crosses, including outcrossing, RILs, F2, and backcrosses. Users should have some knowledge about the R language, as OneMap is developed using R. The analyses are performed using a novel methodology based on multipoint approaches using hidden Markov models.

- *Getting OneMap*: The package source of the latest version can be downloaded from http://cran.r-project.org/web/packages/onemap/index.html
- *Installation*: OneMap can be installed by opening R and typing the command:

  ```
  >install.packages("onemap")
  ```

  Alternatively, it can also be installed using the R console menu, by selecting "Packages > Install packages". After clicking on "Install packages", a pop-up box will ask to choose the CRAN mirror, where the location can be chosen for the installation of the package. After that, select the OneMap package from another box that pops up, and

Table 20.3 A sample input data file for OneMap. A mapping family of 94 samples (hyb220M) was used for the illustration. A total of 17,401 SNPs segregated in the male was used for the linkage analysis. Here, only genotypes for five SNPs are shown. Starting from the second row, the first column contains the SNP IDs starting with the "*"; the second column contains the code indicating segregation types (e.g., B3.7 indicates ABxAB, while D2.15 indicates AA/BBxAB); and the third column contains genotypes of each SNP for the 94 individuals, separated by commas.

94 17401		
*AX-86159488	B3.7	a,ab,a,ab,ab,ab,b,ab,b,ab,ab,a,b,ab,ab,b,ab,b,b,b,b,b,a,ab,ab,ab,a,ab,ab,b,ab,ab,ab,ab,ab,b,ab,ab,b,a,a,ab,b,b,b,ab,ab,ab,ab,b,ab,a,ab,a,ab,a,ab,a,ab,ab,a,ab,b,b,b,b,a,ab,ab,a,a,ab,ab,ab,ab,ab,a,a,ab,ab,ab,ab,ab,a,a,ab,ab,ab,b,b,ab,b,ab,ab
*AX-86159778	B3.7	b,a,b,a,a,a,b,a,b,b,a,ab,a,ab,ab,b,b,b,ab,b,b,a,b,ab,a,b,ab,ab,a,b,ab,b,a,ab,ab,b,a,ab,ab,ab,a,a,a,ab,ab,ab,b,a,ab,ab,ab,ab,a,b,a,ab,a,ab,b,ab,b,b,b,ab,ab,b,a,b,b,b,ab,a,b,b,a,ab,a,ab,b,b,a,b,b,a,ab,ab,ab,ab,b,b,a,b,b,ab
*AX-86160046	B3.7	a,ab,a,b,a,ab,b,a,b,ab,a,ab,ab,a,b,a,ab,ab,a,ab,a,ab,ab,ab,ab,ab,ab,ab,ab,a,b,a,b,ab,ab,ab,ab,ab,ab,ab,ab,ab,ab,a,b,b,b,ab,ab,a,b,ab,ab,b,a,ab,a,ab,b,ab,ab,a,ab,a,b,a,ab,ab,ab,b,a,ab,b,ab,ab,ab,ab,ab,ab,a,ab,ab,a,ab,ab,ab,ab,ab,ab,ab,a,b,b,b
*AX-86160522	B3.7	ab,ab,ab,ab,ab,ab,ab,ab,ab,ab,ab,b,b,b,b,ab,ab,ab,b,ab,ab,ab,ab,a,ab,ab,a,b,ab,ab,a,ab,ab,a,b,ab,b,b,b,a,ab,ab,ab,ab,a,a,ab,ab,a,a,a,b,ab,ab,ab,b,ab,a,ab,ab,ab,ab,ab,b,a,ab,ab,ab,ab,ab,a,ab,ab,ab,ab,b,ab,a,ab,ab,ab,ab,ab,ab,a,a,ab,b,ab,ab,ab,ab,ab,a
*AX-85188597	D2.15	ab,ab,ab,ab,a,ab,a,a,ab,ab,a,a,ab,ab,ab,a,ab,ab,ab,a,a,ab,a,a,ab,a,ab,a,a,ab,ab,ab,a,ab,a,a,a,a,ab,a,ab,a,a,a,a,a,a,ab,ab,a,ab,a,ab,ab,ab,a,ab,a,a,a,ab,a,a,a,ab,ab,a,ab,ab,a,ab,ab,a,a,a,ab,a,ab,a,ab,ab,ab,a,a,a,a,ab,ab,ab,ab,a,ab,a,a

click "OK" to install. The installation will be automatically performed. After installation, OneMap can be loaded in R by typing:

```
library(onemap)
```

- *Input data files*: The input data file is a text file, in which the first line contains two numbers that indicate the number of individuals and the number of markers. The genotype information is included separately for each marker. The character "*" indicates the beginning of information input for a new marker, followed by the marker name. For each marker, a code is assigned to indicate the marker segregation type, according to the notation of Wu *et al.* (2002). A sample input data file is provided in Table 20.3.
- *Importing data*: Once the input file is created, data can be loaded and saved into an R object. The data can be imported using the OneMap function "`read.outcross`":

```
hyb220M<-read.outcross(file="hyb220M.txt")
```

To save the data, type

```
save.image("hyb220M.Rdata")
```

- *Estimating two-point recombination fractions*: To start the analysis, the first step is to estimate the recombination fraction between all pairs of markers, using two-point tests:

```
twopts<-rf.2pts(hyb220M)
```

Different values for the criteria can be chosen using:

```
twopts<-rf.2pts(hyb220M, LOD=6, max.rf=0.35)
```

To save the two-point test results, type:

```
save.image("hyb220M.RFs.Rdata")
```

- *Assigning markers to linkage groups*: Once the recombination fractions and linkage phases for all pairs of markers are determined, markers can be assigned to linkage groups using the function "make.seq" to create a sequence with the markers being analyzed:

```
mark.all<-make.seq(twopts, "all")
```

The grouping step is performed using the function group:

```
LGs<-group(mark.all, LOD=8)
```

To save the linkage group assignment results, type:

```
save.image("hyb220M.LGs.Rdata")
```

- *Genetic mapping of linkage groups*: Once markers are assigned to linkage groups, the genetic mapping step can proceed. First, the mapping function needs to be set up. Two mapping functions (Kosambi or Haldane) can be selected. To use the Kosambi function, type:

```
set.map.fun(type="kosambi")
```

Then, define which linkage group is to be mapped. For instance, if we want to map the linkage group 1, type:

```
LG1<-make.seq(LGs, 1)
```

Similarly, other linkage groups can be mapped.

MergeMap

Traditional software tools are focused on building genetic maps for a single mapping population. In recent years, it is increasingly common to find multiple maps available for the same organism. If the maps can be integrated, a consensus map would result, which can provide a higher density of markers and greater genome coverage (Wu, Close & Lonardi, 2008). MergeMap was developed for constructing consensus genetic maps from a set of individual genetic maps (Wu, Close *et al.*, 2008). The individual genetic maps are given as directed acyclic graphs (DAGs) internally, which are then merged into a consensus graph on the basis of shared vertices. Cycles in the consensus graph indicate ordering conflicts among the individual maps. Conflicts are resolved using a parsimonious approach that takes into account two types of errors that may be present in the individual maps—namely, *local reshuffles* and *global displacements*. Local reshuffles refer to inaccuracies in the order of nearby markers, whereas global displacements refer to the cases where a few markers are placed at wrong positions far from the correct ones. MergeMap tries to resolve conflicts by deleting a minimum set of marker occurrences. To date, MergeMap is the fastest and most accurate software for merging individual maps.

Getting MergeMap

MergeMap can be downloaded free of cost from http://alumni.cs.ucr.edu/~yonghui/mgmap/MergeMap.tar.gz. Then, it should be complied and installed on a Linux machine.

Input Files

Linkage groups can be merged together if they share markers in common. To use MergeMap, you should prepare map files of each individual genetic linkage map and a configuration file in the following format.

```
map1_name map1_weight map1_path
```

In the configuration file, the first column specifies the name of each individual map; the second column specifies the weight of each linkage map; and the last column specifies the path the each individual map. The weight represents the user's confidence in the quality of the map (high weight is associated with high quality). When MergeMap tries to resolve conflict, the minimum-weight set of marker occurrences will be deleted. To construct the consensus map, simply run the following command:

```
$ Consensus_map.exe configuration_file
```

Output Files

Four output files will be generated after running the MergeMap command. There are three graphs in ".dot" format ("lgx.dot", "lgx_consensus.dot", and "lgx_linear.dot"), and one text file showing the final consensus map. The "lgx.dot" graph highlights the conflicts among the individual maps. It also shows the solution by MergeMap as to which marker occurrence is being deleted. The "lgx_consensus.dot" graph shows the simplified consensus DAG, while the "lgx_linear.dot" graph shows the final linear consensus map.

MapChart

After the construction of genetic linkage maps, a quality output of the linkage map is needed for publication and presentation. MapChart is widely used for drawing linkage maps. It is a Microsoft Windows–based computer package that produces charts of genetic linkage and QTL maps. These charts are composed of a sequence of vertical bars representing the linkage groups. MapChart reads the linkage information (i.e., the locus and their positions) from text files, which must be created in advance with genetic mapping software such as JoinMap.

Getting MapChart

MapChart is freeware, and can be obtained from http://www.wageningenur.nl/en/Expertise-Services/Research-Institutes/plant-research-international/About/Organisation/Biometrie/Collaboration/Software-Service/Download-MapChart.htm

- *Input data file*: MapChart reads linkage data from text files, such as the ".map" files from JoinMap. The text files may be opened with an editor program such as TextPad for any desired editing. The text containing the map data specifies the order of linkage

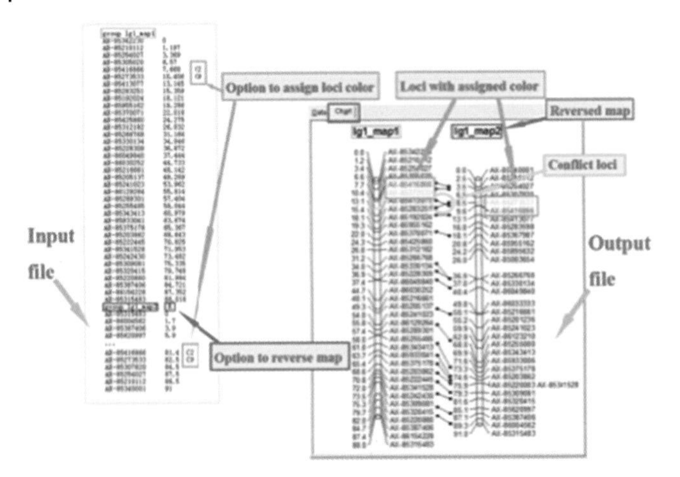

Figure 20.7 An example of the map charts generated by MapChart. The shared markers of two linkage maps are connecting with lines. The two maps are in the same direction after adding the "R" option to one of the maps in the input data file. The markers will be given certain colors when adding the "color" option.

groups. Each linkage group specification consists of a header line followed by a loci section (Figure 20.7).

- *Linkage group header*: The header line of a linkage group starts with the keywords "group" or "chrom", followed by the linkage group name.
- *Loci section*: The names and positions of the loci (or markers) in the linkage group are specified in the Loci section. The loci section starts on the first line after the linkage group header, and continues until the next linkage group header. Each locus contains the following elements in the order of locus name, locus position (increasing order).

- After inputting the data into MapChart, the linkage map chart will be displayed in the "Chart" page. The general appearance of the chart can be adjusted with the "Chart Options" button on the tools menu.
  - "Page size", "Title & Footer", "Page layout", and "Colors & Lines" are the cards used to set the general parameters of the chart, as described by their names.
  - "Bars" are the options to set the linkage bars regarding the length, width, position, and loci indicators.
  - The "Loci" page controls the font and position of locus names.
  - The "Homologs" page highlights the homologues of loci based on their names or homolog numbers between different linkage groups. Loci that are homologous to each other are connected by lines in the chart and given the same color.

Sometimes we can get two maps in opposite directions. In order to reverse the order of one map, we can add the "R" option to the linkage group header line (Figure 20.7). In addition, different colors can be assigned to locus names by adding the color option "C <number>" in the locus section of the input file.

Pros and Cons of Linkage Mapping Software Packages

Each package has its own strengths and weaknesses. For example, OneMap was developed to construct linkage maps of high heterozygous species. It is fairly efficient at assigning markers to linkage groups, but the marker ordering function of OneMap is problematic in both generating the right marker orders and estimating the distances between markers. Similarly, MSTmap can handle very large numbers of markers and generate ultra-dense maps of up to 100,000 markers. However, it tends to inflate map distances due to genotype errors. In contrast, older versions of JoinMap are limited in their ability to handle large numbers of markers, but the use of the maximum likelihood algorithm greatly expedites computational speed in marker ordering. Therefore, it is wise to evaluate several software packages with genotype data before choosing a specific package for linkage analysis. It is also often productive to utilize combinations of software packages—for example, using OneMap to place large numbers of markers into linkage groups, and then using JoinMap to order the markers and estimate their spacing.

References

Baird, N.A., Etter, P.D., Atwood, T.S., *et al.* (2008) Rapid SNP discovery and genetic mapping using sequenced RAD markers. *PloS One*, **3**, e3376.

Bateson, W. and Mendel, G. (1909) *Mendel's principles of heredity*. New York, NY: Putnam's Sons.

Bateson, W., Waunders, E. and Punnett, R.C. (1909) Experimental studies in the physiology of heredity. *Molecular and General Genetics MGG*, **2**, 17–19.

Buetow, K.H. and Chakravarti, A. (1987) Multipoint gene mapping using seriation. I. General methods. *American Journal of Human Genetics*, **41**, 180.

Cervera, M.T., Storme, V., Ivens, B., *et al.* (2001) Dense genetic linkage maps of three Populus species (*Populus deltoides*, *P. nigra* and *P. trichocarpa*) based on AFLP and microsatellite markers. *Genetics*, **158**, 787–809.

Cheema, J. and Dicks, J. (2009) Computational approaches and software tools for genetic linkage map estimation in plants. *Briefings in Bioinformatics*, **10**, 595–608.

Da, Y., VanRaden, P.M., Li, N., *et al.* (1998) Designs of reference families for the construction of genetic linkage maps. *Animal Biotechnology*, **9**, 205–228.

Danzmann, R. (2001) *LINKMFEX: Linkage analysis package for outcrossed mapping families with male or female exchange of the mapping parent*. www.uoguelph.ca/~rdanzman/software.

Davey, J.W., Hohenlohe, P.A., Etter, P.D., *et al.* (2011) Genome-wide genetic marker discovery and genotyping using next-generation sequencing. *Nature Reviews Genetics*, **12**, 499–510.

Doerge, R. (1996) Constructing genetic maps by rapid chain delineation. *Journal of Agricultural Genomics*, **2**, article 6.

Du, Z.Q., Ciobanu, D.C., Onteru, S.K., *et al.* (2010) A gene-based SNP linkage map for pacific white shrimp, *Litopenaeus vannamei*. *Animal Genetics*, **41**, 286–294.

Falk, C. and Chakravarti, A. (1992) Preliminary ordering of multiple linked loci using pairwise linkage data. *Genetic Epidemiology*, **9**, 367–375.

Ferreira, A., Silva, M.F.D. and Cruz, C.D. (2006) Estimating the effects of population size and type on the accuracy of genetic maps. *Genetics and Molecular Biology*, **29**, 187–192.

Gaspin, C. and Schiex, T. (1998) *Genetic algorithms for genetic mapping*. Paper presented at Artificial Evolution, Springer.

Gonen, S., Lowe, N.R., Cezard, T., *et al.* (2014) Linkage maps of the Atlantic salmon (*Salmo salar*) genome derived from RAD sequencing. *BMC Genomics*, **15**, 1.

Green, P., Falls, K. and Crooks, S. (1990) *Documentation for CRI-MAP, version 2.4*, Washington University School of Medicine, St Louis, MO.

Guyomard, R., Boussaha, M., Krieg, F., *et al.* (2012) A synthetic rainbow trout linkage map provides new insights into the salmonid whole genome duplication and the conservation of synteny among teleosts. *BMC Genetics*, **13**, 1.

Hackett, C.A. and Broadfoot, L.B. (2003) Effects of genotyping errors, missing values and segregation distortion in molecular marker data on the construction of linkage maps. *Heredity*, **90**, 33–38.

Haldane, J. (1919) The combination of linkage values, and the calculation of distances between the loci of linked factors. *Journal of Genetics*, **8**, 299–309.

Hollenbeck, C.M., Portnoy, D.S. and Gold, J.R. (2015) A genetic linkage map of red drum (*Sciaenops ocellatus*) and comparison of chromosomal syntenies with four other fish species. *Aquaculture*, **435**, 265–274.

Holliday, R. and Grigg, G.W. (1993) DNA methylation and mutation. *Mutation Research*, **285**, 61–67.

Howe, K., Clark, M.D., Torroja, C.F., *et al.* (2013) The zebrafish reference genome sequence and its relationship to the human genome. *Nature*, **496**, 498–503.

Hubert, S., Higgins, B., Borza, T. and Bowman, S. (2010) Development of a SNP resource and a genetic linkage map for Atlantic cod (*Gadus morhua*). *BMC Genomics*, **11**, 1.

Huehn, M. (2010) Random variability of map distances based on Kosambi's and Haldane's mapping functions. *Journal of Applied Genetics*, **51**, 27–31.

Iwata, H. and Ninomiya, S. (2006) AntMap: constructing genetic linkage maps using an ant colony optimization algorithm. *Breeding Science*, **56**, 371–377.

Jansen, J., De Jong, A. and Van Ooijen, J. (2001) Constructing dense genetic linkage maps. *Theoretical and Applied Genetics*, **102**, 1113–1122.

Jiao, W., Fu, X., Dou, J., *et al.* (2013) High-resolution linkage and quantitative trait locus mapping aided by genome survey sequencing: building up an integrative genomic framework for a bivalve mollusc. *DNA Research*, **21**, 85–101.

Jones, D.B., Jerry, D.R., Khatkar, M.S., *et al.* (2013) A high-density SNP genetic linkage map for the silver-lipped pearl oyster, *Pinctada maxima*: a valuable resource for gene localisation and marker-assisted selection. *BMC Genomics*, **14**, 1.

Kinghorn, B. and van der Werf, J. (2000) Identifying and incorporating genetic markers and major genes in animal breeding programs. *Course notes Belo Horizonte (Brazil), 31 May–5* June 2000..

Kosambi, D. (1943) The estimation of map distances from recombination values. *Annals of Eugenics*, **12**, 172–175.

Kruglyak, L. (1997) The use of a genetic map of biallelic markers in linkage studies. *Nature Genetics*, **17**, 21–24.

Lai, E. (2001) Application of SNP technologies in medicine: lessons learned and future challenges. *Genome Research*, **11**, 927–929.

Lander, E.S., Green, P., Abrahamson, J., *et al.* (1987) MapMaker: an interactive computer package for constructing primary genetic linkage maps of experimental and natural populations. *Genomics*, **1**, 174–181.

Li, Y., Liu, S., Qin, Z., *et al.* (2015) Construction of a high-density, high-resolution genetic map and its integration with BAC-based physical map in channel catfish. *DNA Research*, **22**, 39–52.

Lien, S., Gidskehaug, L., Moen, T., *et al.* (2011) A dense SNP-based linkage map for Atlantic salmon (*Salmo salar*) reveals extended chromosome homeologies and striking differences in sex-specific recombination patterns. *BMC Genomics*, **12**, 615.

Lincoln, S. (1992) Mapping genes controlling quantitative traits with MapMaker/QTL 1.1. *Whitehead Institute Technical Report*.

Liu, D., Ma, C., Hong, W., *et al.* (2014) Construction and analysis of high-density linkage map using high-throughput sequencing data. *PloS One*, **9**, e98855.

Liu, S., Zhou, Z., Lu, J., *et al.* (2011) Generation of genome-scale gene-associated SNPs in catfish for the construction of a high-density SNP array. *BMC Genomics*, **12**, 53.

Lu, H., Romero-Severson, J. and Bernardo, R. (2002) Chromosomal regions associated with segregation distortion in maize. *Theoretical and Applied Genetics* (*TAG, Theoretische und angewandte Genetik*, **105**, 622–628.

Manly, K.F., Cudmore, Jr., R.H. and Meer, J.M. (2001) Map Manager QTX, cross-platform software for genetic mapping. *Mammalian Genome*, **12**, 930–932.

Margarido, G., Souza, A.P. and Garcia, A. (2007a) OneMap: software for genetic mapping in outcrossing species. *Hereditas*, **144**, 78–79.

Margarido, G.R., Souza, A.P. and Garcia, A.A. (2007b) OneMap: software for genetic mapping in outcrossing species. *Hereditas*, **144**, 78–79.

Matsushita, S., Iseki, T., Fukuta, Y., *et al.* (2003) Characterization of segregation distortion on chromosome 3 induced in wide hybridization between indica and japonica type rice varieties. *Euphytica*, **134**, 27–32.

Maxam, A.M. and Gilbert, W. (1977) A new method for sequencing DNA. *Proceedings of the National Academy of Sciences*, **74**, 560–564.

Mester, D., Ronin, Y., Minkov, D., *et al.* (2003) Constructing large-scale genetic maps using an evolutionary strategy algorithm. *Genetics*, **165**, 2269–2282.

Miko, I. (2008) Gregor Mendel and the principles of inheritance. *Nature Education*, **1**, 134.

Miller, M.R., Dunham, J.P., Amores, A., *et al.* (2007) Rapid and cost-effective polymorphism identification and genotyping using restriction site associated DNA (RAD) markers. *Genome Research*, **17**, 240–248.

Morgan, T.H. (1910) Sex limited inheritance in Drosophila. *Science*, **32**, 120–122.

Morgan, T.H. (1911) Random segregation versus coupling in Mendelian inheritance. *Science*, **34**, 384.

Morton, N.E. (1955) Sequential tests for the detection of linkage. *American Journal of Human Genetics*, **7**, 277–318.

Ninwichian, P., Peatman, E., Liu, H., *et al.* (2012) Second-generation genetic linkage map of catfish and its integration with the BAC-based physical map. *G3: Genes| Genomes| Genetics*, **2**, 1233–1241.

Palti, Y., Genet, C., Luo, M.C., *et al.* (2011) A first generation integrated map of the rainbow trout genome. *BMC Genomics*, **12**, 1.

Portnoy, D.S., Renshaw, M.A., Hollenbeck, C.M. and Gold, J.R. (2010) A genetic linkage map of red drum, *Sciaenops ocellatus*. *Animal Genetics*, **41**, 630–641.

Rastas, P., Paulin, L., Hanski, I., *et al.* (2013) Lep-MAP: fast and accurate linkage map construction for large SNP datasets[J]. *Bioinformatics*, **29**, 3128–3134.

Recknagel, H., Elmer, K.R. and Meyer, A. (2013) A hybrid genetic linkage map of two ecologically and morphologically divergent Midas cichlid fishes (Amphilophus spp.) obtained by massively parallel DNA sequencing (ddRADSeq). *G3*, **3**, 65–74.

Ren, B., Robert, F., Wyrick, J.J., *et al.* (2000) Genome-wide location and function of DNA binding proteins. *Science*, **290**, 2306–2309.

Sanger, F., Nicklen, S. and Coulson, A.R. (1977) DNA sequencing with chain-terminating inhibitors. *Proceedings of the National Academy of Sciences*, **74**, 5463–5467.

Schiex, T. and Gaspin, C. (1997) Cartagene: constructing and joining maximum likelihood genetic maps. Paper presented at: Proceedings of the fifth international conference on Intelligent Systems for Molecular Biology.

Semagn, K., Bjørnstad, Å. and Ndjiondjop, M. (2006) Principles, requirements and prospects of genetic mapping in plants. *African Journal of Biotechnology*, **5**, 2569–2587.

Shi, Y., Wang, S., Gu, Z., *et al.* (2014) High-density single nucleotide polymorphisms linkage and quantitative trait locus mapping of the pearl oyster, *Pinctada fucata martensii* Dunker. *Aquaculture*, **434**, 376–384.

Stam, P. (1993) Construction of integrated genetic linkage maps by means of a new computer package: Join Map. *The Plant Journal*, **3**, 739–744.

Tan, Y.-D. and Fu, Y.-X. (2006) A novel method for estimating linkage maps. *Genetics*, **173**, 2383–2390.

Thompson, E. (1984) Information gain in joint linkage analysis. *Mathematical Medicine and Biology*, **1**, 31–49.

van der Beek, S., Groen, A.F. and van Arendonk, J.A. (1993) Evaluation of designs for reference families for livestock linkage mapping experiments. *Animal Biotechnology*, **4**, 163–182.

Van Ooijen, J.W. and Jansen, J. (2013) *Genetic mapping in experimental populations*. Cambridge: Cambridge University Press.

van Os, H., Stam, P., Visser, R.G. and van Eck, H.J. (2005a) RECORD: a novel method for ordering loci on a genetic linkage map. *Theoretical and Applied Genetics*, **112**, 30–40.

van Os, H., Stam, P., Visser, R.G. and van Eck, H.J. (2005b) SMOOTH: a statistical method for successful removal of genotyping errors from high-density genetic linkage data. *Theoretical and Applied Genetics (TAG, Theoretische und angewandte Genetik)*, **112**, 187–194.

Vinod, K. (2011) Kosambi and the genetic mapping function. *Resonance*, **16**, 540–550.

Voorrips, R. (2002) MapChart: software for the graphical presentation of linkage maps and QTLs. *Journal of Heredity*, **93**, 77–78.

Wang, C.M., Bai, Z.Y., He, X.P., *et al.* (2011) A high-resolution linkage map for comparative genome analysis and QTL fine mapping in Asian seabass, *Lates calcarifer*. *BMC Genomics*, **12**, 1.

Watson, J.D. and Crick, F.H. (1953) Molecular structure of nucleic acids. *Nature*, **171**, 737–738.

Weeks, D.E. and Lange, K. (1987) Preliminary ranking procedures for multilocus ordering. *Genomics*, **1**, 236–242.

Whitehouse, H.L. (1982) *Genetic recombination*, John Wiley & Sons, New York, NY.

Wilson, S.R. (1988) A major simplification in the preliminary ordering of linked loci. *Genetic Epidemiology*, **5**, 75–80.

Wu, R., Ma, C.X., Painter, I. and Zeng, Z.B. (2002) Simultaneous maximum likelihood estimation of linkage and linkage phases in outcrossing species. *Theoretical Population Biology*, **61**, 349–363.

Wu, Y., Bhat, P.R., Close, T.J. and Lonardi, S. (2008) Efficient and accurate construction of genetic linkage maps from the minimum spanning tree of a graph. *PLoS Genetics*, **4**, e1000212.

Wu, Y., Close, T.J. and Lonardi, S. (2008) On the accurate construction of consensus genetic maps. *Computational Systems Bioinformatics / Life Sciences Society Computational Systems Bioinformatics Conference*, **7**, 285–296.

Xu, S. (2013) *Principles of statistical genomics*, Springer, New York, NY.

Yue, G.H. (2014) Recent advances of genome mapping and marker-assisted selection in aquaculture. *Fish and Fisheries*, **15**, 376–396.

Zhao, L., Zhang, Y., Ji, P., *et al.* (2013) A dense genetic linkage map for common carp and its integration with a BAC-based physical map. *PLoS One*, **8**, e63928.

Zheng, X., Kuang, Y., Zhang, X., *et al.* (2011) A genetic linkage map and comparative genome analysis of common carp (*Cyprinus carpio* L.) using microsatellites and SNPs. *Molecular genetics and genomics*, **286**, 261–277

21

Genomic Selection in Aquaculture Breeding Programs

Mehar S. Khatkar

Introduction

Genome selection was initially proposed in 2001, and it has now been widely used within livestock species. Thus far its application in aquaculture species has been very limited. In this chapter, the principles of genome selection are introduced, followed by an example of how to conduct genome selection, and, at the end, some perspectives for aquaculture species are provided.

Recent advances in molecular techniques have made it possible to genotype thousands of genetic markers (typically SNPs) even in species in which genome assemblies or genetic maps are not available. The ever-decreasing cost of genotyping and sequencing is now making it possible to use these tools in many aquaculture species. The genotypic data from pedigreed or non-pedigreed populations can provide insights into the population structure, gene association, and information for selective breeding. Indeed, with the availability of molecular data, there is no need to record pedigree information any more, which can greatly simplify breeding schemes, especially for aquaculture species. In this chapter, the concept of *genomic selection* (GS) is first described—that is, the application of genetic markers in selective breeding schemes—and then, its potential in aquaculture breeding programs is explored.

Genomic Selection

The idea of using molecular markers in animal breeding has been around for quite some time. However, only recently has the availability of thousands of genome-wide molecular markers made it practically possible to directly predict the production characteristics of animals solely using their DNA tests. Using simulations, Meuwissen, Hayes, and Goddard (2001) showed that the genetic merit of animals can be accurately estimated, without using information about their phenotype or that of close relatives, but using only genotypes on dense markers covering all the chromosomes. In this approach, known as *genomic selection* (GS) or *whole genome selection*, selection decisions are made on genomic breeding values (GBVs) predicted from high-density markers, typically single-nucleotide polymorphisms (SNPs). GS relies on the assumption that, with sufficiently high density of markers across the genome, most quantitative

Bioinformatics in Aquaculture: Principles and Methods, First Edition. Edited by Zhanjiang (John) Liu.

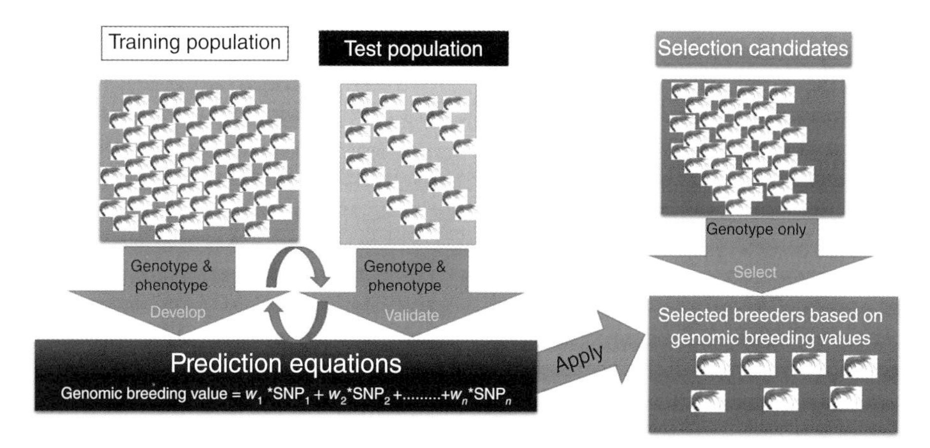

Figure 21.1 Overview of GS (adapted from Goddard & Hayes, 2009). (*See color plate section for the color representation of this figure.*)

trait loci will be in high linkage disequilibrium (LD) with at least one marker. Estimates of marker effects, when combined across the genome, will provide accurate predictions of genetic merit for a trait.

The implementation of GS is simple. It uses a "training population" of individuals, which has been both genotyped and phenotyped, to develop a prediction equation. This equation is then applied to the genotypes of the "test population/selection candidate" to compute molecular or genomic breeding values (GBVs). The GBV is then used to rank and select animals as parents for the next generation (Figure 21.1). GBVs can be further combined, if required, with traditional estimated breeding values (EBVs) to generate genomic estimated breeding values (GEBV).

GS can increase the rate of genetic gain compared to traditional breeding schemes due to a substantial reduction in generation intervals and increased selection intensities (Schaeffer, 2006). For example, in sheep and dairy cattle, the rates of genetic improvement could increase 25% and 100%, respectively. GS is particularly attractive for traits that are costly and difficult to phenotype, measured only by destruction or expressed late in life (Pryce *et al.*, 2010). For instance, GS is likely to increase genetic gain for traits that are difficult to record, such as disease resistance in aquaculture and poultry species, meat quality in pigs, and lifetime wool production and parasite resistance in sheep.

GS technology has been regarded as a huge milestone in animal and plant breeding programs. In domestic animals, SNP chips became first available in bovine species (Khatkar *et al.*, 2007), which led to the first successful application of GS in dairy cattle (Moser *et al.*, 2009); (VanRaden *et al.*, 2009). GS has replaced progeny testing in many countries, or is being used for the pre-selection of young bulls for progeny testing. GS is now being applied in a number of other animal and plant species (Goddard & Hayes, 2009), albeit with varied success. The principles of genomic prediction have even been applied in human studies, particularly for the identification of high-risk individuals for optimal interventions and personalized medicine. For instance, a genomic prediction method was used to predict the possibility of skin cancer in humans with promising results (Vazquez *et al.*, 2012). However, a distinction in the use of the terms "genomic selection" and "genomic prediction" should be noted. "Genomic selection" has been used in animal breeding programs, and this includes the identification of genetically superior animals through genomic prediction, selection, and mating of selected animals to produce the next generation. In contrast, "genomic prediction" only involves the prediction of the genetic worth of individuals from genomic information, and is better suited for human studies.

Steps in GS

An overview of GS is presented in Figure 21.1, consisting of the following main steps.

Preparation of Reference Population

A large sample of animals is required for constructing a reference population. These animals are measured for a trait of interest and genotyped for genome-wide markers, typically a large number of SNP markers using an SNP chip. For statistical analysis, the SNP genotypes are generally coded as numeric variables that take the value of 0, 1, or 2, corresponding to one of the homozygotes, the heterozygote, or the other homozygote. The reference population is generally divided into a training set and a validation set.

Development of Prediction Equations

The statistical analysis of the training set results in the estimates of effect (w) for each SNP marker (coded as 0, 1, and 2, the number or copy of one allele), and the effects of all the marker genotypes are combined to generate a prediction equation to estimate the GBV of each animal.

$$\text{GBV} = w_1{}^* \text{SNP}_1 + w_2{}^* \text{SNP}_2 + w_3{}^* \text{SNP}_3 + \ldots + W_n{}^* \text{SNP}_n$$

Here, W_i is the partial regression coefficient or effect size of the i-th SNP, and SNP_i is the vector of genotypes (numeric codes) for the i-th SNP.

One of the many analytical approaches mentioned in the next section can be used for developing such equations.

Validation of Prediction Equations

The animals in a validation set also have information on both genotypes and phenotypes. The preceding prediction equation is applied to the genotypes of validation animals for estimating GBVs. The accuracy of the prediction equation is assessed by comparing the estimated GBVs with the actual phenotypic information. The predictive accuracy for a continuous trait can be measured by the mean square error or correlation coefficient of the predicted and actual values of the trait. This step is optional, but provides important information on the accuracy of GS, and is hence recommended.

Computing GBVs of Selection/Test Candidates

Selection candidates are required to have only genotypes; no phenotypic information is required. The prediction equation is applied to the genotypes of these animals to compute the GBVs.

Selection and Mating

The animals are ranked based on the GBVs, and the top animals are selected and mated to produce the next generation.

Models for Genomic Prediction

Several analytical approaches have been applied for the whole-genome-based prediction of genetic merits. These can be broadly classified into three main categories:

1) *Regression based approaches*: GS in its simplest form involves estimating the effect of each marker/quantitative trait locus (QTL), and then summing them over all the loci across the entire genome for each individual. It is assumed that the markers that have some effects should be in LD with the QTLs or are the QTLs themselves, and that most of the genetic variance is additive. However, with the use of high-density SNP panels, the number of markers (p) is generally much larger than the number of animals (n) that can be tested in the training set. This makes it challenging to use simple regression models to estimate the effect of all the markers simultaneously. To solve this large-p-with-small-n regression problem, several variable selection and shrinkage estimation procedures have been proposed for the whole genome prediction of phenotypes. Meuwissen *et al.* (2001) proposed three methods—that is, genome-wide BLUP (rrBLUP), Bayes A, and Bayes B—to accommodate the large number of genetic markers in the prediction model. The differences in these, as well as in other subsequent Bayesian methods—Bayes Cπ (Habier *et al.*, 2011), Bayesian LASSO (de Los Campos *et al.*, 2009), and Bayes-R (Erbe *et al.*, 2012)—are in the definition of the prior distribution of SNP/QTL effects. For a detailed review and comparison of these approaches, please see de Los Campos *et al.* (2013). Partial least-squares regression (PLSR) and principal component (PC) regression also reduce dimensionality by computing latent variables, which are then used for predictions (Jannink, Lorenz & Iwata, 2010; Moser *et al.*, 2009).
2) *The genomic relationship approach*: The genomic relationship approach is also called "gBLUP". In this approach, the genomic relationship matrix (GRM) among individuals is computed. The GRM is then used to compute the breeding values for all the genotyped animals. This method is equivalent to the traditional "animal model," and the key difference is in replacing the pedigree-based relational matrix with GRM. The same model can be used to estimate the variance components and genetic parameters. This framework can easily be extended for the analysis of multiple traits. This is very attractive for application in aquaculture, where recording pedigree is very difficult and expensive. gBLUP can be directly applied without the need for recording pedigree information.

   GRM is based on actual genome-wide similarity and estimates realized relationships, and hence is more accurate when compared to pedigree-based expected relationships—for example, random segregation of chromosomes at meiosis can cause variation in actual genomic similarity among full-sibs that are captured by GRM (Nejati-Javaremi, Smith & Gibson, 1997). GRM can be constructed from the matrix of marker genotypes. VanRaden (2008) described three different methods of constructing GRM and evaluated their performance.

   It can be shown that, when a large number of markers have some effects, the gBLUP and other approaches (e.g., rrBLUP) mentioned earlier are similar in accuracy (Habier, Fernando & Dekkers, 2007; VanRaden, 2008). In practice, due to high genome-wide LDs in small populations and high relationships between training and test animals in many breeding schemes, the differences across various methods are generally small (Moser *et al.*, 2009).

3) *Semiparametric and machine-learning approaches*: In most applications, GBVs include only additive effects (genetic merit passed to the next generation). However, in some breeding schemes, it may be desirable to exploit dominance and epistatic effects—for example, to produce sets of (crossbred) progeny with the highest genetic value by exploiting heterosis through mate selection (Falconer & Mackay, 1996). Gianola and van Kaam (2008) proposed that non-parametric methods can account for complex epistatic effects without explicitly modeling them. Semi-parametric and non-parametric procedures for GS—such as reproducing kernel Hilbert spaces regression (Gianola & van Kaam, 2008), radial basis function neural networks (Gonzalez-Camacho *et al.*, 2012), support vector machine (SVM) (Maenhout, De Baets & Haesaert, 2010; Moser *et al.*, 2009), penalized SVM, random forests (Ogutu, Piepho & Schulz-Streeck, 2011), and boosting (Gonzalez-Recio et al., 2010)—can potentially utilize interactions among thousands of markers. Heslot *et al.* (2012) compared performance of 10 different methods, including machine learning algorithms, for GS in plants.

An Example of Implementation of Genomic Prediction

As described in the preceding section, a number of methods have been used for genomic prediction. Here, an example of implementation of genomic prediction is shown using the "rrBLUP" R package (Endelman, 2011). For this demonstration, a dataset of 599 wheat lines genotyped at 1279 DArT markers from another R-package BLR will be used. If you are new to R, there are many online resources available for obtaining a basic introduction to R (Torfs & Brauer, 2014; Venables, Smith & R Development Core Team, 2016).

You need to install the rrBLUP package using "`install.packages('rrBLUP');`", and then,

```
install.packages('BLR');
rm(list=ls())
```

tidy up

```
set.seed(99)
```

for repeatable results

```
library(BLR);
```

for wheat dataset

```
library(rrBLUP);
data(wheat)
```

use a set of 599 wheat lines genotyped at 1279 DArT markers from the BLR package;
we will use M (genotypes), Y (phenotypes) objects from this dataset

```
M <- 2*X-1
```

convert markers to {−1,1} as required by rrBLUP

```
pheno <-Y[,1]
```

grain yield in environment 1

```
geno <- M;
whichTest<-sample(1:length(pheno),100);
```

100 lines randomly selected as test

```
phenoTrain <- pheno[-whichTest];
phenoTest <- pheno[whichTest];
GenoTrain <- as.matrix(geno[-whichTest, ]);
genoTest <- as.matrix(geno[whichTest, ]);
```

Estimate marker effects using the rrBLUP package

```
markerEffects <- mixed.solve(y=phenoTrain, Z=GenoTrain)$u
```

make predictions in validation population

```
predictedGBV <- genoTest%*%markerEffects
```

compute accuracy of genomic prediction in test set

```
(predictionAcc <- cor(predictedGBV, phenoTest))
```

0.5370625; # using gBLUP (using genomic relationships)

```
A1 <-A.mat(M);
```

computing GRM

```
rownames(A1) <- 1:length(pheno);
yNA=pheno;
yNA[whichTest] <- NA;
```

mask trait values for validation set

```
data1 <-data.frame(y=yNA,gid=1:length(pheno));
```

a minimalist data set with two columns, one with the phenotypes, and one with the genotype labels

```
ans1 <- kin.blup(data1,K=A1,geno="gid",pheno="y");
```

compute accuracy of genomic prediction in test set

```
(cor(ans1$g[whichTest],Y[whichTest,1]));
```

0.5370625

Some Important Considerations for GS

How Many Animals Need to Be Genotyped?

The accuracy of GS is most critical to its success, and in itself is a major research area in plant and animal breeding. In addition to the choice of a statistical model, the accuracy of GS depends on many inter-related factors—namely, genome size, marker density,

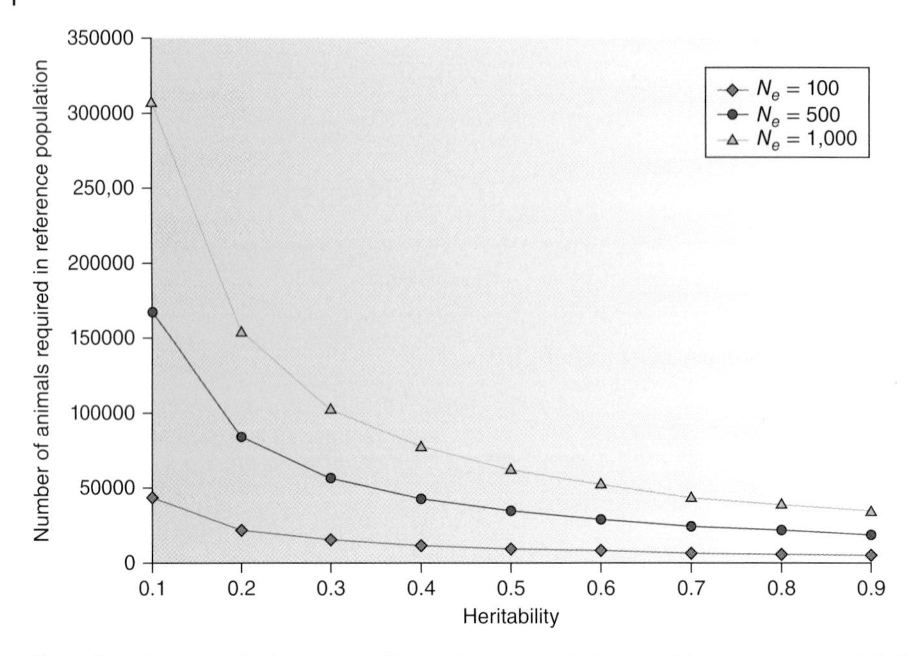

Figure 21.2 Number of animals needed in a reference population to achieve an accuracy of 0.7 for estimated GBVs (Goddard & Hayes, 2009).

LD between markers and QTLs, number and size of the QTL effects, effective population size, size of the training population, relationship between training and test animals, and heritability of the trait. For a required accuracy, the number of animals in a training set will mainly depend on the heritability estimate of a trait and the effective size of the population (*Ne*) (Figure 21.2). Lower heritability estimate of the trait will require a larger training set, also large *Ne* will require correspondingly bigger training population (Figure 21.2). In general, a larger training population will provide more accurate prediction in the test candidates. For example, in Holstein dairy cattle, a reference population of 3,000–10,000 bulls with genotypes and phenotype records (average of their daughters) is required for high accuracy (0.7) of GS for overall profit index.

How Many SNPs Are Enough?

SNP panels should cover the whole genome and be dense enough to ensure that most QTLs are in LD with at least one marker, so that estimates of marker effects capture the most genetic variance. The total number of SNPs required depends mainly on the extent of LD across the genome and genome size. In general, LD between adjacent SNP markers (say $r^2 > 0.2$) provides a good idea of about the adequacy of an SNP panel (Calus et al., 2008).

In practice, it seems that 50 K SNPs provide good prediction accuracies within a breed in dairy cattle. Increasing density further from 50 K to 800 K only adds marginally to the prediction accuracies (Khatkar *et al.*, 2012). However, if the training and test animals are related, the GRM can be used to predict GBVs using gBLUP. GRM can be computed accurately with a much smaller marker panel.

The cost of genotyping with a DNA chip of moderate SNP density (e.g., a 50 K SNP chip in cattle) is around US$40–100 per DNA sample. The cost of genotyping and sequencing is rapidly declining, making it cost-effective to apply this technology in many more species. In addition, genotype imputation can be effectively used to increase the density of genotypic data at lower costs. Genotype imputation involves genotyping a small proportion of the population with a high-density SNP panel and predicting high-density genotypes for the rest of the population genotyped with a low-cost, lower-density SNP panel. These *in silico* genotypes obtained by imputation, *albeit* with some uncertainty, can then be used in GS (Khatkar *et al.*, 2012).

What Is the More Important Factor for Genomic Predictions, the Number of Individuals or the Number of SNPs?

Accuracy improves with more records and closer marker spacing. However, after a certain minimum density of SNPs (say few thousand SNPs), adding more animals in the training set becomes more important for higher accuracies of genomic prediction.

Is Prediction Across Breeds/Population Possible?

Is prediction across breeds/population possible—that is, using training animals of one breed (say Jersey) and doing prediction for another breed (Holsteins)? GS has been successfully applied in within-population prediction of breeding values. However, the success of GS in across-breed prediction has been limited. For example, when genomic prediction was computed for breeds that were not in the training set, the accuracy was close to zero, or very low (Kachman *et al.*, 2013). In such situations, using a few SNPs with moderate/large effects and employing Bayesian methods may be helpful, as this may incorporate only those QTLs segregating across lines/breeds.

Do We Need Knowledge About Genes and Gene Functions?

In practice, most GS methods mainly use SNP effects estimated from (non-)linear models or just GRM, and as such they do not utilize specific knowledge of genes, gene functions, or even exact locations of SNPs in the genome.

Does Accuracy Decline Over Generations?

It is commonly assumed that prediction of GBVs in GS is achieved by capitalizing on LD between markers and quantitative trait loci. However, in practice, and especially for within-population prediction, the reliability of genomic prediction depends on the strength of the genetic relationships of the animals in the candidate/test with the animals in the training/reference set. Hence, accuracy of prediction equations from a fixed training set declines quickly when applied over new generations. This generally means that the prediction equations need to be updated by adding animals from more recent generations.

Would Inbreeding Increase by Using GS?

The shorter generation intervals and higher intensity of selection from GS will lead to higher annual rates of inbreeding, especially inbreeding around selected markers. This would require extra attention in managing inbreeding in the population.

The use of genomic relationship information for careful mate selection, such as minimum-co-ancestry mating and optimum contribution selection, can control inbreeding.

GS in Aquaculture

There are large numbers of species in aquaculture, and the potential of GS in aquaculture will vary across different species owing to the differences in life cycle, fecundity, effective population size, and breeding objectives. Currently, traditional breeding programs with the most aquaculture species use mass selection and family-based selection. Sibling test is conducted for traits that cannot be directly measured on the selection candidates (e.g., traits recorded late in life or at slaughter, disease resistance). Family-based selection uses only a fraction (half) of genetic variance and results in increased inbreeding. GS can also predict genetic differences within families, and hence can utilize all genetic variations. Increases in inbreeding resulting from using only few animals/families is a major concern in aquaculture. In addition, tagging of individual animals, recording of pedigree, and rearing individual families separately, required in traditional selection schemes, are also difficult and expensive in aquaculture species. Information on genetic markers can help address some of these challenges.

The application of genetic markers and GS in aquaculture is becoming attractive with the decreasing costs of sequencing and genotyping. However, the development of SNP panels and application of genomic resources have been undertaken only recently in a few aquaculture species, and hence there is limited information available on the results of practical implementation of GS (Tsai *et al.*, 2015). The published reports are mainly based on the efficiency of GS using simulated data. All simulation studies demonstrate that aquaculture breeding programs can increase the accuracy of selection and genetic gains by using GS, both in production (continuous) and disease (dichotomous) traits (Sonesson & Meuwissen, 2009); (Nielsen, Sonesson & Meuwissen, 2011); (Lillehammer, Meuwissen & Sonesson, 2013), and reduce inbreeding by up to 81% as compared to traditional sib-testing (Sonesson & Meuwissen, 2009). It may also be possible to reduce the cost of genotyping by combining conventional BLUP family breeding values with within-family breeding values based on low-density genotyping without compromising genetic gain (Lillehammer *et al.*, 2013).

In spite of the decreasing cost of genotyping and sequencing, the GS strategy still incurs additional costs due to the logistic and genotyping of a large number of samples. However, part of this additional cost can be recovered from the increased genetic gains and being less reliant on sibling testing as compared to traditional selection schemes. The value of individual animals of most aquaculture species is generally low (e.g., in comparison to cattle). However, due to the high fecundity of most aquaculture species, the impact of genetic gains is amplified by the immediate transfer of genetic gains from the breeding nucleus to commercial ponds. However, further research on optimizing the breeding schemes using GS is required, such as in determining the optimum number of animals in training and selection sets, marker density, minimizing inbreeding for maximizing long-term genetic gains, and comparative economic analysis with traditional schemes. In addition to GS for production traits, genetic markers can offer additional advantages over traditional methods in aquaculture—for example, the use of sex-associated markers would allow the production of mono-sex commercial crops (Robinson *et al.*, 2014); introgression can be used to introduce desirable genes in a population, such as in transferring the resistance to specific diseases from local strains to commercial strains through repeated backcrossing and marker-assisted selection (Odegard *et al.*, 2009); markers can be used for surveillance and breeding out of any deleterious mutation from the population; and for tractability of genetic material to protect breeders.

In summary, with the fast developments in genomic technologies and the ever-decreasing costs of obtaining genomic information, whole genome selection holds great potential to improve the food production potential of aquaculture species while increasing profitability and maintaining genetic variability in improved stocks.

Acknowledgment

I am grateful to Dr. Gerhard Moser for his suggestions on the manuscript.

References

Calus, M.P., Meuwissen, T.H., de Roos, A.P. and Veerkamp, R.F. (2008) Accuracy of genomic selection using different methods to define haplotypes. *Genetics*, **178**, 553–561.

de Los Campos, G., Hickey, J.M., Pong-Wong, R. *et al.* (2013) Whole-genome regression and prediction methods applied to plant and animal breeding. *Genetics*, **193**, 327–345.

de Los Campos, G., Naya, H., Gianola, D. *et al.* (2009) Predicting quantitative traits with regression models for dense molecular markers and pedigree. *Genetics*, **182**, 375–385.

Endelman, J.B. (2011) Ridge regression and other kernels for genomic selection with R package rrBLUP. *Plant Genome*, **4**, 250–255.

Erbe, M., Hayes, B.J., Matukumalli, L.K. *et al.* (2012) Improving accuracy of genomic predictions within and between dairy cattle breeds with imputed high-density single nucleotide polymorphism panels. *Journal of Dairy Science*, **95**, 4114–4129.

Falconer, D.S. and Mackay, T.F.C. (1996) *Introduction to quantitative genetics*, 4th edn, Longman, Essex, UK.

Gianola, D. and van Kaam, J.B. (2008) Reproducing kernel Hilbert spaces regression methods for genomic assisted prediction of quantitative traits. *Genetics*, **178**, 2289–2303.

Goddard, M.E. and Hayes, B.J. (2009) Mapping genes for complex traits in domestic animals and their use in breeding programmes. *Nature Reviews Genetics*, **10**, 381–391.

Gonzalez-Camacho, J.M., de Los Campos, G., Perez, P. *et al.* (2012) Genome-enabled prediction of genetic values using radial basis function neural networks. *Theoretical and Applied Genetics*, **125**, 759–771.

Gonzalez-Recio, O., Weigel, K.A., Gianola, D. *et al.* (2010) L2-Boosting algorithm applied to high-dimensional problems in genomic selection. *Genetics Research (Cambridge)*, **92**, 227–237.

Habier, D., Fernando, R.L. and Dekkers, J.C. (2007) The impact of genetic relationship information on genome-assisted breeding values. *Genetics*, **177**, 2389–2397.

Habier, D., Fernando, R.L., Kizilkaya, K. and Garrick, D.J. (2011) Extension of the Bayesian alphabet for genomic selection. *BMC Bioinformatics*, **12**, 186.

Heslot, N., Yang, H.P., Sorrells, M.E. and Jannink, J.L. (2012) Genomic selection in plant breeding: a comparison of models. *Crop Science*, **52** (1), 146–160.

Jannink, J.L., Lorenz, A.J. and Iwata, H. (2010) Genomic selection in plant breeding: from theory to practice. *Briefings in Functional Genomics*, **9**, 166–177.

Kachman, S.D., Spangler, M.L., Bennett, G.L. *et al.* (2013) Comparison of molecular breeding values based on within- and across-breed training in beef cattle. *Genetics Selection Evolution*, **45**, 30.

Khatkar, M.S., Moser, G., Hayes, B.J. and Raadsma, H.W. (2012) Strategies and utility of imputed SNP genotypes for genomic analysis in dairy cattle. *BMC Genomics*, **13**, 538.

Khatkar, M.S., Zenger, K.R., Hobbs, M. *et al.* (2007) A primary assembly of a bovine haplotype block map based on a 15,036-single-nucleotide polymorphism panel genotyped in Holstein-Friesian cattle. *Genetics*, **176**, 763–772.

Lillehammer, M., Meuwissen, T.H. and Sonesson, A.K. (2013) A low-marker density implementation of genomic selection in aquaculture using within-family genomic breeding values. *Genetics Selection Evolution*, **45**, 39.

Maenhout, S., De Baets, B. and Haesaert, G. (2010) Prediction of maize single-cross hybrid performance: support vector machine regression versus best linear prediction. *Theoretical and Applied Genetics*, **120**, 415–427.

Meuwissen, T.H., Hayes, B.J. and Goddard, M.E. (2001) Prediction of total genetic value using genome-wide dense marker maps. *Genetics*, **157**, 1819–1829.

Moser, G., Tier, B., Crump, R.E. *et al.* (2009) A comparison of five methods to predict genomic breeding values of dairy bulls from genome-wide SNP markers. *Genetics Selection Evolution*, **41**, 56.

Nejati-Javaremi, A., Smith, C. and Gibson, J.P. (1997) Effect of total allelic relationship on accuracy of evaluation and response to selection. *Journal of Animal Science*, **75**, 1738–1745.

Nielsen, H.M., Sonesson, A.K. and Meuwissen, T.H. (2011) Optimum contribution selection using traditional best linear unbiased prediction and genomic breeding values in aquaculture breeding schemes. *Journal of Animal Science*, **89**, 630–638.

Odegard, J., Yazdi, M.H., Sonesson, A.K. and Meuwissen, T.H. (2009) Incorporating desirable genetic characteristics from an inferior into a superior population using genomic selection. *Genetics*, **181**, 737–745.

Ogutu, J.O., Piepho, H.P. and Schulz-Streeck, T. (2011) A comparison of random forests, boosting and support vector machines for genomic selection. *BMC Proceedings*, **5** (3), S11.

Pryce, J.E., Goddard, M.E., Raadsma, H.W. and Hayes, B.J. (2010) Deterministic models of breeding scheme designs that incorporate genomic selection. *Journal of Dairy Science*, **93**, 5455–5466.

Robinson, N.A., Gopikrishna, G., Baranski, M. *et al.* (2014) QTL for white spot syndrome virus resistance and the sex-determining locus in the Indian black tiger shrimp (*Penaeus monodon*). *BMC Genomics*, **15**, 731.

Schaeffer, L.R. (2006) Strategy for applying genome-wide selection in dairy cattle. *Journal of Animal Breeding and Genetics*, **123**, 218–223.

Sonesson, A.K. and Meuwissen, T.H.E. (2009) Testing strategies for genomic selection in aquaculture breeding programs. *Genetics Selection Evolution*, **41**, 1.

Torfs, P. and Brauer, C. (2014). "*A (very) short introduction to R*," Hydrology and Quantitative Water Management Group, Wageningen University, The Netherlands, available at https://cran.r-project.org/doc/contrib/Torfs+Brauer-Short-R-Intro.pdf

Tsai, H.Y., Hamilton, A., Tinch, A.E. *et al.* (2015) Genome wide association and genomic prediction for growth traits in juvenile farmed Atlantic salmon using a high density SNP array. *BMC Genomics*, **16**, 969.

VanRaden, P.M. (2008) Efficient methods to compute genomic predictions. *Journal of Dairy Science*, **91**, 4414–4423.

VanRaden, P.M., Van Tassell, C.P., Wiggans, G.R. *et al.* (2009) Invited review: reliability of genomic predictions for North American Holstein bulls. *Journal of Dairy Science*, **92**, 16–24.

Vazquez, A.I., de los Campos, G., Klimentidis, Y.C. *et al.* (2012) A comprehensive genetic approach for improving prediction of skin cancer risk in humans. *Genetics*, **192**, 1493–1502.

Venables, W.N., Smith, D.M. and the R Core Team. (2016). An introduction to R. Notes on R: A programming environment for data analysis and graphics. *Version 3.2.4 (2016-03-10)* Available: https://cran.r-project.org/doc/manuals/R-intro.pdf. Accessed 15 April 2016.

22

Quantitative Trait Locus Mapping in Aquaculture Species: Principles and Practice

Alejandro P. Gutierrez and Ross D. Houston

Introduction

Understanding the genetic basis of complex traits in aquaculture species is a major research goal. Quantitative trait loci (QTL) mapping is a first step toward identification of the genes and the causal variants underlying these traits. Genetic markers linked to QTL can also be applied in selective breeding schemes to improve the accuracy of selection for economically important traits via marker-assisted selection (MAS) or genomic selection. This has been achieved for a limited number of species, using microsatellite and, latterly, single-nucleotide polymorphism (SNP) markers. In the era of high-throughput sequencing, the genomic resources available for most aquaculture species are rapidly expanding, and now include reference genome assemblies, high-density linkage maps, and SNP arrays for several farmed species. Further, features of the reproductive biology of many aquatic species are amenable to well-powered genetic mapping studies, such as large full-sibling family sizes and flexible mating structures. These tools and features can expedite the discovery and application of QTL, and progress toward the underlying causative mechanisms. In this chapter, the methods and software typically used for QTL mapping using linkage analysis are reviewed, and examples given. The influence of developments in genomic technology and the opportunities it offers to the field are highlighted. Finally, successful examples of QTL mapping and application to improve economically important traits are given, and future directions for QTL research are discussed.

Selective Breeding in Aquaculture

Traditional selective breeding practices achieve genetic improvement in farmed animals via the selection of individuals for desirable phenotypic characteristics. For farmed aquatic species, these traits may include faster growth, disease resistance, robustness, morphology, and sexual maturation, among others. For some aquatic species, programs for improvement can incorporate cross-breeding, or the introduction of strains with desirable traits. However, the main driver of genetic improvement for important production traits in modern breeding schemes is within-strain selection. This is typically achieved within a well-managed, commercial program of family and pedigree tracking combined with extensive trait measurements on selection candidates or their relatives. However, while selective breeding programs have demonstrable benefits to

Bioinformatics in Aquaculture: Principles and Methods, First Edition. Edited by Zhanjiang (John) Liu.

the efficiency and sustainability of production, aquaculture species vary greatly in their level of technical sophistication. These range from production of largely unimproved domestic or wild stocks, to advanced family-based programs incorporating genomic tools—for example, in salmonid species (Gjedrem & Baranski, 2010; Yáñez, Newman & Houston, 2015). It is also important to consider the challenges of implementing a successful breeding program, including the large setup cost, the potentially slow progress for species with long generation intervals, and the complexity of the genetic architecture of the target traits. For example, complex traits are typically a result of many interacting genes and regulatory factors (Doerge, 2002), and care must be taken to avoid unwanted pleiotropic effects (e.g., desirable increases in salmon growth can be associated with unwanted early maturation; Gjerde, 1984).

Selective breeding programs for strain enhancement began in the late 1960s (Gjedrem & Baranski, 2010) with the development of Atlantic salmon breeding programs in Norway. In recent years, selective breeding programs in fish varieties have become more prevalent as fundamental knowledge of reproduction and inheritance has increased, and the aquaculture industry has become more organized. To date, selective breeding has been conducted for over 60 fish and shellfish species (Gjedrem & Baranski, 2010). Advances in sequencing technology and genomics have significantly improved the tools available for the genetic improvement of livestock. In particular, the development of genetic markers and linkage maps has permitted great advances in the quantitative analyses of commercially important traits (discussed in the following text). QTL are genomic loci contributing to variation in a complex trait—that is, one with a continuous distribution. The direct application of QTL in breeding programs via genetic markers is increasingly popular, and MAS has been applied for several species in the last decade. The use of genetic markers has the potential to improve the accuracy of selection and genetic gain by capitalizing on within-family genetic variation in addition to between-family variation (Sonesson, 2007). In recent years, this has been extended to include genomic selection, whereby genome-wide markers are used to calculate genomic breeding values without prior knowledge of the underlying QTL affecting the trait of interest (Meuwissen, Hayes & Goddard, 2001).

QTL Mapping in Aquatic Species

The search for QTL underpinning complex traits in aquatic species is also of interest from the perspective of fundamental genetics and biology. QTL mapping can be viewed as a first step toward identifying the genes and polymorphisms directly contributing to population variation in quantitative traits. There are two main approaches to the detection of QTL—namely, linkage mapping and association mapping (as discussed in detail in the following text). Association mapping is covered in Chapter 23; therefore, this chapter will focus on linkage mapping. Linkage mapping involves tracking of QTL via their genetic linkage to genetic markers within families. Aquaculture species have several advantages for linkage QTL mapping studies when compared to terrestrial livestock species, and in some ways to model organisms. Among the most important is the fact that most farmed aquatic species have high fecundity that provides a great number of progeny from a single mating. This enhances both the statistical power to detect QTL in outbred populations (primarily a function of within-family sample size), and the resolution of mapping (primarily a function of the number of recombination events; Mackay, 2009). These large family sizes and abundant progeny (with relatively low economic value per individual) also make large-scale experimental trials in controlled environments fairly inexpensive when compared to other livestock species. Finally, the external fertilization used for reproduction in many aquatic species has advantages in the ability to create a variety of crosses (e.g., factorial), although for mass-spawning species this also presents drawbacks, as discussed in the following text. To date, QTL for many important traits (e.g., cold and salinity tolerance, sex determination, growth traits, and disease resistance) have been mapped in over 20 aquaculture species (e.g., Laghari *et al.*, 2014; Yue, 2014).

Applications of Genomic Technology

The rapid development of high-throughput sequencing technologies has made it possible to rapidly detect and characterize a large number of DNA markers in species of interest using next-generation sequencing. In particular, thousands of SNPs and microsatellites have been identified for almost all the major aquaculture species (Liu, 2011). The availability of these markers genotyped for animals of known pedigrees allows the construction of dense linkage maps (Goddard & Hayes, 2009). Linkage maps use information on the co-segregation of genetic markers to assign them to groups (which typically correspond to chromosomes), and give measures of the linear order of markers within those groups on the basis of the recombination frequency between them. These maps are generally a prerequisite for QTL mapping, allowing as estimation of QTL position and confidence intervals (Visscher, Thompson & Haley, 1996). Confidence intervals are defined regions of the genome map that are postulated to contain genes associated with the particular quantitative trait of interest. The number, positions, and magnitude of the QTL affecting a trait are assessed by the statistical associations between marker genotypes and the phenotypes of interest (Lynch & Walsh, 1998). While microsatellites were the marker of choice for these studies due to their allelic diversity, the abundance and ease of scoring has meant that SNP markers are increasingly applied for both linkage- and association-based QTL mapping studies.

With the increasing numbers of SNPs and the various SNP genotyping platforms available, genome-wide association studies (GWAS) have gained popularity (see Chapter 23). There are fundamental differences in the principles of GWAS versus linkage-based QTL mapping as a means of detecting QTL underpinning quantitative trait variation. Association mapping typically exploits linkage disequilibrium in a large, heterogeneous population, while linkage mapping is based on tracking the co-segregation of markers and QTL from parents to offspring (Mackay, 2009). In practice, this means that linkage mapping is a powerful approach where clearly defined, large, half- or full-sibling families are available, even if only sparse marker density is available. In contrast, association mapping is suitable for a large outbred population with small family sizes, and requires high marker density for a genome-wide study. The statistical power of GWAS is a function of sample size, QTL effect size, and marker allele frequency (Stranger, Stahl & Raj, 2011). An important difference is that the confidence interval for the position of the QTL is generally smaller for a GWAS than for a linkage analysis (Kemper *et al.*, 2012), essentially meaning that the resolution of GWAS can be higher than linkage mapping. However, linkage QTL mapping (the focus of this chapter) is advantageous for species that are hampered by the lack of reference genome assemblies and genomic tools such as high-density SNP arrays. For aquatic species, this may include recently cultured/domesticate species, or those that have not yet attracted investment in development of the genomic toolbox. In this chapter, we will cover the main components of the linkage analysis of quantitative traits, including the type of markers used, the development of linkage maps, the procedures of QTL analysis, and some recent examples from aquaculture species.

DNA Markers and Genotyping

Genetic markers consist of polymorphisms (where individuals in a population carry different alleles) at certain positions on a chromosome. These genetic markers may be the result of previous mutation events in the evolutionary history of an organism/population that are retained due to selection, genetic drift, or both. This genomic variation is potentially useful for QTL mapping (and other genetics studies), assuming it is heritable and discernable to the researcher (Liu & Cordes, 2004). Genetic markers can comprise single base substitutions (i.e., SNPs), insertions or deletions of one base or more, inversion of a chromosome segment, or copy number variation. To be useful for QTL mapping, a genetic marker must exhibit sufficiently high variation among individuals or populations. Currently, the two types of DNA markers that are usually applied for linkage mapping in aquaculture species are microsatellites and SNPs. This is due to a combination of factors such as their informativeness, abundance, and simplicity of scoring. However, each marker type comes with its relative advantages and drawbacks, as discussed in the following text.

Microsatellites

Microsatellite markers are short tandem repeated (1–6 bp) DNA sequences of nucleotides (e.g., ATATATA) commonly found dispersed throughout a nuclear genome (Weber, 1990). These are abundant in all species studied to date, and are estimated to occur approximately every 10 KB in fishes (Wright, 1993). The repeat units of microsatellites are vulnerable to mutation and, as a result, these markers are typically highly polymorphic and can contain several alleles within a population. Importantly, they are co-dominant markers that exhibit Mendelian inheritance (Chauhan & Kumar, 2010). Further, their small size facilitates ease of detection using PCR, and, partly thanks to the introduction of next-generation sequencing technologies, the abundant microsatellites can be easily identified. However, the use of microsatellites does require up-front effort to sequence the corresponding genomic region and design high-quality PCR primers. Additionally, polymerase slippage during replication can make the discernment of small differences in size between alleles (as small as 2 bp) very difficult (Chauhan & Kumar, 2010). Perhaps the biggest disadvantage of microsatellites for modern genetic studies is that there is not a reliable, high-throughput means of scoring several thousand markers simultaneously. Nonetheless, to date, microsatellites have been utilized in a wide variety of genetic investigations in aquaculture species, and have been the tool used to discover the majority of the QTL published to date.

SNPs

SNPs describe polymorphisms caused by point mutations that give rise to different alleles containing alternative bases at a given nucleotide position within a locus. SNPs are becoming a focal point in molecular marker development since they represent the most abundant polymorphism in any organism's genome (coding and non-coding regions), are adaptable to automation, and reveal hidden polymorphism not detected with other markers (Liu, 2007). An SNP typically results in two alleles (corresponding to alternative base pairs) at a specific nucleotide position in the genome. Therefore, SNPs are regarded as biallelic and, as with microsatellites, are inherited as co-dominant markers. However, SNPs do have drawbacks such as the lack of information per locus as compared to microsatellites, and are less useful on a per-locus basis for several applications, including linkage mapping and parentage assignment. However, these drawbacks are more than offset by the simplicity of rapidly obtaining accurate, genome-wide marker datasets. The development of high-throughput sequencing technologies (especially Illumina sequencing) has made it routinely possible to obtain many thousands of SNP markers for downstream applications. This has been assisted by the release of the whole genome sequence of many aquaculture species such as Atlantic salmon (Davidson *et al.*, 2010), rainbow trout (Berthelot *et al.*, 2014), Atlantic cod (Star *et al.*, 2011), and Pacific oyster (Zhang *et al.*, 2012), together with current efforts to sequence the genome of many other aquaculture species, which is expected to provide a large number of SNPs (Yue, 2014). High-density SNP arrays have also been developed for several aquaculture species, including Atlantic salmon (132K, Houston *et al.*, 2014), rainbow trout (57K, Palti *et al.*, 2015), catfish (250K, Liu *et al.*, 2014), and carp (250K, Xu *et al.*, 2014). These SNP arrays (currently all on the Affymetrix platform) offer researchers the ability to survey many thousand genetic markers on a sample rapidly and accurately, albeit at a relatively high cost.

Genotyping by Sequencing

The direct approach of obtaining genotypes for individual animals using sequence data has become increasingly commonplace and cost-effective in aquaculture species. The advantages of this approach include: (1) SNP discovery and genotyping is essentially simultaneous, and (2) the marker dataset so obtained is specific to the population/families of interest (Houston *et al.*, 2012). These genotyping- by-sequencing techniques are typically based on digestion of the genomic DNA with restriction enzymes, followed by sequencing of the flanking regions of the restriction site. This complexity reduction step results in high read coverage at specific regions dispersed throughout the genome, enabling the accurate calling of SNP genotypes in multiplexed samples carrying a nucleotide barcode. A number of variations on this theme exist, including CRoPS (Van Orsouw *et al.*, 2007), RAD-Seq (Baird *et al.*, 2008; Miller *et al.*, 2007), GBS (Elshire *et al.*, 2011), double-digest RAD-Seq (Peterson *et al.*, 2012), and 2bRAD (see Chapter 19; also Wang *et al.*, 2012). The most commonly applied approach to date in aquaculture species is restriction site associated DNA (RAD) sequencing (RAD-Seq), which uses a single rare-cutting restriction enzyme for complexity reduction (Baird *et al.*, 2008). This approach is designed to sequence short regions surrounding all restriction recognition sites present in the genome. It begins with the digestion of a DNA sample using a chosen restriction enzyme (most commonly the 8 bp cutter SbfI). Barcoded adaptors are then ligated onto fragments with sticky end overhangs produced by the enzyme digestion, which allows the identification of each individual in a population (Davey *et al.*, 2011). Due to its low cost and high efficiency, RAD-Seq is becoming an important source of SNPs for aquaculture species such as Atlantic salmon (Houston *et al.*, 2014) and rainbow trout (Palti *et al.*, 2014). RAD-Seq is particularly useful for exploring genetic variation in non-model species, as it does not require a fully sequenced genome. For example, RAD-Seq has been applied to fine map disease resistance QTL in salmon (Houston *et al.*, 2012), to map growth-related QTL in pearl oysters (Li & He, 2014), and to map sex determination loci in halibut (Palaiokostas *et al.*, 2013a) and tilapia (Palaiokostas *et al.*, 2013b). RAD-Seq and similar techniques have also been widely used in creating genetic maps of the genomes of aquaculture species (see the following text).

Linkage Maps

Many of the details of genetic linkage mapping are covered in Chapter 20. Here, we will provide operational levels of the processes in relation to QTL mapping. A linkage map is an ordered listing of genetic markers along the length of all chromosomes. Since markers that are located physically close to one another on a chromosome tend to be inherited together, one can estimate the recombination distances that separate the markers by determining the frequency at which alleles at different loci are inherited together. Linkage maps are useful as references for determining the position of QTL on a genome, particularly in species without reference genome assemblies (the majority of aquaculture species). However, even where reference genomes are available, the physical map is not as informative as the genetic map for linkage mapping of QTL, since the primary driver of QTL detection and positioning is the recombination rate.

To construct a linkage map, at least one reference population/family is required where the DNA markers segregate and pedigree are known. The sample size required to create a useful map depends on the density of markers genotyped, and on the number of recombination events in the dataset. Traditionally, the creation of linkage maps was performed in reference crosses originating from a limited number of parents, utilizing the recombination within a single generation to order the markers within a chromosome. However, for high-resolution maps of genome-wide SNP data, a large number of individuals are required to sample adequate recombination events. In some cases, the use of multiple families (or indeed populations) could be necessary to map markers that have a low minor allele frequency, and to distinguish map positions for markers that are physically very close together on the genome.

There are several different methods of analyzing genotype data in the reference family/population to obtain a linkage map. At a simple level, to determine if two markers are linked or not, the LOD [logarithm (base 10) of odds] score is used (Risch & Giuffra, 1992), where a score threshold of 3 is traditionally considered as indicating significant linkage between two markers. The family structure and the number of families used in the analysis will determine which software is better suited to construct the map. As explained by Cheema and Dicks (2009), building a linkage map consists of three main tasks: grouping, ordering, and spacing. The grouping step divides the DNA markers into linkage groups, which ideally would illustrate a correspondence between the number of linkage groups and the number of chromosomes (i.e., the karyotype of the organism). The second step, ordering, aims to find the most likely order of the markers within the linkage group. For a linkage group of m markers, there are $m!/2$ possible orders; this requires extensive computing time in large datasets. The final step, spacing, aims to determine the map distances, in cM, between each adjacent pair of marker loci in a previously ordered set of markers in a linkage group, and hence the length of the linkage group as the sum of those distances. There is usually a positive correlation between the length of the linkage group and the physical length of the chromosome, although this clearly depends on the distribution of recombination events across the genome.

The software and pipelines used for constructing linkage maps in aquaculture species have traditionally included Joinmap (Stam, 1993), Mapchart (Voorrips, 2002), Mapmaker (Lincoln, Daly & Lander, 1993), Crimap (Green, Falls & Crooks, 1990), or LINKMFEX (R. Danzmann, University of Guelph, http://www.uoguelph.ca/˜rdanzman/software.htm). More recently, software containing algorithms is developed to cope with increasing marker density, and specifically to build maps for high-density SNP datasets. These include Carthagene (de Givry *et al.*, 2005), Mstmap (Wu *et al.*, 2008), AntMap (Iwata & Ninomiya, 2006), and Lep-MAP (Rastas *et al.*, 2013). The Lep-MAP software has been widely applied to create high-density SNP linkage maps, including those constructed for sea bass (Palaiokostas *et al.*, 2015), large yellow croaker (Ao *et al.*, 2015), chinook salmon (McKinney *et al.*, 2015), sockeye salmon (Larson *et al.*, 2016), and Atlantic salmon (Gonen *et al.*, 2015). The advantages of the Lep-MAP software include its suitability for large SNP datasets measured on multiple outbred families, its ability to account for achiasmatic meiosis, and its amenability to efficient cluster-based computing (Rastas *et al.*, 2013).

A great deal of progress has been made in the last few years in the development of linkage maps for aquaculture species. As of today, linkage maps for more than 40 species have been constructed, as reviewed by Yue (2014), including relatively dense microsatellite or SNP linkage maps for species of importance in aquaculture, including Atlantic salmon (Gonen *et al.*, 2014; Lien *et al.*, 2011), rainbow trout (Guyomard *et al.*, 2012; Palti *et al.*, 2012), Nile tilapia (Guyon *et al.*, 2012), catfish (Li *et al.*, 2015), Japanese flounder (Shao *et al.*, 2015), and Asian sea bass (Wang *et al.*, 2015), among others. First- and second-generation linkage maps of lower density are available for other species such as gilthead sea bream (Tsigenopoulos *et al.*, 2014), bighead carp (Zhu *et al.*, 2014), yellowtail (Aoki *et al.*, 2015), Pacific oyster (Hedgecock *et al.*, 2015), and South African abalone (Vervalle *et al.*, 2013), among others. In addition to providing a framework for subsequent QTL mapping, linkage maps have been valuable assets in the assembly of reference genomes for aquaculture species (e.g., Tine *et al.*, 2014).

Quantitative Trait Loci (QTL) Mapping

As described earlier, QTL are genomic regions (that could harbor genes or regulatory regions) contributing to genetic variation in a complex trait. The individual genes underlying quantitative traits usually have a minor effect, due to the polygenic genetic architecture of most traits. However, genetic architecture varies, and the goals of QTL mapping include assessing the number of putative loci underpinning a target trait, estimating their size of effect, and mapping these loci to a region of the genome. The ultimate goal of QTL mapping is usually to identify the causative genes and polymorphisms that contribute directly to control of the trait. However, a precious few examples exist to date where the causative genes/mutations have been identified. A shorter-term goal of QTL mapping is to develop genetic markers that can assist in breeding programs by incorporating MAS alongside pedigree-based selection (i.e., BLUP). Incorporating MAS into selection is desirable where it can enable within-family selection (particularly in aquaculture species that can have very large full-sibling families), or in cases in which MAS can reduce or replace the need for routine collection of trait data. The former situation is achieved where the marker and QTL are linked, but the latter requires population-wide linkage disequilibrium between the favorable QTL allele and one of the alleles at the marker. It is also worth noting that routine use of genomic selection is partially superseding MAS usage, although it is still very relevant in the case of QTL with major effects, and for species in which genomic selection is not yet implemented.

Traits of Importance

Selective breeding for aquaculture species has focused on traits related to productivity and welfare, such as growth, sexual maturation, and disease resistance. Other potentially important traits that have perhaps had less focus include fillet quality, feed conversion, stress response, and general robustness or survivability.

An important point to consider for selective breeding or QTL mapping is the heritability of the trait. Growth, morphometric, and disease resistance traits related to survival tend to have moderate or even high heritabilities, whereas general robustness, survivability, and stress response tend to be less heritable. Higher heritability means greater opportunity to make genetic gains per generation, and a higher chance of detecting significant QTL in mapping studies.

The initial goal of most selective breeding programs is focused on improving growth and/or body size (Gjedrem & Baranski, 2010), mainly because it is easy to measure and is highly heritable (Gjedrem, 2000). In line with this, growth is arguably the trait that has received the most attention in QTL mapping studies of aquaculture species. QTL affecting growth-related traits have been detected in most aquaculture species (Yue, 2014), including Atlantic salmon (e.g., Baranski, Moen & Vage, 2010; Gutierrez *et al.*, 2012; Reid *et al.*, 2005; Tsai *et al.*, 2015), rainbow trout (e.g., Wringe *et al.*, 2010), turbot (e.g., Sanchez-Molano *et al.*, 2011), Asian and European sea bass (e.g., Massault *et al.*, 2010; Wang *et al.*, 2015), common carp (e.g., Laghari *et al.*, 2013), scallop (e.g., Li, Liu & Zhang, 2012), Pacific oyster (e.g., Guo *et al.*, 2011), and several others. Repeatable major growth QTL have generally not been identified, suggesting that it is likely to be a polygenic trait, and/or that QTL are specific to families, populations, and environments. Genotype–environment interaction has been demonstrated in several studies to be important in the regulation of growth phenotypes in aquaculture species (e.g., Sae-Lim *et al.*, 2014).

Sexual maturation is another important trait for breeders of certain aquaculture species, since it is an energetically expensive process, and precocious maturation results in impaired growth, reduced meat quality, and even increased mortality through susceptibility to pathogens (Küttner *et al.*, 2011; Thorpe, 1994). QTL analyses for this trait have been carried out in many species such as Atlantic salmon (Gutierrez *et al.*, 2014; Pedersen *et al.*, 2013), rainbow trout (Easton *et al.*, 2011; Haidle *et al.*, 2008), and Arctic char (Küttner *et al.*, 2011; Moghadam *et al.*, 2007). Interestingly, a major QTL affecting sexual maturation has recently been identified in Atlantic salmon by two research groups (Ayllon *et al.*, 2015; Barson *et al.*, 2015). This QTL is estimated to explain 33–39% of the phenotypic variation in the trait, with the VGLL3 gene likely to be causative for this QTL (Ayllon *et al.*, 2015; Barson *et al.*, 2015).

Disease outbreaks are a major problem for the culturing of most aquatic species, and therefore disease resistance has also received much attention in aquaculture breeding programs worldwide. Disease resistance is commonly measured during challenge studies, whereby the mortality and survival of animals with known pedigree are recorded. While the heritability of this trait is frequency-dependent (i.e., maximal at a prevalence of 0.5; Bishop & Woolliams, 2010), moderate–high heritabilities have been observed, highlighting the potential for genetic improvement (Ødegård *et al.*, 2011). One of the most successful examples of QTL analyses applied to selective breeding is the case of IPN resistance in Atlantic salmon, where a major QTL explains the majority of the genetic variance for resistance (Houston *et al.*, 2008; Moen *et al.*, 2009), and has been demonstrated as a successful means of controlling the disease (Moen *et al.*, 2015). Other QTL analyses for disease resistance in Atlantic salmon include resistance to salmonid alphavirus, which causes pancreas disease (Gonen *et al.*, 2015), ISAv (Moen *et al.*, 2007), and *Gyrodactylus salaris* parasitic disease (Gilbey *et al.*, 2006). QTL mapping for disease resistance has also been performed for lymphocystis disease in Japanese flounder (Fuji *et al.*, 2006); MSX and Dermo in eastern oyster (Yu & Guo, 2006); Bonamiosis in the European flat oyster (Lallias *et al.*, 2009); and *Flavobacterium psychrophilum* in rainbow trout (Vallejo *et al.*, 2013), among others. Even within the salmonid species alone, dozens of putative QTL identified for disease resistance traits on most chromosomes (Yáñez, Houston & Newman, 2014).

Mapping Populations

QTL mapping studies for aquatic species fall into two broad categories: those based on controlled laboratory experiments on defined families/populations, and those based on data and samples collected in "field" environments (e.g., disease outbreaks in seawater cages, or measurements taken at harvest processing). Each of these comes with its advantages: experimental trials allow much greater control of the environmental conditions and family structure, while data collected from field environments have direct relevance to production under commercial conditions. Considering the case of disease resistance, experimental challenges can facilitate the analysis of crosses between lines (e.g., lines known to be resistant or susceptible to a particular disease), or simply a collection of outbred families where pedigree is known or can be established. Where experimental crosses are established, these may be backcrosses whereby F1 (or subsequent generations) progeny derived from a cross between two lines are backcrossed to the recurrent parent. Alternatively, an F2 intercross design may be applied, whereby these F1 (or subsequent generations) are crossed to each other to establish an F2 mapping population. The idea behind these crosses is to ensure that the mapping population is segregating for QTL that contribute to the phenotypic differences between the founder strains. It is also possible to repeat backcrossing for several generations such that the QTL interval can be narrowed, and this is an approach typically taken with inbred lines. Where outbred families are used for mapping, these tend to be closely related to commercial broodstock, such that the QTL that are segregating in these families are also likely to be segregating in the general breeding population. Finally, for field data collection, it is not always necessary to know family structure going into the experiment because parentage assignment can easily be performed concurrently with QTL mapping, and/or genomic kinship matrices can be used *in lieu* of pedigree data (although this would be better suited to association mapping than linkage mapping). One common theme of linkage mapping is that large family sizes are preferable for adequate statistical power. In typical livestock QTL mapping designs, this is often achieved by a half-sibling design, whereby sires are mated to several dams. However, one of the key advantages of aquaculture species is that large full-sibling families are readily available in the natural course of the breeding cycle.

One important consideration for QTL mapping population design is for mass-spawning species (e.g., sea bass and sea bream), where the effective number of parents contributing to progeny is considerably lower than the actual number of parents used for a given mating structure (Massault *et al.*, 2008). Since some parents tend not to contribute at all, whereas others can dominate the population due to a high fertilization success, the QTL mapping experiments for these species require knowledge of the population structure—that is, family type and size (Vandeputte *et al.*, 2004)—and must account for its study design and analysis (Massault *et al.*, 2008).

Finally, and particularly relevant for salmonid species, the rate and distribution of recombination across the genome of the species of interest need to be considered and accounted for in mapping population design and methods. Recombination events are not randomly distributed across the genome, and tend to cluster into recombination hotspots. As such, the marker density required in areas of high recombination is substantially more than that required across the rest of the genome. In salmonid fish, males have very low recombination across much of the genome, with higher rates typically only observed close to the telomeres of the chromosomes. This means that QTL mapping can be quite efficient by genotyping only a few markers in male parents, and by tracking the male inheritance of marker genotypes for the detection of QTL. Subsequently, for chromosomes demonstrated to harbor QTL from the sire-based analysis, increased marker density can be obtained for the significant chromosomes, and the QTL can be positioned using a dam-based analysis. This strategy has been applied successfully in several studies of salmonid fish (e.g., Hayes, Gjuvsland & Omholt, 2006; Houston *et al.*, 2008; Gonen *et al.*, 2015). The effectiveness of this approach does depend on the genotyping platform used, and it would not be as useful where an SNP chip has been used to obtain marker genotypes, because all animals are already genotyped using a relatively high-density marker panel.

QTL Mapping Methods

At the basic level, the principle of QTL mapping is to assess the difference between the mean phenotype of individuals carrying one genotype at a locus versus those carrying the alternative genotype at a locus. However, since the underlying locus is not known, anonymous genetic markers linked to that locus are used for analysis. Therefore, one of simplest methods used for linkage QTL mapping is the analysis of variance (ANOVA; also known as simple linear regression). However, this method has several fundamental drawbacks; it does not provide reliable estimates of QTL effect or QTL position (Broman *et al.*, 2003), and it does not account for multiple testing of markers. Interval mapping overcomes some of these drawbacks, and is more commonly used for identifying QTLs. It was proposed by Lander and Botstein (1989) as an improvement of the previous methods by considering flanking markers. Interval mapping uses the flanking markers in the context of a genetic map to give a more precise estimate of the location and effect of a QTL. It does, however, require more computational effort and specialized software for analysis (Broman *et al.*, 2003). An approximation of this method was developed by Haley and Knott (1992), which requires much less computing power and is implementable in standard and easy-to-use statistical packages. However, the estimates of locations and effects of QTL under these models can be biased when the QTL are linked (Haley & Knott, 1992). Considering this, composite interval mapping can be applied, whereby multiple QTL can be accounted for (on other intervals or chromosomes), and this increases the precision of QTL detection (Jansen & Stam, 1994; Zeng, 1994). A good alternative to linkage analysis is the joint use of linkage disequilibrium and linkage analysis simultaneously (also called LDLA) (Meuwissen *et al.*, 2002), which permits mapping of QTL more accurately than linkage analysis alone. LDLA uses both the linkage and LD information to determine the probability of a QTL being present in a particular genomic location (Hayes *et al.*, 2006).

Software for QTL Analysis

Many software programs are now available for QTL analysis to utilize the statistical methods described earlier. Most of these are stand-alone software packages, including efficient and user-friendly alternatives for beginners. Depending on the experimental design planned for QTL mapping, GridQTL (Seaton *et al.*, 2006) and MapQTL (Van Ooijen & Kyazma, 2009) are good alternatives that can handle inbred as well as outbred populations, a great number of markers (from hundreds to thousands), and numerous families, and have been frequently used in the QTL analysis of aquaculture species. GridQTL (http://www.gridqtl.org.uk/; Seaton *et al.*, 2006) is a freely available web-based software capable of performing QTL analyses for different family structures, such as sib-pair families, half-sibs, F2, and backcrosses. However, unfortunately, the tool to perform LDLA in tandem with haplotyping is no longer available; however, alternative methods are available (Druet *et al.*, 2008) and have been implemented in other software such as MapQTL (Gilbert *et al.*, 2008). The GridQTL software offers a user-friendly environment without the need for programming skills, and has a relatively short learning curve. Since its release, GridQTL has been used extensively in QTL analysis for a variety of traits in aquaculture species (e.g., Boulton *et al.*, 2011; Gutierrez *et al.*, 2012, 2014; Kuang *et al.*, 2015; Pedersen *et al.*, 2013; Sanchez-Molano *et al.*, 2011; Vallejo *et al.*, 2013). MapQTL is an alternative software package that runs similar models and is user-friendly, but requires a license (Van Ooijen & Kyazma, 2009). Other popular software options for QTL linkage mapping include programs that were developed using the R package—for example, R/qtl (Broman *et al.*, 2003) and R/qtlbim (Yandell *et al.*, 2007)—or others developed to work under the SAS environment, such as the SAS PROC package (Hu & Xu, 2009).

QTL Mapping Example

Since GridQTL appears to be the most popular tool used for linkage-based QTL analysis in aquaculture species, the following is an example of how a QTL mapping study may be performed using this software. Other software tools (e.g., MapQTL or R/QTL) possess similar features as GridQTL, but there are operational differences and nuances in input file formats that need to be considered. The vast majority of software packages have instruction manuals easy to read and easy to follow for creating input files and running the software. In general, good practice in data management and manipulation of file formats is essential to ensure robust and repeatable results of a QTL mapping study. Note that, currently, one must apply for an account to be able to run the GridQTL software.

Assuming a system similar to that used for the testing of salmonid fish, this example will focus on 10 full-sibling families of 100 fish each ($n = 1000$) in which all parents are known. The progeny of these families are reared in a communal environment and measured for a trait (e.g., body size at two specific time points). In trials of salmonid fish (and other finfish species), it is common to tag the fish (e.g., using a Passive Integrated Transponder, or PIT, tag) such that family identification is possible in a mixed family environment. At the same time, a small sample of fin tissue can be taken for DNA extraction and genotyping. Alternatively, if PIT tagging prior to family mixing is not possible (e.g., in a mass spawning system, or with mixing of families at a very young age), then samples can be taken of parents and progeny, and the pedigree constructed using genotyping alone. Assuming a genome size of approximately 3 gigabases, and a linkage map size of approximately 3 Morgans (considering female-based maps), then a few hundred microsatellite markers evenly distributed across the genome is sufficient for a genome scan (but consider the disparity between male and female recombination rates discussed earlier). Alternatively, if SNPs are used, then marker density must be higher due to the lower information content and polymorphism rate per locus. Previously, QTL mapping studies in salmonids have successfully applied an SNP chip of approximately 6K SNPs for this purpose (Baranski *et al.*, 2010; Gutierrez *et al.*, 2012, 2014; Pedersen *et al.*, 2013). GridQTL can deal with both SNP and microsatellite genotype data. Prior to running the QTL analysis, a linkage map that is specific to the families of the study can be created. This can be performed using software such as CriMap (Green *et al.*, 1990) or JoinMap (Stam, 1993), and the output files from the linkage mapping software can be readily transferred into formats suitable for input into the GridQTL software.

The first stage of applying the GridQTL analysis is to choose a suitable model to analyze the data, according to the design of the experiment. For the example population described earlier, either the "sib-pair" analysis or the "half-sibling" analysis (adjusted to account for the full-sibling families, as described in the following text) could be performed. In the sib-pair module of the software, a variance approach is used, expecting that siblings that inherit more QTL alleles identical by descent (IBD) tend to be more phenotypically similar. As such, the differences in their phenotypes tend to be smaller, the more QTL alleles that share IBD (Haseman & Elston, 1972). IBD calculations are made using an approximate algorithm that is fast and requires the parental genotypes to be known (Knott & Haley, 1998). Phenotypes are then adjusted for non-genetic fixed effects and covariates in the model. In the earlier example, the "sex" of the fish could be included as a fixed effect. In the "half-sib" module of the program, the first step uses a multimarker approach for interval mapping, as described by Knott, Elsen, and Haley (1996). Briefly, for each offspring, the probability of inheriting the parent's first haplotype of a linkage group is calculated at fixed intervals (typically 1 cM), conditional on its marker genotype. Subsequently, a putative QTL is fitted and tested at fixed intervals along the linkage group by regression of phenotype on the probability of inheriting the first haplotype of the parent. In the case described earlier with multiple families, the analysis is nested within families, and the residuals are pooled across families to compute a test statistic (de Koning *et al.*, 2001). This test statistic is calculated as an F-ratio for every map position within and across families. Since the family structure is actually a full-sibling one rather than a half-sibling one *per se*, the analyses can be repeated using either the male or female parents. Later, when computing the percentage of variation explained by QTL, the results from the sire- and dam-based analyses can be combined according to the following formula:

$$h^2\text{QTL} = 2 \times \left\{ \left[1 - \left(\frac{MSEfull}{MSEreduced}\text{Sire}\right)\right] + \left[1 - \left(\frac{MSEfull}{MSEreduced}\text{Dam}\right)\right] \right\}$$

Here, *MSEfull* is the mean square error of the model including the QTL, and *MSEreduced* is the equivalent fitting only a family mean (Houston *et al.*, 2008).

There are three input files required for the GridQTL software. The first contains the genotypes of all the analyzed markers in every sample (although missing genotypes are permitted). This file can be produced for each chromosome separately, or combined for the entire genome. It must contain six main lines—line 1 for the number of markers on the chromosome/genome; line 2 for the list of names of the markers; line 3 not used; line 4 to show the identifier of males and females in the file (one per column, e.g., "M" and "F"); line 5 for the identifier of missing genotypes; and line 6 containing genotype data for each individual, but divided into six new columns. The genotype data is itself then formatted—column 1 for the individual ID; column 2 to identify the sire of the progeny (must be present in the file, and use 0 if unknown); column 3 to identify the dam of the progeny (if unknown, use 0); columns 4 and 5 not used in sib-pair analysis; and column 6 contains the genotypes of the markers listed in line 2, showing two alleles per marker (one per cell/tab). An example of an input genotype file is given in Figure 22.1, and it can be prepared in indeed any program capable of producing tabular files, such as Microsoft Excel, Notepad++ or Textpad.

The second input file contains the map location of the markers, as calculated in the linkage mapping section described in the preceding text. As with the genotype file, this map file can be prepared for the entire genome, or for each chromosome separately. This file must contain the location of every marker included in the genotype file. The file consists of four lines. Line 1 contains the number of chromosomes included in the map of the genome. Line 2 contains the interval for the calculation of probabilities and coefficients in cM (1 cM is usually used). Line 3 contains three columns—the first one for the chromosome name (e.g., C1, C2, etc.), the second containing the number of markers on the chromosome, and the third one to indicate if the map is combined for both sexes (=1) or not (=0) (in the case of salmonids with large differences in recombination rates between the sexes, sex-specific maps are usually most appropriate). Line 4 gives the list of IDs of the markers on the aforementioned chromosome. The marker names must correspond to those in the genotype file. Further, it is important to note that the figures between the markers are the distances between the markers, and not the cumulative map position. An example of a map input file is given in Figure 22.2.

500												
M1	M2	M3	M4	M5	M6	M7	M8	M9	M10	M11	M12	MX
THIS LINE IS NOT USED												
M	F											
0												
SIRE	0	0	UNUSED	UNUSED	1	3	1	3	3	3	1	4
DAM	0	0	UNUSED	UNUSED	3	3	1	1	1	3	1	1
ID1	DAM	SIRE	UNUSED	UNUSED	3	3	1	3	1	3	1	1
ID2	DAM	SIRE	UNUSED	UNUSED	1	3	1	3	3	3	1	1
ID3	DAM	SIRE	UNUSED	UNUSED	3	3	1	1	3	3	1	4
ID4	DAM	SIRE	UNUSED	UNUSED	3	3	1	1	3	3	1	1
ID5	DAM	SIRE	UNUSED	UNUSED	3	3	1	1	3	3	1	1
ID6	DAM	SIRE	UNUSED	UNUSED	3	3	1	3	1	3	1	1
ID7	DAM	SIRE	UNUSED	UNUSED	1	3	1	3	1	3	1	4
ID8	DAM	SIRE	UNUSED	UNUSED	3	3	3	3	3	3	1	1
IDx	DAM	SIRE	UNUSED	UNUSED	1	3	1	1	3	3	1	4

Figure 22.1 The example structure of a genotype input file for GridQTL.

```
3
1
C1      9       1
M1 2.5  M2  6.7 M3  1.1 M4  2.8 M5  3.9 M6  3.7 M7  3.8 M8  6.6  M9
C2      5       1
M10  1.3 M11  9.7 M12  1.6  M13  3.5  M14
C3      10      1
M15   2.6  M16  3  M17  10  M18  9.6  M19  5.6  M20  1.8  M21  3.8  M22  3.6  M23  5.1  M24
```

Figure 22.2 The example structure of a map input file for GridQTL, consisting of three chromosomes.

The third and final file contains phenotype values of the trait to be analyzed. Line 1 must contain three columns—one each for the number of fixed effects (e.g., sex in our example), covariates, and the traits of interest. Line 2 contains the name of each fixed effect, covariate, and trait. Line 3 identifies the code for missing values. Line 4 gives the individual IDs and trait values. The number of columns depends on the number of fixed effects, covariates, or traits, but must begin with an individual ID that matches the genotype file. An example of a trait input file is given in Figure 22.3.

Once these three files have been uploaded to the server and checked for errors (e.g., Mendelian inconsistencies will be reported), the software will give you the option to initiate the QTL analysis. The software has the option to obtain appropriate significance thresholds empirically by permutation analysis (Churchill & Doerge, 1994). These thresholds are commonly defined as genome-wide significant (the level at which one false positive is expected in 20 genome scans; Lander & Kruglyak, 1995). In addition, a chromosome-wide threshold can be calculated for each chromosome separately, which is akin to a suggestive threshold but is dependent on the relative size of the chromosome/linkage group. Further, it is possible to run a bootstrapping analysis to calculate confidence intervals for the QTL (Visscher *et al.*, 1996), whereby the top and bottom 2.5% of resampled position estimates define the 95% confidence interval for the QTL. For each of these processes, there are different options for the number of permutations. While higher numbers of permutations are desirable, computing time must be considered, particularly on publicly available servers. The relevant fixed effects can be selected to be tested in the analysis as well as genetic background effects or cofactors (i.e., markers

Figure 22.3 The example structure of a trait input file for GridQTL—in this case, for two traits with one fixed effect.

```
1     0     2
sex   Weight1Weight2
***
ID1   1     35.8   384.2
ID2   0     40     520
ID3   1     41.3   818.7
ID4   0     36     684
ID5   0     30.3   609.7
ID6   1     36.3   693.7
IDx   0     33.3   746.4
```

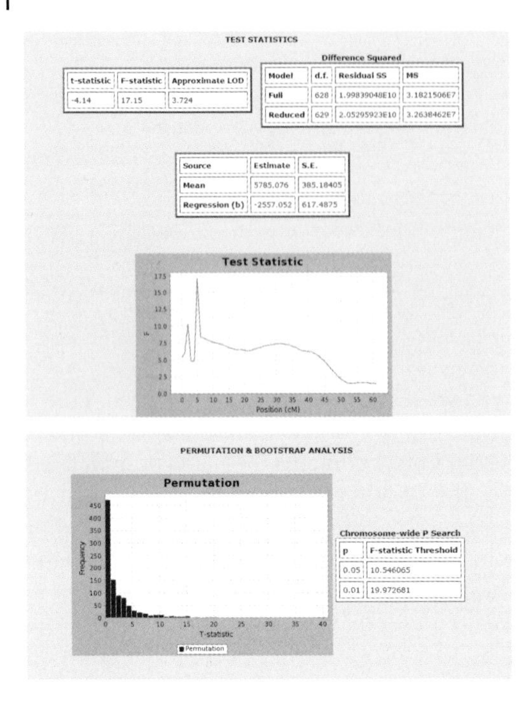

Figure 22.4 An example of the graphical interface to show the results of the sib-pair analysis, showing the detection of a chromosome-wide significant QTL ($p < 0.05$) at 5 cM. The significance level was determined according to the permutation analysis, which shows that the F-statistics threshold for a chromosome-wide significance ($p < 0.05$) is 10.55. The F-statistics value of the QTL at 5 cM is 17.15; therefore, it can be considered chromosome-wide significant.

that have a known association with a trait); however, their use could affect the QTL detection (Jansen, 1994).

Results are shown in a graphical interface, giving the test statistic information associated with the best estimate position of a QTL (if any), particularly the F-statistic value and LOD score. The trait mean, sums of squares, and mean square values with and without the QTL (full and reduced model) are given and can be applied to calculate the percentage of variance explained (described earlier). If a permutation analysis has been performed, the genome-wide or chromosome-wide significance threshold value will also be given (as an F-statistic value). QTL with test statistics exceeding these threshold values are considered significant. An example of the output file is given in Figure 22.4.

Future Directions and Applications of QTL

QTL analyses have been carried out for many aquaculture species, and have been applied to study almost all economically important traits (Laghari *et al.*, 2014; Yue, 2014). However, while a large number of QTL have been identified, most QTL span large regions of the genome, and replication of QTL results for typical complex traits across populations has not been particularly successful. The potential of QTL mapping in aquaculture species is greater than for livestock species, because aquaculture species have been recently domesticated and may face relatively new selection pressures. As such, the standing genetic variation in aquaculture populations may include loci of major effect. A number of quite successful examples of the identification of major QTL have been highlighted, and their implication is discussed in the following text. However, one of the major objectives of QTL identification is to know the genes and causative variants underpinning complex traits, and to use them in MAS to increase high-quality production (Laghari *et al.*, 2014). As such, fine mapping and confirmation of QTL effects in independent populations is important. While several good examples exist, the majority of studies on aquaculture species have not yet progressed to fine mapping of underlying causative genes, or verification in independent populations as a prerequisite to MAS. Arguably, the value of such studies searching for individual loci is of limited commercial value in the era of genomic selection, unless those QTL happen to have major effects. However, it is important to note that a (limited) number of major QTL have already been identified in aquaculture (e.g., IPNV resistance and sexual maturation in salmon), and knowledge of the regulation of a trait at the level of individual genes is a fundamental goal of biology.

Only a few examples exist where QTL mapping has led to the implementation of MAS for favorable alleles in aquaculture breeding schemes. These include selection for Japanese flounder resistant to lymphocystis disease (Fuji *et al.*, 2006), and for Atlantic salmon resistant to infectious pancreatic necrosis virus (IPNV) (Houston *et al.*, 2008, 2010; Moen *et al.*, 2009, 2015). In both cases, major QTL were detected that explain most of the genetic variation in host resistance to the disease. Recently, making use of advances in genomic technologies for the identification of SNPs, major QTL were detected for resistance to pancreas disease virus (Gonen *et al.*, 2015), and for sexual maturation (Ayllon *et al.*, 2015; Barson *et al.*, 2015) in salmon. In the latest studies, association mapping is commonly applied as a means of identifying QTL, and this is likely to continue as the cost of generating genome-wide marker data across large numbers of samples decreases. Studies based on linkage QTL mapping and linkage disequilibrium–based mapping are both useful means of moving toward identification of the individual genes underpinning traits of economic importance. As large-scale genomic and trait datasets become more routinely available for aquaculture species, new opportunities are becoming available to harness these data to identify these causative genes.

References

Ao, J., Li, J., You, X. *et al.* (2015) Construction of the high-density genetic linkage map and chromosome map of large yellow croaker (*Larimichthys crocea*). *International Journal of Molecular Sciences*, **16** (11), 26237–26248.

Aoki, J.-y., Kai, W., Kawabata, Y., *et al.* (2015). Second generation physical and linkage maps of yellowtail (*Seriola quinqueradiata*) and comparison of synteny with four model fish. *BMC Genomics*, **16**, 1–11.

Ayllon, F., Kjærner-Semb, E., Furmanek, T. *et al.* (2015) The vgll3 locus controls age at maturity in wild and domesticated Atlantic salmon (*Salmo salar L.*) males. *PLoS Genetics*, **11** (11), e1005628.

Baird, N.A., Etter, P.D., Atwood, T.S., *et al.* (2008). Rapid SNP discovery and genetic mapping using sequenced RAD markers. *PLoS One*, **3**, e3376.

Baranski, M., Moen, T. and Vage, D. (2010) Mapping of quantitative trait loci for flesh colour and growth traits in Atlantic salmon (*Salmo salar*). *Genetics Selection Evolution*, **42**, 17.

Barson, N.J., Aykanat, T., Hindar, K. *et al.* (2015) Sex-dependent dominance at a single locus maintains variation in age at maturity in salmon. *Nature*, **528**, 405–408.

Berthelot, C., Brunet, F., Chalopin, D. *et al.* (2014) The rainbow trout genome provides novel insights into evolution after whole-genome duplication in vertebrates. *Nature Communications*, **5**, 3657.

Bishop, S.C. and Woolliams, J.A. (2010). On the genetic interpretation of disease data. *PLoS One*, **5**, e8940.

Boulton, K., Massault, C., Houston, R.D. *et al.* (2011) QTL affecting morphometric traits and stress response in the gilthead seabream (*Sparus aurata*). *Aquaculture*, **319**, 58–66.

Broman, K.W., Wu, H., Sen, Ś. and Churchill, G.A. (2003) R/qtl: QTL mapping in experimental crosses. *Bioinformatics*, **19**, 889–890.

Chauhan, T. and Kumar, R. (2010) Molecular markers and their applications in fisheries and aquaculture. *Advances in Bioscience and Biotechnology*, **01**, 281–291.

Cheema, J. and Dicks, J. (2009) Computational approaches and software tools for genetic linkage map estimation in plants. *Briefings in Bioinformatics*, **10**, 595–608.

Churchill, G.A. and Doerge, R.W. (1994) Empirical threshold values for quantitative trait mapping. *Genetics*, **138**, 963–971.

Davey, J.W., Hohenlohe, P.A., Etter, P.D. *et al.* (2011) Genome-wide genetic marker discovery and genotyping using next-generation sequencing. *Nature Reviews Genetics*, **12**, 499–510.

Davidson, W., Koop, B., Jones, S. *et al.* (2010) Sequencing the genome of the Atlantic salmon (*Salmo salar*). *Genome Biology*, **11**, 403.

de Givry, S., Bouchez, M., Chabrier, P. *et al.* (2005) Carh ta Gene: multipopulation integrated genetic and radiation hybrid mapping. *Bioinformatics*, **21**, 1703–1704.

de Koning, D.J., Schulmant, N.F., Elo, K. *et al.* (2001) Mapping of multiple quantitative trait loci by simple regression in half-sib designs. *Journal of Animal Science*, **79**, 616–622.

Doerge, R.W. (2002) Mapping and analysis of quantitative trait loci in experimental populations. *Nature Reviews Genetics*, **3**, 43–52.

Druet, T., Fritz, S., Boussaha, M. *et al.* (2008) Fine mapping of quantitative trait loci affecting female fertility in dairy cattle on BTA03 using a dense single-nucleotide polymorphism map. *Genetics*, **178**, 2227–2235.

Easton, A.A., Moghadam, H.K., Danzmann, R.G. and Ferguson, M.M. (2011) The genetic architecture of embryonic developmental rate and genetic covariation with age at maturation in rainbow trout *Oncorhynchus mykiss*. *Journal of Fish Biology*, **78**, 602–623.

Elshire, R.J., Glaubitz, J.C., Sun, Q., *et al.* (2011). A robust, simple genotyping-by-sequencing (GBS) approach for high diversity species. *PLoS One*, **6**, e19379.

Fuji, K., Kobayashi, K., Hasegawa, O. *et al.* (2006) Identification of a single major genetic locus controlling the resistance to lymphocystis disease in Japanese flounder (*Paralichthys olivaceus*). *Aquaculture*, **254**, 203–210.

Gilbert, H., Le Roy, P., Moreno, C. *et al.* (2008) QTLMAP, a software for QTL detection in outbred population. *Annals of Human Genetics*, **72**, 694.

Gilbey, J., Verspoor, E., Mo, T.A. *et al.* (2006) Identification of genetic markers associated with *Gyrodactylus salaris* resistance in Atlantic salmon *Salmo salar*. *Diseases of Aquatic Organisms*, **71**, 119–129.

Gjedrem, T. (2000) Genetic improvement of cold-water fish species. *Aquaculture Research*, **31**, 25–33.

Gjedrem, T. and Baranski, M. (2010) *Selective breeding in aquaculture: an introduction*, Springer, Dordrecht.

Gjerde, B. (1984) Response to individual selection for age at sexual maturity in Atlantic salmon. *Aquaculture*, **38**, 229–240.
Goddard, M.E. and Hayes, B.J. (2009) Mapping genes for complex traits in domestic animals and their use in breeding programmes. *Nature Reviews Genetics*, **10**, 381–391.
Gonen, S., Lowe, N.R., Cezard, T. *et al.* (2014) Linkage maps of the Atlantic salmon (*Salmo salar*) genome derived from RAD sequencing. *BMC Genomics*, **15**, 1–17.
Gonen, S., Baranski, M., Thorland, I. *et al.* (2015) Mapping and validation of a major QTL affecting resistance to pancreas disease (salmonid alphavirus) in Atlantic salmon (*Salmo salar*). *Heredity*, **115**, 405–414.
Green, P., Falls, K. and Crooks, S. (1990) *Documentation for CRI-MAP, version 2.4*, Washington University School of Medicine, St. Louis, MO.
Guo, X., Li, Q., Wang, Q.Z. and Kong, L.F. (2011) Genetic mapping and QTL analysis of growth-related traits in the Pacific oyster. *Marine Biotechnology*, **14**, 218–226.
Gutierrez, A.P., Lubieniecki, K.P., Davidson, E.A. *et al.* (2012) Genetic mapping of quantitative trait loci (QTL) for body-weight in Atlantic salmon (*Salmo salar*) using a 6.5K SNP array. *Aquaculture*, **358–359**, 61–70.
Gutierrez, A.P., Lubieniecki, K.P., Fukui, S. *et al.* (2014) Detection of quantitative trait loci (QTL) related to grilsing and late sexual maturation in Atlantic salmon (*Salmo salar*). *Marine Biotechnology*, **16**, 103–110.
Guyomard, R., Boussaha, M., Krieg, F. *et al.* (2012) A synthetic rainbow trout linkage map provides new insights into the salmonid whole genome duplication and the conservation of synteny among teleosts. *BMC Genetics*, **13**, 15.
Guyon, R., Rakotomanga, M., Azzouzi, N. *et al.* (2012) A high-resolution map of the Nile tilapia genome: a resource for studying cichlids and other percomorphs. *BMC Genomics*, **13**, 1–17.
Haidle, L., Janssen, J.E., Gharbi, K. *et al.* (2008) Determination of quantitative trait loci (QTL) for early maturation in rainbow trout (*Oncorhynchus mykiss*). *Marine Biotechnology*, **10**, 579–592.
Haley, C.S. and Knott, S.A. (1992) A simple regression method for mapping quantitative trait loci in line crosses using flanking markers. *Heredity*, **69**, 315–324.
Haseman, J.K. and Elston, R.C. (1972) The investigation of linkage between a quantitative trait and a marker locus. *Behavior Genetics*, **2**, 3–19.
Hayes, B.J., Gjuvsland, A. and Omholt, S. (2006) Power of QTL mapping experiments in commercial Atlantic salmon populations, exploiting linkage and linkage disequilibrium and effect of limited recombination in males. *Heredity*, **97**, 19–26.
Hedgecock, D., Shin, G., Gracey, A.Y. *et al.* (2015) Second-generation linkage maps for the Pacific oyster *Crassostrea gigas* reveal errors in assembly of genome scaffolds. *G3: Genes. Genomes, Genetics*, **5**, 2007–2019.
Houston, R.D., Haley, C.S., Hamilton, A. *et al.* (2008) Major quantitative trait loci affect resistance to infectious pancreatic necrosis in Atlantic salmon (*Salmo salar*). *Genetics*, **178**, 1109–1115.
Houston, R.D., Haley, C.S., Hamilton, A. *et al.* (2010) The susceptibility of Atlantic salmon fry to freshwater infectious pancreatic necrosis is largely explained by a major QTL. *Heredity*, **105**, 318–327.
Houston, R.D., Davey, J.W., Bishop, S.C. *et al.* (2012) Characterisation of QTL-linked and genome-wide restriction site-associated DNA (RAD) markers in farmed Atlantic salmon. *BMC Genomics*, **13**, 1–15.

Houston, R.D., Taggart, J.B., Cézard, T. *et al.* (2014) Development and validation of a high density SNP genotyping array for Atlantic salmon (*Salmo salar*). *BMC Genomics*, **15**, 1–13.

Hu, Z. and Xu, S. (2009). PROC QTL—A SAS Procedure for Mapping Quantitative Trait Loci. *International Journal of Plant Genomics*, **2009**, Article ID 141234, 3 pages.

Iwata, H. and Ninomiya, S. (2006) AntMap: constructing genetic linkage maps using an ant colony optimization algorithm. *Breeding Science*, **56**, 371–377.

Jansen, R.C. (1994) Controlling the type I and type II errors in mapping quantitative trait loci. *Genetics*, **138**, 871–881.

Jansen, R.C. and Stam, P. (1994) High resolution of quantitative traits into multiple loci via interval mapping. *Genetics*, **136**, 1447–1455.

Kemper, K.E., Daetwyler, H.D., Visscher, P.M. and Goddard, M.E. (2012) Comparing linkage and association analyses in sheep points to a better way of doing GWAS. *Genetics Research (Cambridge)*, **94**, 191–203.

Knott, S.A. and Haley, C.S. (1998) Simple multiple-marker sib-pair analysis for mapping quantitative trait loci. *Heredity*, **81**, 48–54.

Knott, S.A., Elsen, J.M. and Haley, C.S. (1996) Methods for multiple-marker mapping of quantitative trait loci in half-sib populations. *Theoretical and Applied Genetics*, **93**, 71–80.

Kuang, Y., Zheng, X., Lv, W. *et al.* (2015) Mapping quantitative trait loci for flesh fat content in common carp (*Cyprinus carpio*). *Aquaculture*, **435**, 100–105.

Küttner, E., Moghadam, H.K., Skúlason, S. *et al.* (2011) Genetic architecture of body weight, condition factor and age of sexual maturation in Icelandic Arctic charr (*Salvelinus alpinus*). *Molecular Genetics and Genomics*, **286** (1), 67–79.

Laghari, M.Y., Lashari, P., Zhang, X. *et al.* (2013) Mapping quantitative trait loci (QTL) for body weight, length and condition factor traits in backcross (BC1) family of Common carp (*Cyprinus carpio* L.). *Molecular Biology Reports*, **41**, 721–731.

Laghari, M.Y., Lashari, P., Zhang, Y. and Sun, X. (2014) Identification of quantitative trait loci (QTLs) in aquaculture species. *Reviews in Fisheries Science & Aquaculture*, **22**, 221–238.

Lallias, D., Gomez-Raya, L., Haley, C.S. *et al.* (2009) Combining two-stage testing and interval mapping strategies to detect QTL for resistance to bonamiosis in the European flat oyster *Ostrea edulis*. *Marine Biotechnology*, **11**, 570–584.

Lander, E. and Kruglyak, L. (1995) Genetic dissection of complex traits: guidelines for interpreting and reporting linkage results. *Nature Genetics*, **11**, 241–247.

Lander, E.S. and Botstein, D. (1989) Mapping mendelian factors underlying quantitative traits using RFLP linkage maps. *Genetics*, **121**, 185–199.

Larson, W.A., McKinney, G.J., Limborg, M.T. *et al.* (2016) Identification of multiple QTL hotspots in sockeye salmon (Oncorhynchus nerka) using genotyping-by-sequencing and a dense linkage map. *Journal of Heredity*, **107** (2), 122–133.

Li, Y. and He, M. (2014). Genetic mapping and QTL analysis of growth-related traits in *Pinctada fucata* using restriction-site associated DNA sequencing. *PLoS One*, **9**, e111707.

Li, H., Liu, X. and Zhang, G. (2012). A consensus microsatellite-based linkage map for the hermaphroditic bay scallop (*Argopecten irradians*) and its application in size-related QTL analysis. *PLoS One*, **7**, e46926.

Li, Y., Liu, S., Qin, Z. *et al.* (2015) Construction of a high-density, high-resolution genetic map and its integration with BAC-based physical map in channel catfish. *DNA Research*, **22**, 39–52.

Lien, S., Gidskehaug, L., Moen, T. *et al.* (2011) A dense SNP-based linkage map for Atlantic salmon (*Salmo salar*) reveals extended chromosome homeologies and striking differences in sex-specific recombination patterns. *BMC Genomics*, **12**, 615.

Lincoln, S.E., Daly, M.J. and Lander, E.S. (1993) Mapping genes controlling quantitative traits using MAPMAKER/QTL version 1.1: a tutorial and reference manual, in *Whitehead Institute for Biomedical Research Technical Report*.

Liu, S., Sun, L., Li, Y. *et al.* (2014) Development of the catfish 250K SNP array for genome-wide association studies. *BMC Research Notes*, **7**, 1–12.

Liu, Z. (2007) *Aquaculture genome technologies*, John Wiley & Sons, Oxford, UK.

Liu, Z. (2011) *Genomic variations and marker technologies for genome-based selection, next generation sequencing and whole genome selection in aquaculture*, Wiley-Blackwell, Hoboken, United States, pp. 3–19.

Liu, Z.J. and Cordes, J.F. (2004) DNA marker technologies and their applications in aquaculture genetics. *Aquaculture*, **238**, 1–37.

Lynch, M. and Walsh, B. (1998) *Genetics and analysis of quantitative traits*, Sinauer Associates, Inc., Sunderland, MA.

Mackay, T.F.C. (2009) Q&A: genetic analysis of quantitative traits. *Journal of Biology*, **8**, 1–5.

Massault, C., Bovenhuis, H., Haley, C. and de Koning, D.-J. (2008) QTL mapping designs for aquaculture. *Aquaculture*, **285**, 23–29.

Massault, C., Hellemans, B., Louro, B. *et al.* (2010) QTL for body weight, morphometric traits and stress response in European sea bass *Dicentrarchus labrax*. *Animal Genetics*, **41**, 337–345.

McKinney, G.J., Seeb, L.W., Larson, W.A. *et al.* (2015) An integrated linkage map reveals candidate genes underlying adaptive variation in Chinook salmon (Oncorhynchus tshawytscha). *Molecular Ecology Resources*, **16** (3), 769–783.

Meuwissen, T.H.E., Hayes, B.J. and Goddard, M.E. (2001) Prediction of total genetic value using genome-wide dense marker maps. *Genetics*, **157**, 1819–1829.

Meuwissen, T.H.E., Karlsen, A., Lien, S. *et al.* (2002) Fine mapping of a quantitative trait locus for twinning rate using combined linkage and linkage disequilibrium mapping. *Genetics*, **161**, 373–379.

Miller, M.R., Dunham, J.P., Amores, A. *et al.* (2007) Rapid and cost-effective polymorphism identification and genotyping using restriction site associated DNA (RAD) markers. *Genome Research*, **17**, 240–248.

Moen, T., Torgersen, J., Santi, N. *et al.* (2015) Epithelial cadherin determines resistance to infectious pancreatic necrosis virus in Atlantic salmon. *Genetics*, **200**, 1313–1326.

Moen, T., Baranski, M., Sonesson, A. and Kjoglum, S. (2009) Confirmation and fine-mapping of a major QTL for resistance to infectious pancreatic necrosis in Atlantic salmon (*Salmo salar*): population-level associations between markers and trait. *BMC Genomics*, **10**, 368.

Moen, T., Sonesson, A.K., Hayes, B. *et al.* (2007) Mapping of a quantitative trait locus for resistance against infectious salmon anaemia in Atlantic salmon (*Salmo salar*): comparing survival analysis with analysis on affected/resistant data. *BMC Genetics*, **8**, 1–13.

Moghadam, H.K., Poissant, J., Fotherby, H. *et al.* (2007) Quantitative trait loci for body weight, condition factor and age at sexual maturation in Arctic charr (*Salvelinus alpinus*): comparative analysis with rainbow trout (*Oncorhynchus mykiss*) and Atlantic salmon (*Salmo salar*). *Molecular Genetics and Genomics*, **277**, 647–661.

Ødegård, J., Baranski, M., Gjerde, B. and Gjedrem, T. (2011) Methodology for genetic evaluation of disease resistance in aquaculture species: challenges and future prospects. *Aquaculture Research*, **42**, 103–114.

Palaiokostas, C., Bekaert, M., Davie, A. *et al.* (2013a) Mapping the sex determination locus in the Atlantic halibut (*Hippoglossus hippoglossus*) using RAD sequencing. *BMC Genomics*, **14**, 1–12.

Palaiokostas, C., Bekaert, M., Khan, M.G.Q. *et al.* (2013b) Mapping and validation of the major sex-determining region in Nile tilapia (*Oreochromis niloticus* L.) using RAD sequencing. *PLoS One*, **8**, e68389.

Palaiokostas, C., Bekaert, M., Taggart, J.B. *et al.* (2015) A new SNP-based vision of the genetics of sex determination in European sea bass (Dicentrarchus labrax). *Genetics Selection Evolution*, **47** (1), 1.

Palti, Y., Gao, G., Liu, S. *et al.* (2015) The development and characterization of a 57K single nucleotide polymorphism array for rainbow trout. *Molecular Ecology Resources*, **15**, 662–672.

Palti, Y., Gao, G., Miller, M.R. *et al.* (2014) A resource of single-nucleotide polymorphisms for rainbow trout generated by restriction-site associated DNA sequencing of doubled haploids. *Molecular Ecology Resources*, **14**, 588–596.

Palti, Y., Genet, C., Gao, G. *et al.* (2012) A second generation integrated map of the rainbow trout (*Oncorhynchus mykiss*) genome: analysis of conserved synteny with model fish genomes. *Marine Biotechnology*, **14**, 343–357.

Pedersen, S., Berg, P.R., Culling, M. *et al.* (2013) Quantitative trait loci for precocious parr maturation, early smoltification, and adult maturation in double-backcrossed trans-Atlantic salmon (*Salmo salar*). *Aquaculture*, **410–411**, 164–171.

Peterson, B.K., Weber, J.N., Kay, E.H., *et al.* (2012). Double digest RADseq: an inexpensive method for de novo SNP discovery and genotyping in model and non-model species. *PLoS One*, 7, e37135.

Rastas, P., Paulin, L., Hanski, I. *et al.* (2013) Lep-MAP: fast and accurate linkage map construction for large SNP datasets. *Bioinformatics*, **29**, 3128–3134.

Reid, D.P., Szanto, A., Glebe, B. *et al.* (2005) QTL for body weight and condition factor in Atlantic salmon (*Salmo salar*): comparative analysis with rainbow trout (*Oncorhynchus mykiss*) and Arctic charr (*Salvelinus alpinus*). *Heredity*, **94**, 166–172.

Risch, N. and Giuffra, L. (1992) Model misspecification and multipoint linkage analysis. *Human Heredity*, **42**, 77–92.

Sae-Lim, P., Komen, H., Kause, A. and Mulder, H.A. (2014) Identifying environmental variables explaining genotype-by-environment interaction for body weight of rainbow trout (*Onchorynchus mykiss*): reaction norm and factor analytic models. *Genetics Selection Evolution*, **46**, 1–11.

Sanchez-Molano, E., Cerna, A., Toro, M. *et al.* (2011) Detection of growth-related QTL in turbot (*Scophthalmus maximus*). *BMC Genomics*, **12**, 473.

Seaton, G., Hernandez, J., Grunchec, J.A. *et al.* (2006) GridQTL: a grid portal for QTL mapping of compute intensive datasets, 8th World Congress on Genetics Applied to Livestock Production, Belo Horizonte, MG, Brasil.

Shao, C., Niu, Y., Rastas, P. *et al.* (2015) Genome-wide SNP identification for the construction of a high-resolution genetic map of Japanese flounder (*Paralichthys olivaceus*): applications to QTL mapping of *Vibrio anguillarum* disease resistance and comparative genomic analysis. *DNA Research*, **22**, 161–170.
Sonesson, A.K. (2007) Possibilities for marker-assisted selection in aquaculture breeding schemes, in *Marker-assisted selection: current status and future perspectives in crops, livestock, forestry and fish* (eds E. Guimarães, J. Ruane, B. Scherf *et al.*), Rome, FAO, pp. 309–328.
Stam, P. (1993) Construction of integrated genetic linkage maps by means of a new computer package: Join map. *The Plant Journal*, **3**, 739–744.
Star, B., Nederbragt, A.J., Jentoft, S. *et al.* (2011) The genome sequence of Atlantic cod reveals a unique immune system. *Nature*, **477**, 207–210.
Stranger, B.E., Stahl, E.A. and Raj, T. (2011) Progress and promise of genome-wide association studies for human complex trait genetics. *Genetics*, **187**, 367–383.
Thorpe, J.E. (1994) Reproductive strategies in Atlantic salmon, *Salmo salar* L. *Aquaculture Research*, **25**, 77–87.
Tine, M., Kuhl, H., Gagnaire, P.-A. *et al.* (2014) *European sea bass genome and its variation provide insights into adaptation to euryhalinity and speciation*, Nature Communications, p. 5.
Tsai, H.Y., Hamilton, A., Guy, D.R. *et al.* (2015) The genetic architecture of growth and fillet traits in farmed Atlantic salmon (*Salmo salar*). *BMC Genetics*, **16**, 1–11.
Tsigenopoulos, C.S., Louro, B., Chatziplis, D., *et al.* (2014). Second generation genetic linkage map for the gilthead sea bream *Sparus aurata* L. *Marine Genomics*, **18**(Part A), 77–82.
Vallejo, R.L., Palti, Y., Liu, S. *et al.* (2013) Detection of QTL in rainbow trout affecting survival when challenged with *Flavobacterium psychrophilum*. *Marine Biotechnology*, **16**, 349–360.
Vandeputte, M., Kocour, M., Mauger, S. *et al.* (2004) Heritability estimates for growth-related traits using microsatellite parentage assignment in juvenile common carp (*Cyprinus carpio* L.). *Aquaculture*, **235**, 223–236.
Van Ooijen, J. and Kyazma, B. (2009) *MapQTL 6. Software for the mapping of quantitative trait loci in experimental populations of diploid species*, Kyazma BV, Wageningen, Netherlands.
Van Orsouw, N.J., Hogers, R.C.J., Janssen, A., *et al.* (2007). Complexity reduction of polymorphic sequences (CRoPSTM): a novel approach for large-scale polymorphism discovery in complex genomes. *PLoS One*, **2**, e1172.
Vervalle, J., Hepple, J.-A., Jansen, S. *et al.* (2013) Integrated linkage map of *Haliotis midae* Linnaeus based on microsatellite and SNP markers. *Journal of Shellfish Research*, **32**, 89–103.
Visscher, P.M., Thompson, R. and Haley, C.S. (1996) Confidence intervals in QTL mapping by bootstrapping. *Genetics*, **143**, 1013–1020.
Voorrips, R.E. (2002) MapChart: software for the graphical presentation of linkage maps and QTLs. *Journal of Heredity*, **93**, 77–78.
Wang, L., Wan, Z.Y., Bai, B. *et al.* (2015) Construction of a high-density linkage map and fine mapping of QTL for growth in Asian seabass. *Scientific Reports*, **5**, 16358.
Wang, S., Meyer, E., McKay, J.K. and Matz, M.V. (2012) 2b-RAD: a simple and flexible method for genome-wide genotyping. *Nature Methods*, **9**, 808–810.

Weber, J.L. (1990) Informativeness of human (dC-dA) n•(dG-dT) n polymorphisms. *Genomics*, 7 (4), 524–530.

Wright, J.M. (1993) DNA fingerprinting in fishes, in *Biochemistry and molecular biology of fishes* (eds P. Hochachka and T. Mommsen), Elsevier, Amsterdam, pp. 58–91.

Wringe, B.F., Devlin, R.H., Ferguson, M.M. *et al.* (2010) Growth-related quantitative trait loci in domestic and wild rainbow trout (*Oncorhynchus mykiss*). *BMC Genetics*, **11**, 63.

Wu, Y., Bhat, P.R., Close, T.J. and Lonardi, S. (2008). Efficient and accurate construction of genetic linkage maps from the minimum spanning tree of a graph. *PLoS Genetics*, **4**, e1000212.

Xu, J., Zhao, Z., Zhang, X. *et al.* (2014) Development and evaluation of the first high-throughput SNP array for common carp (*Cyprinus carpio*). *BMC Genomics*, **15**, 1–10.

Yandell, B.S., Mehta, T., Banerjee, S. *et al.* (2007) R/qtlbim: QTL with Bayesian interval mapping in experimental crosses. *Bioinformatics*, **23**, 641–643.

Yáñez, J.M., Houston, R. and Newman, S. (2014) Genetics and genomics of disease resistance in salmonid species. *Frontiers in Genetics*, **5**.

Yáñez, J.M., Newman, S. and Houston, R.D. (2015) Genomics in aquaculture to better understand species biology and accelerate genetic progress. *Frontiers in Genetics*, **6**, 1–3.

Yu, Z. and Guo, X. (2006) Identification and mapping of disease-resistance QTLs in the eastern oyster, *Crassostrea virginica* Gmelin. *Aquaculture*, **254**, 160–170.

Yue, G.H. (2014) Recent advances of genome mapping and marker-assisted selection in aquaculture. *Fish and Fisheries*, **15**, 376–396.

Zeng, Z.B. (1994) Precision mapping of quantitative trait loci. *Genetics*, **136**, 1457–1468.

Zhang, G., Fang, X., Guo, X. *et al.* (2012) The oyster genome reveals stress adaptation and complexity of shell formation. *Nature*, **490**, 49–54.

Zhu, C., Tong, J., Yu, X. *et al.* (2014) A second-generation genetic linkage map for bighead carp (*Aristichthys nobilis*) based on microsatellite markers. *Animal Genetics*, **45**, 699–708.

23

Genome-wide Association Studies of Performance Traits

Xin Geng, Degui Zhi and Zhanjiang Liu

Introduction

Genome-wide association study (GWAS) is becoming a powerful tool for investigating the genetic architecture of important traits of humans, crops, and livestock. However, it has not been widely used in aquaculture species to uncover the genetic variants underlying the complex traits. In this chapter, we briefly introduce the key concepts, study designs, and statistical analysis methods for GWAS for aquaculture species. To make the chapter also practically more useful, a step-by-step section is included as an example of GWAS processes. While we understand that there are many possibilities for the combined use of various statistical methods, demonstration using some popular methods is practically useful for beginners. We also briefly discuss the challenges and future research avenues of GWAS.

Traditional selection breeding has been conducted for decades with aquaculture species, and major progress has been made with various traits, especially growth (Smitherman, Dunham & Tave, 1983). However, with some traits such as disease resistance, many genes are involved, and accurate selection using traditional selection is difficult. Whole genome marker–assisted selection allows increased selection accuracy and efficiency, which should be conducted to develop superior brood stocks for the aquaculture industry, but genetics work must be done first to dissect the genomic architecture controlling the traits of interest. More importantly, there are many important traits. Selection in one trait may adversely affect other traits, especially when the traits are closely linked. Therefore, understanding of various important performance and production traits is a prerequisite for the application of genome-based selection programs.

Agricultural genetics is about the inheritance of agriculturally important traits, that is, understanding the genetic basis of phenotypes of economic importance. The central goal of genetic stock enhancement is to discover the relationship between genetic polymorphism and the phenotypic variances observed among individuals. The phenotype of an organism is the measurement of observable characteristics or traits, while the genotype is the inherited genetic information. Qualitative traits, where the phenotypes can be assigned into different categories, are controlled by a single gene, or by a limited number of genes, such that the segregation of the traits can be followed by classical Mendelian genetics. However, quantitative traits have continuous variation, which is attributable to the combination of segregation of alleles at the multiple loci controlling

Bioinformatics in Aquaculture: Principles and Methods, First Edition. Edited by Zhanjiang (John) Liu.

the trait, environment, and genotype–environment interactions. Different quantitative traits have different levels of sensitivity to genetic, sexual, and external environmental effects (Mackay, 2001a). Moreover, because each causal gene may only have a small contribution to overall heritability, identifying the genes related to quantitative traits can be difficult (Hirschhorn & Daly, 2005). Most aquaculture performance and production traits of economic importance are quantitative in nature, such as growth rate, feed conversion efficiency, disease resistance against many different diseases, low oxygen tolerance, body shape, carcass and fillet yield, and behavioral traits (e.g., aggressiveness of feeding, and seinability).

Although many types of molecular markers can be used for marker-assisted selection, single-nucleotide polymorphism (SNP) markers are becoming the markers of choice for two reasons. First, SNPs are abundant and widespread throughout the genomes of most species. In most aquaculture species studied to date, SNP rates are 0.5–5% among species. For instance, one SNP exists within approximately 116 bp in catfish genome on average (Sun *et al.*, 2014). Such a polymorphic rate provides no limitation for a dense genome coverage for GWAS. Although not perfectly evenly distributed, SNPs are far superior to any other types of molecular markers in this respect. Secondly, SNPs are biallelic in most cases and codominantly inherited, making them more amenable to automation with reduced complexity for genotyping and analysis.

SNPs can be readily discovered in a cost-effective fashion using next-generation sequencing technology (Liu *et al.*, 2011; Sun *et al.*, 2014). After SNPs are defined, SNP arrays can be developed that provide high efficiency for high-throughput genotyping. Several SNP arrays have been developed for aquaculture species, including catfish, carp, rainbow trout, and Atlantic salmon (Liu *et al.*, 2014; Palti *et al.*, 2014; Xu *et al.*, 2014). Because of these advantages, SNPs have rapidly become the markers of choice for genome-wide marker assisted selection (Morin, Luikart & Wayne, 2004). Of course, it is the choice of researchers to possibly use other types of markers such as microsatellites, but the genome coverage will be low, and the involved labor and cost will be high.

Association study, also known as linkage disequilibrium (LD) mapping, detects and locates quantitative trait loci (QTLs) based on the strength of the correlation between mapped markers and the trait in question. GWAS, that is, conducting association studies using genome-wide genotyping data, has evolved into a powerful tool for investigating the genetic architecture of important traits in human beings, crops, and livestock during the last decade. However, in aquaculture, GWAS has been seldom utilized. In this chapter, we will describe the considerations for genome-wide association studies, and focus on the bioinformatics methods and procedures for analysis of the genetic loci controlling the traits of interest. We will provide a general introduction to the procedures to conduct GWAS in aquaculture (Figure 23.1). Specially, we will focus on the unique strategies and solutions considering the distinct situations in aquaculture as compared to crops, livestock, and human beings. Readers with greater interest in the detailed and step-by-step descriptions of GWAS procedures are referred to other books such as Gondro, Van der Werf, and Hayes (2013) and Stram (2014).

Study Population

The ideal samples should be homogenous in genetic background without population stratification, highly contrasted in phenotype, and highly intercrossed to provide high

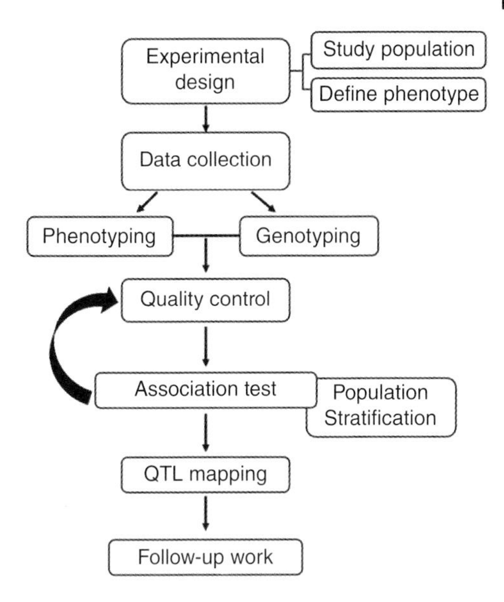

Figure 23.1 Flowchart of procedures to conduct GWAS in aquaculture.

mapping resolution. Population stratification is generated from the different allele frequencies among subpopulations, and it always confounds association tests in practical situations. If phenotypic variation exists among different subpopulations, imbalanced sampling from the subpopulations will generate false positive results. For example, if fish from strain A are more resistant to one disease than those from strain B intrinsically, it is then possible that most resistant fish are sampled from strain A. As a consequence, the identified "associated" loci could be more associated with the strain difference than with disease resistance. Because researchers are faced with various biological or economic limitations, the most appropriate samples may not be available in practical situations, which may lead to false positive results. To eliminate the effect of population stratification, a number of experimental population structures and corresponding statistical methods have been designed for GWAS. In this section, we describe a few of the most popular designs for population structure.

Samples from Natural Population

Existing samples from non-manipulated natural populations with known phenotypes can be used in GWAS. Obviously, using this kind of samples is more cost-efficient and time-effective as compared to using samples from family-based populations, because the latter requires additional time to generate individuals of higher generations. While it is easy to assume that natural populations are unrelated, this may or may not be true. Population stratifications could be more problematic with aquaculture species than livestock because, in many cases, a large number of individuals could be derived from a very limited number of founders, forming subpopulations in a natural population. Therefore, it must be noted that population stratification in random natural samples could cause false positive results. Recent developments in GWAS methodologies for random natural samples have offered mature software packages for association analysis to control population stratifications, so GWAS with samples from natural populations could be widely performed in aquaculture species, considering the relatively abundant natural population resources as compared to livestock and human beings.

Samples from Family-Based Population

In family-based association tests, families with one or more offspring are used as the subjects rather than unrelated samples. Family-based population designs are more immune to population stratification, which cannot be efficiently addressed in natural population design. Moreover, many aquaculture species have high fecundities, with thousands of progenies per spawn, saving tremendous labor for reproduction compared with livestock. Thus, samples for a GWAS in aquaculture can be produced by a few parents, and the progenies with homogenous genetic background are suitable for GWAS, which cannot be realized in humans and mammals. This situation makes aquaculture species unique for GWAS using family-based samples. Even though the samples consist of more than one full-sibling family in most practical experiments, the clear pedigree information of family-based population design makes correction of population stratification much easier as compared to the natural population design. This and the other prominent advantages—including highly contrasted phenotypes (sometimes realized by interspecific hybrid), allowing investigation of specific questions such as parent-of-origin effects, and the power to detect rare variants—make it popular to use family-based populations for GWAS in aquaculture (Mott *et al.*, 2000). For example, the interspecific hybrid catfish, obtained from mating female channel catfish (*Ictalurus punctatus*) with male blue catfish (*I. furcatus*), serves as a great model to detect the major QTLs involved in columnaris disease resistance, because channel catfish is generally resistant to the disease while blue catfish is generally susceptible (Geng *et al.*, 2015).

However, family-based population design has some disadvantages. Due to limited founders, family-based population design is not powerful enough to detect the causal alleles that are homozygous in the subpopulation used in the association test but heterozygous in the whole population. Moreover, compared with the natural population, which has more rounds of historical recombination, the limited numbers of recombination events in family-based population design make its mapping resolution low. In addition, for aquaculture, using family-based sample requires an additional breeding period. For example, the generation time of catfish is long (3 years). Therefore, family-based samples, especially higher generations, for association mapping are time-consuming and costly. Last, the between-family stratification in both phenotypes and genotypes still needs to be addressed by statistical methods.

Mott *et al.* (2000) proposed that higher generations of intercross hybrids can be produced by intermating F_2 individuals for several generations, and that such higher generations of hybrids can provide a higher resolution for association mapping. They applied multi-parent advanced generation intercross (MAGIC), and this approach provides ideal samples with a highly diverse and a no-population-stratification structure, which is suitable for fine mapping. The idea of higher generations of intercross hybrids is very simple: basically, the haplotype blocks become shorter and shorter surrounding the gene of interest to allow the identification of the candidate genes within a small chromosome region. However, in spite of the advantages and theoretical attractiveness, this approach has limited application potential for many important aquaculture species, simply because of the long generation time of aquaculture species. It takes too long to produce high enough generations of progenies to effectively reduce the LD block sizes.

Phenotype Design

A good understanding of the observational data and a correct adjustment for the phenotype are key prerequisite steps for further analysis. Based on the phenotypes, that is, qualitative and quantitative traits, two types of study design can be made: qualitative trait design (case-control design) and quantitative trait design. In some cases, if the trait does not have well-established quantitative measures, the samples can be classified with categorical variables. For instance, in the case of disease resistance, quantitative measurements may not be available in many species. Then, the disease resistance trait can be classified as "resistant" versus "susceptible" as a binary variable (Geng *et al.*, 2015). If the traits are binary, the data could be analyzed with logistic regression models. Although some methods for association tests were developed with quantitative traits, they can also be used to analyze case-control datasets by using dummy variables (i.e., coding case phenotypes as 1 and control phenotypes as 0) (Kang *et al.*, 2010). From a statistical perspective, genetic effect size (the proportion of phenotypic variance explained by two alleles at a locus) can be easily calculated with quantitative traits, since quantitative traits are measured by continuous numbers.

Factors that influence the trait should be adjusted in or before the association tests to exclude spurious associations caused by confounding factors. These factors, which describe the circumstances under which the data were collected and the characteristics of the samples, may include gender, age, experimental batch, body weight, known family structure, etc. The adjustment procedure could be conducted by linear models, with the factors as the explanatory variables and the observational data of interest as the response variables, and the residuals could be used as the phenotype for further analysis, which just consist of the genetic component, other unaccounted effects, and random effects (Dominik, 2013). In addition, outliers should be flagged because they will affect the fitting of the model.

Power of Association Test and Sample Size

The power of a study is the probability that a true association between a marker and the trait of interest is found significant by the designed study. Calculating the power before conducting an experiment is a central element in study design. The power depends on the significance level α set by the experimenter, design of the experiment, statistical test, effect size of QTL, allele frequency of the causal allele, the LD between the causal allele and the genotyped markers on the array, and sample size (Hayes, 2013). Increasing the sample size is an obvious method to improve the power to detect associations.

For quantitative trait designs, selective genotyping is often utilized as a cost-effective strategy, which just genotypes individuals from the extremes of the phenotypic distribution (Lynch & Walsh, 1998). It requires less sample size but keeps the high power to detect QTLs (Van Gestel *et al.*, 2000). However, the tradeoff is that it may cause a potential overestimation of the effect size.

Some software tools are available to calculate the sample size for unrelated individuals to ensure sufficient power (Gauderman & Morrison, 2006). However, including a statistician during the planning phase is often recommended to ensure a solid and powerful design.

Quality Control Procedures

After the genotype-calling procedure based on signal intensities generated by the SNP assay for the alleles (Ziegler, König & Thompson, 2008), quality control (QC) should be performed for genotypes to avoid false results. QC for GWAS data includes sample-level QC and SNP-level QC.

Samples with low genotyping quality or a low call rate should be excluded from analysis. The "outliers" with different ancestry may cause false positive loci. For example, in the study conducted by Gudbjartsson *et al.* (2008), individuals having large deviations in terms of genetic background were removed to keep the samples homogenous. Principal component analysis (PCA) or cluster analysis based on identity by state (IBS) kinship matrix with the genotypes of all samples can be used to detect outliers. After visualizing these structures, the outliers can be identified and removed (Geng *et al.*, 2015).

For markers, SNPs with low genotyping quality should also be excluded if they have any Mendelian inheritance errors or low calling rates. Rare SNPs, possibly generated by genotyping errors or population stratification, may lead to spurious results. Moreover, GWAS is not powerful to detect the effect of rare SNPs. Therefore, SNPs with low minor allele frequencies (MAFs) should always be discarded. The Hardy–Weinberg equilibrium test compares the observed proportion of the marker versus the expected proportion. If unrelated samples are used, the SNPs that are severely out of Hardy–Weinberg equilibrium should be flagged before further analysis, because disequilibrium can result from a true association, a potential genotyping error, or population stratification (Turner *et al.*, 2011).

LD Analysis

In population genetics, the LD describes the correlations of alleles at two or more neighboring loci (Reich *et al.*, 2001). If one locus has alleles "A" and "a" with frequencies p_A and p_a, and a second has alleles "B" and "b" with frequencies p_B and p_b, then the expected haplotype frequencies at equilibrium are the product of the two component allele frequencies. For example, $p_{AB} = p_A \times p_B$, where p_{AB} is the frequency of the AB haplotype. The deviation of the observed frequency of a haplotype from the expected under equilibrium is the LD, which is denoted by D, that is, $D = p_{AB} - p_A \times p_B$ (Table 23.1). The D statistic is dependent on the frequencies of the individual alleles (p_A and p_B), so D is not useful in describing the LD on different pairs of loci. Two alternative methods D' and r^2 are used to normalize D (Lewontin, 1964; Pritchard & Przeworski, 2001):

Table 23.1 The relationship between the haplotype frequencies, allele frequencies, and D.

	A	a	Total
B	$p_{AB} = p_A p_B + D$	$p_{aB} = p_a p_B - D$	p_B
b	$p_{Ab} = p_A p_b - D$	$p_{ab} = p_a p_b + D$	p_b
Total	p_A	p_a	1

1) $D' = \frac{D}{D_{max}}$, where $D_{max} = \begin{cases} \min(p_A p_B, p_a p_b) & \text{when } D < 0 \\ \min(p_A p_b, p_a p_B) & \text{when } D > 0 \end{cases}$

2) $r^2 = \frac{D^2}{p_A p_B p_a p_b}$

LD is caused by the lack of recombinations breaking the linkage of nearby loci. Therefore, LD decays with increasing distance between loci. LD decay is also influenced by several other factors, such as population size, the number of founders in the population, and the number of generations of the populations (Bush & Moore, 2012). A significantly associated SNP detected from association mapping could be a causal variant, but, in most cases, the identified SNPs are in high LD with the causal variants.

There are several reasons why LD is interesting. First, the resolution of the association mapping depends on the decaying extent of LD. Second, we could generate independent SNPs that are not correlated with the surrounding SNPs (LD pruning). Using the number of independent SNPs, we can conduct the Bonferroni correction (Geng *et al.*, 2015).

Association Test

There are different kinds of association test models. The proper statistical test method should be chosen carefully according to specific situations—for example, quantitative trait studies versus qualitative trait studies, samples from family-based population versus samples from natural population, and different genetic effects including dominant, additive, and recessive. If no population stratification exists in the samples, it is simple to evaluate the association between markers and traits by common methods, including linear model, Cochran–Armitage trend test, etc. However, population stratification almost always exists within the sample population and, therefore, correction of population stratification is the key issue in association tests.

Various software packages have been developed for statistical analysis of GWAS in different situations. Most are free and can be downloaded from the Internet. For example, Purcell *et al.* (2007) developed PLINK to conduct association test with free access, and it has been widely used in GWAS. Some commercial software packages assemble popular methods into an easy-to-use toolsets with user-friendly interfaces. Table 23.2 summarizes the software packages that are commonly used for GWAS.

In the following, we will elucidate the strategies to detect population stratification and infer genetic ancestry. Genomic control, principal component analysis, mixed linear model, and transmission disequilibrium test are introduced.

Genomic Control

Genomic control is useful in detecting and correcting population stratification. Population stratification could generate association between a trait and markers distributed over the whole genome (Mackay & Powell, 2007). The distribution of the test statistics for association of markers could be inflated from the expected null distribution owing to population stratification. The inflation factor λ, which is assumed as constant for all the SNPs, can be estimated from a set of markers unlinked to the traits, and can be used to evaluate population stratification (Devlin & Roeder, 1999; Price *et al.*, 2006). Genomic control is the practice that corrects the inflation of test statistics by dividing

Table 23.2 Examples of commonly used software packages for GWAS.

Software package	Application	Source
PLINK	Multipurpose GWAS toolset	http://pngu.mgh.harvard.edu/~purcell/plink/ (Purcell *et al.*, 2007)
Haploview	Haplotype analysis with graphical interface	http://www.broadinstitute.org/scientific-community/science/programs/medical-and-population-genetics/haploview/haploview (Barrett *et al.*, 2005)
Quanto	Sample size calculations	http://biostats.usc.edu/software (Gauderman & Morrison, 2006)
EMMAX	Mixed model analysis	http://genetics.cs.ucla.edu/emmax/ (Kang *et al.*, 2010)
GCTA	Multipurpose GWAS toolset	http://cnsgenomics.com/software/gcta/ (Yang *et al.*, 2011)
STRUCTURE	Population structure inference based on structured association	http://pritchardlab.stanford.edu/structure.html (Pritchard, Stephens & Donnelly, 2000)
EIGENSTRAT	Population structure inference based on PCA	http://genetics.med.harvard.edu/reich/Reich_Lab/Welcome.html (Price *et al.*, 2006)

the test statistics of all markers by λ (Devlin & Roeder, 1999). However, genomic control suffers from weak power when the effect of population stratification is large (Price *et al.*, 2006). Moreover, genomic control always decreases the power and underestimates test statistics of all markers rather than only those associated with population stratification. The uniform adjustment of genomic control will not change the rank of the SNPs according to p-value, and it also cannot adjust the errors in test results caused by different allele frequencies among subpopulations (Price *et al.*, 2010).

PCA

PCA is a dimensionality reduction method that is powerful in representing genetic relationships. PCA summarizes the variations among different samples across all independent markers into a smaller number of principal components, which indicates the relationship of the individuals. When using PCA to correct population stratification, the regression of genotypes at a candidate SNP to phenotypes is adjusted by including the loadings of top principal components to remove all correlations to ancestry (Price *et al.*, 2006). This method assumes a small number of ancestral populations and simple admixture, and it cannot correct stratification due to complex relationships (Yu *et al.*, 2005). Among PCA-based software packages that have been proposed, EIGENSTRAT is the most widely used (Price *et al.*, 2006).

Linear Mixed Models

Linear mixed model can be used to model population structure, family structure, and cryptic relatedness (Yu *et al.*, 2005). It has the ability to capture multiple levels of population structure of the samples, even from several families or inbred lines (Kang *et al.*, 2008).

The model is listed as follows:

$$\mathbf{Y} = \mathbf{Xb} + \mathbf{Za} + \mathbf{e}$$

Here, **Y** is the vector of phenotype; **X** is the matrix of fixed effects and **b** is the coefficient vector; **Z** is a matrix relating the instances of the random effect to the phenotypes and **a** is the vector representing the coefficients with covariance structure based on kinship matrix **G**; **e** is the vector of random residuals. This method models phenotypes using a mixture of fixed and random effects. Fixed effects (**Xb**) include the SNPs and optional covariates, and random effects include heritable (**Za**) and non-heritable (**e**) random variation (Price *et al.*, 2010). An efficient mixed-model association method, EMMAX, was developed that markedly reduced the computational cost and have been widely used in GWAS (Kang *et al.*, 2010). First, EMMAX computes a genetic relatedness matrix representing the sample structure, whose entries are the genetic relationships between every pair of individuals. Second, using a variance component model, the contribution of the sample structure to the covariance of phenotype is estimated, generating an estimated covariance matrix of phenotypes that models the effect of genetic relatedness on the phenotypes. Third, a generalized least square F-test or a score test is applied at each marker to detect associations accounting for the sample structure using the covariance matrix (Kang *et al.*, 2010). Although EMMAX was designed preferably for quantitative traits that follow a normal distribution, the association test for qualitative traits can be approximately conducted using 0–1 quantitative response variable to represent the case-control status (Kang *et al.*, 2010). There are some similar methods—such as GCTA, TASSEL, and GEMMA—that are proven to be effective in correcting complex structure stratification (Yang *et al.*, 2011; Zhang *et al.*, 2010; Zhou & Stephens, 2012). Despite the advantage that linear mixed model could help eliminate false positives caused by complex population stratification, it is not guaranteed to adjust for all possible confounding population structures. A recommended practice is both using principal components as fixed effects and using estimated kinship matrix as a variance–covariance matrix in the random effects (Price *et al.*, 2010). Considering imperfect adjustments, the samples with less population stratification are still preferred to avoid spurious results.

Transmission Disequilibrium Test and Derivatives

The transmission disequilibrium test (TDT), in which family pedigrees of samples are ascertained, is robust to the effects of population stratification (Laird & Lange, 2006). The TDT was proposed by Spielman, McGinnis, and Ewens (1993) with family-based populations for the association test between a genetic marker and a trait. When conducting TDT, the progenies in each family with a certain extreme phenotype of interest are selected—for example, fish with albino or with an exceptionally high growth rate (Lange, DeMeo & Laird, 2002). Parents and progenies are genotyped, and the loci where parents are heterozygous will contribute to the analysis. From each parent, one allele must be transmitted to the progeny, and the other one not. Over all families, the ratio of transmission to non-transmission will be compared with the expected value of 1:1 (Mackay & Powell, 2007).

Various extensions of the TDT have been developed, of which the family-based association test for quantitative traits is widely used (Abecasis, Cardon & Cookson, 2000).

It can accommodate nuclear families of any size, with or without parental information. It breaks down the genotypes into between-family and within-family components, and the latter is free of population structure.

The major drawback of TDT is its extreme susceptibility to genotype errors of parents. TDT, which needs additional genotype information of parents, is of lower power as it only uses the allele transmission information within pedigrees. Moreover, the family based studies still need to incorporate between-family information, which may be confounded from stratification (Lasky-Su *et al.*, 2010; Won *et al.*, 2009).

However, with a large number of progenies, aquaculture species may be ideally suited for the TDT design. Compared with the trio-design in humans, larger family design is possible with most aquaculture species, and this advantage greatly reduces the efficiency penalty for genotyping parents. Moreover, parental genotypes could be validated to correct the genotype errors by the genotypes of numerous offspring based on the Mendelian laws of inheritance. Considering the immunity to population stratification, family-based design will perform efficiently for most aquaculture species.

Significance Level for Multiple Testing

Using a strict significance level for GWAS is important, because GWAS typically tests a very large number of hypotheses, and spurious false positive results may arise by chance. In GWAS, the null hypothesis refers to the statement that no association exists between the markers and the trait. Thus, rejecting the null hypothesis means an association. Under the null hypothesis, low p-value indicates that the chance for obtaining the observed sample results is small. When p-value falls below a predetermined alpha value (significance level), which is usually 0.05 for single marker testing, the null hypothesis will be disproved. This also means that the null hypothesis will be disproved with a probability of 5% when it is true (type 1 error), so the probability for a false positive in one single test will be 5%. However, when we conduct a multiple test in GWAS, hundreds of thousands of SNPs are tested simultaneously. Therefore, the cumulative likelihood of false positive results will increase. To control the false positive results, Bonferroni correction converts $\alpha = 0.05$ to $\alpha = 0.05/n$, where n equals the number of independent tests. Because of LD among GWAS markers, each association test of all the markers is not independent. Duggal *et al.* (2008) proposed that the threshold p-value (α value) for genome-wide significance could be calculated based on Bonferroni correction with the estimated number of independent markers and LD blocks. For instance, if the probability of one type 1 error should be controlled at 0.05 with a total of 15,000 haplotype blocks, the genome-wide significance level now is at $0.05/15{,}000 = 0.0000033$. Apart from Bonferroni correction, an alternative method to adjust α value is by using false discovery rate (FDR), which is widely used in multiple hypotheses testing but less common in the GWAS context (Hochberg & Benjamini, 1990).

Step-by-Step Procedures: A Case Study in Catfish

For many aquaculture students, the challenges of simultaneously understanding genetics, genomics, and informatics processes involving various different software packages are huge. To ease such a steep learning curve, in this section, we provide a step-by-step procedure using a case study for the analysis of QTLs for the head length trait of catfish (Geng *et al.*, 2016).

Description of the Experiment

The goal of this study is to identify the QTLs controlling the head length of hybrid catfish. The head lengths and body weights are recorded. Genotyping is performed using the catfish 250 K SNP array (Liu *et al.*, 2014).

Phenotyping

Quantitative design is chosen in this study, since the head length is a quantitative measurement, with incremental values in millimeters. To eliminate the effect of body weight and between-family phenotypic stratification, phenotypic data need to be adjusted with the cubic root of body weight by simple linear regression within each family. The residuals could be used as adjusted phenotypes to carry out GWAS. In other words, all the fish individuals should be first brought to the "same" body weight. To obtain the adjusted phenotypes, the R software (https://www.r-project.org/) (Team, 2014) can be utilized. After inputting original records (body weight and head length) into R, the command for simple linear model could be run as follows:

```
phenotypicdata$head_length_residual=
  resid(lm(phenotypicdata$head_length~
  phenotypicdata$cubic_root_body_weight)
```

The residuals are given in the output file.

Data Input for PLINK

Subsequent statistical analysis could be carried out with PLINK (v1.07), a command-line program (Purcell *et al.*, 2007). Two files should be input into PLINK, in this case, "data.ped" and "data.map". The ".ped" file contains information such as family ID, individual ID, paternal ID, maternal ID, sex, phenotype, and genotype. The ".map" file describes the information on markers, including chromosome ID and their physical positions on the chromosome. The format requirements of ".ped" and ".map" files are described in detail on the PLINK website (http://pngu.mgh.harvard.edu/~purcell/plink/).

QC

QC can be performed in PLINK, by applying filters both for samples and for SNPs. Samples with too much missing genotype data should be excluded using commands as follows:

```
plink --file data --mind 0.05 --nofounders --allow-no-sex
   --recode --out cleaned1
```

Here: "`--file data`" instructs to read the *data* file; "`--mind 0.05`" excludes samples with more than 5% missing genotypes; "`--nofounders`" ensures founder genotypes are not required; "`--allow-no-sex`" ensures the phenotype will not be set as missing if no gender information is recorded; "`--recode`" generates a new dataset from the result; and "`--out cleaned1`" instructs to name the new dataset as "cleaned1".

After samples with missing genotype data (at the set threshold, here 5%) are excluded, SNPs with low MAF (<0.05) too should be excluded:

```
plink --file cleaned1 --maf 0.05 --nofounders --allow-no-sex
   --recode --out cleaned2
```

Subsequent analyses should exclude SNPs with high missing genotyping rate (more than 10% in this case) with the "`--geno`" option:

```
plink --file cleaned2 --geno 0.1 --nofounders --allow-no-sex
   --recode --out cleaned3
```

LD-based SNP Pruning

To estimate the number of independent SNPs, LD-based SNP pruning should be conducted in PLINK. In this case, the LD block is defined as a set of contiguous SNPs with the minimum pairwise r^2 value exceeding 0.50. In order to generate a set of independent SNPs, pairwise LD for the backcross progeny population will be calculated according to the r^2 value. LD pruning will be conducted with a window size of 50 SNPs, a step of five SNPs, and an r^2 threshold of 0.5. This procedure could be performed in PLINK by running:

```
plink --file cleaned3 --indep-pairwise 50 5 0.5
```

The parameters for "`--indep-pairwise`" are: window size in SNPs (50); the number of SNPs to shift the window at each step (5); and the r^2 threshold (0.5). The command would consider a window size of 50 SNPs; calculate LD between each pair of SNPs in the window; prune out one of a pair of SNPs if r^2 of the LD is greater than 0.5; shift the window five SNPs forward; and repeat the procedure (http://pngu.mgh.harvard.edu/~purcell/plink/summary.shtml#prune). The output files of this command are two lists of SNPs: those that are pruned out, and those that are not. The remaining SNPs are considered as the independent markers. The number of the independent markers (n) could be used to adjust the significance level for multiple testing. The threshold for genome-wide significance level could be set as $0.05/n$ according to Bonferroni correction.

Family-based Association Tests for Quantitative Traits (QFAM)

QFAM is applicable for family-based population to control population stratification. QFAM partitions the genotypes into between- and within-family components (Abecasis *et al.*, 2000; Fulker *et al.*, 1999). The within-family components could control stratification. The command is described as follows:

```
plink --file cleaned3 --qfam --mperm 100000 --allow-no-sex
```

Here, "`--qfam`" uses a within-family test, and "`--mperm`" specifies permutation. In the result, the p-value of each maker could be generated to construct Manhattan plots.

Follow-up Work after GWAS

Validation

If genomic regions are identified with the potential effects on a trait, it is important to confirm and verify whether the association is true across other populations. Since different populations have differences in genetic architectures, the influence of each associated locus on the trait in question depends on the populations. Thus, the association results may not be the same if using different populations, because of population-specific or family-specific loci (Brachi, Morris & Borevitz, 2011). However, it is important to test whether the identified markers are commonly associated with the traits of interest across various populations by using different populations.

Fine Mapping

GWAS can narrow down the loci associated with the trait in question into regions ranging from hundreds of KB to several MB. Especially in the family-based study, the resolution is drastically limited. One of the reasons limiting the resolution is that the sample size of genotyped individuals may not be sufficient. Another reason is that the coverage of markers in the local region may not be dense enough. In addition, multiple causal genes within a "functional hub" could lead to a long-associated region, which requires additional evidence by fine mapping the region (Geng *et al.*, 2015). Theoretically, very large numbers of individuals and very-high-density SNPs can be used, but the involved costs would be tremendously high. Practically, to get a higher resolution, saturated SNPs from the local regions can be genotyped with a very large number of individuals. Genotyping SNPs from a local region for sub–whole genome study applications is more economical when performing fine mapping studies than genotyping SNPs over the whole genome.

Functional Confirmation of Implicated Molecular Mechanisms

After the associated loci are identified, the underlying molecular mechanisms can be explored. However, phenotypic variations caused by genetic architecture could be generated at different levels from the composite of genes, transcriptional regulation, post-transcriptional modification and regulation, translational regulation, and post-translational modification and regulation—along with environmental impact and genotype–environment interactions (Wang *et al.*, 2013). Therefore, functional exploration is a very complex task. Methods of functional exploration include investigating gene expression with real-time PCR; determination of genomic variations that may cause changes in protein structures, or those that may cause changes in the level of expression of the involved genes; investigating deletions of the syntenic region; knockdown and knockout of candidate genes; gene editing; *in situ* hybridization; and *in vitro* expression studies with constructs of candidate genes. For instance, Sun *et al.* (2013) evaluated candidate genes located in the regions associated with meat quality by real-time quantitative PCR (Q-PCR) in subsets of six chickens with lowest or highest trait phenotypic values. Schoenebeck *et al.* (2012) investigated the role of the candidate gene *BMP3* for head shape by *in situ* hybridization. Liu *et al.* (2013) performed knockdown of newly identified genes impacting liver development to validate the candidate genes identified by GWAS. Nonetheless, functional assignments from associate markers to associated genes, from associated genes to causal genes, and from causal genes to the mechanisms behind the cause of phenotypic variations involve a series of very complex and difficult tasks that, in many cases, may become too difficult and too expensive to resolve with aquaculture species. Functional inference from other species could be a choice, considering the cost and technical limitations of aquaculture species.

Pitfalls of GWAS with Aquaculture Species

Although GWAS allows whole-genome scans to locate small haplotype blocks associated with the traits in question, some challenges still hinder the identification of QTLs. First, the investment is high for assembling genomes, and for designing and manufacturing SNP arrays. Second, the prevalent association tests with random population samples are underpowered, owing to low allele frequencies and allelic heterogeneity when testing the association of rare variants with the traits in question (Brachi *et al.*, 2011). The rare variants, especially those with modest effects, are difficult to be identified, unless the family-based populations that contain these rare variants are used. Third, GWAS cannot estimate the effect size of one gene correctly if multiple genes are involved in the same trait and located in LD. Fourth, multiple variants at a locus, where various functional alleles of the same gene exist and are associated with different phenotypes, will impede the identification of the associated variants (Wood *et al.*, 2011). Finally, the quantitative traits usually have a complex genetic architecture (e.g., epistasis effect; Cordell, 2009). The existing statistical tools do not have the full power to detect these loci with complex gene interactions by considering massive numbers of tests (Brachi *et al.*, 2011).

Comparison of GWAS with Alternative Designs

Apart from GWAS, alternative statistical methods are available to investigate the genetic basis of variation causing different phenotypes. Here, we describe the advantages and disadvantages of these methods compared with GWAS.

Linkage-based QTL Mapping

Chapter 22 is fully devoted to QTL linkage analysis. Here, we will provide the perspectives of their similarities and differences. Similar to GWAS, QTL mapping is also a statistical method that links phenotypic and genotypic data to explain the genetic basis that causes phenotypic variations. QTL mapping can be regarded as a special case of GWAS where LD is derived from the small number of founders who established the population in the recent past (Mackay & Powell, 2007). Different from GWAS, QTL analysis requires two or more strains of organisms as the parental population that differ genetically with regard to the trait of interest. Moreover, genetic markers that are different in the parental lines should segregate with the contrasted phenotype.

QTL mapping has been a powerful traditional method used to identify loci co-segregating with a given trait. Without significant investment in the development of large genotyping platforms such as SNP arrays, QTL mapping is still widely used. However, QTL mapping suffers from some fundamental limitations. First, the mapping resolution of QTL is limited by the amount of recombination events within the pedigrees, although it can be improved by several generations of intercrossing (Darvasi & Soller, 1995). Linkage analyses thus have a lower level of resolution than association studies, which also leverage all historic recombination events among founders (Mackay, 2001a). In the natural populations that are utilized by GWAS, LD often decays more rapidly with increasing physical distances than in controlled crosses (Mackay & Powell, 2007). Second, some loci will remain undetected if the analyzed families contain no segregating alleles at the loci. Third, linkage analyses have less power to identify common genetic variants with modest effects (Risch & Merikangas, 1996). Fourth, important quantitative traits usually have complex genetic architectures, such that the phenotype is determined by multiple factors (Wang *et al.*, 2005), such as genotype-by-sex, genotype-by-environment, and epistatic interactions between QTLs. However, not all QTL studies were designed to detect such interactions (Mackay, 2001b). Moreover, the allele frequencies and combinations present in the sampled families may differ from those in other populations (Korte & Farlow, 2013). Because of these reasons, the number of times that individual genes have been identified by utilizing QTL mapping remains very small. In Table 23.3, the advantages and disadvantages of GWAS and linkage mapping are compared.

Bulk Segregant Analysis

The basic idea of bulk segregant analysis (BSA) is that phenotypic extremes should have drastic differences in the loci associated with the phenotype when samples are selected from phenotypic extremes and their genotypes are analyzed in bulk. Although it may be difficult to detect the associated loci by comparing individuals with different performances in phenotypes, the pooled samples (bulk) with the phenotypic extremes should reveal the contrast in the genotype (Michelmore, Paran & Kesseli, 1991; Wang *et al.*, 2013). In other words, if samples are grouped according to the contrasted traits,

Table 23.3 A comparison of association analysis with linkage analysis.

Property of mapping approach	Association mapping[a)]	Linkage mapping
Localization of the detected region	Short (<1 MB)	Long (~5 MB)
Number of markers for genome-wide coverage	Large	Moderate
Cost	High[#]	Low
Pedigree required	No	Yes
Susceptible to stratification	Yes	No
Power to detect rare alleles	No	Yes
Detection of variants with modest effect	Suitable	Not suitable

a) refers to association mapping with samples from natural population but not with family-based samples. # indicates significant upfront investment for the development of SNP arrays, although the cost for genotyping SNPs is low on a per-marker basis.

the frequencies of the two marker alleles present within each of the two bulks should deviate significantly from the expected ratio in their specific populations (Quarrie *et al.*, 1999). Thus, the correlation between genotype and phenotype can be identified. The major drawback of BSA is the imprecision caused by the genotype generated from the pooled sample. Moreover, the effect of family stratification, if existing, is impossible to be eliminated, owing to the bulk analysis. Furthermore, BSA will only be able to detect the genetic effects of single locus, precluding any analysis of haplotype or gene-by-gene interaction effects. However, because of high efficiency, low cost, and analytical simplicity, it is still broadly used, especially with plant species. The high fecundities of aquaculture species make BSA potentially a useful tool to provide preliminary results for aquaculture species (Wang *et al.*, 2013).

Conclusions

Undoubtedly, GWAS has accelerated the field of human, plant, and livestock genetics. Using GWAS, numerous genetic risk factors for many common human diseases have been identified, and many genetic regions controlling important economical traits have been located in plants and livestock. Genome-wide association studies could open new frontiers in our understanding of the relations between traits and the underlying genetic architecture in aquaculture. With the development of genotyping technologies, especially high-density SNP arrays, GWAS could be widely used for the analysis of aquaculture traits to improve the brood stocks of aquaculture species, with lower costs in the long term.

References

Abecasis, G., Cardon, L. and Cookson, W. (2000) A general test of association for quantitative traits in nuclear families. *The American Journal of Human Genetics*, **66**, 279–292.

Barrett, J.C., Fry, B., Maller, J. and Daly, M.J. (2005) Haploview: analysis and visualization of LD and haplotype maps. *Bioinformatics*, **21**, 263–265.

Brachi, B., Morris, G.P. and Borevitz, J.O. (2011) Genome-wide association studies in plants: the missing heritability is in the field. *Genome Biology*, **12**, 232.

Bush, W.S. and Moore, J.H. (2012). Genome-wide association studies. *PLoS Computational Biology*, **8**, e1002822.

Cordell, H.J. (2009) Detecting gene–gene interactions that underlie human diseases. *Nature Reviews Genetics*, **10**, 392–404.

Darvasi, A. and Soller, M. (1995) Advanced intercross lines, an experimental population for fine genetic mapping. *Genetics*, **141**, 1199–1207.

Devlin, B. and Roeder, K. (1999) Genomic control for association studies. *Biometrics*, **55**, 997–1004.

Dominik, S. (2013) Descriptive statistics of data: Understanding the data set and phenotypes of interest, in *Genome-wide association studies and genomic prediction* (eds C. Gondro, J. Van der Werf and B. Hayes), Humana Press, New York, pp. 19–35.

Duggal, P., Gillanders, E.M., Holmes, T.N. and Bailey-Wilson, J.E. (2008) Establishing an adjusted p-value threshold to control the family-wide type 1 error in genome wide association studies. *BMC Genomics*, **9**, 516.

Fulker, D., Cherny, S., Sham, P. and Hewitt, J. (1999) Combined linkage and association sib-pair analysis for quantitative traits. *The American Journal of Human Genetics*, **64**, 259–267.

Gauderman, W. and Morrison, J. (2006). QUANTO 1.1: A computer program for power and sample size calculations for genetic-epidemiology studies. (Available at http://hydra.usc.edu/gxe/.)

Geng, X., Liu, S., Yao, J. *et al.* (2016) A genome wide association study identifies multiple regions associated with head size in catfish. *G3: Genes| Genomes| Genetics*, **6**, 3389–3398.

Geng, X., Sha, J., Liu, S. *et al.* (2015) A genome-wide association study in catfish reveals the presence of functional hubs of related genes within QTLs for Columnaris disease resistance. *BMC Genomics*, **16**, 196.

Gondro, C., Van der Werf, J. and Hayes, B. (2013) *Genome-wide association studies and genomic prediction*, Humana Press, New York.

Gudbjartsson, D.F., Walters, G.B., Thorleifsson, G. *et al.* (2008) Many sequence variants affecting diversity of adult human height. *Nature Genetics*, **40**, 609–615.

Hayes, B. (2013) Overview of statistical methods for genome-wide association studies (GWAS), in *Genome-wide association studies and genomic prediction* (eds C. Gondro, J. van der Werf and B. Hayes), Humana Press, New York, pp. 149–169.

Hirschhorn, J.N. and Daly, M.J. (2005) Genome-wide association studies for common diseases and complex traits. *Nature Reviews Genetics*, **6**, 95–108.

Hochberg, Y. and Benjamini, Y. (1990) More powerful procedures for multiple significance testing. *Statistics in Medicine*, **9**, 811–818.

Kang, H.M., Sul, J.H., Service, S.K., Zaitlen, N.A., Kong, S.-y. *et al.* (2010). Variance component model to account for sample structure in genome-wide association studies. *Nature Genetics*, **42**, pp. 348–354.

Kang, H.M., Zaitlen, N.A., Wade, C.M. *et al.* (2008) Efficient control of population structure in model organism association mapping. *Genetics*, **178**, 1709–1723.

Korte, A. and Farlow, A. (2013) The advantages and limitations of trait analysis with GWAS: a review. *Plant Methods*, **9**, 29.

Laird, N.M. and Lange, C. (2006) Family-based designs in the age of large-scale gene-association studies. *Nature Reviews Genetics*, **7**, 385–394.

Lange, C., DeMeo, D.L. and Laird, N.M. (2002) Power and design considerations for a general class of family-based association tests: quantitative traits. *The American Journal of Human Genetics*, **71**, 1330–1341.

Lasky-Su, J., Won, S., Mick, E. *et al.* (2010) On genome-wide association studies for family-based designs: an integrative analysis approach combining ascertained family samples with unselected controls. *The American Journal of Human Genetics*, **86**, 573–580.

Lewontin, R. (1964) The interaction of selection and linkage. I. General considerations; heterotic models. *Genetics*, **49**, 49.

Liu, L.Y., Fox, C.S., North, T.E. and Goessling, W. (2013) Functional validation of GWAS gene candidates for abnormal liver function during zebrafish liver development. *Disease Models & Mechanisms*, **6**, 1271–1278.

Liu, S., Sun, L., Li, Y. *et al.* (2014) Development of the catfish 250 K SNP array for genome-wide association studies. *BMC Research Notes*, 7, 135.

Liu, S., Zhou, Z., Lu, J. *et al.* (2011) Generation of genome-scale gene-associated SNPs in catfish for the construction of a high-density SNP array. *BMC Genomics*, **12**, 53.

Lynch, M. and Walsh, B. (1998) *Genetics and analysis of quantitative traits*, Sinauer Associates, Sunderland, MA.

Mackay, I. and Powell, W. (2007) Methods for linkage disequilibrium mapping in crops. *Trends in Plant Science*, **12**, 57–63.

Mackay, T.F. (2001a) The genetic architecture of quantitative traits. *Annual Review of Genetics*, **35**, 303–339.

Mackay, T.F. (2001b) Quantitative trait loci in Drosophila. *Nature Reviews Genetics*, **2**, 11–20.

Michelmore, R.W., Paran, I. and Kesseli, R. (1991) Identification of markers linked to disease-resistance genes by bulked segregant analysis: a rapid method to detect markers in specific genomic regions by using segregating populations. *Proceedings of the National Academy of Sciences*, **88**, 9828–9832.

Morin, P.A., Luikart, G. and Wayne, R.K. (2004) SNPs in ecology, evolution and conservation. *Trends in Ecology & Evolution*, **19**, 208–216.

Mott, R., Talbot, C.J., Turri, M.G. *et al.* (2000) A method for fine mapping quantitative trait loci in outbred animal stocks. *Proceedings of the National Academy of Sciences*, **97**, 12649–12654.

Palti, Y., Gao, G., Liu, S. *et al.* (2014) The development and characterization of a 57 K single nucleotide polymorphism array for rainbow trout. *Molecular Ecology Resources*, **15** (3), 662–672.

Price, A.L., Patterson, N.J., Plenge, R.M. *et al.* (2006) Principal components analysis corrects for stratification in genome-wide association studies. *Nature Genetics*, **38**, 904–909.

Price, A.L., Zaitlen, N.A., Reich, D. and Patterson, N. (2010) New approaches to population stratification in genome-wide association studies. *Nature Reviews Genetics*, **11**, 459–463.

Pritchard, J.K. and Przeworski, M. (2001) Linkage disequilibrium in humans: models and data. *The American Journal of Human Genetics*, **69**, 1–14.

Pritchard, J.K., Stephens, M. and Donnelly, P. (2000) Inference of population structure using multilocus genotype data. *Genetics*, **155**, 945–959.

Purcell, S., Neale, B., Todd-Brown, K. *et al.* (2007) PLINK: a tool set for whole-genome association and population-based linkage analyses. *The American Journal of Human Genetics*, **81**, 559–575.

Quarrie, S.A., Lazić-Jančić, V., Kovačević, D. *et al.* (1999) Bulk segregant analysis with molecular markers and its use for improving drought resistance in maize. *Journal of Experimental Botany*, **50**, 1299–1306.

Reich, D.E., Cargill, M., Bolk, S. *et al.* (2001) Linkage disequilibrium in the human genome. *Nature*, **411**, 199–204.

Risch, N. and Merikangas, K. (1996) The future of genetic studies of complex human diseases. *Science*, **273**, 1516–1517.

Schoenebeck, J.J., Hutchinson, S.A., Byers, A. *et al.* (2012). Variation of BMP3 contributes to dog breed skull diversity. *PLoS Genetics*, **8**, e1002849.

Smitherman, R.O., Dunham, R.A. and Tave, D. (1983) Review of catfish breeding research 1969–1981 at Auburn University. *Aquaculture*, **33**, 197–205.

Spielman, R.S., McGinnis, R.E. and Ewens, W.J. (1993) Transmission test for linkage disequilibrium: the insulin gene region and insulin-dependent diabetes mellitus (IDDM). *American Journal of Human Genetics*, **52**, 506.

Stram, D.O. (2014) *Design, analysis, and interpretation of genome-wide association scans*, Springer, New York.

Sun, L., Liu, S., Wang, R., Jiang, Y. *et al.* (2014). Identification and analysis of genome-wide SNPs provide insight into signatures of selection and domestication in channel catfish (*Ictalurus punctatus*). *PloS One*, **9**, e109666.

Sun, Y., Zhao, G., Liu, R. *et al.* (2013) The identification of 14 new genes for meat quality traits in chicken using a genome-wide association study. *BMC Genomics*, **14**, 458.

Team, R.C. (2014). *R: A language and environment for statistical computing*. Vienna, Austria: R Foundation for Statistical Computing. ISBN 3-900051-07-0.

Turner, S., Armstrong, L.L., Bradford, Y. *et al.* (2011) Quality control procedures for genome-wide association studies. *Current Protocols in Human Genetics*, **1** (19), 11–11.19.18.

Van Gestel, S., Houwing-Duistermaat, J.J., Adolfsson, R. *et al.* (2000) Power of selective genotyping in genetic association analyses of quantitative traits. *Behavior Genetics*, **30**, 141–146.

Wang, R., Sun, L., Bao, L. *et al.* (2013) Bulk segregant RNA-seq reveals expression and positional candidate genes and allele-specific expression for disease resistance against enteric septicemia of catfish. *BMC Genomics*, **14**, 929.

Wang, W.Y., Barratt, B.J., Clayton, D.G. and Todd, J.A. (2005) Genome-wide association studies: theoretical and practical concerns. *Nature Reviews Genetics*, **6**, 109–118.

Won, S., Wilk, J.B., Mathias, R.A. *et al.* (2009). On the analysis of genome-wide association studies in family-based designs: a universal, robust analysis approach and an application to four genome-wide association studies. *PLoS Genetics*, **5**, e1000741.

Wood, A.R., Hernandez, D.G., Nalls, M.A. *et al.* (2011) Allelic heterogeneity and more detailed analyses of known loci explain additional phenotypic variation and reveal complex patterns of association. *Human Molecular Genetics*, **20**, 4082–4092.

Xu, J., Zhao, Z., Zhang, X. *et al.* (2014) Development and evaluation of the first high-throughput SNP array for common carp (*Cyprinus carpio*). *BMC Genomics*, **15**, 307.

Yang, J., Lee, S.H., Goddard, M.E. and Visscher, P.M. (2011) GCTA: a tool for genome-wide complex trait analysis. *The American Journal of Human Genetics*, **88**, 76–82.

Yu, J., Pressoir, G., Briggs, W.H. *et al.* (2005) A unified mixed-model method for association mapping that accounts for multiple levels of relatedness. *Nature Genetics*, **38**, 203–208.

Zhang, Z., Ersoz, E., Lai, C.-Q. *et al.* (2010) Mixed linear model approach adapted for genome-wide association studies. *Nature Genetics*, **42**, 355–360.

Zhou, X. and Stephens, M. (2012) Genome-wide efficient mixed-model analysis for association studies. *Nature Genetics*, **44**, 821–824.

Ziegler, A., König, I.R. and Thompson, J.R. (2008) Biostatistical aspects of genome-wide association studies. *Biometrical Journal*, **50**, 8–28.

24

Gene Set Analysis of SNP Data from Genome-wide Association Studies

Shikai Liu, Peng Zeng and Zhanjiang Liu

Introduction

Genome-wide association studies (GWAS) have been widely conducted to identify genetic variants underlying complex diseases in humans. However, the identified genetic variations usually accounted for only a small proportion of the heritability of the diseases, suggesting that real but perhaps weaker associations were missed in typical GWAS data analyses due to adjustments in the massive number of tests. In recent years, gene set or pathway analysis was introduced to GWAS analysis to discover associations of gene sets that share common biological functions with the genetic variants. In such analyses, SNPs were assigned into subsets based on genes or pathways, and thereby reducing the number of tests. By analyzing SNPs at the gene set level, GWA studies are able to reveal numerous coordinated associations that were ignored previously. Although gene set analysis (GSA) of GWAS data is being widely used, the resulting significant gene sets often vary substantially with different analysis methods. Comparisons of various GSA methods are needed to understand the pros and cons of different analysis methods, and are useful to draw significant conclusions. In this chapter, we provide a brief review of the GSA in GWAS situations. The procedures of different methods are illustrated using a dataset from a GWAS of Alzheimer's disease (AD): a typical GWAS analysis was first conducted using SNPs, and then the results were compared with GWAS using alternative approaches, including GWAS using the gene set-based approach. The example demonstrated that GSA in GWAS could improve performance and enable the identification of important genes with mild SNP "signals," which are essentially missed when using the standard GWAS approach. Detailed comparisons indicated that a combination use of several methods could provide even greater power to the analysis, allowing better conclusions to be drawn.

GWAS, or GWA study, is currently one of the most widely used approaches for the identification of genes underlying complex traits (Hirschhorn & Daly, 2005; Stranger, Stahl & Raj, 2011). A typical GWAS analysis either tests SNP allele frequency differences between case and control groups, or tests differences in the mean phenotypes associated with each SNP genotype. In either case, the analysis requires a relatively large association of a single marker with the phenotype to achieve statistical significance. GWAS have been widely conducted to detect the associations of human diseases with hundreds of thousands to millions of single-nucleotide polymorphisms (SNPs) across the

Bioinformatics in Aquaculture: Principles and Methods, First Edition. Edited by Zhanjiang (John) Liu.

human genome by genotyping hundreds to thousands of individuals. Many genetic variants underlying the susceptibilities of complex diseases have been identified, but only a small proportion of the heritability were accounted for in these diseases. One of the possible reasons could be the missing of some real but weak associations after multiple comparison adjustments in GWAS data analyses, due to the large number of SNPs examined and massive number of tests conducted (Wang *et al.*, 2011).

In recent years, GSA (also referred to as *pathway analysis*) methods were introduced to GWAS to determine the associations of gene sets that share common biological functions with the genetic variants. GSA methods were originally developed to improve the interpretation of the results of genome-wide expression analysis for identifying the differential expression profiles of pre-defined gene sets. One of the implementations of this method is the gene set enrichment analysis (GSEA; see Chapter 10), which focuses on groups of genes that share biomedical or cellular functions, and regulation or chromosomal locations (Subramanian *et al.*, 2005). GSA exhibits higher statistical power than single-gene analysis, and have revealed many novel gene sets with minor but coordinated expression patterns (Subramanian *et al.*, 2005). As the basic goal of GWAS is to identify the gene networks or biological pathways associated with the trait of interest, it is reasonable to regard the pre-defined gene sets or pathways as the units for the association analysis in GWA studies (Nam *et al.*, 2010). By analyzing SNPs at the gene set level, GWA studies are able to reveal numerous coordinated association patterns that are missed in single-SNP analysis. For instance, Wang, Li and Bucan (2007) applied the GSEA framework into GWAS SNP data analysis, where the most highly associated SNP (best SNP) was assigned to each gene to summarize the association of multiple SNPs. Using this approach, they successfully identified the pathways of Parkinson's disease susceptibility. Since then, the same approach was also used to clarify the molecular mechanisms of autism beyond individual genes (Glessner *et al.*, 2009) and used in a number of other GWA studies (Deelen *et al.*, 2013; Holden *et al.*, 2008; Kar *et al.*, 2013; Lee *et al.*, 2013; Nam *et al.*, 2010; Song & Lee, 2013; Wang *et al.*, 2007, 2011; Zhang *et al.*, 2011, 2013; Zhao *et al.*, 2011). Several comprehensive reviews on GSA of GWAS were provided by Wang (Wang, Li & Hakonarson, 2010), Cantor *et al.* (Cantor, Lange & Sinsheimer, 2010), and Wang *et al.* (Wang *et al.*, 2011).

Although GSA of GWAS data is being widely used, the current approaches for GSA are still in an early stage (Wang *et al.*, 2011). The resulting significant gene sets often vary substantially with different analysis methods (Elbers *et al.*, 2009). One possible reason might be the lack of statistical power in the tests (Wang *et al.*, 2011). In a recent simulation study, all three GSA methods—GSEA, Fisher's exact test, and SNP ratio test—were found lacking in statistical power for detecting disease-associated gene sets (Jia *et al.*, 2011). Studies also indicated that current GSA results are prone to biases related to different gene set sizes, linkage disequilibrium (LD) patterns, and overlapping genes (Cantor *et al.*, 2010; Hong *et al.*, 2009; Wang *et al.*, 2007, 2011).

GSA may become more and more popular for the analysis of genetic and molecular data with the accumulation of knowledge on genes and biological pathways responsible for complex traits (Cantor *et al.*, 2010; Fridley & Biernacka, 2011; Wang *et al.*, 2010). With the increasing application of GSA in GWAS data analysis, it is important to examine different GSA methods, determine the advantages and limitations, and address challenges in these analyses. Here, several widely used GSA approaches are compared, including a standard set-based test from Plink (hereafter referred to as

"Plink-set") (Purcell *et al.*, 2007), logistic kernel-machine-based test ("LKM") (Wu *et al.*, 2010), adaptive rank truncated product (ARTP)–based test ("ARTP") (Yu *et al.*, 2009), and GSEA (Wang *et al.*, 2007). Among these, Plink-set, LKM, and ARTP are gene based, while GSEA is based on pathways. Because of the large demand for GWAS analyses, numerous novel GSA methods are expected to be developed. The objective of this chapter is not to provide a comprehensive survey of GSA methods; instead, we will focus on several of the most popular approaches to provide an introduction to the GSA approach and a summary of the statistical methods, and compare the results obtained by applying these methods using a real GWAS dataset.

GSA in GWAS

A typical GSA of GWAS data (Wang *et al.*, 2011) involves the following steps: (1) preprocessing of data and defining the SNP sets, (2) formulating a hypothesis, (3) constructing the corresponding statistical test, and (4) assessing the statistical significance of the results.

Preprocessing Data and Defining the Gene Sets

The first step of GSA in GWAS is to define the gene sets. In this process, the SNPs involved in GWAS need first to be "assigned" to genes, and then the genes need to be "assigned" to gene pathways, which are a set of genes with similar functions, or play complementary, additive, or synergistic roles in the same biological process. To do this, first, gene pathway databases need to be selected. There are several widely used pathway databases, including the Kyoto Encyclopedia of Genes and Genomes (KEGG) (Ogata *et al.*, 1999), Gene Ontology (Ashburner *et al.*, 2000), and Molecular Signature Database (MSigDB) (Subramanian *et al.*, 2005). Second, SNPs are associated with genes. SNPs are assigned to genes based on SNP coordinates in the genome. Various criteria are used for this purpose, but the most widely used approach is to assign SNPs within a certain distance (e.g., 20 KB) to a gene, most often upstream of the first exon or downstream of the last exon of a gene into a gene-based SNP set. Then, genes are assigned to gene pathways based on information extracted from pathway databases. Genes are generally linked to pathways according to gene symbols, such as HUGO gene symbols for human genes.

Formulating a Hypothesis

In the analysis of gene expression data, two statistical hypotheses were formulated for testing coordinated associations between a set of genes with a phenotype of interest (Tian *et al.*, 2005): (1) the genes in a gene set showing the same magnitude of association with the disease phenotype are compared with the remaining genes in the genome; and (2) the genes in a gene set are not associated with the disease phenotype.

To test an individual gene set, Goeman and Buhlmann (2007) classified tests according to null hypotheses (1) and (2) as competitive and self-contained tests, respectively. While a competitive test compares disease association test statistics for genes in the gene set versus that for the remaining genes in the genome, a self-contained test directly tests gene set associations with disease, and does not depend on genes not included in the gene set. When the causal SNPs are fully contained in the gene set, testing the two null hypotheses is approximately the same. However, when causal genes are in multiple gene sets, or the disease is associated with multiple gene sets, using competitive tests that compare gene set association signals with the remaining genes in the genome may result in loss of power (Hong *et al.*, 2009).

Constructing Corresponding Statistical Tests

A test statistic can be constructed with units based on either gene or SNP, which are classified as gene-based and SNP-based methods, respectively. In the gene-based method, SNPs in each gene are first used to evaluate their associations with the gene, followed by aggregation of the gene-level tests to test for the association of the phenotype with the gene sets. In the SNP-based method, all SNPs in a gene set are used in the analysis without consideration of gene-level effects. Several studies reported that gene-based methods may have higher power because only a few SNPs within different genes usually contribute to disease risk or in LD with causal variants (Yu *et al.*, 2009). However, in gene-based methods, a consensus opinion on the best strategy for SNP information reduction within each gene has not been reached. Strategies include using the most significant SNP, using a summary measure of all SNPs within a gene, or simultaneously modeling the effects of all SNPs in the gene on the phenotype. A common and simple approach is to use the most significant SNP to represent each gene. However, this strategy would result in bias for larger genes because larger genes have more SNPs, and are likely to have more significant SNPs (smaller minimum p-value) as compared to smaller genes with fewer SNPs. Multiple testing corrections need to be made to adjust the most significant p-value for the number of SNPs on the gene to assess the false discovery rate (FDR) (Benjamini, 2010). In addition, this strategy may not be the most powerful approach when each of several SNPs in a gene has a modest effect on the phenotype.

Assessing the Statistical Significance of the Results

To preserve LD patterns, permutations of sample labels are typically used to establish null distribution of gene set scores. Several difficulties exist for the application of permutation tests to GWAS. For a typical GWAS with half a million or more SNPs on hundreds or even thousands of samples, the recalculation of a gene set score for each permutation is extremely computationally intensive, especially for the competitive tests based on markers from the entire genome. In contrast, permutation tests for the GSA of GWAS would require much less computing time and resources due to the much smaller number of statistical tests.

Statistical Methods

Single-SNP Analysis

Single-SNP analysis has been widely used in standard GWA studies. Typically, for a population-based case–control GWAS, a simple logistic regression model is used for assessing SNP effects by individual SNPs as well as for accounting for other covariates. Take a population-based case–control GWAS as an example, where n independent subjects are genotyped. For a given SNP set containing p SNPs, let z_{i1} , z_{i2} , … , z_{ip}, be genotypes for the SNPs in the SNP set for the i-th subject ($i = 1, \ldots, n$). The SNPs are coded as $z_{ij} = 0, 1, 2$, corresponding to homozygotes for the major allele (AA), heterozygotes (Aa), and homozygotes (aa) for the minor allele, respectively. For each individual, m confounding variables, such as demography, age, gender, and environmental factors, are included. For the i-th subject, let $x_{i1}, x_{i2}, \ldots, x_{im}$ denote the values of the covariates. The case–control status for the i-th subject is denoted by y_i ($y_i = 1$ for cases, and $y_i = 0$ for controls). The logistic regression model is presented as:

$$\text{logit } P(y_i = 1) = \alpha_0 + \alpha_1 x_{i1} + \ldots + \alpha_m x_{im} + \beta_1 z_{i1} + \beta_2 z_{i2} + \ldots + \beta_p z_{ip}$$

Here, α_0 is the intercept term; $\alpha_1, \alpha_2, \ldots, \alpha_m$ are the regression coefficients of covariates; and β_j is the regression coefficient corresponding to the j-th SNP.

Set-based Tests

For most practical purposes, the analytical unit of interest for a set-based test is the gene. The tests estimate the significance of each SNP and then calculate the average χ^2 statistic for the most significant SNPs per gene. Permutation tests are used to obtain empirical significance levels (p-values) of the gene-based test, while accounting for the number of tests conducted within a gene and the lack of independence of SNPs within the gene. Sets could comprise up to five SNPs, and gene-wide statistical significance was estimated with 50000 permutations.

The set-based SNP tests in Plink v1.07 (Purcell *et al.*, 2007) generally work as follows:

1) For each SNP set, each SNP is determined in LDs with other SNPs above a certain threshold.
2) Standard single-SNP analysis (e.g., case–control association) is performed.
3) For each SNP set, select the “independent” SNPs with p-values below a certain threshold. The best SNP is selected first; subsequent SNPs are selected in the order of decreasing statistical significance, after removing SNPs in LDs with previously selected SNPs.
4) From these subsets of SNPs, the statistic for each set is calculated as the mean of these single-SNP statistics.
5) Permute the dataset a large number of times, keeping the LD between SNPs constant (i.e., permute phenotype labels).
6) For each permutation dataset, repeat steps 2–4.
7) The empirical p-value for the set (EMP1) is the number of times the permuted set-statistic exceeds the original one for that set. Empirical p-values are corrected for the multiple SNPs within a set (taking account of the LD between these SNPs).

LKM Regression Approach

As described in Wu *et al.* (2010), this approach is illustrated in the following text using the same example as in a typical GWAS using single-SNP analysis. Given that n independent subjects are genotyped, and, for a given SNP set containing p SNPs, let $z_{i1}, z_{i2}, \ldots, z_{ip}$ be genotypes for the SNPs in the SNP set for the i-th subject ($i = 1, \ldots, n$). The SNPs are coded as $z_{ij} = 0, 1, 2$, corresponding to homozygotes for the major allele, heterozygotes, and homozygotes for the minor allele, respectively. For each individual, m indicates confounding variables such as demography, age, gender, and environmental factors. For the i-th subject, let $x_{i1}, x_{i2}, \ldots, x_{im}$ denote the values of the covariates. The case–control status for the i-th subject is denoted by y_i ($ji = 1$ for cases, and $ji = 0$ for controls). The LKM-based regression model is presented as (Wu *et al.*, 2010):

$$\text{logit } P(y_i = 1) = \alpha_0 + \alpha_1 x_{i1} + \cdots + \alpha_m x_{im} + h(z_{i1,} \ z_{i2} , \ldots, z_{ip})$$

The SNPs, $z_{i1} , \ldots, z_{ip}$, influence y_i through a general function $h(.)$, which is an arbitrary function that has a form defined only by a positive, semi-definite kernel function $K(.,.)$. The $h(.)$ is the only model component to evaluate the SNP effects. Using the representer theorem (Kimeldorf & Wahba, 1971), $h(z_{i1,} \ z_{i2} , \ldots, z_{ip})$ is equal to $h_i = h(\boldsymbol{Z}_i) = \sum_{i'}^{n} \gamma_{i'} K(\boldsymbol{Z}_i, \boldsymbol{Z}_{i'})$ for some $\gamma_1 , \ldots, \gamma_n$. This shows that $h(.)$ is fully defined by the kernel function $K(.,.)$. By choosing different kernel functions, different bases and corresponding models can be specified. It is apparent that the choice of kernel changes the underlying basis for the nonparametric function governing the relationship between case–control status and the SNPs in the SNP set. $K(\boldsymbol{Z}_i, \boldsymbol{Z}_{i'})$ can be viewed as a function that measures the similarity between two individuals, the i-th and i'-th subject, on the basis of the genotypes of the SNPs in the SNP set. For this perspective, $K(.,.)$ can be linear, Gaussian, identical-by-state (IBS), and weighted IBS kernels.

The linear kernel is $K(\boldsymbol{Z}_i , \boldsymbol{Z}_{i'}) = \sum_{j=1}^{p} z_{ij} z_{i'j}$.

The Gaussian kernel is $K(\boldsymbol{Z}_i, \boldsymbol{Z}_{i'}; d) = exp\left\{-\sum_{j=1}^{p} (z_{ij} - z_{i'j})^2/d\right\}$, where d is a parameter that approximately controls curvature of the kernel function.

The IBS kernel is $K(\boldsymbol{Z}_i, \boldsymbol{Z}_{i'}) = \sum_{j=1}^{p} \{2I(z_{ij} = z_{i'j}) + I(|z_{ij} - z_{i'j}| = 1)\}/2p$.

The weighted IBS kernel is $K(\boldsymbol{Z}_i, \boldsymbol{Z}_{i'}; \boldsymbol{w}) = \sum_{j=1}^{p} w_j\{2I(z_{ij} = z_{i'j}) + I(|z_{ij} - z_{i'j}| = 1)\}/2p$, where $w_j = 1/\sqrt{q_i}$ and q_i is the minor allele frequency for the j-th SNP in the SNP set.

To test whether there is a true SNP-set effect: it is noted that the probability of the i-th subject is a case depending on the SNPs only through the function $h(\boldsymbol{Z}_i)$; therefore, the null hypothesis can be:

$$H_0: \quad \boldsymbol{h}(\boldsymbol{Z}) = 0$$

To test the hypothesis, the connection between the LKM framework and generalized linear mixed models is exploited (Liu, Ghosh & Lin, 2008). In brief, let $\mathbf{K}$ be the n × n matrix with the $(\boldsymbol{ii'})$-th element equal to $K(\boldsymbol{Z}_i, \boldsymbol{Z}_{i'})$. It is straightforward to see that $\boldsymbol{h} = \mathbf{K}\gamma$, where $\boldsymbol{h} = [h_1, \ldots, h_n]'$. Treat $\boldsymbol{h}$ as a subject-specific random effect; then, through the generalized linear mixed model, h follows an arbitrary distribution F with a mean of 0 and a variance of $\tau\mathbf{K}$. τ measures the effect of the SNPs in the SNP set such that

$$H_0: \quad \boldsymbol{h}(\boldsymbol{Z}) = 0 \rightarrow H_0: \quad \tau = 0$$

Then it is only needed to test whether the indexing parameter τ is significantly different from 0. This can be done with the variance-component score test of Zhang and Lin (2003) using the statistic:

$$Q = \frac{(\boldsymbol{y} - \hat{\boldsymbol{p}}_0)'\boldsymbol{K}(\boldsymbol{y} - \hat{\boldsymbol{p}}_0)}{2}, \text{ where logit } \hat{p}_{0i} = \hat{\alpha}_0 + \hat{\alpha}_1 x_{i1} + \cdots + \hat{\alpha}_m x_{im}.$$

In this score test, since $\hat{\alpha}_0$ and $\hat{\alpha}_j$ are estimated under the null model, which does not contain $\boldsymbol{h}$, the standard estimate from the logistic-regression model without the genotypes can be used. To compute the p-value for significance, Q is compared to a scaled χ^2 distribution with the scale parameter κ and degree of freedomν (Lin, 1997; Zhang & Lin, 2003).

ARTP Method

The ARTP method was proposed by Yu *et al.* (2009). In this approach (Yu *et al.*, 2009), to test the null hypothesis (H_0) that a pathway consisting of L SNPs is not associated with a disease outcome, the tests on individual SNPs are performed. Let the resulting p-values be $p_1, \ldots, p_L$; then, the ordered statistics of those p-values are $p_{(1)} \leq \ldots \leq p_{(L)}$, with $p_{(l)}$ being the l-th smallest p-value. To test for a global null hypothesis, the RTP statistic (Dudbridge & Koeleman, 2003) has been proposed:

$$W(K) = \prod_{i=1}^{K} p_{(i)}$$

Here, K is a predefined integer that is the truncation point, and $1 \leq K \leq L$. $W(K)$ is the product of the K smallest p-values. When all tests are independent, the p-value associated with $W(K)$ can be calculated. However, if the p-values are correlated (due to the LD among the SNPs with a gene), a permutation procedure is generally needed to obtain the significance level of $W(K)$.

To use this RTP statistic, a prior truncation point K should be selected. When the number of individual tests is large, it is difficult to make a choice of K. To solve this problem, the optimized association evidence can be obtained on each of the J candidate truncation points $K_1 \leq \ldots \leq K_J$. Specifically, let $\hat{s}\ (K_j)$ be the estimated p-value for $W(K_j)$, $1 \leq j \leq J$; then, the statistic based on minimum p-value are defined (Dudbridge & Koeleman, 2004; Hoh, Wille & Ott, 2001; Yu *et al.*, 2009) as:

$$MinP = \min_{1 \leq j \leq J} \hat{s}(K_j)$$

To get the adjusted p-value for $MinP$, a two-level permutation procedure is generally needed (Hoh *et al.*, 2001), with the inner level for estimating $\hat{s}\ (K_j)$ and the outer level for the adjustment needed to account for multiple testing over different truncation points. This type of permutation procedures, however, can become computationally infeasible if the number of tests for L is relatively large. Yu *et al.* (2009) proposed to use a single layer of permutation for determining the significance level of the ARTP statistic by borrowing techniques originally introduced for gene-expression data analysis (Ge, Dudoit & Speed, 2003). The p-values for each test on the null hypothesis based on the observed data were first obtained, and denoted as $p_1^{(0)}, \ldots, p_L^{(0)}$. Then, the B datasets were generated using the appropriate permutation (or re-sampling) procedure under the null hypothesis H_0. Based on the b-th permuted dataset, $1 \leq b \leq B$, L individual tests were performed, and the resulting p-values were denoted as $p_1^{(b)}, \ldots, p_L^{(b)}$. Based on these p-values, the following steps are applied to obtain the adjusted p-value for $MinP$ (Yu *et al.*, 2009):

1) Based on $p_1^{(b)}, \ldots, p_L^{(b)}$, $0 \leq b \leq B$, calculate the RTP statistics for each candidate truncation point, and denote them as:

$$W_j^{(b)} \prod_{i=1}^{K_j} p_{(i)}^{(b)}, 1 \leq j \leq J,\ \ 0 \leq b \leq B$$

2) Based on $W_j^{(b)}, 1 \leq j \leq J, 0 \leq b \leq B$, the estimated p-value $\hat{s}_j^{(b)}$ corresponding to $W_j^{(b)}$ was obtained using Ge's algorithm (Ge *et al.*, 2003).

3) Let $MinP^{(b)} = \min_{1 \le j \le J} \hat{s}_j^{(b)}, 0 \le b \le B$; then, the adjusted p-value for the ARTP statistic $MinP^{(0)}$ is estimated as:

$$\frac{\sum_{b=0}^{B} I(MinP^{(b)} \le MinP^{(0)})}{B+1}$$

In step 2, the estimated p-values $\hat{s}_j^{(b)}$ for $W_j^{(b)}$ from the observed data ($b = 0$) and from the b-th (≥ 1) generated dataset need to be calculated. Generally, the standard approach would apply another level of permutations specifically based on the b-th (≥ 1) permuted dataset. To avoid this, the set $\{W_j^{(b)},\ b = 0,\ \ldots,\ B\}$, itself was used to form a common reference distribution for the evaluation of the significance level of each $W_j^{(b)}$ (Ge *et al.*, 2003). Specifically, the p-value for $W_j^{(b)}$ is estimated as:

$$s_j^{(b)} = \frac{\sum_{b*=0}^{B} I(W_j^{(b*)} \le W_j^{(b)})}{B+1}$$

In step 3, the adjusted p-value from the observed ARTP statistic $MinP^{(0)}$ is estimated. In a calculation similar to that in step 2, the adjusted p-value for $MinP^{(b)}$, the ARTP statistic from the b-th generated dataset $1 \le b \le B$, can be estimated by:

$$\frac{\sum_{b*=0}^{B} I(MinP^{(b*)} \le MinP^{(b)})}{B+1}$$

GSEA-based Approach

The GSEA method is proposed by Subramanian *et al.* (2005) to detect significant "enrichment" of disease association in a gene set from gene expression analysis. GSEA is then extended to GWAS analyses to jointly incorporate SNP data with known gene pathways (Holden *et al.*, 2008; Wang *et al.*, 2007).

To illustrate the principle of this method, here we introduce the original GSEA analysis. Let $L_0 = \{G_1,\ G_2,\ \ldots,\ G_N\}$ be the list of all genes quantified with expression. Sorting their association test statistic values (e.g., t-test score) from largest to smallest, we obtain $r_{(1)},\ r_{(2)},\ \ldots,\ r_{(N)}$ and a ranked gene list $L = \{G_{(1)},\ G_{(2)},\ \ldots,\ G_{(N)}\}$. Let $S \in L_0$ be the gene set of interest. The enrichment in S is evaluated in three steps:

1) *Calculate the enrichment score (ES)*: A weighted Kolmogorov–Smirnov statistic is used to measure over-representation of S at the top of the ranked list L. It is calculated by examining the ranked gene list L, to increase a running sum statistic when encountering a gene in S, and to decrease it when not. The ES is defined by maximization of the running sum statistics:

$$ES_{(SD)} = \max_{1 \le j \le N} \left\{ \sum_{G_{(j*)} \in S,\ j* \le j} \frac{|r_{(j*)}|^P}{N_R} - \sum_{G_{(j*)} \notin S,\ j* \le j} \frac{1}{N - |S|} \right\}$$

Here, D denotes observed data; $|S|$ is the number of genes in S; $N_R = \sum_{G_{(j*)} \in S} |r_{(j*)}|^P$; and P is a tuning parameter giving higher weights to highly associated genes.

2) *Evaluate the significance of ES by permutation*: To estimate the significance level of ES, permutation procedures were used to create the null distribution of ES for each pathway/gene set. ES is calculated by repeatedly shuffling the disease status

of samples to generate an empirical null distribution of ES. An empirical p-value is estimated by the proportion of permutations that results in larger ES than originally observed.

3) *Correct for multiple testing*: The FDR and family-wise error rate ($FWER$) are used to correct for multiple testing when a large number of gene sets are tested simultaneously. The FDR is a procedure used to control the portion of expected false-positive findings to stay below a certain threshold, while $FWER$ is considered to be a highly conservative procedure seeking to ensure that the list of reported results does not include even a single false-positive gene set (Holden *et al.*, 2008; Wang *et al.*, 2007). The ES is normalized first to minimize the effects of varying gene set size and within-gene-set corrections. The normalized ES is defined as:

$$NES_{(SD)} = \frac{ES_{(SD)}}{mean(ES_{(S,\pi)})},\ \text{where } ES_{(S,\pi)} \text{ is the ES for permutation } \pi.$$

4) *GSEA of SNP data*: One of the key steps to extend GSEA to GWAS is to derive a summary score that combines signals from individual SNPs into each gene. To denote the m SNPs in the gene G_k as $V_{k1}, V_{k2}, \ldots, V_{km}$, and their association test statistic as $t_{k1}, t_{k2}, \ldots, t_{km}$, the gene score for G_k is usually assigned with the highest test statistic value among all the SNPs, $\max(t_{kj})$ (Holden *et al.*, 2008; Wang *et al.*, 2007). Following this, the ESs are calculated as in the original GSEA. To compare the significance among pathways with different numbers of genes (different gene set size), the normalized ES is first calculated as follows:

$$NES_{(S,D)} \frac{ES_{(SD)} - mean(ES_{(S,\pi)})}{SD(ES_{(S,\pi)})}$$

To adjust the multiple-hypothesis testing, FDR and FWER are calculated. For a pathway/gene set, $NES_{(S,D)}$ denotes the normalized ES in the observed data. The FDR q value (q_{FDR}) was calculated as the ratio of the fraction of all permutations with $NES \geq NES_{(S,D)}$ to the fraction of observed pathways/gene sets with $NES \geq NES_{(S,D)}$.

$$q_{\text{FDR}} = \frac{\%\text{ of permutations with } NES_{(S,\pi)} \geq NES_{(S,D)}}{\%\text{ of observed } S \text{ with } NES \geq NES_{(S,D)}}$$

FWER p-value (p_{FWER}) is calculated as the fraction of pathways/gene sets whose largest NES among all permutations is greater than $NES_{(S,D)}$.

$$p_{\text{FWER}} = \%\ \text{of } S \text{ with } \max(NES_{(S,\pi)}) \geq NES_{(S,D)}$$

Demonstration Using Alzheimer's Disease Neuroimaging Initiative's AD Data

Data Information

To demonstrate the application of GSA methods, the aforementioned methods were applied to a real GWAS dataset, and the results were compared with those found using standard single-SNP analysis. Data used for this analysis were obtained through the Alzheimer's Disease Neuroimaging Initiative (ADNI, http://adni.loni.ucla.edu/) database. ADNI was initiated in 2003 by Michael W. Weiner, from VA Medical Center and University of California, San Francisco, with the goal to evaluate the biomarkers of AD–related neuropathology in patients with mild cognitive impairment (MCI) and early AD. This study was supported by the National Institute on Aging (NIA), the National Institute of Biomedical Imaging and Bioengineering (NIBIB), the Food and Drug Administration (FDA), private pharmaceutical companies, and non-profit organizations.

Participants in ADNI include older individuals, aged 55–90 years, recruited from 59 sites across the United States and Canada. These subjects include approximately 200 cognitively normal patients (CN), 400 patients diagnosed with MCI, and 200 patients diagnosed with early probable AD. As described previously, diagnoses of participants were made on a clinical basis via neuropsychological assessment data and patient and informant reports of cognitive performance and functioning in activities of daily living at consensus conferences involving neurologists, neuropsychologists, and study coordinators (Jack *et al.*, 2009; Petersen *et al.*, 2010). Written informed consent was obtained for all participants, and prior Institutional Review Board approval was obtained at each participating institution. All demographic information, neuropsychological and clinical assessment data, and diagnostic information used in this study were downloaded from the ADNI clinical data repository (http://adni.loni.ucla.edu/).

Genotype data used in ADNI included SNP data obtained from a GWAS on the full ADNI sample (Saykin *et al.*, 2010). The acquisition and processing of genotype data for the ADNI sample was previously described (Saykin *et al.*, 2010). Briefly, all participants analyzed were genotyped using the Human610-Quad BeadChip (Illumina, Inc., San Diego, CA), which included 620901 SNPs and copy number variant markers. The ADNI genotype data was subjected to standard quality control procedures (Shen *et al.*, 2010). Genotype markers were excluded if they had a call rate of $<95\%$, Hardy–Weinberg equilibrium test score of $<10^{-6}$, or minor allele frequency of $<5\%$. Samples were excluded if they had a call rate of $<95\%$. Finally, a total of 521971 SNPs passed quality control and were used for subsequent analyses.

Data Analysis: General Strategy

In the present analysis, we only concentrated on the participants diagnosed with early AD and the CN samples. A total of 392 participants—178 AD and 214 CN—were used, while 364 participants with MCI were excluded for further analysis. Three covariates—age, gender, and education—were included in the analysis.

The SNP data from GWA studies are generally organized as an $n \times p$ matrix, with the inclusion of phenotype variable and covariates. As shown in Figure 24.1, a partial set of data was used for illustrating the data structure. It should be noted that, in the full dataset of a typical GWA study, the number of samples is greatly smaller than the number of SNPs (i.e., $n << p$)

We conducted the following data analysis: (1) standard single-SNP analysis using the logistic model; (2) basic set-based SNP analysis using Plink; (3) the LKM method, which uses gene-based SNP sets and assesses the SNP effects using the kernel machine; (4) the ARTP method, which uses gene-based SNP sets and considers the combination of p-values; and (5) the GSEA method, which uses pathway-based SNP sets (Table 24.1). The results from GSA-based GWAS were compared with each other, and compared with the results from standard analysis. The description of input data and the codes used for these analyses are provided in great detail in the following section.

Sample Covariates Phenotype Genotype SNP ID

SNP code

AD

CN

	A	B	C	D	E	F	G	H	I	J	K
1	Patient ID	Gender	age	Education	Affection status	rs3094315	rs12562034	rs12124819	rs4475691	rs28705211	rs13303118
2	123_S_0094	1	70.7	14	1	Aa 1	aa 2	AA 0	aa	aa	AA
3	126_S_0606	1	68.5	10	1	Aa	aa	AA	Aa	aa	Aa
4	022_S_0219	0	84.9	12	1	Aa	aa	Aa	Aa	Aa	Aa
5	109_S_1157	1	70.7	13	1	Aa	aa	AA	aa	aa	aa
6	002_S_1018	1	74.7	8	1	Aa	[illegible]	[illegible]	aa	aa	Aa
7	041_S_1435	0	86.6	13	1	AA	[illegible]	[illegible]	aa	Aa	Aa
8	123_S_0091	0	84.7	12	1	AA	aa	AA	Aa	Aa	aa
9	099_S_0372	0	71.6	12	1	AA	aa	AA	Aa	aa	Aa
10	006_S_0547	0	72.4	13	1	AA	aa	aa	Aa	Aa	aa
11	007_S_1304	1	72.7	16	1	AA	aa	AA	Aa	Aa	Aa
12	133_S_1170	0	72.6	18	1	AA	AA	AA	Aa	Aa	AA
13	128_S_0266	0	82.7	16	1	aa	aa	AA	Aa	AA	Aa
14	099_S_0470	1	76.1	7	1	AA	aa	Aa	aa	aa	aa
15	012_S_0803	1	76	12	1	AA	aa	AA	aa	aa	AA
16	005_S_1341	1	76.9	18	1	AA	aa	AA	aa	aa	Aa
17	129_S_0778	0	76.9	20	0	AA	aa	Aa	Aa	Aa	Aa
18	057_S_0934	1	74.1	17	0	AA	Aa	AA	aa	Aa	aa
19	007_S_1222	1	76.6	14	0	AA	AA	AA	aa	aa	aa
20	036_S_0813	0	77.5	18	0	Aa	Aa	AA	aa	Aa	aa
21	068_S_0473	0	70.2	16	0	Aa	Aa	Aa	aa	aa	aa
22	136_S_0086	1	75.8	18	0	Aa	aa	aa	aa	aa	AA
23	016_S_0538	0	80	20	0	AA	aa	Aa	aa	aa	AA
24	068_S_0210	1	79.1	15	0	Aa	aa	AA	aa	Aa	aa
25	126_S_0405	0	74.3	16	0	AA	aa	aa	aa	aa	AA
26	141_S_0717	0	73.7	20	0	Aa	aa	AA	AA	AA	Aa
27	128_S_0229	1	83.3	15	0	AA	Aa	AA	aa	aa	Aa
28	130_S_0886	1	72	20	0	AA	aa	Aa	aa	aa	Aa
29	100_S_0035	0	77.5	20	0	Aa	aa	AA	Aa	Aa	Aa
30	941_S_1195	0	76.2	16	0	AA	aa	Aa	Aa	Aa	aa

Figure 24.1 A presentation of GWAS data using several samples and SNP markers. The rows are samples ($n = 29$), columns two–four are covariates, the fifth column is phenotype data (AD or CN), and the last six columns are SNP genotypes (AA, Aa, and aa, which are commonly coded into 0, 1, 2 in analysis).

Table 24.1 Methods used for GSA of GWAS data.

Method	SNP set	Input data	Gene set test statistic	Significance assessment	References
Plink set	Gene-based	Genotype	Mean of significant SNPs	Adaptive permutation	Purcell *et al.* (2007)
GSEA	Gene-based	Genotype	Modified KS	Sample permutation	Wang, Li & Bucan (2007)
LKM	Gene-based	Genotype	Kernel function	Empirical null distribution	Wu *et al.* (2010)
ARTP	Gene-based	Genotype	Adaptive RTP	Permutation	Yu *et al.* (2009)

Data Analysis: Protocol and Codes

Standard GWAS Using Plink

1) *Input data files*: binary PED files

   ADNI392.bed: containing genotype information

   ADNI392.bim: containing phenotype information including six mandatory columns: "Family ID", "Individual ID", "Paternal ID", "Maternal ID", "Sex" (1 = male; 2 = female; "other" = unknown), and "Phenotype"

   ADNI392.fam: an extended MAP file containing the marker information including chromosome#, rs#/SNP identifier, genetic distance (cM), and physical position (bp)

2) *Run association test*:

```
plink --noweb --bfile ANDI392 -logistic -covar
  covariates.txt
```

The covariates are loaded using the option "-covar", and the "covariates.txt" file has three columns containing information on age, gender, and education status.

Set-based Test Using Plink

1) *Make set file with gene list* (hg18) 20 KB upstream and downstream of a gene (assign SNPs into genes):

```
plink --noweb --bfile ../ADNI392 --make-set ../glist-hg18
   --make-set-border 20 --write-set
```

"glist-hg18" is a gene coordinate file (Chromosome, Start position, Stop position, Gene name), retrieved from http://pngu.mgh.harvard.edu/~purcell/plink/res.shtml#sets.

2) *Perform LD-based set test* (default: "--set-r2"=0.5; "--set-p"=0.05; "--set-max"=5):

```
plink --noweb --bfile ../ADNI392 --set-test --set plink
  .set --mperm 10000 -logistic
```

LKM Method Using SKAT, an R Package

1) *Input data files*: It can read plink binary format, but we need to generate the SSD file first:
   Three plink binary files: ADNI.bed, ADNI.bim, ADNI.fam
   SNP_set file: ADNI.SetID
   Create two empty files: ADNI.SSD and ADNI.SSD.info
2) *R Commands*:
3) #Load the package

```
library(SKAT)
```

#Import data files

```
File.Bed<-"F:/ADNI/ADNI392.bed"
File.Bim<-"F:/ADNI/ADNI392.bim"
File.Fam<-"F:/ADNI/ADNI392.fam"
File.SetID<-"F:/ADNI/ADNI392.SetID"
File.SSD<-"F:/ADNI/ADNI392.SSD"
File.Info<-"F:/ADNI/ADNI392.SSD.info"
```

To generate SSD file:

```
Generate_SSD_SetID(File.Bed, File.Bim, File.Fam,
   File.SetID, File.SSD, File.Info)
FAM<-Read_Plink_FAM(File.Fam, Is.binary=FALSE)
y<-FAM$Phenotype
```

```
# To use an SSD file, open it first. After using, if must be closed.
SSD.INFO<-Open_SSD(File.SSD, File.Info)
# Number of samples
SSD.INFO$nSample
# Number of Sets
SSD.INFO$nSets
obj<-SKAT_Null_Model(y ~ 1, out_type="C")
#The default kernel is "linear.weighted"
out<-SKAT.SSD.All(SSD.INFO, obj)
# save results "out" to a file
write.csv(out$results, file="F:/ADNI/ADNI392.gs.output
  .csv")
#Using linear kernel
out.linear<-SKAT.SSD.All(SSD.INFO, obj, kernel="linear")
write.csv(out.linear$results, file="F:/ADNI/ADNI392.gs
  .linear.csv")
#Using IBS kernel
out.ibs<-SKAT.SSD.All(SSD.INFO, obj, kernel="IBS")
write.csv(out.ibs$results, file="F:/ADNI/ADNI392.gs.ibs
  .csv")
#Using IBS.weighted kernel
out.ibs.wt<-SKAT.SSD.All(SSD.INFO, obj, kernel="IBS
  .weighted")
write.csv(out.ibs.wt$results, file="F:/ADNI/ADNI392.gs
  .ibs.wt.csv")
out.qd.wt<-SKAT.SSD.All(SSD.INFO, obj, kernel="quadratic")
write.csv(out.qd$results, file="F:/ADNI/ADNI392.gs.qd
  .csv")
```

Adaptive Rank Truncated Product Method Using an R Package, ARTP

1) *Input data files*:
   "geno_data.txt": first column is SNPid; column header is "ldat"; and the other columns are SNP genotypes, with sample IDs as column headers. The SNPs are coded as "AA, AG, GG", which is file type 2.
   "phenol_data.txt": first column is sample name, with ID as header; second column is phenotype (case: 1, control: 0); header is Y; and the third and fourth columns are continuous values with X1 and X2 as headers.
   "gene_SNP_data.txt": first column is SNPs with the header of "SNP"; second column is genes with the header of "Gene"

2) *R Commands*:

#Load the package

```
library(ARTP)
```

#Import data files

```
geno_file <- ("/home/sl/ARTP/ADNI392.AB2AG.txt")
pheno_file <- ("/home/sl/ARTP/ADNI_pheno_data.txt")
gs_file <- ("/home/sl/ARTP/ADNI_SNP-Gene_data.txt")
print(geno_file)
print(pheno_file)
print(gs_file)
```

#Define the three files

```
geno.list <- list(file = geno_file, delimiter = "\t",
   file.type = 2, in.miss= "NA")
pheno.list <- list(file = pheno_file, delimiter = "\t",
   header = 1,id.var = "ID", response.var = "Y")
gs.list <- list(file = gs_file, snp.var = "SNP",
   gene.var = "Gene", delimiter = "\t", header = 1)
```

#Define an output directory

```
out.dir <- ("/home/sl/ARTP/out.dir")
print(out.dir)
```

#Define the two output files: observed p-values and permutated p-values

```
obs.outfile <- paste(out.dir, "/", "obs.txt", sep = "")
perm.outfile <- paste(out.dir, "/", "perm.txt", sep = "")
```

#Compute p-values

```
nperm <- 500
op.list <- list(nperm = nperm, obs.outfile = obs.outfile,
   perm.outfile = perm.outfile, perm.method = 2)
runPermutations(geno.list, pheno.list, 1, op = op.list)
```

#Compute the gene and pathway p-values

```
set.seed(76523)
ret <- ARTP_pathway(obs.outfile, perm.outfile, nperm,
   out.dir,gene.list = gs.list)
print(ret)
```

#Write out the running results

```
write.csv(ret$gene.table, file="/home/sl/ARTP/out.dir/
  gene.table.csv")
```

Results of Single-SNP Analysis

A regular GWAS analysis using the logistic model was performed in Plink (plink – logistic) for single-SNP associations. The top 30 SNPs from this analysis are shown in Table 24.2.

The top SNP-associated gene, APOE, was reported previously by several independent studies as the highly significant association locus with AD (Grupe *et al.*, 2007; Li *et al.*, 2008; Reiman *et al.*, 2007). However, even the SNP in APOE, the top-ranked

Table 24.2 A list of top 30 significant SNPs from the GWAS analysis of ADNI's AD data based on single-SNP association using the logistic model in Plink.

Rank	SNP ID	*p*-value	Chromosome	Gene name
1	rs2075650	1.60E−07	19	*APOE*
2	rs12822144	2.62E−06	12	*CCND2*
3	rs6116375	2.63E−06	20	*PRNP*
4	rs11253696	4.49E−06	10	*C10orf97*
5	rs4953672	1.08E−05	2	*HAAO*
6	rs12691813	2.22E−05	2	*KIF5C*
7	rs11930385	2.31E−05	4	*ZNF827*
8	rs4972310	2.95E−05	2	*KIF5C*
9	rs1975545	3.33E−05	2	*KIF5C*
10	rs17246639	3.48E−05	11	*UBASH3B*
11	rs3815501	3.60E−05	2	*BZW1*
12	rs7661978	3.68E−05	4	*ADH1C*
13	rs10892831	4.10E−05	11	*UBASH3B*
14	rs6585082	4.22E−05	10	*ADRA2A*
15	rs2883782	4.86E−05	2	*MYO3B*
16	rs7544111	5.05E−05	1	*SLC2A7*
17	rs4795895	5.15E−05	17	*CCL11*
18	rs6505403	5.29E−05	17	*CCL11*
19	rs10491109	6.05E−05	17	*CCL2*
20	rs288496	6.23E−05	3	*MITF*
21	rs1320914	6.49E−05	3	*PLOD2*
22	rs6816078	6.71E−05	4	*ZNF827*
23	rs11706690	6.98E−05	3	*CHL1*
24	rs7985759	6.99E−05	13	*MYO16*
25	rs530652	7.20E−05	20	*RSPO4*
26	rs1498853	7.47E−05	3	*FRMD4B*
27	rs8141950	7.92E−05	22	*PARVB*
28	rs10804857	9.12E−05	3	*NAALADL2*
29	rs11620374	9.33E−05	13	*C13orf36*
30	rs1461707	9.46E−05	18	*PIK3C3*

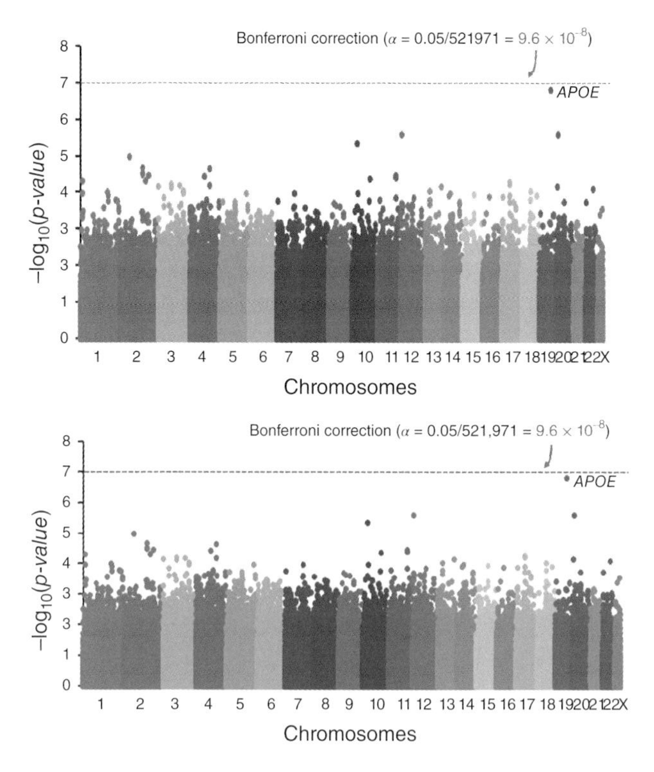

Figure 24.2 Manhattan plot of the standard GWAS analysis of ADNI's AD data.

variant, does not reach genome-wide significance with the use of either a Bonferroni correction ($\alpha = 0.05 / 521971 = 9.6 \times 10^{-8}$) or an FDR correction in the genome scan (Figure 24.2).

Results of Set-based SNP Tests

Set-based SNP tests were conducted in Plink. The set file was made with the human genome annotation (hg18). SNPs were mapped to genes if they were located in a genomic region within 20 KB 5′-upstream and 3′-downstream of the gene. Only the best SNP was selected from each set, and the permutation was set with 10000 times. A total of 18657 gene-based SNP sets were formed and used for further analysis, which were represented by 337538 of the 521965 SNPs that were genotyped in the GWAS. Among them, 179 SNP sets were significant with association signals at $p < 0.01$. The top 37 significant SNP sets are shown in Table 24.3.

Results of LKM-based Test

To evaluate the performance of SNP-set analysis with the LKM-based test, the SNP sets were formed by grouping SNPs that lie within the same gene, from 20 KB upstream of a gene to 20 KB downstream of a gene. Using these criteria, a total of 15464 SNP sets were formed that consisted of 521965 unique SNPs. Each of the gene-based SNP sets was tested by using the LKM test under the linear kernel, the IBS kernel, and the quadratic

Table 24.3 A list of top 37 significant SNP sets (***p*** < **2.00E − 03**) resulted from set-based tests using Plink.

Rank	SNP set	Number of SNPs	Empirical *p*-value	Gene name
1	APOE	5	2.00E−03	*APOE*
2	ADH1C	18	2.00E−03	*ADH1C*
3	ALG5	7	2.00E−03	*ALG5*
4	ATP6V0E1	13	2.00E−03	*ATP6V0E1*
5	BZW1	4	2.00E−03	*BZW1*
6	CCL11	14	2.00E−03	*CCL11*
7	CCL2	8	2.00E−03	*CCL2*
8	CCL7	11	2.00E−03	*CCL7*
9	CCL8	16	2.00E−03	*CCL8*
10	CCNI	5	2.00E−03	*CCNI*
11	CD99L2	13	2.00E−03	*CD99L2*
12	CHL1	104	2.00E−03	*CHL1*
13	CHST7	9	2.00E−03	*CHST7*
14	CST9	8	2.00E−03	*CST9*
15	CST9L	5	2.00E−03	*CST9L*
16	EXOSC8	4	2.00E−03	*EXOSC8*
17	FAM48A	5	2.00E−03	*FAM48A*
18	FSD1L	13	2.00E−03	*FSD1L*
19	HIST1H2BI	5	2.00E−03	*HIST1H2BI*
20	HIST1H4H	4	2.00E−03	*HIST1H4H*
21	KIF5C	11	2.00E−03	*KIF5C*
22	KMO	30	2.00E−03	*KMO*
23	LOC653147	9	2.00E−03	*LOC653147*
24	OPN3	16	2.00E−03	*OPN3*
25	PML	14	2.00E−03	*PML*
26	PSMC6	4	2.00E−03	*PSMC6*
27	PVRL2	16	2.00E−03	*PVRL2*
28	RPL26L1	9	2.00E−03	*RPL26L1*
29	SAMM50	32	2.00E−03	*SAMM50*
30	SEP11	17	2.00E−03	*SEP11*
31	SLC2A7	19	2.00E−03	*SLC2A7*
32	STYX	2	2.00E−03	*STYX*
33	TOMM40	9	2.00E−03	*TOMM40*
34	ZNF600	11	2.00E−03	*ZNF600*
35	ZNF673	3	2.00E−03	*ZNF673*
36	ZNF674	3	2.00E−03	*ZNF674*
37	ZNF827	32	2.00E−03	*ZNF827*

Table 24.4 A list of genes from the LKM-based SNP-set analysis of the ADNI's AD GWAS data ($p < 0.001$).

Rank	Gene name	p-value		
		Linear kernel	IBS kernel	Quadratic kernel
1	*APOE*	2.33E−07	1.23E−07	2.14E−06
2	*ADH1C*	8.80E−05	9.53E−05	4.79E−04
3	*GAK*	1.19E−04	2.82E−04	7.69E−05
4	*UBASH3B*	1.47E−04	7.87E−05	1.96E−04
5	*IQUB*	1.66E−04	1.61E−04	4.76E−04
6	*OR1E1*	2.03E−04	2.37E−04	2.60E−04
7	*F2RL1*	2.04E−04	2.15E−04	2.31E−04
8	*APOC1*	2.93E−04	3.86E−04	8.43E−04
9	*SHISA2*	3.06E−04	1.06E−04	8.99E−05
10	*CNO*	3.14E−04	4.42E−04	1.52E−04
11	*KCNC2*	4.22E−04	7.87E−04	4.29E−04
12	*CCL11*	4.45E−04	3.28E−04	2.30E−04
13	*PEX26*	4.88E−04	1.28E−03	2.87E−03
14	*RHOBTB3*	4.90E−04	6.79E−04	7.18E−04
15	*ADAMTS12*	5.83E−04	6.74E−04	3.15E−04
16	*CCND2*	6.99E−04	7.45E−04	1.90E−03
17	*ATP6V0E1*	7.19E−04	3.44E−04	1.40E−03
18	*PML*	7.38E−04	7.06E−04	8.12E−04
19	*LAMC1*	7.49E−04	7.66E−04	1.92E−03
20	*SLC2A7*	9.54E−04	1.16E−03	3.76E−04
21	*KIF5C*	9.70E−04	6.54E−04	8.45E−04
22	*GEMIN8*	9.93E−04	1.04E−03	2.05E−03

kernel. SNPs were coded in the additive mode, and the parametric effects of the age group, and whether the individual had hormone therapy, were adjusted, and the first four principal components of genetic variation were included to control for population stratification.

The results of the SNP-set analysis based on the LKM-based test are shown in Table 24.4. With the linear kernel, the SNP set formed by the *APOE* gene is the most highly ranked SNP set, with a p-value of 2.3×10^{-7}. At this significance level, it reaches genome-wide significance if we apply a Bonferroni correction ($\alpha = 0.05/15464 = 3.2 \times 10^{-6}$), or if we control the FDR.

Results of ARTP Statistic

The performance of SNP-set analysis with the ARTP method was evaluated with the SNP sets formed as mentioned earlier, by grouping SNPs that lie within the same gene, from 20 KB upstream of a gene to 20 KB downstream of a gene. Using these criteria, a total of 15464 SNP sets were formed that consisted of 521965 unique typed SNPs. Each

Table 24.5 A list of genes from the ARTP-based SNP-set analysis of the ADNI's AD GWAS data.

Rank	SNP set	Number of SNPs	Empirical *p*-value	Gene name
1	ADH1C	8	2.00E−03	*ADH1C*
2	ADRA2A	127	2.00E−03	*ADRA2A*
3	ALG5	7	2.00E−03	*ALG5*
4	APOE	2	2.00E−03	*APOE*
5	ATP6V0E1	13	2.00E−03	*ATP6V0E1*
6	C10orf97	54	2.00E−03	*C10orf97*
7	CCL11	14	2.00E−03	*CCL11*
8	CCL2	12	2.00E−03	*CCL2*
9	CCND2	45	2.00E−03	*CCND2*
10	CCNI	8	2.00E−03	*CCNI*
11	CST9	8	2.00E−03	*CST9*
12	DCUN1D1	5	2.00E−03	*DCUN1D1*
13	FSD1L	11	2.00E−03	*FSD1L*
14	GPR89A	1	2.00E−03	*GPR89A*
15	HAAO	78	2.00E−03	*HAAO*
16	HIST1H2BH	2	2.00E−03	*HIST1H2BH*
17	KIF5C	16	2.00E−03	*KIF5C*
18	KMO	23	2.00E−03	*KMO*
19	MNAT1	9	2.00E−03	*MNAT1*
20	MYCT1	16	2.00E−03	*MYCT1*
21	NANOS3	2	2.00E−03	*NANOS3*
22	OR1E1	9	2.00E−03	*OR1E1*
23	ORMDL1	2	2.00E−03	*ORMDL1*
24	PML	13	2.00E−03	*PML*
25	PRNP	65	2.00E−03	*PRNP*
26	RAD51AP1	5	2.00E−03	*RAD51AP1*
27	TCF3	9	2.00E−03	*TCF3*
28	UBASH3B	113	2.00E−03	*UBASH3B*
29	ZBTB9	4	2.00E−03	*ZBTB9*
30	ZNF827	46	2.00E−03	*ZNF827*

of the gene-based SNP sets was then tested by ARTP (Yu *et al.*, 2009). A total of 158 SNP sets were identified with a significance level of $p < 0.01$. The top 30 SNP sets are shown in Table 24.5 ($p < 2.00E{-}03$).

Results of GSEA-based SNP-Set Analysis

The modified GSEA of Wang *et al.* (2007) was used to identify additional genotype–phenotype association signals in pathways from different resources (Wang *et al.*, 2007).

The pathway data were collected from KEGG. Genes belonging to these pathways were associated with SNPs that were used in the ADNI GWA studies. SNPs were mapped to genes if they were located within a genomic region within 20 KB 5′-upstream and 3'-downstream of the gene. The analysis was restricted to pathways with 5–200 genes to alleviate multiple testing problems by avoiding testing too narrowly or too broadly defined functional categories.

A total of 194 pathways involving 18250 genes were included. These were represented by 521902 of the 521965 SNPs that were genotyped in the GWAS. Of the 194 examined pathways, 23 were significantly enriched with association signals at the $p < 0.01$ level (Table 24.6). The top-ranked pathway is glycolysis/gluconeogenesis, followed by neurodegenerative diseases, tyrosine metabolism, notch signaling pathway,

Table 24.6 A list of pathways associated with AD identified by SNP-set analysis based on GSEA ($p < 0.01$); p-values are computed for the pathway ES using 10000 permutations.

Rank	KEGG ID	Pathway name	Number of genes	Normalized ES	p-value
1	hsa00010	Glycolysis/gluconeogenesis	61	3.996	0
2	hsa01510	Neurodegenerative diseases	40	3.949	0
3	hsa00350	Tyrosine metabolism	57	3.802	0
4	hsa04330	Notch signaling pathway	46	3.731	0
5	hsa00380	Tryptophan metabolism	57	3.076	0
6	hsa05010	AD	28	2.858	0
7	hsa00624	1- and 2-methylnaphthalene degradation	19	2.059	0
8	hsa04110	Cell cycle	110	3.773	0.001
9	hsa04520	Adherens junction	77	2.875	0.001
10	hsa05215	Prostate cancer	89	2.949	0.002
11	hsa05216	Thyroid cancer	31	2.65	0.003
12	hsa01430	Cell communication	134	2.979	0.003
13	hsa04310	Wnt signaling pathway	147	3.044	0.006
14	hsa04612	Antigen processing and presentation	81	2.913	0.006
15	hsa00310	Lysine degradation	42	2.466	0.006
16	hsa00190	Oxidative phosphorylation	114	2.605	0.006
17	hsa05218	Melanoma	70	2.484	0.007
18	hsa00641	3-Chloroacrylic acid degradation	14	2.073	0.007
19	hsa05217	Basal cell carcinoma	56	2.432	0.007
20	hsa04950	Maturity onset diabetes of the young	25	2.531	0.007
21	hsa00340	Histidine metabolism	40	2.384	0.009
22	hsa03050	Proteasome	22	1.389	0.009
23	hsa04120	Ubiquitin-mediated proteolysis	128	2.259	0.01

tryptophan metabolism, AD, and 1- and 2-methylnaphthalene degradation. Obviously, the identification of significant pathways of neurodegenerative diseases and AD supported the viability of this method. Other pathways were also involved in neurodegenerative disorders according to extensive literature search. For instance, it has been reported that the glycolysis/gluconeogenesis pathway is closely associated with AD and Parkinson's disease (Zabel *et al.*, 2010). The altered tryptophan and tyrosine metabolism is associated with the chronic low-grade inflammation in elderly persons, suggesting their potential roles in neuropsychiatric symptoms (Capuron *et al.*, 2011).

Comparison of the GSA Methods

The results from the three gene-based GSA methods, that is, Plink set, LKM, and ARTP, were compared. As shown in Figure 24.3, it can be seen that nine genes were detected by all the three methods. Furthermore, five genes were detected by LKM and ARTP, three genes were detected by both LKM and Plink set, and four genes were detected by both ARTP and Plink set methods. Taken together, a total of 21 genes were identified as significant genes by at least two methods. It is speculated that such genes would be more likely involved in AD progression than genes identified only by one method. Therefore, it appears that the use of a combination of these methods could potentially provide greater power for the detection of associated genes.

The 21 genes identified by at least two gene-based SNP-set analyses were then compared with the significant pathways discovered by the GSEA-based methods. As summarized in Table 24.7, five genes involved in seven pathways were identified by the gene/pathway-based SNP-set analysis, including *APOE*, *ADH1C*, *HAAO*, *PRNP*, and *KMO*. Four of these five genes (*APOE*, *ADH1C*, *HAAO*, *PRNP*) were in the top 30 significant SNP list, but not all of them were among the most significant gene candidates like the *APOE* gene. Notably, the *ADH1C* gene, which was identified by all the GSA-based methods, only ranked at 12th place when using the standard single-SNP analysis. This gene may be ignored in a typical GWAS analysis, while it could play important roles in the progression of AD. Moreover, the gene *KMO*, which was detected by ARTP, Plink set, and GSEA-based methods, was not in the top 30 significant genes, indicating that this gene is highly likely to be ignored if using the standard single-SNP analysis alone.

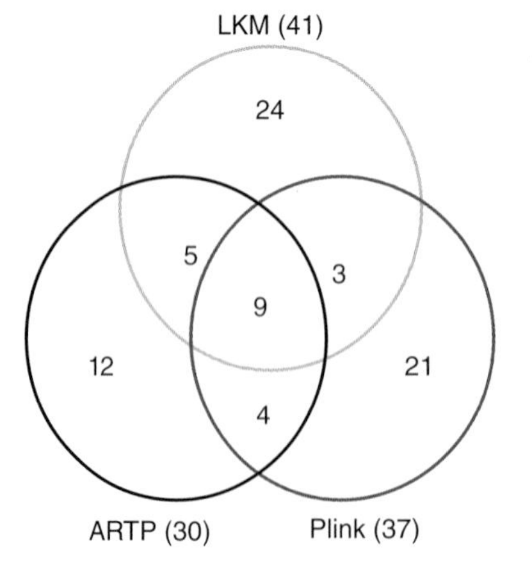

Figure 24.3 Comparison of the number of significant genes identified using three gene-based GSA methods.

Table 24.7 The comparison of results from GSA methods.

Pathway	Gene name	ARTP method	LKM method	Plink set method	Rank in top 30 SNPs
AD	*APOE*	+	+	+	1
Neurodegenerative diseases	*APOE*	+	+	+	1
Glycolysis/ gluconeogenesis	*ADH1C*	+	+	+	12
Tyrosine metabolism	*ADH1C*	+	+	+	12
1- and 2-methylnaphthalene degradation	*ADH1C*	+	+	+	12
Tryptophan metabolism	*HAAO*	+	−	−	5
Tryptophan metabolism	*KMO*	+	−	+	NA
Neurodegenerative diseases	*PRNP*	+	−	−	3

The *ADH1C* gene encodes the alcohol dehydrogenase (ADH) enzymes, which are thought to play important roles in the metabolism of ethanol and in the synthesis of retinoic acid, and may also be important in the detoxification of reactive substances such as 4-hydroxynonenal. The dehydrogenases may, therefore, constitute a first line of defense in the gastrointestinal tract against exogenous toxic agents, and the genetic defects in these enzymes may lead to increased uptake of potentially harmful substances that may eventually reach the dopamine system of the brain (Buervenich *et al.*, 2005). Studies on genetic variation in *ADH1C* revealed its involvement in several neurodegenerative diseases (Buervenich *et al.*, 2005; Das & Mukhopadhyay, 2011).

The *KMO* (kynurenine 3-monooxygenase) gene encodes a mitochondrion outer-membrane protein that catalyzes the hydroxylation of L-tryptophan metabolite, L-kynurenine, to form L-3-hydroxykynurenine. Numerous studies have reported that the kynurenine pathway, in which KMO is involved, is the promising new target with therapeutic potential for neurodegenerative diseases (Amaral *et al.*, 2013; Crunkhorn, 2011; Malpass, 2011; Pellicciari *et al.*, 2003; Zwilling *et al.*, 2011).

Conclusion

In this chapter, a brief review of the gene-set-based analysis of SNP data from GWAS was provided. With the objective to resolve multiple testing issues, the SNPs were assigned into groups based on genes or pathways. To assign SNPs into genes, the approaches are generally the same; however, the choice of representative SNPs and the use of p-values from single-SNP tests varied among different methods. Accordingly, the results of GSA of GWAS are comparatively provided. To demonstrate the powers for the detection of associated genes, we applied these methods with a real GWAS dataset for AD from ADNI. The results supported that GSA of GWAS could improve the analysis and enable the identification of important genes with mild SNP "signals," which are usually missed using the standard GWAS analysis protocol. The comparison of the results from the four GSA methods indicate that statistical power can be insufficient in some specific GSA methods owing to sources of bias—including gene set size, LD patterns, and overlapping genes. Although the combined use of these methods could provide a greater power for the detection of associated genes, the methods are yet to be integrated for improved performance.

GSA analysis is gaining popularity with human studies, but its application in aquaculture settings are yet to be explored. One major obstacle for the use of GSA in aquaculture is the lack of whole genome sequences, gene location coordinates, and pathway databases in aquaculture species. However, it is possible to use comparative genomic tools to overcome these deficiencies. In addition, it is possible to use human gene orthologs as proxies for aquaculture species for pathway analysis, assuming that gene pathways are well conserved through evolution.

References

Amaral, M., Outeiro, T.F., Scrutton, N.S. and Giorgini, F. (2013) The causative role and therapeutic potential of the kynurenine pathway in neurodegenerative disease. *Journal of Molecular Medicine (Berlin)*, **91**, 705–713.

Ashburner, M., Ball, C.A., Blake, J.A. *et al.* (2000) Gene Ontology: Tool for the Unification of Biology. *Nature Genetics*, **25**, 25–29.

Benjamini, Y. (2010) Discovering the false discovery rate. *Journal of the Royal Statistical Society: Series B*, **72**, 405–416.

Buervenich, S., Carmine, A., Galter, D. *et al.* (2005) A rare truncating mutation in *ADH1C* (G78Stop) shows significant association with Parkinson disease in a large international sample. *Archives of Neurology*, **62**, 74–78.

Cantor, R.M., Lange, K. and Sinsheimer, J.S. (2010) Prioritizing GWAS results: a review of statistical methods and recommendations for their application. *The American Journal of Human Genetics*, **86**, 6–22.

Capuron, L., Schroecksnadel, S., Feart, C. *et al.* (2011) Chronic low-grade inflammation in elderly persons is associated with altered tryptophan and tyrosine metabolism: role in neuropsychiatric symptoms. *Biological Psychiatry*, **70**, 175–182.

Crunkhorn, S. (2011) Neurodegenerative disorders: restoring the balance. *Nature Reviews Drug Discovery*, **10**, 576.

Das, S. and Mukhopadhyay, D. (2011) Intrinsically unstructured proteins and neurodegenerative diseases: conformational promiscuity at its best. *IUBMB Life*, **63**, 478–488.

Deelen, J., Uh, H.W., Monajemi, R. *et al.* (2013) Gene set analysis of GWAS data for human longevity highlights the relevance of the insulin/IGF-1 signaling and telomere maintenance pathways. *Age*, **35**, 235–249.

Dudbridge, F. and Koeleman, B.P.C. (2003) Rank truncated product of *p*-values, with application to genomewide association scans. *Genetic Epidemiology*, **25**, 360–366.

Dudbridge, F. and Koeleman, B.P.C. (2004) Efficient computation of significance levels for multiple associations in large studies of correlated data, including genomewide association studies. *The American Journal of Human Genetics*, **75**, 424–435.

Elbers, C.C., van Eijk, K.R., Franke, L. *et al.* (2009) Using genome-wide pathway analysis to unravel the etiology of complex diseases. *Genetic Epidemiology*, **33**, 419–431.

Fridley, B.L. and Biernacka, J.M. (2011) Gene set analysis of SNP data: benefits, challenges, and future directions. *European Journal of Human Genetics*, **19**, 837–843.

Ge, Y.C., Dudoit, S. and Speed, T.P. (2003) Resampling-based multiple testing for microarray data analysis. *Test*, **12**, 1–77.

Glessner, J.T., Wang, K., Cai, G.Q. *et al.* (2009) Autism genome-wide copy number variation reveals ubiquitin and neuronal genes. *Nature*, **459**, 569–573.

Goeman, J.J. and Bühlmann, P. (2007) Analyzing gene expression data in terms of gene sets: methodological issues. *Bioinformatics*, **23** (8), 980–987.

Grupe, A., Abraham, R., Li, Y. *et al.* (2007) Evidence for novel susceptibility genes for late-onset Alzheimer's disease from a genome-wide association study of putative functional variants. *Human Molecular Genetics*, **16**, 865–873.

Hirschhorn, J.N. and Daly, M.J. (2005) Genome-wide association studies for common diseases and complex traits. *Nature Reviews Genetics*, **6**, 95–108.

Hoh, J., Wille, A. and Ott, J. (2001) Trimming, weighting, and grouping SNPs in human case-control association studies. *Genome Research*, **11**, 2115–2119.

Holden, M., Deng, S.W., Wojnowski, L. and Kulle, B. (2008) GSEA–SNP: applying gene set enrichment analysis to SNP data from genome-wide association studies. *Bioinformatics*, **24**, 2784–2785.

Hong, M.G., Pawitan, Y., Magnusson, P.K.E. and Prince, J.A. (2009) Strategies and issues in the detection of pathway enrichment in genome-wide association studies. *Human Genetics*, **126**, 289–301.

Jack, C.R., Lowe, V.J., Weigand, S.D. *et al.* (2009) Serial PIB and MRI in normal, mild cognitive impairment and Alzheimer's disease: implications for sequence of pathological events in Alzheimer's disease. *Brain*, **132**, 1355–1365.

Jia, P.L., Wang, L., Meltzer, H.Y. and Zhao, Z.M. (2011) Pathway-based analysis of GWAS datasets: effective but caution required. *International Journal of Neuropsychopharmacology*, **14**, 567–572.

Kar, S.P., Seldin, M.F., Chen, W. *et al.* (2013) Pathway-based analysis of primary biliary cirrhosis genome-wide association studies. *Genes & Immunity*, **14**, 179–186.

Kimeldorf, G. and Wahba, G. (1971) Some results on Tchebycheffian spline functions. *Journal of Mathematical Analysis and Applications*, **33** (1), 82–95.

Lee, D., Lee, G.K., Yoon, K.A. and Lee, J.S. (2013) Pathway-based analysis using genome-wide association data from a Korean non-small cell lung cancer study. *PLoS One*, **8**, e65396.

Li, H., Wetten, S., Li, L. *et al.* (2008) Candidate single-nucleotide polymorphisms from a genomewide association study of Alzheimer's disease. *Archives of Neurology (Chicago)*, **65**, 45–53.

Lin, X.H. (1997) Variance component testing in generalised linear models with random effects. *Biometrika*, **84**, 309–326.

Liu, D., Ghosh, D. and Lin, X. (2008) Estimation and testing for the effect of a genetic pathway on a disease outcome using logistic kernel machine regression via logistic mixed models. *BMC Bioinformatics*, **9** (1), 292.

Malpass, K. (2011) Neurodegenerative disease: the kynurenine pathway – promising new targets and therapies for neurodegenerative disease. *Nature Reviews Neurology*, **7**, 417.

Nam, D., Kim, J., Kim, S.Y. and Kim, S. (2010) GSA-SNP: a general approach for gene set analysis of polymorphisms. *Nucleic Acids Research*, **38**, W749–W754.

Ogata, H., Goto, S., Sato, K. *et al.* (1999) KEGG: Kyoto Encyclopedia of Genes and Genomes. *Nucleic Acids Research*, **27**, 29–34.

Pellicciari, R., Amori, L., Costantino, G. *et al.* (2003) Modulation of the kynurine pathway of tryptophan metabolism in search for neuroprotective agents. Focus on kynurenine-3-hydroxylase. *Advances in Experimental Medicine and Biology*, **527**, 621–628.

Petersen, R.C., Aisen, P.S., Beckett, L.A. *et al.* (2010) Alzheimer's Disease Neuroimaging Initiative (ADNI) Clinical characterization. *Neurology*, **74**, 201–209.

Purcell, S., Neale, B., Todd-Brown, K. *et al.* (2007) PLINK: A tool set for whole-genome association and population-based linkage analyses. *The American Journal of Human Genetics*, **81**, 559–575.

Reiman, E.M., Webster, J.A., Myers, A.J. *et al.* (2007) GAB2 alleles modify Alzheimer's risk in APOE epsilon 4 carriers. *Neuron*, **54**, 713–720.

Saykin, A.J., Shen, L., Foroud, T.M. *et al.* (2010) Alzheimer's disease neuroimaging initiative biomarkers as quantitative phenotypes: genetics core aims, progress, and plans. *Alzheimer's & Dementia*, **6**, 265–273.

Shen, L., Kim, S., Risacher, S.L. *et al.* (2010) Whole genome association study of brain-wide imaging phenotypes for identifying quantitative trait loci in MCI and AD: A study of the ADNI cohort. *NeuroImage*, **53**, 1051–1063.

Song, G.G. and Lee, Y.H. (2013) Pathway analysis of genome-wide association studies for Parkinson's disease. *Molecular Biology Reports*, **40**, 2599–2607.

Stranger, B.E., Stahl, E.A. and Raj, T. (2011) Progress and promise of genome-wide association studies for human complex trait genetics. *Genetics*, **187**, 367–383.

Subramanian, A., Tamayo, P., Mootha, V.K. *et al.* (2005) Gene set enrichment analysis: A knowledge-based approach for interpreting genome-wide expression profiles. *Proceedings of the National Academy of Sciences USA*, **102**, 15545–15550.

Tian, L., Greenberg, S.A., Kong, S.W. *et al.* (2005) Discovering statistically significant pathways in expression profiling studies. *Proceedings of the National Academy of Sciences USA*, **102**, 13544–13549.

Wang, K., Li, M.Y. and Bucan, M. (2007) Pathway-based approaches for analysis of genomewide association studies. *The American Journal of Human Genetics*, **81**, 1278–1283.

Wang, K., Li, M.Y. and Hakonarson, H. (2010) Analysing biological pathways in genome-wide association studies. *Nature Reviews Genetics*, **11**, 843–854.

Wang, L.L., Jia, P.L., Wolfinger, R.D. *et al.* (2011) Gene set analysis of genome-wide association studies: Methodological issues and perspectives. *Genomics*, **98**, 1–8.

Wu, M.C., Kraft, P., Epstein, M.P. *et al.* (2010) Powerful SNP-set analysis for case-control genome-wide association studies. *The American Journal of Human Genetics*, **86**, 929–942.

Yu, K., Li, Q.Z., Bergen, A.W. *et al.* (2009) Pathway analysis by adaptive combination of *p*-values. *Genetic Epidemiology*, **33**, 700–709.

Zabel, C., Nguyen, H.P., Hin, S.C. *et al.* (2010) Proteasome and oxidative phosphorylation changes may explain why aging is a risk factor for neurodegenerative disorders. *Journal of Proteomics*, **73**, 2230–2238.

Zhang, D.W. and Lin, X.H. (2003) Hypothesis testing in semiparametric additive mixed models. *Biostatistics*, **4**, 57–74.

Zhang, K.L., Chang, S.H., Cui, S.J. *et al.* (2011) ICSNPathway: identify candidate causal SNPs and pathways from genome-wide association study by one analytical framework. *Nucleic Acids Research*, **39**, W437–W443.

Zhang, R.Y., Zhao, Y., Chu, M.J. *et al.* (2013) Pathway analysis for genome-wide association study of lung cancer in Han Chinese population. *PLoS One*, **8**, e57763.

Zhao, J., Gupta, S., Seielstad, M. *et al.* (2011) Pathway-based analysis using reduced gene subsets in genome-wide association studies. *BMC Bioinformatics*, **12**, 17.

Zwilling, D., Huang, S.Y., Sathyasaikumar, K.V. *et al.* (2011) Kynurenine 3-monooxygenase inhibition in blood ameliorates neurodegeneration. *Cell*, **145**, 863–874.

Part IV

Comparative Genome Analysis

25

Comparative Genomics Using CoGe, Hook, Line, and Sinker

Using CoGe Tools for Catching Fish Genome Evolution

Blake Joyce, Asher Baltzell, Matt Bomhoff and Eric Lyons

Introduction

The comparative genomics platform named CoGe is designed to manage, analyze, and compare genomes within or between species in a user-friendly graphical interface. CoGe also removes the need to have specialty computer science knowledge, and expensive computational software and hardware to carry out these kinds of comparative analyses. Several tools exist to find syntenic regions, provide evidence of whole genome duplications (WGDs), and create phylogenetic trees of genes or syntenic genome regions. These analyses are especially important in fish species, as there is evidence for genome duplications throughout their lineage. Identifying WGDs sheds light on how polyploidization affects speciation in organisms. Here, CoGe tools and analysis workflows are presented and explained from a bench-top scientist's perspective. The ultimate goal of this chapter is to help researchers with no previous background in comparative genomics or bioinformatics understand how CoGe's tools may be used to easily carry out broadscale analyses within and between species to gain insights into their genetics and unique evolutionary histories.

There are few certainties in biology, but genetics has proven to be strikingly universal. All organisms from the most rudimentary bacteria to the most complex plant or animal carry their self-replicating biological data in the form of adenine, guanine, cytosine, and thymine/uracil. This simple fact underpins the understanding we have about all of the fundamental processes inherent in biology: evolution, inheritance, population and community ecology, molecular biology, cell cycle, genetic improvement, and many others. Our maturing understanding of DNA, RNA, and the proteins responsible for cellular processes has led to a revolution in biology, in turn drastically altering the landscape of hypotheses that can be tested. Cellular biologists can now test hypotheses about entire regulatory gene networks (Gama-Castro *et al.*, 2011), metabolic cascades, gene expression profiles, and carbon flux between biosynthetic pathways (Dandekar *et al.*, 2014; Toubiana *et al.*, 2013; Weitzel *et al.*, 2013). Ecologists can investigate the interactions between changes in environmental conditions and molecular processes (Feil & Fraga, 2012; Schilmiller, Pichersky & Last, 2012). Evolutionary biologists have begun to rewrite our understanding of common descent between organisms (Miettinen *et al.*, 2012) and the evolutionary milestones along the way (Christoffels *et al.*, 2004; Dehal & Boore, 2005;

Bioinformatics in Aquaculture: Principles and Methods, First Edition. Edited by Zhanjiang (John) Liu.

Freeling, 2009)—for example, WGDs. Notably, testing broad hypotheses such as these now require years rather than decades, given the correct tools and expertise.

Perhaps the most striking change to biology in the last 15 years is the wide availability of genetic sequences for model and non-model organisms. DNA sequencing technology has increased the ease of generating datasets that contain a genome or the entire nuclear and/or organelle DNA present in an organism, metagenomes, or multiple organisms' genomes in a particular place or environment. In addition, DNA sequencing technology may be leveraged for generating transcriptomes, which are mRNA expression profiles of whole tissues, single cells, or organisms under certain conditions. Figure 25.1A details the rate of growth in the International Nucleotide Sequence Database Collaboration

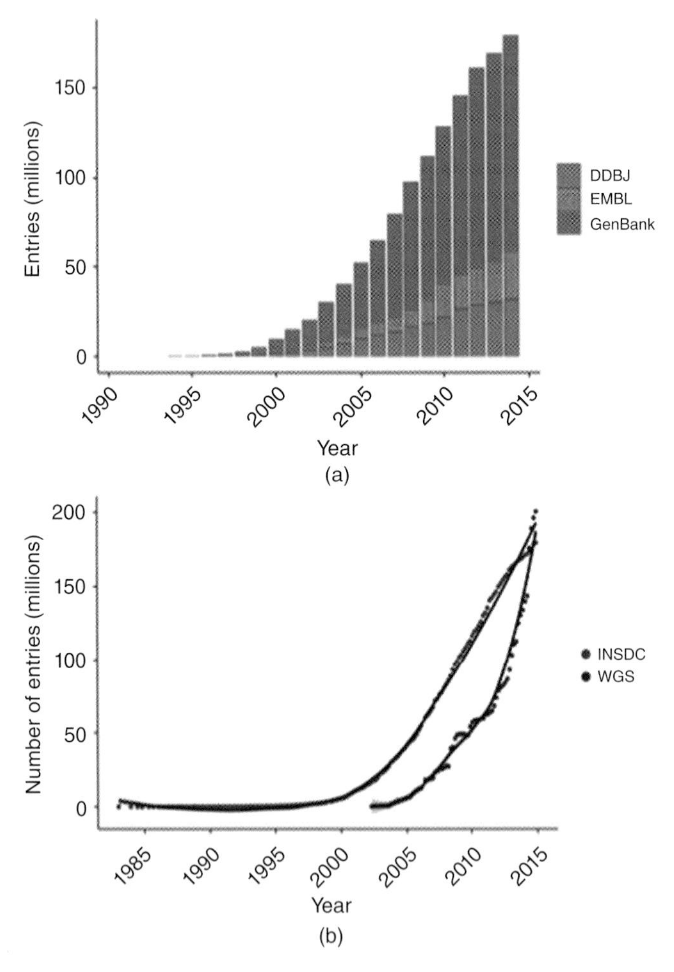

Figure 25.1 The number of entries in the three INSDC databases (NCBI, EMBL, and DDBJ) and the growth WGS sequencing projects in GenBank (NCBI). (a) Entries available in the three major international databases from 1992 to present. The number of entries available in the INSDC has displayed exponential growth since 1999. (b) The number of entries in the GenBank database and WGS sequence database of NCBI. The WGS database matches the number of entries in GenBank in only 11 years (2002–2013). Data taken from NCBI (http://www.ncbi.nlm.nih.gov/genbank/statistics) and DDBJ (http://www.ddbj.nig.ac.jp/documents-e.html). (*See color plate section for the color representation of this figure.*)

(INSDC), which is comprised of three databases: the National Center for Biotechnology Information (NCBI), the European Molecular Biology Laboratory (EMBL), and the DNA Data Bank of Japan (DDBJ). Currently, around 180 million entries of single genes or proteins exist across all three databases. These entries have been accrued since the early 1980s. However, whole genome shotgun (WGS) sequencing has developed this same number of entries in only 11 years, or half the time needed for the first 180 million entries (Figure 25.1B). This technological revolution has turned biology into a data-driven science and has created a need for infrastructure and expertise to process these large and complex datasets to yield biological information. To this end, platforms such as CyVerse (Goff *et al.*, 2011), iAnimal, and CoGe (Lyons, 2008) have been created to allow scientists with expertise in wet-bench techniques to process large genome or transcriptome datasets. This eliminates the need to invest funds in expensive computing resources, or spend time to develop highly specialized informatics knowledge. In short, these platforms are designed to enable researchers with no or minimal bioinformatics expertise to begin running analyses on supercomputers without any financial investment.

Generally, scientists have turned to bioinformatics in an effort to leverage large datasets to investigate large-scale processes in biology. In particular, investigating the process of evolution through genome-wide comparison across taxa has yielded deeper understanding about the events that occur to create new molecular processes and, ultimately, new species. As a specific example, WGDs have been observed and are traditionally well documented in plants (Freeling, 2009). These WGDs have been implicated in metabolic evolution and adaptation to land in early plants (Rensing *et al.*, 2007), but they have also occurred in animals.

Fish genomes have been investigated to determine vertebrate genome duplication events in an effort to understand the genome structure of tetrapods (i.e., mammals, reptiles, birds, and amphibians). Genome-wide comparisons have identified two hypothesized WGDs that are common to all vertebrates (1R and 2R; Dehal & Boore, 2005), one that occurred in ray-finned fish before teleosts (3R; Christoffels *et al.*, 2004), and several others specific to salmonids (4R), carp, and sturgeons (Davidson *et al.*, 2010; Meyer & Van de Peer, 2005). Figure 25.2 displays a phylogenetic tree of the lamprey, shark, and fish genomes available in CoGe along with the WGDs (stars) that are thought to occur in these taxons (Berthelot *et al.*, 2014; Betancur-R. *et al.*, 2013; Braasch *et al.*, 2014; Near *et al.*, 2012). We do not have a clear understanding about when the first vertebrate WGD (1R) occurred, but is believed to have occurred before cyclostomes (jawless fish), because of the paralogs present in lampreys and hagfish (Steinke *et al.*, 2006). The second vertebrate WGD (2R) is thought to have occurred around the same time as the division between jawless fish and cartilaginous/bony fish (Meyer & Van de Peer, 2005). The fish genomes available to the public are positioned at key points in fish evolutionary history to identify WGD events, with the hope that researchers can eventually discern the impact that WGDs have on genome structure and the fate of gene families during evolutionary processes.

WGDs are thought to create duplicated genetic material that may increase a variety of genetic and phenotypic variability in a relatively short evolutionary time (Dehal & Boore, 2005; Meyer & Van de Peer, 2005). WGDs cause polyploidization and usually are followed by diploidization, where rapid gene loss restores meiotic viability (Renny-Byfield *et al.*, 2013). Detecting WGDs over large spans of evolutionary history can be difficult because diploidization results in fractionation of gene content (homeologous gene loss

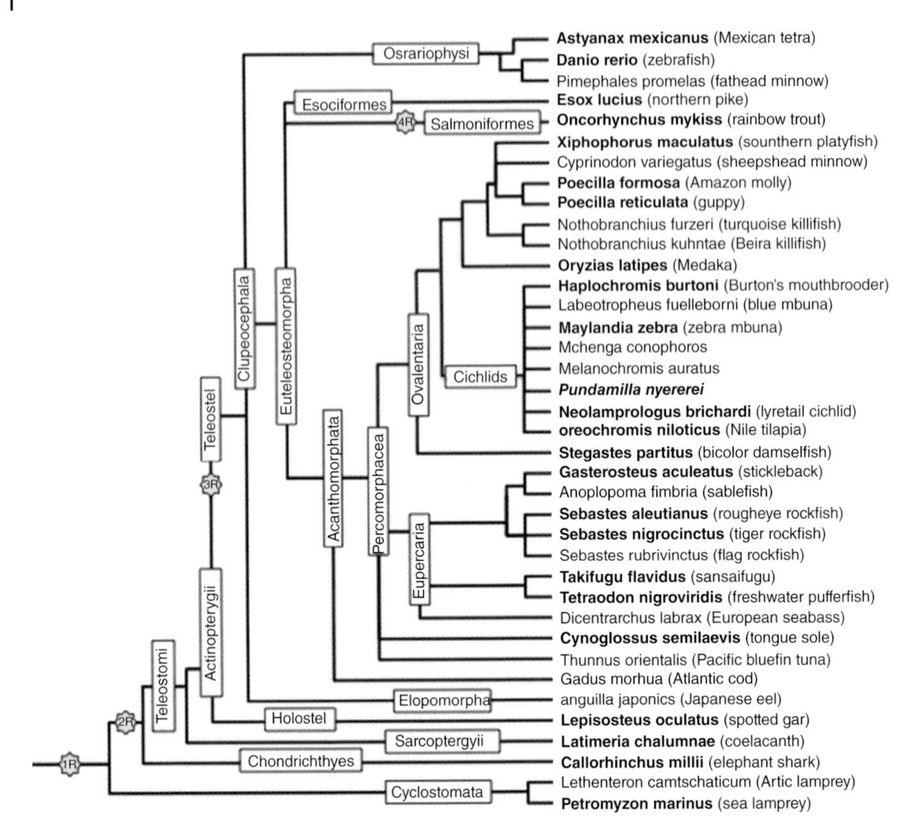

Figure 25.2 Phylogenetic relationships of fish genomes available in CoGe (March 2015; https://genomevolution.org/CoGe/NotebookView.pl?nid=). WGDs are displayed as stars, and bolded species have annotations available in CoGe. Tree based on Uniprot taxonomy (http://www.uniprot.org); Berthelot *et al.* (2014); Betancur-R. *et al.* (2013); Braasch *et al.* (2014); and Near *et al.* (2012).

of duplicated genes) in syntenic regions (Dehal & Boore, 2005), obfuscating the ability to infer their evolutionary history.

Describing the complex relationships of genes across species caused by speciation events from common ancestors and duplications or deletions requires specific terminology. *Homology,* in a genomic context, refers to genes derived from a common ancestor. Homologous genes in two species that are derived from the divergence of their lineages are called *orthologs.* Homologous genes that are derived from gene duplication within a specific lineage are called *paralogs,* but this terminology is further divided based on when the duplication occurred in evolutionary history, or how the duplication occurred. *In-paralogs* (https://genomevolution.org/wiki/index.php/In-Paralog) are duplicated genes that happen after a speciation event, whereas *out-paralogs* (https://genomevolution.org/wiki/index.php/Out-Paralog) are duplicated genes that happen before a speciation event. These terms are coined after two traditional phylogeny terms: *in-groups*—recently diverged species within a clade; and *out-groups*—an anciently diverged group that serves as a comparison (Sonnhammer & Koonin, 2002). *Ohnologs,* or *syntelogs,* are derived from WGD events, and are named after Susumu Ohno, who hypothesized in his 1970 book *Evolution by Gene Duplication* that gene duplication from polyploidization leads to speciation (Wolfe, 2000).

Identifying evidence of WGDs in fish genomes using CoGe is a simple process that can be summed up in three major steps: hook, line, and sinker. The first step is to "hook" into the CoGe platform by loading genome(s) for analysis into notebooks or analyzing pre-loaded genomes directly with CoGe's tools. Next, the genomes are analyzed, and then they are visualized to draw "lines" of comparison to highlight regions of structural similarity. CoGe's tools *SynMap* (Lyons *et al.*, 2008) visualizes whole genome comparisons using syntenic dot plots; *GEvo* (Lyons & Freeling, 2008) is used for in-depth microsynteny analysis; and *SynFind* finds all syntenic regions among genomes. Finally, in the sinker step, weight can be added to the analysis by uploading additional datasets such as annotations; transcriptomes; quantitative measurements such as gene expression fragments per kilobase of exon per million fragments mapped (FPKM); and single-nucleotide polymorphisms (SNPs). These associated datasets are referred to as "experiments" in CoGe, and may be visualized in GenomeView, which uses the JBrowse viewer (Skinner *et al.*, 2009) to stack multiple levels of data for drawing comparisons.

Getting Hooked into the CoGe Platform

It is best to log in to CoGe before starting any analysis, so that you can create notebooks to organize your loaded genomes, publicly available genomes, and any associated experiments. CoGe uses the cyber-infrastructure provided by CyVerse (Goff *et al.*, 2011) for user account management and computational scalability. If you are logged into CoGe, you can easily import and export data from your CyVerse Data Store (http://data.iplantcollaborative.org).

Logging in to CoGe

1) Logging in to CoGe requires an CyVerse/iAnimal username and password. Registering is free; simply visit the registration page (https://user.iplantcollaborative.org/register/).
2) Once you register, or if you already have an CyVerse/iAnimal username, select "Sign in" at the top-right of the CoGe page. If you are logged in, CoGe will automatically keep track of any data loaded and analyses run, and give access to any data present in your CyVerse/iAnimal Data Store folder "coge_data" (https://genomevolution.org).

Navigating CoGe Using MyProfile

A brief explanation of the layout of CoGe is necessary before continuing to load data and running analyses. This section is meant to serve as a broad explanation on how to navigate the CoGe platform. CoGe's tools and methods for loading data will be explained in further detail later in this chapter. The toolbar at the top-right next to the "Sign in" tab is available for navigation on every page in CoGe. The "My Profile" button brings users to their profile once they are signed in (https://genomevolution.org/CoGe/User.pl). "My Profile" organizes data, notebooks, user groups, past activity, and any deleted genomes or experiments. The "My Data" section tracks any genomes or experiments that have been loaded by the user or placed into a user's notebook for easy reference. A list of genomes or experiments can be viewed separately by using the arrow next to the "My Data" link. In "My Profile", there are four buttons at the top of any list of data. From left to right, they are: (1) the bust icon ("Share Selected Items"), (2) the folder icon ("Add Selected Items to a Notebook"), (3) the trashcan icon ("Delete Selected Items"), and (4) the arrow icon ("Send Selected Items to").

Privately Sharing Genomes or Experiments in CoGe

Genomes and experiments can be privately shared within CoGe by navigating to the "My Profile" page and selecting "My Data" on the left-hand side (https://genomevolution.org/CoGe/User.pl).

1) Select either "Genomes" or "Experiments" in "My Data".
2) Find the data to be shared and check the box next to each name.
3) Select the bust icon at the top of the list ("Share Selected Items").
4) A "Share Items" window will open. Type the person's name or username in the search box and select the name from the drop-down menu.
5) Click the "Add" button. This will allow those user(s) to have access to the data even if the data is private.
6) The owner of the data can remove access by selecting the "X" next to a user's name.

Making a Genome or Experiment Publicly Available

A genome or experiment can be made publicly available by viewing it in GenomeView. All data in CoGe are considered private until made public by the owner.

1) Navigate to the "My Data" list by selecting "My Profile" in the top-right toolbar.
2) Select the genome or experiment of interest by double-clicking the name.
3) This opens a new pop-up window called "GenomeView". All of the information for a genome is displayed in this window. Under the info section, select the "Edit Info" button.
4) Find the "Restricted?" button and click the box to remove the checkmark.
5) The data is now public and can be discovered using the OrganismView search tool. The data can be made private again by rechecking the box and saving the changes.

Data Management in CoGe

Data, genomes, and experiments can also be organized from the "My Data" list in "My Profile". Data can be sent to notebooks for organization by checking the boxes next to the data and selecting the folder icon ("Add Selected Items to a Notebook"). Notebooks will be described in more detail later in the chapter.

Checking the boxes next to the data names and then selecting the trashcan icon ("Delete Selected Items") will delete the data and remove it from "My Data". However, deleted data can be recovered. To recover data, navigate to "My Profile" and select the "Trash" link in the list. Deleted data will be displayed in a list; select the data to recover, and select the asterisk icon ("Undelete Selected Item") at the top of the list.

Tracking History and Progress in CoGe

Once a user signs in, CoGe tracks the history of analyses and any data loaded in the "Activity" link in "My Profile". Each time a CoGe tool is run, it will deposit a log in "Analyses" (https://genomevolution.org/CoGe/User.pl). Each logged event will contain a link to revisit the event. For example, if an analysis was logged, following the link will regenerate the analysis exactly as previously configured (selected data, algorithms, and visualization). Any genomes or experiments loaded will deposit a log into the "Data Loading" link. CoGe will track analyses and data upload events that are currently running, and report those that have failed or those that have been successfully completed. The date and time when the analyses are completed are also displayed along with the amount of time the analysis ran. Selecting any of the names will send the user to the data or analysis without the need to search for the analysis using OrganismView or the "My Data" link. This is an easier way to access more recent data and analyses instead of searching through a list of new and old in the "My Data" link.

The "Tools" toolbar button generates a drop-down list of CoGe tools. This can be used to easily navigate between tools in CoGe. The "Home" toolbar button will return the user to the CoGe homepage. The "Help" toolbar button will lead to a drop-down list of additional resources to help users use CoGe. CoGePedia contains wiki pages dedicated to explaining terminology and CoGe tools in further detail. The "Fish Comparative Genomics" CoGePedia page is dedicated to this book chapter and contains:

1) All of the URLs present in this chapter
2) Details on each CoGe tool
3) A link to every fish genome discussed in this chapter
4) Sample datasets that can be used to practice or follow along with this chapter
5) A glossary of terminology (https://genomevolution.org/wiki/index.php/Fish_Comparative_Genomics).
6) Video tutorials with step-by-step directions for how to use nearly every aspect of CoGe
7) Past workshop outlines available in the tutorial section, providing more information about the workflows available in CoGe and what results can be expected from using CoGe

Finding Genomes Already Present in CoGe

CoGe contains over 23500 genomes covering 16700 species. Public genomes may be added into a new notebook. Table 25.1 lists fish genomes currently available (March 2015) in CoGe according taxonomic class. Genomes can also be analyzed directly by searching and selecting them using the OrganismView tool.

1) Open OrganismView (https://genomevolution.org/CoGe/OrganismView.pl).
2) Type in a species name or general taxonomic search terms to yield all of the genomes (nuclear or mitochondrial) available for that species.
3) Select the organism of interest (many may match a given name).
4) Select the genome of interest for analysis by clicking a CoGe tool (CodeOn, SynMap, or CoGeBlast) located under "Genome Information" next to "Tools", or by clicking on "GenomeInfo" to view more information about the genome.

Loading Genome(s) into CoGe

If a genome does not already exist in CoGe, or if you have a new genome, you may add it to CoGe. Genomes can always be kept private to ensure the safety/privacy of

Table 25.1 A list of fish genomes currently available in CoGe (as of March 2015).

Class	Order	Genus species	Common name	Notes	NCBI taxonomy
Cephalaspidomorphi	Petromyzontiformes	*Lethenteron camtschaticum*	Artic lamprey	One of few extant jawless fish	980415
		Petromyzon marinus	Sea lamprey	One of few extant jawless fish	7757
Chondrichthyes	Chimaeriformes	*Callorhinchus milii*	Australian ghostshark	Proposed model cartilaginous fish	7868
Sarcopterygii	Coelacanthiformes	*Latimeria chalumnae*	Coelacanth	Oldest known extant Sarcopterygii	7897
Actinopterygii	Anguilliformes	*Anguilla japonica*	Japanese eel	Ecological model, Industry species	7937
	Beloniformes	*Oryzias latipes*	Medaka	Model fish, Japanese pet fish	8090
	Characiformes	*Astyanax mexicanus*	Mexican tetra	Two forms: seeing, and cave-dwelling (blind)	7994
	Cichliformes	*Haplochromis burtoni*	Burton's mouthbrooder	Ecological model species, aquarium species	8153
		Labeotropheus fuelleborni	Blue mbuna		57307
		Maylandia zebra	Zebra mbuna	Ecological model species, aquarium fish	106582
		Mchenga conophoros			35575
		Melanochromis auratus	Auratus cichlid	Phenotypic and ecological diversity	27751
		Neolamprologus brichardi	Princess cichlid	Aquarium fish	32507
		Oreochromis niloticus	Nile tilapia	Industrial species	8128
		Pundamilia nyererei			303518
	Cypriniformes	*Danio rerio*	Zebrafish	Model species	7955
		Pimephales promelas	Fathead minnow	Industrial baitfish	90988

Cyprinodontiformes	*Cyprinodon variegatus*	Sheepshead minnow	Toxicology model	28743
	Nothobranchius furzeri	Turquoise killifish	Short-life-span model species, metabolic diapause	105023
	Nothobranchius kuhntae	Beira killifish	Short-life-model species	321403
	Poecilia formosa	Amazon molly	Gynogenesis: all female populations	48698
	Poecilia reticulata	Guppy	Model species, aquarium fish	8081
	Xiphophorus maculatus	Southern platyfish	Gives live birth	8083
Esociformes	*Esox lucius*	Northern pike	Angling species	8010
Gadiformes	*Gadus morhua*	Atlantic cod	Industrial species	8049
Perciformes	*Anoplopoma fimbria*	Sablefish	Industrial species	229290
	Dicentrarchus labrax	European seabass	Industrial species	13489
	Gasterosteus aculeatus	Three-spined stickleback	Model species	69293
	Sebastes nigrocinctus	Tiger rockfish	Angling species, long-life model, live bearer	72089
	Sebastes rubrivinctus	Flag rockfish	Angling species, industrial species	72099
	Stegastes partitus	Bicolor damselfish	Medical model species	144197
Pleuronectiformes	*Cynoglossus semilaevis*	Tongue sole	Industrial species	244447
Salmoniformes	*Oncorhynchus mykiss*	Rainbow trout	Angling and industrial species	8022
Semionotiformes	*Lepisosteus oculatus*	Spotted gar		7918
Scombriformes	*Thunnus orientalis*	Pacific bluefin tuna	Industrial species	8238
Tetraodontiformes	*Takifugu flavidus*	Sansaifugu		433684
	Tetraodon nigroviridis	Spotted green puffer fish	Model species, low amount of repetitive sequence	99883
	Takifugu rubripes	Puffer fish	Shortest vertebrate genome	31033

your data, shared with collaborators, or made fully public. Genomes may be added through the CyVerse Data Store, retrieved from a remote server by an FTP/HTTPS URL, uploaded from your computer, or retrieved from NCBI using the accession number of that genome. To load a genome, the file must be in the FASTA file format.

Metadata can and should be associated with the genome to assist in tracking the genome. Metadata includes the genome version, individual specimen or biological samples used to generate the dataset, provenance for where the genome was obtained, and any other user-specified aspects that could be relevant to genome analysis. The only metadata that is required when uploading a genome is the organism's scientific name, the version of the genome being uploaded, the type of genome (i.e., unmasked or masked), and the source of the sequencing data. However, including a link about where the genome was obtained and entering a brief description of the genome is highly recommended.

The *Takifugu rubripes* genome and annotation are available for download as an example dataset. Directions for downloading the *T. rubripes* genome FASTA file into CoGe can be found in the FTP/HTTP upload section in the following text. An annotation file (in GFF format) is also available and will be detailed in the "LoadAnnotation" section of this chapter.

1) You must be logged into CoGe to add new genomes.
2) Start by clicking "My Profile" on the top-right toolbar.
3) Then, click the "Create" button on the left side toolbar.
4) Choose "New Genome", and then input the required metadata in each field.
   a) Organism scientific name (searched and selected from NCBI with the option of modifying it as required).
   b) Version of the genome assembly. If using a publicly available genome assembly, be sure to synchronize the version name that is published.
   c) Type of genome assembly. Usually an unmasked genome, but, if masked, be sure to select the appropriate type.
   d) Source of the genome assembly. Most public repositories are discoverable by typing in the name in this field, but, if the genome was assembled by your lab, be sure to credit yourself appropriately.
   e) By default, genomes are set as restricted, meaning they are only visible and available to you and the users you share them with. If you would like to make your data publicly available (highly recommended), unselect the restricted option.
   f) A description and a link to the genome can be input by selecting the "More" option.
   g) Finally, select the genome FASTA file by clicking the "Add Data" button. This can be accomplished through the methods described in the following text.

Adding Data from the CyVerse Data Store

1) Direct from the CyVerse Data Store: CoGe will automatically access the directory named "coge_data" in your CyVerse Data Store (http://data.iplantcollaborative.org).
2) If a genome is already loaded into the Data Store, simply move that genome to the folder "coge_data". It will now be viewable in CoGe.
3) If a genome is on the computer from which you are accessing CoGe, you can add it to the CyVerse Data Store

a) From a separate browser tab, go to the CyVerse Data Store (https://data.iplant collaborative.org).
b) Click on the data icon to open the data window.
c) Click on the "Upload" tab on the top-left and select "Simple Upload from Desktop".
d) Find the FASTA file of the genome(s) and select to upload.
e) Once the upload is complete, be sure to move the file to the folder named "coge_data", so that it is accessible from the CoGe web site.

Uploading from an FTP/HTTPS Site

An FTP/HTTP link can also be used for uploading genomes into CoGe from a remote server.

1) In the top-left of the screen, click on the "FTP/HTTP" button.
2) Find the URL of the genome FASTA file from an FTP site or web page.
3) Paste the URL into a blank field on the "Add Data" page. For example, to add the *T. rubripes* genome, paste the following URL: http://de.cyverse.org/dl/d/D8201F18-6D99-4970-BFFF-4E668768F4EA/Takifugu_rubripes_genome_NCBIv1.faa

Uploading Directly to CoGe from Local Storage

This option allows you to upload a local file from your computer directly to CoGe.

1) Click the "Browse" button.
2) Select the FASTA file from your computer.

Using the NCBI Loader

The NCBI loader accesses NCBI directly to download genomes. The NCBI loader will choose the first genome containing scaffolds (WGS_SCAFLD) if it exists or the first whole genome sequence (WGS) associated with a genome accession number. However, these genomes should be checked to ensure that the correct genome is uploaded, since the first WGS or WGS_SCAFLD are not always the desired assemblies. For example, the first WGS_SCAFLD for *T. rubripes* is the genome that contains chromosome assemblies (https://genomevolution.org/r/en3t), whereas the first WGS_SCAFLD for *Anguilla japonica* is a single accession for the mitochondrial sequence, and the second WGS_SCAFLD has scaffolds (https://genomevolution.org/r/en3r). The NCBI loader will also access any annotations that are associated with the genomic sequence. However, annotations uploaded from NCBI must also be checked to ensure that the annotations were properly associated with the genomic sequence.

1) Select the NCBI tab.
2) Open the NCBI genome database (http://www.ncbi.nlm.nih.gov/genome/) and search for the organism's scientific name.
3) Copy the accession number, and then paste it into the CoGe NCBI loader.
4) Once the genome is loaded, check to make sure the latest assembly and annotation files have been loaded by viewing the genome with CoGe's GenomeView (available from GenomeInfo).

Each of these data import methods has advantages and disadvantages, and ranges from essentially hands-free to hands-on. The NCBI loader is an automated process and requires the least amount of hands-on time; however, this loader assumes that NCBI uses standard formats and is uniform in presenting data. This is not always the case with NCBI repositories. Therefore, care should be taken to double-check the data loaded before using genomes or annotations imported through the automated NCBI loader.

Uploading a genome from an FTP/HTTPS web site only requires copying and pasting the URL into the loader. Generally, this method is minimally hands-on, uploads quickly, and can be more precise than using the fully automatic NCBI loader, because the researcher handpicks which files to upload. Of course, this comes with the added responsibility of finding the appropriate file and being sure that you have a set of genome and annotation files that match.

The most hands-on way to load a genome and annotation is by uploading through the CyVerse Data Store. However, this method is private and has the fastest upload speeds to CoGe. This method assumes that you have an iPlant/iAnimal account and have previously uploaded data to the Data Store. Uploading a genome from a local computer is private, but the most hands-on, and ultimately has the slowest upload speeds that are based on a local machine's connection speed, and therefore is the most time consuming. Consider using the Data Store instead of a direct upload from a local computer.

Organizing Genomes Using Notebooks

Once genomes have been uploaded into CoGe, they can be used directly in different CoGe tools, or they can be organized into notebooks. Notebooks are a useful way to group genomes for analysis, and many CoGe tools can be launched directly from the notebook view. Additionally, the genomes in a notebook can be sent to tools in CoGe that analyze multiple genomes. For example, GenomeList creates a table view of metadata associated with all genomes in that notebook, whereas CoGeBlast lets you blast a sequence against any set of genomes in CoGe. Information about individual genomes in a notebook can be viewed by double-clicking the genome to bring up the GenomeInfo tool. All of the fish genomes currently available through CoGe are organized into notebooks. The links to find the notebooks are given in Table 25.2. The notebooks are also available as GenomeLists that display metadata about each genome in a table for comparison (see Table 25.2).

Table 25.2 Links of fish genomes organized as notebooks in CoGe.

Notebook	Description	GenomeList link	Notebook link
Fish genomes	Fish genomes publicly available in CoGe	https://genomevolution.org/r/egv5	https://genomevolution.org/r/eb5w
Annotated unmasked fish genomes	Fish genomes publicly available with annotations and not masked	https://genomevolution.org/r/egv4	https://genomevolution.org/r/edoh
Annotated hard-masked fish genomes	Fish genomes with annotations and hard-masked by CoGe	https://genomevolution.org/r/en40	https://genomevolution.org/r/egvo

On the GenomeInfo tool page, information about the uploaded genomes are presented in four sections: (1) the "Info" section displays the organism name for the genome and all of the metadata associated with the genome available for that organism; (2) the "Sequence & Gene Annotation" and "Experiments" sections display data files that have been uploaded and associated with the organism; (3) the "Statistics" section describes the content of the genome; and (4) the "Features" section generates a table of all annotated structural features of the genome (e.g., genes). Essentially, the GenomeInfo page displays all of the metadata and possible analyses that can be done to the genome in one easy-to-access window. Once all genomes of interest are loaded or organized into notebooks, they are ready for comparison. The last consideration before analyzing a newly added genome should be loading structural annotations of genomes if they are available.

Loading Genome Annotation

Genome annotations can be loaded in GFF3 or GTF file formats, which are the popular formats for NCBI and EMSEBML, respectively. Directions for downloading the sample *T. rubripes* genome annotation are listed in the following text.

1) Bring up the GenomeInfo tool by double-clicking on the genome of interest in "My Profile", or search for it in OrganismView.
2) Find the "Load Gene Annotation" button on the bottom-left of the screen. If a genome was loaded from NCBI directly, then the annotation will be found automatically; in this case, the "Add Annotation" button will not be present.
3) This brings up the LoadAnnotation tool. The genome name will be automatically populated in the metadata. Input the version number and data source of the annotation file.
4) Load the annotation file by clicking "Select Data File" and following the same steps as in loading a genome. The *T. rubripes* sample annotation can be downloaded by copying and pasting this URL into the FTP/HTTP loader: http://de.iplantcollaborative.org/dl/d/B6359141-500D-43BA-988B-33A0FCA925B3/T_rubripes_annotation_GCF_000180615.1.gff.gz.
5) Click the "Load Annotation" button when all the required fields are filled in.
6) Once the genome annotations have been loaded, clicking the "Click for Features" button in the GenomeInfo tool will produce a summary table of what has been annotated (e.g., counts of genes and coding DNA sequence [CDA]) for the genome.

Casting the Line: Analyses for Comparing Genomes

Genes are arranged on linear chromosomes at specific loci. As chromosomes are passed down through generations, they sometimes go through rearrangements or experience mutations during the meiotic phase, which change their linear order. If two genomes contain an area of genes that are arranged in the same order, they are considered *colinear*. This colinear arrangement of genes is used to infer *synteny*, which is defined as genomic regions sharing a common ancestry (homologus genomic regions.) Synteny observed within the genome of the same species is evidence of WGD or polyploidization, or large-scale segmental duplications of chromosomes. Following WGDs, genomes reduce to a diploid state over time by loss of duplicated genes through a process known as *fractionation* (Freeling, 2009). The process of fractionation disrupts the pattern of colinear gene order, and distantly related species will have fewer areas of identifiable synteny.

CoGe's tool SynMap is used for whole genome comparisons of two genomes, with the results viewable in a syntenic dot plot. In this plot, each genome is laid along the X- or Y-axis, with neighboring chromosomes or contigs separated by vertical or horizontal lines, respectively. The pair of genomes are compared to one another, either CDS or whole genome sequences, using a blast-like algorithm, and all regions of sequence similarity are fed into an algorithm to identify patterns of colinearity (DAGChainer) (Haas *et al.*, 2004). Colinear pairs of genes are drawn as dots on SynMap's syntenic dot plot.

Figure 25.3 displays the results from three genome self-comparisons: *Lepisosteus oculatus* (regenerate analysis at https://genomevolution.org/r/ewrn), *T. rubripes* (regenerate analysis at https://genomevolution.org/r/ewqm), and *Oncorhynchus mykiss* (regenerate analysis at https://genomevolution.org/r/ewc5). *L. oculatus* has not experienced a WGD event recently (Figure 25.3A), and therefore the genome self-comparison displays only noise from the algorithm used to identify synteny (Figures 25.3A and 25.3B). The *T. rubripes* self-comparison yields evidence for an older WGD (3R) common to all Teleostei fish (Figures 25.3A and 25.3C). The *O. mykiss* self-comparison shows evidence for both a newer WGD (4R) and some areas of synteny from the 3R Teleostei WGD (Figures 25.3A and 25.3D).

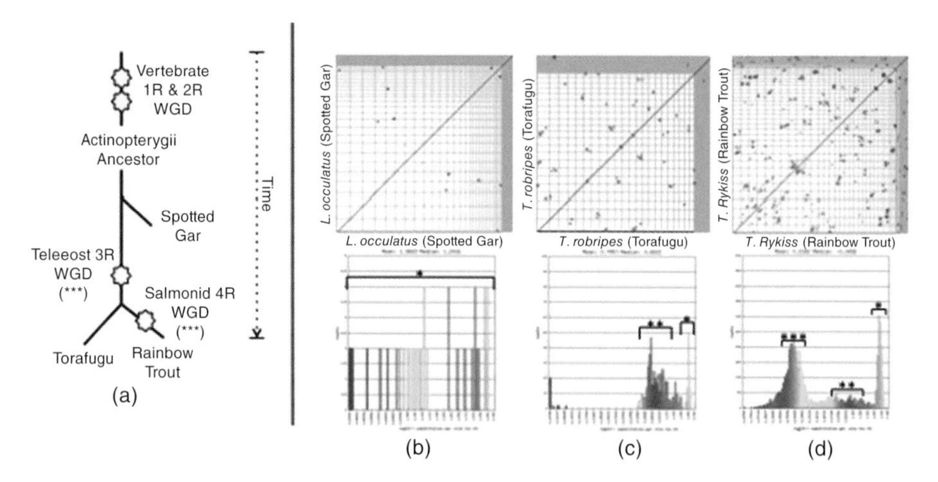

Figure 25.3 Intergenomic comparative dot plots and synonymous mutation rate distributions of *L. oculatus, T. rubripes*, and *O. mykiss* show evidence of Teleost and salmonid lineage WGDs. (a) Ancestral evolutionary relationship of *L. oculatus* (Lo), *T. rubripes* (Tr), and *O. mykiss* (Om) overlaid with hypothesized WGD events (gold stars). (b–d) Syntenic dot plots comparing genomes of three species against themselves (upper). Dots represent syntenic windows of genes, and are colored based on synonymous mutation rates (*Ks*). Distribution of *Ks* values are illustrated in histograms (lower). (b) Intragenomic syntenic dot plot of *L. oculatus*, an outgroup for the Teleost WGD. Although *L. oculatus* experienced an ancient WGD in the early vertebrate linage, evidence is harder to visualize due to fractionation. (c) Syntenic dot plot of *T. rubripes*. Single peak in *Ks* distribution indicative of a single, Teleost lineage WGD. (d) Dot plot of synteny in the genome of *O. mykiss*. Two peaks in *Ks* distribution indicative of two recent WGDs, the Teleost as well as a salmonid-lineage specific duplication. * Noise from the SynMap algorithm (Tang & Lyons, 2012); ** Teleost WGD; *** salmonid-lineage WGD. All analyses can be regenerated for *L. oculatus, T. rubripes*, and *O. mykiss* at: https://genomevolution.org/r/ewrn, https://genomevolution.org/r/ewqm, and https://genomevolution.org/r/ewrm, respectively. (*See color plate section for the color representation of this figure.*)

Running SynMap to Generate Syntenic Dot Plots

1) Select two genomes from a notebook for comparison, or select SynMap from the "Tools" drop-down menu on the top-right. Using genomes that are annotated and masked will result in faster analysis. Unmasked genomes can be masked in CoGe by opening the GenomeInfo tool and clicking on the "Copy & Mask" button in the "Tools" section.
2) Once both genomes are selected, you can run the analysis, or select from a group of optional functions.
   a) *Optional*: SynMap will calculate the synonymous substitution value, *Ks*, if the option is selected on the "Analysis Options" tab under the CodeML heading. Selecting to "Calculate syntenic CDS pairs and color dots" from the drop-down menu will generate a histogram of *Ks* values of syntenic gene pairs and color the dots in the dot plot according to their *Ks* values. The syntenic dots can be colored using different schemes for better visualization.
   b) *Optional*: The "Syntenic Path Assembly" (SPA) option will arrange and orient the contigs in one genome based on synteny to the other genome. This option is useful for visualizing synteny between a reference genome and a shotgun-sequenced genome that contains many contigs. Additionally, it is useful for visualizing synteny across distantly related organisms.

SynMap will generate an image when the analysis completes. The image will represent syntenic gene pairs as individual dots inside the graph. If *Ks* was calculated, a histogram showing the distribution of synonymous substitutions will also be presented below the syntenic dot plot. Once the syntenic dot plots have been generated, clicking on a chromosome–chromosome comparison will magnify that part for closer investigation. A single-click on a syntenic dot will bring up information about the paired genes, including their designated names, length of syntenic region identified, and several links that can be used to send the syntenic regions to different CoGe tools. A double-click will send the gene pair to CoGe's microsynteny analysis tool, GEvo.

SynMap will also compare across species. These comparisons can show genome structure changes across speciation events. SynMap is designed to use genomes that have been assembled to either the scaffold or chromosome level. Many newly sequenced species have a highly fragmented genome due to incomplete sequencing, or areas of tandem repeats that are difficult to resolve. SynMap can still utilize these fragmented genomes to develop genome structure understanding. When the fragmented *Pundamilia nyererei* genome, containing 7236 scaffolds, is initially compared to *Oreochromis niloticus*, it results in a random distribution of syntenic gene pair dots (Figure 25.4A). Rerunning the assembly by selecting the SPA option organizes the syntenic pairs from *P. nyererei* according to the chromosome-level assembly of *O. niloticus* (Figure 25.4B). The pseudo-assembly can also be exported for future use as a less fragmented assembly of *P. nyererei*.

Using the Syntenic Path Assembly Algorithm in SynMap

1) Load two genomes of interest into the SynMap tool as described earlier.
2) Under the "Display Options" tab, select the box next to "Syntenic Path Assembly (SPA)".

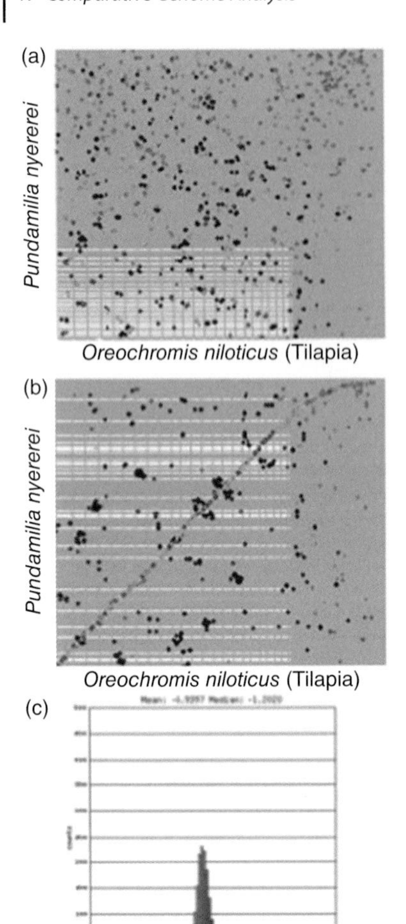

Figure 25.4 Syntenic dot plots generated using SynMap to compare the genomes of *Oreochromis niloticus* and *Pundamilia nyererei* without and with the SPA algorithm. (a) Syntenic dot plot of *O. niloticus* and *P. nyererei* unmasked and annotated CDSs without the SPA algorithm. The colors of the histogram syntenic dots based on synonymous substitution rate (*Ks*) are randomly distributed, and their pattern is difficult to discern. (b) Syntenic dot plot of *O. niloticus* and *P. nyererei* unmasked and annotated CDSs using the SPA algorithm. The SPA algorithm pseudo-assembles syntenic pairs using a chromosome-level genome assembly (*O. niloticus*) as the reference for a genome made of only contigs (*P. nyererei*). The pseudo-assembly mapping can be exported for use as a more complete genome assembly. (c) The SynMap histogram displays syntenic orthologs in yellow and paralogs from the Teleost WGD (3R) in blue. The syntenic dot plot also shows a recent WGD. Results can be regenerated for *O. niloticus* and *P. nyererei* at https://genomevolution.org/r/f74x and https://genomevolution.org/r/f74z, respectively. (*See color plate section for the color representation of this figure.*)

3) Confirm that the reference genome that will be used to map the fragmented genome to has more or less pieces (i.e., contigs or scaffolds).
4) Run SynMap.
5) To export the mapped pseudo-assembly, find and click on the "Links and Downloads" link in the SynMap output screen underneath the images.
6) Under the "Results" section in "Links and Downloads", select the "Syntenic Path Assembly Mapping" link to open a window with the mapped pieces' chromosome positions and orientations.

At this point, the typical workflow is to identify interesting genomic regions and send them to GEvo for microsynteny analysis. In addition, there are two workflows that can be carried out by clicking on a gene-pair to view information about the genes. The annotation box for genes contains links to additional analytical tools in CoGe. First, a gene from one organism can be sent to SynFind to be used as a query to identify syntenic genes from other genomes. SynFind operates on genes within a genome, and therefore can only be used for genomes that have CDS annotations. SynFind will calculate how many times the query genome region shows up the target genomes and the syntenic depth, and reports these in a table with how much of each level of duplication exists in the target genome in a percentage. These regions can then be sent to the GEvo tool to investigate the regions of microsynteny between the genomes. Additionally, a gene may be sent to the CoGe basic local alignment search tool (CoGeBLAST), which will identify BLAST hits (putative homologous genes and sequences) in any number of other genomes in CoGe. These results may then be used to extract sequences of putative homologous genes, sent to Phylogeny.fr to build phylogenic trees, or sent to other tools in CoGe.

Seek Synteny and Ye Shall SynFind

SynFind can be used to identify syntenic regions for a given gene in one genome across a user-defined selection of genomes. Note that SynFind only operates on genomes with CDS structural annotations. The strength of SynFind is that it will identify syntenic regions in the target genes even if a homolog to the query gene is not present. This is accomplished by using a window of neighboring genes to query for synteny.

1) You can directly select SynFind from the "Tools" drop-down menu located in the top-right of all pages. However, it is more common to use a link from a gene to launch SynFind, with that gene pre-loaded as the query gene.
2) From the results dot plot of SynMap, select a syntenic dot. A popup window will show that area in more detail.
3) Click on a specific syntenic dot within the dot plot. An information table about the syntenic gene pair for that region will be displayed. Each gene's information box will have a link to send the gene to SynFind, which can be found to the right of the CoGe tools row.
4) The SynFind tool will open and have the gene already selected below for the organism you chose. This will be the query sequence that will be used to search other genomes.
5) At the top, select a list of other genomes by typing in the "Organism" box. Selecting "Add" will put the genome(s) of interest into the "Selected Genomes" table. These are now referred to as *target genomes*.
6) Once all genomes of interest are selected, the "Run SynFind" button will begin the analysis.

Once the analysis is finished, the SynFind tool will display the syntenic regions that have been found in the target genomes using the query gene. This table displays the number of syntelogs (ohnologs) found in each target genome, their name(s), their positions on scaffolds or chromosomes, and their synteny score. Below this, there are several links to regenerate the analysis for later investigation or publication, to send the SynFind hits to GEvo for additional microsynteny analysis, or to generate tables containing nucleotide information for the SynFind hits. Below the "Links" section is the "Syntenic Depth" tables. These tables show the percentage of regions in each target genome compared to the genes in the query genome broken into number of duplications—that is, depth zero is no synteny, depth 1 has genome regions that match once to the target genome, depth 2 has genome regions that match twice to the target genome, etc. Selecting the "Compare and visualize region in GEvo" link will send the syntelogs to the GEvo tool for microsynteny comparison.

Visualizing Genome Evolution Using GEVo

Once the global whole genome comparison has been carried out, for example, using SynMap or SynFind, areas of microsynteny can be investigated to provide further evidence for WGDs. Linking to GEvo from SynFind begins the visualization automatically, whereas SynMap will require you to click the red "Run GEvo" button located below the sequence submission boxes.

GEvo will run pair-wise sequence similarity comparison of the genomic regions selected using any number of BLAST-like algorithms (LastZ being the default; Harris, 2007). The results of GEvo are displayed graphically in a set of multiple panels. Each panel represents a genomic region, with the dashed line separating the top and bottom strands of DNA. Examples of results from GEvo microsynteny analysis between *T. rubripes* and *O. mykiss* (regenerate results at https://genomevolution.org/r/eoln) and between *L. oculatus* and *O. mykiss* (regenerate results at https://genomevolution.org/r/f5jo) are displayed in Figure 25.5. Gene models are drawn as composite arrows, with grey regions being the extent of the gene, blue being mRNA, and green being the protein coding (CDS) sequences. Located above and below the gene models are colored boxes representing regions of sequence in the same and opposite orientations, respectively. Clicking on a gene will cause a dialog box to pop up with information about that gene; clicking on a region of sequence similarity will draw a transparent wedge connecting it to its partner region, and also cause a dialog box to pop up with information about the alignment. Microsynteny comparisons between *T. rubripes* and *O. mykiss* reveal a 1:2 syntenic relationship, which supports the proposed 4R salmonid lineage WGD (Figure 25.5A). The *L. oculatus* and *O. mykiss* microsynteny analysis has a 1:4 syntenic relationship, suggesting that two WGDs, Teleost WGD (3R), and salmonid WGD (4R), have occurred since the divergence of these two species (Figure 25.5B).

1) Linking to GEvo from SynFind begins the visualization automatically, whereas SynMap will require you to click the red "Run GEvo" button located below the sequence submission boxes.
2) Once the results are generated, they will be automatically displayed.
3) Clicking the "Shift" button on the keyboard while selecting the syntenic sections will connect all the syntenic lines for that track. Selecting each syntenic block in a track will show the syntenic lines for only that block.
4) The alignments, sequence files, and a URL to regenerate the analysis can be found under the image.

This analysis workflow is useful for validating synteny to confirm genomic regions that arise from a common ancestor. Analysis of regions in GEvo also helps determine tandem gene duplications, inversions, or fractionation of gene content.

CoGe Phylogenetics: Tree Branches and Lines for Something Other than Making a Rod

Comparing many species or genes that are spread throughout several chromosomes often leads to having too many tracks to visualize in one window in GEvo. In this case, phylogenetic trees are better for comparing gene families across multiple species that are

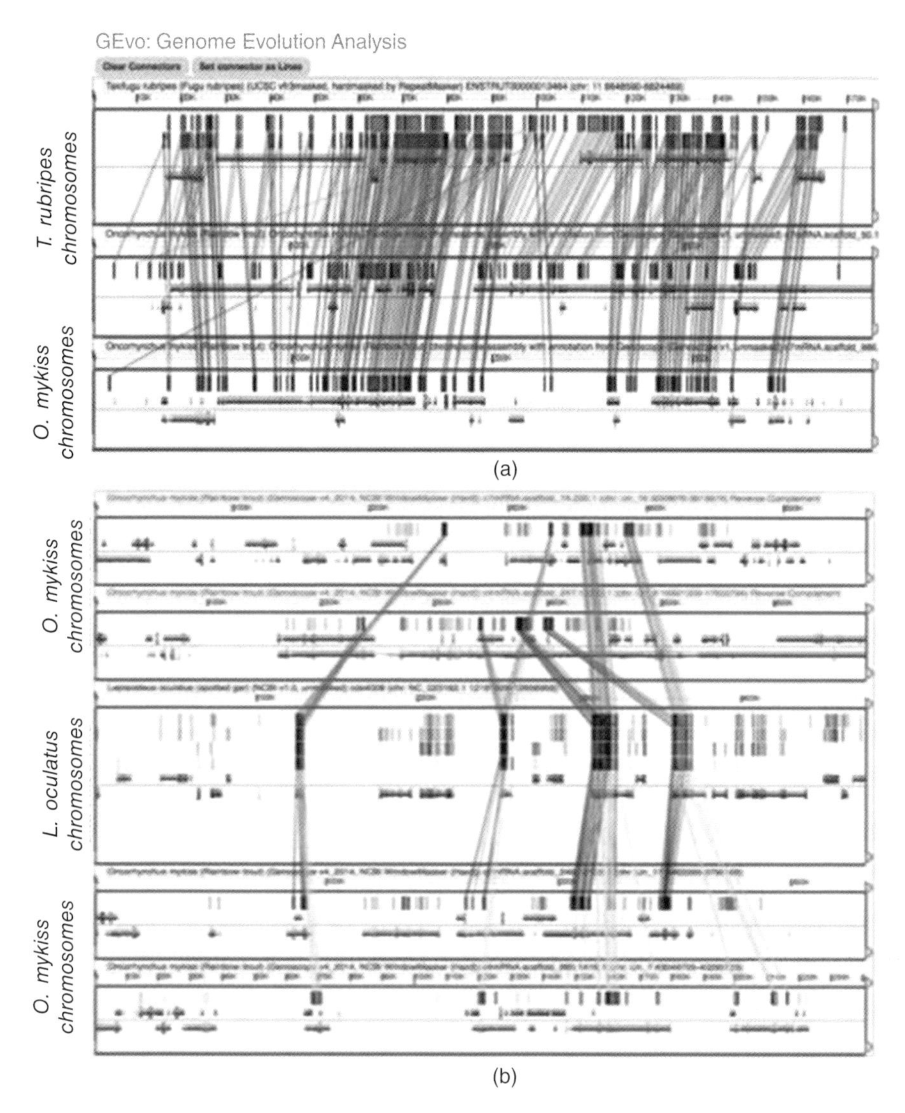

Figure 25.5 Regions of microsynteny from SynMap comparisons of Figure 25.4 displayed using the CoGe tool GEvo. (a) Comparison of a syntenic region in the *T. rubripes* genome on chromosome 11 (top track) with two separate scaffolds of the *O. mykiss* genome (second and third tracks). This genome region has a 1:2 ratio with two duplications in *O. mykiss*, suggesting that these may have resulted from the salmonid WGD. (b) A 1:4 *L. oculatus* to *O. mykiss* syntenic genome region. Five genes in this region display a 1:4 syntenic pattern, suggesting two duplications (Teleost 3R and salmonid 4R). The other genes in this region display a 1:2 or 1:3, suggesting fragmentation of the genome. Results can be regenerated at https://genomevolution.org/r/eoln and https://genomevolution.org/r/f5jo, respectively. (*See color plate section for the color representation of this figure.*)

not linked in a region of the chromosome. There are two steps to creating phylogenetic trees in CoGe: (1) use the CoGeBLAST tool to find homologous genes of interest across species, and (2) send genes identified by CoGeBLAST to phylogeny.fr to build a phylogenetic tree.

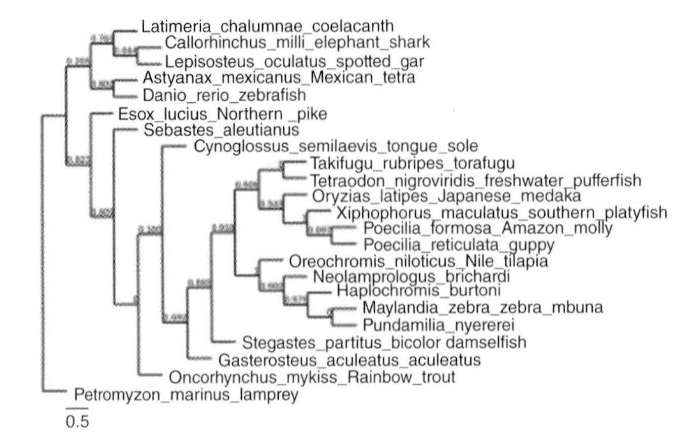

Figure 25.6 Phylogenetic relationships of annotated, masked fish genomes available in CoGe (https://genomevolution.org/r/en40) using an *O. mykiss* polyunsaturated fatty acid elongase (GenBank AY605100.1) query sequence in CoGeBLAST. Most fish species group into their clades. The tree was rooted using the jawless fish *Petromyzon marinus* as an outgroup for the gnathostomata fish species. Results can be regenerated at https://genomevolution.org/r/f5vl.

Figure 25.6 is a sample phylogenetic tree created using the annotated, masked fish genomes notebook (https://genomevolution.org/r/en40), by using the *O. mykiss* polyunsaturated fatty acid elongase (GenBank AY605100.1) as a query sequence in CoGeBLAST (results regenerated at https://genomevolution.org/r/f5vl). The tree was rooted using the jawless fish *Petromyzon marinus* as an outgroup for the gnathostomata fish species.

BLAST at CoGe

The CoGeBLAST tool can be accessed directly from the drop-down menu on the CoGe toolbar, or syntenic sequences can be exported from SynMap or SynFind by selecting the appropriate links from those tools.

1) Once the CoGeBLAST tool is opened, a list of target genomes needs to be populated. Simply type the species name into the "Organism:" search box, and add them to the "Selected Genomes" list in the same manner as in SynFind. A list of genomes inside a notebook can also be imported in bulk by searching for the notebook's name.
2) Next, select whether to submit protein or nucleotide sequence to the BLAST search. Parameters can be modified, but the defaults are usually sufficient.
3) If CoGeBLAST was opened from SynFind or SynMap, then the selected query sequence will automatically populate in "Query Sequence(s)".
4) Otherwise, copy and paste a query sequence to be used in the BLAST search. Afterward, select the "Run CoGeBLAST" button to start the analysis.

Once the analysis finishes, a list of BLAST hits is given at the top of the screen, with species names arranged left to right. The top row shows how many hits are displayed in the table below, and the bottom row displays the total BLAST hits found. The "HSP Table" displays the information about BLAST hits, such as organism name, chromosome number, chromosome position, and closest genomic feature to the BLAST hit. CoGeBLAST will identify the overlapping gene to the blast hit, and those sequences may be selected and sent to other tools in CoGe. To do this, from the drop-down menu next to the text "Send selected to:", select the "Go" button. *Note*: Be sure to wait for the "Closest Genomic Feature" column to fill in before sending the sequences to FastaView, or the BLAST hits will not transfer properly.

Select the "Phylogenetics" option in the drop-down menu, and select the "Go" button to send genes overlapping BLAST hits to the FastaView tool to retrieve their sequences. In FastaView, you can select whether to translate those sequences, download them, or send them to other tools for downstream analysis. One place you can send these sequences to is the Phylogeny.fr platform. If genes are missing from the phylogenetic tree, simply return to the FastaView table and export the sequences to Phylogeny.fr again to rerun the tree-building tool.

Phylogeny.fr: The Trees in France Are Lovely This Time of Year

Sequences sent to Phylogeny.fr (Dereeper *et al.*, 2008) are automatically input into their "One Click Mode" phylogenetic tree builder. The FASTA sequences will automatically be aligned to one another using MUSCLE (Edgar, 2004), then sent to PhyML (Guindon *et al.*, 2010) to build the phylogenetic tree, and then the tree is rendered in TreeDyn (Chevenet *et al.*, 2006). Once the automatic analysis is complete, the tree can be rooted and modified in a number of different ways. The tools for changing the tree are underneath the "Dynamic Tree Edition" section. After a tree has been modified, it can be exported as an image for publication, or in Newick format for use in other tree-building applications.

Sinkers to Cast Further and Deeper: Adding Weight to Genomes with Additional Data Types

After initial genome comparisons have been carried out, there are additional data types, such as diversity and functional genomics data, which can be associated with genomes. These data sets are called "experiments", and they add to the number of analyses and visualizations that CoGe can provide. Loading multiple experiments allows for comparisons across each experiment and to the associated genome, as well as visualization in GenomeView. Stacking multiple tracks of data in GenomeView allows researchers to visually compare sections of the genome. The CoGe tool LoadExperiment imports a variety of data types, but has a set file format that can be used for each type of data: RNA-Seq (FASTQ); quantitative measurements such as FPKM (CSV, BED, WIG); markers (GFF and GTF); and SNPs (VCF). Once loaded, the experiments can be visualized in GenomeView as tracks. SNP variant data has been loaded for *Tetraodon nigroviridis* (https://genomevolution.org/coge/GenomeView.pl?embed=&gid=24926) as an example. Selecting the experiment tracks in GenomeView brings up the SNP data for the *T. nigroviridis* genome. Areas of the genome can be magnified to view the SNP variants (e.g., G > T). GenomeView allows for visualization of many tracks and fluid movement across chromosomes to ease comparisons between many different experiment datasets.

Using LoadExperiment to Include Other Data Types

1) View a genome of interest in GenomeInfo.
2) Click the "Load Experiment" button located at the bottom-left of the GenomeInfo page.
3) Fill in the required metadata fields, and select "Add Data".
4) Follow the same steps to upload an experiment as previously in genome or annotation uploads.
5) Note that you can have a private experiment (which can be shared) on a public genome. This is a convenient way to mix and match public and private data.

Viewing Genomes and Experiments in GenomeView

GenomeView is the tool used to visualize the sequence of a genome in CoGe, and any experiments associated with the genome. There are no limits to how many experiments can be associated with a genome, and many tracks can be stacked vertically in GenomeView for comparison.

1) View a genome of interest in GenomeInfo.
2) Under the "Info" section, select the "View" button.
3) This brings up GenomeView. On the right is a list of tracks that are available for that genome. This is where the genome sequence and experiments can be selected for visualization.
4) GenomeView allows movement over chromosomes by scrolling or by selecting the left and right arrows in the top toolbar. Next to the left and right arrows are the zoom-in and zoom-out icons.
5) Genomes with labeled chromosomes can be navigated to by selecting the chromosome of interest from the drop-down menu next to the zoom icons.

Loading experiments into CoGe associates those datasets with their corresponding genomes. Some data sets will generate additional analyses automatically. For example, when RNA-Seq data (FASTQ) is loaded into CoGe, CoGe will run an RNA-Seq processing pipeline to clean, map, and quantitate the FASTQ reads. This pipeline will generate three data sets for visualization in GenomeView: reads quantified to each nucleotide position, reads normalized to transcripts as FPKM values, and an alignment of the individual reads to the genome (similar pipelines are available to detect and quantitate SNP variants). These tracks allow researchers to easily compare areas of upregulation and/or downregulation expression and discover SNPs across different experiments. SNPs can also be overlaid with marker data to help discover the underlying biological changes associated with breeding markers.

SNP Analysis of BAM Alignment Experiment Datasets

1) Navigate to the GenomeInfo page of the genome containing the BAM alignment experiment of interest. *T. rubripes* BAM alignment files have been loaded to demonstrate analyses that can be conducted with experiments (https://genomevolution.org/coge/GenomeInfo.pl?gid=25444).
2) In the "Experiments" section, select the experiment to analyze.

3) This will bring the user to the ExperimentView page containing metadata, and a "Tools" section.
4) Select "Identify SNPs".
5) Select an algorithm to use to identify the SNPs in the experiment. The analysis will run, and, when completed, it will be viewable as an experiment track in GenomeView.

CoGe also supports visualization of epigenetic data sets in a collaboration called EPIC-CoGe. Epigenetic data can be uploaded into the experiments section using the CSV, TSV, BED, or WIG file formats, and then visualized in GenomeView along with the genome and other experiments. Visualizing methylation patterns in addition to transcript expression and genome structure can help to correlate expression patterns to epigenetic alterations between samples.

Conclusions

The CoGe platform is dedicated to providing genome management and comparative genomics analysis and visualization tools to the scientific community. Any number of genomes and experiments can be loaded onto the CoGe platform from the CyVerse Data Store, using FTP/HTTP URLs, from NCBI, or directly from your local computer. Data loaded into CoGe is initially assumed to be private, but the owner of data can choose to share it privately with collaborators or make it public for the entire research community. Notebooks can be used to manage datasets, analyses, and permissions of collaborators associated with a research project.

Tools are available for the comparison of 40 fish genomes (https://genomevolution.org/CoGe/NotebookView.pl?nid=890) on both broad scales using SynMap and SynFind as well as smaller scales using GEvo. Genes of interest can be used as query sequences to rapidly find related genes in other genomes. BLAST hits can then be exported to the Phylogeny.fr platform to build phylogenetic trees without having to do any hands-on analysis.

A host of additional datasets can be associated with genomes by loading them as experiments into CoGe. Quantitative data such as transcript expression, markers, polymorphisms, and RNA-Seq alignments can be loaded and easily visualized using GenomeView for comparisons of genome regions or genes of interest. GenomeView allows for visualization of multiple data sets that range from the initial genome to methylation patterns, SNPs, markers, and transcript expression profiles.

The Fish Comparative Genomics CoGePedia page contains supplemental information, all of the URLs presented in this chapter, and step-by-step directions for analyses (https://genomevolution.org/wiki/index.php/Fish_Comparative_Genomics). Additional help such as video tutorials and workshop outlines can be found on CoGePedia or by accessing the "Help" button in the main toolbar. Data and analyses are tracked in your profile and can easily be discovered by name or by time and date. Each analysis can easily be rerun using permanent, unique URLs generated by the analysis. This is useful for sharing results in publications as they can always be regenerated by anyone with an Internet connection. Taken together, the CoGe platform tools allow scientists without expertise in comparative genomics to run analyses within minutes of registering for access, and we believe it is one of the ideal platforms for the aquaculture research community.

References

Berthelot, C., Brunet, F., Chalopin, D. *et al.* (2014) The rainbow trout genome provides novel insights into evolution after whole-genome duplication in vertebrates. *Nature Communications*, **5**, 3657.

Betancur-R.R., Broughton, R.E., Wiley, E.O., *et al.* (2013). The tree of life and a new classification of bony fishes. *PLoS Currents*, doi: 10.1371/currents.tol .53ba26640df0ccaee75bb165c8c26288.

Braasch, I., Peterson, S.M., Desvignes, T. *et al.* (2014) A new model army: emerging fish models to study the genomics of vertebrate evo-devo. *Journal of Experimental Zoology. Part B, Molecular and Developmental Evolution*, **9999B**, 1–26.

Chevenet, F., Brun, C., Banuls, A.-L. *et al.* (2006) TreeDyn: towards dynamic graphics and annotations for analyses of trees. *BMC Bioinformatics*, **7**, 439.

Christoffels, A., Koh, E.G., Chia, J. *et al.* (2004) Fugu genome analysis provides evidence for a whole-genome duplication early during the evolution of ray-finned fishes. *Molecular Biology and Evolution*, **21**, 1146–1151.

Dandekar, T., Fieselmann, A., Majeed, S. and Ahmed, Z. (2014) Software applications toward quantitative metabolic flux analysis and modeling. *Briefings in Bioinformatics*, **15**, 91–107.

Davidson, W.S., Koop, B.F., Jones, S.J. *et al.* (2010) Sequencing the genome of the Atlantic salmon (*Salmo salar*). *Genome Biology*, **11**, 403.

Dehal, P. and Boore, J.L. (2005). Two rounds of whole genome duplication in the ancestral vertebrate. *PLoS Biology*, **3**, e314.

Dereeper, A., Guignon, V., Blanc, G. *et al.* (2008) Phylogeny.fr: robust phylogenetic analysis for the non-specialist. *Nucleic Acids Research*, **36**, W465–W469.

Edgar, R.C. (2004) MUSCLE: multiple sequence alignment with high accuracy and high throughput. *Nucleic Acids Research*, **32**, 1792–1797.

Feil, R. and Fraga, M.F. (2012) Epigenetics and the environment: emerging patterns and implications. *Nature Reviews Genetics*, **13**, 97–109.

Freeling, M. (2009) Bias in plant gene content following different sorts of duplication: tandem, whole-genome, segmental, or by transposition. *Annual Review of Plant Biology*, **60**, 433–453.

Gama-Castro, S., Salgado, H., Peralta-Gil, M. *et al.* (2011) RegulonDB version 7.0: transcriptional regulation of Escherichia coli K-12 integrated within genetic sensory response units (Gensor Units). *Nucleic Acids Research*, **39**, D98–D105.

Goff, S.A., Vaughn, M., McKay, S. *et al.* (2011) The iPlant collaborative: cyberinfrastructure for plant biology. *Frontiers in Plant Science*, **2**, 34.

Guindon, S., Dufayard, J.-F., Lefort, V. *et al.* (2010) New algorithms and methods to estimate maximum-likelihood phylogenies: assessing the performance of phyml 3.0. *Systematic Biology*, **59**, 307–321.

Haas, B.J., Delcher, A.L., Wortman, J.R. and Salzberg, S.L. (2004) DAGchainer: a tool for mining segmental genome duplications and synteny. *Bioinformatics*, **20**, 3643–3646.

Harris, R.S. (2007) *Improved pairwise alignment of genomic DNA*, The Pennsylvania State University, Ph.D..

Lyons, E. and Freeling, M. (2008) How to usefully compare homologous plant genes and chromosomes as DNA sequences. *The Plant Journal*, **53**, 661–673.

Lyons, E., Pedersen, B., Kane, J. and Freeling, M. (2008) The value of nonmodel genomes and an example using SynMap within CoGe to dissect the hexaploidy that predates the rosids. *Tropical Plant Biology*, **1**, 181–190.

Lyons, E.H. (2008) *CoGe, a new kind of comparative genomics platform: Insights into the evolution of plant genomes*, University of California, Berkeley, Ph.D..

Meyer, A. and Van de Peer, Y. (2005) From 2R to 3R: evidence for a fish-specific genome duplication (FSGD). *Bioessays*, **27**, 937–945.

Miettinen, O., Larsson, E., Sjökvist, E. and Larsson, K. (2012) Comprehensive taxon sampling reveals unaccounted diversity and morphological plasticity in a group of dimitic polypores (Polyporales, Basidiomycota). *Cladistics*, **28**, 251–270.

Near, T.J., Eytan, R.I., Dornburg, A. *et al.* (2012) Resolution of ray-finned fish phylogeny and timing of diversification. *Proceedings of the National Academy of Sciences*, **109**, 13698–13703.

Renny-Byfield, S., Kovarik, A., Kelly, L.J. *et al.* (2013) Diploidization and genome size change in allopolyploids is associated with differential dynamics of low- and high-copy sequences. *The Plant Journal*, **74**, 829–839.

Rensing, S.A., Ick, J., Fawcett, J.A. *et al.* (2007) An ancient genome duplication contributed to the abundance of metabolic genes in the moss Physcomitrella patens. *BMC Evolutionary Biology*, **7**, 130.

Schilmiller, A.L., Pichersky, E. and Last, R.L. (2012) Taming the hydra of specialized metabolism: how systems biology and comparative approaches are revolutionizing plant biochemistry. *Current Opinion in Plant Biology*, **15**, 338–344.

Skinner, M.E., Uzilov, A.V., Stein, L.D. *et al.* (2009) JBrowse: a next-generation genome browser. *Genome Research*, **19**, 1630–1638.

Sonnhammer, E.L. and Koonin, E.V. (2002) Orthology, paralogy and proposed classification for paralog subtypes. *Trends in Genetics*, **18**, 619–620.

Steinke, D., Hoegg, S., Brinkmann, H. and Meyer, A. (2006) Three rounds (1R/2R/3R) of genome duplications and the evolution of the glycolytic pathway in vertebrates. *BMC Biology*, **4**, 16.

Tang, H. and Lyons, E. (2012) Unleashing the genome of Brassica rapa. *Frontiers in Plant Science*, **3**, 172.

Toubiana, D., Fernie, A.R., Nikoloski, Z. and Fait, A. (2013) Network analysis: tackling complex data to study plant metabolism. *Trends in Biotechnology*, **31**, 29–36.

Weitzel, M., Nöh, K., Dalman, T. *et al.* (2013) 13CFLUX2 – high-performance software suite for 13C-metabolic flux analysis. *Bioinformatics*, **29**, 143–145.

Wolfe, K. (2000) Robustness – it's not where you think it is. *Nature Genetics*, **25**, 3–4.

Part V

Bioinformatics Resources, Databases, and Genome Browsers

26

NCBI Resources Useful for Informatics Issues in Aquaculture

Zihao Yuan, Yujia Yang, Shikai Liu and Zhanjiang Liu

Introduction

The National Center for Biotechnology Information (NCBI) hosts very comprehensive bioinformatics databases that are extremely useful for molecular biologists, providing them access to genomic and biomedical information. Two major databases that NCBI houses are PubMed for biomedical literature and GenBank for nucleotide sequences. NCBI also harbors online software tools that can be used via web browsers—for example, NCBI's Basic Local Alignment Search Tool (BLAST) is a popular tool for searching sequence similarities. An NCBI user can not only download data but also deposit data to the databases, and such data is curated by the NCBI staff. NCBI is of vital importance for molecular biology research. Many, if not most, of the tasks of aquaculture researchers can probably be executed by using NCBI tools if one knows the NCBI resources and tools well. In this chapter, we will sequentially introduce: how to search and download data from popular NCBI databases; how to apply frequently used NCBI tools to solve biological problems; and how to submit data to the NCBI. Readers who are familiar with NCBI resources and tools can skip this chapter.

NCBI is a branch of the national library of medicine, part of the National Institutes of Health (NIH). Established in 1988, it is one of the world's most popular resources for bioinformatics and biomedical research. The NCBI resources comprise multiple databases and tools offering information on the analysis of molecular and genetic processes. Briefly, the NCBI offers the following functions: it accepts submissions of primary data from sequencing centers or individual researchers; and it provides tools for the analysis of genomic data. NCBI provides free links to these databases and tools through its Entrez search engine (Geer & Sayers, 2003; Wheeler *et al.*, 2004).

In a nutshell, the missions of NCBI can be summarized as: developing software at the molecular level based on mathematical methods; maintaining and supporting databases for the scientific and medical research communities; fostering international collaboration in biological and medical research; and developing and promoting standards in the field of biological nomenclatures.

Bioinformatics in Aquaculture: Principles and Methods, First Edition. Edited by Zhanjiang (John) Liu.

Popularly Used Databases in NCBI

PubMed

PubMed (http://www.ncbi.nlm.nih.gov/pubmed) is a free search engine that primarily indexes the life sciences and biomedical literatures. As of August 2016, PubMed covers more than 26 million literature records, and this number is continuously growing and updating. On the top of PubMed's search web page is a search box. The user can retrieve data using standard searches by simply entering key words of a subject into the search window—similar to searching in Google—and PubMed will retrieve related literatures with similar keywords or titles (Clarke & Wentz, 2000). For example, if we are interested in the *BCL2* gene in humans, simply typing "bcl2 *Homo Sapiens*" in the search box will prompt PubMed to present literature related to the human *BCL2* gene. Below the search box, there are options to set up the search preferences. The search preferences contain the "Format" (Summary, Abstract, MEDLINE, XML, PMID List); "Items per page", and "Sort by" options. The user can sort the search results by time of publication, relevance, publication date, authorship, and journal titles by using the "Sort by" option. On the left of the web page is the "Manage Filters" option, and the user can narrow down the scope of searching to the articles that he or she is interested by using this option. The commonly used filter options are "Article types", "Text availability", "PubMed Commons", "Publication dates", "Species", etc. For example, if we set the parameters for the search of the *BCL2* genes in humans to "Free full text" in "recent 5 years", then the number of items will narrow down from 4327 to 920 hits. If the user is interested in a certain article, he or she can click on the link; also, on the upper right corner of the web page are provided with the full text links to the database where the user can access the articles.

Entrez Nucleotide

The GenBank is the primary database of NCBI that stores nucleotide sequences (Benson *et al.*, 2005). However, the user can retrieve nucleotide information not only from GenBank but also from other nucleotide databases—such as NCBI Reference Sequence Database and Third-Party Annotation (TPA) Sequences via the Entrez Nucleotide database (http://www.ncbi.nlm.nih.gov/gene). The main goal of the Entrez Nucleotide database is to supply key connections between maps, sequences, expression profiles, structures, functions, homology data, and the scientific literature.

To conduct a search, the user needs to enter into the search box the name of the nucleotide and the Latin name of the species. For example, entering "Beta-actin *Ictalurus punctatus*" in the search box will display a list of all the nucleotide sequences of channel catfish Beta-actins that have been submitted to NCBI. Each item has a link to the web page with a detailed description. For example, when the user clicks on "*Ictalurus punctatus* beta-actin mRNA, partial cds" with accession number of "AY555575.1", it will link to the web page http://www.ncbi.nlm.nih.gov/nuccore/AY555575.1, which describes the beta-actin mRNA in channel catfish with the features and references for this sequence. The user can also view the sequence in FASTA format and graphics views. On the right side of the web page are the commonly used tools that can analyze this sequence. The user can run BLAST on the sequence, design primers of the sequence, and highlight and search in the sequence. On the right side of the web page are provided links on the gene, protein, literature, taxonomy, and UniGene of the sequence. The user can always download the sequence using the "Send" button in the upper right corner. For example, by choosing "Complete Record", "File" in "Choose Destination", and "FASTA" in format, the sequence can be exported to a text file in FASTA format.

Entrez Genome

The user can retrieve genome information via the Entrez system and Entrez genome (http://www.ncbi.nlm.nih.gov/genome/) to search for organisms from prokaryotes, eukaryotes, viruses, plasmids, to organelles. The Entrez genome contains the genomes that have been completely sequenced, as well as the genome sequencing projects that are currently in progress. The Entrez genome provides a user-friendly interface to search for genomes. For example, when we type "*Danio rerio*" in the search box and click "Search", it will present a summary of statistics describing all the available zebrafish genomes. The web page also contains publications, reference genomes, chromosomes, related BioProjects, and external databases that are related to the zebrafish genome. For the current version (GRCz10) of the reference genome assembly of zebrafish, all the statistics about the 25 chromosomes of zebrafish, mitochondrial genome, and unassembled sequences are presented in a table in the "Representative" section. On the top-most part of the web page is a column with links to download the genomic data in FASTA format for genome, transcript, and protein; links to download genome annotation in GFF, GenBank, or tabular format; and links to BLAST sequences against zebrafish genome, transcript, and protein databases. If the user is interested in other versions of zebrafish genome assemblies, he or she can click on the link presented under the title of "All 2 genomes for species" or the "Genome assembly and Annotation report" in the summary to get the table containing the two assemblies of the genome of zebrafish (GCA_000002035.3, GCA_000767325.1), and download the assembly sequence through the FTP link provided at the end of the table.

The Entrez genome can be used to download the assembly of the nuclear genome, as well as mitochondrial genomes for most of the non-model organisms. For instance, when searching for *Ictalurus furcatus* (blue catfish) in the Entrez genome search column, the system will provide a link to the blue catfish mitochondrial genome sequence. There is a table under the category link, and it contains basic information on the blue catfish mitochondrial genome—such as genome size, GC content, number of genes and their coded proteins, and number of rRNAs and tRNAs. Below the column is a graph showing the genome regions of the genome. On the upper right corner of the web page is a link "Send to". From this link, the user can choose to download the sequence in complete sequence, coding sequence, or gene sequences. If the user wishes to download the complete mitochondrial genome sequence, he or she can click on "Complete Record"; choose "File" in "Choose Destination"; and choose "FASTA" as the format. After clicking on the "Create File" button, a sequence file in FASTA format will be generated.

Sequence Read Archive

The Sequence Read Archive (SRA) (http://www.ncbi.nlm.nih.gov/sra) stores raw sequence data from the vast resources of "next-generation" sequencing platforms, including 454, Illumina, SOLiD, Ion Torrent, and Helicos. The archives in the SRA database have grown rapidly in recent years. SRA is NIH's primary archive of high-throughput sequencing data, and is part of the international partnership of archives named International Nucleotide Sequence Database Collaboration (INSDC) cooperated by NCBI, European Bioinformatics Institute (EMBL-EBI), and DNA Data Bank of Japan (DDBJ). Data submitted to any of the three organizations are shared among them (Kodama, Shumway & Leinonen, 2012).

The raw data stored in the SRA database are stored in four levels: Bioproject, experiment, BioSample, and run. For each project, there can be several experiments; for each experiment, there can be multiple BioSamples; and every BioSample can have multiple runs. The raw data are stored in each run. For example, if we want to download the short reads to identify SNPs from common carp (*Cyprinus carpio*), simply type in "*Cyprinus carpio*" in the search box, and it will list all of the raw reads that are related to common carp. Among the listed experiments, there is an item with Accession: "SRX317022"; click on that link, and we can see the details; it lists sections such as "Design", "Submitter", "Study abstract", "Sample", "Library" information, and "Spot descriptor". From the description, one can learn detailed information about the raw data. For instance, the dataset "SRR924334" contains pair-end whole genome shotgun sequences generated by Illumina HiSeq 2000. If the user wants to have a more detailed view of the sequence and its quality, he or she can click on the link "SRR924334" under the "Run" category, and the browser will jump to a web page describing the metadata and the sequencing base quality. The user can also provide a preview of the reads in the link "Reads".

In order to download SRA data, the "SRA toolkit" can be used. The user can download the toolkit from http://trace.ncbi.nlm.nih.gov/Traces/sra/sra.cgi?view=software. In Linux and macOS systems, after installation, the following command line can be used to download the SRA data:

```
$ ~/path/to/sratoolkit.2.5.5-mac64/bin/prefetch SRR924334
```

In this command line, "prefetch" is the command to conjunct the computer with the Hypertext Transfer Protocol (HTTP). "SRR924334" is the accession number of the SRA file.

This can be a simpler way for many users to download SRA data. However, the download file is in SRA format; if the user wants to convert an SRA-format file to FASTQ format, the "`fastq-dump`" command is needed:

```
$ ~/path/to/sratoolkit.2.5.5-mac64/bin/fastq-dump SRR924334
```

This command line will convert "SRR924334.sra" to "SRR924334.fastq" in the current working directory.

dbSNP Database

The dbSNP database (http://www.ncbi.nlm.nih.gov/SNP) is a public repository for a broad range of single-nucleotide variations, including large numbers of sequence variants identified by next-generation sequencing (Day, 2010; Sherry, Ward & Sirotkin, 1999; Sherry *et al.*, 2001). Genetic variations in the dbSNP database include both germline and somatic variations (NCBI, 2014). Additionally, the dbSNP database contains population-specific allele frequencies and genotypes, which presents the validation state of each variant (NCBI, 2014). The dbSNP database is essential for integrating biological information into genetic association experiments on a genome-wide scale, such as genome-wide association studies (GWAS; Saccone *et al.*, 2011).

The dbSNP database can be searched from the its homepage (http://www.ncbi.nlm.nih.gov/SNP/) by using Entrez SNP, or by using the links to the other basic dbSNP search options located in the "Search" column of the dbSNP homepage (Kitts & Sherry, 2002). The other basic dbSNP search options include BLAST SNP, Batch Query, Genotype Query, etc.

Using Entrez SNP, small variations or large structural variations can be searched in dbSNP or dbVar. Input organism names in the search box, then click on the "Search" button, and the related data found will be displayed. dbSNP also provides searches by IDs on all assemblies, including Reference cluster ID (rs#), Gene ID or symbol, etc. For instance, input "zebrafish rs3727524" in the search box, then click "search", and the reference SNP cluster report of rs3727524 will be displayed on a new web page. The refSNP cluster details are organized in the following sections: "Integrated Maps", "GeneView", etc. In the "Integrated Maps" section, click "Chr Pos" or "Contig Pos" to display the position of rs3727524 on the chromosome or the contig. In addition, the "GeneView" section provides analysis of contig annotation by checking the location of reference SNP clusters in the genes.

Gene Expression Omnibus

The NCBI Gene Expression Omnibus (GEO; http://www.ncbi.nlm.nih.gov/geo/) is the largest data repository and retrieval system for high-throughput molecular abundance gene expression data generated by microarray and next-generation sequencing technologies (Barrett *et al.*, 2005; NCBI, 2014). Currently, the GEO database hosts about 1,900,000 samples and 70 billion individual abundance measurements for around 1600 organisms. GEO data are housed in two Entrez databases: GEO DataSets (http://www.ncbi.nlm.nih.gov/gds) and GEO Profiles (http://www.ncbi.nlm.nih.gov/geoprofiles/).

GEO DataSets stores descriptions of all original submitter-supplied records (Wheeler *et al.*, 2004). Here, we use catfish as an example to illustrate GEO DataSets usage. First, input the text terms of interest, for example, "catfish" and "liver", in the search box of the GEO DataSets Advanced Search Builder web site (http://www.ncbi.nlm.nih.gov/gds/advanced), and then click the "Search" button for the closest match in the GEO database. A list of 53 related results of the DataSet (GDS), Series (GSE), or Platform (GPL) accession number will be displayed on a new web page. After checking its summary, type, and platform, click on any link of interest in GEO DataSet Records. For each GEO DataSet Record, a link of "Analyze with GEO2R" is provided to compare two or more groups of samples in order to identify genes that are differentially expressed across experimental conditions. The GEO DataSet Record can be downloaded by selecting one of the several download options, including Series family SOFT file, Series family MINiML file, etc. In addition, raw data, processed data in a sample table, and processed data provided as a supplementary file can also be downloaded.

GEO Profiles is a gene-level database used for storing quantitative gene expression measurements for one gene across one experiment (Barrett *et al.*, 2005; NCBI, 2014). Here, we use zebrafish Interleukin-2 as an example to illustrate GEO Profiles usage. When entering the terms "zebrafish" and "Interleukin-2" in the search box of GEO Profiles Advanced Search Builder (http://www.ncbi.nlm.nih.gov/geoprofiles/advanced), 120 GEO Profiles are detected as matches in the NCBI database. First, click on any link of interest in GEO Profiles Records, and then click on annotated gene names in the

"Annotation" section to present gene symbol, full name, and aliases in the NCBI Gene, UniGene, or Nucleotide databases and gain more information on the related RefSeqs, maps, pathways, variations, phenotypes resources, etc. In addition, click on Profile Neighbors, Chromosome Neighbors, Sequence Neighbors, or Homologene Neighbors, and a list of GEO Profiles in similar expression patterns that pass the user-selected criteria will be displayed. For instance, to retrieve other genes with similar expression patterns in that DataSet, click on the "Profile Neighbors" button on the web page of GEO Profiles ID 88058162, and 200 similar expression patterns to GEO Profiles ID 88058162 will be detected within the same DataSet. This use of GEO Profiles will help in identifying functionally related genes and investigating gene expression neighborhoods.

Notably, to display the list of pathways in which gene expression profiles participate, the interface of GEO Profiles provides a "Profile pathways" module. For instance, click on the "Find pathways" button, and the list of pathways in which GEO Profiles ID 88058162 participates will be shown on a new web page. This process of searching for pathways can be particularly useful for characterizing lists of profiles that have been determined as differentially expressed genes during experiments.

UniGene

UniGene (http://www.ncbi.nlm.nih.gov/UniGene/) is a largely automated analytical system for producing an organized view of the transcriptome (Pontius, Wagner & Schuler, 2003). UniGene provides information on protein similarities, gene expression, cDNA clone reagents, and genomic locations (Romiti & Cooper, 2005). Additionally, the UniGene database provides a source of unique sequences for the fabrication of microarrays for the large-scale study of gene expression (Wheeler *et al.*, 2004). Further, the UniGene database computationally identifies transcript sequences (including ESTs) into a non-redundant set of clusters from the same locus (NCBI, 2014). Currently, UniGene clusters are created for 142 eukaryotes, which contain more than 70000 ESTs in the UniGene database (NCBI, 2014).

The interface of UniGene provides UniGene clusters information on protein similarities, gene expression, cDNA clone reagents, and genomic location. For instance, if we are interested in zebrafish protein-coding gene Rhodopsin (*RHO*), input the terms "zebrafish" and "Rhodopsin" in the search box of UniGene Advanced Search Builder (http://www.ncbi.nlm.nih.gov/unigene/advanced), then click the "Search" button, and the links to zebrafish protein-coding gene *RHO* will be displayed on the search result page. UniGene provides several sections, such as "Selected protein similarities", "Gene expression", "Mapping position", and "Sequences". In the module of "Selected protein similarities", the best hits and hits from model organisms to the cluster transcripts will be displayed, which provides a comparison of the cluster transcripts with RefSeq proteins. In this process, the alignments can suggest function of the cluster. In addition, the "Gene expression" section provides information on gene expression in tissues and development stages, cDNA resources, etc. The "Mapping position" section helps specify genomic location by transcript mapping, radiation hybrid mapping, genetic mapping, or cytogenetic mapping. Click on the "Map Viewer" button in "Mapping position", and the genomic location of zebrafish protein-coding gene *RHO* will be displayed in an integrated zebrafish genome map using NCBI Map Viewer. Further, the UniGene interface also supports downloading sequences representing this gene, mRNAs, ESTs, and gene predictions supported by transcribed sequences. All mRNA sequences and EST sequences representing the zebrafish protein-coding gene *RHO* can be downloaded by clicking the “Download sequences” button.

Probe

Probe (http://www.ncbi.nlm.nih.gov/probe) is a public database of nucleic acid reagents designed for use in a wide variety of biomedical research applications, including genotyping, gene expression studies, SNP discovery, genome mapping, and gene silencing (NCBI, 2015; Romiti & Cooper, 2005; Tatusova, Smith-White & Ostell, 2007). Probe records contain information on reagent distributors, Probe effectiveness, and computed sequence similarities (NCBI, 2015; Romiti & Cooper, 2005). Currently, the Probe database contains more than 14 million Probe sequences for all types of biomedical researches (NCBI, 2014).

In the following example, we use channel catfish primer set Probe rRNA 18S for demonstration. First, input the terms “channel catfish” and “rRNA 18S” in the search box of Probe Advanced Search Builder (http://www.ncbi.nlm.nih.gov/probe/advanced); click the “Search” button; and then “*Ictalurus punctatus* primer set Probe rRNA 18S” will be displayed on a new web page. The Probe interface provides basic information on the name, type, function, and organism of Probe sequences. For channel catfish rRNA 18S primer set, the application type is “gene expression”. In addition, primer set sequences of channel catfish rRNA 18S can be downloaded in the “Sequence” section. Further, computational maps for Probe are also available for some of the Probe sequences.

Conserved Domain Database

Conserved Domain Database (CDD) (http://www.ncbi.nlm.nih.gov/cdd/) is a resource for annotation of functional units in proteins with a collection of well-annotated multiple-sequence alignment models (Marchler-Bauer *et al.*, 2011). The application of CDD is to refine domain models of 3D protein structure and provide insights into the relationships among sequence, structure, and function (Marchler-Bauer *et al.*, 2011). Additionally, CDD provides annotation of protein sequences based on the location of conserved domain and functional sites (Marchler-Bauer *et al.*, 2012). Currently, CDD contains more than 46000 domains imported from several external source databases (Pfam, SMART, COG, PRK, TIGRFAM, etc.) (Marchler-Bauer *et al.*, 2011; NCBI, 2014). In addition, CDD provides 3300 superfamily records, each of which possesses a set of conserved domains (CDs) (NCBI, 2014).

Here, we use zebrafish Pleckstrin homology-like superfamily (PH-like) as an example. First, launch a new search in CDD; input the terms “zebrafish” and “PH-like” in the search box of Conserved Domains Advanced Search Builder (http://www.ncbi.nlm.nih.gov/cdd/advanced); click the “Search” button; and zebrafish PH-like domain will be displayed at the top in the list of search results. Several links to selected types of data that are related to zebrafish PH-like domain are available just below each conserved domain models record on the search results page. Next, click on the thumbnail image or title for zebrafish PH-like domain model, and the summary page of the conserved protein domain model provides a brief summary, links to related resources in Entrez, etc. Click on the thumbnail image of zebrafish PH-like domain, and the 3D structure will be downloaded in CN3 format. On the summary page, the "Conserved features/sites" section provides the location and biological function in the domain family. Typically, *common conserved features* refer to catalytic residues, binding sites, or motifs. Click on the "Scroll to sequence alignment display" button at the bottom of the "Conserved features/sites" section, and multiple sequence alignments of proteins that contain the domain model will be displayed in the viewer. The sequence alignment can be switched to a new format using the pull-down list of "Format", "Row display", "Color bits", and "Type selection" just above the multiple sequence alignment. The selection in "Type selection" includes "The most diverse members" and "Top listed sequences". At last, click on the "reformat" button, and the viewer will be refreshed to display the new multiple alignment.

In order to provide insights into residue conservation and functional properties, the domain family hierarchy is provided in CDD, which presents a common ancestor, a common set of conserved residues, and a common general function. Further, click on the "Interactive display with CDTree" button just below the title "Sub-family Hierarchy"; select "Whole hierarchy" or any branch of the whole hierarchy; and the interactive display of domain hierarchy will be downloaded in CN4 format.

Popularly Used Tools in NCBI

BLAST

BLAST is based on the similarity of nucleotide or protein sequences (see Chapter 2). NCBI provides several BLAST methods (http://blast.ncbi.nlm.nih.gov/Blast.cgi), including those optimized for highly similar sequences (MegaBLAST), optimized for more dissimilar sequences (Discontiguous MegaBLAST), and optimized for somewhat similar sequences (BLASTN).

MegaBLAST (Morgulis *et al.*, 2008) is the default BLAST program selection provided by NCBI. It is designed for the rapid identification of novel sequences. MegaBLAST works best if the query-target identity is 95% or more. Discontiguous MegaBLAST (Baltimore *et al.*, 2015; Johnson *et al.*, 2008) uses noncontiguous word match and an initial seed that can align the sequences with mismatches. The method is fast and is intended for cross-species comparisons. BLASTN (Johnson *et al.*, 2008) is an earlier version of BLAST. It allows shorter-word-size queries, as short as seven bases. It can be applied to align sequences with low similarities, but the program itself is slow in speed.

NCBI provides basic BLAST tools (Table 26.1) for identification of sequence similarities. Besides finding sequences based on similarities, NCBI-BLAST also supports other forms of specialized BLAST tools for different aims (Table 26.2).

As the detailed uses of BLAST searches are covered in Chapter 2, here we focus on the usage of NCBI-BLAST—for instance, to identify the homologous nucleotide sequences of "*Ictalurus punctatus* beta-actin mRNA, partial cds" (Accession: AY555575). We need to conduct the nucleotide BLAST program using BLASTN (http://blast.ncbi.nlm.nih.gov/Blast.cgi?PROGRAM=blastn&PAGE_TYPE=BlastSearch&LINK_LOC=blasthomes). In the query box, we either enter the sequence in FASTA format or the accession number, or upload the sequence file in FASTA format. The default search option is to search the database "Nucleotide collection (nr/nt)" using MegaBLAST. In

Table 26.1 Basic NCBI-BLAST tools.

BLAST type	Description
BLASTN	BLAST nucleotide databases using a nucleotide query
BLASTP	BLAST protein databases using a protein query
BLASTX	BLAST protein databases using a translated nucleotide query
TBLASTN	BLAST translated nucleotide databases using a protein query
TBLASTX	BLAST translated nucleotide databases using a translated nucleotide query

Table 26.2 A list of all the specialized BLAST tools provided by NCBI.

BLAST type	Description
SmartBLAST	Generate fast protein BLAST results with graphical views
Primer-BLAST	BLAST method to design primers, ensuring full primer–target alignment, and to detect targets with mismatches to primers
MOLE-BLAST	BLAST multiple sequences together with their database neighbors
Conserved Domains	BLAST for conserved domains within a protein or coding nucleotide sequence
Conserved Domain Architecture Retrieval Tool	Find sequences with similar conserved domain architecture
GEO Nucleotide BLAST	BLAST sequences with their expression profiles
IGBLAST	Search immunoglobulin (IG) and T-cell receptor (TR) V domain sequences
VecScreen	Find nucleotide sequences that may originate from vectors; check sequences for vector contamination
bl2seq	Align two or more sequences using BLAST
PubChem BioAssay BLAST	Search protein or nucleotide targets in PubChem BioAssay
SRA Nucleotide BLAST	BLAST sequences in the SRA nucleotide database
Constraint-based Multiple Protein Alignment Tool (COBALT)	Align multiple protein sequences based on conserved domain and sequence similarities
Needleman–Wunsch Global Align Nucleotide Sequences	Align two nucleotide sequences based on the Needleman–Wunsch alignment algorithm
RefSeqGene Nucleotide BLAST	BBLAST RefSeqGene databases using a nucleotide query
Trace Archive Nucleotide BLAST	BLAST Trace Archive databases using a nucleotide query
Targeted Loci Nucleotide BLAST	BLAST targeted loci databases using a nucleotide query

the "Choose Search Set" section, the user can set the database and organism for the BLASTN search. In "Program Selection", the user can choose the BLAST search scheme from three options—the MegaBLAST, Discontiguous MegaBLAST, or BLASTN methods. The default setting is recommended. After conducting the BLASTN search, the output can be presented in several formats, such as HTML, XML, and plain text. HTML is the default and most popular output format for web page NCBI-BLAST. The results are presented in a graph with all the hits, and displayed as a table with similar sequences, sorted by scoring, along with query cover, E-values, identity, accession number, etc. (Johnson *et al.*, 2008; McEntyre & Ostell, 2002). From the resulting table, one can see that the sequence with the highest similarity is "*Ictalurus punctatus* beta-actin mRNA, partial cds" itself, with an identity of 100%. The second-most homologous sequence is the *Rhamdia quelen* (Rhamdia) beta-actin mRNA, complete cds, with an identity of 96%.

ORF Finder

ORF Finder (https://www.ncbi.nlm.nih.gov/orffinder/) is a simple and efficient web interface for detecting potential protein coding sequences or open reading frames (ORFs) from genomic DNA fragments (Rombel *et al.*, 2002). To use this program, the user needs to provide either an accession number or a nucleotide sequence in FASTA format. For example, we copy and paste the sequence of "*Ictalurus punctatus* beta-actin mRNA, partial cds" (Accession: AY555575) in FASTA format, or just type "AY555575" into the search box labeled "Enter Query Sequence", and then click the button "Submit". A graphical view of all the potential ORFs will be generated. The predicted ORFs are shown as colored boxes in the graphical view in ORF Finder. The colored box can be chosen and, once selected, the chosen ORF's sequence and features will appear in the columns under the graph. Users can choose to conduct SmartBLAST or conventional BLAST analysis of the predicted ORFs against either UniProtKB/Swiss-Prot database, Reference proteins or Non-redundant protein sequences databases to confirm the ORF predictions. This web interface facilitates analysis of nucleic acid sequences containing genes, and is quite helpful in preparing accurate sequences for submissions (Wheeler *et al.*, 2000).

Splign

Splign (https://www.ncbi.nlm.nih.gov/sutils/splign/splign.cgi) is a cDNA-to-genomic alignment program that relies heavily on the alignment algorithm developed by NCBI (Kapustin *et al.*, 2008). The basic goal of the Splign program is to identify the gene duplications, introns and splicing signals of a gene based on the alignment of cDNA to genomic DNA. For a large number of sequences, a stand-alone application for Linux 64 can be downloaded from ftp://ftp.ncbi.nlm.nih.gov/genomes/TOOLS/splign/linux-i64/, For small number of sequences, online job is available. For example, consider two sequences:

1) Sequence1, *Homo sapiens* amylase, alpha 2B (pancreatic) (AMY2B), mRNA (Accession NM_020978.4)
2) Sequence 2, *Homo sapiens* amylase, alpha gene cluster (AMY@) on chromosome 1 (Accession NG_004750.1)

The two sequences are from human chromosome 1. In the Splign user interface, for the cDNA sequence, we add the accession number of the sequence1: NM_020978.4; and, for the genomic sequence, we add the access numbers of sequence 2: NG_004750.1. There are several options available for different sequences alignment such as "lower quality query sequences", "reverse and complement the queries", "more partial alignments", and "using discontinuous megablast for cross-species analysis", the user can choose different modifications from case to case. Click the "Align" button, and the cDNA-genomic sequence alignment results will be displayed. The best alignment (model 1) shows sequence 1 spans from 29189bp to 54079bp on sequence 2 in 12 segments with exact matches. The results illustrate that sequence 1 *Homo sapiens* amylase, alpha 2B is transcribed from sequence 2 *Homo sapiens* amylase, alpha gene cluster (AMY@) on chromosome 1 in 12 exons.

Map Viewer

Map Viewer (http://www.ncbi.nlm.nih.gov/mapview/) allows the user to view and search for an organism's complete genome, display chromosome maps, and zoom into the sequence of interest in detail. Also, it is a powerful tool for exploring the genes on the genome. For example, the *FMR1* gene (fragile X mental retardation 1) is associated with various human diseases. To view the neighboring genome context of *FMR1* in the human genome, we enter "*FMR1*" in the search column, and choose "*Homo sapiens*" as the species. On clicking the "Go" button, the Map Viewer will present the search results in a new web page. These search results can be filtered by choosing from the options of "Gene", "RefSeq", and/or "UniGene" in the "Quick Filter" column on the right. If we are interested in certain sequence, the "Genes_seq" link can be accessed, which will lead to a graphic view of the *FMR1* gene on the human chromosome. The user can choose the display region of the maps in the "Region Shown" option on the left column of the web page. The user can also choose to view the sequence in detail by clicking the link that is associated with the gene name.

GEO DataSets Data Analysis Tools

The GEO DataSets homepage provides several data analysis tools, of which "Find genes" and "Compare 2 sets of samples" are widely used.

"Find genes" provides two methods for finding specific genes within the same dataset—"Find gene name or symbol" and "Find genes that are up/down for this condition(s)". In the "Find gene name or symbol" section, the name or symbol of the required gene will be located in the dataset, and the relevant profiles will also be located. To search for related genes of interest, for example, interleukin 7 (IL7), input the term "interleukin 7" in the "Find genes" search box within GDS4268; only four related GEO Profiles are found in GEO Datasets GDS4268. Based on gene annotation, only two genes, interleukin 7 receptor (IL7R) and interleukin 7 (IL7), are detected in *Homo sapiens* GEO Profiles. The "Find genes that are up/down for this condition(s)" section helps identify genes that are differentially expressed according to experimental subsets. For instance, select "Disease state" within GEO Datasets GDS4268, and 1353 potentially differentially expressed genes related to disease states in *Homo sapiens* are detected.

"Compare 2 sets of samples" is a tool for identifying genes that display marked differences in expression levels between two sets of samples. Three steps are involved:

Step 1: Select the statistics test to perform, and a significance level. Two-tailed *t*-test (A vs. B), one-tailed *t*-test (A > B) or (A < B), value, or rank means fold differences are available.

Step 2: Select the samples to put in Group A and Group B by clicking on accessions individually. Then click "OK".

Step 3: Query Group A vs. Group B. Each group is calculated for *t*-test scores or means fold differences. Genes that pass the user-selected criteria are displayed in GEO Profiles.

Conserved Domain Search Service (CD-Search)

The NCBI Conserved Domain Search (CD-Search) (http://www.ncbi.nlm.nih.gov/Structure/cdd/wrpsb.cgi) is a web interface for identifying structural and functional domains, annotation, and classification of domain models (Derbyshire *et al.*, 2015; Marchler-Bauer & Bryant, 2004). CD-Search uses reverse position-specific BLAST (RPS-BLAST) to identify a query sequence with a matching CDD model (Marchler-Bauer *et al.*, 2007). CD-Search enables searching for conserved domains in several databases, including CDD, Pfam, SMART, PRK, TIGRFAM, COG, and KOG (Marchler-Bauer & Bryant, 2004).

To search for conserved domains within a protein or coding nucleotide sequence, enter the protein or nucleotide query as accession or GI number, or sequence in FASTA format. Then, set the several available options, such as "expect value", "maximum number of hits", "Apply Low Complexity Filter", etc. For most cases, the default options work well. For instance, if gi|157830769 is entered in the search box and default options are set for CD-Search, clicking the "Submit" button will display a graphical summary to report the details of the list of domain hits, such as accessions, description, and E-values for each individual hit (Marchler-Bauer & Bryant, 2004). Click on any of the specific hits, superfamilies, or multi-domains colored boxes in the graphical summary. The viewer will be directly taken to the links of the conserved domains. To retrieve proteins that contain one or more of the domains present in the query sequence, click the "Search for similar domain architectures" button, and a list of domain architectures similar to your query will be displayed in the viewer using Conserved Domain Architecture Retrieval Tool (CDART). In order to modify the query to search against a different database or use advanced search options, click the "Refine search" button, and your query can be resubmitted on the CD-Search homepage.

CDART

CDART (http://www.ncbi.nlm.nih.gov/Structure/lexington/lexington.cgi) searches for functional domains in a given protein sequence and lists proteins with a similar annotated functional domain architecture across significant evolutionary distances (Geer *et al.*, 2002; NCBI, 2014).

To launch a new search in CDART, input a protein query as a sequence identifier (GI or accession number) or as sequence data, for example, "NP_001229488.1", in the search box (Geer *et al.*, 2002), and then click the "Submit" button. A list of similar conserved domain architecture will be displayed in a graphical summary. The list of conserved domain architectures can be narrowed down by using the "Filter by Taxonomy" and "Filter by Superfamilies" options from the pull-down list of "Filter your results" just above the graphical summary. Click on any of conserved domain architectures in the graphical summary, and the user will be taken directly to the links to the conserved domains architecture. The current search results can be downloaded in a comma-delimited table in XML format by clicking the "Download" icon below the graphical summary.

Submit Data to NCBI

Submission of Nucleotide Sequences Using BankIt

The most frequently used tool to submit nucleotide sequences to GenBank is a web-based tool named BankIt. The specific kinds of data that can be submitted via BankIt includes: sequences >200 nt long, protein-coding genes, ribosomal RNA genes, internal transcribed spacers (ITS), microsatellite markers, complete viral or phage genomes, complete mitochondrial genomes, and complete chloroplast or other plastid genomes.

BankIt is a web-based submission tool using which all submission steps can be completed online (Benson *et al.*, 2005). The user needs a computer with an Internet connection and a web browser such as Internet Explorer 9 or above, Firefox, Chrome, or Safari. Also, to submit a newly identified gene sequence to NCBI, an NCBI account is required to sign in to BankIt. The account can be created at: https://www.ncbi.nlm.nih.gov/account/. BankIt can be accessed at: http://www.ncbi.nlm.nih.gov/WebSub/?tool=genbank.

Once logged in, the first step of submission is to fill out the contact information of the submitter. Then, BankIt will require the reference information of the sequence, such as the sequence authors, and the publication status of the sequence. Second, the user needs to provide information on how the sequence is produced: the sequencing technology and assembly information (whether the sequence is assembled or not, and what program was used to assemble the sequences). Last, the sequence data can be input for submission. The user can define the release date of the data, and also needs to specify the sequence type, topology, and genomic completeness. All sequences must be entered in the FASTA format.

The user needs to clarify the features of the sequence: whether it is a "Coding Region (CDS) of Gene/mRNA", "RNA (rRNA, tRNA, non-coding RNA, misc_RNA, etc.)", "Repeat region (for sequence repeats, mobile elements, and satellites)", "Regulatory feature (promoter, TATA_signal, RBS, etc.)", or other sequences. In most cases, the features are added by completing input forms. If the submitted sequence contains an incomplete coding sequence, the user also needs to define the 5′ or 3′ end of the coding sequence that is partial. If the user does not provide any features, the submitted sequences will be considered as unverified data.

Before completing submission, the user needs to review all the submission information. BankIt will ask him or her to check the "Additional email addresses", "Resubmission", "Submission title", "Additional information", and "Updates of the gene sequence". Once the user clicks on "Finish submission", NCBI will send a confirmation email to confirm the submission.

NCBI staff will review the sequence and contact the submitter if anything needs to be modified. Once the sequence passes the review process, the user will receive an email that provides an accession number for the submitted sequence.

Submission of Short Reads to SRA

For short reads generated from next-generation sequencing platforms, the user can submit raw data to the SRA database. In general, the submission of data to SRA is involved in five steps:

1) Create a BioProject for your project
2) Create a BioSample submission for the biological samples
3) Prepare the sequence data files
4) Enter the metadata on the SRA web site: (a) create SRA submission; (b) create experiment(s) and link to BioProject and BioSample; and (c) create run(s)
5) Transfer data files to SRA

To demonstrate the submission process, we use a set of RNA-Seq raw reads sequenced from the gill tissue of a single 1-year-old male channel catfish as an example. The data file is named as "transcriptome.fastq".

Create a BioProject

After logging into the NCBI account to submit raw data, a BioProject needs to be created at https://submit.ncbi.nlm.nih.gov/subs/bioproject/. Clicking on the "New Submission" button will take the user to the submission web page. On this web page, the submitter needs to fill out his or her contact information such as email, address, and the institution. This is followed by the "Project Type" page; choose the "Project data type" to illustrate the primary goal of the study, and "Sample scope" to provide a general overview of the biological sample used for the study.

The next step is to provide target information for the sequencing data by entering the organism's name, strain/breed information, and description. The information on biological properties is optional, but can be provided for lesser-known organisms. The next step is to provide general information. In this section, the user needs to identify "When this submission should be released to the public", "Project title", "Public description", and "Relevance". For instance, the "Project title" can be "*Ictalurus punctatus* raw sequence reads"; the "Public description" would be "Channel catfish raw sequence reads"; and the "Relevance" can be "Agricultural". This is followed by the "BioSample" step, in which the user can provide the accession number of the existing BioSample. If there is no BioSample available, the user can skip this step and fill the blank after the BioSample has been created as described in the next section. The next step is the "Publications" web page, where the user needs to identify the publications that are related to this submission, if available. The final step is the review of the input information. The user can make any necessary changes by going back before clicking on the "Submit" button for submission.

Once submitted, NCBI staff will review the project. Once processed, a BioProject will be created, and an accession number will be provided through email.

Create a BioSample Submission

BioSample submission can be done at https://submit.ncbi.nlm.nih.gov/subs/biosample/. The first step is to identify the submitter's information. The basic steps for this are similar to the BioProject submission process. The user needs to fill out the general information section. This part is to identify when the submission should be released to the public, and to specify if he or she is submitting a single sample, or multiple. For this demonstration, we have chosen the "Release immediately following curation" and "Single BioSample" options.

The next step is to specify the sample type that describes the samples used for the study. In the next page, the attribute part is to specify the "Sample name", "Organism", "Sex", and "tissue". From the "Strain", "Isolate", "Breed", "Cultivar", and "Ecotype" options, the user must specify at least one. Also, in the "Age" and "Developmental stage" section, at least one choice is required from the user.

The next part is the "BioProject" section. The user needs to provide the project accession number to link the BioProject with the BioSample. In our case, we will select the BioProject we created in the last section.

After that is the "Comments" section, where the user needs to identify the title and comments, public description, and private comments to NCBI staff. The sample title is auto-generated, but the user can always modify it if necessary. The public description comments in this field will appear in the publicly released record. After all these steps, the web page will provide an overview of the BioSample, and the user can check it before submission by clicking on the "Submit" button. Again, once reviewed by NCBI staff, an accession number will be provided through email.

Prepare Sequence Data Files

The user can prepare his or her data files in several formats, including ".bam", ".fastq", ".qseq", or ".srf". Most often, the short reads generated from next-generation sequencers are in the ".fatstq" format.

Enter Metadata on SRA Web Site

Create SRA Submission Go to the SRA submission web page (http://www.ncbi.nlm.nih.gov/Traces/sra_sub/sub.cgi?view=submissions), and log in via the "NCBI PDA" link for NCBI primary data archive submitters.

To start an SRA submission, click on the "Create new submission" button. The system requires submission "Alias" and "Submission Comment" to be input. "Alias" should be something that is used internally to refer to the project and makes sense to the submitter. In the "Release date" column, the user can also set the date on which he or she wishes to release the data to the public. By clicking on the "Save" button, the submission will be created.

Create Experiment and Link to BioProject and BioSample An experiment is essential for the submission, which is linked to one BioProject and one BioSample. On the confirmation page of the SRA submission, click on the "New Experiments" button in the web page to create a new experiment. After that, a description of the experiment needs to be provided. The following are the descriptions of all the required options.

Platform—The sequencing platform used in the experiment
Alias—Similar to "Alias" in the previous section, used as a reference for the user and the archive
Title—A publicly viewable and formal title used to describe the experiment
BioProject Accession—Links this experiment to a BioProject
BioSample Accession—Links this experiment to a BioSample
Strategy—Sequencing strategy used in the experiment
Source—Type of genetic source material sequenced
Selection—Method of selection or enrichment used in the experiment
Layout—Configuration of the read layout
Nominal Size—Size of the insert for paired reads
Nominal Standard Deviation—Standard deviation of insert size (typically ~10% of *Nominal Size*)

By clicking "Save", a new experiment with an ID is created.

Create Run(s) *Runs* describe the files that belong to the previously created experiments. They specify the data files for a specific sample to be processed by SRA. Each experiment may contain more than one run. To establish a new run for the BioProject, click on the "New Run" button on the BioSample confirmation page. Then, add the alias as described earlier. For the "Run data file type" option, the user needs to choose a proper format such as ".bam", ".fastq", ".qseq", or ".srf", according to the format of the data being submitted. Next, the data file information is provided by entering the file name and the MD5 checksum. The file name is the full file name with extension, and the MD5 checksum is a 32-character alphanumeric string that is used to identify each file, and can be computed with the native command "md5" (macOS) or "md5sum" (Linux). For instance, to generate such an MD5 value for the example data file, the following command line can be run in the terminal window:

```
$ md5 transcriptome.fastq (macOS)
$ md5sum transcriptome.fastq (Linux)
```

Last, click on the "Save" button to create the new run.

Transfer Data Files to SRA

After providing the metadata, the information on BioProject and BioSample, and the information on the data files being submitted, the last step is to transfer the data files to the SRA. This can be achieved by using File Transfer Protocol (FTP). The FTP site can be accessed by command lines and various freeware FTP clients. For instance, the user can log into the following address using a popular open-source FTP client named FileZilla (available from https://filezilla-project.org/), and the appropriate login ID and password.

Address: ftp://ftp-private.ncbi.nlm.nih.gov
Login ID: sra
Password: as provided

After accessing the SRA location, the user can create a folder with the name of the project and upload the raw data files into the folder. After processing, the data files will be automatically linked to the metadata in the SRA database.

References

Barrett, T., Suzek, T.O., Troup, D.B. *et al.* (2005) NCBI GEO: mining millions of expression profiles—database and tools. *Nucleic Acids Research*, **33**, D562–D566.

Baltimore, B.D., Berg, P., Botchan, M. *et al.* (2015) A prudent path forward for genomic engineering and germline gene modification. *Science*, **348**, 36–38.

Benson, D.A., Karsch-Mizrachi, I., Lipman, D.J. *et al.* (2005) GenBank. *Nucleic Acids Research*, **33**, D34–D38.

Clarke, J.M. and Wentz, R. (2000) Pragmatic approach is effective in evidence based health care. *BMJ*, **320**, 954–955.

Day, I.N. (2010) dbSNP in the detail and copy number complexities. *Human Mutation*, **31**, 2–4.

Derbyshire, M.K., Gonzales, N.R., Lu, S. *et al.* (2015) Improving the consistency of domain annotation within the Conserved Domain Database. *Database*, **2015**, bav012.

Geer, L.Y., Domrachev, M., Lipman, D.J. and Bryant, S.H. (2002) CDART: protein homology by domain architecture. *Genome Research*, **12**, 1619–1623.

Geer, R.C. and Sayers, E.W. (2003) Entrez: making use of its power. *Briefings in Bioinformatics*, **4**, 179–184.

Johnson, M., Zaretskaya, I., Raytselis, Y. *et al.* (2008) NCBI BLAST: a better web interface. *Nucleic Acids Research*, **36**, W5–W9.

Kapustin, Y., Souvorov, A., Tatusova, T. *et al.* (2008) Splign: Algorithms for computing spliced alignments with identification of paralogs. *Biology Direct*, **3**, 1.

Kodama, Y., Shumway, M. and Leinonen, R. (2012) The Sequence Read Archive: explosive growth of sequencing data. *Nucleic Acids Research*, **40**, D54–D56.

Marchler-Bauer, A., Anderson, J.B., Derbyshire, M.K. *et al.* (2007) CDD: a conserved domain database for interactive domain family analysis. *Nucleic Acids Research*, **35**, D237–D240.

Marchler-Bauer, A. and Bryant, S.H. (2004) CD-Search: protein domain annotations on the fly. *Nucleic Acids Research*, **32**, W327–W331.

Marchler-Bauer, A., Lu, S., Anderson, J.B. *et al.* (2011) CDD: a Conserved Domain Database for the functional annotation of proteins. *Nucleic Acids Research*, **39**, D225–D229.

Marchler-Bauer, A., Zheng, C., Chitsaz, F. *et al.* (2012) CDD: conserved domains and protein three-dimensional structure. *Nucleic Acids Research*, **41**, D348–D352.

McEntyre, J. and Ostell, J. (2002) *The NCBI handbook*, National Library of Medicine (US), NCBI, Bethesda, MD.

McEntyre, J., Ostell, J., Kitts, A. and Sherry, S. (2011) *The Single Nucleotide Polymorphism database (dbSNP) of nucleotide sequence variation.*

Morgulis, A., Coulouris, G., Raytselis, Y. *et al.* (2008) Database indexing for production MegaBLAST searches. *Bioinformatics*, **24**, 1757–1764.

NCBI, RC. (2014) Database resources of the National Center for Biotechnology Information. *Nucleic Acids Research*, **42**, D7.

NCBI, RC. (2015) Database resources of the National Center for Biotechnology Information. *Nucleic Acids Research*, **43**, D6.

Pontius, J.U., Wagner, L. and Schuler, G.D. (2003) 21. UniGene: a unified view of the transcriptome, in *The NCBI handbook*, National Library of Medicine (US), NCBI, Bethesda, MD.

Rombel, I.T., Sykes, K.F., Rayner, S. and Johnston, S.A. (2002) ORF-FINDER: a vector for high-throughput gene identification. *Gene*, **282**, 33–41.

Romiti, M. and Cooper, P. (2005) *Entrez help*, National Library of Medicine (US), NCBI, Bethesda (MD); Saccone, S.F., Quan, J., Mehta, G. *et al.* (2011) New tools and methods for direct programmatic access to the dbSNP relational database. *Nucleic Acids Research*, **39**, D901–D907.

Sherry, S.T., Ward, M. and Sirotkin, K. (1999) dbSNP—database for single nucleotide polymorphisms and other classes of minor genetic variation. *Genome Research*, **9**, 677–679.

Sherry, S.T., Ward, M.-H., Kholodov, M. *et al.* (2001) dbSNP: the NCBI database of genetic variation. *Nucleic Acids Research*, **29**, 308–311.

Tatusova, T., Smith-White, B. and Ostell, J. (2007) A collection of plant-specific genomic data and resources at NCBI. *Plant Bioinformatics*, **406**, 61–87.

Wheeler, D.L., Chappey, C., Lash, A.E. *et al.* (2000) Database resources of the national center for biotechnology information. *Nucleic Acids Research*, **28**, 10–14.

Wheeler, D.L., Church, D.M., Edgar, R. *et al.* (2004) Database resources of the National Center for Biotechnology Information: update. *Nucleic Acids Research*, **32**, D35–D40.

27

Resources and Bioinformatics Tools in Ensembl

Yulin Jin, Suxu Tan, Jun Yao and Zhanjiang Liu

Introduction

The Ensembl project (http://www.ensembl.org/) produces comprehensive genome databases for chordate and key model organisms. The project makes the information freely available online, providing valuable resources for both the research community and industries. Since its start in 1999, the Ensembl project has gradually evolved as an accurate and reliable genomic interpretation system, providing the most up-to-date gene set with annotations, querying tools, and access methods for chordates and key model organisms (Cunningham *et al.*, 2015). The range of available data has expanded to include comparative genomics, genomic variation, and regulatory data. Meanwhile, the project provides a number of powerful and reusable tools for processing data in Ensembl as well as those from the user, such as BLAST and BioMart. In this chapter, we describe genome resources in Ensembl, and introduce a number of user-friendly tools for the analysis of genomic data, especially those useful for aquaculture genomics researchers.

Ensembl is a joint project between the European Bioinformatics Institute (EBI) and the Wellcome Trust Sanger Institute. The project was launched in 1999 in response to the completion of the Human Genome Project (Flicek *et al.*, 2010). Since then, it has developed and gradually evolved into a highly reliable resource for genome research communities. Its genome browser includes a number of species including human, mouse, zebrafish, and a total of over 80 species in various taxa (Table 27.1). The Ensembl is updated periodically, and, in some cases, such as for human genome, the earlier genome assembly has been kept as a stable assembly to allow researchers to continue their analysis as they did with the earlier assembly.

Ensembl provides a number of useful tool packages including Variant Effect Predictor, BLAST/BLAT, Assembly Converter, ID History Converter, and BioMart. They also offer many genomic resources such as genome browsers, single-nucleotide polymorphisms (SNPs) and other genomic variants, RNA-Seq datasets, comparative genomic analysis resources, etc. In this chapter, we will first introduce some Ensembl resources, and then discuss several tools useful for genomic analysis, especially for researchers working with aquaculture species.

Bioinformatics in Aquaculture: Principles and Methods, First Edition. Edited by Zhanjiang (John) Liu.

Table 27.1 Genomes available in Ensembl as of December 2015 (Version 83), along with data and source of the "gene build" and the assembly used. Species marked by the asterisk are those genomes in the process of being annotated.

Scientific name	Common name	Assembly	Gene build	Gene build date
Ailuropoda melanoleuca	Giant panda	ailMel1	Ensembl	December 2011
Anas platyrhynchos	Mallard (wild duck)	BGI_duck_1.0	Ensembl	February 2010
Anolis carolinensis	Carolina anole lizard	AnoCar2.0	Ensembl	February 2013
Astyanax mexicanus	Mexican tetra (blind cave fish)	AstMex102	Ensembl	December 2013
Bos taurus	Cattle	UMD3.1	Ensembl	September 2011
Caenorhabditis elegans	*C. elegans*	WBcel235	WormBase	October 2014
Callithrix jacchus	Common marmoset	C_jacchus3.2.1	Ensembl	February 2014
Canis lupus familiaris	Dog	CanFam3.1	Ensembl	July 2015
Cavia porcellus	Guinea pig	cavPor3	Ensembl	May 2010
*Ceratotherium simum simum**	White rhinoceros	CerSimSim1		
Chlorocebus sabaeus	Green monkey	ChlSab1.1	Ensembl	February 2015
Choloepus hoffmanni	Hoffmann's two-toed sloth	choHof1	Ensembl	May 2010
*Chrysemys picta bellii**	Western painted turtle	ChrPicBel3.0.1		
Ciona intestinalis	*C. intestinalis*	KH	Ensembl	March 2013
Ciona savignyi	Pacific transparent sea squirt	CSAV 2.0	Ensembl	April 2013
*Cricetulus griseus**	Chinese hamster	CriGri_1.0		
Danio rerio	Zebrafish	GRCz10	Ensembl	May 2015
Dasypus novemcinctus	Nine-banded armadillo	Dasnov3.0	Ensembl	December 2013
Dipodomys ordii	Ord's kangaroo rat	dipOrd1	Ensembl	May 2010
Drosophila melanogaster	Common fruit fly	BDGP6	FlyBase	September 2014
Echinops telfairi	Lesser hedgehog tenrec	TENREC	Ensembl	April 2013
Equus feruscaballus	Horse	Equ Cab 2	Ensembl	November 2012
Erinaceus europaeus	European hedgehog	eriEur1	Ensembl	April 2013
Felis catus	Cat	Felis_catus_6.2	Ensembl	January 2013
Ficedula albicollis	Collared flycatcher	FicAlb_1.4	Ensembl	September 2013
Gadus morhua	Atlantic cod	gadMor1	Ensembl	August 2011
Gallus gallus	Red junglefowl	Galgal4	Ensembl	December 2014

Table 27.1 (Continued)

Scientific name	Common name	Assembly	Gene build	Gene build date
Gasterosteus aculeatus	Three-spined stickleback	BROAD S1	Ensembl	May 2010
Gorilla gorilla gorilla	Western lowland gorilla	gorGor3.1	Ensembl	July 2011
*Heterocephalus glaber**	Naked mole-rat	HetGla_female_1.0		
Homo sapiens	Human	GRCh38.p3	Ensembl	July 2015
Ictidomys tridecemlineatus	Thirteen-lined ground squirrel	spetri2	Ensembl	May 2012
Latimeria chalumnae	West Indian Ocean coelacanth	LatCha1	Ensembl	November 2012
Lepisosteus oculatus	Spotted gar	LepOcu1	Ensembl	December 2013
Loxodonta africana	African bush elephant	Loxafr3.0	Ensembl	December 2011
*Macaca fascicularis**	Crab-eating macaque	MacFas5.0		
Macaca mulatta	Rhesus macaque	MMUL 1.0	Ensembl	May 2010
Macropus eugenii	Tammar wallaby	Meug_1.0	Ensembl	May 2010
Meleagris gallopavo	Wild turkey	Turkey_2.01	Ensembl	April 2014
*Melopsittacus undulatus**	Budgerigar	MelUnd6.3		
Microcebus murinus	Gray mouse lemur	micMur1	Ensembl	May 2010
*Microtus ochrogaster**	Prairie vole	MicOch1.0		
Monodelphis domestica	Gray short-tailed opossum	monDom5	Ensembl	August 2012
Mus musculus	House mouse	GRCm38.p4	Ensembl	September 2015
Mustela putorius furo	Ferret	MusPutFur1.0	Ensembl	August 2012
Myotis lucifugus	Little brown bat	Myoluc2.0	Ensembl	June 2011
Nomascus leucogenys	Northern white-cheeked gibbon	Nleu1.0	Ensembl	October 2012
Ochotona princeps	American pika	OchPri2.0	Ensembl	April 2013
Oreochromis niloticus	Nile tilapia	Orenil1.0	Ensembl	March 2012
Ornithorhynchus anatinus	Platypus	OANA5	Ensembl	August 2012
*Orycteropus afer afer**	Aardvark	OryAfe1.0		
Oryctolagus cuniculus	European rabbit	OryCun2.0	Ensembl	June 2013

(continued)

Table 27.1 (Continued)

Scientific name	Common name	Assembly	Gene build	Gene build date
Oryzias latipes	Japanese rice fish (medaka)	HdrR	Ensembl	April 2013
Otolemur garnettii	Northern greater galago	OtoGar3	Ensembl	December 2011
Ovis aries	Sheep	Oar_v3.1	Ensembl	May 2015
Pan troglodytes	Common chimpanzee	CHIMP2.1.4	Ensembl	November 2012
Papio anubis	Olive baboon	PapAnu2.0	Ensembl	July 2014
*Papio hamadryas**	Hamadryas baboon	Pham		
Pelodiscus sinensis	Chinese softshell turtle	PelSin_1.0	Ensembl	February 2014
Petromyzon marinus	Sea lamprey	Pmarinus_7.0	Ensembl	April 2013
*Physeter macrocephalus**	Sperm whale	PhyMac_2.0.2		
Poecilia formosa	Amazon molly	Poecilia_ formosa-5.1.2	Ensembl	August 2014
Pongo abelii	Sumatran orangutan	PPYG2	Ensembl	August 2012
Procavia capensis	Rock hyrax	proCap1	Ensembl	April 2013
Pteropus vampyrus	Large flying fox	pteVam1	Ensembl	May 2010
Rattus norvegicus	Brown rat	Rnor_6.0	Ensembl	June 2015
Saccharomyces cerevisiae	Baker's yeast	R64-1-1	SGD	December 2011
*Saimiri boliviensis**	Black-capped squirrel monkey	SaiBol1.0		
Sarcophilus harrisii	Tasmanian devil	Devil_ref v7.0	Ensembl	December 2011
Sorex araneus	Common shrew	sorAra1	Ensembl	May 2010
Sus scrofa	Wild boar	Sscrofa10.2	Ensembl	February 2014
Taeniopygia guttata	Zebra finch	taeGut3.2.4	Ensembl	May 2012
Takifugu rubripes	Japanese puffer	FUGU 4.0	Ensembl	May 2010
Tarsius syrichta	Philippine tarsier	tarSyr1	Ensembl	April 2013
Tetraodon nigroviridis	Green spotted puffer	TETRAODON 8.0	Ensembl	May 2010
Tupaia belangeri	Northern treeshrew	tupBel1	Ensembl	April 2013
Tursiops truncatus	Common bottlenose dolphin	turTru1	Ensembl	April 2013
Vicugna pacos	Alpaca	vicPac1	Ensembl	April 2013
Xenopus tropicalis	Western clawed frog	JGI 4.2	Ensembl	April 2013
Xiphophorus maculatus	Southern platyfish	Xipmac4.4.2	Ensembl	April 2013

Ensembl Resources

Genome Browsers

Ensembl (http://useast.ensembl.org/index.html) displays genomic resource information, and provides tools useful for genome research (Spudich & Fernández-Suárez, 2010). Its genome browser allows intuitive examination of most relevant genome information. As of December 2015, 80 species are included in Ensembl, of which information for 69 species is mostly complete. Of all the species included, nine have been included for a long time, and their information could be more comprehensive than that of others. These nine species are: human, mouse, rat, zebrafish, pufferfish, fruit fly, mosquito, and two nematode worms (*Caenorhabditis elegans* and *C. briggsae*). The user can get detailed information and statistics for genome assembly, gene annotation, comparative genomics, variation, and regulation through various entries on the genome web page of each species. The Pre-Ensembl site (http://pre.ensembl.org/index.html) lists partially supported species whose genomes are in the process of being annotated. Resources for metazoan genomes are available from the EnsemblMetazoa database; plant and fungal genomes can be found at the EnsemblPlants and EnsemblFungi databases, respectively; and the unicellular eukaryotic and prokaryotic genomes are available at the EnsemblProtists and EnsemblBacteria databases, respectively. Besides these databases, the user can make use of these data from Ensembl's web sites, public MySQL databases, the Ensembl application programming interface (API), and FTP site (Cunningham *et al.*, 2015).

Gene Annotation

More and more genomes are being sequenced. However, the biological information contained within the ATGC sequences is not easily accessible without genome annotation. Manual annotation, determined by experts on a case-by-case basis, is accurate but time-consuming and labor-intensive. Automatic annotation can rapidly be performed using software pipelines. Ensembl utilizes both automatic and manual annotation to provide accurate annotation information. The Ensembl automatic gene annotation system, named *Ensembl genebuild*, is evidence-based, using protein, cDNA, EST, RNA-Seq, and other relevant data sources.

The main steps in a standard gene build process are: align available species-specific and orthologous transcripts plus their protein sequences to the genome assembly to build a gene set and transcripts. Manual annotation for several species is provided by Havana (Wilming *et al.*, 2008). In every Ensembl release, an automatically produced Ensembl gene set is merged with Havana manual annotation to produce the GENCODE gene set for human and made available in Ensembl (Cunningham *et al.*, 2015). Zebrafish, mouse, as well as pig gene sets are also regularly updated to provide combined manual and automatic annotations (Flicek *et al.*, 2013). Ensembl also continues to collaborate with the Consensus Coding Sequence (CCDS) project (Pruitt *et al.*, 2009) to produce consistent annotation among genome resources, with the aim of creating high-quality annotation for human and mouse protein coding regions (Cunningham *et al.*, 2015).

Ensembl makes continuous and considerable progress on annotation for chordates and model organisms. Approximately five times a year, Ensembl updates the annotation of already supported species, and includes newly supported species when significant new data become available. Recent updates in Ensembl (Cunningham *et al.*, 2015; Flicek *et al.*, 2012, 2013) have focused on two aspects: (1) the inclusion of RNA-Seq data in gene annotation, and (2) the newly released human assembly, GRCh38. In recent years, RNA-Seq data has been introduced to update existing gene annotation, or to be one part of the gene-build process (Flicek *et al.*, 2013). Ensembl has built RNA-Seq-based transcript models for many species. These transcript models can be used to add new genes or transcripts into the gene set, and to provide the accurate structure of protein-coding models. In addition to the use of RNA-Seq datasets, the new human genome assembly, GRCh38, contains major improvements, particularly in annotations. For instance, some genes that were previously annotated as non-coding in GRCh37.p13 assembly are now protein-coding in the new assembly.

For protein-coding gene annotation, Ensembl also considers it on low-coverage genomes (2x genomes). Low-coverage genomes exhibit various problems such as missing sequences, incorrect sequences, and fragmentation. Standard gene build pipeline application is not suited for low-coverage genomes because the standard Ensembl genebuild pipeline is based on the alignment of complete proteins and transcripts to the genome sequence. Therefore, Ensembl has developed a new pipeline that relies on a whole genome alignment (WGA) to an annotated reference genome.

As other algorithms are developed, the range of gene annotation is expanding. In addition to protein-coding gene annotation, Ensembl also automatically annotates pseudogenes and some RNA genes such as long non-coding RNAs (Birney *et al.*, 2004). All the improved annotations are available from the Ensembl databases and can be accessed via the Ensembl web site, BioMart, and Ensembl Perl API. For annotation of any specific gene, the gene name and species are selected; for example, for annotation of the human RHOA gene, in Ensembl home page, enter "RHOA" in the search box, choose "Human" as species, and then click "Go". All the available annotation information is displayed in the output.

Variation Annotation

Because of the exponential growth of genome sequencing and re-sequencing projects, a large amount of variation data is simultaneously generated. The variation information has not only expanded our understanding of the genomic diversity of species (Chen *et al.*, 2010), but also provided invaluable resources for disease diagnosis and treatment, and for the analysis of genotype–phenotype associations. Ensembl has developed browsers and other methods to archive and display variation data.

Variation data in Ensembl is mostly imported from other sources. Ensembl imports two levels of variation data: (1) nucleotide level, such as SNPs, short nucleotide insertions and/or deletions (InDels), and short tandem repeats; (2) structural level, such as copy number variations (CNVs), large-scale inversion, and translocation. Where available, other associated data (allele frequencies, genotypes, phenotypes, etc.) are also imported from a variety of external sources. The majority of sequence variants in Ensembl (SNPs, insertions, and deletions) are imported from NCBI's SNP database (dbSNP) (Foelo & Sherry, 2007). For human SNPs particularly, Ensembl updates the variation data in every release, with the aim to keep up with dbSNP. Structural variations are from the Database of Genomic Variants archive (DGVa) (http://www.ebi.ac.uk/dgva), and additional structural variants for some species are generated from uniform processing and variation discovery using sequencing reads (Chen *et al.*, 2010). Phenotype data in Ensembl are currently imported from two sources: the National Human Genome Research Institute's (NHGRI) Office of Population Genomics, and the European Genome–phenome Archive (EGA) database.

Once variation data is stored in the Ensembl database, a series of processing steps is performed. Ensembl filters doubtful data using quality control processes, categorize different classes of variant data, and then predict their consequences of variants. Ensembl provides calculations for four different sets of variant consequences:

1) *Variant consequences in transcripts.* Ensembl identifies stored transcripts that overlap imported variants. Then, the effects that each allele of the variant may have on the transcript are predicted based on a rule-based approach.
2) *Protein function prediction.* For variants that may lead to an amino acid substitution, Ensembl provides PolyPhen and SIFT predictions for the effect of this substitution on protein function.
3) *Linkage disequilibrium information.* For each population with a large sample size, variants are arranged by their positions, and pairwise LD values (r^2) are calculated. Linkage disequilibrium with r^2 values more than 0.05 are retained for the calculation of tagged SNPs (Chen *et al.*, 2010).
4) *Calculation of tagged SNPs.* A tagged SNP is a representative SNP in a region of the genome with high linkage disequilibrium. In Ensembl, it is produced for each of the HapMap and Perlegen populations and used for haplotype analysis. The representative SNPs are very useful for genetic analysis, reducing computational demands without losing any genetic analysis powers.

All the variation datasets are accessible online using the Ensembl genome browser visualization tools and the Ensembl BioMart tool, or through the Ensembl Perl API. Ensembl provides displays at four levels: variation, gene, transcript, and genomic location.

Gene-based Views

To show variation data based on the gene-based view, let us use the zebrafish gene *dyrk1b* as an example. *dyrk1b* is crucial for endoderm formation (Mazmanian *et al.*, 2010). On the Ensembl home page, select "Zebrafish" as the species, type gene name "*dyrk1b*" in the search box, and click on "Go". By clicking on the most matched result, the user will be guided to the *dyrk1b* gene summary page. First, the user can view variation within the gene sequences by following these steps. Click the "Sequence" link at the left of the page. The gene sequence will show up, with exons highlighted in red. In order to turn on the variation, click on "Configure this page", and change the "Show variation" option to "Yes and show links". The page will be reloaded, and the user can now see the short sequence variations (SNPs and small insertions and deletions) as highlighted along the sequence. The color keys at the top allow the user to know the variation consequence such as frameshift, missense, and synonymous. The user can click any highlighted variation on the sequence or click any links other than the variation tab for more information about specific SNPs or InDels. Another way to view short sequence variants is to click the "Variation table" at the left of the page. Here, the user can see a summary of all the variations in this gene containing variation ID, chromosome coordinates, alleles, sources, consequences on the transcript, etc. Most variation IDs are imported from dbSNP, and begin with "rs"—for example, "rs499678744". The "Variation image" link allows the user to view variation information graphically. The protein domains from different sources are mapped to the genome and combined with variants for the convenient estimation of variant effect on protein function. In order to view large structural variants, the user can click the "Structural variation" link at the left. This link displays both a table and an image, which show detailed information containing chromosome positions, genomic sizes, classes, sources, and study descriptions.

Transcript-based Views

The transcript tab can be accessed by clicking on a transcript ID of a gene. In this sample, let us choose one transcript (ID: ENSDART00000038748) of the zebrafish *dyrk1b* gene. Different from variation data in the gene tab, the transcript tab provides a focused view of one splice variant. First, as we did with the gene tab, let us look at the variation data on the sequence. The user can click on the "Exon" link at the left side to achieve this. This view shows exons (blue), introns (grey), untranslated regions (UTRs, colored in red), and flanking sequences (purple). To show variants, the user can click on the "Configure this page" link. Variants are highlighted according to consequence types. To view variations on cDNA sequence, the user can click on the "cDNA" link. UTR is highlighted in yellow, and amino acid sequences are displayed under protein-coding sequences. Similarly, variants are highlighted with different colors, based on the consequence type.

Location-based Views

To explore the variation in region, the user can click on the location tab after searching for a gene on the Ensembl home page. The user can also get to the location-based view by clicking on the location link on the gene tab or variation tab. Let us display the location-based view by clicking on the location link on any transcript tab of the *dyrk1b* gene. "Region in detail" shows the gene list with the *dyrk1b* gene in the center. At the bottom, there is a more detailed view. The user can change the region by using the location box or by searching for another gene in the gene box below. This detailed view is highly customizable, and the user can add or delete variation tracks by configuring this page.

Variation-based View

Searching for a variant brings us to the variation tab containing specific information about one SNP. Let us search for "rs499678744", which was mentioned earlier. In the variation tab, clicking on the available icons or pictures below the page can lead the user to specific information about the variation. In this example, information including genomic content, gene regulation, population genetics, and in the phylogenetic context can be explored. The user can also discover related information by clicking on the links at the left of the page. The summary on this page also presents basic information, including source, alleles, location, consequence, and synonyms.

Currently, variation databases in Ensembl contain information for 22 species (http://www.ensembl.org/info/genome/variation/sources_documentation.html#danio_rerio). Huge amounts of new biological variation data are continuously being added. With the comprehensive variation resources and clear visualization interface in Ensembl, researchers can perform a variety of studies.

Comparative Genomics

The Ensembl Compara database provides extensive cross-species resources and analyses focusing on: (1) homologous gene pairs between genomes, and (2) the alignments of long-range multiple sequences. Data can be easily accessed through the comparative genomics pages on the browser, or by using BioMart and Compara Perl API. By utilizing the gene-based resources such as genomic alignments, phylogenetic trees, Ensembl families, orthologs, and paralogs, Ensembl can automatically integrate the gene annotations produced by the Ensembl genebuild team across all the available species. Sequence-based resources allow whole genome alignments either between or among species. Additional analyses such as "Ancestral sequences", "Age of base", "Conservation scores and constrained elements", and "Syntenies" can be applied on the whole genome alignments. However, such analyses are only available for species whose genomes are hosted by Ensembl.

Phylogenetic trees provide most helpful information to infer evolutionary history and relationships among various biological species. In Ensembl, two kinds of phylogenetic trees (protein trees and ncRNA trees) across the whole set of protein-coding genes and non-coding RNA genes can be computed with two different pipelines. Both the results will be displayed in a set of trees or branching diagrams. Homology is determined by using all the genes in all the species in Ensembl. Protein trees are constructed using the longest reliable protein as a representative protein for each gene in Ensembl. The main steps to build a gene tree are: using "hcluster_sg", which is based on NCBI BLAST+ e-values to cluster proteins; aligning each cluster of proteins by using "M-Coffee" or "Mafft"; and, finally, using "TreeBeST" to produce a gene tree from each multiple alignment, reconciling it with the species tree to call duplication events (Vilella *et al.*, 2009).

Non-coding RNA genes are also valuable resources for phylogenetic studies, since they are abundant in the genome, and can produce functional RNAs rather than protein-coding RNAs (Eddy, 2001). They have specific secondary structures, with pairs of residues being matched to form loops and other structures. Thus, a similar strategy as for protein trees is utilized for the analysis of orthologies and paralogies for short ncRNA genes, but with specific modifications.

Various statistics are displayed on a gene tree, such as the number of homologs, the size of the gene tree (split by their root taxon), gene-tree nodes, and the inference of speciation/duplication events. On the other hand, from both sets of gene trees, the gene gain/loss tree can be generated and used to summarize the phylogenetic history of an Ensembl gene family by showing gene gain events (expansions) and gene loss events (contractions) over evolutionary time.

Homologs refer to proteins that share a common ancestor. They are extracted and separated into orthologs (across different species) and paralogs (within a species). *Orthologs* are genes that have common ancestors and have split due to speciation events, while *paralogs* are defined as genes for which the most common ancestors have undergone a duplication event within the species. Homologs are inferred from gene trees, which are determined by using all species in that particular database, such as all the (mostly) chordates in Ensembl, all the metazoa in Ensembl Metazoa, all the protists in Ensembl Protists, all the fungi in Ensembl Fungi, all the plants in Ensembl plants, or all the species in the Pan-Compara set for Pan-Compara orthologs in Ensembl Genomes. Orthologs and paralogs determined with the trees are listed in the tables of their corresponding gene pages.

Another resource for a target gene is Ensembl Families, which provides a platform to explore orthologs and closely related homologs across a range of animal species. The pre-calculated multiple sequence alignments of Ensembl or all family members (including UniProtKB) for each cluster can be viewed and exported. Based on the high sequence similarity of proteins, the families are determined by the clustering of all Ensembl proteins, along with all the metazoan sequences from UniProtKB, SwissProt, and SPTREMBl. Generally, the pipeline builds similarity clusters by running the Markov Cluster Algorithm based on the set of all Ensembl proteins (potentially several per gene) and the set proteins from UniProt. Clusters are then aligned using the MAFFT program (Enright, Van Dongen & Ouzounis, 2002).

Sequence-based resources within the Ensembl Compara database provide information on pre-calculated whole genome alignments either between or among vertebrate species. Because of the differences in sequencing depth and completeness, all genome assemblies of species are grouped into two tiers for differential alignment processing. The Enredo-Pecan-Ortheus (EPO) pipeline (Enright *et al.*, 2002; Flicek *et al.*, 2013) is used to generate a progressive multiple sequence alignment from high-quality and high-coverage genomes of 13 species. The low-coverage genomes of an additional 23 species are first inserted into the previous alignment, and the genome sequences are aligned pairwise by mapping them onto the human assembly with BlastZ-net (Flicek *et al.*, 2013). The clade-specific multiple sequence alignments for primates, birds, fish, and amniotes can also be generated. Particularly, the multiple alignments for fish have been extended to 11 species, of which five teleost fishes (Zebrafish, Medaka, Tetraodon, Stickleback, and Spotted gar), whose genome assemblies were generated through the EPO pipeline, have available genomes with relatively high quality and coverage.

Here, we will use a zebrafish gene, *dmd*, as an example and explore its datasets, and analyze the graphics features of the comparative genomics tools provided by Ensembl.

Genomic Alignments

Enter the gene name of interest—here "*dmd*"—and input "zebrafish" as the species in the search box on the Ensembl home page. Click on "Go", and the results will show all the related categories of this gene. Usually, the first hit is the gene of interest; click on it to go to this gene's summary page. On the left panel of this page, there is a "comparative genomics" tab, including categories of genomic alignments, gene trees, gene gain/loss trees, orthologs, paralogs, and the Ensembl protein families for this gene. In order to see how this gene compares to the same locus in other species, click on "Genomic alignments". Then, select the type of alignment to display from the "Alignment" search box. There are multiple pre-calculated whole genomic alignments available for certain species. If you are in a zebrafish gene web page, alignment options such as "5 teleost fish EPO" and "11 fish EPO LOW COVERAGE" for fish alignments are available. If you search for a human gene, the first option is the multiple alignments for certain taxonomic groups—for example, the primate, eutherian mammals, or vertebrates. "Pecan" and "EPO" refer to the computation analyses used to determine alignments. Further down the list, there are whole genome alignments done on a pairwise basis using BlastZ-net for comparison on the nucleotide level or translated BLAT on the amino acid level to compare two genomes. The "?" icon provides more information about genomic alignments. Choose the alignment of "5 teleost fish EPO" in this case, and then click "Go". Three alignment blocks are displayed; select a block from the alignment column to view the result. In this way, the regions where alignments can be generated from multiple species are listed, and information such as the scaffold, chromosome, base pair range, and strand of DNA are indicated. For example, "Zebrafish chromosome:GRCz10:9:24213558:24215135:1" indicates that the zebrafish alignment is on chromosome 10 on the forward strand, as indicated by 1 (reverse strand is indicated by −1). Aligned sequences are shown under these coordinates. By default, the exons are highlighted in red, and dashes indicate gaps in the alignments. Multiple customization options such as conserved residues, variations, codons, and line numbering can be managed by clicking on the "Configure this page" link on the left display panel. The alignments can be exported as a file by clicking on the "Download alignments" link on this page.

Phylogenetic Trees

Gene Tree

If you click on "Gene tree" on the left display panel of the *dmd* gene summary page, a branching diagram for this gene will be displayed on the left side of the viewer, with the protein alignments on the right side. This provides an intuitive display of phylogenetic relationships of known *dmd* genes. In this gene tree, the zebrafish *dmd* gene is marked in red, and the within-species paralogs are shown in blue if the option to view paralogs is selected (below the tree diagram). Collapsed nodes are fan-shaped: the blue notes indicate speciation events, and the red nodes indicate duplication events giving rise to orthologs and paralogs. Clicking on each note opens a pop-up box with specific information for each event, providing links to expand a collapsed set of branches into a full tree and links to view alignments or sub-trees using Jalview. Light blue nodes represent unsupported duplication, which can be considered as a speciation event with a confidence core of 0. In the protein alignments view: the light green bars correspond to matches in the individual protein; the dark green bars are consensus sequences corresponding to the collapse nodes of the tree; and the white spaces are alignment gaps.

Gene Gain/Loss Tree

If you click on the "Gene gain/loss tree" on the left panel, the gene gain/loss tree for the *dmd* gene will be displayed, which is helpful for understanding the phylogenetic history of this gene family by showing gene gain events (expansions) and gene loss events (contractions) over evolutionary time. The significance level for the gene gain/loss events of this gene family is presented by the p-value on the top of the tree. The user can click the tree nodes or image icons to interact with the tree. The species for the queried gene is labeled in red, and the other species that lost this gene in this tree are labeled in gray. A red branch on the tree indicates a significant expansion of the gene, and a green branch represents a significant contraction at that point in its history. A grey branch indicates that there was no significant change. The numbers at each node reflects the number of members in the ancestral species. Clicking on any node opens a pop-up box with specific information on the expansion or contraction event.

Orthologs and Paralogs

If you click on "Orthologs" on the left panel of the gene summary page, several tables will be displayed. The top table is the summary of orthologs of this gene, which provides the number of species for each ortholog type. Species are grouped by clades in this table, such as "Primate", "Rodents", and "Fish". Ortholog types are assigned as one-to-one orthologs, many-to-many orthologs, and multiple orthologs by comparing two species. Click on "Show details" to list only the orthologs for species in one clade underneath the top table. In this case, choose "Fish" to explore the selected orthologs. The list of orthologs contains the species, ortholog type, dN/dS value (if calculated), Ensembl identifier and name, and links to view the location of the ortholog (either on a chromosome or scaffold) in the browser. The "Compare" column in this table provides links to protein and cDNA alignments, as well as links to a graphical display of the alignment ("Region Comparison") and the gene tree in the image. In addition, the "Target %ID" column gives the numbers that indicate the percentage of identical amino acids in the ortholog as compared with the gene of interest. The numbers in the "Query %ID" column refer to the identity at the amino acid level of the gene of interest when compared with the ortholog. For example, if you are searching for a gene in zebrafish, and looking for its homolog in another species such as tilapia, the "Query % ID" refers to the percentage of the query sequence (zebrafish) that matches to the homolog (the tilapia protein). "Target % ID" refers to the percentage of the target sequence (tilapia) that matches with the query sequence (zebrafish). Further down the list, a table displays the species without orthologs, which are not shown in the table of selected orthologs.

Next, click on "Paralogs" on the left panel, and a table will be displayed that describes the paralogs of this gene. It includes columns of "Ancestral taxonomy", which indicates the taxonomic level of the ancestor duplication node, the Ensembl identifier and gene name, the location of the paralog, the "Target %ID", and the "Query %ID". The "Compare" column also provides links to view the graphical display of the alignment ("Region Comparison"), and links for protein and cDNA alignments of the paralogs.

Ensembl Families

Click on "Ensembl protein families" on the left panel of the *dmd* gene summary page. A summary table will be displayed that lists the contents of protein families, with details such as family ID, consensus annotation, and other proteins in this family. JalView is also available for cross-species alignments. Clicking on the family ID opens links that helps explore both UniProt and Ensembl proteins in the family, which allows the user to compare data between the two datasets.

Whole Genome Alignments

In addition to the constrained exploration of comparative genomics for one gene, whole genome alignments and synteny can also be explored in the zoomable region. It includes pairwise sequence alignments between two species, and multi-species alignments using genomes of more than two species. Go to the "Location" tab of this gene, and the genomic region of the *dmd* gene will be displayed in the "Region in detail" view. On the top is the region of interest on chromosome 1 (of zebrafish), with the target gene (*dmd*) indicated by a red box. Below is an image centered on this gene, with other genes indicated in the ~250 KB region, and the third image is a zoomable region. Comparative genomics tracks can be turned on in this view by clicking on "Configure this page" in the left panel. Choose the appropriate options under the "Comparative genomics" menu, such as selecting constrained elements and conservation score for "11 fish EPO_LOW_COVERAGE". In this way, the view is reloaded, and the pink histograms indicate the GERP scores, which represent how conserved the area is between the 11 fishes. The pinkish brown bars can be considered as a summary of the GERP scores. The highlighted regions are high GERP scores, which are known as "constrained elements".

There are other views in the location tab dedicated to comparative genomics—to view the alignments as text or as image; to view regions in the browser; or by synteny, to view the syntenic regions between two species' genomes.

1) First, click on "Alignments (image)" on the left side of the "Location" page. Then, choose the type of alignment to view the multi-species alignments across more than two species, or a pairwise alignment between two species—for example, select the "5 teleost fish EPO". At the top is the zebrafish genome region, and the other species are shown underneath. In this image, the vertical peach-colored stripes show alignments, and the white stripes are gaps in the alignment; the blue horizontal bars represent contigs. Contigs are the blocks of sequences that make up the whole genome assembly. If a bar is filled, the forward strand of the chromosome was aligned, whereas hollow bars represent alignments using the reverse strand.
2) The second option to view the text alignments is by clicking on "Alignments (text)". It is the same display as the genomic alignment option from the gene tab, but not confined to one gene, and the user can look into the genomic regions. The "5 teleost fish EPO" option is already selected. Again, the chromosomal regions for each species are listed above, and alignments are listed below, with exons in red. Click on "Configure this page"; there are customizable view options to add or change the display—for example, select "Show variations and links" to view any variants in this region for any species.
3) The third option is to click on "Region Comparison", and similar graphics as the "Region in detail" view will be displayed. In order to compare against other species, the species needs to be selected or removed by using the "Select species" button on the left side. Multiple species can be added to the view—for instance, select fugu to compare with the zebrafish genome region. The zoomed-out view of the region on the top shows the *dmd* gene in zebrafish and fugu, which is indicated by a red box. The view below shows how the regions are aligned. The pink bars show alignments, with the green links connecting them between two species. The white spaces are unaligned sequences. Gaps in the blue bar indicate that regions have not been sequenced yet. Fortunately, there are no gaps in the assembly of zebrafish or fugu genomes in this region. Features can also be added in this view by changing options in the "Configure this page" link, such as adding CpG islands in this region.
4) The last option is the "Synteny" view, which is similar to "Region Comparison", but it is large in scale, and has less detailed information. Click on "Synteny" on the left panel, and the view will show a graphic display of the synteny. Syntenic regions are displayed as 100 KB or more of conserved sequences across two species in Ensembl. Species and chromosomes of interest can be changed on the right side. In this case,

zebrafish chromosome 1 is shown in the center, and the gene of interest is marked in red. Chromosomes of the second species with syntenic blocks are scattered alongside. Colored blocks indicate synteny on the chromosomes of the second species, which are connected by lines. The table below these blocks is the gene list, from which the upstream gene, downstream gene, and center of the gene of interest can be viewed.

Ensembl Regulation

Ensembl Regulation contains abundant resources to provide the most up-to-date and comprehensive mechanisms of gene regulation in human and mouse cells (Flicek *et al.*, 2012). It particularly focuses on transcriptional and post-transcriptional mechanisms. This annotation work provides a solid framework for ongoing epigenomic research, with continuous datasets and methodology improvement (Zerbino *et al.*, 2015). Currently, the data covers 18 human cell types and five mouse cell types (Cunningham *et al.*, 2015). A process known as Ensembl Regulatory Build annotates the putative regulatory regions from the public experimental data of several large-scale projects (e.g., ENCODE, BLUEPRINT, Roadmap Epigenomics, and HipSci), and associates these regions with regulatory functions (Zerbino *et al.*, 2015).

Regulatory features in Ensembl refer to regions in the human and mouse genomes that could be involved in gene transcription or post-transcription regulation. The Ensembl Regulation resources help the user characterize the dynamics of regulatory features across genome and cell types, based on various experimental techniques. Regions of open chromatin can be mapped using nuclease digestion by DNaseI, coupled with high-throughput sequencing (DNase-seq) (Song & Crawford, 2010) or formaldehyde-assisted isolation of regulatory elements (FAIRE) (Giresi *et al.*, 2007). High-throughput chromatin immunoprecipitation sequencing (ChIP-Seq) (Barski *et al.*, 2007) and microarrays (ChIP-chip) (Ren *et al.*, 2000) are utilized to detect a range of histone modifications, as well as transcription factor binding regions. Methylation analysis is another aspect of epigenetic regulation. For instance, methylation was detected with 47 reduced representations in bisulfite sequencing (RRBS) assays and whole genome bisulfite sequencing datasets in humans (Cunningham *et al.*, 2015). In addition, the Ensembl Regulation resources store external annotations, such as the Diana TarBase microRNA target predictions and VISTA enhancers for human and mouse.

From these types of datasets, Ensembl Regulatory Build computes MultiCell features and assigns functions to each regulatory feature by its biochemical activity, independently of cell type. Each feature is annotated into one of the predicted categories: promoters, promoter flanking regions, enhancers, CTCF binding sites, unannotated transcription factor binding sites, and unannotated open chromatin regions (Zerbino *et al.*, 2015).

To find the regulation features from the Ensembl web site, search the gene name of interest, for example, "breast cancer 2", and the results will show all the related genes from human and mouse. Click any of the links for your gene, then click "Regulation" on the left, and a graphic view is then provided to configure the region to display tracks linked to regulation features. For instance, the upstream region, downstream region, and exons for seven BRCA2 transcripts are shown in this regulation view. Regulation features such as the open chromatin and transcription factor binding sites (TFBSs), histones, and polymerases are drawn and color-coded alongside the human genome together with BRCA2 transcripts. Detailed description for the gene related to each regulatory feature of interest is provided by clicking the colored bars. Also, the pop-up window can display information about the evidence supporting that regulatory feature, as well as cell-specific activity estimates. In another way, regulatory features are shown in a table with trackable ENSR IDs or blocks to obtain more detailed information. In addition to the browser, the regulation data in Ensembl can also be accessed through BioMart (Ensembl Regulation database), or the Ensembl Funcgen database by using Perl APIs.

Ensembl Tools

Variant Effect Predictor (VEP)

The Ensembl VEP is a powerful tool for the user to annotate his or her own variation data. In recent years, sequencing and genotyping technologies have produced a large amount of variation data. Of all the information that a variant carries, the effect of the observed variation on transcripts is the most attractive and useful part. Variations with a visible effect can be used to discover biological and medical loci associated with drug treatment or disease susceptibility. In order to predict the effect that variants have on transcripts, considerable computational resources and genomic annotation databases should be available (McLaren *et al.*, 2010). Currently, three major tools can annotate genetic variants: Annovar (Wang, Li & Hakonarson, 2010), SnpEff (Cingolani *et al.*, 2012), and Variant Effect Predictor (VEP) (McLaren *et al.*, 2010). Each application has its advantages. VEP can be applied to all species within the Ensembl Genomes database, including species without variation data (McLaren *et al.*, 2010). For species not in Ensembl, the user can still use VEP by providing a genome FASTA file and a transcript annotation GTF file.

The researcher can use the Ensembl default format, VCF, variant identifiers, HGVS notations, and Pileup formats for input in both the web and script versions of VEP. Three options (file upload, paste file, and file URL) are available for uploading data. VEP offers rich annotation for species based on three transcript databases, including Ensembl transcripts, RefSeq transcripts, and GENCODE basic transcripts. Let us use the zebrafish SNP (ID: rs499678744) discussed earlier to demonstrate how to use this tool. In the VEP web interface, select "Zebrafish" as the species, paste "rs499678744" in the search box, leave the transcript database as default (Ensembl transcripts database), and then click on "Run". On the VEP results page, results are displayed as pie charts and tables. In the pie chart view, all consequences of this variant on transcripts (e.g., "regulatory region variant" and "frameshift variant") and their proportions are displayed. In the results table, various bits of information about this variant on different transcripts are shown, including locations, alleles, consequences, and the related genes. The user can filter the columns of the table to display the required data combinations and download the results in VCF, VEP, or TXT format.

BLAST/BLAT

Ensembl provides BLAST and BLAT to allow the alignment of sequences against reference genome sequences. The position of the gene on chromosomes and its genomic neighborhood is provided after the BLAST/BLAT searches. This can be achieved by providing the query in the search box (or by uploading), and then running BLAST/BLAT. For instance, CTCCGCACTGCTCACTCCCGCGCAGTGAGGTTGGCACAGCCA-CCGCTCTGTGGCTCGCTTGGTTCCCTTAGTCCCGAGCGCTCGCCCACTGC-AGATTCCTT TCCCGTGCAGACATGGCCT is a part of the human *MTAP4* gene. The search can be carried out in the following four steps.

1) Click on the BLAST/BLAT link at the top of the Ensembl home page.
2) Copy and paste the sequence into the "Sequence data" box, or upload the sequence file on the input page.
3) Choose all the correct options, including species, databases, and the search tool. In this example, we have selected "Homo sapiens" and use the BLAT search tool to quickly look for identical matches in the DNA database.
4) Click "Run" to perform the search.

The results of a BLAT search show that this gene is located on human chromosome 9. By clicking the links in the results table, the information available for this gene can be obtained. Since much about BLAST/BLAT is presented under NCBI resources (Chapter 26) as well as in Chapter 2, we will not repeat the discussion here.

Assembly Converter

Genome assemblies, especially those for humans or model organisms, are continuously updated with different versions. Therefore, information and coordinates from one version in relation to their location on the updated version is needed. The "Assembly converter" in Ensembl Tools offers a software tool to perform these functions. Currently, using an efficient conversion tool named CrossMap (Zhao *et al.*, 2014), Assembly converter can convert genome coordinates between different assemblies (GRCh37, GRCh38, NCBI 34, NCBI 35, and NCBI 36). Assembly converter supports various frequently used file formats such as BED, GFF, GTF, WIG, and VCF, and includes data for eight species (human, mouse, zebrafish, cow, pig, dog, rat, and *Saccharomyces cerevisiae*). Data can be directly pasted in the search box, or uploaded as a file. In order to view the result, the user has to download the result file. This tool is likely not very useful for aquaculture genome workers as updates are slow in coming. However, it may be useful for zebrafish, which is a useful reference to various fish species.

ID History Converter

Generally speaking, Ensembl identifiers (IDs) are highly stable. A gene or transcript ID can only change if the gene structure changes dramatically—for instance, if an old gene is found to consist of two genes, or if two genes are now treated as one. ID History Converter is designed to display Ensembl IDs in the current version by using old IDs of the previous release. Researchers can use this tool to find out which ID is still used, which one is out of usage, and which one is changed to a different ID. The input form allows the user to choose his or her species and input a list of Ensembl IDs. IDs can be pasted in the search box directly, uploaded as a file, or attached as a web file. Click on

"Run" to carry out the search job. In order to describe this process efficiently, we use the human *RHOA* gene ID (ENSG00000067560) as an example. After choosing "Human" as species, enter the ID, and submit the job. The match ID is the same as the query ID, and only one ID is found. The results table also shows the links that can guide the user to view the gene information with this ID in different Ensembl releases. Once again, this tool may not be very useful for aquaculture researchers.

BioMart

Ensembl BioMart (http://useast.ensembl.org/biomart/martview) is a highly customizable web-based tool that allows accessing and querying complex data without any programming knowledge or understanding of the underlying database structure. It has been proved to be a valuable resource for mining and extracting the growing volume of genomic data available in the Ensembl project. Generally, tables of Ensembl data for certain species can be downloaded by directly accessing the BioMart interface in Ensembl (http://useast.ensembl.org/index.html), or by using the Ensembl Genomes project (http://www.ensemblgenomes.org), where five distinct domains of life (bacteria, fungi, metazoa, plants, and protists) are presented with links to separate web sites.

Ensembl BioMart is a search engine that can find multiple terms from the various components of the Ensembl project, and then put them into a table format. Each BioMart release incorporates the addition of new species, updated assemblies, updates to the germline, updates to the somatic variation and structural variation datasets, as well as updates to the regulation data (Flicek *et al.*, 2012). It includes four visible mart databases on the BioMart interface. They are: Ensembl Genes, Ensembl Variation, Ensembl Regulation, and Vega (Kinsella *et al.*, 2011). The user can first select a mart database that corresponds to the type of data that he or she is interested in. Then, select the corresponding species whose data you want to retrieve from the "Choose Dataset" entry. Currently, there are 69 species entries in the Ensembl Genes database, and 22 species in the Ensembl Variation database. The regulation database is comprised of only human, mouse, and fruit fly. The Vega database includes only human, mouse, rat, pig, and zebrafish. Next, the mart query (query restriction and input data) can be defined flexibly by clicking on the "Filters" button, and the desired output columns can be selected with "Attributes" button. Finally, a mart result is generated as per the user's requirement, and he or she can retrieve the query using the "Results" button. Results can be displayed in HTML format or be sent as an email with an attached compressed file.

For instance, in order to export the homology information of the zebrafish *dmd* gene using BioMart, first click on "BioMart" on the top panel of the Ensemble home page. Then, choose the database of Ensembl Genes, and the dataset of *Danio rerio* genes. Now, the user can choose many or all the genes in the zebrafish genome, or only the required gene, as in this case. Going to "Filters", open "Gene", and narrow it down to the "ZFIN symbol" and enter "*dmd*"; then, click "Count". It shows that only one gene passes the filter. Next, pick "Attributes" to ask for output information about this gene. Select "Homologs" in this page; you can now choose items in the "ORTHOLOGS" or "PARALOGS" menu, or choose both of them. In this case, select the Ensembl Gene ID for Atlantic cod, cave fish, fugu, and platyfish in the "ORTHOLOGS" menu, and the zebrafish paralog

Ensembl Gene ID in the "PARALOGS" menu. If you do not need separate results for each Ensembl transcript, this can be turn off in the "GENE" menu. Results can be pre-viewed by clicking on the "Results" button. The table shows the orthologs of Atlantic cod, cave fish, fugu, and platyfish for this zebrafish gene, as well as some zebrafish paralogs. Each matched item is clickable, in order to link to the Ensembl genome browser for more information. All the results can be exported as a file, or viewed in rows in HTML format. More options and advanced queries for the gene can be selected in the "Filters" and "Attributes" lists according to needs of the user, such as the chromosomal regions of a set of genes, the associated GO terms, variations, or sequences. More information about utilizing BioMart for genomic research is available at http://useast.ensembl.org/info/data/biomart/index.html.

References

Barski, A., Cuddapah, S., Cui, K. *et al.* (2007) High-resolution profiling of histone methylations in the human genome. *Cell*, **129**, 823–837.

Birney, E., Andrews, T.D., Bevan, P. *et al.* (2004) An overview of Ensembl. *Genome Research*, **14**, 925–928.

Chen, Y., Cunningham, F., Rios, D. *et al.* (2010) Ensembl variation resources. *BMC Genomics*, **11**, 293.

Cingolani, P., Platts, A., Wang, L.L. *et al.* (2012) A program for annotating and predicting the effects of single nucleotide polymorphisms, SnpEff: SNPs in the genome of Drosophila melanogaster strain w1118; iso-2; iso-3. *Fly*, **6**, 80–92.

Cunningham, F., Amode, M.R., Barrell, D. *et al.* (2015) Ensembl 2015. *Nucleic Acids Research*, **43**, D662–D669.

Eddy, S.R. (2001) Non-coding RNA genes and the modern RNA world. *Nature Reviews Genetics*, **2**, 919–929.

Enright, A.J., Van Dongen, S. and Ouzounis, C.A. (2002) An efficient algorithm for large-scale detection of protein families. *Nucleic Acids Research*, **30**, 1575–1584.

Flicek, P., Ahmed, I., Amode, M.R. *et al.* (2012) Ensembl 2013. *Nucleic Acids Research*, **41**, D48–55.

Flicek, P., Aken, B.L., Ballester, B. *et al.* (2010) Ensembl's 10th year. *Nucleic Acids Research*, **38**, D557–D562.

Flicek, P., Amode, M.R., Barrell, D. *et al.* (2013) Ensembl 2014. *Nucleic Acids Research*, **42**, D749–755.

Foelo, M. and Sherry, S.T. (2007) NCBI dbSNP database: content and searching, in *Genetic variation: a laboratory manual*, Cold Spring Harbor Laboratory Press, Cold Spring Harbor, NY, pp. 41–61.

Giresi, P.G., Kim, J., McDaniell, R.M. *et al.* (2007) FAIRE (Formaldehyde-Assisted Isolation of Regulatory Elements) isolates active regulatory elements from human chromatin. *Genome Research*, **17**, 877–885.

Kinsella, R.J., Kähäri, A., Haider, S. *et al.* (2011) Ensembl BioMarts: a hub for data retrieval across taxonomic space. *Database*, **2011**, bar030.

Mazmanian, G., Kovshilovsky, M., Yen, D. *et al.* (2010) The zebrafish *dyrk1b* gene is important for endoderm formation. *Genesis*, **48**, 20.

McLaren, W., Pritchard, B., Rios, D. *et al.* (2010) Deriving the consequences of genomic variants with the Ensembl API and SNP Effect Predictor. *Bioinformatics*, **26**, 2069–2070.

Pruitt, K.D., Harrow, J., Harte, R.A. *et al.* (2009) The consensus coding sequence (CCDS) project: Identifying a common protein-coding gene set for the human and mouse genomes. *Genome Research*, **19**, 1316–1323.

Ren, B., Robert, F., Wyrick, J.J. *et al.* (2000) Genome-wide location and function of DNA binding proteins. *Science*, **290**, 2306–2309.

Song, L. and Crawford, G.E. (2010). DNase-seq: a high-resolution technique for mapping active gene regulatory elements across the genome from mammalian cells. *Cold Spring Harbor Protocols*, **2010**, pdb.prot5384.

Spudich, G.M. and Fernández-Suárez, X.M. (2010) Touring Ensembl: a practical guide to genome browsing. *BMC Genomics*, **11**, 295.

Vilella, A.J., Severin, J., Ureta-Vidal, A. *et al.* (2009) EnsemblCompara GeneTrees: Complete, duplication-aware phylogenetic trees in vertebrates. *Genome Research*, **19**, 327–335.

Wang, K., Li, M. and Hakonarson, H. (2010). ANNOVAR: functional annotation of genetic variants from high-throughput sequencing data. *Nucleic Acids Research*, **38**, e164.

Wilming, L.G., Gilbert, J.G., Howe, K. *et al.* (2008) The vertebrate genome annotation (Vega) database. *Nucleic Acids Research*, **36**, D753–D760.

Zerbino, D.R., Wilder, S.P., Johnson, N. *et al.* (2015) The Ensembl regulatory build. *Genome Biology*, **16**, 56.

Zhao, H., Sun, Z., Wang, J. *et al.* (2014) CrossMap: a versatile tool for coordinate conversion between genome assemblies. *Bioinformatics*, **30**, 1006–1007.

28

iAnimal: Cyberinfrastructure to Support Data-driven Science

Blake Joyce, Asher Baltzell, Fiona McCarthy, Matt Bomhoff and Eric Lyons

Introduction

Next-generation sequencing technologies produce large-scale datasets that have outpaced the computational resources available to the majority of wet-bench scientists. The CyVerse cyberinfrastructure was created to address the need to store, process, analyze, annotate, submit to public repositories, curate, and manage in perpetuity these large-scale datasets. CyVerse itself is cyberinfrastructure that provides the computational resources needed to analyze "big data" sets for any kind of science, from astronomy to the life sciences. Associated with CyVerse are domain-specific platforms such as iAnimal, iMicrobe, etc., that leverage the same computational resources. In this chapter, we give an overview of the "Discovery Environment" (DE), "Atmosphere", and third-party federated platforms that are associated with CyVerse, as well as outline how to use some aspects of each of these resources.

High-throughput (HT) sequencing of DNA, RNA, and proteins has revolutionized how researchers conduct experiments to unravel the dynamics of the central dogma of biology. While every discipline in biology can now generate massive quantitative datasets through next-generation sequencing, animal science is at the forefront of this endeavor. This trend is perhaps made most clear by data from the International Nucleotide Sequence Database Collaboration (INSDC), which is comprised of three molecular biology databases: the National Center for Biotechnology Information (NCBI), the European Molecular Biology Laboratory (EMBL), and the DNA Data Bank of Japan (DDBJ). The INSDC has ranked the top 100 most sequenced organisms, and animal species make up 47 of those positions (Table 28.1). Animal species have more than twice the amount of nucleotides sequenced, and also twice the number of sequence entries, than the next closest kingdom in the top 100 list, plants.

The sheer amount of sequences publicly available to scientists underscores the inherent need for bioinformatics tools and analyses. The need for large-scale analysis comes hand-in-hand with large sequence-based datasets. While the need for large-scale analysis may seem obvious, many of the other needs that come with large datasets are less apparent. Large-scale datasets, sometimes several gigabytes or possibly even terabytes in size, need to be stored, processed, annotated with metadata, submitted to public repositories, curated, analyzed, and managed in perpetuity to be available for future reuse and referencing. These tasks may seem simple at first, but upon closer inspection

Bioinformatics in Aquaculture: Principles and Methods, First Edition. Edited by Zhanjiang (John) Liu.

Table 28.1 Distribution of publicly available sequencing data according to taxons. Data accessed February 2015 (http://www.ddbj.nig.ac.jp/breakdown_stats/org1000/top100-e.html).

Kingdom	Portion of top 100 organisms (%)	Median rank in top 100	Nucleotides (billion)	Entries (million)
Animal	47	44	67.53	61.96
Plant	43	60	25.45	30.05
Metagenomes	4	43	2.14	7.35
Bacteria	3	36	5.04	6.63
Chromalveolate	2	86	0.39	0.31
Fungi	1	79	0.21	0.28

they can become a Herculean task for wet-bench scientists, who are accustomed to producing written and graphic data that is managed by scientific publishers. Additionally, investment in cyberinfrastructure to support large-scale data processing within a lab group is often outside the focus for many members of the biological research community. Therefore, the need for publicly available community resources, training materials, cyberinfrastructure, and analysis tools that enable all animal scientists to be data scientists has become commonplace. Here, we describe various computing platforms and resources available to animal researchers under the moniker of "iAnimal".

Background

Many of the services for iAnimal were originally developed by the iPlant Collaborative, a 10-year NSF-funded project that started in 2008 (NSF #DBI-0735191, NSF #DBI-1265383). The specific goal for iPlant was to develop cyberinfrastructure to assist plant scientists in assembling, annotating, and comparing large datasets (e.g., genomes, transcriptomes, phenomes, and proteomes) (Goff et al., 2011). After its success in enabling many plant science research communities, iPlant was charged with an expanded mandate to serve all life science research as well as physical scientists. This expanded project has been rebranded as "CyVerse" and maintains the fundamental resources that power iPlant. These resources are leveraged to create platforms for other research communities such as iAnimal, iMicrobe, etc.

iAnimal is a United States Department of Agriculture (USDA)–funded project (USDA # 2013-67015-21231) to federate bioinformatics platforms used by animal researchers with the CyVerse infrastructure, specifically "AgBase", "Comparative Genomics" (CoGe), and "VCMap" (Koltes, 2012; Lyons, 2008; McCarthy et al., 2006). Federation provides several advantages over independent platforms. First, CyVerse shares cyberinfrastructure between the aforementioned platforms, which allows for increased scalability of computing and storage resources. Federated platforms also benefit from interoperability, which enables data and computational results to be more easily passed between them. This provides researchers the ability to analyze larger datasets, and move data more easily between bioinformatics tools for more varied analyses. In short, the federated platforms are comparable to a "digital ecosystem" of data storage, data management, analysis platforms, and data repositories that all provide specific "ecosystem services." Federation also allows iAnimal to share services such as user identity management and authentication with other CyVerse platforms. This means that each platform shares basic infrastructure, and researchers, known as platform "end users," do not need to repeat authentication or maintain several user identities. In addition, federation permits fluid navigation between platforms and analyses because the requirements to download datasets to a personal computer, convert data types, and then upload them to another platform are largely removed. Lastly, federation reduces the amount of time that platform developers need to spend on creating, maintaining, and updating hardware and security infrastructure, which translates to more time invested in delivering scientific value to their communities. This also means end users experience a smooth transition, both in the (1) interface, and (2) movement of data and analyses between platforms. This is important because no single platform is capable of all of the analyses that a scientist may want to perform on a dataset, whether privately or with collaborators.

Third-party platforms that are "Powered by CyVerse" maintain their own branding. As such, iAnimal refers not to a single entity, but to the broad category of tools and platforms federated with iPlant. We will use the names "CyVerse" and "iAnimal" as appropriate and sometimes interchangeably.

Signing Up for a Free iAnimal User Account

To sign up for an account with iAnimal, one needs to:

1) Go to CyVerse's (used to be iPlant) user portal (https://user.iplantcollaborative.org).
2) Click on "Register".
3) After registration, you will have access to most platforms federated with iPlant (CyVerse), which includes iAnimal's tools and resources

iAnimal Resources

The resources that will be covered in this chapter are the "Data Store", "Discovery Environment" (DE), and "Atmosphere". These three resources form the core of iAnimal's cyberinfrastructure, but additional third-party resources such as AgBase, VCmap, and CoGe are federated with iAnimal to extend the range of analyses that can be performed on data. Data is the core of all scientific endeavors, and likewise data is the core of the iAnimal platform. The Data Store (http://www.iplantcollaborative.org/ci/data-store) is the central data repository where datasets can be managed and then accessed for analysis.

Each user is allocated 100 GB of storage space in the Data Store, free of charge, which can be increased upon request. Using the Data Store reduces the need for local storage on a laboratory computer. Maintaining data in a single location also minimizes the time required to upload data for future analyses. The Data Store can collect many biological data types ranging from molecular sequences to micrographs by uploading them directly from a local computer, retrieving data from online databases, or using data shared between collaborators. The Data Store is connected to all of CyVerse's analytical resources and federated platforms. These data are considered sacred, and no direct change is made to the original data files. When end users run analyses, a dataset is sent out from the Data Store to one or more of the iAnimal resources. Computation occurs on those resources, and then results come back and are saved in the Data Store.

The Data Store is integrated directly into CyVerse's DE, which is one of the primary user interfaces for data management and analysis (http://www.iplantcollaborative.org/ci/discovery-environment). DE has a web-based interface, and therefore can be accessed to manage data or conduct analyses from any computer connected to the Internet. DE leverages large computing clusters at the University of Arizona's (UA; located in Tucson, Arizona, United States) CyVerse and supercomputers at the Texas Advanced Computing Center (TACC; located in Austin, Texas, United States) to conduct analyses. Therefore, data can often be analyzed more quickly than on most standard laboratory computers.

Atmosphere is a cloud computing service that allows researchers to request a custom, virtual computer with pre-installed software (also known as a virtual machine, abbreviated as "VM") that is accessible through the web. A VM is analogous to a desktop computer and comes complete with random access memory (RAM), hard drive (HD) space, an operating system (OS), and software. The software on a VM is flexible and could be used for a specific bioinformatics analysis. In addition, software may be installed to optimize the VM for the end user's needs. In short, Atmosphere provides flexible servers that are dedicated to running various scientific analyses to researchers.

The Data Store

The Data Store is a central repository for uploading and managing data such as genomes, transcriptomes, and images. The Data Store is a high-performance cloud storage hub that connects your data to cloud computing, third-party platforms, and remote services. All registered users get an initial 100 GB of storage that can be increased if requested.

The Data Store provides secure and mirrored storage for data (data are replicated between supercomputers at the UA and the TACC), whereas storing precious data on lab-owned desktops, laptops, external HDs, or flash drives is a short-term solution prone to loss of data through hardware failures, computer viruses, software incompatibilities, thefts, or being misplaced. The Data Store tracks metadata and data about your datasets, and also allows you to tag datasets with metadata. Methodical use of metadata tags is analogous to keeping a good lab notebook. Metadata tags will simplify understanding key aspects about your datasets, such as how they were created and what analyses were done to them. Therefore, discovering the datasets, interpreting them, analyzing them, and citing them becomes easier for collaborators, current and future members of your lab, manuscript reviewers, as well as other scientists in your community. Metadata tags can help identify specific analyses and software settings that were used on a dataset, and mark data with information such as what organisms and tissues samples were sequenced. This information can also simplify downstream publication writing, and make your data searchable and discoverable.

In the world of arts and antiques, the idea of *provenance*, that is, a clear record of creation and ownership, is well established, and is often used to vouch for the authenticity and quality of an art work or antique piece. Likewise, provenance should be an integral aspect of data generation and curation. Metadata can be used to create data provenance by tagging data with the list of investigators involved in data generation, algorithms involved in analyses, and each step of transforming raw data to high-value results for publication. Metadata tags are also used as search terms to discover data. These tags can be placed on all datasets that are associated with a single project, which effectively groups the related datasets for future researchers to access.

Lastly, reproducibility can be achieved using metadata and automated pipelines of analyses within the iAnimal ecosystem of resources. Thorough metadata tagging will record what software versions were used to create the analyses, so that those same analyses can be rerun with newer software versions, or if reviewers request portions of an analyses be rerun in other programs. Researchers can replicate analyses with new data, so that data analysis best practices can be maintained in the scientific community.

Uploading Data to the Data Store: Choosing the Right Tool

The Data Store uses a software system called iRODS (www.irods.org) to manage hardware resources, and presents them to users in a single directory. There are several ways to access the Data Store, from web-based tools to application programming interfaces (APIs) (https://pods.iplantcollaborative.org/wiki/display/DS/Data+Store+Quick+Start). Keep in mind that there is no one "best" way to access the Data Store; each method has its own strengths and weakness, depending on the needs of the user. For example, DE provides a web-based graphical user interface (GUI) to the Data Store and some easy-to-use tools for sharing data and adding metadata. Although a standard HTTP upload from a local computer is the simplest way to move data to the Data Store, it is also the slowest method. Alternatively, a stand-alone application called iDrop is available to upload larger datasets (https://pods.iplantcollaborative.org/wiki/display/DS/Using+iDrop+Desktop). The iDrop interface is similar to "File Explorer" in Microsoft Windows and "Finder" in macOS. iDrop can create parallel threads to transfer datasets faster than a simple upload, provided the necessary bandwidth is available. The final option, iCommands (https://pods.iplantcollaborative.org/wiki/display/DS/Using+iCommands), requires the ability to use the command line. While it is the most flexible and offers the best performance, some researchers may find it rather difficult to use. For a summary of the strengths and weakness of the various means to access the Data Store, please see https://pods.iplantcollaborative.org/wiki/display/DS/Data+Store+Quick+Start

Uploading Data to the Data Store from a Local Computer

DE (http://data.iplantcollaborative.org) provides a web-based graphical way to upload data to the CyVerse Data Store. This resource also provides tools for easily sharing your data, tagging your data with metadata, and running analyses with the data.

Simple Upload

This is the easiest to use, but does not use parallel file transfers and may fail when transferring data files larger than 2 GB.

1) When logged into DE, click on the icon labeled "Data".
2) Click on the "Upload" tab on the top-left, and select "Simple upload from desktop".

3) Select the file on your computer that you wish to upload.
4) Click on "Upload".

Using FTP/HTTP Links for Uploading Genomes to the Data Store

1) Click on the "Upload" button at the top-left of the screen.
2) Select "Import using URL".
3) This will display several blank fields where you may paste URLs that point to data residing on remote servers.
4) Copy the link location file (e.g., in another browser window or tab).
5) Paste the URL into a blank field on the Data Store page.
6) Click on "Import from URL" to upload.

Bulk Upload

This link launches a Java program and supports parallel transfer of files. However, it may have compatibility issues with some browsers and OSs.

Uploading Data to the Data Store from iDrop

iDrop is stand-alone software that facilitates HT parallel data transfer to the Data Store. This way of transferring data is much faster than traditional methods, but your Internet connection bandwidth should be considered. iDrop will use most of your available bandwidth, so using newer, higher-capacity commercial connections such as those found at institutes of higher learning are preferable to using home connections.

1) All documentation about downloading and installing iDrop can be found on the CyVerse Collaborative Wiki (http://goo.gl/0FtTBO)
2) Once iDrop is installed, files from a local computer can be dragged and dropped into the Data Store

Command Line: iCommands

The Data Store has a suite of command-line tools for researchers who are comfortable with using the command line. They generally follow the Linux/Unix style of common commands for navigation, such as "`icd`" to change directories (instead of "`cd`") and "`imkdir`" to create a directory (instead of "`mkdir`"). Details for using these commands, along with some tips for transferring very large sets of data, are available at http://goo.gl/qMxUah.

Managing and Sharing Data

The easiest way to manage and share data is through DE. From there, it is simple to share folders and files with other users privately, as well as generate public links that can be used by anyone to download a file. Please note that public links are not secure, and anyone with a link can download that file.

1) Go to http://data.iplantcollaborative.org.
2) Click on the "Data" icon.
3) Data can be organized into traditional folders in the Data Store. To create a new folder, click on "File" in the menu bar, and then click on "New Folder".

4) Data can be shared in several ways, as listed in the "Share" dropdown list in the Data Store toolbar. Select a file or folder to share, and then click on "Share" in the menu bar.
5) The "Share with Collaborators" button will allow you to search and associate data with other iPlant/iAnimal–registered scientists within Data Store. You will have to search and select people by their user names or real names. You can also select the level of access they have on the file (read, write, own).
6) The "Create a Public Link" option will generate a URL that can be sent to collaborators to access data. However, caution should be exercised, as this link does not require a login or password to access, and therefore can be copied and shared by your collaborators at any time.
7) The "Send to" option will link the data to other third-party platforms such as CoGe, Genome Browser, and Tree Viewer.
8) To search for data in the Data Store, input search terms in the search field in the menu bar. A list of matching files and folders (by name) will appear. To search for metadata, click on the icon of a funnel to the right of the search bar.

DE

In addition to the data management tools discussed in the preceding text, DE provides an easy way to run complex workflows and analyses that require more compute resources than a laboratory computer can provide (Oliver et al., 2013). When you open DE, you will see three buttons in the upper left of the screen: "Data", "Apps", and "Analyses". The Data button, which was covered in the previous section, will open a window to the Data Store to manage, transfer, upload, download, or share datasets. The Apps button brings up a list of bioinformatics applications and workflows that can be run on data in your data store. The Analyses button will take you to a list of your previously run analyses to monitor their progress and retrieve their results.

Each application available through DE provides a GUI built on top of Linux command-line programs, called "tools", preloaded into the DE. This saves the time of finding and loading software into a local computer, an especially valuable feature for users lacking experience in running command-line software. Also, the computing resources on which these apps run are often more powerful than what is available to researchers on a desktop computer.

To run an analysis, simply find the application of interest—for example, SOAPdenovo next-generation sequence aligner—in the dropdown menu or by using the search bar. This will bring up the GUI that allows you to configure and run the program. DE runs on high-performance supercomputers located at UA and TACC. Combining the simple user interface and powerful computer resources results in fast analysis times and parallel processing batches of data rather than running single datasets in a linear progression. Running batches of genome assemblies, annotations, or other analyses reduces technical variation in datasets by ensuring uniform software settings. This is particularly powerful when you wish to repeat an analysis weeks, months, or years later using the same settings, software version, etc. However, users may experience wait times for running a job, depending on how many jobs have been submitted to the queue and how much resources are being requested to run the job. Please note that other researchers integrate the apps available through DE, and you have the opportunity to modify the interfaces that they have designed, as well as to add new apps to DE. Each app that has been integrated will have a documentation page on how to use it, and will also provide sample input and output files.

Finding Practice Data in DE

DE contains practice data for each app, so that researchers can run analyses and see the results from each before using their own data. The practice data comes in two folders: (1) assembly test data, which contains several assembled genomes and transcriptomes as well as unassembled reads (/iplant/home/shared/iplant_assembly_test_data), and (2) practice data for each of the apps present in DE (/iplant/home/shared/iplant collaborative/example_data).

Accessing Practice Data in DE

1) Pressing the "Data" button in DE (https://de.iplantcollaborative.org/de/) will open the user's data folder.
2) In the "Viewing" bar, paste the assembly practice data folder location (/iplant/home/shared/iplant_assembly_test_data), or paste the app's practice data folder location (/iplant/home/shared/iplantcollaborative/example_data).
3) *Alternatively*, in the "Navigation" panel on the left side of the Data window, find the "Community Data" folder and open it. Next, find the "iplantcollaborative" folder and open it. Finally, open the "example_data" folder.

Inside the "example_data" folder is a list of apps available in DE. Each app's folder contains practice data that can be used in that app.

How to Run "Apps" in DE

There are three major groups of applications available in DE. The "Workspace Apps" are your available applications, including any applications you have created, your favorite applications, and any applications that have been shared with you. The "Public Apps" section contains any applications that have been shared with all users. The "High-Performance Computing Apps" section shows all applications that are submitted directly through the Agave API, which will be discussed in a later section. To view and access the "High-Performance Computing Apps" section, you will need to authenticate a second time by clicking on that category.

To Run an App in DE

1) Pressing the "Apps" button in DE (https://de.iplantcollaborative.org/de/) will open the available apps for data analysis.
2) Browse and select the app that you wish to use, or search for the name of an app in the search toolbar.
3) Once the app window opens, follow the documentation within the app to run the analysis.
4) Once you run an app, you will receive notifications on its submission process. You can monitor its progress by clicking on the "Analyses" icon. This will tell you of its status (submitted, running, failed, completed) and provide a link to the folder in which the data files are deposited. Each analysis will get its own folder under the "Analyses" directory in your data store.

Using Workflows to Run a Pipeline of Analyses

Often, there are analyses that must be performed on a dataset in a set order. For example, genome sequencing datasets must be filtered to remove low-quality reads, trimmed to remove adapter sequences, assembled, and annotated before they are useful for testing biological hypotheses. Ideally, each of these analyses would be run in a linear pipeline, or workflow, that would uniformly carry out each of these steps across all datasets in a project or experiment. This reduces technical variation in datasets, increases reproducibility, and increases the clarity of data analysis, which all serve to instill confidence in manuscript reviewers and the scientific community. Optimized workflows can be described as "set it and forget it," focusing hands-on time and energy away from complicated, obtuse computations and debugging, and toward assessing results and testing biological hypotheses.

Currently, only linear workflows can be created in DE. Therefore, the output file of one app must be the input file of the next app. This also means that the output file format of an app must match the format that the next app can use as input. Last, any app included in a workflow must allow the output file to be renamed, so that it does not contain any part of the input file name. In DE, you can run previously generated workflows as well as create your own.

Creating a Workflow in DE

1) Navigate to DE (https://de.iplantcollaborative.org/de/) and select the "Apps" button to bring up the apps window.
2) Select the "Workflow" dropdown list and then select "Create New...".
3) Add all apps that will be in the workflow by clicking on the "Add" button at the top-left of the workflow window. The order of the apps does not matter at this step because the apps can be rearranged throughout the workflow process.
4) Once all apps have been added, they should be ordered from top to bottom for a linear workflow. Use the "Move Up" and "Move down" functions to reorder them.
5) Synchronize the output of each app to the input of the next app in the workflow. The first app in the workflow will show the output file format. The second app in the workflow will have an input dropdown menu; synchronize the two types, and repeat for each set of apps in the workflow.
6) When finished, test the workflow by going to the "Workspace" portion of the DE apps list. Note that workflows can be shared and made public.

Building workflows are more complex than running a single app in DE. It is likely that individual apps will have to be copied and modified to allow certain output or input formats. Documentation and support can be found on the CyVerse Wiki workflow creation page (http://goo.gl/t6O6iS).

Batching: Running an App Across all Data in a Directory

Apps can also be run across a group of files inside of a directory. To do this, a special HT analysis path list file must be created. The HT list file contains the list of data files to be analyzed in a batch, which then can be uploaded into apps like a single file. The HT list can be used as input for any app that is not run on the Agave API platform.

1) From DE (https://de.iplantcollaborative.org/de/), click on the "Data" icon to open the "Data" window.
2) Navigate to the Data Store and select the "File" dropdown list in the Data Store toolbar.
3) Place the cursor over "Create" in the dropdown list, and then select "New HT Analysis Path List file..." in the submenu.
4) Drag and drop a folder or a group of individual files to be analyzed in a batch into the HT list file window, and then select "Save" in the toolbar. Save the HT list file somewhere convenient in your Data Store.
5) This creates an HT list file that can be uploaded into DE apps to be run. HT list files also have a different icon—a green list that will let you know that the list was generated successfully. An app in DE will run each file in the HT list file in order, and output analyses as it would for individual files.

How to Modify Existing Apps

Apps that are already available in DE can be modified to suit your particular needs. Modifications can expose more options in the software package, or change the default settings to match accepted/published program settings in your community or organism.

1) From DE (https://de.iplantcollaborative.org/de/), click on the "Apps" icon to open the "Apps" window.
2) Select the app you wish to copy and modify in DE. The app name will be highlighted in blue.
3) Select the "Apps" dropdown list in the toolbar. Select "Copy".
4) This brings up the interface for modifying an app.

Requesting Apps that Are Not in DE

If an application is not already available in DE, you can request a new "tool" for installation. The tool will then be installed on CyVerse's computing cluster and made available to you to create an interface that is available through the DE "Apps" window. You can then create private or public apps based on that tool. It may be noted that only certain programs can be installed on CyVerse's computing cluster, namely Linux command-line tools that do not require user interactions while the program runs.

1) Click the "Apps" button in DE (https://de.iplantcollaborative.org/de/) to bring up the "Apps" window.
2) In the toolbar, click the "Apps" dropdown list and select the "Request tool" option.
3) Input the information on whatever bioinformatics software you would like to be installed, and submit. An email will be sent to you when the tool is installed.
4) Once the tool is installed, you will have to create an interface for it by opening the "Apps" window, clicking the "Apps" dropdown menu, and selecting "Create New".
5) For details on creating and testing the interface, visit http://goo.gl/by5aii.

Atmosphere: Accessible Cloud Computing for Researchers

Atmosphere is a cloud-computing platform that provides configurable VMs containing a full OS and pre-installed software for performing tasks such as sequence analysis, genome assembly, and image processing (Skidmore et al., 2011). The interface for Atmosphere is more advanced than DE, and is designed for intermediate users who are somewhat comfortable at the command line and in using remote servers. Atmosphere allows you to select and launch VMs of different sizes (number of CPUs, RAM, and HD space) in order to match the computing power that you need for specific projects. Atmosphere's underlying cyberinfrastructure is structurally different from traditional computing hardware, but the end user's experience is not fundamentally different.

End users start by "spawning" one or more cloud-computing instances. From your perspective, an instance is analogous to a physical running computer with an OS, preinstalled software, RAM, CPUs, and an HD. An instance is spawned from a template called a VM image. Most images are loaded with specific analysis tools to accomplish a set of bioinformatics analyses. Images are powerful for iAnimal end users because they can package and compartmentalize all necessary analysis software, and their dependencies, in one easy-to-access package. Once an instance is spawned, it can be modified with additional software or workflows. Atmosphere allows instance sharing with collaborators or the general public so that they can access analyses and workflows, datasets, and results. Once an instance is modified and optimized for a set of analyses, an image can be taken of that instance to provide the optimized instance to other groups.

However, the need for greater computational savvy is the counterbalance to having the enhanced computational power and freedom that Atmosphere provides. First, the user must request for access to Atmosphere because it is not automatically provided when registering for an account.

Requesting for Access to Atmosphere

1) Go to your user management dashboard (https://user.iplantcollaborative.org/dashboard/).
2) Under the "Apps & Services" tab, scroll down to the bottom and find the Atmosphere section.
3) Select the "Request Access" button.
4) The request will be processed, and access granted.

To Launch an Atmosphere Instance

1) Log into Atmosphere (https://atmosphere.iplantcollaborative.org).
2) Select "Launch New Instance".
3) Search and select an image to use from the list on the right (e.g., "Biolinux").
4) Select the size of your instance and click on "Launch Instance".

Please note that you have "credits." The larger the instance, the more credits it will consume. Also, instances that have been spawned will be turned off after a set period of time or if they are unused. You can monitor, suspend, resume, and shutdown instances from the Atmosphere control panel, shown on the left of the webpage.

There are two general ways to access an instance: through the command line, or through specialty programs that virtualize the image's graphical desktop on your computer. To access the graphical desktop, you will need to install specialized software, such as "VNC Viewer", on your computer. Connecting through the command line requires SSH, which is available on Linux and macOS, and available to Microsoft Windows through programs such as PuTTY.

Also, with great computational power comes great computational responsibility. Atmosphere is a shared resource, and therefore provisioning resources must be considered. Each user is allocated a limited amount of computational power and usage time. It is best to ask only for the minimum server size needed for your analyses. Ultimately, deciding how much computational power each of the specific analyses requires will come down to experience. Allotments of CPU hours refreshes monthly, and additional resources can be requested by users who need to practice with Atmosphere (https://pods.iplantcollaborative.org/wiki/display/atmman/Requesting+More+Atmosphere+Resources).

And finally, using Atmosphere can be a great introduction to using the command line for bioinformatics analyses. Atmosphere provides an environment where researchers can leverage greater computational flexibility available through a full OS to run data-intensive analyses.

The Agave API

The Agave API is an advanced tool for bioinformaticians and software developers to build applications or entire platforms using portions of the existing CyVerse infrastructure (Dooley et al., 2012). Software development kits are available for many languages, but designing new platforms or tools is beyond the scope of this chapter. To learn more about Agave, visit http://agaveapi.co.

Bio-Image Semantic Query User Environment: Analysis of Images

This type of work is primarily focused on how to process genetic data; however, the CyVerse is dedicated to the analysis of many data types. For example, image files can be analyzed using the Bio-Image Semantic Query User Environment (BisQue) CyVerse resource (http://www.iplantcollaborative.org/ci/bisque-image-analysis-environment). Image data such as 2D micrographs, medical scans, and even 3D z-stack confocal micrographs can be uploaded and analyzed. Analyses can be performed on images to generate additional data to test hypotheses. For instance, the BisQue platform can annotate images; count pixels and voxels for comparison between samples; create movies from 3D stacks of images; count and track nuclei and organelles; and even create interactive figures or images. More information can be found on the iPlant BisQue wiki page (https://pods.iplantcollaborative.org/wiki/display/BIS/Getting+Help+with+Bisque).

Selecting the Right Tool for the Job

There are several platforms in iAnimal that can be used to perform your bioinformatics analyses. In general, moving from DE to Atmosphere to the Agave API enhances the computational power and control that a researcher has over running an analysis; however, this also increases the complexity and necessity for understanding computer science. A list of pros and cons associated with each resource is discussed in Table 28.2 to help end users select the right tool for their analysis.

Reaching Across the Aisle: Federated Third-Party Platforms

Several third-party platforms are federated with iAnimal. As discussed earlier, this eases the movement of data and simplifies the user interface between different platforms that specialize in certain data types or analyses. Federating these third-party resources, which are already familiar to animal researchers, creates an ecosystem that can more easily move data and analyses between platforms by using resources such as a common user-identity management system and the Data Store. Federation also provides cyberinfrastructure and backend resources that support the federated platforms, allowing them to leverage computational and data storage scalability without large investments of time and capital. This permits developers of third-party platforms to focus on providing scientific value to users, rather than on system administration. Federated platforms fall under the "Powered by CyVerse" program (http://www.iplantcollaborative.org/about-iplant/powered-iplant). Figure 28.1 shows a set of commonly used bioinformatics platforms being federated on this common cyberinfrastructure. Each federated platform in the figure specializes in a type of data analysis, and so is most useful when included in an overall workflow.

AgBase

AgBase (http://www.agbase.msstate.edu/) provides curated functional information for agriculture plant and animal species and, more recently, for agricultural pathogens (McCarthy et al., 2010). AgBase provides computationally inferred function for a broad range of species, targeted manual biocuration, and support and training (as requested) on how to functionally model large-scale datasets. AgBase biocurators use Gene Ontology (Ashburner et al., 2000; Eilbeck et al., 2005) for providing functional annotations, as well as for providing interaction data to the EBI IntAct database. The data provided by AgBase can be used to answer questions such as:

- What metabolic pathways are expressed in different populations of crops or livestock?
- What major biological functions are represented in a transcriptome from a non-model species?

GO annotations are typically used to perform functional enrichment analyses, and many bioinformatics tools exist for this purpose. Rather than re-inventing bioinformatics tools, AgBase focuses on allowing agricultural researchers to more easily generate the functional information they require to support functional enrichment analyses.

AgBase's Genome2Seq tool will generate a FASTA file of genome sequences from a list of coding DNA sequence (CDS) coordinates. This is useful if the researcher has a list of CDS coordinates of interest, as it is typically generated by standard RNA-Seq assembly and annotation software. Genome2Seq will find the corresponding genome sequence

Table 28.2 A comparison of iAnimal resources' strengths and considerations.

Resource	Strength	Things to consider
CyVerse DE (All skill levels)	• Web-based system for running analyses and managing data • GUIs for command-line programs (hundreds are available) • Linear workflow wizard • Batch file processing • GUI for installing and creating interfaces for new apps • Easy data management and sharing	• Limited to command-line apps • Cannot interact with apps as they run
CyVerse Atmosphere (Intermediate)	• Easy-to-use web system for requesting, provisioning, and managing VMs, and for running instances • Full Linux OS • Ability to install your own programs • Specialized software stacks of pre-installed scientific software • Can have GUIs or command line • Highly interactive visualization of data possible • Extend personal software to a greater community by creating new VM images	• Familiarity with Linux helps • Provisioning of VMs dependent on usage of the system • Knowledge of using VMs helps
CyVerse Agave API (Advanced)	• Science-as-a-service platform • A grid and virtualization environment • Access to high-performance computational systems to run analyses • RESTful interface	• Entirely command-line based; mainly for use in programs • Must understand RESTful interfaces and JSON
CyVerse Data Store	• Scalable cloud storage for data • Accessible through web (DE), command line (iCommands), stand-alone software (iDrop), and API • Ability to easily share data • Ability to easily connect data to various iPlant computing resources	• Different means to access data depending on user needs • Free registration before use
CoGe/EPIC-CoGe	• Web-based platform for the management and analysis of genomic and functional genomics data • Keep data private or share it with collaborators • Mix and match public and private data • Thousands of genomes available	• Figuring out which version of a genome is the correct one to use
AgBase	• Annotation of agricultural plant, animal, microbe, and parasite species • Curated database for dependable data • Produces biological meaning from genetic data	• Limited to 12 curated animal species and seven curated plant species • Requires experimental evidence or bioinformatics inference to annotate

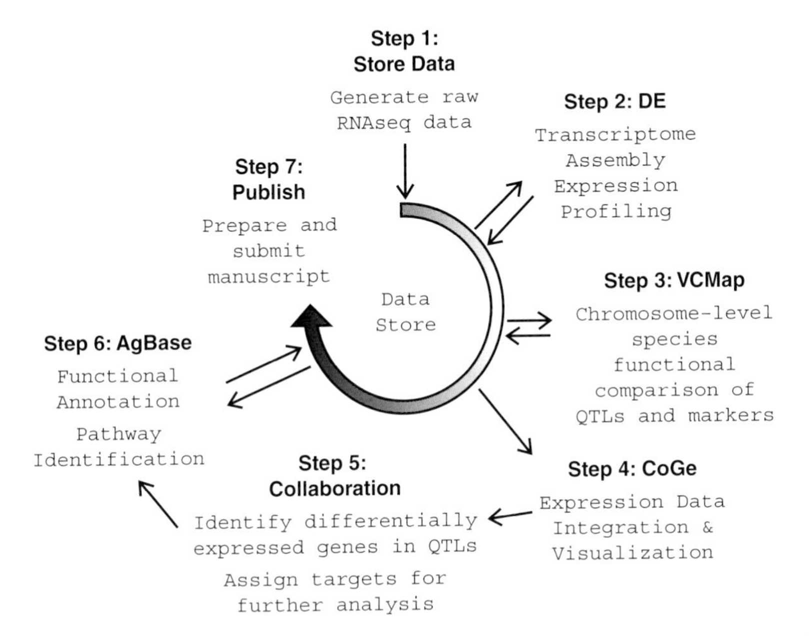

Figure 28.1 A common RNA-Seq analysis workflow across the federated iAnimal platform. (1) First, data is generated and uploaded to the CyVerse Data Store. (2) Transcriptomes are then assembled, and broad expression patterns are profiled using the CyVerse DE. (3) Transcriptomes can then be compared to a chromosome map available for seven animal species using the VCMap platform. This comparison matches common breeding markers such as quantitative trait loci (QTLs) to genes from the transcriptome. (4) Afterward, the genetic mapping data (QTLs) and sequencing data are imported into CoGe to be visualized and compared to the expression profile. (5) Visualization of the genome regions near QTLs and the differential expression of genes in those regions result in researchers selecting areas of biological interest for further analysis. (6) The differentially expressed genes are then annotated using AgBase to determine the structural and functional information about the genes that are differentially expressed at the QTL sites. (7) Once the biological function is determined, this leads to an understanding of the genes associated with the QTLs, and the results are prepared for publication.

and any GO annotations associated with those CDS coordinates. A common use case is to generate a list of functional annotations associated with the genome regions identified from RNA-Seq data that have been mapped to a genome. However, RNA-Seq data from organisms that do not have annotated genomes available can also be analyzed using the AgBase GO annotation tools.

Any transcripts that do not have GO annotations can be analyzed using GOanna to run a BLAST against annotated genomes (McCarthy et al., 2006). If the alignments are valid, then the annotations associated with the BLAST hits will be transferred to the transcripts in the RNA-Seq data. Once a set of annotations is created for the RNA-Seq data, they can be summarized and compared to similar genomes using GOSlimViewer (McCarthy et al., 2010). The quality of GO annotations can be assessed using the GO Annotation Quality (GAQ) Score tool.

The InterProScan tool produces "inferred from electronic annotation" (IEA) annotations for genes (Hunter et al., 2009). This tool is available in DE (https://de.iplantcollaborative.org/de/), and complements GoAnna by annotating newly assembled genomes with conserved protein domains and functional sites. InterProScan searches a curated database of protein domains and gene families to assign annotations.

InterProScan can be used to complete initial annotations for newly sequenced genomes or transcriptomes.

VCmap

The VCmap (http://www.animalgenome.org/VCmap/) tool specializes in discovering the underlying genetic sequences and elements responsible for QTL (Koltes, 2012). QTLs are used by breeders to identify areas of a genome associated with beneficial or deleterious phenotypes in breeding populations. Comparing genetic sequences with QTL maps allows researchers to identify the underlying biological basis for previously identified QTL markers. VCmap supports seven model animal species currently: rat (*Rattus norvegicus*), wild boar (*Sus scrofa*), red junglefowl (*Gallus gallus*), cattle (*Bos taurus*), house mouse (*Mus musculus*), horse (*Equus ferus caballus*), and humans (*Homo sapiens*).

QTL and genetic map comparisons can be carried out on the chromosome level between sets of individuals or across species. VCmap supports genomic sequence maps along with linkage maps and radiation hybrid maps. Visualization of this data can be modified to identify areas of synteny between individuals in a population or across different species. Further documentation about VCmap can be found on the webpage support manual (http://www.animalgenome.org/VCmap/manual/v3/).

CoGe

CoGe (https://genomevolution.org/CoGe/) is a platform that enables scientists to easily manage, analyze, visualize, and compare genomic data from the level of nucleotides to entire genomes (Lyons, 2008). CoGe allows users to either upload genome assemblies and annotations, or use preexisting genomes that are publicly available (over 23000 genomes from 16000 organisms), and then associate experimental data such as function genomics data (RNA-Seq, alignments, ChIP-Seq), variation data (polymorphisms, single-nucleotide polymorphisms (SNPs)), and/or markers with assembled genomes. Once genomes and experimental data have been integrated, CoGe has tools to investigate broad areas of synteny in genomes, visualize areas of microsynteny across genomes, search sets of genomes for homologs, search for blast sequences against any set of genomes in CoGe, and visualize genomes and experiments using JBrowse (Skinner et al., 2009). In addition, CoGe has tools to let researchers organize genomes and experiments, add new genomes and keep them private, share data with collaborators, and make data fully public.

While there are over 20 tools in CoGe to compare and analyze genomes, some of the more frequently used are summarized. SynMap allows for whole genome comparisons between two genomes or within one genome to find syntenic gene pairs and regions of genomes (Lyons et al., 2008). These are displayed as dot plots, and synonymous (K_s) and non-synonymous (K_a) substitution rates may also be calculated for each pair. Selecting an area of the dot plot will bring up another window with that area magnified. Selecting a single dot will allow you to analyze the region's microsynteny with a tool called "GEvo".

SynFind will identify syntenic regions across multiple genomes to a given gene in one genome. Results from SynFind can also be sent to GEvo for microsynteny analysis

(Tang & Lyons, 2012). CoGeBLAST will search any number of genomes in CoGe by sequence.

GenomeView (based on JBrowse) allows researchers to visualize and navigate the genome(s) and all associated experiments (Lyons, 2008; Skinner et al., 2009). Each experiment is represented as tracks that can be stacked vertically for comparison.

CoGe also has workflows for processing raw data such as FASTQ data from genome resequencing or transcriptome projects and quantifying/mapping them to genomes in CoGe. These workflows rely on CyVerse's cyberinfrastructure for data management and analysis. This is an example of how leveraging CyVerse's cyberinfrastructure resources enables on-demand computational scalability. For specific examples on how to use CoGe, or for a more in-depth explanation of CoGe tools, please refer to Chapter 25 of this book.

How to Find Help Using iAnimal

There are many ways to find help with iAnimal services, platforms, or general questions about analyses. The fastest way to get additional help is to visit the Learning Center (http://www.iplantcollaborative.org/learning-center). The Learning Center has video tours of various platforms and links to YouTube tutorials for specific applications. The Learning Center also has links to upcoming events and workshops that have hands-on training. The CyVerse Wiki (http://goo.gl/0FtTBO) contains detailed descriptions and links for using many of the platforms and tools. This resource also describes advanced options not discussed here, and maintains an up-to-date list of applications (http://goo.gl/ZStcyA). Forums are also available to discuss specific topics with other users and with support staff (http://ask.iplantcollaborative.org/questions/). Additionally, support staff can be contacted directly for technical questions about the cyberinfrastructure—for example, about app installation, file format synchronization, issue reporting when a service is not working, and/or scientific help (http://www.iplantcollaborative.org/about-iplant/science/science-support).

Each third-party platform has its own help and tutorials. AgBase has an educational resource with workshop examples, presentations with specific examples, and online courses (http://www.agbase.msstate.edu/cgi-bin/education.cgi). VCmap has an extensive manual to support first-time users (http://www.animalgenome.org/VCmap/manual/v3/). CoGepedia is the wiki page for CoGe that contains descriptions of CoGe tools, video tutorials, information about databases linked to CoGe, and outlines for past workshops that provide specific examples and datasets for learning how to use the CoGe platform (https://genomevolution.org/wiki/index.php/Main_Page).

Coming Soon to CyVerse

The Data Store currently provides researchers with an area to upload and manage data. However, further organizational tools are being developed to supplement the current Data Store. The Data Commons will offer a higher layer of organization and control over data in the Data Store. The Data Commons is designed to organize data into projects that allow for tracking data, analyses, and ultimately provide an automated output format to NCBI, SRA, Dryad, or other public repositories for publication. Additionally, datasets or results from projects will be discoverable in future projects or attached to publication information for easy reference in future publications. This will be streamlined in domain-centered research groups—such as iAnimal, iMicrobe, etc., within CyVerse—which can organize projects, bring in data from either Data Store or publicly available projects, and find analyses commonly performed by that research community (e.g., animal phylogenetics). Each dataset in the Data Commons has a set of metadata associated with it to facilitate data discovery and reuse. This includes metadata about the specimens that were used to generate data; analyses performed on the dataset; and administrative metadata such as the person who entered it, or when it was uploaded. The end result is a set of metadata that can be tracked in an analysis pipeline from raw data that is imported all the way to a final dataset that can be easily deposited into repositories such as NCBI.

The Data Commons is designed to be an extension of the Data Store, allowing greater collaboration and organization of data, and simplifying publication by providing a data submission pathway to the major international sequence databases—for example, NCBI, DDBJ, and ENSEMBL. This is all driven by project organization and metadata that is automatically generated or added by researchers to aid discovery and integration in a community of scientists. Additionally, all metadata and datasets are accessible from anywhere in CyVerse, since it is rooted in the central Data Store. The metadata tags can be added at any time through the lifetime of the project, leading to a "metadata workflow" that can be modified at a time that is most appropriate.

Projects and associated data can be curated and refined, or packaged, by users for multiple targeted uses such as publication or reuse by other communities and future projects. Because the metadata tags move with the datasets and any analyses that those datasets experience, they are always available, and are the stable, searchable, discoverable, and citable identifiers for the datasets, and ultimately for downstream publications.

Conclusion

Biology has become a data-driven science backed by large-scale datasets. Next-generation sequencing technologies is only one example of how nearly all biological researchers can generate large sets of data to connect fundamental genetic aspects to broad-scale processes such as evolution, population dynamics, interspecies competition, adaptation, and ecology. The CyVerse infrastructure has been developed by funding from NSF to provide data management and analysis solutions to researchers, which is being augmented by additional funding from sources such as USDA to build specialized systems for use by animal researchers.

iAnimal is an example of this extension, enabling animal researchers to focus on biological questions, removing the need for each scientist to maintain computational expertise and expensive computational hardware and software. Popular and trusted genetic data analysis software and tools are preloaded into the iAnimal platform, and once data is loaded into the Data Store there is no further requirement to move it for analysis.

While iAnimal aims to provide a comprehensive set of tools and services, the ecosystem of data analysis is always changing. However, by participating in an ecosystem, new platforms may be added to enrich the computational environment. Access to this infrastructure removes the barriers associated with large-scale datasets and allows researchers to go back to doing what they do best: testing hypotheses and deriving knowledge from experiments to understand the living world around us.

Acknowledgments

This work was funded by a grant from the USDA (# 2013-67015-21231). We would also like to thank the United States of America National Science Foundation for funding the iPlant Collaborative (#DBI-0735191, NSF #DBI-1265383).

References

Ashburner, M., Ball, C.A., Blake, J.A. *et al.* (2000) Gene ontology: tool for the unification of biology. *Nature Genetics*, **25**, 25–29.

Dooley, R., Vaughn, M., Stanzione, D. *et al.* (2012) Software-as-a-service: the iPlant foundation API. *5th IEEE Workshop on Many-Task Computing on Grids and Supercomputers (MTAGS).*

Eilbeck, K., Lewis, S., Mungall, C. *et al.* (2005) The sequence ontology: a tool for the unification of genome annotations. *Genome Biology*, **6**, R44.

Goff, S.A., Vaughn, M., McKay, S. *et al.* (2011) The iPlant collaborative: cyberinfrastructure for plant biology. *Frontiers in Plant Science*, **2**, 34.

Hunter, S., Apweiler, R., Attwood, T.K. *et al.* (2009) InterPro: the integrative protein signature database. *Nucleic Acids Research*, **37**, D211–D215.

Koltes, J. (2012). VCMap3. 0: a comparative genetics viewer designed to transfer annotation, QTL and biological information across species. *Plant and Animal Genome XX Conference* (January 14–18, 2012).

Lyons, E.H. (2008) *CoGe, a new kind of comparative genomics platform: Insights into the evolution of plant genomes*, PhD, University of California, Berkeley.

Lyons, E., Pedersen, B., Kane, J. and Freeling, M. (2008) The value of nonmodel genomes and an example using SynMap within CoGe to dissect the hexaploidy that predates the rosids. *Tropical Plant Biology*, **1**, 181–190.

McCarthy, F.M., Gresham, C.R., Buza, T.J., Chouvarine, P., Pillai, L.R., Kumar, R. et al. (2010). AgBase: supporting functional modeling in agricultural organisms. *Nucleic Acids Research*, **39**, D497–D506.

McCarthy, F.M., Wang, N., Magee, G.B. *et al.* (2006) AgBase: a functional genomics resource for agriculture. *BMC Genomics*, **7**, 229.

Oliver, S.L., Lenards, A.J., Barthelson, R.A. *et al.* (2013) Using the iPlant collaborative discovery environment. *Current Protocols in Bioinformatics*, **42**, 1–22.

Skidmore, E., Kim, S., Kuchimanchi, S. *et al.* (2011) iPlant atmosphere: a gateway to cloud infrastructure for the plant sciences. *Proceedings of the 2011 ACM workshop on Gateway computing environments*, 59–64.

Skinner, M.E., Uzilov, A.V., Stein, L.D. *et al.* (2009) JBrowse: a next-generation genome browser. *Genome Research*, **19**, 1630–1638.

Tang, H. and Lyons, E. (2012) Unleashing the genome of *Brassica rapa*. *Frontiers in Plant Science*, **3**, 172.

Index

Bioinformatics in Aquaculture: Principles and Methods, First Edition. Edited by Zhanjiang (John) Liu.

n

o

p

q

r

s

t

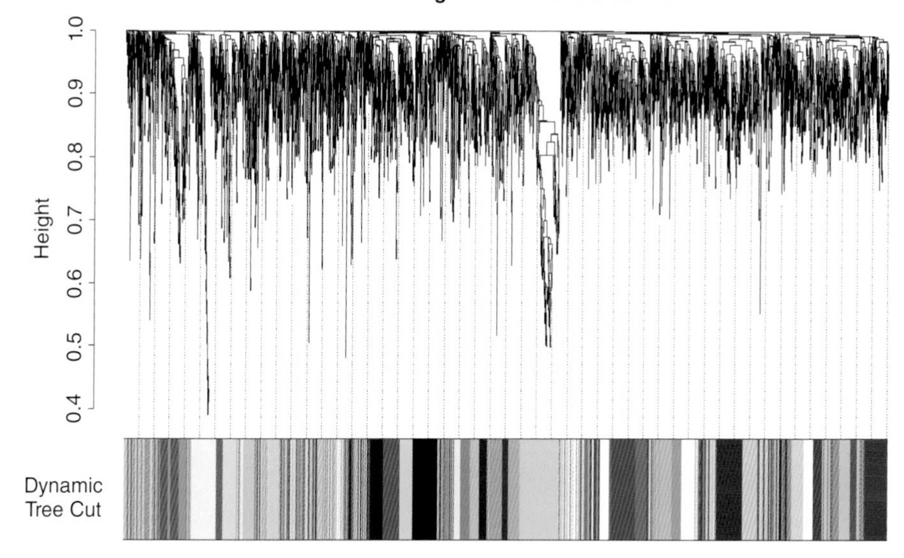

Figure 9.3 A dendrogram for the genes and detected modules.

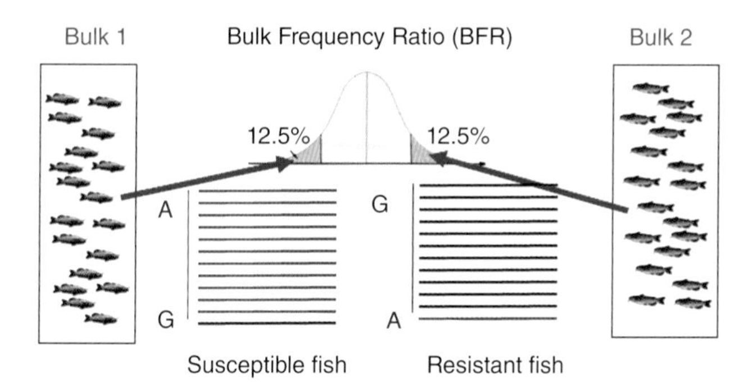

Figure 11.1 Schematic presentation of bulk frequency ratio (BFR). Phenotypic extremes are pooled into bulks (bulk 1 for susceptible fish and bulk 2 for resistant fish). A allele is indicated by blue lines, and G allele is indicated by black lines. In this example, BFR = (11/12)/(1/12) = 11.

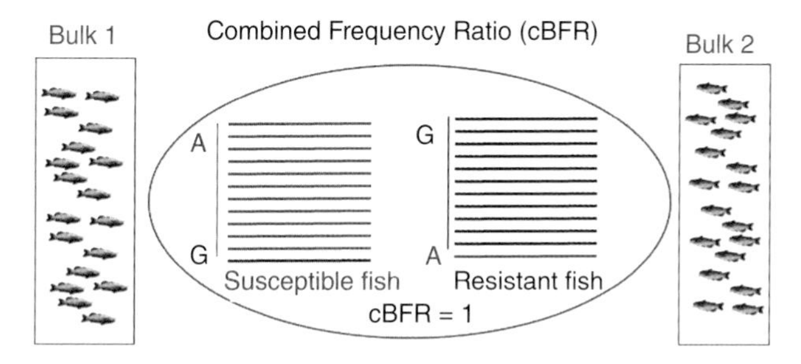

Figure 11.2 Schematic presentation of combined frequency ratio (cBFR). Allele frequency of A in bulk 1 is 11/12, and in bulk 2 is 1/12. However, when the bulks are combined, the overall allele frequency ratio is 12A/12G = 1.

Bioinformatics in Aquaculture: Principles and Methods, First Edition. Edited by Zhanjiang (John) Liu.

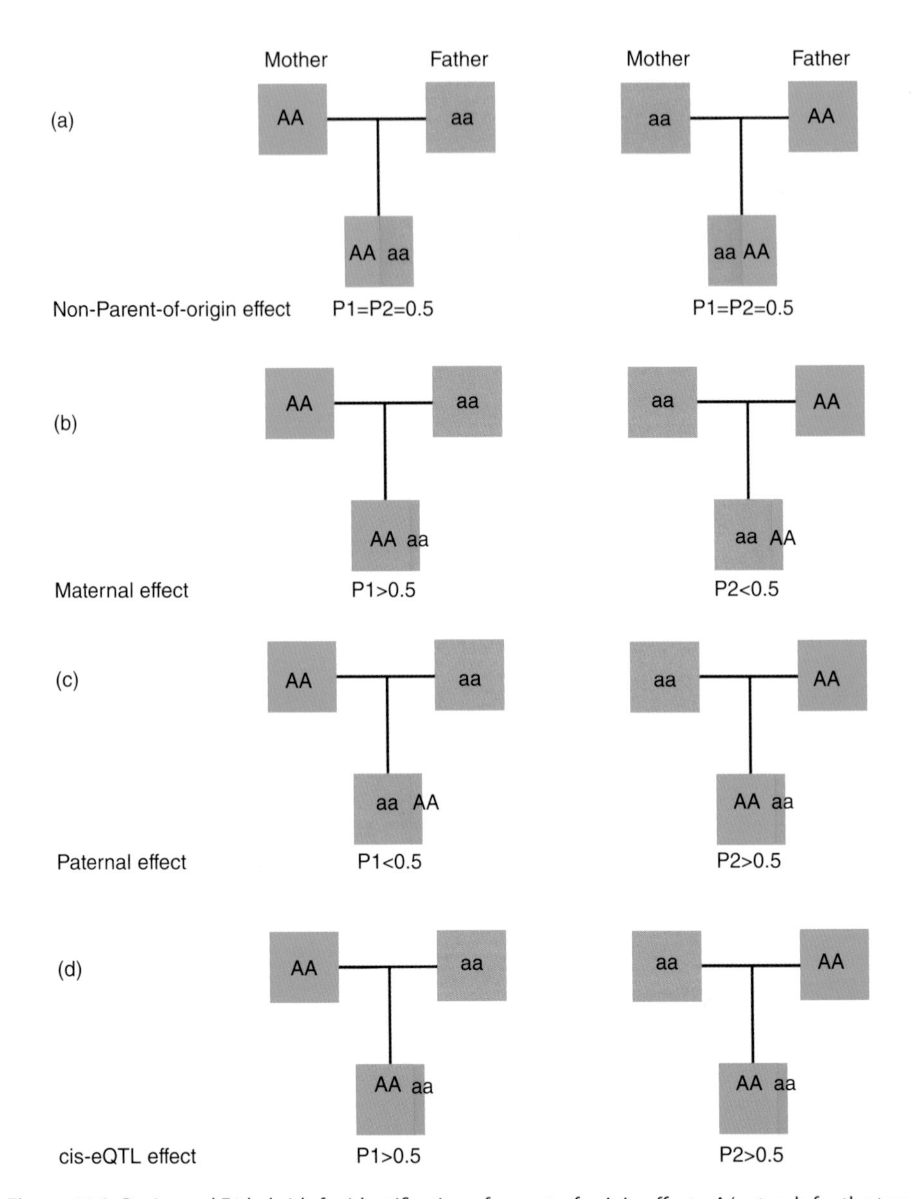

Figure 14.3 Reciprocal F1 hybrids for identification of parent-of-origin effects. A/a stands for the two alleles of informative SNP site in polymorphic parents; (a) alleles are expressed equally, exhibiting with a pattern of non-parent-of-origin effect; (b) maternal alleles are expressed more abundantly, exhibiting a pattern of maternal effect; (c) paternal alleles are expressed more abundantly, exhibiting a pattern of paternal effect; (d) a specific allele is preferentially expressed, exhibiting a pattern of *cis*-eQTL effect.

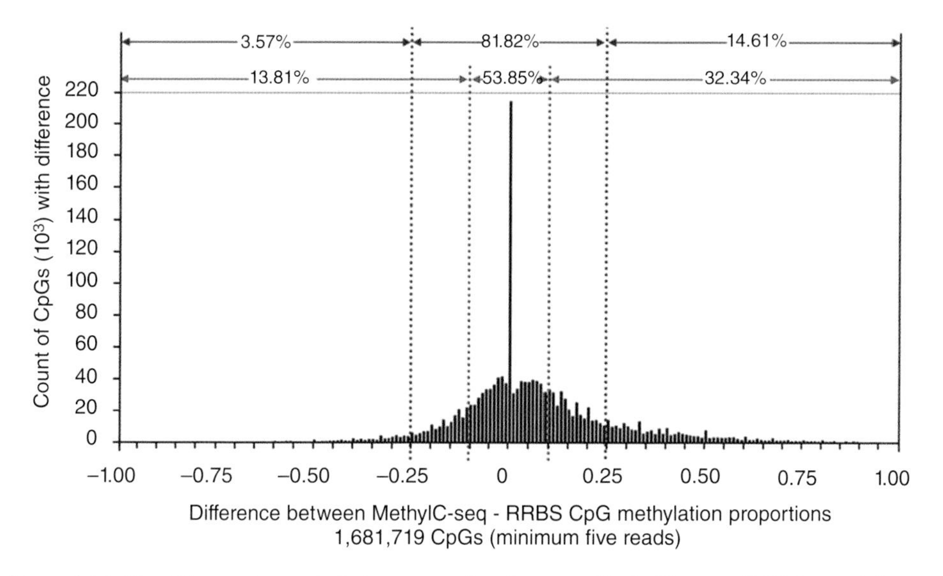

Figure 15.3 The difference in methylation proportions between MethylC-seq and RRBS at a minimum read depth of 5 was calculated for individual CpGs. Of the CpGs compared between MethylC-seq and RRBS, only 12.75% displayed identical methylation levels with a difference threshold of zero. If the difference threshold is relaxed to 0.1 (green dashed lines) or 0.25 (red dashed lines), the concordance increased to 53.85% or 81.82%, respectively (Meissner *et al.*, 2008).

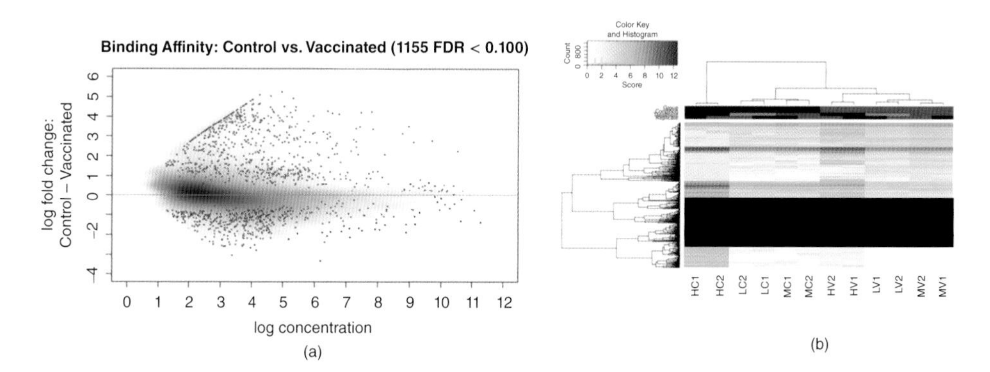

Figure 15.6 Visualization and cluster of differentially methylated region (DMR) samples considering their methylation levels (Eck *et al.*, 2009): (a) the MA plot shows in red the DMRs between chickens immunized with infectious laryngotracheitis (ILT) vaccine and control individuals with FDR < 0.1; (b) heat-map of samples demonstrates a perfect classification of the condition of the individuals based on methylation levels in the DMRs. Each condition (control and vaccinated) has three elute concentrations (high, medium, and low), with two replications for MBD sequencing.

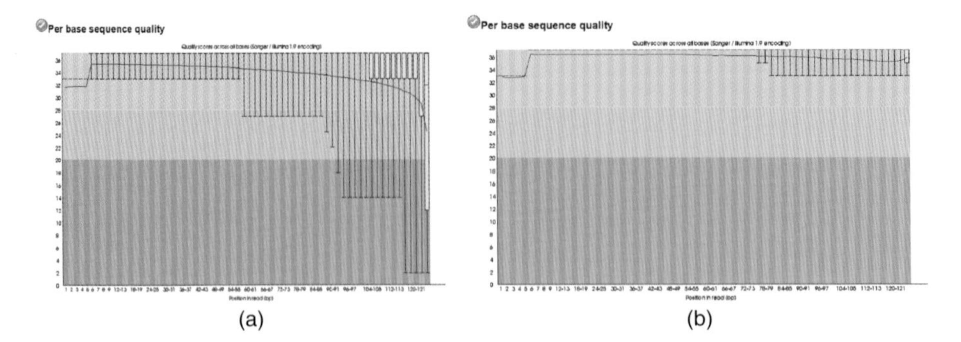

Figure 17.1 Summary of reads quality generated by FastQC: an overview of the range of base quality values at each position before (a) and after (b) trimming.

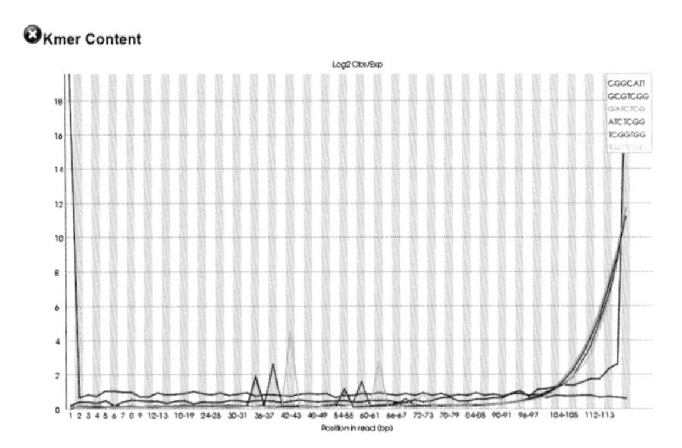

Figure 17.3 Summary of reads quality generated by FastQC. Analysis of overrepresented sequences. The underlined part of the TruSeq universal adapter sequence "AATGATACGGCGACCACCGAGATC TACACTCTTT CCCTACACGACGCTCTTCCGATCT" was identified with positional biases enrichment.

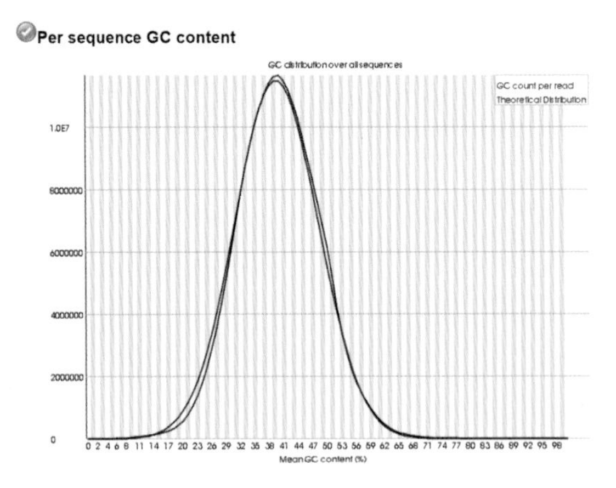

Figure 17.4 An example of the output file for GC content analysis: well-aligned distribution between the theoretical and observed G/C content (two colored distributions) indicate good quality of the data.

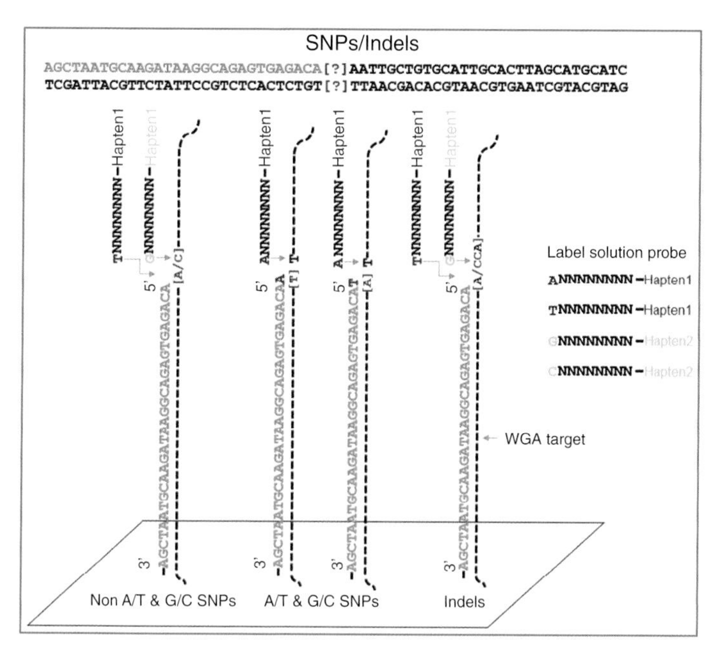

Figure 18.1 Probe design for interrogation of SNPs and indels on Affymetrix SNP array.

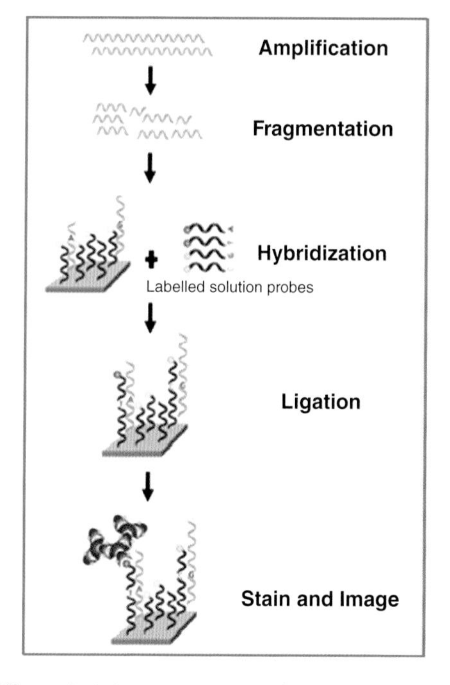

Figure 18.3 Overview of Affymetrix Axiom genotyping: chemistry and workflow.

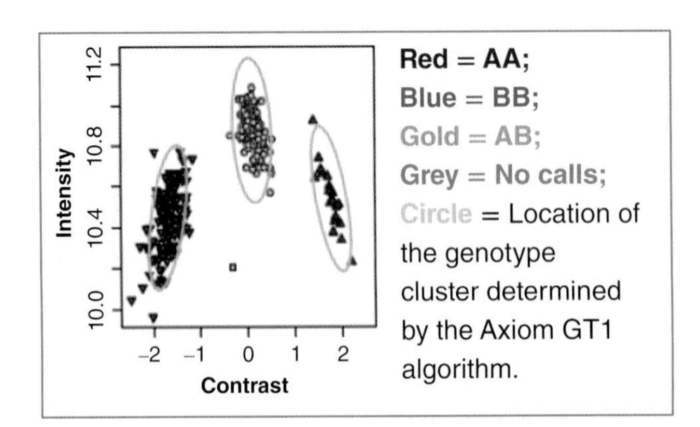

Figure 18.5 GTC cluster graph for an example SNP. In this figure, each cluster plot is for one SNP, and each point in the cluster plot is for one sample. For each sample, "A" indicates the summarized intensity of the A allele, and "B" indicates the summarized intensity of the B allele. The values are transformed into the cluster space of the genotyping algorithm, referred to as *contrast* and *size*. Contrast = $\text{Log}_2(\text{A/B})$, and size = $[\text{Log}_2(\text{A}) = \text{Log}_2(\text{B})]/2$. The samples are colored by the genotype called by the algorithm, where AA calls are red, BB calls are blue, AB calls are gold, and no calls are grey.

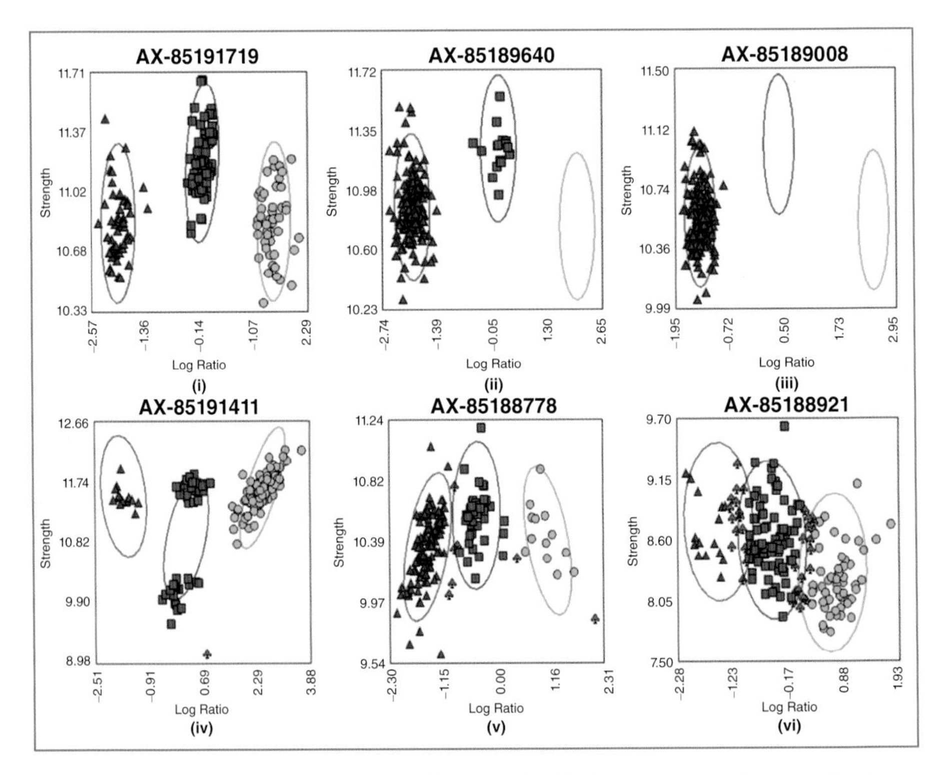

Figure 18.6 Examples of six SNP categories. SNPs were classified into six categories according to cluster properties: **(i)** "PolyHighResolution"; **(ii)** "NoMinorHom"; **(iii)** "MonoHighResolution"; **(iv)**, "OTV" off-target variants; **(v)** "CallRateBelowThreshold"; and **(vi)** "Other". The figure is adapted from Liu *et al.* (2014).

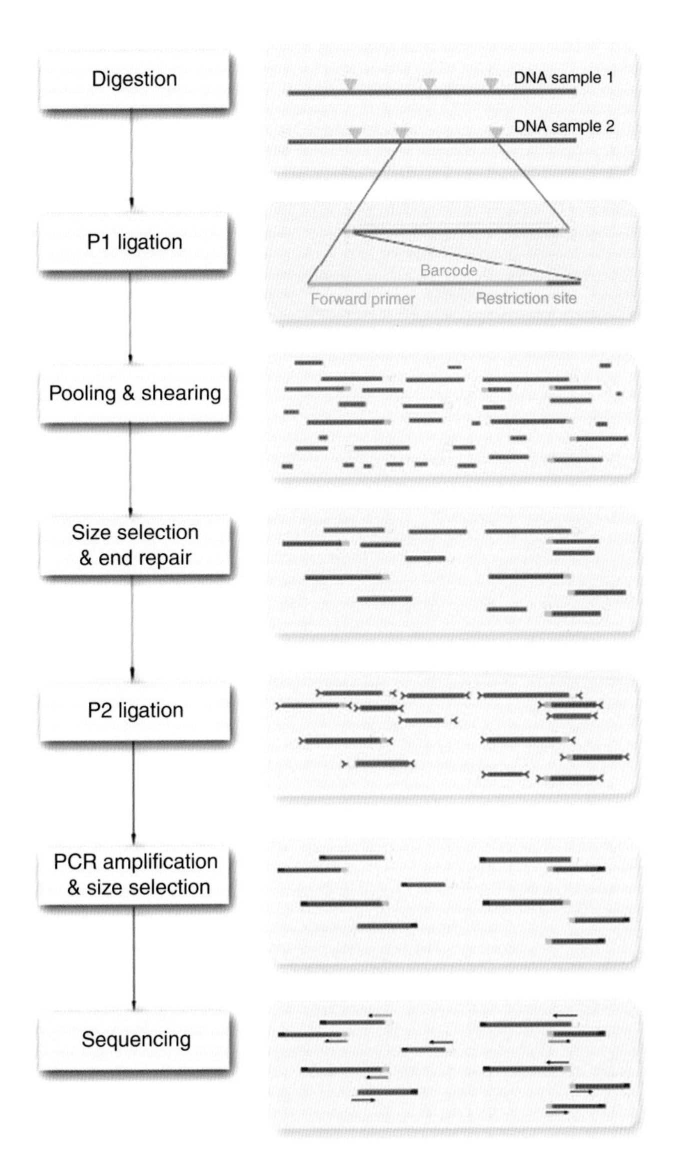

Figure 19.2 Schematic overview of RAD library preparation.

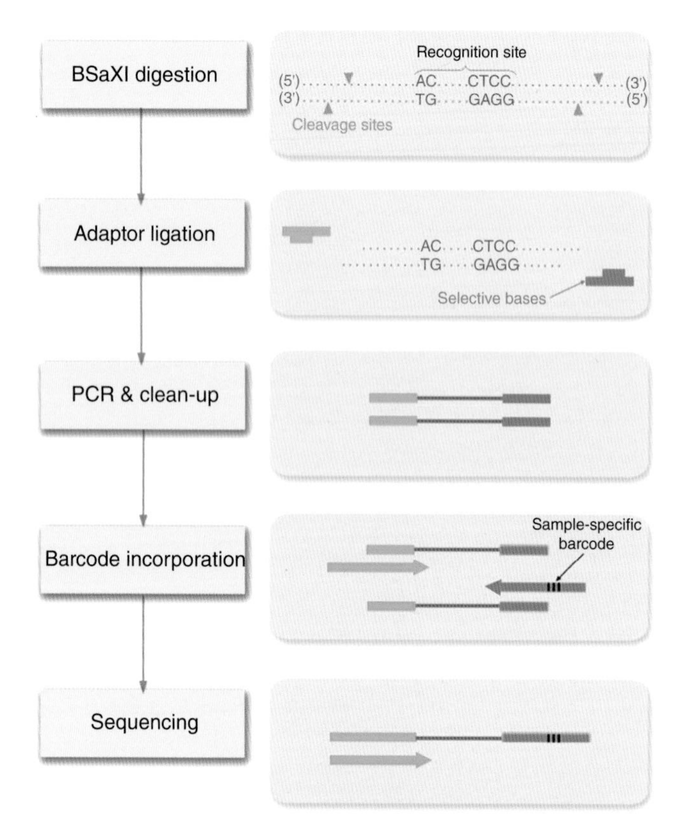

Figure 19.3 Schematic overview of 2b-RAD library preparation.

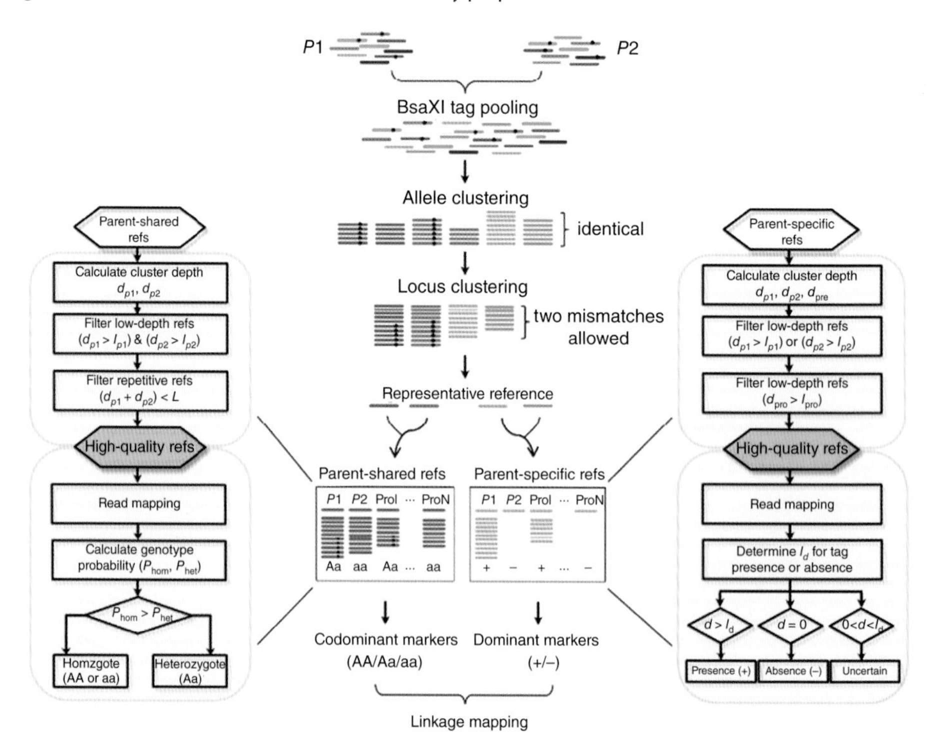

Figure 19.7 An overview of the RADtyping analytical approach for *de novo* codominant and dominant genotyping in a mapping population (adopted from Fu *et al.*, 2013). The main principles of codominant and dominant genotyping are shown in flowcharts.

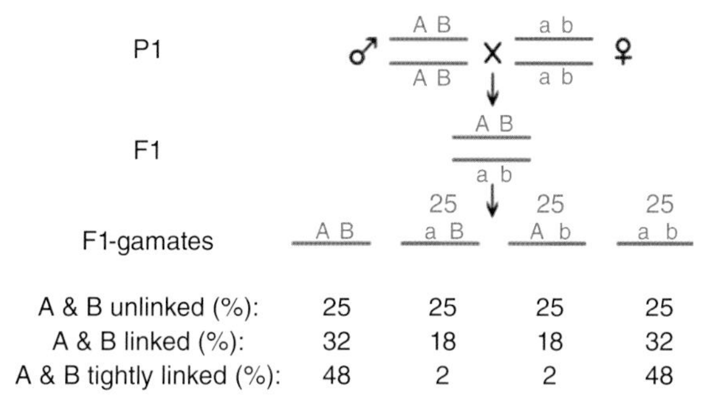

Figure 20.1 An example of the percentage of gametes generated by independent segregation and linkage.

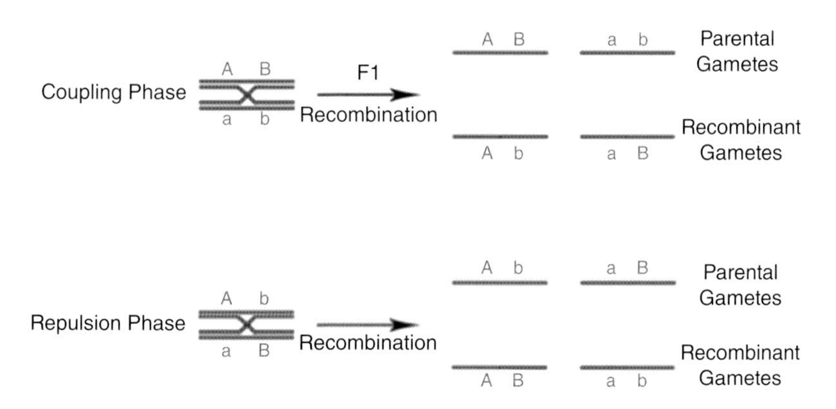

Figure 20.2 An example of the gamete composition for linked markers or genes for coupling and repulsion crosses.

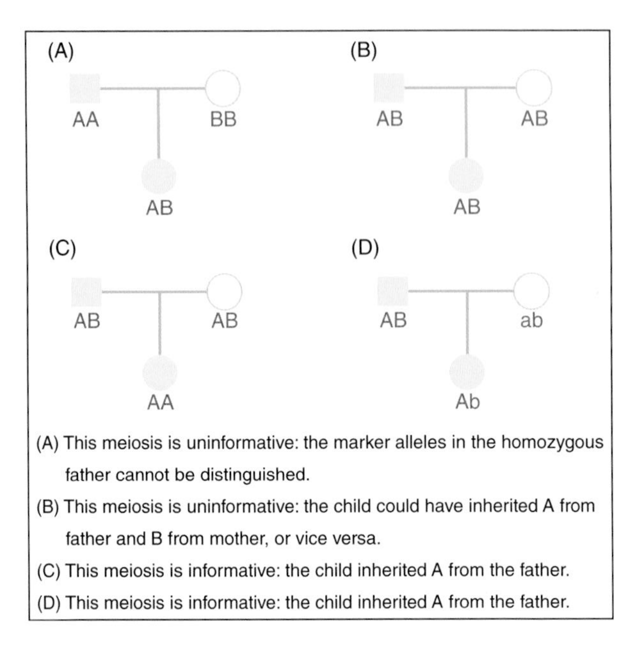
(A)
AA
BB
AB
(B)
AB
AB
AB
(C)
AB
AB
AA
(D)
AB
ab
Ab
(A) This meiosis is uninformative: the marker alleles in the homozygous father cannot be distinguished.
(B) This meiosis is uninformative: the child could have inherited A from father and B from mother, or vice versa.
(C) This meiosis is informative: the child inherited A from the father.
(D) This meiosis is informative: the child inherited A from the father.

Figure 20.3 An example of informative and non-informative meiosis.

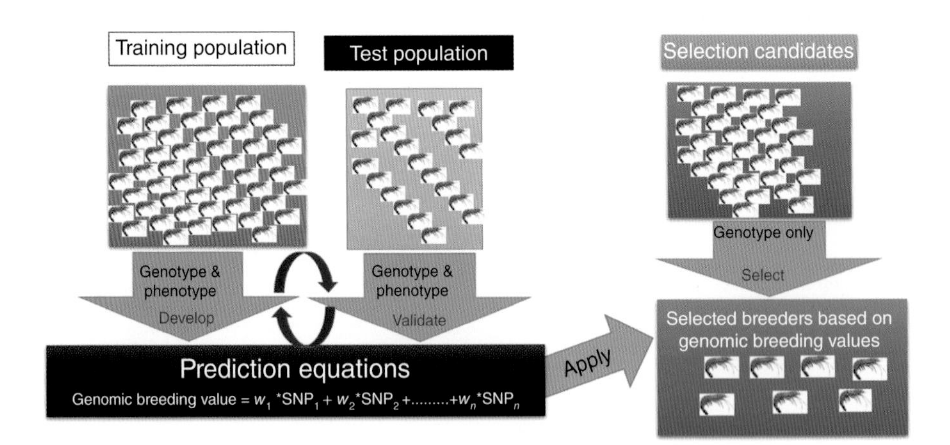
Training population
Test population
Selection candidates
Genotype & phenotype
Develop
Genotype & phenotype
Validate
Genotype only
Select
Prediction equations
Genomic breeding value = w_1 *SNP$_1$ + w_2*SNP$_2$ +.........+w_n*SNP$_n$
Apply
Selected breeders based on genomic breeding values

Figure 21.1 Overview of GS (adapted from Goddard & Hayes, 2009).

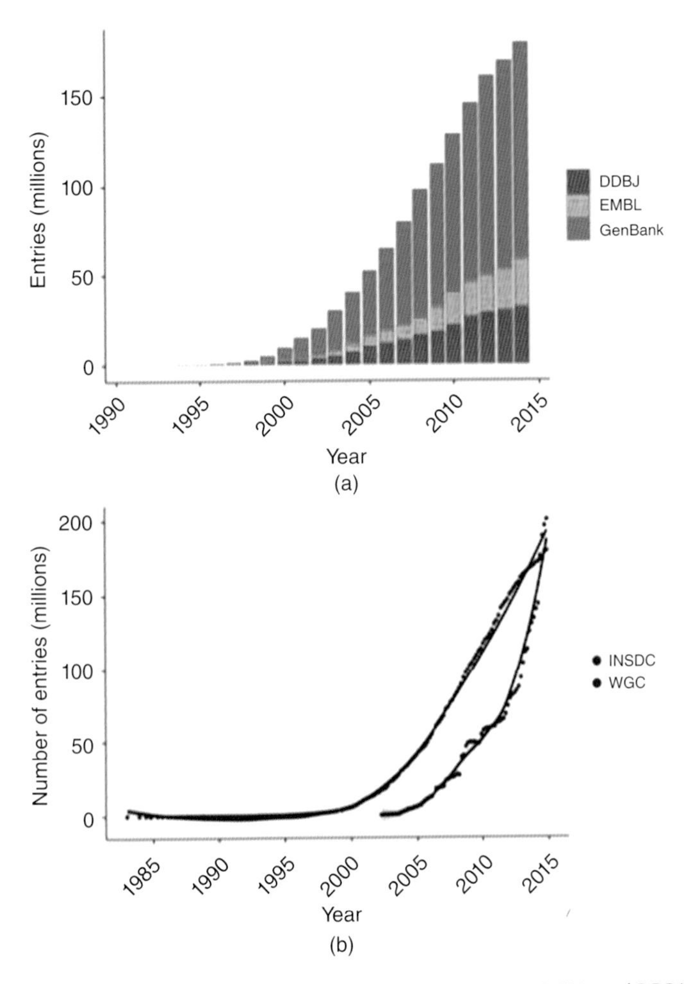

Figure 25.1 The number of entries in the three INSDC databases (NCBI, EMBL, and DDBJ) and the growth WGS sequencing projects in GenBank (NCBI). (a) Entries available in the three major international databases from 1992 to present. The number of entries available in the INSDC has displayed exponential growth since 1999. (b) The number of entries in the GenBank database and WGS sequence database of NCBI. The WGS database matches the number of entries in GenBank in only 11 years (2002–2013). Data taken from NCBI (http://www.ncbi.nlm.nih.gov/genbank/statistics) and DDBJ (http://www.ddbj.nig.ac.jp/documents-e.html).

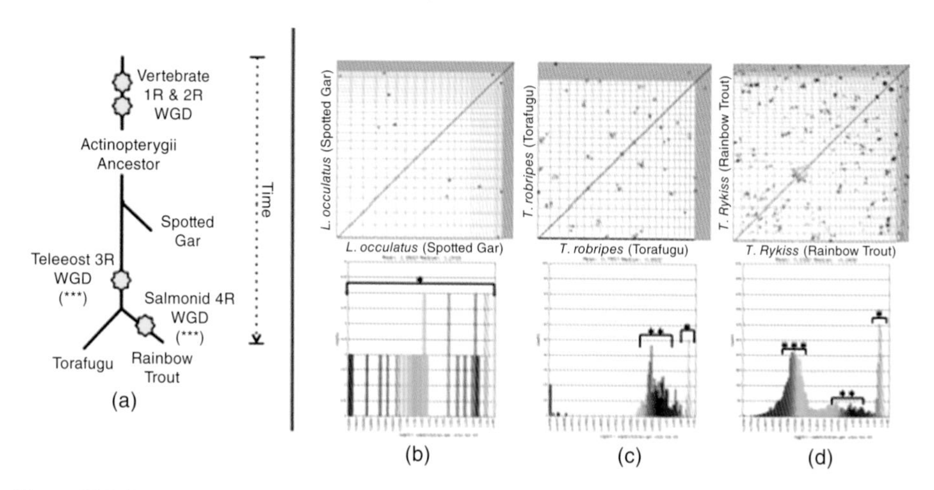

Figure 25.3 Intergenomic comparative dot plots and synonymous mutation rate distributions of *L. oculatus*, *T. rubripes*, and *O. mykiss* show evidence of Teleost and salmonid lineage WGDs. (a) Ancestral evolutionary relationship of *L. oculatus* (Lo), *T. rubripes* (Tr), and *O. mykiss* (Om) overlaid with hypothesized WGD events (gold stars). (b–d) Syntenic dot plots comparing genomes of three species against themselves (upper). Dots represent syntenic windows of genes, and are colored based on synonymous mutation rates (*Ks*). Distribution of *Ks* values are illustrated in histograms (lower). (b) Intragenomic syntenic dot plot of *L. oculatus*, an outgroup for the Teleost WGD. Although *L. oculatus* experienced an ancient WGD in the early vertebrate linage, evidence is harder to visualize due to fractionation. (c) Syntenic dot plot of *T. rubripes*. Single peak in *Ks* distribution indicative of a single, Teleost lineage WGD. (d) Dot plot of synteny in the genome of *O. mykiss*. Two peaks in *Ks* distribution indicative of two recent WGDs, the Teleost as well as a salmonid-lineage specific duplication. * Noise from the SynMap algorithm (Tang & Lyons, 2012); ** Teleost WGD; *** salmonid-lineage WGD. All analyses can be regenerated for *L. oculatus*, *T. rubripes*, and *O. mykiss* at: https://genomevolution.org/r/ewrn, https://genomevolution.org/r/ewqm, and https://genomevolution.org/r/ewrm, respectively.

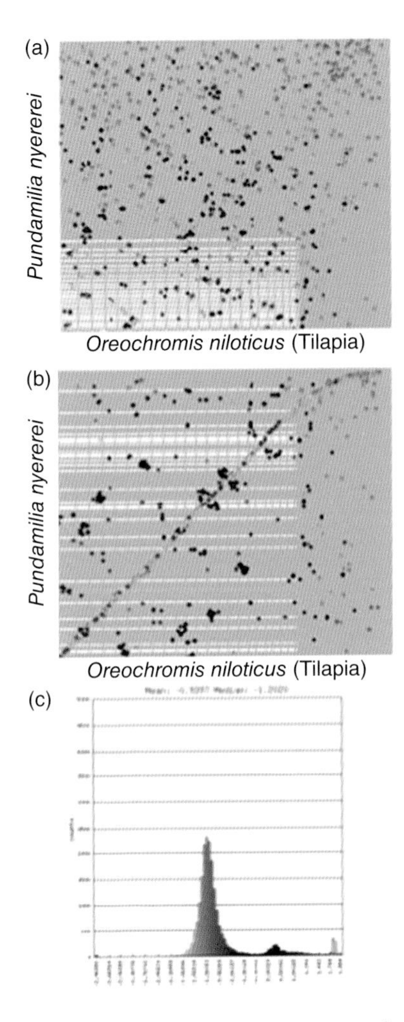

Figure 25.4 Syntenic dot plots generated using SynMap to compare the genomes of *Oreochromis niloticus* and *Pundamilia nyererei* without and with the SPA algorithm. (a) Syntenic dot plot of *O. niloticus* and *P. nyererei* unmasked and annotated CDSs without the SPA algorithm. The colors of the histogram syntenic dots based on synonymous substitution rate (*Ks*) are randomly distributed, and their pattern is difficult to discern. (b) Syntenic dot plot of *O. niloticus* and *P. nyererei* unmasked and annotated CDSs using the SPA algorithm. The SPA algorithm pseudo-assembles syntenic pairs using a chromosome-level genome assembly (*O. niloticus*) as the reference for a genome made of only contigs (*P. nyererei*). The pseudo-assembly mapping can be exported for use as a more complete genome assembly. (c) The SynMap histogram displays syntenic orthologs in yellow and paralogs from the Teleost WGD (3R) in blue. The syntenic dot plot also shows a recent WGD. Results can be regenerated for *O. niloticus* and *P. nyererei* at https://genomevolution.org/r/f74x and https://genomevolution.org/r/f74z, respectively.

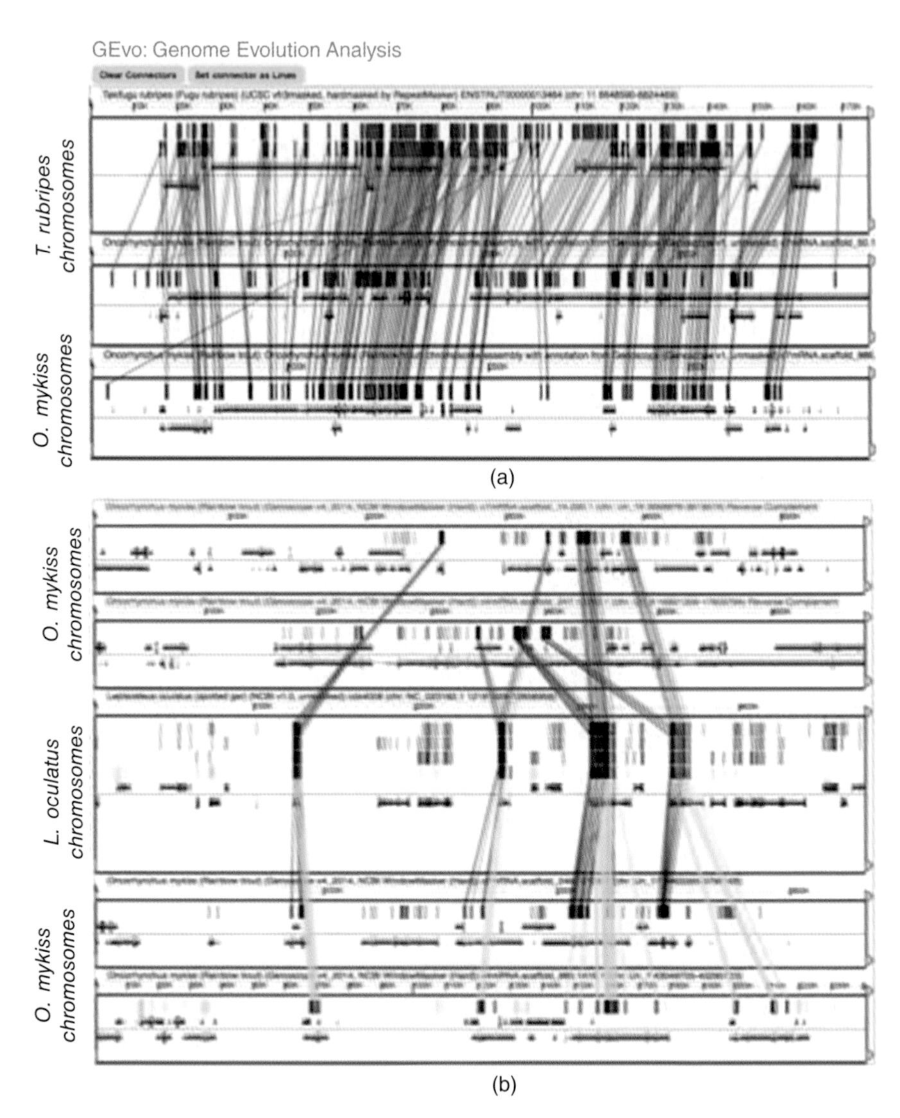

Figure 25.5 Regions of microsynteny from SynMap comparisons of Figure 25.4 displayed using the CoGe tool GEvo. (a) Comparison of a syntenic region in the *T. rubripes* genome on chromosome 11 (top track) with two separate scaffolds of the *O. mykiss* genome (second and third tracks). This genome region has a 1:2 ratio with two duplications in *O. mykiss*, suggesting that these may have resulted from the salmonid WGD. (b) A 1:4 *L. oculatus* to *O. mykiss* syntenic genome region. Five genes in this region display a 1:4 syntenic pattern, suggesting two duplications (Teleost 3R and salmonid 4R). The other genes in this region display a 1:2 or 1:3, suggesting fragmentation of the genome. Results can be regenerated at https://genomevolution.org/r/eoln and https://genomevolution.org/r/f5jo, respectively.